PROBLEMS
ON
PHYSICAL CHEMISTRY

PROBLEMS
ON
PHYSICAL CHEMISTRY

[For BSc (Honours) and Postgraduate students of all Indian universities]

Dola Pahari MSc, PhD
Lecturer, Department of Chemistry
Kharagpur College, Kharagpur, West Bengal

Dr S Pahari MSc, PhD
Retd. Reader, Department of Chemistry
Midnapore College, Midnapore, West Bengal

New Central Book Agency (P) Ltd

Co-Published by

Bijay Publications

E-mail : bijaypublications@gmail.com

Exclusively Distributed by

Bijay Pustak Bhandar (P) Ltd

90/6A M. G. Road, Bhawani Dutta Lane, Kolkata-700007

✆ : (033) 4605-7856, 8777898576, 8274916104

E-mail : bijaypustakbhandar@gmail.com

NCBA

REGD OFFICE
8/1 Chintamoni Das Lane, Kolkata 700 009, India
email: ncbakolkata@gmail.com

OVERSEAS
NCBA (UK) Ltd, 149 Park Avenue North, Northampton, NN3 2HY, UK
email: ncbauk@yahoo.com

CORPORATE OFFICE
212 Shahpur Jat, New Delhi 110 049
email: ncbaexports@ncbapvtltd.com

BIJAY PUBLICATIONS
90/6A M. G. Road, Bhawani Dutta Lane
Kolkata-700 007
E-mail : bijaypublications@gmail.com

BIJAY PUSTAK BHANDAR (P) LTD
90/6A M. G. Road, Bhawani Dutta Lane
Kolkata-700 007
E-mail : bijaypustakbhandar@gmail.com

PROBLEMS ON PHYSICAL CHEMISTRY • Dola Pahari and S Pahari

First Published: January 2009
Reprinted: January 2011, July 2012
Thoroughly Revised, Enlarged and Updated Second Edition: February 2015
Reprinted: July 2016, October 2018, October 2019, September 2021, **June 2023**

PUBLISHER AND TYPESETTER
New Central Book Agency (P) Ltd
8/1 Chintamoni Das Lane, Kolkata 700 009

PRINTER
New Central Book Agency (P) Ltd
Web-Offset Division, Dhulagarh, Sankrail, Howrah

TECHNICAL EDITOR
Dipan Roy

PROJECT TEAM
Partha Pratim Biswas and Pradip Biswas

ISBN: 978 81 7381 958 2

Price ₹ 765.00

Contents

Preface to the Second Edition

The book first appeared in 2009 with the object of helping the students to learn the subject. Acceptance of the book by the students as well as the teachers inspired us to revise it. While revising, we felt that—to give the students the flavour of the subject through short questions and numerical problems—some more questions and problems of different varieties should be added. We have increased the number of problems in many chapters. Few short questions have also been added. Although we had plan to add few more short questions, we could not do it to meet the deadline.

We have followed IUPAC recommendations as to symbols of physical quantities so as to make the students familiar with it. Hence, symbols (e.g., density, velocity etc.) have been changed. Only few symbols would be found unaltered in this book. Use of molar conductivity, instead of equivalent conductivity, is strongly advocated by IUPAC, but both the conductivities have been kept here since value of equivalent conductivity, instead of molar conductivity, is often asked in many numerical problems. Molar conductivity, in the book has been denoted by $\wedge_m$ instead of $\wedge$ as recommended by IUPAC. IUPAC recommends using molarity (c) or molality (M) as to defining concentration of a solution. Since we also use normality, it is kept in the book. We could not strictly adhere to the SI system as to unit of any quantity. The problems given by many universities in UG exams do not always follow the SI system. So we have answered in the unit as required by the problem. Hence, the units of physical quantities in the numerical problems have been kept unaltered. However, we have added few problems in SI unit and promise to give more so in the next edition.

In thermodynamics, work has been defined according to IUPAC convention. Hence its sign has been changed. Although we also had the intention to change the working formulae required to work out numerical problems in other chapters to comply with the IUPAC recommendations, we could not do it for lack of time. We have tried our level best to eliminate all the errors and incorporate the alterations uniformly. Since there were many alterations and corrections, new errors might have crept in while some older ones may have remained unnoticed despite our sinceremost effort. We deeply regret all such errors and will try to correct them in the next edition.

Only by solving out different types of problems, students can get a clear conception of physical chemistry. The more multifarious problems a student tries to solve, the deeper he/she will go into the subject. So our advice is that students should look up the solution to any problem only after giving substantial effort to solve the problem themselves. This will help them to understand the subject better. Then only our purpose will be fulfilled. Students should thorougly read Chapter 1 wherein we have tried to give them the idea as to how to tackle any numerical problem in Physical Chemistry.

We are grateful to Sri Amitabha Sen, Director, New Central Book Agency (P) Ltd, to agree to publish the Second Edition. We remain truly indebted to our family members—Gita Pahari, Sri Shubhashis Acharya, Subhajit Pahari—for their constant care, support, and encouragement. We are truly thankful to Shri Prabhat Jas and other members of the Project Team of NCBA for their sincere effort. Shri Dipan Roy (Technical Editor) helped us a lot in proof correction. We express our gratitude to him. The Printer, Cover Printer, Cover Designer and others have done their jobs quite well. We may not know them personally but without their help our effort would not have been fruitful.

Last but not the least, we are grateful to all those teachers and students whose valuable comments helped us a lot in the revision of the book.

Raspurnima
7 November 2014
Midnapore

D Pahari
S Pahari

Preface to the First Edition

A study of physical chemistry is never complete without solving problems—either theoretical or numerical. It enables the student to understand the subject clearly. We have written this book on problems to help the students to learn how to solve the problems.

Numerous books are available either on numerical problems or theoretical problems. But this book on problems includes both numerical problems as well as theoretical problems.

This book consists of three sections. In the first section, we have introduced various formulae necessary for solving the numerical problems. Second section is contributed to theoretical problems, and the last section is for numerical problems. Besides this, few chapters have *Miscellaneous* sections at the end with some conceptual questions and tricky derivations. Major portion of problems have been answered. Few problems have been left for the students to work out by themselves but hints have been given to guide them. At the end of the book there is a Miscellaneous chapter having various types of problems consisting of multiple choice questions, short questions, numerical problems covering the whole subject of the BSc syllabus. Some short questions as well as all the multiple choice questions have been answered. Few numerical problems have been solved. Some short questions have been left unanswered. For most numerical problems in this chapter, only the answers have been given. They are left for the students to solve to check their grasp of the knowledge of the subject. It will help them to grow confidence in themselves.

Many different techniques can be employed to solve a problem. In this book we have tried to stick to one method as far as possible to solve similar type of problems. We have chosen that method which the students will find very easy to understand, and can use it to solve other problems. However, sometimes, for some problems, another alternative and simpler method of solution suitable for students has been shown and students can choose either one.

We express our gratitude to Shri Gouri Sankar Maity for providing us with IIT and other Entrance Examination question papers. We are grateful to Late Jogendra Nath Sen and the publisher Shri Amitabha Sen. We owe much to Shri Prabhat Jas and others of New Central Book Agency (P) Ltd. We are grateful to Shri Dipan Roy (Technical Editor) who helped us by offering suggestions at various stages of composing of the manuscript. The editorial staff and the printers, who work behind the scenes, do a thankless job; but without them the project cannot be completed. We are indebted to all of them.

We invite suggestions and criticisms from the teachers as well as the students which will help us to make the book more purposeful and error-free in the subsequent editions.

15 August 2008 **S Pahari**
Midnapore **D Pahari**

CHAPTER 1

INTRODUCTION

Long back Prof. W.V. Evans defined a Physical Chemist as a mathematician sitting behind a mahogany desk with a platinum slide rule in his hand. A popular science and novel writer Sinclair Lewis held a very high opinion about Physical Chemistry. In one of his novels Published in 1924, Arrowsmith, the character Max Gotlieb, a teacher of a medical school asks Arrowsmith, the hero of the novel—"How can you know Physical Chemistry without much mathematics?" In the same novel Gotlieb utterd, "Physical Chemistry is power, it is life, it is exactness." In this line we tell you to learn Physical Chemistry by heart. For that you must be able to solve Problems to learn a basic theory by heart.

For solving a problem at the start :

(i) Write what has been given.

You may write in short in boxes. For example, suppose that a problem is like this—

A. 2 moles of an ideal gas at 27 °C and 10 atm is allowed to expand to 1 atm at 327 °C expanding against a constant pressure of 1 atm. Calculate the work done, ΔU, ΔH and Q. Comment.

Thus it has been given :

$n = 2$		$n = 2$
$T_1 = 300$ K $P_1 = 10$ atm	Expansion against $P_{ex} = 1$ atm	$T_2 = 600$ K $P_2 = 1$ atm

And comment on the result.

You may not draw such boxes always. But, however tough the problem may appear, if you can write a simple summary of what has been given, you will find that the problem is not so tough.

Then,

(ii) Write what has to be calculated.

In the given problem A, the quantities to be calculated are W, Q, ΔU and ΔH. In this given problem, no extra quantity has to be calculated. The states mentioned in the boxes are equilibrium states. This may not be always so. For example, let the exercise be :

B. 2 moles of an ideal gas at 10 atm and 27 °C are allowed to expand against a constant pressure of 1 atm to double the volume at a temperature of 327 °C. Calculate W, ΔU, ΔH and Q. Comment.

Here it has been given :

$n = 2$		$n = 2$
$T_1 = 300$ K $P_1 = 10$ atm V_1	Expansion against 1 atm →	$T = 600$ K $P_2 = ?$ $2V_1$

In this problem, the extra quantity to be calculated is volume V_1 at 300 K and 10 atm. And then

(iii) Write formulae connecting known to unknown.

In problem A given,

$$W = -P_{ex}(V_2 - V_1)$$

$$= -P_2\left(\frac{nRT_2}{P_2} - \frac{nRT_1}{P_1}\right)$$

$$\Delta U = nC_v(T_2 - T_1)$$

$$\Delta H = nC_p(T_2 - T_1)$$

$$\Delta U = Q + W.$$

Solve it yourself and comment if you can (see the chapter on first law of Thermodynamics).

In problem B :

$$W = -P_{ex}(V_2 - V_1)$$

$$= -P_{ex}(2V_1 - V_1)$$

$$= -P_{ex} \cdot V_1 = -P_{ex}\frac{nRT_1}{P_1}$$

ΔU, ΔH are same as problem A.

$$\Delta U = Q + W.$$

To comment, calculate the pressure of the final state.

Do not use an equation where it does not apply. Be sure of the fact that you are using an equation which is applicable.

If you are in difficulty, then do backwards from unknown to known. Then come from known to unknown.

(iv) Calculate using formulae and modifications. Use a calculator. Round off the final result to the desired place of decimal.

Rounding off : In calculations, round up if the last digit is above 5 and round down if the last digit is below 5. Thus, if there has to be 4 significant figures, then report 2·4637 as 2.464 and 2.4634 as 2.463. But for numbers ending in 5 always round to the nearest even number. Thus 2.4635 is 2.464 but 2.4645 is 2.464.

Writing and reporting results : A number should be written as A × 10^a : A is a decimal number before decimal point and a is a whole number, positive or negative. Thus 333 is 3.33×10^2 and 0.043 is 4.3×10^{-2}

PROBLEMS
ON
PHYSICAL CHEMISTRY

[For BSc (Honours) and Postgraduate students of all Indian universities]

Dola Pahari MSc, PhD
Lecturer, Department of Chemistry
Kharagpur College, Kharagpur, West Bengal

Dr S Pahari MSc, PhD
Retd. Reader, Department of Chemistry
Midnapore College, Midnapore, West Bengal

New Central Book Agency (P) Ltd

Co-Published by

Bijay Publications

E-mail : bijaypublications@gmail.com

Exclusively Distributed by

Bijay Pustak Bhandar (P) Ltd

90/6A M. G. Road, Bhawani Dutta Lane, Kolkata-700007

✆ : (033) 4605-7856, 8777898576, 8274916104

E-mail : bijaypustakbhandar@gmail.com

NCBA

REGD OFFICE
8/1 Chintamoni Das Lane, Kolkata 700 009, India
email: ncbakolkata@gmail.com

OVERSEAS
NCBA (UK) Ltd, 149 Park Avenue North, Northampton, NN3 2HY, UK
email: ncbauk@yahoo.com

CORPORATE OFFICE
212 Shahpur Jat, New Delhi 110 049
email: ncbaexports@ncbapvtltd.com

BIJAY PUBLICATIONS
90/6A M. G. Road, Bhawani Dutta Lane
Kolkata-700 007
E-mail : bijaypublications@gmail.com

BIJAY PUSTAK BHANDAR (P) LTD
90/6A M. G. Road, Bhawani Dutta Lane
Kolkata-700 007
E-mail : bijaypustakbhandar@gmail.com

PROBLEMS ON PHYSICAL CHEMISTRY • Dola Pahari and S Pahari

First Published: January 2009
Reprinted: January 2011, July 2012
Thoroughly Revised, Enlarged and Updated Second Edition: February 2015
Reprinted: July 2016, October 2018, October 2019, September 2021, **June 2023**

PUBLISHER AND TYPESETTER
New Central Book Agency (P) Ltd
8/1 Chintamoni Das Lane, Kolkata 700 009

PRINTER
New Central Book Agency (P) Ltd
Web-Offset Division, Dhulagarh, Sankrail, Howrah

TECHNICAL EDITOR
Dipan Roy

PROJECT TEAM
Partha Pratim Biswas and Pradip Biswas

ISBN: 978 81 7381 958 2

Price ₹ 765.00

Contents

Preface to the Second Edition

The book first appeared in 2009 with the object of helping the students to learn the subject. Acceptance of the book by the students as well as the teachers inspired us to revise it. While revising, we felt that—to give the students the flavour of the subject through short questions and numerical problems—some more questions and problems of different varieties should be added. We have increased the number of problems in many chapters. Few short questions have also been added. Although we had plan to add few more short questions, we could not do it to meet the deadline.

We have followed IUPAC recommendations as to symbols of physical quantities so as to make the students familiar with it. Hence, symbols (e.g., density, velocity etc.) have been changed. Only few symbols would be found unaltered in this book. Use of molar conductivity, instead of equivalent conductivity, is strongly advocated by IUPAC, but both the conductivities have been kept here since value of equivalent conductivity, instead of molar conductivity, is often asked in many numerical problems. Molar conductivity, in the book has been denoted by $\wedge_m$ instead of $\wedge$ as recommended by IUPAC. IUPAC recommends using molarity (c) or molality (M) as to defining concentration of a solution. Since we also use normality, it is kept in the book. We could not strictly adhere to the SI system as to unit of any quantity. The problems given by many universities in UG exams do not always follow the SI system. So we have answered in the unit as required by the problem. Hence, the units of physical quantities in the numerical problems have been kept unaltered. However, we have added few problems in SI unit and promise to give more so in the next edition.

In thermodynamics, work has been defined according to IUPAC convention. Hence its sign has been changed. Although we also had the intention to change the working formulae required to work out numerical problems in other chapters to comply with the IUPAC recommendations, we could not do it for lack of time. We have tried our level best to eliminate all the errors and incorporate the alterations uniformly. Since there were many alterations and corrections, new errors might have crept in while some older ones may have remained unnoticed despite our sinceremost effort. We deeply regret all such errors and will try to correct them in the next edition.

Only by solving out different types of problems, students can get a clear conception of physical chemistry. The more multifarious problems a student tries to solve, the deeper he/she wili go into the subject. So our advice is that students should look up the solution to any problem only after giving substantial effort to solve the problem themselves. This will help them to understand the subject better. Then only our purpose will be fulfilled. Students should thorougly read Chapter 1 wherein we have tried to give them the idea as to how to tackle any numerical problem in Physical Chemistry.

We are grateful to Sri Amitabha Sen, Director, New Central Book Agency (P) Ltd, to agree to publish the Second Edition. We remain truly indebted to our family members—Gita Pahari, Sri Shubhashis Acharya, Subhajit Pahari—for their constant care, support, and encouragement. We are truly thankful to Shri Prabhat Jas and other members of the Project Team of NCBA for their sincere effort. Shri Dipan Roy (Technical Editor) helped us a lot in proof correction. We express our gratitude to him. The Printer, Cover Printer, Cover Designer and others have done their jobs quite well. We may not know them personally but without their help our effort would not have been fruitful.

Last but not the least, we are grateful to all those teachers and students whose valuable comments helped us a lot in the revision of the book.

Raspurnima
7 November 2014
Midnapore

D Pahari
S Pahari

Preface to the First Edition

A study of physical chemistry is never complete without solving problems—either theoretical or numerical. It enables the student to understand the subject clearly. We have written this book on problems to help the students to learn how to solve the problems.

Numerous books are available either on numerical problems or theoretical problems. But this book on problems includes both numerical problems as well as theoretical problems.

This book consists of three sections. In the first section, we have introduced various formulae necessary for solving the numerical problems. Second section is contributed to theoretical problems, and the last section is for numerical problems. Besides this, few chapters have *Miscellaneous* sections at the end with some conceptual questions and tricky derivations. Major portion of problems have been answered. Few problems have been left for the students to work out by themselves but hints have been given to guide them. At the end of the book there is a Miscellaneous chapter having various types of problems consisting of multiple choice questions, short questions, numerical problems covering the whole subject of the BSc syllabus. Some short questions as well as all the multiple choice questions have been answered. Few numerical problems have been solved. Some short questions have been left unanswered. For most numerical problems in this chapter, only the answers have been given. They are left for the students to solve to check their grasp of the knowledge of the subject. It will help them to grow confidence in themselves.

Many different techniques can be employed to solve a problem. In this book we have tried to stick to one method as far as possible to solve similar type of problems. We have chosen that method which the students will find very easy to understand, and can use it to solve other problems. However, sometimes, for some problems, another alternative and simpler method of solution suitable for students has been shown and students can choose either one.

We express our gratitude to Shri Gouri Sankar Maity for providing us with IIT and other Entrance Examination question papers. We are grateful to Late Jogendra Nath Sen and the publisher Shri Amitabha Sen. We owe much to Shri Prabhat Jas and others of New Central Book Agency (P) Ltd. We are grateful to Shri Dipan Roy (Technical Editor) who helped us by offering suggestions at various stages of composing of the manuscript. The editorial staff and the printers, who work behind the scenes, do a thankless job; but without them the project cannot be completed. We are indebted to all of them.

We invite suggestions and criticisms from the teachers as well as the students which will help us to make the book more purposeful and error-free in the subsequent editions.

15 August 2008 **S Pahari**
Midnapore **D Pahari**

CHAPTER 1

INTRODUCTION

Long back Prof. W.V. Evans defined a Physical Chemist as a mathematician sitting behind a mahogany desk with a platinum slide rule in his hand. A popular science and novel writer Sinclair Lewis held a very high opinion about Physical Chemistry. In one of his novels Published in 1924, Arrowsmith, the character Max Gotlieb, a teacher of a medical school asks Arrowsmith, the hero of the novel—"How can you know Physical Chemistry without much mathematics?" In the same novel Gotlieb utterd, "Physical Chemistry is power, it is life, it is exactness." In this line we tell you to learn Physical Chemistry by heart. For that you must be able to solve Problems to learn a basic theory by heart.

For solving a problem at the start :

(i) Write what has been given.

You may write in short in boxes. For example, suppose that a problem is like this—

A. 2 moles of an ideal gas at 27 °C and 10 atm is allowed to expand to 1 atm at 327 °C expanding against a constant pressure of 1 atm. Calculate the work done, ΔU, ΔH and Q. Comment.

Thus it has been given :

$n = 2$		$n = 2$
$T_1 = 300$ K $P_1 = 10$ atm	Expansion against $P_{ex} = 1$ atm	$T_2 = 600$ K $P_2 = 1$ atm

And comment on the result.

You may not draw such boxes always. But, however tough the problem may appear, if you can write a simple summary of what has been given, you will find that the problem is not so tough.

Then,

(ii) Write what has to be calculated.

In the given problem A, the quantities to be calculated are W, Q, ΔU and ΔH. In this given problem, no extra quantity has to be calculated. The states mentioned in the boxes are equilibrium states. This may not be always so. For example, let the exercise be :

B. 2 moles of an ideal gas at 10 atm and 27 °C are allowed to expand against a constant pressure of 1 atm to double the volume at a temperature of 327 °C. Calculate W, ΔU, ΔH and Q. Comment.

Here it has been given :

$n = 2$		$n = 2$
$T_1 = 300$ K, $P_1 = 10$ atm, V_1	Expansion against 1 atm	$T = 600$ K, $P_2 = ?$, $2V_1$

In this problem, the extra quantity to be calculated is volume V_1 at 300 K and 10 atm.

And then

(iii) Write formulae connecting known to unknown.

In problem A given,

$$W = -P_{ex}(V_2 - V_1)$$

$$= -P_2\left(\frac{nRT_2}{P_2} - \frac{nRT_1}{P_1}\right)$$

$$\Delta U = nC_v(T_2 - T_1)$$

$$\Delta H = nC_p(T_2 - T_1)$$

$$\Delta U = Q + W.$$

Solve it yourself and comment if you can (see the chapter on first law of Thermodynamics).

In problem B :

$$W = -P_{ex}(V_2 - V_1)$$

$$= -P_{ex}(2V_1 - V_1)$$

$$= -P_{ex} \cdot V_1 = -P_{ex}\frac{nRT_1}{P_1}$$

ΔU, ΔH are same as problem A.

$$\Delta U = Q + W.$$

To comment, calculate the pressure of the final state.

Do not use an equation where it does not apply. Be sure of the fact that you are using an equation which is applicable.

If you are in difficulty, then do backwards from unknown to known. Then come from known to unknown.

(iv) Calculate using formulae and modifications. Use a calculator. Round off the final result to the desired place of decimal.

Rounding off : In calculations, round up if the last digit is above 5 and round down if the last digit is below 5. Thus, if there has to be 4 significant figures, then report 2·4637 as 2.464 and 2.4634 as 2.463. But for numbers ending in 5 always round to the nearest even number. Thus 2.4635 is 2.464 but 2.4645 is 2.464.

Writing and reporting results : A number should be written as A × 10^a : A is a decimal number before decimal point and a is a whole number, positive or negative. Thus 333 is 3.33 × 10^2 and 0.043 is 4.3 × 10^{-2}

Significant figures : In reported measurement, the digits, the report of which depends on the accuracy of measurement, are called the significant figures (SF). Thus there are two significant figures (with 2SF) in 2.4 and three significant figures (with 3SF) in 3.40. Trailing zero and zero in middle of a number are significant, but leading zeros are not included in significant figures. Thus there are 3SF in 0.120 (which is 1.20×10^{-1}) and 2SF in 0.0017 (which is 1.7×10^{-3}).

When adding or subtracting, the number of decimal places in the result should be equal to the smallest number of decimal places in the data. Thus, 12.1 added with 7.22 is 19.3, not 19.32.

When multiplying or dividing, the number of significant figures in the result should be the same as the smallest number of significant figures in the data (least exact factor). Hence, $12.14 \times 7.11 = 86.32$ (and not 86.3154).

Similarly, $\frac{12.14}{3.21} = 3.78$ (and not 3.781931464 or 3.782).

But if the first significant figure is 1, the number should contain one more significant figure. Thus $\frac{12.14}{7.11} = 1.707$ (and not 1.71 or 1.70745429).

However, while using calculator the rounding off should be done at the last stage of calculation and all digits are to be carried in the memory of the calculator until that stage.

Using different formulae : Let the collision frequency Z be given as

$$Z = \frac{\pi\sigma^2}{\sqrt{2}}\sqrt{\frac{8RT}{\pi M}}\left(\frac{N_AP}{RT}\right)^2.$$

All the data including σ is given. You can calculate Z. But let σ be not given—λ, the mean free path, is given :

Then $\lambda = \frac{RT}{\sqrt{2}\pi PN_A\sigma^2}$.

$$\therefore \quad Z = \left(\frac{N_AP}{RT}\right)\frac{1}{2\lambda}\sqrt{\frac{8RT}{\pi M}}.$$

Let λ be not also given, but viscosity η and density d be given. Then

$$\eta = \frac{1}{3}d\bar{v}\lambda$$

$$\lambda = \frac{3\eta}{d\bar{v}}.$$

$$\text{Thus,} \quad Z = \left(\frac{N_AP}{RT}\right)\frac{d\bar{v}}{6\eta}\bar{v}$$

$$= \left(\frac{N_AP}{RT}\right)\frac{d}{6\eta}\bar{v}^2.$$

This is a simple illustration. You have to be free in such uses. These require practice.

The more you practise, the more confident you become in solving. And note the key-word : Confidence!

Solve similar problems by the same method. Start from simpler problems. Tackle tougher problems later.

Be proficient in using formulae. Then you may be able to tackle "Prove that $w = \ldots$), etc" problems easily.

To solve and write answers to short questions a thorough study and deep thinking on the chapter are needed.

Be confident that if there is a problem, then there is a solution. 'Problems are solvable, and by study and practice, I can solve problems, however tough they may appear'. The same thing applies for short questions.

Method of solving theoretical questions

A student of physical chemistry who cannot solve problems has not understood this subject. Numerical problems as already outlined can be solved by the given method. Test it slowly finding a problem appearing tough to you. Then, follow the method outlined blindly. See what happens.

Similarly, derivations and theoretical problems can be solved if you have understood the subject by heart. Let us check. It is given that

(i) Cooling due to expansion of an ideal gas occurs due to decrease in internal energy per unit volume. Is it O.K?

Hints to help you : This is from first law of thermodynamics. Adiabatic expansion leads to cooling of ideal gas since it is done at the expense of internal energy.

(ii) Cooling due to expansion of real gas occurs because of reduction in the intermolecular energy, since it depends on intermolecular attractive forces between molecules. Is this right?

Hints to help you : Cooling of real gases also take place during adiabatic expansion even during free expansion. However, it is to be remembered that increase in volume does not necessarily mean decrease in intermolecular attraction energy.

Suppose, you are asked to derive $\left(\frac{\partial H}{\partial P}\right)_T$ for van der Waals gas which is often set in university examination. It is needed to calculate $\mu_{J.T}$ (Joule-Thomson Coefficient) which is $\mu_{J.T} = -\frac{1}{C_P}\left(\frac{\partial H}{\partial P}\right)_T$.

This can be solved at least by three different ways. These are the followings :

Method A :

van der Waals equation is $\left(P+\frac{a}{V^2}\right)(V-b) = RT.$ A(1)

Expanding and neglecting $\frac{ab}{V^2}$ in comparison to other terms,

$$PV + \frac{a}{V} - Pb = RT. \qquad \text{A(2)}$$

Replacing V in the denominator by the ideal gas value $\frac{RT}{P}$ (which will not incur huge error),

$$PV + \frac{aP}{RT} - Pb = RT \qquad \text{A(3)}$$

or, $PV = Pb - \dfrac{aP}{RT} + RT.$ A(4)

$$\therefore\ V = \frac{RT}{P} + \left(b - \frac{a}{RT}\right). \quad A(5)$$

Hence, $\left(\dfrac{\partial V}{\partial T}\right)_P = \dfrac{R}{P} + \dfrac{a}{RT^2}$ A(6)

$$\therefore\ T\left(\frac{\partial V}{\partial T}\right)_P = \frac{RT}{P} - \frac{a}{RT}.$$

So $V - T\left(\dfrac{\partial V}{\partial T}\right)_P = \left(b - \dfrac{2a}{RT}\right)$ A(7)

Again, we know that $dH = T\,dS + V\,dP$ A(8)

$$\therefore\ \left(\frac{\partial H}{\partial P}\right)_T = T\left(\frac{\partial S}{\partial P}\right)_T + V. \quad A(9)$$

By Maxwell's relation, $\left(\dfrac{\partial S}{\partial P}\right)_T = -\left(\dfrac{\partial V}{\partial T}\right)_P$ A(10)

$$\left(\frac{\partial H}{\partial P}\right)_T = V - T\left(\frac{\partial V}{\partial T}\right)_P$$

From equation A(7),

$$\left(\frac{\partial H}{\partial P}\right)_T = V - T\left(\frac{\partial V}{\partial T}\right)_P = \left(b - \frac{2a}{RT}\right). \quad A(11)$$

$$\therefore\ \mu_{J.T} = -\frac{1}{C_P}\left(b - \frac{2a}{RT}\right). \quad A(12)$$

$$\therefore\ \mu = \frac{1}{C_P}\left(\frac{2a}{RT} - b\right). \quad A(13)$$

Related Derivation :

We know that $C_P - C_V = T\left(\dfrac{\partial P}{\partial T}\right)_V \left(\dfrac{\partial V}{\partial T}\right)_P$ A(14)

van der Waals equation :

$$P = \frac{RT}{V-b} - \frac{a}{V^2} \quad A(15)$$

$$V = \frac{RT}{P} + \left(b - \frac{a}{RT}\right) \quad A(16)$$

$$\therefore\ \left(\frac{\partial P}{\partial T}\right)_V = \frac{R}{V-b}. \quad A(17)$$

Again, $\left(\dfrac{\partial V}{\partial T}\right)_P = \dfrac{R}{P} + \dfrac{a}{RT^2}.$ A(18)

Now from the expanded van der Waals equation neglecting $\frac{ab}{V^2}$ and V in denominator of the term $\frac{a}{V}$ by $\frac{RT}{P}$ (the ideal gas value),

$$PV = Pb - \frac{aP}{RT} + RT \qquad \text{A(19)}$$

$$\text{or, } P(V - b) = \frac{RT}{P} - \frac{a}{RT}. \qquad \text{A(20)}$$

$$\text{Hence, } \frac{R}{P} = \frac{V-b}{T} + \frac{a}{RT^2}. \qquad \text{A(21)}$$

$$\text{Thus, } C_P - C_V = T\left(\frac{R}{V-b}\right)\left(\frac{R}{P} + \frac{a}{RT^2}\right). \qquad \text{A(22)}$$

Putting A(21) in Equation A(22),

$$C_P - C_V = \frac{RT}{V-b}\left(\frac{V-b}{T} + \frac{2a}{RT^2}\right)$$

$$= R\left(1 + \frac{2a}{RTV}\right) \; [V >> b] = R + \frac{2aP}{RT^2} \; \left[\text{putting } V = \frac{RT}{P}\right] \qquad \text{A(23)}$$

Method B :

$$\mu_{JT} = -\frac{1}{C_P}\left(\frac{\partial H}{\partial P}\right)_T \qquad \text{B(1)}$$

Now van der Waals equation :

$$\left(P + \frac{a}{V^2}\right)(V - b) = RT. \qquad \text{B(2)}$$

Neglecting $\frac{ab}{V^2}$ term in comparison to others and expanding,

$$PV + \frac{a}{V} - Pb = RT. \qquad \text{B(3)}$$

$$\therefore \frac{\partial}{\partial T}\left(PV + \frac{a}{V} - Pb\right) = \frac{\partial}{\partial T}(RT) \qquad \text{B(4)}$$

$$\text{or, } \left(\frac{\partial V}{\partial T}\right)_P \left\{\frac{\partial}{\partial V}\left(PV + \frac{a}{V} - Pb\right)\right\}_P = R. \qquad \text{B(5)}$$

$$\therefore \left(\frac{\partial V}{\partial T}\right)_P \left[P - \frac{a}{V^2}\right] = R \qquad \text{B(6)}$$

$$\text{or, } \left(\frac{\partial V}{\partial T}\right)_P = \frac{R}{\left(P - \frac{a}{V^2}\right)}. \qquad \text{B(7)}$$

$$\text{So, } T\left(\frac{\partial V}{\partial T}\right)_P = \frac{RT}{P+\dfrac{a}{V^2}-\dfrac{2a}{V^2}} \quad \text{B(8)}$$

$$= \frac{RT}{\dfrac{RT}{V-b}-\dfrac{2a}{V^2}} \quad \text{B(9)}$$

$$= \frac{RT}{\dfrac{RT}{V-b}\left(1-\dfrac{2a(V-b)}{RTV^2}\right)}. \quad \text{B(10)}$$

$$\therefore\ T\left(\frac{\partial V}{\partial T}\right)_P = (V-b)\left(1-\frac{2a(V-b)}{RTV^2}\right)^{-1} \quad \text{B(11)}$$

$$= (V-b)\left(1+\frac{2a(V-b)}{RTV^2}\right) \quad \text{B(12)}$$

$$\text{or, } T\left(\frac{\partial V}{\partial T}\right)_P - V = (V-b)\left(1+\frac{2a(V-b)}{RTV^2}\right) - V \quad \text{B(13)}$$

$$= \left[V-b+\frac{2a}{RT}\left(\frac{V-b}{V}\right)^2\right] - V \quad \text{B(14)}$$

$$\text{or, } T\left(\frac{\partial V}{\partial T}\right)_P - V = V - b + \frac{2a}{RT} - V - V \quad \text{B(15)}$$

Neglecting b in comparison to V,

$$V - T\left(\frac{\partial V}{\partial T}\right)_P = \left(b-\frac{2a}{RT}\right) \quad \text{B(16)}$$

$$\mu_{J.T} = -\frac{1}{C_P}\left(\frac{\partial H}{\partial P}\right)_T \quad \text{B(17)}$$

$$dH = TdS + VdP. \quad \text{B(18)}$$

$$\text{Hence, } \left(\frac{\partial H}{\partial P}\right)_T = T\left(\frac{\partial S}{\partial P}\right)_T + V. \quad \text{B(19)}$$

But by Maxwell's relation,

$$\left(\frac{\partial S}{\partial P}\right)_T = -\left(\frac{\partial V}{\partial T}\right)_P. \quad \text{B(20)}$$

$$\text{Thus, } \left(\frac{\partial H}{\partial P}\right)_T = -T\left(\frac{\partial V}{\partial T}\right)_P + V. \quad \text{B(21)}$$

$$\text{Hence, } \left(\frac{\partial H}{\partial P}\right)_T = V - V + b - \frac{2a}{RT}. \quad \text{B(22)}$$

$$\mu_{JT} = -\frac{1}{C_P}\left(\frac{\partial H}{\partial P}\right)_T \quad B(23)$$

$$= -\frac{1}{C_P}\left(b - \frac{2a}{RT}\right) \quad B(24)$$

$$\text{or, } \mu_{JT} = \frac{1}{C_P}\left(\frac{2a}{RT} - b\right). \quad B(25)$$

Related Derivation :

$$C_P - C_V = T\left(\frac{\partial P}{\partial T}\right)_V\left(\frac{\partial V}{\partial T}\right)_P.$$

By van der Waals equation,

$$P = \frac{RT}{V-b} - \frac{a}{V^2}. \quad B(26)$$

$$\text{Hence, } \left(\frac{\partial P}{\partial T}\right)_V = \frac{R}{V-b}. \quad B(27)$$

From Equation B(12),

$$T\left(\frac{\partial V}{\partial T}\right)_P = (V-b)\left[1 + \frac{2a(V-b)}{RTV^2}\right]. \quad B(28)$$

$$\text{Hence, } C_P - C_V = \frac{R}{V-b} \times (V-b)\left[1 + \frac{2a(V-b)}{RTV^2}\right] \quad B(29)$$

$$= R\left[1 + \frac{2a(V-b)}{RTV^2}\right] \quad B(30)$$

$$= R\left(1 + \frac{2aV}{RTV^2}\right) \quad B(31)$$

$$= R\left(1 + \frac{2a}{RTV}\right) \quad B(32)$$

$$= R + \frac{2aP}{RT^2} \text{ (putting } V = \frac{RT}{P}\text{)}. \quad B(33)$$

Method C :

$$\mu_{JT} = -\frac{1}{C_P}\left(\frac{\partial H}{\partial P}\right)_T \quad C(1)$$

$$= -\frac{1}{C_P}\left[\left(\frac{\partial U}{\partial P}\right)_T + \left\{\frac{\partial (PV)}{\partial P}\right\}_T\right] \quad C(2)$$

$$= -\frac{1}{C_P}\left[\left(\frac{\partial U}{\partial V}\right)_T\left(\frac{\partial V}{\partial P}\right)_T + \left\{\frac{\partial (PV)}{\partial P}\right\}_T\right] \quad C(3)$$

For a van der Waals gas,

$$\left(\frac{\partial U}{\partial V}\right)_T = \frac{a}{V^2} \quad \text{C(4)}$$

Since, $dU = T\,dS - P\,dV$, C(5)

$$\therefore \quad \left(\frac{\partial U}{\partial V}\right)_T = T\left(\frac{\partial S}{\partial V}\right)_T - P. \quad \text{C(6)}$$

By Maxwell's relation,

$$\left(\frac{\partial S}{\partial V}\right)_T = \left(\frac{\partial P}{\partial T}\right)_V. \quad \text{C(7)}$$

$$\text{Thus,} \left(\frac{\partial U}{\partial V}\right)_T = T\left(\frac{\partial P}{\partial T}\right)_V - P. \quad \text{C(8)}$$

$$\text{Now,} \left(\frac{\partial P}{\partial T}\right)_V = \frac{R}{V-b}. \quad \text{C(9)}$$

$$\text{Thus,} \left(\frac{\partial U}{\partial V}\right)_T = \frac{R}{V-b} - \left(\frac{RT}{V-b} - \frac{a}{V^2}\right) \quad \text{C(10)}$$

$$= \frac{a}{V^2}. \quad \text{C(11)}$$

$$\text{Hence,} \left(\frac{\partial H}{\partial P}\right)_T = \frac{a}{V^2}\left(\frac{\partial V}{\partial P}\right)_T + \left[\frac{\partial(PV)}{\partial P}\right]_T. \quad \text{C(12)}$$

We know that

$$\left(P + \frac{a}{V^2}\right)(V - b) = RT \text{ [For van der Waals gas].}$$

Negleeting $\frac{ab}{V^2}$ in comparison to other terms and expanding,

$$PV - Pb + \frac{a}{V} = RT \quad \text{C(13)}$$

$$\text{or,} \left\{\frac{\partial(PV)}{\partial P}\right\}_T - b - \frac{a}{V^2}\left(\frac{\partial V}{\partial P}\right)_T = \left\{\frac{\partial(RT)}{\partial P}\right\}_T. \quad \text{C(14)}$$

$$\therefore \quad \left\{\frac{\partial(PV)}{\partial P}\right\}_T = b + \frac{a}{V^2}\left(\frac{\partial V}{\partial P}\right)_T. \quad \text{C(15)}$$

$$\text{Now} \quad \frac{\partial}{\partial P}\left(P + \frac{a}{V^2}\right) = \frac{\partial}{\partial P}\left(\frac{RT}{V-b}\right) \quad \text{C(16)}$$

$$\text{or, } 1 + \frac{\partial}{\partial P}\left(\frac{a}{V^2}\right) = \frac{\partial}{\partial P}\left(\frac{RT}{V-b}\right) \quad \text{C(17)}$$

$$\text{or, } 1 + \frac{\partial}{\partial V}\left(\frac{a}{V^2}\right)\left(\frac{\partial V}{\partial P}\right)_T = \frac{\partial}{\partial V}\left(\frac{RT}{V-b}\right)\left(\frac{\partial V}{\partial P}\right)_T \quad \text{C(18)}$$

$$\text{or, } 1 + \left(-\frac{2a}{V^3}\right)\left(\frac{\partial V}{\partial P}\right)_T = -\frac{RT}{(V-b)^2}\cdot\left(\frac{\partial V}{\partial P}\right)_T. \quad \text{C(19)}$$

$$\text{Thus, } \left(\frac{\partial V}{\partial P}\right)_T = \frac{1}{\dfrac{2a}{V^3} - \dfrac{RT}{(V-b)^2}} \quad \text{C(20)}$$

$$= \frac{1}{\dfrac{2a}{V^3}\left\{1 - \dfrac{RTV^3}{2a(V-b)^2}\right\}} \quad \text{C(21)}$$

$$= \frac{1}{\dfrac{2a}{V^3}\left(1 - \dfrac{RTV}{2a}\right)}. \quad \text{C(22)}$$

$$\text{Now, } \left(\frac{\partial H}{\partial P}\right)_T = \frac{a}{V^2}\left(\frac{\partial V}{\partial P}\right)_T + \left\{\frac{\partial (PV)}{\partial P}\right\}_T \quad \text{C(23)}$$

$$= \frac{a}{V^2}\left(\frac{\partial V}{\partial P}\right)_T + \left(b + \frac{a}{V^2}\right)\left(\frac{\partial V}{\partial P}\right)_T \quad \text{C(24)}$$

$$= \left(b + \frac{2a}{V^2}\right)\left(\frac{\partial V}{\partial P}\right)_T \quad \text{C(25)}$$

$$= \left(b + \frac{2a}{V^2}\right)\frac{1}{\dfrac{2a}{V^3}\left[1 - \dfrac{RTV^3}{2a(V-b)^2}\right]}. \quad \text{C(26)}$$

Since $V >> b$, $\dfrac{V^3}{(V-b)^2} = V.$

$$\text{Thus, } \left(\frac{\partial H}{\partial P}\right)_T = b + \frac{V}{1 - \dfrac{RTV}{2a}}. \quad \text{C(27)}$$

$$\text{Now } \frac{RTV}{2a} >> 1. \quad \text{C(28)}$$

$$\text{Hence, } \left(\frac{\partial H}{\partial P}\right)_T = b - \frac{V}{\dfrac{RTV}{2a}} \quad \text{C(29)}$$

$$= \left(b - \frac{2a}{RT}\right). \quad \text{C(30)}$$

Thus, $\mu_{JT} = -\frac{1}{C_P}\left(\frac{\partial H}{\partial P}\right)_T = -\frac{1}{C_P}\left(b - \frac{2a}{RT}\right)$ C(31)

$$= \frac{1}{C_P}\left(\frac{2a}{RT} - b\right). \quad \text{C(32)}$$

We cannot proceed further to derive $(C_P - C_V)$ along this path, since both $\left(\frac{\partial P}{\partial T}\right)_V$ and $\left(\frac{\partial V}{\partial T}\right)_P$ have to be derived anew.

However, $(C_P - C_V)$ can also be derived in a separate method employing another formula.

Derivation of $(C_P - C_V)$:

We know that,

$$C_P - C_V = \left[\left(\frac{\partial U}{\partial V}\right)_T + P\right]\left(\frac{\partial V}{\partial T}\right)_P. \quad \text{C(33)}$$

Now, $dU = T\,dS - P\,dV$.

$$\therefore \left(\frac{\partial U}{\partial V}\right)_T = T\left(\frac{\partial S}{\partial V}\right)_T - P. \quad \text{C(34)}$$

By Maxwell relation,

$$\left(\frac{\partial S}{\partial V}\right)_T = \left(\frac{\partial P}{\partial T}\right)_V. \quad \text{C(35)}$$

Hence, $\left(\frac{\partial U}{\partial V}\right)_T = T\left(\frac{\partial P}{\partial T}\right)_V - P.$ C(36)

For a van der Waals gas,

$$P = \frac{RT}{V-b} - \frac{a}{V^2}. \quad \text{C(37)}$$

$$\therefore \left(\frac{\partial P}{\partial T}\right)_V = \frac{R}{V-b}. \quad \text{C(38)}$$

Putting Equation C(37) and Equation C(38) in Equation C(36),

$$\left(\frac{\partial U}{\partial V}\right)_T = \frac{RT}{V-b} - \left[\frac{RT}{V-b} - \frac{a}{V^2}\right] \quad \text{C(39)}$$

$$= \frac{a}{V^2}. \quad \text{C(40)}$$

Putting Equation C(40) in Equation C(33),

$$C_P - C_V = \left(P + \frac{a}{V^2}\right)\left(\frac{\partial V}{\partial T}\right)_P. \quad \text{C(41)}$$

Again expanding van der Waals equation and neglecting $\frac{ab}{V^2}$ in comparison to other terms,

$$PV - Pb + \frac{a}{V} = RT \tag{C(42)}$$

$$\text{or, } P\left(\frac{\partial V}{\partial T}\right)_P - \frac{a}{V^2}\left(\frac{\partial V}{\partial T}\right)_P = R \tag{C(43)}$$

$$\text{or, } \left(\frac{\partial V}{\partial T}\right)_P = \frac{R}{P - \frac{a}{V^2}} \tag{C(44)}$$

$$= \frac{R}{\left(P + \frac{a}{V^2}\right) - \frac{2a}{V^2}} \tag{C(45)}$$

$$= \frac{R}{\frac{RT}{V-b} - \frac{2a}{V^2}} \tag{C(46)}$$

$$= \frac{R}{\frac{RT}{V-b}\left[1 - \frac{2a(V-b)}{RTV^2}\right]}. \tag{C(47)}$$

Since $V >> b$,

$$\left(\frac{\partial V}{\partial T}\right)_P = \frac{V-b}{T\left(1 - \frac{2a}{RTV}\right)} \tag{C(48)}$$

$$= \frac{V-b}{T}\left(1 - \frac{2a}{RTV}\right)^{-1}. \tag{C(49)}$$

$$\text{Thus, } \left(\frac{\partial V}{\partial T}\right)_P = \frac{V-b}{T}\left(1 + \frac{2a}{RTV}\right). \tag{C(50)}$$

$$\text{Hence, } C_P - C_V = \frac{RT}{V-b}\left[\frac{V-b}{T}\left(1 + \frac{2a}{RTV}\right)\right] \tag{C(51)}$$

$$= R\left(1 + \frac{2a}{RTV}\right). \tag{C(52)}$$

Replacing V in denominator by ideal gas value,

$$C_P - C_V = R + \frac{2aP}{RT^2}. \tag{C(53)}$$

Which one of the above methods will you choose for deriving μ_{JT} and why? Ask it yourself.

Units and Dimensions

For details see Physical Chemistry—Dr S. Pahari—Chapter 1

Quantities are to be reported with units consistent with the dimension of the quantity. In International System of Units (SI) there are 7 base quantities which have their own dimensions. Out of these we will never use in this book the dimension, J, and unit, candela, of the base quantity luminous intensity (Iv). So, the remaining 6 base quantities with symbols to denote them and their dimensions along with symbols and definitions of their units are given below :

Physical quantities	*Symbol for quantities*	*Symbol for dimension*	*Name of unit*	*Symbol for unit*	*Definition*
Length	l	L	metre	m	length of the path travelled by light in vacuum during a time interval of 1/299792458 of a sec
Mass	m	M	kilogram	kg	mass of the cylinder, kept in Paris by International Bureau (BIPM) of Weights and Measures, made of Pt-Iralloy
Time	t	T	second	s	duration of 9,192,631,770 periods of the radiation corresponding to the specified transition of the caesium 133 atom at rest at a temperature of 0° K
Electric current	I	I	ampere	A	magnitude of the constant current which, if maintained between two parallel plates of infinite length and negligible circular crosssection and placed one metre apart in vacuum, would produce will exert a force of 2×10^{-7} newton per metre-length.
Thermodynamic temperature	T	Θ	kelvin	K	fraction of 1/273.16 of the thermodynamic temperature of triple point (273.16 K) of water.
Amount of substance	n	N	mole	mol	amount contains as many atoms as there are in 0.012 kg of ^{12}C.

All other quantities are expressed in these units or in terms of derived units obtained algebraically by multiplication or division. Some derived units have their special symbols. The principal derived units are shown below (to be used in problems) :

Quantity	*Unit*	*Symbol*	*Definition*
Force	newton	N	kg. $m.sec^{-2}$
Work, energy, heat	joule	J	$N.m = kg.m^2sec^{-2}$
Power	watt	W	$J. sec^{-1}$
Pressure	pascal	Pa	$N.m^{-2} = kg.m^{-1}.sec^{-2}$
Electric charge	coulomb	C	A.s
Electric potential difference	volt	V	$kg.m^2.sec^{-3}.A^{-1}$ $= J.A^{-1}.sec^{-1} = J.C^{-1}$
Electric resistance	ohm	Ω	kg $m^2.sec^{-3}.A^{-2}$ $= V.A^{-1} = J.C^{-1}.A^{-1}$
Electric capacitance	farad	F	$A.s.V^{-1}$ $= kg^{-1}.m^{-2}.sec^4.A^2$
Magnetic flux density	tesla	T	$kg.sec^{-2}.A^{-1}$ $= N.A^{-1}.m^{-1}$
Frequency	hertz (cycle/second)	Hz	sec^{-1}
Megnetic flux	weber	Wb	V.s $= kg.m^2.sec^{-2}.A^{-1}$
Inductance	henry	H	$V.sec.A^{-1}$ $= kg.m^2.sec^{-2}.A^{-2}$ $= J.A^{-2}$

SI PREFIXES

Submultiple	*Symbol*	*Prefix*	*Multiple*	*Prefix*	*Symbol*
10^{-1}	d	deci	10	deca	da
10^{-2}	c	centi	10^2	hecto	h
10^{-3}	m	milli	10^3	kilo	k
10^{-6}	μ	micro	10^6	mega	M
10^{-9}	n	nano	10^9	giga	G
10^{-12}	p	pico	10^{12}	tera	T
10^{-15}	f	femto			
10^{-18}	a	atto			
10^{-21}	z	zepto			

Writing Symbols

Abbreviated units are never plural and also do not end with dots.

Thus, 10 kg, and not 10 kg. or 10 kgs.

In case of derived units, there are three methods of writing, e.g., joule is N.m or Nm or N × m. The first method is followed here.

Conversion of Units

C.G.S. and practical system of units are commonly used. The C.G.S. unit of energy is erg: $1 J = 10^7$ erg. The conversion factors for mechanical quantities in the two units may be done by writing dimensions. The dimension of energy is $F.l = ML^2T^{-2}$.

$$1 \text{ joule} = 1 \text{kg.m}^2.\text{s}^{-2} = x \text{ erg} = x \text{ gm.cm}^2.\text{s}^{-2}$$

$$\therefore x = 1\frac{\text{kg}}{\text{gm}}.\left(\frac{\text{m}}{\text{cm}}\right)^2 = 10^3\ \frac{\text{gm}}{\text{gm}}\left(\frac{10^2\text{cm}}{\text{cm}}\right)^2 = 10^3 \times 10^4 = 10^7.$$

There are many practical units. As for example, pressure is expressed in atm—it is the weight of 76 cm high Hg column.

$$P = d\,g\,h$$

$$= 13.5951\ \frac{\text{gm}}{\text{cm}^3} \times 980.665\ \frac{\text{cm}}{\text{sec}^2} \times 76 \text{ cm}$$

$$= 1.01325 \times 10^6\ \frac{\text{dyne}}{\text{cm}^2}$$

$$= 1.01325 \times 10^5 \text{ Pa}.$$

$$1\text{Pa} = 1\ \frac{\text{N}}{\text{m}^2} \text{ (dimension of pressure} = ML^{-1}T^{-2})$$

$$= \text{kg.m}^{-1}.\text{s}^{-2}$$

$$= x\ \frac{\text{dyne}}{\text{cm}^2} = x \text{ gm.cm}^{-1}.\text{sec}^{-2}.$$

$$\therefore x = \frac{1\text{kg}}{\text{g}}.\frac{\text{cm}}{\text{m}} = 10^3\frac{\text{gm}}{\text{gm}} \times \frac{1\text{cm}}{10^2\text{cm}} = 10.$$

$$P = 1.01325 \times 10^2 \text{ kPa (kilo pascal)}$$

$$= 1.01325 \text{ bar (1bar} = 10^6\ \frac{\text{dyne}}{\text{cm}^2} = 10^5 \text{ Pa}).$$

$$1 \text{ torr} = 1\text{mm of Hg} = \frac{1.01325 \times 10^5}{760} \text{ Pa} = 133.32 \text{ Pa}.$$

Energy : $1 \text{ l-atmosphere} = 10^{-3}\text{m}^3 \times 1.01325 \times 10^5 \frac{\text{N}}{\text{m}^2}$

$$= 1.01325 \times 10^2 \text{ J}.$$

1 bar-litre = 100 J (exactly).

S.I. system is coherent in the sense that if in an equation there are 2 variables or more and all are expressed in S.I., then the resulting quantity will be in S.I.

Some of the conversion factors of some special units are indicated in the following table :

Physical quantity	*Name of Unit*	*Symbol*	*Equivalent in SI unit*
Length	Angstorm	Å	10^{-10} m(10^{-1}nm)
Energy	electron-volt	eV	1.602189×10^{-19} J
			= 8065.5 cm^{-1}
			= 96.485 kJ.mol^{-1}
	wave number	cm^{-1}	1.986×10^{-23}J
	calorie	cal	4.1845 J
	erg	erg	10^{-7} J
	l-atmosphere	l-atm	101.325 J
	bar-litre	b-litre	100·0 J
Force	dyne	gm.cm.s^{-2}	10^{-7}N
Pressure	atmosphere	atm	101.325 kPa
	bar	bar	100 kPa
	torr	tor	133.322 Pa
Electric charge	e.s.u. (electrostatic unit)		~3.334×10^{-10} C
Dipole moment	debye	10^{-18} esu-cm	~3.334×10^{-30} C.m
Magnetic flux density	gauss	G	10^{-4} T
Magnetic flux	maxwell	Mx	10^{-8} Wb
Viscocity	poise	P	10^{-1} Pl

The values of standard quantities that will be used in the book with symbols are :

Name	*Symbol*	*Values*
1. Velocity of light	c	2.9979×10^{8}m.sec^{-1}
2. Electronic charge	e	1.6022×10^{-19} C (coulomb)
3. Planck constant	h	6.626×10^{-34}J.sec
	$\hbar = \dfrac{h}{2\pi}$	1.0546×10^{-34}J.sec
4. Avogadro number	N_A	$6.022\times10^{23}$$mol^{-1}$
5. Mass of one atom of $\frac{1}{12}$ of C^{12} isotope	1 a.m.u.	1.6606×10^{-27} kg
6. Mass of electron	m_e	9.109×10^{-31} kg
7. Mass of proton	m_p	1.6726×10^{-27} kg
8. Mass of neutron	m_n	1.6750×10^{-27} kg
9. Faraday	F	9.6484×10^{4} C.mol^{-1}

Contd.

	Name	Symbol	Values
10.	Boltzmann constant	k	1.3807×10^{-23} J.K^{-1}
11.	Gas constant	R	8.314 J.K^{-1}.mol^{-1}
			0.0821 l-atm.K^{-1}.mol^{-1}
			1.987 cal.K^{-1}.mol^{-1}
12.	Acceleration due to gravity	g	9.80665 m.sec^{-2}
13.	Gravitational constant	G	6.673×10^{-11} N.m^2.kg^{-2}
14.	Standard pressure	P^o	1 bar
	Standard concentration	C^o	1 mol.l^{-1} of solution
	Standard molality	m^o	1 mol.kg^{-1}
15.	NTP, STP		1 atm, 273.15 K (old)
			1 bar, 273.15 K (new)
16.	SATP		1 bar, 298.15 K

The symbols for quantities commonly used in the book are listed below :

Quantity	Symbol	Quantity	Symbol	
Number of molecules	N	Time	t	
Number of molecules per unit volume	N'	Temperature °C	θ	
Root mean square (r.m.s.) velocity	$\bar{c}$	Number of molecules in zero energy level	N_0	
Concentration in		E.M.F.	E	
molarity	C	Internal energy	U	per mole U_m
molality	m	Enthalpy	H	per mole H_m
normality	N	Entropy	S	per mole S_m
Equivalent conductance	Λ	Helmholtz	F	per mole F_m
Ion conductance	$\lambda_\pm$	Gibbs free energy	G	per mole G_m
Average velocity	$\bar{v}$	Number of moles	n	
Most probable speed (mps)	$\bar{u}$	Viscosity coefficient	η	
Ionic mobility	$u_\pm$	Mean free path	λ	
Ionic velocity	$v_\pm$	Density	ρ	
Volume	V	Molar mass	M	
Any velocity	v	Transport number	$t_i(t_\pm)$	
Molar volume	V_m	Charge	Q	
Conductance	L	Coulomb	C	
Resistance	R	Wavelength	λ	
Pressure	P	Frequency	ν	
Momentum	$\bar{p}$	Wave number	$\bar{\nu}$	
Temperature	T	C_p/C_v	γ	
		Avogadro Number	N_A	

CHAPTER 2

GAS

A Gas Laws

A1. Required Formulae

1. Ideal Gas Laws

1a. $PV = nRT = \frac{W}{M}RT,$ where W is the weight of the gas taken and M its molecular weight.

1b. $PM = \frac{W}{V}RT = \rho RT,$ where ρ is the density.

1c. $\frac{P_1}{P_2} = \frac{V_2}{V_1} \cdot \frac{T_1}{T_2}$

2. Real Gas Laws

2a. van der Waals equation : $\left(P + \frac{an^2}{V^2}\right)(V - nb) = nRT.$

2b. $\frac{RT_c}{P_c V_c} = \frac{8}{3}$ for van der Waals gas, where T_c, P_c, V_c, etc. are critical temperature, pressure and volume respectively.

2c. Compressibility factor is Z.

$$Z = \frac{PV}{nRT}, \; n = \frac{PV}{ZRT}$$

$Z = 1$, ideal gas

$Z > 1$, less compressible than ideal gas

$Z < 1$, more compressible than ideal gas.

3. Dieterici equation of state : $P = \frac{nRT}{V - nb} e^{-\frac{nA}{RTV}}$

4. Virial equation of state :

$$Z = 1 + \frac{A}{V} + \frac{B}{V^2} + \frac{C}{V^3} + \dots$$

. **A, B, C, etc are 2nd, 3rd, and 4th Virial coefficients.**

5. Redlich-Kwong Equation :

$$\left(P + \frac{An^2}{\sqrt{T}V(V+nB)}\right)(V - nB) = nRT.$$

6. Law of corresponding states :

$$Z = \frac{PV}{nRT} = F(T_r, P_r),$$

where $T_r = \frac{T}{T_c}$, $P_r = \frac{P}{P_c}$, $V_r = \frac{V}{V_c}$.

At the same T_r and P_r, all gases have same V_r.

7. Near Boyle temperature, T_B, a real gas behaves more ideally. However, in the high temperature range the major deviation from ideality is closer to $2T_B$. For a van der Waals or Dicterici gas,

$$T_B = \frac{a}{Rb}.$$

8. *a, b* of van der Waals equation in terms of T_c, P_c,

$$b = \frac{RT_c}{8P_c};\quad a = \frac{27R^2T_c^2}{64P_c}$$

$$V_c = 3b,\quad P_c = \frac{a}{27b^2},\ T_c = \frac{8a}{27Rb}.$$

A2. Short Questions

1. What is the dimension of change?

Ans. : Change is dimensionless. If the change in x is Δx, the dimension of this change is the dimension of x.

2. When does a real gas behave like an ideal gas?

Ans. : In a real gas there is attraction and repulsion between molecules. When these two are same, the real gas behaves like an ideal gas. Boyle temperature T_B of a real gas is the temperature at which a real gas behaves like an ideal gas up to certain range of pressures. At any temperature, every real gas tries to behave like an ideal gas as $P \to 0$. At very high temperatures also a real gas may behave like an ideal gas.

3. Define Compressibility factor. State its significance.

Ans. : Compressibility factor, Z, is defined as the ratio of observed molar volume of a gas to the molar volume that an ideal gas would occupy at the same temperature and pressure. It may also be defined as the ratio of pressure that one mole of gas exerts to the pressure that one mole of an ideal gas would exert at the same temperature and volume.

$$\therefore\quad Z = \frac{V_{\text{real}}}{V_{\text{ideal}}} = \frac{PV_{\text{real}}}{nRT} = \frac{P_{\text{real}}}{P_{\text{ideal}}} = \frac{P_{\text{real}}V}{nRT}.$$

Significance : Z is measure of the deviation from ideal gas behaviour. For an ideal gas, Z is equal to one at any temperature or pressure. A real gas can have negative or positive deviation. Negative deviation ($Z < 1$) signifies that the real gas is more compressible than ideal gas or $V_{real} < V_{ideal}$ whereas positive deviation signifies less compressibility of real gas compared to ideal gas, i.e., $V_{real} > V_{ideal}$. However, at Boyle temperature, T_B, Z becomes 1 for a certain range of pressure when $V_{real} = V_{ideal}$.

4. If P and V are the pressure and molar volume of a real gas and P_{id}, V_{id} are those for an ideal gas, then can we write $P_{id}V = PV_{id}$ = constant?

Ans. : Acording to the definition of Z (See Q. 3)

$$Z = \frac{V}{V_{ideal}} = \frac{P}{P_{ideal}}$$

or, $$Z = \frac{P_{ideal}V}{nRT} = \frac{PV_{ideal}}{nRT}.$$

Thus, $P_{ideal}V = PV_{ideal} = ZnRT$ = constant as T is constant.

5. Is Z in $PV = ZnRT$ extensive?

Ans. : For a particular gas under a particular set of conditions Z is fixed, and in case of Z parts are not added to give the whole, thus, Z is intensive.

6. What is the basic difference between volume and molar volume?

Ans. : Volume is an extensive property whereas molar volume is an intensive property.

7. Is T_B higher than T_C?

Ans. : Critical temperature is the temperature above which a gas cannot be liquified by application of pressure alone. But attractive forces still remain stronger than repulsive forces. A gas above T_C is a gas, below T_C it is vapour. Gases cannot be liquified because K. E. is so high that they cannot be brought closer. At Boyle temperature, T_B, attractive force is balanced by repulsive force—a gas behaves like an ideal gas. Since repulsion increases its importance with increase in temperature, T_B is certainly higher than T_C. Thus T_B has to be higher than T_C.

8. What are the dimensions and S.I. units of a, b and R?

Ans. : The van der Waals equation is $\left(P + \frac{an^2}{V^2}\right)(V - nb) = nRT$. Thus, $\frac{an^2}{V^2}$ has the dimension of pressure, $ML^{-1}T^{-2}$. Hence, a has the dimension of ML^5T^{-2} mole^{-2} and hence, SI unit is J.m^3.mol^{-2}.

nb has the dimension of volume and so the dimension of b is L^3 mole^{-1} and S1 unit is m^3.mol^{-1}.

R has the dimension of energy per deg per mole, so it is $ML^2T^{-2}K^{-1}$mole^{-1}. In SI unit it is J.K^{-1}.mol^{-1}.

9. Can we liquify a van der Waals gas for which $a = 0$?

Ans. : There is no attraction between the molecules of a gas for which $a = 0$, and thus it cannot be liquified.

10. "A real gas in a container starts behaving ideally as temperature is raised"—Comment.

Ans. : In a real gas molecule, there are attractive as well as repulsive forces between the molecules. With increase in temperature, the kinetic energy of the molecules increases and hence chaos increases. At a particular temperature, T_B, called Boyle temperature, the intermolecular attraction becomes equal to repulsive forces. The gases behave ideally over a wide range of pressures. As temperature still increases, only intermolecular repulsive forces are important up to twice the Boyle temperature. Above that temperature, intermolecular repulsive forces also start losing importance. Only the chaos of motion remains. Thus, as temperature increases above that, the gas starts behaving ideally.

11. "van der Waals equation takes into account only the attractive forces between molecules, and repulsive interactions are not considered".—Comment.

Ans. : The above assertion is incorrect. No doubt, the van der Waals constant '*a*' takes into account the attractive forces. '*b*' in van der Waals equation is correction factor for volume of gas molecules—the molecules cannot come closer than *2r*, where *r* is the radius. There is infinite repulsion at touch. Thus both attractive and repulsive forces are considered.

12. Can you suggest a condition under which a van der Waals gas will behave like an ideal gas at all temperatures?

Ans. : A van der Waals gas behaves like an ideal gas when $b - \dfrac{a}{RT} = 0$. Thus if a van der Waals gas has variables *a* and *b* so that at all temperatures, $b = \dfrac{a}{RT}$, then the van der Waals gas will behave like an ideal gas at all temperatures.

13. How do you characterize the deviation from ideality of a real gas?

Ans. : We define the compressibility factor, $Z = \dfrac{V_{\text{real}}}{V_{\text{ideal}}} = \dfrac{PV}{nRT}$. For ideal gas, $Z = 1$. If for a gas $Z > 1$, then it is less compressible than ideal gas. If, on the other hand, $Z < 1$, the gas is more compressible than ideal gas. Thus, the more closer is *Z* to 1, the more nearly ideal is the gas.

At the Boyle temperature, T_B, all gases behave like an ideal gas within a certain range of pressure. Thus, the closer is the temperature to T_B, the more ideal is the gas.

Z goes on increasing from T_B to $2T_B$. Above $2T_B$, *Z* starts decreasing reaching $Z = 1$ as $T \to \infty$. Thus, at very high temperatures every gas behaves ideally.

Thus, we characterise deviation by calculating *Z*, and/or by noting the temperature.

14. Express the compressibility factor *Z* of a van der Waals gas in terms of reduced temperature (T_r) and pressure (P_r).

Ans. : *Z* of a van der Waals gas,

$$= \frac{P\overline{V}}{RT} = \frac{V}{V-b} - \frac{a}{RTV} = \left(1 - \frac{b}{V}\right)^{-1} - \frac{a}{RTV}.$$

Replacing *V* in the denominator by ideal gas value $V = \dfrac{RT}{P}$

$$Z = \left(1 - \frac{bP}{RT}\right)^{-1} - \frac{aP}{(RT)^2}.$$

But $b = \dfrac{RT_c}{8P_c}$ and $a = \dfrac{27R^2T_c^2}{64P_c}$.

$$\therefore \; Z = \left(1 - \frac{RT_c P}{RT 8 P_c}\right)^{-1} - \frac{27 R^2 T_c^2 P}{64 P_c R^2 T^2}$$

$$= \left(1 - \frac{P_r}{8T_r}\right)^{-1} - \frac{27 P_r}{64 T_r^2}.$$

15. Is it possible to have gaseous dissociation without the gas showing any abnormal vapour density?

Ans. : In majority of cases of dissociation, the number of molecules increases, e.g., $COCl_2 = CO + Cl_2$, $PCl_5 = PCl_3 + Cl_2$, etc. and thus the volume increases at constant pressure, and hence the density, vapour density and apparent molecular weight decrease. However, if dissociation takes place without change in the number of molecules, then there will be no decrease of density or vapour density—there will be no abnormal vapour density. Some examples are $2HI = H_2 + I_2$, $2NO = N_2 + O_2$

16. "The compressibility factor of a real gas is greater than unity at high pressure and temperature".—Comment.

Ans. : There is a force of attraction and repulsion between molecules in a real gas. The repulsive force is of very short range and acts when the molecules come closer at high pressure, $Z = \frac{PV}{RT} > 1$. At high temperature the molecules move with high velocity and hence, attraction between gas molecules becomes less important. They collide more often, and hence, only repulsive forces are important and $Z > 1$. At a very high temperature, however, intermoleculer forces lose importance.

17. C_v of Cl_2 is greater than N_2 at room temperature, but at high temperature both the gases have same C_v. Explain.

Ans. : Heat capacity, C_v, is defined as the amount of heat required to raise the temperature of a system through 1°K at constant volume and the heat absorbed by the system increases its internal energy, U. Hence, $C_v = \frac{dQ_v}{dT} = \left(\frac{\partial U}{\partial T}\right)_v$. Internal energy is composed of kinetic energy and potential energy of constituent molecules. Intermolecular attractions and repulsions contribute to the potential energy. At room temperature attractions are predominant. For non-polar molecules like N_2 and Cl_2 dispersion energy mainly contribute to potential energy. Being larger in size, Cl_2 molecule is more polarizable than N_2 and thus, former has more negative dispersion energy. However both the gas molecules have the same positive contribution of kinetic energy to internal energy since both are diatomic molecules. Hence, at ordinary temperature, internal energy is more negative i.e., has greater absolute value in case of Cl_2. For the same increase in temperature at room temperature increase in U for Cl_2 is greater than N_2 and it absorbs more heat. Thus C_v of Cl_2 is greater than N_2 at room temperature. At high temperature, however, intermolecular atractions of both the gases become negligibly small while they have, still, the same kinetic energy. Hence, they have same internal energy at high temperature and, thus their change in internal energy is same for the same change in temperature. So, now they have same C_v.

18. The experimental value of $\frac{RT}{V}$ is 2.0 for 1 mole of a non-ideal gas. The gas is at 1 atm and below its critical temperature. If the pressure is now halved at constant temperature, would the new volume be twice of original valume?

Ans. : For the given A gas, $Z = \frac{PV}{RT} = 0.5$. at $P = 1$ atm below critical temperature T_c.

Below critical temperature at 1 atm pressure, Z increases with decrease in pressure. Thus, the new volume will be more than twice the original volume.

19. A non-ideal gas is at T_c but above P_c. Would its volume be halved of the original volume if pressure is doubled at constant T?

Ans. : At critical temperature above P_c, Z increases with increase in pressure. So its volume will be less than halved, if P is doubled.

20. Two separate bulbs are filled with an ideal gas A and a non-ideal gas B respectively in such a way that PV remains same. B is below its Boyle temperature. Which gas has greater temperature?

Ans. : For gas A, $Z = \frac{PV}{RT_A} = 1$ and $PV = RT_A$.

For gas B, $Z = \frac{PV}{RT_B} < 1$ (since gas is below its Boyle temperature)

or, $\frac{RT_A}{RT_B} < 1$ (since $PV = RT_A$).

$\therefore\ T_A < T_B$.

Thus, B has greater temperature than A.

21. Doubling the temperature at constant P will be more than double the volume if the gas is below critical temperature and the volume will be less than double if it is above its Boyle temperatur.—Explain. Can the gas be ideal?

Ans. : For an ideal gas, Z, the compressibility factor, is always one. Now, $Z = \frac{PV}{RT}$

So if T is doubled at constant P, V would exactly doubles. It could not be more or less than double. So the gas is certainly not an ideal gas.

Below critical temperature (T_C), at constant pressure Z will increase with increasing T. Let Z_1 be the corresponding compressibility factor at temperature, T_1 and volume, V_1. If now temperature is doubled to T_2 at constant pressure to attain new volume V_2 and Z_2 is the new compressibility factor, then $\frac{Z_1}{Z_2} = \frac{T_1 V_2}{T_2 V_1} = \frac{V_2}{2V_1}$ and $Z_2 > Z_1$. Hence $V_2 > V_1$.

Thus V should have more than doubled at T is doubled.

Above temperature twice as high as Boyle temperature, T_B, Z decreases with increases in temperature at constant P, since then Z approaches the ideal gas value. Thus, at constant pressure, above T_B at higher T, $Z_2 < Z_1$. Hence, $V_2 < 2V_1$. Thus, V should have less than doubled as T is doubled.

A3. Numercial Problems

1. What is 200 giga pascal in atmosphere and bar?

Soln. : We know that

1 atm = 1.01325×10^5 Pa

1 Pa = $9.86679822 \times 10^{-6}$ atm.

Thus, 200 giga Pa = 200×10^9 Pa

= $9.87 \times 10^{-6} \times 200 \times 10^9$ atm = **1.97×10^6 atm.**

Again, 1 atm = 1.01325 bar

Hence, 200 giga Pa = $1.97 \times 10^6 \times 1.01325$ bar.

= 1.996×10^6 bar = 199.6×10^3 bar = **199.6 kilo bar.**

2. Convert 1013.25 millibar to atm, Pa and torr.

Soln. : 1 atm = 1.01325 bar = 1013.25×10^{-3} bar

Thus, 1013.25 millibar = 1 atm = 760 torr = 1.01325×10^5 Pa.

3. At what temperature are the Celsius and the Fahrenheit scales equal?

Soln. : We know that °C scale is divided into 100 degrees (0-100) and °F scale is divided into 180 degrees from 32 to 212.

Thus, $\frac{t\,°C}{100} = \frac{t\,°F - 32}{180}$

or, $9t\,°C = 5t\,°C - 160$

By the problem, $t\,°C = t\,°F = x$

Thus, $9x = 5x - 160$ or, $4x = -160$, $\therefore x = -40°$

4. Air is approximately 21% O_2 and 79% N_2. If 8.0 gm H_2 is now added at 0 °C in a 22.4 litre flask initially filled with air at 1 atm, what is the average molar weight of the mixture?

Soln. : At NTP volume occupied by 1 gm-mole of any gas = 22.4 litres.

Now, by the problem, contents of 1 gm-mole of air

= (0.79×28) gm of N_2 + (0.21×32) gm of O_2

= 22.12 gm N_2 + 6.72 gm O_2

Hence, 1 gm-mole of air = (22.12 + 6.72) gm = 28.84 gm.

Now, 8 gm H_2 i.e. 4 moles of H_2 are added.

Now, mass of air = (28.84 + 8) gm = 36.84 gm.

Total number of moles of gases in air = 4 mole H_2 + 1 mole air = 5 moles.

Hence, average molar mass = $\frac{36.84\text{ g}}{5\text{ mol}}$ = 7.36 gm.mol^{-1}.

5. Calculate the number of molecules in a container of 0.1 cc volume at 25 °C and at a pressure of 10^{-13} torr.

Soln. : $P = 10^{-13}$ torr = $\frac{10^{-13}}{760}$ atm

Assuming the gas to be ideal, $n = \frac{PV}{RT}$.

Here, V = 1 cc, R = 0.08206 l.atm.deg^{-1}.mol^{-1}, T = (273.15 + 23)K = 298.15 K

Thus, $n = \frac{10^{-13}\text{ atm} \times 0.1\text{ cc} \times 10^{-13}\text{ l/cc}}{760 \times 0.08206\text{ l.atm.K}^{-1}.\text{mol}^{-1} \times 298.15\text{K}}$

= $5.37799112 \times 10^{-22}$ mole.

Total number of molecules present = $6.022 \times 10^{23} \times 5.38 \times 10^{-22}$

= 323.86 ≈ **323 molecules.**

6. A mixture of H_2 and N_2 has a density of 0.2 kg.m^{-3} at 27 °C and a pressure of 500 cm of Hg. The total volume of the mixture is one cubic metre. Calculate the mole fraction of N_2 and H_2.

Soln. : Let the total pressure be P.

$$\text{Thus, } P = \frac{500\,\text{cm of Hg}}{760\,\text{cm of Hg.atm}^{-1}}$$

$$= \frac{500}{760}\,\text{atm} \times 1.01325 \times 10^5 \frac{\text{Pa}}{\text{atm}}$$

$$T = (27 + 273)\text{ K} = 300\text{ K}$$

$$V = 1\text{ m}^3$$

$$\text{Now, } n = \frac{PV}{RT} = \frac{500 \times 101325\,\text{Pa} \times 1\,\text{m}^3}{760 \times 8.314\,\text{J.K}^{-1}.\text{mol}^{-1} \times 300\,\text{K}}$$

$$= \frac{500 \times 101325\,\text{N.m}^{-2} \times 1\,\text{m}^3}{760 \times (8.314 \times 300)\,\text{J.mol}^{-1}} = 26.72648\text{ mol..}$$

Let number of moles of N_2 be n_{N_2}.

Thus, number of moles of $H_2 = n_{H_2} = 26.7 - n_{N_2}$

$M_{N_2} = 0.028$ kg.mol^{-1}

$M_{H_2} = 0.002$ kg.mol^{-1}

By the problem,

$$0.2\text{ kg} = 0.028\text{ kg.mol}^{-1}\, n_{N_2}\text{ mol} + 0.002\text{ kg.mol}^{-1}(26.7 - n_{N_2})\text{ mol}$$

$$\text{or,}\quad 0.2 - 0.0534 = 0.026\; n_{N_2}$$

$$\therefore\quad n_{N_2} = \frac{0.1466}{0.026} = 5.6$$

$$\text{Thus, } X_{N_2} = \frac{5.6}{26.7} = \mathbf{0.2}$$

$$\therefore\; X_{H_2} = 1 - 0.2 = \mathbf{0.8}.$$

7. 15.0 l of nitrogen measured at 300 torr is compressed to 1.00 at a particular temperature. Calculate the pressure of the gas.

Soln. : $300\text{ torr} = \frac{300}{760} \times 1.01325 \times 10^5\text{ Pa}$

$$= 3.99967 \times 10^4\text{ Pa.}$$

Now at a particular temperature, PV = constant.

$$\therefore\quad P_1V_1 = P_2V_2$$

$$\text{i.e.,}\quad P_2 = \frac{P_1V_1}{V_2} = \frac{3.99967 \times 10^4\,\text{Pa} \times 151}{11}$$

$$= 5.999506579 \times 10^5\text{ Pa} = 6{\cdot}0000\ (\pm 0{\cdot}001) \times 10^5\text{ Pa}$$

$$= \mathbf{600\ kPa.}$$

8. A faulty barometer reads a pressure of 750 mm of Hg when the actual pressure is 760 mm of Hg. The evacuated part of the barometer is 20.0 mm long. Calculate the actual pressure when the barometer reads 730 mm.

Soln. : The barometer may be faulty if there is some gas in the open space. The pressure exerted by the gas is (760 – 750) = 10 mm of Hg when the open space has a volume of 20 A mm^3 (area of barometric tube is A sq. mm). When the barometer reads 730 mm, then the open volume is $[(750 + 20) - 730]A = 40A$ mm^3.

Thus, by Boyle's law,

$$P \times 40A = 10 \times 20A$$

$$\text{or, } P = 5.$$

Hence, the real pressure is 730 + 5 = 735 mm of Hg

$$\text{or, } 735 \text{ torr} = \frac{735}{760} \times 1.01325 \times 10^5 \text{ Pa}$$

$$= 97991.94079 \text{ Pa} = \mathbf{97.99\ kPa.}$$

9. (a) A faulty barometer reads 740 torr when the pressure is 750 torr. The barometric tube is 790 mm long. What pressure will the barometer read when the real pressure is 760 mm of Hg?

Soln. : By the problem, the gas occupies a volume of 50 A mm^3 and exerts 10 torr pressure. Let the barometer read x torr when the real pressure is 760 torr. The open space then $(790 - x)$ mm and the gas pressure is P'.

$$\text{So, } P'(790 - x)A = 10 \times 50A$$

$$\text{i.e., } P' = \frac{500}{790 - x} \text{ torr.}$$

$$\text{But } P' + x = 760.$$

$$\therefore \quad \frac{500}{790 - x} + x = 760$$

$$\text{i.e., } x^2 - 1550x + 599900 = 0.$$

$$\text{Solving, } x = \frac{1550 - \sqrt{1550^2 - 4 \times 599900}}{2} = 748.07418 \text{ torr} = \mathbf{748\ torr.}$$

(b) A faulty barometer reads 760 mm of Hg when the real pressure is 768 mm of Hg. There is an open space of 32.0 mm in the barometer. What will the barometer read when the real pressure is 745 mm of Hg? **[Ans.** 740 mm]

10. A one-litre bulb is filled at 300 K with acetylene at 1.00 bar and oxygen at 5.00 bar, and the mixture is exploded. Calculate the final pressure if the final temperature is (i) 300 K, (ii) 353 K and (iii) 600 K. Aqueous tension at 300 K is 26.7 mm of Hg and at 353 K is 353 mm of Hg.

Soln. : When explosion takes place, C_2H_2 reacts with O_2 as follows :

$$2C_2H_2 + 5O_2 = 4CO_2 + 2H_2O\ (l)$$

Under given condition, $P \propto n \left(n = \frac{PV}{RT}\right)$. 1 mole of C_2H_2 reacts with 2.5 moles of O_2 to give 2 moles of CO_2. Thus, 1 bar of C_2H_2 is completely consumed. The remaining will be 2.5 bar of O_2 and 2 bar of CO_2. Therefore, the total pressure is now

$P_{CO_2} + P_{O_2}$ + aq. tension at 300 K

$$= 2 + 2.5 + \left(\frac{26.7}{760} \times 1.01325\right) \text{bar}$$

$$= 4.5 + 0.03559 = 4.536 \text{ bar} = \mathbf{4.54 \text{ bar.}}$$

At 353 K,

$P_{CO_2} + P_{O_2}$ = 4.5 bar at 300 K.

$\therefore$ at 353K, $P_{CO_2} + P_{O_2} = \frac{4.5 \times 353}{300} = 5.29$ bar.

$P_{CO_2} + P_{H_2O}$ + aq. tension = **5.76 bar.**

At 600 K, H_2O is gaseous. As the gas at 300 K, there is a pressure of 4.5 + 1 = 5.5 bar which at 600 K will be **11 bar.**

11. **A one-litre bulb at 27 °C is filled with ethylene at 1 bar and oxygen at 5 bar. The mixture is exploded. Calculate the final pressure if the final temperature is (i) 27 °C and (ii) 327 °C. Use data of problem 5.** [**Ans.** 4.04 bar, 12 bar]

12. **1l of N_2 at a pressure of 2.1 bar in a 4l flask mixed with 2l of O_2 at a pressure of 3.4 bar at a constant temperature without any reaction taking place. Suppose that initial temperature of O_2 is 304 K and that of N_2 is 402 K. If final temperature is 377 K, what wlll be final pressure?**

Soln. : For an ideal gas, $n = \frac{PV}{RT}$.

Initial $P_{N_2} = 2.1 \times 10^5$ Pa (1 bar = 10^5 Pa)

$V_{N_2} = 1l = 1 \times 10^{-3}\,m^3$

$P_{O_2} = 3.4$ bar $= 3.4 \times 10^5$ Pa

$V_{O_2} = 2l = 2 \times 10^{-3}\,m^3$

$$\text{Thus, } n_{N_2} = \frac{2.1\times10^5\,\text{Pa}\times(1\times10^{-3}\,\text{m}^3)}{R(\text{J.K}^{-1}.\text{mol}^{-1})T_{N_2}\,\text{K}} = \frac{210\,\text{N.m}^{-2}.\text{m}^3}{RT_{N_2}\,\text{J.mol}^{-1}}.$$

$$= \frac{201}{RT_{N_2}} \frac{\text{N.m}}{\text{J.mol}^{-1}} = \frac{210}{RT_{N_2}} \text{mol} \quad (\because \text{N.m} = \text{J})$$

$$n_{O_2} = \frac{3.4\times10^5\times2\times10^{-3}\,\text{N.m}}{RT_{O_2}\,\text{J.mol}^{-1}} = \frac{680}{RT_{O_2}} \text{mol.}$$

Given, $T_{N_2} = 304$ K, $T_{O_2} = 402$ K

Thus, total number of moles $= n_T = n_{N_2} + n_{O_2} = \frac{210\text{J}}{R(304\text{k})} + \frac{680\text{J}}{R(402\text{k})}$

Let P_f be final pressure.

Thus, $P_f = \frac{n_T R T_f}{V_f}$, $T_f = 377$ k, $V_f = 4l = 0.004\,m^3$

Hence, $P_f = \left[\dfrac{210\,J}{R304K} + \dfrac{680\,J}{R402K}\right]\dfrac{R377K}{0.0004\,m^3}$

$= 224534.77$ Pa $= 224.53$ kPa

13. It takes 0.37 gm of N_2 to fill a glass tube at 298.2 K and 0.01 bar pressure. Under the same set of condition it takes 0.92 gm of an unknown gas (homonuclear diatomic) to fill the same tube. Find out what the unknown gas was.

Soln. : Let M = molar mass of unknown gas,

W = mass of unknown gas.

Thus, number of moles of unknown gas $= n = \dfrac{W}{M}$

Here, $W = 0.92$ gm.

By Avogadro's law, same volume of all gases contains same number of molecules at same temperature and pressure.

Given : $W_{N_2} = 0.37$ gm.

$M_{N_2} = 28$ gm.-mol^{-1}

Hence, $n = \dfrac{0.37\,gm}{28\,gm.mol^{-1}} = \dfrac{0.92\,gm}{M}$

or, $M = \dfrac{0.92\,gm \times 28\,gm.mol^{-1}}{0.37\,gm} = 70.7$ gm.mol^{-1} ≈ 71 gm.mol^{-1}

Therefore, the unknown gas was **chlorine (Cl_2)**

14. Find the density of fluorine in S.I. unit at 27 °C and 288 cm of Hg pressure assuming that it behaves ideally.

Soln. : We know that $M_{F_2} = 38$ gm.mol^{-1} $= 0.038$ kg.mol^{-1}

Given that $T = (273.15 + 27)$ K $= 300.15$ K

$P = 288$ cm of Hg

$$= \frac{2880\,mm\ of\ Hg \times (1.01325 \times 10^5)\ Pa.atm^{-1}}{760\,mm\ of\ Hg.atm^{-1}} = 383968.4211\ Pa.$$

Now, density, $\rho = \dfrac{MP}{RT} = \dfrac{0.038\,kg.mol^{-1} \times 3.83968 \times 10^5\,Pa}{8.314\,N.m.mol^{-1}.K^{-1} \times 300.15\,K}$

$= 5.8469$ kg.m^{-3} (1 Pa = 1 N.m^{-2})

In C.G.S. unit, ρ = **0.0584 gm.cm^{-3}**.

15. 2 glass bulbs of equal volume are connected by a narrow tube and are filled with a gas at 273.15 K and 1.00 bar. One of the bulbs is then placed in melting ice and the other in a water bath at 62 °C. What is the new value of P inside the bulb.

[I.I.T. Entrance, 1985]

Soln. : Initially the volume of the two bulbs was $2V$ and there were

$$n = \frac{1 \times 2V}{R \times 273.15} = \frac{2V}{0.08314 \times 273.15}\ \text{mole.}$$

Under new condition, let the pressure be P. Then

$$n_1 = \frac{PV}{0.08314 \times 273.15};\ n_2 = \frac{PV}{0.08314 \times 335.15}.$$

So, $n_1 + n_2 = n$

or, $$\frac{PV}{0.08314 \times 273.15} + \frac{PV}{0.08314 \times 335.15} = \frac{2V}{0.08314 \times 273.15}$$

$$\therefore \quad \frac{P}{273.15} + \frac{P}{335.15} = \frac{2}{273.15}$$

or, $$\frac{608.3P}{273.15 \times 335.15} = \frac{2}{273.15}.$$

$$\therefore\ P = \frac{2 \times 335.15}{608.3} = 1.10923392 \text{ bar} = \mathbf{1.109\ bar.}$$

16. 2 glass bulbs of volume 3 litres and 1 litre respectively are connected by a narrow tube of negligible volume. Air at a pressure of 1 bar at 30 °C is contained in the apparatus which is then hermetically sealed. If the 3-litre bulb is now suddenly immersed in steam at 100 °C, what would be the pressure of air in the bulbs? **[J.E.E. 1981]**

[Ans. 1.16 bar]

17. 2 bulbs of volume 600 cc and 300 cc are connected by a short tube of negligible volume at 27 °C. The pressure is 760 torr. Now, the temperature of the larger bulb is quickly changed to 100 °C. What will be the final pressure? **[Ans.** 1.16 bar]

18. Spherical baloons of 15 cm diameter are to be filled with oxygen from a cylinder containing the gas at 25 bar and 0 °C. If the cylinder can hold 2 litres of water at 1 bar pressure, calculate the number of baloons that can be filled up. How much oxygen will go out?

Soln. : Volume of the baloon is $\frac{4}{3}\pi\left(\frac{15}{2}\right)^3 = 1767.375 \text{ cc} \doteq 1.767 \times 10^{-3}\ m^3$.

The cylinder will hold 2 litres of O_2 at 1 bar.

Thus at 25 bar it can hold 50 litres = $50 \times 10^{-3}\ m^3$.

Consequently the baloons will get $(50 - 2)\ 10^{-3}\ m^3 = 48 \times 10^{-3}\ m^3$, since 2 litres will remain in the cylinder.

Hence, it can fill $\frac{48 \times 10^{-3}}{1.767 \times 10^{-3}} = 27.1646859 = 27$ baloons.

Naturally, $0.1646859 \times 1.767 \times 10^{-3}\ m^3 = 2.91 \times 10^{-4}\ m^3$ = **0.291l O_2** will go out.

19. A spherical balloon of 21 cm diameter is to be filled up with hydrogen at STP from a cylinder containing the gas at 20 bar at 27 °C. If the cylinder can hold 2.82 litres of water, calculate the number of balloons that can be filled up. **[Ans.** 10]

20. 0.540 gm of an organic liquid displaces 71.52 cc of dry air at 99.80 °C. The barometer pressure was 691.4 mm of Hg. The compound is saturated and contains only carbon, fluorine and chlorine. There is 57% Cl_2 by weight. Determine its molecular formula.

Soln. : 99.8 °C = 372.95 °K

$$PV = nRT = \frac{W}{M}RT$$

$$M = \frac{WRT}{PV} = \frac{0.54\times0.08206\times372.95}{\frac{691.4}{760}\times\frac{71.52}{1000}} \text{ gm}$$

$$= \frac{0.54\times0.08206\times372.95\times760\times1000}{691.4\times71.52} \text{ gm}$$

$$= 253.999 \text{ gm} = 254 \text{ gm}.$$

Let the formula of the compound be $C_nCl_xF_{2n+2-x}$.

So, $12n + 19(2n + 2 - x) + 35.5x = 254$.

$$\% \text{ of } Cl_2 = 57 = \frac{x\times35.5}{254}\times100.$$

$$\therefore x = \frac{57\times254}{35.5\times100} = 4.078309859 = 4.$$

Thus, $12n + 19(2n - 2) + 142 = 254$

or, $50n = 112 + 38 = 150$.

$\therefore \quad n = 3$.

The formula of the compound is $\mathbf{C_3Cl_4F_4}$.

21. A meteorological balloon had a radius of 1 m when raised from sea level and expanded to a radius of 3 m when it has risen to its maximum altitude. When the temperature is –20 °C, what is the pressure inside this balloon at this altitude? **[B.U.(H) 1996]**

Soln. : Assuming sea level to be at SATP,

$T_i = 298.15$ K $\qquad T_F = 253.15$ K

$P_i = 1$ bar $\qquad P_F = ?$

$r_i = 1$ m $\qquad r_F = 3$ m

$$\frac{P_FV_F}{T_F} = \frac{P_iV_i}{T_i}$$

$$\therefore P_F = \frac{V_i}{V_F}\cdot\frac{T_F}{T_i}P_i = \frac{\frac{4}{3}\pi r_i^3}{\frac{4}{3}\pi r_F^3}\times\frac{253.15}{298.15}\times 1 \text{ bar}$$

$$= \frac{1^3}{3^3}\times\frac{253.15}{298.15} \text{ bar} = \mathbf{0.0314 \text{ bar.}}$$

22. The Celsius scale is based on 0 °C as the melting point of ice and 100 °C as the normal boiling point of water, the respective limiting PV values are 22.71 bar-l.mol^{-1} and 31.02 bar-l.mole^{-1}.

A Chemistry Honours student Sandip proposed a special scale on an unknown liquid taking 0 °SM as the freezing pt. of that liquid and 100 °SM as the normal boiling pt. of that liquid. He also measured that corresponding limiting PV values and found these to be 20 and 33 bar-l.mole^{-1} respectively.

(a) Calculate the m. pt. and b. pt. of the unknown liquid in °C.

(b) Calculate absolute zero in 0_{SM} and the value of the gas constant in bar.l. T_{SM}^{-1} where T_{SM} is absolute temp. in SM Scale.

Soln. : Now 100 °C = 31.02 − 22.71 = 8.31 in PV units.

Thus, temperature in °C may be given by

$$t\,°C = \frac{100(PV - 22.71)}{8.31}.$$

Melting point of the unknown liquid is then

$$100 \times \frac{20 - 22.71}{8.31} = -\frac{100 \times 2.71}{8.31} = \mathbf{-32.61\ °C.}$$

Boiling point of the unknown liquid is

$$100 \times \frac{33 - 22.71}{8.31} = -\frac{100 \times 10.29}{8.31} = \mathbf{123.83\ °C.}$$

Now, $\dfrac{P_1V_1}{T} = \dfrac{P_2V_2}{T+100} = R.$

$$\therefore \quad \frac{20}{T} = \frac{33}{100+T}$$

or, $13T = 2000$, $\quad\therefore T = 153.85.$

$\therefore \quad 0_{SM} = \mathbf{-153.85\ °SM.}$

Hence, $T_{SM} = 0_{SM} + 153.85$

R will be $\dfrac{20}{153.85} = 0.1299$ l-bar.T_{SM}^{-1}.mol^{-1}

$= \mathbf{0.13}$ **l-bar.T_{SM}^{-1}.mol^{-1}.**

23. With the data of the Celsius scale in hand as in problem 14, another student found 0 °D as the f. pt. of another liquid and 100 °D as the b. pt. of that liquid and found PV as 10 bar.litre.mol^{-1} and 25 bar.litre.mol^{-1} respectively.

Calculate the m. pt. and b. pt. of the unknown liquid in °C and absolute zero in °D and gas constant R. [**Ans.** −152.9 °C, 27.5 °C, −66.7 °D, 0.15 bar.litre.mol^{-1}.T_D^{-1}]

24. The inhabitants of a planet have a centigrade scale, but based on 0 °P as the m. pt. of a substance X in that planet and t = 100 °P as the b. pt. of that substance. They know about ideal gas behaviour and find limiting PV product to be 20 at 0 °P and 25 at 100 °P. Calculate the temperature of absolute zero in degree P.

Hints : A diff. of 100 °P reduces PV by 5 units. So 20 will be reduced to zero at − 400 °P.

25. The space above mercury column in a thermometer is ordinarily evacuated, but due to faulty manufacture, a particular thermometer has a pressure of 0.665 kPa in this space when the whole thermometer is immersed in a bath at 0 °C. Calculate the pressure when the whole thermometer is immersed at 100 °C. At 0 °C, the length of the air space is 0.012 m, and at 100 °C it is 0.002 m.

Hints : $\dfrac{V_1}{V_2} = \dfrac{0.012A}{0.002A} = 6$

$$P_2 = \frac{T_2}{T_1} \times \frac{V_1}{V_2} \times P_1 = \frac{373.15}{273.15} \times 6 \times 0.665 = \mathbf{5.45\ kPa}\ (40.98\ \text{mm of Hg}).$$

26. A gaseous compound X contained 44.4% carbon, 51.9% nitrogen and 3.7% hydrogen. Under like conditions 50 cc of X is diffused through a porous plug in 25 sec and the same volume of hydrogen diffused in 6.8 sec. Deduce the molecular formula of X.

Soln. : Let the vapour density of X be V_{d_x}; that of hydrogen being $V_{d_{H2}}$. Let the velocity of diffusion of X be V_x and that of H_2 be V_{H_2}.

$$\therefore \quad \text{by Graham's Law : } V_x \propto \frac{1}{25} \propto \sqrt{\frac{1}{V_{d_x}}}.$$

$$\text{Since } V_{H_2} \propto \frac{1}{6.8} \propto \sqrt{\frac{1}{V_{d_{H_2}}}},$$

$$\therefore \quad \frac{6.8}{25} = \frac{1}{\sqrt{V_{d_x}}} \text{ (since, } V_{d_{H2}} = 1).$$

$$\therefore \quad V_{d_x} = \frac{(25)^2}{(6.8)^2} = 13.51643599 = 13.5.$$

Thus, molecular weight of X in C.G.S. scale is 27.

In the compound. atomic %.

$$\text{of} \quad C = \frac{44.4}{12} = 3.7$$

$$\text{of} \quad N = \frac{51.9}{15555} = 3.7$$

$$\text{of} \quad H = \frac{3.7}{1} = 3.7.$$

$\therefore \quad C : N : H = 1 : 1 : 1$

Formula : $(CNH)_x = 27.$

$\therefore \quad x = 1.$

The compound is **HCN**.

27. The time taken for a certain volume of gas to stream through a small hole is 0.96 min while an equal volume of oxygen took 1.2 min to pass through. Calculate the approximate molecular weight of the gas. [**Ans.** 20.5]

28. If 5 gm of ammonium carbamate dissociates to $2NH_3$ and CO_2 at 200 °C, the substance occupies a volume of 7.66 litres at a pressure of 740 mm of Hg. Calculate the degree of dissociation.

Hints : $NH_2COONH_4 \rightarrow 2NH_3 + CO_2$

78 gm occupies 3 × 22.414 = 67.242 l at NTP. Hence, 5 gm occupies 4.3108 l at NTP.

$$\therefore V = \frac{P_0V_0}{T_0} \times \frac{T}{P} = \frac{760 \times 4.318}{273.15} \times \frac{473.15}{740} = 7.68 \text{ litres.}$$

$$\therefore \quad \frac{7.66}{7.68} \times 100 = 99.74, \therefore \alpha = \mathbf{0.99}.$$

29. Calculate the density of moist air in gm-litre^{-1} of moist air at 298.15 K and 1 bar pressure when dry air contains 79% nitrogen and 21% oxygen by volume. The relative humidity of moist air is 60%. The aqueous tension at 298.15 K is 0.032 bar. Given, R = 0.0821 l.atm.K^{-1}.mole^{-1}.

Soln. : Volume fraction is proportional to mol-fraction which again is proportional to total pressure.

$$\therefore \quad \frac{P_{N_2}}{P_{O_2}} = \frac{0.79}{0.21} = 3.7619 \qquad (1)$$

$$R = 0.0821 \text{ l-atm.K}^{-1}.\text{mol}^{-1} = 0.0821 \times 1.01325 \text{ l-bar.K}^{-1}.\text{mol}^{-1}$$

$$= 0.0831 \text{ l-bar.K}^{-1}.\text{mol}^{-1}. \qquad (2)$$

$$\rho = \frac{W \text{ gram}}{V \text{ litre}} = \frac{PM \text{ bar. gm. mol}^{-1}}{RT \text{ bar. litre. mol}^{-1}} = \frac{PM}{RT} \text{ gm.l}^{-1}. \qquad (3)$$

But, $W = W_{N_2} + W_{O_2} + W_{H_2O}$

$$\therefore \quad \rho = \frac{W_{N_2} + W_{O_2} + W_{H_2O}}{V} = \frac{P_{N_2}M_{N_2} + P_{O_2}M_{O_2} + P_{H_2O}M_{H_2O}}{RT}. \qquad (4)$$

Now, $P_{N_2} + P_{O_2}$ = 1 bar – 0.032 × 0.6 bar (since, relative humidity = 60%)

= (1 – 0.0192) bar = 0.9808 bar.

Now, $\frac{P_{N_2}}{P_{O_2}} + 1 = 4.7619$ [from eqn. (1)].

$$\therefore \quad \frac{P_{N_2} + P_{O_2}}{P_{O_2}} = \frac{0.9808}{P_{O_2}} = 4.7619.$$

$$\therefore \quad P_{O_2} = \frac{0.9808 \text{ bar}}{4.7619} = 0.2059 \text{ bar}.$$

$$P_{N_2} = (0.9808 - 0.2059) \text{ bar} = 0.7748 \text{ bar}.$$

$$\text{From eqn. (3), } \rho = \frac{0.7748 \times 28 + 0.2059 \times 32 + 0.0192 \times 18}{0.08314 \times 298.15}$$

$$= \frac{28.629}{0.08314 \times 298.15} \text{ gm/litre} = \mathbf{1.155 \text{ gm/litre.}}$$

30. Solve problem 29 if the total pressure is 1 atm and aqueous tension at 25 °C is 23.76 mm of Hg. [**Ans.** d = 1.171 gm/l]

31. At 300 K, 1000 cc of H_2 measured under a pressure of 600 mm of Hg and 1000 cc of N_2 measured under a pressure of 400 mm of Hg are introduced into an evacuated vessel having a volume of 4 litres. Calculate the resulting pressure.

Hints : $P = P_{H_2} + P_{N_2} = \frac{1}{4} \times 600 + \frac{1}{4} \times 400 = 150 + 100 =$ **250 mm of Hg.**

32. The compressibility factor of O_2 is 0.927 at 0 °C, and 100 atm pressure. Calculate the weight of O_2 necessary to fill a cylinder of 100 litres capacity at the given condition.

Hints : Given, T °K = 273.15 K

P = 100 atm, Z = 0.927

$$n = \frac{PV}{ZRT} = \frac{100 \times 100 \text{ l.atm}}{0.927 \times 0.08206 \text{ l.atm.K}^{-1}.\text{mol}^{-1} \times 273.15°\text{K}}$$

$$= 481.26233 = 481.3.$$

$\therefore$ wt. in kg = 481.3 × 32 gm = 15401.6 gm = **15.4 kg.**

33. The highest allowed presssure of a 20-litre flask containing 1.6 kg of oxygen is 150 atm. To what temperature can it be heated? Assume that O_2 obeys van der Waals equation with a = 1.360 litre^2.atm.mole^{-2} and b = 31.83 cm^3.mole^{-1}.

Hints : $T = \dfrac{\left(P + \dfrac{an^2}{V^2}\right)(V - nb)}{nR}$ **[Ans.** 711.95 K ≈ 712 K]

34. Calculate the value of R in S.I. unit from the following data. A 20.00 l container was filled with 0.25132 gm of He gas and heated up to 500 °C. A student doing this experiment measured the pressure which he found to be 206 cm of water in a manometer at 27 °C. If that density of water is 0.9965 gm.cm^{-3} at 25 °C and molecular wt. of He is 4.002 gm, did he measure the pressure correctly?

Soln. : Given that V = 20.00 l = 0.02 m^3

T = (500 + 273.15) K = 773.15 K

Mol. wt. of He = M_{He} = 4.0026 gm

Mass of He = 0.25132 gm

Hence, number of moles of He = $n = \dfrac{0.25132 \text{ gm}}{4.0026 \text{ gm.mol}^{-1}}$ = 0.06279 mol.

By the problem, P = 206 cm of H_2O = 206 × 10^{-2} m of H_2O, density of water = ρ = 0.9965 gm.cm^{-3}.

Thus, $P = h\rho g$

= 206 × 10^{-2} m × 0.9965 × 10^3 kg.m^{-3} × 9.80665 m.s^{-2}

= 20130.9931 N.m^{-2} = 20130.9931 Pa.

Assuming the gas to be ideal, $R = \dfrac{PV}{nT} = \dfrac{20130.9931 \text{ Pa} \times 0.02 \text{ m}^3}{0.06279 \text{ mol} \times 773.15 \text{ k}}$

= **8.2936 J.K^{-1}.mol^{-1}.**

The reading of the student was obviously inaccurate. Since R = 8.3146 J.k^{-1}.mol^{-1}, he should have read the pressure as 206.5 cm of water to get the correct value of R.

35. Pressure exerted by a sample of nitrogen is 202 k Pa at 300 K. Volume occupied by it under this condition is 1 m^3. If it is now compressed to a volume of 400 l, calculate its new pressure.

Soln. : By the problem,

P_1 = 202 kPa P_2 = ?

V_1 = 1 m^3 V_2 = 400 × 10^{-3} m^3

T_1 = 300 K T_2 = 400 K

Now, $\dfrac{P_1V_1}{T_1} = \dfrac{P_2V_2}{T_2}$.

$$\therefore \quad P_2 = \frac{P_1V_1}{T_1}\times\frac{T_2}{V_2} = \frac{202\,\text{k Pa}\times 1\,\text{m}^3}{300\text{K}}\times\frac{400\,\text{K}}{0.4\,\text{m}^3}$$

$= \mathbf{673.32\ kPa.}$

36. The air space in a driving bell is 4.00 m³ on the deck of a boat moving in the sea. What will be volume of the air space when the bell is taken to a depth of 60 m. Given that density of sea water is 1.026 gm.cm^{-3}?

Soln. : By the problem, $V_1 = 4\ \text{m}^3$

$P_1 = 1$ atm (say) $= 101.325$ kPa

$\rho = 1.026$ gm.cm^{-3}

$h = 60$ m

$V_2 = ?$

$P_2 = P_1 + h\rho g$

$P_2 = 101.325\ \text{kPa} + (60\ \text{m} \times 1.026 \times 10^3\ \text{kg.m}^{-3} \times 9.80665\ \text{m.s}^{-2})$

$= 101.325\ \text{kPa} + 603.69737\ \text{kPa} = 705.02237\ \text{kPa} = 705\ \text{kPa}$

If we assume temperature remains constant,

$$V_2 = \frac{P_1V_1}{P_2} = \frac{101.325\,\text{kPa}\times 4m^3}{705\,\text{kPa}} = \mathbf{0.5748\ m^3}.$$

37. In a glass globe of a particular size, chemical nitrogen prepared by heating NH_4NO_2 (a mixture of $NaNO_2$ and NH_4Cl) is found to have a mass of 2.2991 gm. In the same globe nitrogen, made from atmosphere by removing O_2, CO_2 and water vapour, had a mass of 2.31018 gm. Suppose that atmospheric nitrogen and argon have molar masses 28.02 gm and 39.952 gm respectively. Calculate the mole fraction of nitrogen in the mixture.

Soln. : Let T and P be same for both cases. Then same volume of gases will contain same number of moles.

Mole number of $N_2 = \dfrac{2.2991}{28.02} = 0.0820521.$

Let there be A mole of N_2 and B mole of Ar in the nitrogen made from air.

Thus, $28.02A + 39.952B = 2.31018$ (1)

and $A + B = 0.0820521$

$\therefore \quad B = 0.0820521 - A$ (2)

Putting the value of B in Eqn. (1), we get $A = 0.081123$ mole.

Thus, $X_A = \dfrac{0.081123}{0.082052} = 0.9886$

and $X_B = 1 - 0.9886 = 0.0114.$

Alternative method :

Let there be A gm of Ar and B gm of N_2, then $A + B = 2.31018$ g

and $\dfrac{A}{28.02} + \dfrac{B}{39.952} = \dfrac{2.2991}{28.02} = 0.0820521$

or, $\dfrac{A}{28.02} + \left(\dfrac{2.31018 - A}{39.952}\right) = 0.082052.$

A can be found out from the above equation and $n_A = \dfrac{A}{28.02}$, $X_A = \dfrac{n_A}{n_A + n_B}$.

38. At a pressure of 1 atm and a temperature 296 K, density of air is 1.19 gm/*l*. Calculate the mole fraction of nitrogen in this mixture. There are only oxygen and nitrogen in air.

Soln. : We know, for an ideal gas, $M = \dfrac{\rho RT}{P}$.

By the problem, ρ = 1.19 gm/l, T = 296 K, P = 1 atm, R = 0.08206 l.atm.K^{-1}.mol^{-1}

$$\therefore \quad M = \frac{1.19\,\text{gm.l}^{-1} \times 0.08206\,\text{l.atm.K}^{-1}.\text{mol}^{-1} \times 296\text{K}}{1\,\text{atm}} = 28.90481\ \text{gm. mol}^{-1}$$

Let x be the mole fraction of O_2 and then $(1 - x)$ be the mole fraction of N_2. If we assume there is one mole of air, then

$$x_{O_2} + (1 - x)_{N_2} = 1.$$

$\therefore \quad x \times 32 + (1 - x) \times 28 = 28.90481$

or, $\quad 4x = 0.90481 \quad \therefore \quad x = 0.22.$

Thus, mole fraction of nitrogen = $(1 - x)$ = **0.78**.

39. One mole of gas A at 66 K and 3.2 atm and gas B at 608 K and 18.2 atm occupy a volume of 1718 cc and 2782 cc respectively. Their critical constant values are shown below. Find the missing volume in the table and state which gas is more easily liquefiable and which is more close to ideal behaviour at STP?

Gas	T_c	P_c (atm)	V_c(l)
A	33 K	12.8	0.065
B	304 K	72.8	—

Soln. : For gas A, the reduced temperature and pressure under the given condition are $T_r = \dfrac{66}{33} = 2$, $P_r = \dfrac{3.2}{12.8} = 0.25$, and for gas B the reduced temperature and pressure are $T_r = \dfrac{608}{304} = 2$, $P_r = \dfrac{18.2}{72.8} = 0.25$. Since these are same, they will have same reduced volumes. V_r of A is $\dfrac{1.718}{0.065} = 26.43$.

$\therefore$ V_r of B will be 26.43.

Hence, V_c for B is $\dfrac{V}{V_r}$.

$$\therefore (V_c)_B = \frac{2.782}{26.43}\ \text{litre} = \mathbf{0.1052\ litre.}$$

(Assuming that they have the same Z_c, $(V_c)_B$ may also be calculated.)

T_c of B is higher—it can be liquified at 304 K. Thus, in comparison to A (liquefiable by application of pressure up to 33 K only), $\mathbf{B}$ is more easily liquefiable.

Boyle temperature for A, $(T_B)_A = \dfrac{27}{8} \times 33 = 111.375$ °K

Boyle temperature for B, $(T_B)_B = \dfrac{27}{8} \times 304 = 1026$ °K.

At STP, A is above twice its Boyle temperature while B is much below its Boyle temperature.

Naturally, at STP, A will be more close to ideal behaviour.

40. Gases A and B obeying van der Waals equation have T_c and P_c values given below:

	T_c (°K)	P_c (atm)
A	44	26
B	304	72

Which gas (i) has higher V_c value and (ii) shows more nearly ideal behaviour at 25 °C and 10 atm pressure? Give reasons for your answer. **[C.U.(H) 1992, V.U. 1998]**

Soln. : (i) For a van der Waals gas,

$$\frac{RT_c}{P_cV_c} = \frac{8}{3}.$$

$$\therefore \quad V_c = \frac{3RT_c}{8P_c}$$

$$\therefore \quad (V_c)_A = \frac{3\times 0.08206\ \text{l.atm.K}^{-1}.\text{mol}^{-1}\times 44\ \text{K}}{8\times 26\ \text{atm}} = 0.052\ \text{l.mol}^{-1}.$$

$$(V_c)_B = \frac{3\times 0.08206\ \text{l.atm.K}^{-1}.\text{mol}^{-1}\times 304\ \text{K}}{8\times 72\ \text{atm}} = 0.1298\ \text{l.mol}^{-1} = 0.13\ \text{l.mol}^{-1}.$$

Thus, **B with very high T_c has higher V_c value.**

(ii) Boyle temperature for A, $(T_B)_A = \frac{27}{8}(T_c)_A$

$$= \frac{27}{8}\times 44\ \text{°K} = 148.5\ \text{°K} = -124.5\ \text{°C}.$$

Boyle temperature of B, $(T_B)_B = \frac{27}{8}\times (T_c)_B = \frac{27}{8}\times 304\ \text{°K} = 1026\ \text{°K} = 753\ \text{°C}.$

Boyle temperature of B is much above 25 °C than that of A below it. A is close to $2T_B$ at 25 °C. Thus, **A will show more nearly ideal behaviour than B** at 25 °C and 10 atm.

41. The following table gives the values of a, b and experimentally measured volume at 273 K and 200 atm of two van der Waals gases A, B :

	a atm.l^2.mol^{-2}	b litre.$mole^{-1}$	Volume litre.$mole^{-1}$
A	0.24	0.026	0.112
B	1.36	0.032	0.111

Calculate (i) the compressibility factor for 1 mole of each at 273 K and 200 atm and hence state which one is more compressible than predicted by the ideal gas law, (ii) the temperature at which B obeys Boyle's law over a wide range of pressure and (iii) the temperature below which A would be subjected to Joule-Thomson expansion for liquefaction.

Soln. : (i) Compressibility factor $Z = \frac{PV}{RT}$.

$$\therefore\ Z_A = \frac{200 \times 0.112}{0.082 \times 273} = 1.000625391 = 1.0006 = 1.00$$

$$Z_B = \frac{200 \times 0.111}{0.082 \times 273} = 0.99169 = 0.992.$$

Thus, ***B* is more compressible than an ideal gas.**

(ii) The temperature is the Boyle temperature : $T_B = \dfrac{a}{Rb}$.

$$(T_B)_B = \frac{1.36}{R \times 0.032} = \frac{1.36}{0.082 \times 0.032} = 518.29 \text{ K} = \mathbf{578\ K.}$$

(iii) The temperature is inversion temperature.

$$T_i = \frac{2a}{Rb}.$$

$$\therefore\ (T_i)_A = \frac{2 \times 0.24}{0.082 \times 0.026} = 225.14 \text{ K} = \mathbf{225\ K.}$$

In this problem R is chosen as 0.082 l.atm.K^{-1}.mole^{-1} rather than 0.0821 since choice of R = 0.08206 will lead to both Z_A and Z_B less than one. What is more, there are only 2-3 significant figures in the reported data.

42. Gases *A, B, C* and *D* obey van der Waals equation with *a* and *b* values in S1 units given below :

	A	*B*	*C*	*D*
a	0.6	0.6	0.2	0.005
$10^3 b$	0.025	0.15	0.10	0.02

Which of the gases has (i) the highest T_c (ii) the largest molecules and (iii) the most ideal bahaviour at STP? **[C.U.(H) 2003]**

Hints : Calculate T_c and T_B.

T_B of D = 30.1K, C = 240.5, B = 481K, A = 2886 K

D is most ideal at STP (above $2T_B$).

43. Gases NO and CH_4 obeying van der Waals equation have T_c and P_c values given below :

	T_c K	P_c (atm)
NO	177	64
CH_4	550	45

Which gas has (i) smaller value of *a* (ii) smaller value of *b* (iii) larger value of V_c and (iv) which gas is nearly ideal in behaviour at 300 K and 10 atm pressure? Give reason. **[C.U. 1998]**

Hints :

$$b_{NO} = \frac{RT_c}{8P_c} = 0.028 \text{ l.mol}^{-1}$$

$$a_{NO} = \frac{27R^2T_c^2}{64P_c} = 1.392 \text{ atm.l}^2\text{.mole}^{-2}$$

$$b_{CH_4} = 0.125 \text{ l.mole}^{-1}$$

$$a_{CH_4} = 19.115 \text{ atm.l}^2\text{.mole}^{-2}$$

$$V_c = 3b$$

$$(T_B)_{NO} = \frac{a_{NO}}{Rb_{NO}} = \frac{1.392}{0.0821 \times 0.028} = 605.53 \text{ °K}$$

$$(T_B)_{CH_4} = \frac{a_{CH_4}}{Rb_{CH_4}} = \frac{19.115}{0.0821 \times 0.125} = 1862.61 \text{ °K.}$$

T_B of NO is more close to 300 K than that of CH_4. So NO is more nearly ideal at 300 K and 10 atm.

44. At 300 °C, one mole of ether occupied a volume of 740 cc. It is a van der Waals gas with $a = 17.4$ atm.l^2.mole^{-2} and $b = 13.4 \times 10^{-2}$ l.mol^{-1}. Calculate the pressure. What would have been the pressure if it were an ideal gas? Hence, calculate Z.

Soln. : For 1 mole of a van der Waals gas,

$$T = 573.15 \text{ K}$$

$$V = 0.74 \text{ litre}$$

$$b = 0.134 \text{ l.mole}^{-1}$$

$$a = 17.4 \text{ atm.l}^2\text{.mole}^{-2}$$

$$R = 0.08206 \text{ l.atm.deg}^{-1}\text{.mole}^{-1}$$

$$P = \frac{RT}{V-b} - \frac{a}{V^2} = \frac{0.08206 \times 573.14}{0.74 - 0.134} - \frac{17.4}{0.74^2}$$

$$= 77.61169802 - 31.775018$$

$$= 45.83667976 = \mathbf{45.8 \text{ atm.}}$$

$$P_{\text{ideal}} = \frac{RT}{V} = \frac{0.08206 \times 573.15}{0.74} = 63.55768784 = \mathbf{63.6 \text{ atm.}}$$

$$\therefore \quad Z = \frac{P}{P_{\text{ideal}}} = \frac{45.8}{63.6} = 0.72012579 = \mathbf{0.72.}$$

Under the given condition, the gas is more compressible than ideal gas.

45. The pressure exerted by 3.02×10^{20} molecules of a gas in a 2 litre vessel is 1.52 cm of Hg. Calculate the temperature of the gas if the gas is ideal. What will be the temperature if the gas is van der Waals or Dieterici with same $a = 1.38$ atm. l^2.mole^{-2} and $b = 0.039$ l/mole.

Soln. : For ideal gas, $T = \dfrac{PV}{nR}$

$$= \frac{\frac{1.52}{76} \times 2}{\frac{3.02 \times 10^{20}}{6.022 \times 10^{23}} \times 0.08206} \text{ K}$$

$$= \frac{1.52 \times 2 \times 6.022 \times 10^{23}}{3.02 \times 10^{20} \times 0.08206 \times 76} \text{K} = 971.99 \text{ K} = 972 \text{ K.}$$

$$n = \frac{3.02\times10^{20}}{6.022\times10^{23}} = 5.01 \times 10^{-4}.$$

For van der Waals gas,

$$\frac{\left(P+\frac{an^2}{V^2}\right)(V-nb)}{nR} = T.$$

$$\therefore \quad T = \frac{\left(\frac{1.52}{76}+\frac{1.38\times\left(5.01\times10^{-4}\right)^2}{2^2}\right)\left(2-5.01\times10^{-4}\times0.0394\right)}{5.01\times10^{-4}\times0.08206}$$

$$= \frac{\left(0.02+8.65\times10^{-8}\right)\left(2-1.97394\times10^{-5}\right)}{5.01\times10^{-4}\times0.08206}$$

$$= \frac{0.020000086\times1.999980261}{5.01\times10^{-4}\times0.08206} = 972.9451217 \text{ K} = 973 \text{ K}.$$

For Dieterici gas,

$$T = \frac{P(V-nb)}{nR}e^{\frac{-nA}{RTV}}.$$

$$\ln T = \ln 972.94 - \frac{4.210649281\times10^{-3}}{T}$$

Neglecting $\frac{1}{T}$ terms, $T = 972.94 =$ **973 K.**

46. For a particular gas, $b = 4.42 \times 10^{-2}$ l.mole^{-1}. How near can the centres of two molecules approach? **[C.U.(H) 1987]**

Soln. : $b = 44.2$ cm^3.mole^{-1}

$$= 4N_A \times \frac{4}{3}\pi r^3.$$

$$\therefore \quad r^3 = \frac{3b}{16N_A\pi} = \frac{3\times44.2\,\text{cm}^3}{16\times6.022\times10^{23}\times3.142}$$

$$= 4.38002 \times 10^{-24} \text{ cm}^3$$

$$r = 1.63615 \times 10^{-8} \text{ cm}.$$

The centres can approach as close as $2r$ (σ).

$$\therefore \quad \sigma = 3.2723 \times 10^{-8} \text{ cm}$$

$$= 3.27 \times 10^{-10} \text{ m}$$

$$= \mathbf{327\ pm.}$$

47. Given that $P_c = 48$ atm and $T_c = 155$ K for Argon. Calculate the radius of Argon atoms.

Hints : $b = \frac{RT_c}{8P_c} = 33.123$ cm^3

$\therefore \quad r = 1.49$ Å $=$ **149 pm.**

48. Using van der Waals equation, calculate the pressure for 2 moles of N_2 occupying a volume of 5 l at 40 °C.

Given, $a = 1.38$ atm.l^2.mole^{-2}, $b = 0.0399$ l.mole^{-1}, $R = 0.0821$ l.atm.deg^{-1}.mol^{-1}.

[C.U.(H) 1974]

[**Ans.** $P = 10.2$ atm, $P_{id} = 10.3$ atm]

49. For CO_2, $T_c = 31$ °C and critical density is 0.341 gm/cc. Its molecular weight is 44. Calculate a and b and radius of CO_2. **[C.U.(H) 1977]**

[**Ans.** $V_c = 129.03$ cc, $b = 43.01$ cc/mole,
$a = 3.6$ atm.l^2.mole^{-2}, $r = 162$ pm]

50. Calculate the radius of a van der Waals gas molecule if $b = 3.2 \times 10^{-2}$ l.mol^{-1} and $b = 4.42 \times 10^{-3}$ l.mole^{-1}. **[C.U.(H) 1997]**

[**Ans.** $r = 147$ pm, $r = 75.9$ pm]

51. 3 gm-moles of CO_2 occupy 10 l at 15 bar. Calculate its temperature if it is (i) a van der Waals gas, (ii) Dieterici gas and (iii) ideal gas. Given, $a = 3.59$ bar.l^2.mole^{-2} and $b = 0.043$ l.mole^{-1}. [**Ans.** (i) 606.5 K, (ii) 607.2 K, (iii) 601.4 K]

52. Assuming the intermolecular attraction to be negligible in case of Helium gas, and taking its molar volume at 0 °C and 100 atm to be 1.107×10^{-2} times the volume at NTP, calculate the diameter of He atom. **[C.U.(H) 1983]**

Hints : $a = 0$

$\therefore \quad P(V - b) = RT.$

By the problem, $100\,(1.107 \times 10^{-2}\,V - b) = 273.15\,R$

$$1(V - b) = 273.15\text{ R}$$

$$\text{or,} \quad \frac{1.107V - 100b}{V - b} = 1$$

$$\text{or,} \quad b = \frac{0.107\,V}{99} = 0.0242251 \text{ l.mole}^{-1}.$$

$\therefore \quad \sigma = $ **268 pm.**

53. If the compressibility factor Z for a van der Waals gas be 1.000054 at 0 °C and 1 atm, and the Boyle temperature be 107 K, calculate neglecting the higher terms of P, the values of a, b and molecule diameter. **[C.U.(H) 1994]**

Soln. : For a van der Waals gas, neglecting the higher terms in Virial form—

$$Z = 1 + \frac{P}{RT}\left(b - \frac{a}{RT}\right) = 1.000054$$

$$\therefore \quad b - \frac{a}{RT} = 0.000054 \times \frac{RT}{P}$$

$$= 0.000054 \times 0.08206 \times 273.15$$

$$= 1.2103932206 \times 10^{-3}$$

$$= 1.21 \times 10^{-3}.$$

$$T_B = \frac{a}{Rb} = 107$$

$$a = 107\,Rb.$$

$\therefore \quad b - \dfrac{107 \cdot Rb}{R \times 273.15} = 1.21 \times 10^{-3}$

or, $\quad b - 0.391726157b = 1.21 \times 10^{-3}$

or, $\quad 0.608273842b = 1.21 \times 10^{-3}$.

$\therefore \quad b = \mathbf{1.988 \times 10^{-3}\ l.mole^{-1}}$.

$\therefore \quad a = 107 \times 0.08206 \times 1.988 \times 10^{-3}$

$= 0.017466324$ atm.l^2.mole^{-2}

$= \mathbf{0.0175\ atm.l^2.mole^{-2}}$.

$$\therefore \quad r^3 = \frac{3 \times b}{16 N_A . \pi} = \frac{3 \times 1.988}{16 \times 6.022 \times 10^{-23} \times 3.142}$$

$= 1.97002 \times 10^{-25}$ cc.

$\therefore \quad r = 5.81868 \times 10^{-9}$ cm $= 0.58$ Å $= 58$ pm.

$\sigma = \mathbf{116\ pm}$.

54. A certain vapour obeys the van der Waals equation with $a = 0.52$ m^6.Pa.mole^{-2}, Its volume is 4.99×10^{-4} m^3.mole^{-1} at 300 K and 3.20×10^3 kPa. Calculate the value of van der Waals constant. Find the temperature at this volume, when $P = 0$.

Soln. : van der Waals equation for 1 mole is

$$\left(P + \frac{a}{V^2}\right)(V - b) = RT.$$

$$\therefore V - b = \frac{RT}{P + \dfrac{a}{V^2}}$$

By the problem,

$V = 4.99 \times 10^{-4}$ m^3

$a = 0.52$ m^6. Pa. mole^{-2}

$T = 300$ K

$R = 0.08206$ l-atm.deg^{-1}.mole^{-1} = **8.31473 m^3.Pa.deg^{-1}.mole^{-1}.**

$P = 3.2 \times 10^6$ Pa

$\therefore \quad V - b = 4.7164 \times 10^{-4}$m^3

$\therefore \quad b = 4.99 \times 10^{-4} - 4.7164 \times 10^{-4}$

$= 2.7359 \times 10^{-5}$ m^3.mole^{-1} = 2.74 m^3.mole^{-1}.

Now $P = \dfrac{RT}{V - b} - \dfrac{a}{V^2}$.

When $P = 0$,

$$0 = \frac{RT}{4.7164 \times 10^{-4}} - \frac{0.52}{\left(4.99 \times 10^{-4}\right)^2}.$$

$$\therefore \quad T = \frac{V - b}{R} \times \frac{a}{V^2} = \frac{4.7164 \times 10^{-4} \times 0.52}{8.31473 \times \left(4.99 \times 10^{-4}\right)^2}$$

$= 118.45809$ K $\approx$ **118 K.**

Since there is attraction, P turns out to be zero above absolute zero. This only indicates the existence of molecular attraction.

55. One gm-mole of a gas PQ which is dissociated to an extent of 10% at 127 °C according to PQ = P + Q occupies a volume of 4×10^4 cc. Calculate the total pressure at this temperature assuming the gas to be perfect. **[C.U. 1975; B.U. 1974]**

Soln. : Let there be a gm-mole of PQ.

PQ = P + Q

$(1 - \cdot 1)a \quad \cdot 1a \quad \cdot 1a$

$V = 40$ litres.

$n = 1\cdot 1a$ mole = 1.1 mole (since, $a = 1$)

$R = 0.08206$ l.atm.deg^{-1}.mole^{-1}

$$\therefore \quad P = \frac{1.1 \times 0.08206 \times 400}{40} = 0.90266 \text{ atm} = \mathbf{0.903\ atm.}$$

56. If the temperature above which a van der Waals gas cannot be liquefied be 32.3 °C and minimum pressure to be applied at that temperature for liquefaction be 48.2 atm,

(i) find the diameter of gas molecule.

(ii) calculate a and hence pressure of 60 gm of the gas at 27 °C with a volume of 2 litres if its molecular weight be 30.

Soln. : (i) For van der Waals gas, $\dfrac{RT_c}{P_cV_c} = \dfrac{8}{3}$.

$T_c = 32.3 + 273 = 305.3$ K

$V_c = 3b$

$P_c = 48.2$ atm

$$\text{Thus, } b = \frac{RT_c}{8P_c} = \frac{305.3\text{K} \times 0.08206 \text{ l.atm.deg}^{-1}\text{.mole}^{-1}}{8 \times 48.2 \text{ atm}} = 0.065 \text{ litre.}$$

$$\text{Now, } b = 4N_A\left(\frac{4}{3}\pi r^3\right)$$

$$\text{or, } r^3 = \frac{3b}{16N_A\pi} = \frac{3 \times 0.065 \times 10^3 \text{ cc}}{16 \times 6.022 \times 10^{23} \times \pi} = 6.44 \times 10^{-24} \text{ cc}$$

or, $r = 1.86$ Å .

Thus, σ = **3.72 Å** .

$$P_c = \frac{a}{27b^2}.$$

$\therefore a = 27 \times 48.2 \times (0.065)^2 =$ **5.498 atm.l^2.mol^{-2}.**

$$n = \frac{60}{30} = 2.$$

$$\therefore P = \frac{nRT}{V - nb} - \frac{an^2}{V^2} = \frac{2 \times 0.08206 \times 300}{2 - 2 \times 0.065} - \frac{5.498 \times 4}{4} = \mathbf{20.83\ atm.}$$

57. A real gas obeys $Z = a_1 + a_2P + a_3P^2$, Z = compressibility factor. It shows minimum at some pressure P_0 and temperature T_1. Find (i) a_1 (ii) sign of a_2 and a_3 (iii) P_0 in terms of constants. (iv) If the temperature variation of a_2 is given by $a_2 = C_1 + \frac{C_2}{T}$ and the Boyle temperature T_B obeys $T_B > T_1$, remark on the signs of C_1 and C_2 and express T_B in terms of these constants. **[B.U. 2002]**

Soln. : (i) As $P \to 0$, $Z = a_1$.

Now as P tends to zero, Z = 1. $\therefore$ a_1 = **1.**

Thus, $Z = 1 + a_2P + a_3P^2$.

(ii) and (iii) : $\left(\frac{\partial Z}{\partial P}\right) = a_2 + 2a_3P$

When $P = P_0$, $a_2 + 2a_3P_0 = 0$.

$$\therefore \quad P_0 = -\frac{a_2}{2a_3},$$

Pressure P_0 can never be negative, so a_2 and a_3 should have opposite signs.

Now at P_0, $Z < 1$ and $Z = 1 + a_2P_0 + a_3P_0{}^2 = 1 - \frac{a_2^2}{4a_3}$.

a_2^2 is always (+)ve, Thus, a_3 **should be positive** so that $Z < 1$ at temperature T_1 and pressure P_0. Thus, $\boldsymbol{a_2}$ ***is*** **negative.**

(iv) At T_B, $Z = 1$

As P is very low, a_3P^2 can be neglected. Thus $a_2 = 0$.

By the problem, $C_1 + \frac{C_2}{T_B} = 0$.

$$\therefore \quad \boldsymbol{T_B = -\frac{C_2}{C_1}}.$$

T_B is positive, hence $\boldsymbol{C_1}$ **and** $\boldsymbol{C_2}$ **should have opposite signs.**

a_2 increases with increase in temperature from its negative value at T_1, becoming zero at Boyle temperature T_B, $\frac{d}{dT}[a_2(T)] = -\frac{C_2}{T^2} > 0$. Thus $\boldsymbol{C_2}$ **should have negative sign while** $\boldsymbol{C_1}$ **sould have positive sign.**

58. Two gases N_2 and C_2H_6 obeying van der Waals equation have following T_c and P_c :

	T_c	P_c
N_2	126 K	34 atm
C_2H_6	305 K	48 atm

Which gas has (i) smaller value of a (ii) larger value of b (iii) which gas shows nearly ideal behaviour at 25 °C and 10 atm? **[Ans.** N_2,C_2H_6,N_2]

59. Calculate the temperature at which slope of Z vs P curve ($P = 0$) has a maximum value for van der Waals gas. Calculate also the value of the slope.

Soln. : At $P = 0$, for van der Waals gas Z can be written as

$$Z = 1 + \left(b - \frac{a}{RT}\right)\frac{P}{RT}.$$

Hence, slope $m = \left(\frac{\partial Z}{\partial P}\right)_{P=0} = \frac{1}{RT}\left(b - \frac{a}{RT}\right).$

Thus, the slope of Z vs P curve (m) is a function of temperature. This will be maximum at a temperature when $\frac{dm}{dT} = 0.$

$$\text{Now, } \frac{dm}{dT} = \frac{1}{RT}\left(\frac{a}{RT^2}\right) - \frac{1}{RT^2}\left(b - \frac{a}{RT}\right).$$

$$= \frac{2a}{R^2T^3} - \frac{b}{RT^2}.$$

$$\text{Thus, } \frac{2a}{R^2T^3} - \frac{b}{RT^2} = 0, \qquad \therefore\ T = \frac{\mathbf{2a}}{\mathbf{Rb}}.$$

The slope will be maximum at $T = \frac{2a}{Rb}$.

The value of maximum slope,

$$m_{\max} = \left[b - \frac{a}{R\frac{2a}{Rb}}\right] \times \frac{1}{R.\frac{2a}{2b}} = \frac{\mathbf{b^2}}{\mathbf{4a}}.$$

Hence, after reaching the temperature twice the Boxle temperature of the van der Waals gas, its Z value starts decreasing and approaches ideal gas value ($Z = 1$) with increasing temperature.

60. The behaviour of two gases A and B can be approximated by van der Waals' equation. The critical constants of these gases are given below :

Gas	P_c/atm	V_c/ccm^3.mol^{-1}	T_c/K
A	81.5	81.0	324.7
B	2.26	57.76	5.21

Explain (i) which gas has greater intermolecular force of attraction, (ii) which gas obeys van der Waals' equation more closely at the critical state. **[C.U. 2010]**

Soln. : Hints : $a = \frac{27R^2T_c^2}{64P_c}$

$a_A = 3.68$ atm.Al^2.mol^{-2}

$a_B = 0.03$ atm.Al^2.mol^{-2}

Hence, **A has greater intermolecular force of attraction.**

(ii) For A, $\dfrac{RT_c}{P_cV_c} = 4.04$

For B, $\dfrac{RT_c}{P_cV_c} = 3.28$

For van der Waals gas, $\dfrac{RT_c}{P_cV_c} = 2.66.$

So B obeys van der Waals' equation more closely at the critical state.

61. For a van der Waals gas, show that the approximate molar mass of a gas

$$M = \frac{\rho RT}{P}\left[1+\frac{PT_c}{8P_cT}\left(1-\frac{27}{8}\frac{T_c}{T}\right)\right].$$

Soln. : For n moles of van der Waals gas, the equation can be written as

$$PV = nRT + n\left(b-\frac{a}{RT}\right)P$$

or, $$PV = \frac{W}{M}RT + \frac{W}{M}\left(b-\frac{a}{RT}\right)P$$

or, $$\frac{M}{\rho} = \frac{RT}{P} + \left(b-\frac{a}{RT}\right) \quad \left[\text{Since, } \rho = \frac{W}{V}\right]$$

or, $$\frac{M}{\rho} = \frac{RT}{P} + \left(b-\frac{27bRT_c}{8RT}\right) \quad \left[T_c = \frac{8a}{27Rb}\right]$$

or, $$\frac{M}{\rho} = \frac{RT}{P}\left[1+\frac{bP}{RT}\left(1-\frac{27}{8}\frac{T_c}{T}\right)\right]$$

or, $$\frac{M}{\rho} = \frac{RT}{P}\left[1+\frac{RT_c}{8P_c}\cdot\frac{P}{RT}\left(1-\frac{27}{8}\frac{T_c}{T}\right)\right].$$

$$\therefore \quad M = \frac{\rho RT}{P}\left[1+\frac{PT_c}{8P_cT}\left(1-\frac{27}{8}\frac{T_c}{T}\right)\right] \textbf{(Proved).}$$

B. Different velocities, energy, Maxwell distribution, equipartition, mean free path, collision frequency and viscosity

B1. Required Formulae

(1) Most probable velocity $\bar{u} = \sqrt{\dfrac{2RT}{M}} = \sqrt{\dfrac{2kT}{m}}.$

Root mean square velocity $\bar{c} = \sqrt{\frac{3RT}{M}}$.

Average velocity $\bar{v} = \sqrt{\frac{8RT}{\pi M}}$.

(2) Fraction of molecules having velocity between c and $c + dc$:

$$\frac{dN_c}{N} = 4\pi\left(\frac{m}{2\pi kT}\right)^{\frac{3}{2}} c^2 e^{-\frac{mc^2}{2kT}} dc$$

(3) Barometric formula : $P = P_0 e^{-\frac{mgh}{kT}}$.

(4) Fraction of molecules having velocity in any direction between c_x and $c_x + c_x$

$$= F(c_x)dc_x = \left(\frac{m}{2\pi kT}\right)^{\frac{1}{2}} e^{-\frac{mc_x^2}{2kT}}.$$

(5) Absolute value of c_x in one direction $\sqrt{\frac{kT}{2\pi m}}$.

(6) Number of molecules leaking through area A in time $t = \frac{1}{\sqrt{6\pi}} \frac{N}{V} \bar{c}\, A\, t = \frac{1}{4} \frac{N}{V} \bar{v} A t$

(7) Number of molecules leaking through unit area in unit time $\frac{1}{\sqrt{6\pi}} \frac{N}{V} \bar{c}$.

(8) Number of molecules colliding in unit time with wall in unit area :

$$Z = \frac{1}{\sqrt{6\pi}} \frac{N}{V} \sqrt{\frac{3kT}{m}} = \frac{N}{V} \sqrt{\frac{kT}{2\pi m}}$$

or, $\frac{N}{V} = \frac{P}{kT}$.

$$\therefore Z = \frac{P}{\sqrt{2\pi mkT}}.$$

(9) Mass striking with the wall in unit area in unit time

$$= \mu = Zm = \sqrt{\frac{m}{2\pi KT}} P = \sqrt{\frac{M}{2\pi RT}} P.$$

(10) Number of molecules having energy between ε and $\varepsilon + d\varepsilon$:

$$dN_\varepsilon = 2\pi N \left(\frac{1}{\pi kT}\right)^{\frac{3}{2}} \varepsilon^{\frac{1}{2}} e^{\frac{-\varepsilon}{kT}} d\varepsilon.$$

(11) Fraction of molecules having kinetic energy greater than ε:

$$\frac{N_{>\varepsilon}}{N} = \frac{2}{\sqrt{\pi}} \sqrt{\left(\frac{\varepsilon}{kT}\right)} e^{-\frac{\varepsilon}{kT}}.$$

(12) Equipartition of energy :

energy per degree of freedom : $\frac{1}{2}kT$

energy per vibrational degree of freedom : kT.

(13) Mean free path :

$$\lambda = \frac{RT}{\sqrt{2}\pi PN_A\sigma^2} = \frac{kT}{\sqrt{2}\pi P.\sigma^2} = \frac{1}{\sqrt{2}\pi N\sigma^2},$$

where $N' = \frac{N}{V}$ = number of molecules per cc.

(14) Collision frequency :

$$Z_1 = \frac{\bar{v}}{\lambda} = \sqrt{\frac{8RT}{\pi m}} \times \frac{\sqrt{2}\pi\sigma^2 N}{V} = \frac{4\sigma^2 N}{V}\sqrt{\left(\frac{\pi RT}{M}\right)}.$$

Frequency of bimolecular collision per unit volume

$$= Z_{AA} = \frac{1}{2} \times \sqrt{\frac{8RT}{\pi M}} \times \sqrt{2}\pi\sigma^2\left(\frac{N}{V}\right)^2$$

$$= 2\sigma^2\left(\frac{N}{V}\right)^2\sqrt{\left(\frac{\pi RT}{M}\right)}.$$

(15) Viscosity coefficient $\eta = \frac{1}{3}\rho\bar{v}\lambda$,

where $\rho = \frac{Nm}{V}$ = density.

(16) Poiseuille equation for liquid :

$$\eta = \frac{\pi R^4(P_1 - P_2)}{8Vl},$$ where V = volume of gas flowing in unit time.

$$V_g = \frac{\pi R^4(P_1^2 - P_2^2)}{16\eta P_0 l}.$$

(17) Some Important Integrals :

(a) $$\int_0^\infty x^{2n+1}e^{-\lambda x^2}dx = \frac{n!}{2\lambda^{n+1}}$$

(b) $$\int_0^\infty x^{2n+1}e^{-\lambda x^2}dx = \frac{1 \ldots 3 \ldots 5 \ldots (2n-1)}{2^{n+1}}\sqrt{\frac{\pi}{\lambda^{2n+1}}}$$

(c) $$\int_{-\infty}^{+\infty} e^{-\lambda x^2}dx = \sqrt{\frac{\pi}{\lambda}}$$

18. Error Function (erf)

$$\text{erf}(x) = \frac{2}{\sqrt{\pi}}\int_0^x e^{-y^2}dy$$

$$\text{erf}c(x) = 1 - \text{erf}(x) = \frac{2}{\sqrt{\pi}}\int_x^\infty e^{-y^2}dy$$

erf$c(x)$ is co-error function.

Table 1 : Values of erf(x) for different x

x	erf(x)	x	erf(x)
0.00	0.00	0.1	0.11
0.2	0.22	0.3	0.33
		0.4	0.43
0.5	0.52	0.6	0.60
0.7	0.68	0.8	0.74
		0.9	0.79
0.100	0.84		
1.10	0.88	1.20	0.91
		1.30	0.93
1.40	0.95	1.5	0.96

B2. Short Questions

1. Should an ideal gas have viscosity? Why?

Ans. : Viscosity is the internal friction opposing the relative motion of adjacent layers. In liquids this arises because of attraction between molecules in adjacent layers. In a gas the reason of viscosity is intermolecular collision between molecules of adjacent layers. Assuming hard sphere for molecules having collision diameter σ, it can be shown that viscosity coefficient (η) is

$$\eta = \frac{1}{3}\rho\bar{v}\lambda,$$

where ρ is density, $\bar{v}$ is average velocity and λ is mean free path. However, if we strongly adhere to the ideal gas kinetic model and assume molecules as geometric points, then since points, cannot collide, η = 0.

2. Viscosity and thermal conductivity must be related to one another. Why?

Ans. : Viscosity can be interpreted in terms of momentum transfer by molecular collisions. Thermal conductivity is the phenomenon of transfer of kinetic energy through a gas by molecular collision. Thus they should be interrelated. The relation is shown below in brief :

If Q is energy transport, the energy is transported due to collision. Hence,

$$Q = \frac{1}{3}\frac{N}{V}\bar{v}A\lambda\frac{dE}{dy},$$

where $\frac{dE}{dy}$ is K.E transported per unit length through area A.

But $Q = K\,A\dfrac{dT}{dy}$.

Amount of heat flow per unit area in unit time has to be proportional to temperature gradient. K is thermal conductivity coefficient in joule per metre per degree in SI unit.

Thus,

$$KA\frac{dT}{dy} = \frac{1}{3}\frac{N}{V}\bar{v}\lambda\frac{dE}{dT}\frac{dT}{dy}$$

$$K = \frac{1}{3}\frac{N}{V}\bar{v}\lambda\frac{dE}{dT}.$$

But for flow, $E = U$.

$$\therefore \quad \frac{dE}{dT} = mC_v,$$

where C_v is heat capacity per gram.

$$\therefore \quad K = \frac{1}{3}\frac{Nm}{V}\bar{v}\lambda C_v$$

$$= \frac{1}{3}\rho\bar{v}\lambda C_v = \eta C_v,$$

where η is viscosity coefficient.

3. "Viscosity and diffusion coefficient are related."—Explain and show the relation.

Ans. : Diffusion of a gas is a transport of mass across a concentration gradient. The rate of flow of molecules should be directly proportional to concentration gradient $\dfrac{dC}{dy}$.

$$\eta = D\frac{dC}{dy},$$

where D is diffusion coefficient. It can be shown that

$$D = \frac{1}{3}\bar{v}\lambda = \frac{\eta}{\rho},$$

where ρ is density and η is viscosity coefficient which arises due to momentum transfer and hence, they are related.

4. Why does collision diameter decrease with temperature?

Ans. : The collision diameter σ_0 is the sum of radius of two molecules when they collide. Because of intermolecular attraction collisions may take place when they are at a distance greater than σ_0 so that apparent collision diameter is greater than σ_0. As temperature T increases, the velocity increases so that attraction may be felt only at a shorter distance and the apparent collision diameter σ' decreases being equal to σ_0 only at infinite temperature. The empirical Sutherland equation expresses this relation as

$$\sigma^2 = \sigma_0^2\left(1 + \frac{C}{T}\right),$$

where C is a constant.

5. (i) Arrange in decreasing order : most probable velocity $(\bar{u})$, root mean square velocity $(\bar{c})$, average velocity $(\bar{v})$. (ii) What would happen to the differences and the ratio between them if temperature is increased and if molecular weight (*M*) is decreased?

Soln. : (i) $\bar{c} : \bar{v} : \bar{u} = \sqrt{\frac{3RT}{M}} : \sqrt{\frac{8RT}{pM}} : \sqrt{\frac{2RT}{M}}$

$$= \sqrt{3} : \sqrt{\frac{8}{\pi}} : \sqrt{2}$$

$$= 1 : 0{\cdot}92 : 0{\cdot}82.$$

Hence, $\bar{c} > \bar{v} > \bar{u}$.

Ans. : (ii) Difference between them will increase with incease in temperature but the ratio of them will remain constant. The same will happen to the difference and ratio between them if molecular weight is decreased.

B3. Numercial Problems

1. At what temperature will H_2 molecules have the same K.E. and r.m.s. velocity as N_2 molecules at 35 °C ? **[M.Sc. Admission Test, IIT, 1993]**

Soln. : r.m.s. of H_2 at T°K $\sqrt{\frac{3RT}{2}}$.

r.m.s. of N_2 at 35 °C = 308 K is $\sqrt{\frac{3R\times 308}{28}}$.

By the problem, $\frac{3RT}{2} = \frac{3R\times 308}{28}$.

$$\therefore \quad T = \frac{2\times 308}{28} = \mathbf{22\ K.}$$

2. In one litre container 10^{23} gas molecules each of mass 10^{-25} kg are present. The r.m.s. velocity of the molecules is 1000 ms^{-1}.

(i) What is the total K.E. of the molecules?

(ii) What is the temperature of the gas? **[GATE 1996]**

Soln. : By the problem,

$$\text{K.E.} = \frac{1}{2}m\overline{c^2}.xN = \frac{1}{2}\times 10^{-25}\times (10^3)^2 \times 10^{23} = \frac{1}{2}\times 10^4 = \mathbf{5000\ J.}$$

$$\text{Again, } PV = \frac{1}{3}mN\overline{c^2} = \frac{2}{3}\times N \times \frac{1}{2}m\overline{c^2} = \frac{2}{3}\text{K.E.}$$

$$\text{K.E.} = \frac{3}{2}PV = \frac{3}{2}nRT = \frac{3}{2}\frac{N}{N_A}RT$$

$$\text{or,} \quad 5000 = \frac{3}{2}\times\frac{10^{10}}{6.022\times 10^{23}} \times 8.314 \times T.$$

$$\therefore \quad T = \frac{5000\times 6.022\times 2}{3\times 8.314} = 2414{\cdot}40\text{K} = \mathbf{2414\ K.}$$

3. The density of oxygen at 273 °K and 1 atm is 1.4290 gm/l. Calculate the r.m.s. velocity of oxygen at 273 °K.

Soln. : $\bar{c} = \sqrt{\frac{3RT}{M}} = \sqrt{\frac{3PV}{M}} = \sqrt{\frac{3P}{d}}$

$P = 1 \text{ atm} = 1.01325 \times 10^5 \text{N.m}^{-2}$

$\rho = 1.4290 \text{ gm/l} = 1.4290 \text{ kg.m}^{-3}$

$\therefore \quad \bar{c} = \sqrt{\frac{3 \times 1.01325 \times 10^5}{1.4290}} = 461.2143589 \text{ m.s}^{-1} = 461.21 \text{m.s}^{-1}.$

4. It was found that at 2800 K tungsten ($M = 183.80$) loses a weight of 2.15×10^{-6} gm per second from a heated filament of 5 sq cm area. Calculate the vapour pressure of tungsten at 2800 K.

Soln. : We know that mass leaking per area A in t sec is

$$\mu = \sqrt{\frac{M}{2\pi RT}} P.At.$$

By the problem,

$\mu = 2.15 \times 10^{-9}$ kg

$M = 0.1838$ kg.mole^{-1}

$R = 8.314$ J.K.mole^{-1}

$T = 2800$ K

$A = 5 \times 10^{-4}$ sq. metre.

$$\therefore P = \frac{\mu\sqrt{2\pi RT}}{A\sqrt{M}} = \frac{2.15 \times 10^{-9} \sqrt{2 \times 3.142 \times 8.314 \times 2800}}{5 \times 10^{-4} \sqrt{0.1838}}$$

$$= \frac{8.223194714 \times 10^{-7}}{5 \times 10^{-4} \sqrt{0.1838}} = 3.84 \times 10^{-3} \text{ Pa}$$

$\approx$ **2.9×10^{-5} mm of Hg** (760 mm of Hg = 1.01325×10^5 Pa).

5. The vapour pressure of a solid at 2600 K was found to be 5.24×10^{-7} atm. It was found that 6.48×10^{-4} gm of the substance passes through an opening of 3.25 mm^2 in 3.5 hours. Calculate its atomic weight.

Soln. : Mass leaking $\mu = \sqrt{\frac{M}{2\pi RT}} P.A.t$

$\mu = 6.48 \times 10^{-7}$ kg

$A = 3.25 \times 10^{-6}$ m^2

$P = 0.0530943$ Pa

$t = 3.5 \times 60 \times 60$ sec.

$$\sqrt{M} = \frac{\mu\sqrt{2\pi RT}}{P.At}$$

or, $M = 0.012066072,$

$\therefore$ **$M = 12$.**

6. Two flasks (A) and (B) have equal volumes. (A) contains H_2 gas and (B) contains an equal mass of CH_4. (A) is maintained at 300 K while (B) is mantained at 600 K. Assuming ideal behaviour for the gases, answer the following :

(i) In which flask is pressure greater? How many times one is greater than the other?
(ii) In which flask is the mean free path of molecules greater? How many times one is greater than the other? (The collision diameter of CH_4 is twice that of H_2.)

[C.U.(H) 2002]

Soln. : By the problem : Let w gm H_2 be taken, then

(i) $$P_{H_2} = \frac{w \times R \times 300}{2 \times V}$$

$$P_{CH_4} = \frac{w \times R \times 600}{16 \times V}$$

$$\therefore \quad \frac{P_{H_2}}{P_{CH_4}} = \frac{300}{2} \times \frac{16}{600} = 4.$$

Hence P_{H_2} is greater and it is 4 times greater than P_{CH_4}.

(ii) $$\lambda_{H_2} = \frac{RT}{\sqrt{2}\pi P_{H_2} N_A \sigma^2_{H_2}} = \frac{300R}{\sqrt{2}\pi P_{H_2} N_A \sigma^2_{H_2}}.$$

$$\lambda_{CH_4} = \frac{600R}{\sqrt{2}\pi P_{CH_4}\left(4\sigma^2_{H_2}\right)} \quad (\text{since, } \sigma_{CH_4} = 2\sigma_{H_2}).$$

$$\therefore \quad \frac{\lambda_{CH_4}}{\lambda_{H_2}} = \frac{2 \times 4}{4} = 2.$$

$\therefore$ **λ_{CH_4} is 2 times greater than λ_{H_2}.**

7. The viscosity coefficient η of H_2 is 8.41×10^{-5} poise at 0 °C and 1 atmosphere pressure. Determine λ.

Soln. : $$\eta = \frac{1}{3}\rho\bar{v}\lambda$$

$$= \frac{1}{3}\frac{MP}{RT}\sqrt{\frac{8RT}{\pi M}}\lambda$$

$$= \frac{1}{3} \times \frac{2\,\text{gm} - \text{mole}^{-1} \times 1\,\text{atm}}{82.06\,\text{cc.atm.K}^{-1}.\text{mole}^{-1} \times 273.15\text{K}} \times \sqrt{\frac{8 \times 8.314 \times 10^7\,\text{erg.K}^{-1}.\text{mol}^{-1} \times 273.15\,\text{K}}{3.142 \times 2\,\text{gm} - \text{mole}^{-1}}}$$

$$= \frac{1}{3} \times \frac{2 \times 1}{82.06 \times 272.15}\sqrt{\frac{8 \times 8.314 \times 10^7 \times 273.15}{3.142 \times 2}}\lambda$$

$$= 5.05708894\ \text{gm.cm}^{-2}.\text{sec}^{-1}\ \lambda = 5.06\ \text{gm.cm}^{-2}.\text{sec}^{-1}\lambda.$$

$$\therefore \quad \lambda = \frac{8.41 \times 10^{-5}}{5.06} = \mathbf{1.66 \times 10^{-5}\ cm.}$$

8. Calculate η (viscosity coefficient) of nitrogen at 300 K ($\sigma = 4\times10^{-8}$ cm).

Soln. : $$\eta = \frac{1}{3}\rho\bar{v}\lambda = \frac{MP}{RT}\bar{v} = \sqrt{\frac{8RT}{\pi M}}$$

$$\therefore \quad \eta = \frac{1}{3}\frac{MP}{RT}\sqrt{\frac{8\times8.314\times10^7\times300}{3.142\times28}}.\frac{RT}{\sqrt{2}\pi PN_A\sigma^2}$$

$$= \frac{1}{3}28\sqrt{\frac{8.314\times10^7\times2400}{3.142\times28}}\times\frac{1}{\sqrt{2}\times3.142\times\left(4\times10^8\right)^2 N_A}$$

$$= 1.038205653\times10^{-4} \quad = \mathbf{1.04\times10^{-4}\ poise.}$$

9. Molecular diameter of CO is 3.19 Å. At 27 °C and 100 mm of Hg pressure what could be (i) collision frequency and (ii) mean free path.

Hints : $$Z = \frac{\pi\sigma^2}{\sqrt{2}}.\bar{v}.\left(\frac{N_AP}{RT}\right)$$

By the problem,

$$Z = \frac{3.142\times\left(3.19\times10^{-8}\right)^2}{\sqrt{2}}\times\sqrt{\frac{8\times8.314\times10^7\times300}{3.142\times28}}\times\left(\frac{6.022\times10^{23}\times100}{760\times0.08206\times10^3\times300}\right)$$

$$\lambda = \frac{RT}{\sqrt{2}\pi PN_A\sigma^2}.$$ **[Ans.** 3.46×10^8 sec^{-1}, 6.87×10^{-5} cm]

10. A gas contains 10^{15} molecules/l. The mean free path of the molecules is $\frac{1}{\sqrt{2}\pi}$ metre. Calculate σ. **[B.U. 1991]**

Soln. : We know that $\lambda = \frac{1}{\sqrt{2}\pi\sigma^2N'}$,

where σ^2 is the square of collision diameter and N' is the number of molecules per unit volume

By the problem, $N' = 10^{18}$ molecules/m^3

$$\frac{1}{\sigma^2N'} = 1 \text{ metre}$$

$$\sigma^2 = \frac{1}{N'}m^2$$

or, $$\sigma = \sqrt{\frac{1}{10^{18}}} = 10^{-9}\text{ m.}$$

$$\therefore \quad \sigma = \mathbf{1.0\ Å.}$$

11. The viscosity of hydrogen at NTP is 8.4×10^{-5} poise and average velocity of gas moles is 1.7×10^5 cm/sec. Calculate the mean free path. **[B.U. 1997]**

Soln. : We know that $$\eta = \frac{1}{3}\rho\bar{v}\lambda = \frac{1}{3}\frac{MP}{RT}\bar{v}\lambda$$

$$= \frac{1}{3}\frac{2\times1}{82.06\times298.15}\times1.7\times10^5\lambda$$

$R = 82.06$ cc.atm.deg^{-1}.mole^{-1}

$$\lambda = \frac{\eta\times82.06\times273.15\times3}{2\times1\times1.7\times10^5}.$$

$$= \frac{8.4\times10^{-5}\times82.06\times273.15\times3}{2\times1.7\times10^5}$$

$= 1.661324\times10^5$ cm $\approx$ **1.7×10^5 cm.**

12. At NTP, the viscosity coefficient of oxygen is 0.2 millipoise. Find the collision diameter of oxygen. **[C.U. 1996]**

Soln. : $\eta = \frac{1}{3}\rho\bar{v}\lambda = \frac{1}{3}\frac{MP}{RT}\sqrt{\frac{8RT}{\pi M}}\frac{RT}{\sqrt{2}\pi PN_A\sigma^2}$

$$\sigma^2 = \frac{1}{3}\sqrt{\frac{8RTM}{\pi}}\frac{1}{\sqrt{2}\pi N_A\eta}$$

$\eta = 0.2\times10^{-3}$ poise $= 2\times10^{-4}$ poise

$$\sigma^2 = \frac{1}{3}\sqrt{\frac{8\times8.314\times10^7\times273.15\times32}{3.142}}\times\frac{1}{\sqrt{2}\times3.142\times6.022\times10^{23}\times2\times10^{-4}}$$

$= 8.473735682\times10^{-16}$ cm^2

$\therefore\ \sigma$ = **2.91×10^{-8} cm.**

13. In a carefully designed vacuum tube it is possible to reach a pressure of 10^{-10} mm of Hg. Calculate the mean free path of He atoms at this pressure and at 25 °C. Given, $\sigma = 2.5$ Å. **[C.U. 1987]**

Hints : $\lambda = \frac{RT}{\sqrt{2}\pi PN_A\sigma^2}$ = **1.11117×10^8 cm.**

14. The molecular diameter of a gas is 2.85 Å. What will be the mean free path of the molecules at 25 °C and 1 atm. **[C.U. 1989]**

[Ans. 1.125×10^{-5}cm]

15. The coefficient of viscosity of He at 500 K is 279×10^{-4} g.cm^{-1}.sec^{-1}. Calculate the collision diameter of He. Given that $k = 1.38\times10^{-16}$ erg.K^{-1} **[C.U. 1991]**

Hints : $\eta = \frac{1}{3}\rho\bar{v}\lambda$

$R = 1.38\times10^{-16}\times N_A$. **[Ans.** 1.7×10^{-8} cm]

$= 8.3103\times10^7$ erg (Use this value of R).

16. The van der Waals constant b for CO_2 is 42.9×10^{-3} l per mole. Calculate η for CO_2 at 27 °C. **[B.U.(H) 1990]**

Hints : $\sigma = 3.329\times10^{-8}$ cm

$\eta = \frac{1}{3}\rho\bar{v}\lambda$ = **1.97269×10^{-4} gm.cm^{-1}.sec^{-1}.**

17. At 15 °C, the visocity coefficient of CO_2 is 172×10^{-4} poise. Calculate its mean free path and collision diameter. **[C.U. 1984]**

Hints : $\eta = \frac{1}{3}\rho\bar{v}\lambda$

$\sigma = 3.47$ Å

$\lambda = \frac{7.67}{P}$ cm (P in dyne/cm^2).

18. One mole of HI is present in 1 dm^3 at 700 K. Find out the number of collision in 1×10^{-3} dm^3 and the mean free path of the gas. **[V.U. 1989]**

Hints : $Z_{AA} = \frac{\pi\sigma^2}{\sqrt{2}}\sqrt{\frac{8RT}{\pi M}}(N_A \times 10^{-3})^2$ (since 10^{-3} dm^3 = 1 cc)

$= 2.74 \times 10^{46}\sigma^2$

$\lambda = \frac{3.7376 \times 10^{-22}}{\sigma^2}$.

19. λ for N_2 at 0 °C and 1 atm is 10^{-5} cm. Calculate the collision diameter of N_2. What will be the value of λ at a pressure of 1 torr. Calculate the average time between collisions at 0 °C and 1 atm. **[C.U. 1997]**

Hints : $\lambda = \frac{RT}{\sqrt{2}\pi P N_A \sigma^2}$

$\therefore \sigma^2 = \frac{RT}{\sqrt{2}\pi P N_A \lambda}$.

$\therefore \sigma = 2.89 \times 10^{-8}$ cm = 2.89 Å

760 torr = 1 atm

$\therefore \lambda = 7.6 \times 10^{-3}$ cm at 1 torr

Average time $= \frac{\pi}{\bar{v}} = \frac{10^{-5}}{\sqrt{\frac{8RT}{\pi M}}} = 10^{-5} \times \sqrt{\frac{\pi M}{8RT}} = $ **2.20×10^{-10} sec.**

20. Calculate the number of binary collision per cc of N_2(g) per sec at 2 atm pressure and 30 °C. The bond length of the gas molecules is 1.87Å . **[C.U. 1998]**

Hints : By the problem,

$\sigma = 1.87 \times 10^{-8}$ cm

$T = 303$ K

$M = 28$ gm

$P = 2$ atm

$Z_{AA} = \frac{\pi\sigma^2}{\sqrt{2}}\bar{v}\left(\frac{N_A P}{RT}\right)^2 = 8.72 \times 10^{28}$ per cc per sec.

21. Calculate the total kinetic energy of all the molecules of 1 mole of an ideal gas present in 3-l vessel under 2 atm pressure at a certain temperature. **[C.U.(H) 1999]**

Soln. : K.E. $= \frac{3}{2}PV = $ **911.92 J.**

22. A 10-litre capacity metallic container with 0.1 gm-mole of $H_2(g)$ in it has a leakage hole of radius 0.02 mm. The r.m.s. velocity of H_2 molecule in the container is 400 m/sec. How many hydrogen molecules escape through the hole per minute? **[C.U. 1980]**

Soln. : A hole of radius 0.002 cm means an area A of $\pi r^2 = 3.142(2\times10^{-3})^2$ cm^2

Molecules striking area A of the container per minute will escape. Z, the number of escaping molecules

$$= \frac{1}{\sqrt{6\pi}}\frac{N}{V}\bar{c}\ At$$

$$= \frac{1}{\sqrt{6\pi}}\times\frac{0.1\text{mole}\times6.022\times10^{23}\text{ molecules.mole}^{-1}}{10\times10^{3}\text{cc}}\times 400\times10^2\text{ cm.sec}^{-1}$$

$$\times\ 3.142\times(2\times10^{-3})^2\text{ cm}^2\times 60\text{ sec}$$

$= 4.18349\times10^{19} = \mathbf{4.183\times10^{19}}$.

23. At 2500 K, 0.5 mg of carbon passed through an opening of 3 mm^2 in 2.5 hours. Calculate the v. p. of carbon at 2500 K.

Hints : $\mu = \sqrt{\dfrac{M}{2\pi RT}}\ P.\ A.t.$

$$\therefore\ P = \frac{\mu\sqrt{2\pi RT}}{\sqrt{MA.t}} = \mathbf{0.6109\ dyne/cm^2}.$$

24. The van der Waals constant b for CO_2 is 42.9×10^{-3} l.mole^{-1}. Calculate η for CO_2 at 27 °C. **[B.U. 1990]**

Hints : σ from $b = 3.239\times10^{-8}$ cm.

$$\eta = \frac{1}{3}\rho\bar{v}\lambda$$

$$= \frac{1}{3}\frac{MP}{RT}\sqrt{\frac{8RT}{\pi M}}\cdot\frac{RT}{\sqrt{2}\pi PN_A\sigma^2} = \frac{2}{3}\frac{\sqrt{MRT}}{\sqrt{\pi^3}N_A\sigma^2} = \mathbf{1.99\times10^{-4}\ poise.}$$

25. A gas whose viscosity is 200 micropoise flows through a capillary tube 2 mm in diameter and 2 metres long. If 5 litres of a gas pass through the tube every 10 sec, what must be the pressure head in atmosphere unit under which the gas is flowing? **[C.U. 1988]**

Soln. : For any fluid,

$$V = \frac{\pi R^4(P_1 - P_2)}{8\eta l}\lambda.$$

$$\therefore\ (P_1 - P_2) = \frac{8V\eta l}{\pi R^4},$$

where V is volume passing per unit time, η is viscosity, λ is length, R is the radius of the tube.

By the problem,

$\eta \quad = 200\times10^{-6}$ poise

$$V = \frac{5000}{10} = 500 \text{ cc/sec}$$

$P_1 - P_2 = P$, the pressure head

$$R = \frac{0.2}{2} \text{ cm} = 0.1 \text{ cm}$$

l = length of the tube = 2 m = 200 cm

$$P = \frac{8 \times 500 \text{ cc.sec}^{-1} \times 200 \times 10^{-6} \text{ poise} \times 200 \text{ cm}}{3.142 \times (0.1)^4 \text{ cm}^4}$$

$= 509229.7899$ dynes/cm^2

= 0.50257 atm.

26. What will be the ratio of final to initial Z values for a gas, if P is halved at constant gas density?

Soln. : Z = wall collision frequency $= \dfrac{P}{\sqrt{2\pi mkT}}$

Now, $P = \dfrac{N}{V}kT.$

New pressure, $P_1 = \dfrac{P}{2}.$

Since $\rho = \dfrac{Nm}{V}$ = constant, $\dfrac{N}{V}$ = constant. Hence, $T_1 = \dfrac{T}{2}.$

Thus, $\dfrac{Z_1}{Z} = \dfrac{P_1}{\sqrt{T_1}} \times \dfrac{\sqrt{T}}{P} = \dfrac{P_1}{\sqrt{T_1}} \times \dfrac{\sqrt{2T_1}}{2P_1} = \dfrac{1}{\sqrt{2}}$ **= 0.71.**

So wall collision frequency will decrease.

27. Two separate bulbs are filled with two ideal gases A and B respectively. The molecular weight of B is twice that of A. If B is at twice the absolute temperature and has half the density of A, what is the ratio of wall collision frequencies?

Soln. : By the problem, $T_B = 2T_A,$

$M_B = 2M_A$

$\rho_B = \dfrac{\rho_A}{2}.$

Now, $P = \dfrac{\rho}{M}RT.$

Thus, $\dfrac{P_A}{P_B} = \dfrac{\rho_A T_A}{M_A} \cdot \dfrac{M_B}{\rho_B T_B} = 2.$

Now, $Z = \dfrac{PN_A}{\sqrt{2AMRT}}.$

Hence, $\dfrac{Z_A}{Z_B} = \dfrac{P_A\sqrt{M_B T_B}}{P_B\sqrt{M_A T_A}}$ **= 4.**

28. Determine the ratio $\bar{v}_{old}/\bar{v}_{new}$ of average velocity and Z_{old}/Z_{new} of wall collision frequencies of an ideal gas, if (a) T is doubled at constant P (b) P is doubled at constant T.

[**Ans.** (a) $\frac{1}{\sqrt{2}}, \sqrt{2}$, (b) 1, $\frac{1}{2}$]

29. A bulb contains a mixture of H_2 and O_2 at 0 °C and 1 atm total pressure. The bulb contains 68% by weight of H_2. Calculate the ratio Z_{H_2}/Z_{O_2}.

Soln. : In 100 gm mixture,

weight of H_2, $w_{H_2} = 68$ gm

weight of O_2, $w_{O_2} = 32$ gm.

Hence, $P_{O_2} = n_{O_2}\frac{RT}{V} = \frac{RT}{V} (M_{O_2} = 32)$

$P_{H_2} = n_{H_2}\frac{RT}{V} = 34\frac{RT}{V}$ $(M_{H_2} = 2)$.

Hence, $\frac{Z_{H_2}}{Z_{O_2}} = \frac{P_{H_2}}{P_{O_2}}\sqrt{\frac{M_{O_2}}{M_{H_2}}} = 34\sqrt{\frac{32}{2}} = \mathbf{136.}$

30. What would be the molecular weight of a gas if pressure of the gas falls to one half its value at a height of 1 m at 25 °C ?

Hints : $P_h = P_0 e^{-\frac{Mgh}{RT}}$

By problem, $P_h = \frac{P_0}{2}$

$M = \mathbf{175.23\ kg.}$

31. What will be the percentage change in barometric pressure at 2000 m above the sea level at 300 K, if it is assumed that air is roughly a mixture of 20% O_2 and 80% N_2 ? **[B.U. 2004]**

Soln. : At sea level, total pressure $P = 1$ atm.

$P_0^{O_2} = \frac{1}{5}$ atm, $P_0^{N_2} = \frac{4}{5}$ atm

$P_h = P_h^{N_2} + P_h^{O_2} = P_0^{N_2} e^{-\frac{M_{N_2}gh}{RT}} + P_0^{O_2} e^{-\frac{M_{O_2}gh}{RT}}$

$M_{O_2} = 32 \times 10^{-3}$ kg, $M_{N_2} = 28 \times 10^{-3}$ kg, g = 9.8 m/sec^{-2}

$R = 8.314$ J.K^{-1}.mole^{-1}, $h = 2000$ m, $T = 300$ K.

$\therefore$ $P_h = 0.797$.

Thus, pressure decreases by **20.3%**.

32. At what temperature will N_2 molecules have same momentum as H_2 at 25 °C?

Soln. : Momentum of gas molecules = molar mass × average velocity

Momentum of N_2 molecules $= M_{N_2}\bar{v} = 28$ gm $\times \sqrt{\frac{8RT}{\pi 28}}$ cm.sec^{-1}

Momentum of H_2 molecules $= M_{H_2}\bar{v} = 2 \text{ gm} \times \sqrt{\dfrac{8RT_{H_2}}{\pi 2}}$ cm.sec^{-1}

By the problem, $28\sqrt{\dfrac{8RT}{28\pi}} = 2 \times \sqrt{\dfrac{8RT_{H_2}}{2\pi}}$

Given $T_{H_2} = 25\ °C = (273.15 + 25)\ K = 298.15\ K$

$$28\sqrt{\frac{T}{28}} = 2 \times \sqrt{\frac{298.15}{2}}$$

$\therefore$ T = **21.29 K.**

At **21.29 K** N_2 molecules will have same momentum as H_2 at 25 °C.

33. An ideal gas, kept in a thin walled vessel of volume V at a temperature T, is slowly escaping into vacuum through a small hole of area A. How much time is required for the pressure in the vessel to decrease to $\frac{1}{e}$ of its original value?

Soln. : Let N be the number of gas molecules in the vessel.

By the problem, volume of the vessel = V

Temperature = T

Area of the hole = A.

If P is the pressure, $PV = NkT$.

Let dN number of molecules escape through the hole in time dt.

Here $dN = \dfrac{1}{4}\dfrac{N}{V}\bar{v}Adt$,

where $\bar{v}$ = average velocity.

Thus, the leakage reduces the number of gas molecules in the vessel by dN, thereby reducing the pressure by dP in time dt.

Hence, $dP = -\dfrac{1}{4}\dfrac{P}{V}\bar{v}Adt$

or, $\displaystyle\int_P^{P/e}\frac{dP}{P} = -\int_0^t \frac{\bar{v}A}{4V}dt$ or, $\ln\dfrac{1}{e} = -\dfrac{\bar{v}At}{4V}$.

Hence, $t = \dfrac{4V}{\bar{v}A}$.

Thus, the gas requires $\dfrac{4V}{\bar{v}A}$ time to decrease the pressure to $\dfrac{1}{e}$ of its orginal value.

34. A container with a small hole in its thin wall contains an ideal gas A at temperature T and pressure P. The pressure is found to be halved in one hour due to effusion of the gas into vacuum.

If the same container is filled with an equi-molecular mixture of ideal gases A and B at temperature T and pressure P, find out the ratio $\dfrac{P_A}{P_B}$ after one hour.

Soln. : Initially, let N be the number of A gas molecules, and if V be the volume of the container, then at temperature T and pressure P,

$PV = N.KT.$

If dN gas molecules effuse in dt time, then

$$-dN = \frac{P}{kT}\bar{v}_A A dt, \tag{1}$$

where $\bar{v}_A$ = average velocity of A

A = area of the hole.

$$\text{Again, } dN = \frac{VdP}{kT}. \tag{2}$$

$$\text{Thus, } \frac{dP}{P} = -\frac{\bar{v}_A}{4V}Adt \text{ [from eqn. (1)]}$$

$$\text{or, } \int_P^{\frac{P}{2}} d\ln P = -\frac{\bar{v}_A}{4V}A\int_0^1 dt.$$

$$\text{Thus, } \ln 2 = \frac{\bar{v}_A}{4V}A. \tag{3}$$

If N_A and N_B be the number of molecules of A and B respectively,

$$P_AV = N_AkT \tag{4a}$$

$$P_BV = N_BkT. \tag{4b}$$

By the problem, $N_A = N_B = \frac{N}{2}$.

where N = total number of molecules = $N_A + N_B$.

By the problem, $PV = NkT = 2N_AkT = 2P_AV$ [from eqns. (4a) and (4b)]

$$\text{Hence, } P_A = P_B = \frac{P}{2}. \tag{5}$$

Let $P_A(f)$ and $P_B(f)$ be the final pressures of A and B after one hour respectively.

$$\text{Thus, } \ln\frac{P_A(f)}{P_A} = -\frac{\bar{v}_A A}{4V} = -\ln 2 = \ln\frac{1}{2}.$$

$$\text{Hence, } P_A(f) = \frac{P_A}{2}. \tag{6}$$

$$\text{Again, } \ln\frac{P_B(f)}{P_B} = -\frac{1}{4}\frac{\bar{v}_B A}{V} \quad [\bar{v}_B = \text{average velocity of } B]$$

$$= -\frac{\bar{v}_B \ln 2}{\bar{v}_A} \text{ [from eqn. (3)]}$$

$$\text{or, } \ln\left(\frac{P_B}{P_B(f)}\right) = \ln(2)^{\frac{\bar{v}_B}{\bar{v}_A}}.$$

or, $\dfrac{P_B}{P_B(f)} = 2^{\frac{\bar{v}_B}{\bar{v}_A}}$.

$\therefore \quad P_B(f) = P_B 2^{-\frac{\bar{v}_B}{\bar{v}_A}}$. (7)

Hence, $\dfrac{P_A(f)}{P_B(f)} = \dfrac{P_A}{2.P_B.2^{-\frac{\bar{v}_B}{\bar{v}_A}}}$ [from eqn. (6) and eqn. (7)]

$= 2^{\left(\frac{\bar{v}_B}{\bar{v}_A}-1\right)}$ [since $P_A = P_B$].

35. Find the value of most probable kinetic energy of gas molecules.

Soln. : Fraction of molecules having kinetic energy between ε and $\varepsilon + d\varepsilon$ is given by $\rho_\varepsilon = \dfrac{dN_\varepsilon}{N} = 2\pi\sqrt{\left(\dfrac{1}{\pi kT}\right)^3}\sqrt{\varepsilon}e^{-\frac{\varepsilon}{kT}}d\varepsilon$.

Most probable kinetic energy corresponds to maximum of energy distribution curve. At the maximum.

$$\frac{d\rho_\varepsilon}{d\varepsilon} = 0$$

or, $\dfrac{d}{d\varepsilon}\left(\varepsilon^{\frac{1}{2}}e^{-\frac{\varepsilon}{kT}}\right) = 0$

or, $\dfrac{1}{2}\varepsilon^{-\frac{1}{2}}e^{-\frac{\varepsilon}{kT}} - \varepsilon^{\frac{1}{2}}e^{-\frac{\varepsilon}{kT}} \times \dfrac{1}{kT} = 0$ or, $\dfrac{1}{2}\varepsilon^{-\frac{1}{2}} - \dfrac{\varepsilon^{\frac{1}{2}}}{kT} = 0$

or, $\dfrac{1}{2} = \dfrac{\varepsilon}{kT}$, $\therefore \varepsilon = \dfrac{1}{2}kT$

Miscellaneous

1. If 5 gm of ammonium carbamate dissociates to ammonia and carbon dioxide at 200 °C, the products occupying a volume of 7.66 l at a pressure of 740 mm of Hg. Calculate the degree of dissociation of NH_2COONH_4.

Soln. : The reaction is

$NH_4COONH_2 \rightarrow 2NH_3 + CO_2$

78 gm

$1 - \alpha \qquad 2\alpha \qquad \alpha$

NH_3 occupies 44.828 l at NTP and CO_2 occupies 22.414 l—a total of 67.242 l at NTP from decomposition of 78 gm of NH_2COONH_4.

Thus at NTP, 5 gm NH_4COONH_2 on complete decomposition will occupy

$$\frac{5}{78} \times 67.2 = 4.31 \text{ l},$$

which under the given condition is

$$V = \frac{P_o V_o}{T_o} \times \frac{T}{P} = \frac{760 \times 4.31}{273.15} \times \frac{473.15}{740}$$

$= 7.6675$ l $= 7.67$ l.

So the decomposition is not complete and $\frac{7.66}{7.67} \times 100 = 99.869\%$ has decomposed.

$\therefore\ \alpha =$ **0.998.**

2. An iron pipe 3 m long closed at one end is lowered vertically into water until the close end is flushed with water. Calculate the height of the water level in the pipe. Given diameter of the pipe = 2 inches.

density of water = 1×10^3 kg.m^{-3}

pressure = 1×10^5 Pa = 10 m (hydrostatic head of water), temp. is kept constant.

Soln. : Since temperature is constant,

$$P_f V_f = P_{in} V_i$$

$$P_f = P_{in} \frac{3 \times \text{area of the pipe}}{h \times \text{area of the pipe}} = \frac{3P_{in}}{h}$$

But by the problem,

$P_{in} = 10$ m and $P_f = h_{in} + 10$ m

$$\therefore \quad h + 10 = \frac{3 \times 10}{h}$$

i.e., $h^2 + 10h - 30 = 0$.

$$\therefore \quad h = \frac{-10 + \sqrt{10^2 + 120}}{2} = 2.4161985 \text{ m} = \mathbf{2.4\ m.}$$

The pipe is 3 m long. Hence, water level would rise up **0.6 m** in the pipe (3 – 2.4 = 0.6)

3. A sealed vessel of 1 litre capacity contains air and a small quantity of water. At 25 °C, the pressure in the vessel is 1 atm, and it is found that at 200 °C the measured pressure is 4.6 atm. Assume that the vapour pressure of water at 25 °C is negligible and that of steam at 200 °C behaves like an ideal gas and that the vessel does not expand on heating. Calculate the volume of water in the vessel at 25 °C. Given that density of water at 25 °C 10^3 kg/m^3, 0 °C = 273 °K, $M_{H_2O} = 18$ and $R = 0.08206$ l.atm.K^{-1}.mole^{-1}.

Soln. : Let n be the number of moles of air and n' be the number of moles of water at 200 °C, then

$$n + n' = \frac{PV}{RT} = \frac{4.6 \text{ atm} \times 1 \text{ litre}}{0.08206 \text{ l.atm.deg}^{-1}.\text{mole}^{-1} \times 473\text{K}} = 0.118585.$$

Now, at 25 °C, n' mole of water occupies

$18n'$cc $= 18(0.118585 - n)$ cc $= 0.018(0.118585 - n)$ l

$$\therefore \quad n = \frac{1 \text{ atm} \times [1 - 0.018(0.118585 - n)]1}{0.08206 \text{ l.atm.deg}^{-1}.\text{mole}^{-1} \times 298\text{K}} = 0.0408609 = 0.04.$$

$n' = 0.077719 = 0.08$

$V' = 18n' = 1.398942$ cc $=$ **1.40 cc.**

4. Hydrogen will dissociate into atoms at high enough temperature, that is $H_2 = 2H$. Assuming ideal gas behaviour of H_2 and H atom, what would be the density of hydrogen at 2000 °C, if 35% dissociates into atoms? The pressure is 1 atm.

Soln. : The reaction is $H_2 = 2H$

At 2000 °C and $1 - 0.35 \quad 2 \times 0.35$

1 atm, $\alpha = 0.35$ $= 0.65 \quad = 0.70$

Thus total number of moles is 1.35.

Average molecular weight is $\frac{\text{total weight}}{\text{toal number of mole}} = \frac{2}{1.35} = 1.481.$

$$\therefore \quad \rho = \frac{PMa}{RT} = \frac{1 \times 1.481}{0.08206 \times 2273.15} = \mathbf{7.94 \times 10^{-3}\ gm/l.}$$

5. Calculate the value of mean square deviation $\left\langle (c - \langle \bar{c} \rangle)^2 \right\rangle$ for O_2 at 27 °C.

Soln. : $\left\langle (c-\langle\bar{c}\rangle)^2\right\rangle = \left\langle c^2 - 2c\langle\bar{c}\rangle + \langle\bar{c}\rangle^2\right\rangle$

$$= \left\langle \overline{c^2} \right\rangle - 2\langle\bar{c}\rangle\langle\bar{c}\rangle + \left\langle \bar{c}^2 \right\rangle = \left\langle \bar{c}^2 \right\rangle - \langle\bar{c}\rangle^2$$

$$\therefore \quad \left\langle (c-\langle\bar{c}\rangle)^2\right\rangle = \frac{3RT}{M} - \frac{8RT}{\pi M} = \left(3 - \frac{8}{\pi}\right)\frac{RT}{M}$$

$$= \left(3 - \frac{8}{\pi}\right)\frac{8.314\,\text{J.K}^{-1}.\text{mole}^{-1} \times 300\,\text{K}}{0.032\ \text{kg.mole}^{-1}}$$

$$= \left(3 - \frac{8}{\pi}\right) 7.794375 \times 10^4\ \text{m}^2.\text{sec}^{-2}$$

$$= \mathbf{3.53748828 \times 10^4\ m^2.sec^{-2}.}$$

6. A 12-litre flask is filled with 20 gm neon and an unknown weight of hydrogen. The gas density is 0.002 gm/cc at 0 °C. Calculate the average molecular weight and number of grams of H_2 added. Calculate the pressure.

Soln. : Given density is 2 gm/l. Thus, there is 24 gm materials out of which 20 gm of neon has been taken. Automatically 4 gm of H_2, i.e., 2 moles of H_2 has to be taken.

Total weight 24 gm and there are 3 moles (1 mole neon + 2 moles of H_2).

$$\therefore\ Mav = \frac{24}{3} = 8 \text{ gm/mole}$$

$$P = \frac{nRT}{V} = \frac{3 \times 0.08206 \times 273.15}{12} = \mathbf{5.603\ atm.}$$

7. An ideal gas at NTP has a most probable velocity $\bar{u} = 4.20 \times 10^4$ cm/sec and a mean free path of 7.90×10^{-8} cm. Calculate the mean time between collision.

Soln. : We know that $\bar{u} = \sqrt{\frac{2RT}{M}}$ and average velocity $\bar{v}$ is $\sqrt{\frac{8RT}{\pi M}}$.

$$\therefore \quad \frac{\bar{v}}{\bar{u}} = \sqrt{\frac{8}{2\pi}}.$$

$$\therefore \quad \bar{v} = \frac{2}{\sqrt{\pi}}\bar{u} = \frac{2}{\sqrt{\pi}} \times 4.20 \times 10^4 \text{ cm.sec}^{-1} = 4.7392 \times 10^4 \text{ cm/sec.}$$

Now average time between collision is $\tau = \frac{1}{Z} = \frac{\lambda}{\bar{v}}$, where λ is mean free path, and $\bar{v}$ is average velocity.

$$\therefore \quad \tau = \frac{7.90 \times 10^{-8}}{4.732 \times 10^4} = 1.666948 \times 10^{-12} \text{ sec} = \mathbf{1.67 \times 10^{-12} \text{ sec.}}$$

8. Calculate the average time of collision made by a N_2 molecule with successive oxygen molecules in air at 298.15 K at a total pressure of 1 atm. Given, $\sigma_{O_2} = 3.41$ Å and σ_{N_2} = 3.71 Å.

Soln. : Average time between collision is $\tau = \frac{\lambda}{\bar{v}}$ (mean free path divided by average velocity). It is also inverse of $Z_{N_2(O_2)}$.

$$Z_{N_2(O_2)} = \pi\sigma_{AB}^2\sqrt{\frac{8RT}{\pi\mu}}\,C_{O_2}$$

$$\frac{1}{\mu} = \frac{1}{M_A} + \frac{1}{M_B} = \frac{M_A + M_B}{M_A M_B}.$$

$$\therefore \quad \mu = \frac{M_A M_B}{M_A + M_B} = \frac{28 \times 32}{32 + 28} = 14.93 \text{ gm.}$$

$$\sigma_{AB} = \frac{\sigma_A + \sigma_B}{2} = \frac{3.41 + 3.71}{2} \times 10^{-8} \text{ cm} = 3.56 \times 10^{-8} \text{ cm.}$$

$$\text{Concentration } \frac{N}{V} = \frac{P \times N_A}{R \times 298.15}$$

$$= \frac{6.022 \times 10^{23} \text{ molecules. mol}^{-1} \times 1 \text{ atm}}{0.08206 \text{ l.atm.K}^{-1}.\text{mole}^{-1} \times 298.15 \text{ K}}$$

$$= 2.46 \times 10^{22} \text{ molecules per l}$$

$$= 2.46 \times 10^{19} \text{ molecules per cc.}$$

Air contains 20 per cent of O_2 and 80 per cent of N_2.

$$C_{O_2} = 2.46 \times 10^{19} \times 0.2 = 4.92 \times 10^{18} \text{ molecules/cc.}$$

$$Z_{N_2(O_2)} = 3.142(3.56 \times 10^{-8})^2 \sqrt{\frac{8 \times 8.314 \times 10^7 \times 298.15}{3.142 \times 14.93}} \times 4.92 \times 10^{18} \text{ collisions/sec.}$$

$$= 3.9820 \times 10^{-15} \times 4.92 \times 10^{18} \sqrt{\frac{8 \times 8.314 \times 10^7 \times 298.15}{3.142 \times 14.93}} \text{ collisions/sec}$$

$$= 1.9751 \times 10^4 \times 6.5018 \times 10^4 \text{ collisions/sec}$$

$$= 1.2737 \times 10^9 \text{ collisions/sec}$$

$$\therefore \tau = \frac{1}{Z} = \mathbf{7.85 \times 10^{-10} \text{ sec.}}$$

9. Show that to a first approximation the equation of state of a gas that dimerizes to a small extent is given by $\frac{PV}{RT} = 1 - \frac{K_c}{V}$, where K_c is the equilibrium constant for the formation of dimer, $2A = A_2$, $K_c = \frac{[A_2]}{[A^2]}$. **[V.U. 2002]**

Soln. : Let α be the degree of dissociation for the reaction $2A = A_2$.

Thus, $2A = A_2$

$1 - \alpha \quad \frac{\alpha}{2}$

Hence, $[A_2] = \frac{\alpha}{2V}$ and $[A] = \frac{1-\alpha}{V}$, where V is the total volume.

Thus, $K_c = \frac{\alpha V}{2(1-\alpha)^2}$.

Now, total number of moles $= 1 - \alpha + \frac{\alpha}{2} = 1 - \frac{\alpha}{2}$.

If P is the total pressure, then

$$PV = \left(1 - \frac{\alpha}{2}\right)RT$$

or, $\frac{\alpha}{2} = 1 - \frac{PV}{RT}$.

Hence, $\frac{K_c}{V} = \frac{\alpha}{2(1-\alpha)^2} = \frac{\alpha}{2}$ [since, $1 >> \alpha$] $= 1 - \frac{PV}{RT}$.

$\therefore \quad \frac{PV}{RT} = 1 - \frac{K_c}{V}$ **(Proved).**

10. A 2 m long tube is provided with inlets at both the ends so that HCl and NH_3 gases can be admitted simultaneously. Calculate the distance from HCl inlet end to the tube at which NH_4Cl will first appear if two gases are admitted at the same time, one from one end and the other from the other end. **[B.U. 2006]**

Hints : Let distance be x m from HCl inlet end, where NH_4Cl first appears.

By the problem, $t_{NH_3} = t_{HCl}$

or, $\frac{2-x}{\sqrt{\frac{8RT}{\pi M_{NH_3}}}} = \frac{x}{\sqrt{\frac{8RT}{\pi M_{HCl}}}}$ $(x = 0.81 \text{ m})$

11. A flask contains 10^{20} molecules of He at 27 °C. Find out the number of molecules having (i) average kinetic energy (ii) 100 times average kinetic energy.

Soln. : According to Maxwell velocity distribution,

$$\frac{1}{N}\frac{dN_c}{dc} = 4\pi\left(\frac{M}{2\pi RT}\right)^{\frac{3}{2}} c^2 e^{-Mc^2/2RT},$$

where $\frac{1}{N}\frac{dN_c}{dc}$ = fraction of molecules having velocity c,

M = molar mass.

Now, $\frac{1}{2}Mc^2 = E.$

Hence, $dc = \frac{dE}{\sqrt{2EM}}$ and $c = \sqrt{\frac{2E}{M}}.$

Hence, $\frac{1}{N}\frac{dN_E}{dE} = \frac{2\pi}{\sqrt{(\pi RT)^3}}\sqrt{E}\, e^{-\frac{E}{RT}}$

where E = energy per mole.

(i) Average kinetic energy, $E = \frac{3}{2}RT.$

Thus, $\frac{1}{N}\frac{dN_E}{dE} = \frac{2\pi}{\sqrt{(\pi RT)^3}}\sqrt{\left(\frac{3}{2}RT\right)}\, e^{-\frac{3}{2}}$

$= \sqrt{\frac{6}{\pi}}\left(\frac{1}{8.314\times 10^7\times 300}\right)e^{-\frac{3}{2}}$ $(T = 273 + 27 = 300\text{ K})$

$= \mathbf{1.236\times 10^{-11}}.$

This is the fraction of molecules having average energy. Thus, the required number of He molecules having average kinetic energy = $1.236\times 10^{-11}\times 10^{20}$ = $\mathbf{1.236\times 10^{9}}$.

(ii) 100 times average kinetic energy = $100\times\frac{3}{2}RT.$

Fraction of molecules having 100 times average kinetic energy

$= \frac{1}{N}\frac{dN_E}{dE} = \sqrt{\frac{600}{\pi}}\times\left(\frac{1}{8.314\times 10^7\times 300}\right)e^{-300/2}$

$= 3.976\times 10^{-75}.$

So, number of the molecules having 100 times average kinetic energy

$= 3.976\times 10^{-75}\times 10^{20} = 3.976\times 10^{-55} \approx \mathbf{0}.$

12. A flask contains 6.022×10^{23} neon molecules at 300 K. Calculate the number of molecules having (i) average kinetic energy (ii) 20 times of average kinetic energy.

Hints : Energy distribution of mass is independent.. [**Ans.** : 7.45×10^{12}, 13.96]

13. Consider a cubic, 10 cm on a side, of He gas at STP. Estimate the number of times one wall is struck by molecule in one second. (Answer should be given in S.I. unit).

[PhD Entrance Examination, Columbia University]

Soln. : We know that, $Z = \frac{P}{\sqrt{2\pi mkT}}$ A.t

$= \frac{PN_A}{\sqrt{2\pi MRT}}$ A.t.

Given that $P = 1.01325 \times 10^5$ Pa (1 atm)

$T = 273$ K

$A = 10^2 \text{ cm}^2 = 10^2 \times 10^{-4} \text{ m}^2$

$t = 1$ sec

∴ number of coallisions/sec,

$$Z = \frac{1.01325 \times 10^5 \text{ Pa} \times 6.022 \times 10^{23} \times 10^{-2} \text{ m}^2 \times 1 \text{ sec}}{\sqrt{2 \times \pi \times 4 \times 10^{-3} \text{ kg.mole}^{-1} \times 8.314 \text{ J.K}^{-1}.\text{mol}^{-1} \times 273 \text{ K}}}$$

$= 8.07 \times 10^{25}$.

14. If c m/sec is velocity of gas molecules, then (a) calculate the average time required to travel one metre. (b) Calculate the root mean square deviation of the time from average time. (c) Calculate fraction of molecules taking more than average time to travel one metre.

Soln. : (a) Velocity of gas molecules = c m/sec

So they take $\frac{1}{c}$ sec to travel 1 metre.

Average time required by gas molecules to travel 1 metre $= \bar{t} = \overline{\left(\frac{1}{c}\right)}$.

$$\text{Thus } \bar{t} = \overline{\left(\frac{1}{c}\right)} = \int_0^\infty \frac{dN_c}{N} \cdot \frac{1}{c} = \int_0^\infty f(c) \frac{1}{c}\, dc$$

$$= \int_0^\infty 4\pi \left(\frac{m}{2\pi kT}\right)^{\frac{3}{2}} c^2 e^{-\frac{mc^2}{2kT}} \cdot \frac{1}{c}\, dc = 4\pi \cdot \left(\frac{m}{2\pi kT}\right)^{\frac{3}{2}} \int_0^\infty c e^{-\frac{mc^2}{2kT}}\, dc.$$

$$\therefore \quad \bar{t} = \overline{\left(\frac{1}{c}\right)} = 4\pi \left(\frac{m}{2\pi kT}\right)^{\frac{3}{2}} \cdot \frac{1}{2\left(\frac{m}{2kT}\right)} = \sqrt{\left(\frac{2m}{\pi kT}\right)}.$$ (See page no 48)

(b) Root mean square deviation

$$= \sqrt{\overline{t^2} - \bar{t}^2} = \sqrt{\overline{\left(\frac{1}{c^2}\right)} - \overline{\left(\frac{1}{c}\right)}^2}.$$

$$\text{Now, } \overline{\left(\frac{1}{c^2}\right)} = \int_0^\infty f(c) \frac{1}{c^2}\, dc$$

$$= \int_0^\infty 4\pi \left(\frac{m}{2\pi kT}\right)^{\frac{3}{2}} c^2 e^{-\frac{nc^2}{2kT}} \frac{1}{c^2} dc$$

$$= 4\pi \left(\frac{m}{2\pi kT}\right)^{\frac{3}{2}} \int_0^\infty e^{-\frac{mc^2}{2kT}} dc \sqrt{\overline{t^2} - \bar{t}^2}$$

$$= 4\pi \left(\frac{m}{2\pi kT}\right)^{\frac{3}{2}} \cdot \left(\frac{2\pi kT}{m}\right)^{\frac{1}{2}} \cdot \frac{1}{2}$$.[See page no 49]

$$= \frac{m}{kT}.$$

$$\therefore \quad \sqrt{\overline{t^2} - \bar{t}^2} = \sqrt{\frac{m}{kT} - \frac{2m}{\pi kT}} = \left(\frac{m}{kT}\right)^{\frac{1}{2}}\left(1 - \frac{2}{\pi}\right)^{\frac{1}{2}}.$$

(c) Molecules moving with less than average velocity will take more than average time to cross one metre. Since average time $= \overline{\left(\frac{1}{c}\right)}$, average velocity $= \frac{1}{\overline{\left(\frac{1}{c}\right)}}$.

Let $\frac{1}{\overline{\left(\frac{1}{c}\right)}} = u.$

Fraction of molecules with velocity less than $u = \frac{N'}{N}$,

$$= 4\pi\left(\frac{m}{2\pi kT}\right)^{\frac{3}{2}} \int_0^u e^{-\frac{mc^2}{2kT}} c^2 dc = -4\pi \left(\frac{m}{2\pi kT}\right)^{\frac{3}{2}} \frac{kT}{m} \int_0^u c d\left(e^{-\frac{mc^2}{2kT}}\right)$$

$$= -4\pi \left(\frac{m}{kT}\right)^{\frac{1}{2}} \left(\frac{1}{2\pi}\right)^{\frac{3}{2}} \left[\left(ce^{-\frac{mc^2}{2kT}}\right)\right]_0^u - \int_0^u e^{-\frac{mc^2}{2kT}} dc$$

$$= -4\pi \left(\frac{m}{kT}\right)^{\frac{1}{2}} \left(\frac{1}{2\pi}\right)^{\frac{3}{2}} ue^{-\frac{mu^2}{2kT}} + B.$$

Now, $u = \frac{1}{\overline{\left(\frac{1}{c}\right)}} = \left(\frac{\pi kT}{2m}\right)^{\frac{1}{2}}.$

$$\therefore \quad \frac{N'}{N} = e^{-\frac{\pi}{4}} + B,$$

where $B = 4\pi\left(\frac{m}{kT}\right)^{\frac{1}{2}}\left(\frac{1}{2\pi}\right)^{\frac{3}{2}} \int_0^u e^{-\frac{mc^2}{2kT}} dc.$

Let $\frac{mc^2}{2kT} = x^2.$

$$\therefore \quad dc = \frac{2kTx}{mc}\, dx.$$

Now, $x = \sqrt{\frac{m}{2kT}}\, c$

Thus, $dc = \sqrt{\left(\frac{2kT}{m}\right)}\, dx.$

When $c = 0$, $x = 0$.

$c = u$, $x = \sqrt{\frac{m}{2kT}}$, $u = \frac{\sqrt{\pi}}{2}$,

since $u = \frac{1}{\left(\overline{\frac{1}{c}}\right)} = \sqrt{\left(\frac{\pi kT}{2m}\right)}$.

$$\therefore\ B = 4\pi\left(\frac{m}{kT}\right)^{\frac{1}{2}}\left(\frac{1}{2\pi}\right)^{\frac{3}{2}}\left(\frac{2kT}{m}\right)^{\frac{1}{2}}\int_0^{\frac{\sqrt{\pi}}{2}} e^{-x^2}dx = \frac{2}{\sqrt{\pi}}\int_0^{\frac{\sqrt{\pi}}{2}} e^{-x^2}dx$$

$$= \text{erf}\left(\frac{\sqrt{\pi}}{2}\right) \text{ (See formula 18 in B1)}$$

$$= \text{erf}\ (0.886) = 0.79 \text{ (See page no 49)}$$

$$\therefore\quad \frac{N'}{N} = -\,0.4559 + 0.79 = \mathbf{0.334}.$$

0.334 fraction of molecules takes more than average time to travel one metre.

15. Calculate the fraction of molecules moving with velocity between $\bar{u}$ and $\bar{u}$ + $0.1\,\bar{u}$. Does it depend on temperature? Explain.

Soln : Fraction of molecules between $\bar{u}$ and $\bar{u} + 0.1\,\bar{u}$

$$= 4\pi\left(\frac{m}{2\pi kT}\right)^{\frac{3}{2}}\int_{\bar{u}}^{\bar{u}+0.1\bar{u}} e^{-\frac{mc^2}{2kT}}c^2dc$$

$$= 4\pi\left(\frac{m}{2\pi kT}\right)^{\frac{3}{2}}\int_0^{1.1\bar{u}} e^{-\frac{mc^2}{2kT}}c^2dc - 4\pi\left(\frac{m}{2\pi kT}\right)^{\frac{3}{2}}\int_0^{\bar{u}} e^{-\frac{mc^2}{2kT}}c^2dc = I_1 - I_2$$

Proceeding like previous problem and putting $\bar{u} = \sqrt{\frac{2kT}{m}}$,

$$I_1 = -4\pi\left(\frac{m}{2\pi kT}\right)^{\frac{3}{2}} \times 1.1 \times \frac{kT}{m}\sqrt{\frac{2kT}{m}}\,e^{-1.21} + \text{erf}\ (1.1)$$

$$= -\frac{2.2}{\sqrt{\pi}}e^{-1.21} + 0.88.$$

$$I_2 = -4\pi\left(\frac{m}{2\pi kT}\right)^{\frac{3}{2}} \times 1.0 \times \frac{kT}{m}\sqrt{\frac{2kT}{m}}e^{-1.0} + \text{erf}(1.0)$$

$$\text{Hence, } I_1 - I_2 = -\frac{2.0}{\sqrt{\pi}}e^{-1.0} + 0.84 = \mathbf{0.0849}.$$

Thus, 0.0849 fraction of molecules moving with velocity between $\bar{u}$ and 1.1 $\bar{u}$.

This number is independent of temperature. At any temperature, this number will remain same. Since $\bar{u}$ itself depends on temperature (increases with increasing temperature), the given range ($\bar{u}$ to 1.1 $\bar{u}$) becomes temperature dependent. However, curve flatters with increasing temperature, Hence, number of molecules in that range temperature independent.

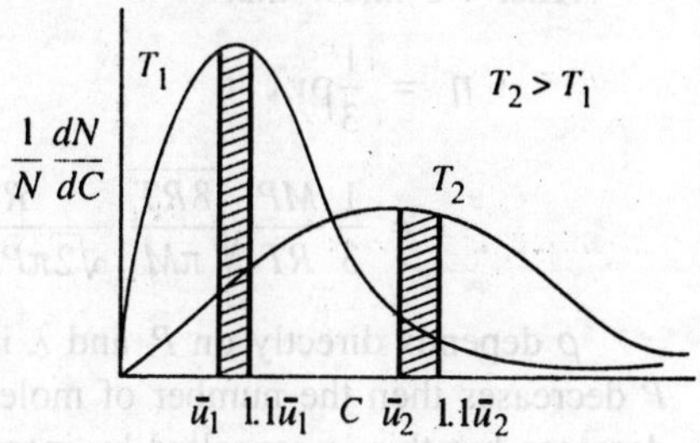

16. Compare the fraction of oxygen molecules to that of Hydrogen molecules moving with velocity between $\bar{u}$ and 1.1 $\bar{u}$.

Soln. : Follow Prob. 15

The value will remain same for both gases which is 0.0849. Compare the temperatures when the Maxwell distribution curve of a gas will look like that of another gas with twice of its molar mass.

Soln. : Two Maxwell distribution curves will look alike only when most probability velocities ($\bar{u}$) match with each other.

Since area under the distribution curve is always one, height of the peaks will be equal if two ($\bar{u}$)'s remain same. Let gas *A* with molar mass M_A at temperature T_A have same $\bar{u}$ as that of gas *B* with molar mass M_B at temperature T_B.

By the problem, $M_B = 2M_A$

$$\bar{u}_A = \sqrt{\frac{2RT_A}{M_A}} \quad \text{(at temp } T_A)$$

$$\bar{u}_B = \sqrt{\frac{2RT_B}{M_B}} \quad \text{(at temp } T_B)$$

Now, $U_A = U_B$

$$\text{or, } \frac{T_A}{M_A} = \frac{T_B}{M_B} \quad \text{or, } \frac{T_A}{T_B} = \frac{M_A}{M_B} = \frac{1}{2}$$

$\therefore \quad T_B = 2T_A$.

Short Questions with Hints

1. According to kinetic theory how will the temperature of an ideal gas be affected by adiabatic expansion into vacuum? Explain briefly.

Hints : No work is done since external force is zero. Energy remains the same because there is neither attraction, nor repulsion.

2. Explain, in terms of kinetic theory, why an automobile pump gets heated while pumping air into a tyre.

Hints : The pumping is done at a higher pressure. So velocity of molecules increases. Hence temp. increases, while the pump is heated.

3. Show that viscosity coefficient is independent of pressure and give a physical explanation of why this should be so.

Ans. We know that

$$\eta = \frac{1}{3}\rho v \lambda$$

$$= \frac{1}{3}\frac{MP}{RT}\sqrt{\frac{8RT}{\pi M}}.\frac{RT}{\sqrt{2}\pi P N_A \sigma^2}.$$

ρ depends directly on P, and λ inverses on P. So the pressure terms cancels out. As P decreases then the number of molecules that cross adjacent layers carrying momentum decrease but this is cancelled by increase of λ as now they have to travel longer distance (so that $\lambda \times \rho$ remains unchanged).

4. Why is the thermal conductivity coefficient independent of pressure?

Ans. Thermal conductivity coefficient $K = \eta \times C_v$, where η is viscosity coefficient and C_v is molar heat capacity at constant volume. Neither η nor C_v depends on P. The flux of heat carrying molecules across unit area in the gas decreases with decrease of P but the mean free path increases. The two effects just cancel out.

5. $PV = \frac{1}{3}mN\overline{c^2}$. If there are two gases having the same volume and same number of moles, will they exert same pressure?

Hints : Not necessarily as molecular weight is different.

6. Under what conditions does real gas approach ideal gas?

Hints : At high temperature and as $P \to 0$.

7. In normal practice aqueous tension is subtracted from the observed pressure of a gas at a particular temperature Why is this done?

Hints : Aq. tension is the vapour pressure of water present in atmosphere.

8. For a fixed temperature indicate graphically how PV of a van der Waals gas depends upon P when (i) $a = 0$ (i) $b = 0$ and (iii) $a = b = 0$.

Hints :

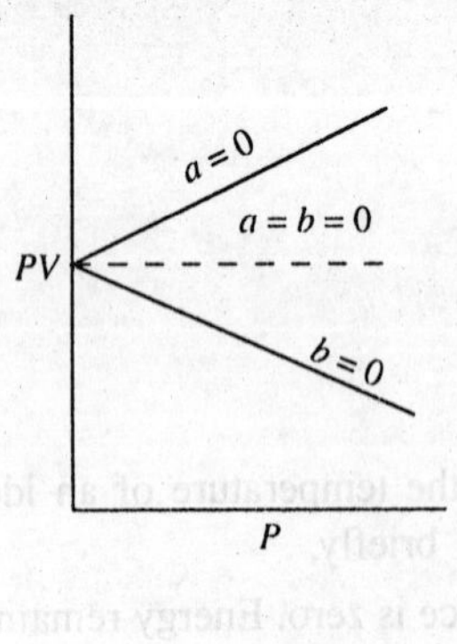

9. Establish the concept of Avogadro number from the expression $PV = \frac{1}{3} mn\overline{c^2}$.

Hints : $\frac{N}{N_A} = n$, where N is the number of molecules in volume V and n is the number of moles. N_A is then ratio.

10. Establish the relation between vapour density (V_d) and density (ρ) of a gas.

Hints : $\frac{\rho_1}{\rho_2} = \frac{V_{d_1}}{V_{d_2}}$.

Gas densities are proportional to moleculer weight and hence vapour densities.

$$\left(V_d = \frac{M}{2}\right)$$

11. What is meant by the term : relative humidity ?

Hints : It is defined as the partial pressure of water in atmosphere to the vapour pressure of water at that temperature.

12. Explain why spark plugs are not needed in a diesel engine.

Hints : The compression of the gas heats the diesel above their ignition temperature.

13. When a tire is rapidly pumped, its temperature rises. Would this be expected if air is an ideal gas?

Hints : Even if air is an ideal gas, compression would cause a temperature rise because energy is expended on the gas in the process. The energy converted into K.E. of the molecules is heat.

CHAPTER 3

LIQUID STATE (SURFACE TENSION AND VISCOSITY)

Required Formulae

1. Excess pressure on the concave side of a bubble $P = \frac{4\gamma}{r}$

In a cavity $P = \frac{2\gamma}{r}$.

2. Rise and fall of height in a capillary of radius r.

$\gamma \cos\theta = \frac{1}{2} h\rho g r$, where h = height, ρ = density.

3. In stalagmometer, $\frac{\gamma_1}{\gamma_2} = \frac{m_1}{m_2} = \frac{\rho_1 n_2}{\rho_2 n_1}$,

where m_1 and m_2 are masses of one drop of two liquids respectively and d and n are respective density and number of drops.

4. Poiseuille equation, $\eta = \frac{\pi P r^4 t}{8Vl}$.

η = viscosity coefficient, V = volume of the liquid flowing through a tube of length l and radius r in time t under a pressure difference P.

5. In viscometer, $\frac{\eta_1}{\eta_2} = \frac{\rho_1 t_1}{\rho_2 t_2}$.

6. Stokes laws : F (force) $= 6\pi r \eta v$ (where v is velocity).

7. $\eta = Ae^{\frac{E_{vis}}{RT}}$, where, E_{vis} = activation energy for flow.

Short Questions

1. It is difficult to blow a balloon initially but becomes easier as time passes.

Ans. : A bubble is formed many times and an inflated balloon is comparable to it. There is excess pressure in the concave side given by $\Delta P = \frac{2\gamma}{r}$. Since γ is approximately constant, when r is small, P is very high. As the balloon is filled up, r increases, thus the effective excess pressure decreases. Then it becomes easier to inflate.

2. At what temperature surface tension will be zero?

Ans. : At critical temperature, the liquid surface is not formed. Surface tension arises due to liquid surface. Thus at critical temperature, it should be zero. However, Ramsay and Shield have given an equation explaining the dependence of γ on temperature. The eqn. is $\gamma \times (MV)^{\frac{2}{3}} = K(T_c - T - 6)$.

Thus γ becomes zero (for many reasons) at $T = T_c - 6$.

3. Why some materials are surface active and others are not?

Ans. : The materials which are surface active are usually a long chain carbon compounds with a soluble part and an insoluble part. Thus they will be thrown back from bulk to surface, thereby reducing surface tension. Non-surface active materials are generally highly soluble in bulk.

4. Why does a piece of blotting paper soak water?

Ans. : Blotting paper consists of fine capillaries. When put over water, it soaks water because water wets paper and there is capillary rise.

5. When pure liquids are boiled in clean and smooth vessels, there is occasional bumping. Why?

Ans. : Bubbles can grow in presence of rough surfaces. In the absence of rough surfaces, bubbles are very small. The pressure required to form such small bubbles is very high ($P = \frac{2\gamma}{r}$, where P is the pressure, γ is the surface tension and r is the radius). The temperature thus rises above normal boiling point (boiling point of a liquid rises with increase in pressure). This again serves to increase the pressure sufficiently so that small bubbles are produced. As these bubbles grow the pressure which has become greater than necessary for their formation leads to rapid expansion. This leads to occasional bumping.

6. Do electrolytes lead to increase in surface tension of water with increase in concentration always?

Ans. : Liquid molecules in the surface are attracted strongly to the interior (cohesion). This is the reason of surface energy and surface tension. Electrolytes attract polar solvents very strongly and this increased attraction leads to increase in surface tension. In a solution, ions are solvated. As suggested by M. Dole (1938), a special orientation of hydrated ions (in water) of the interface may reduce the attractions of surface molecules to the bulk. This may lead to decrease in surface tension at low concentrations with increase in concentration of electrolytes.

7. What should be the latent heat of evaporation of mercury? Will it be higher or lower than water? Do you expect $\frac{Le}{T_b}$ of mercury is as per Trouton's rule?

Ans. : The latent heat of evaporation of mercury will be lower, since it is non-associated. That is why, Trouton's rule will be obeyed.

8. Show that ST of 1 dyne is equal to mN/m.

Ans. : Let $x\dfrac{\text{dyne}}{\text{cm}} = 1\dfrac{\text{mN}}{\text{m}}$.

$$\therefore\ x = \frac{\text{dyne}}{\text{mN}} \times \frac{\text{m}}{\text{cm}}$$

$$= \frac{\text{dyne}}{10^{-3} \times 10^{5}\,\text{dyne}} \times \frac{100\,\text{cm}}{\text{cm}} = 1.$$

$$\therefore\ 1\ \text{dyne/cm} = 1\ \text{mN/m}.$$

9. "Highly viscous liquids are less volatile."—Explain.

Ans. : Attractive forces are operating between liquid molecules. Only those molecule which have sufficient energy to overcome these forces can get into vapour state. Energ distribution of liquid molecules is governed by Maxwell-Boltzmann distribution functio which is characteristic of only temperature as long as temperature is sufficiently high. Thu the less is cohesive energy of liquid, the greater is number of molecules in vapour sta at certain temperature. Highly viscous liquids have large cohesive energy. Thus, lowe number of molucules will vaporize compared to less viscous liquids at certain temperatur Hence, highly viscous liquids are less volatile.

10. "Volatile liquids must have lower surface tension and viscosity."—Explain.

Hints : Volatile liquids have small cohesive energy and thus lower surface tension an viscosity, since they arise due to cohesive forces. See Q. 9.

11. What are bubbles and cavities?

Ans. : A bubble is a region in which vapour is trapped by thin film, probably som air is also trapped. A cavity is vapour-filled hole in a liquid. What are widely called bubble in a liquid, are actually cavities. True bubbles have two surfaces, one on each side of th film. The cavities have only one surface.

12. At critical temperature surface tension of liquid is zero.—Criticize.

Ans. : Surface tension will exist if there is a surface separating two phases . At critic temperature, interface between liquid and vapour disappears as both have same densit So, surface tension of liquid is zero at critical temperature.

13. Why are the raindrops spherical?

Ans. : Raindrops are spherical due to surface tension which liquid tends to reduc by minimizing surface area and sphere has minimum surface area to volume ratio. Althoug force of gravity opposes this tendency of liquid, it has little effect on the small size c raindrops.

14. What is Newtonian fluid? What is laminar flow?

Ans. : If velocity of fluid is always same in magnitude and direction at a particul point, the fluid is Newtonian fluid, and the flow is called laminar flow.

Numerical Problems

1. At 25 °C water rose in a capillary of diameter 0.08 cm to a height of 3.72 cn Calculate the surface tension of water if the density of water at 25 °C be 0.998 gm/cc an g = 980.665 cm/sec^2.

Soln. : Water wets glass. Therefore,

$\gamma = \frac{1}{2} r.h.\rho.g$, where r = radius; h = height of water column;

ρ = density; g = acceleration due to gravity.

In this expression θ has been taken as zero.

Given, $r = 0.04$ cm

$h = 3.72$ cm

$\rho = 0.998$ gm/cc

$g = 980.665$ cm/sec^2.

$$\therefore \gamma = \frac{1}{2} r.h.\rho.g = \frac{1}{2} \times 0.04 \times 3.72 \times 0.998 \times 980.665 = \mathbf{72.8\ dyne.cm^{-1}}.$$

2. How high will water rise in a capillary of radius 3×10^{-4} metre at 25 °C. Given, $\gamma_{H_2O} = 72.75 \times 10^{-3}$ N.m^{-1}

$g = 9.80665$ metre.sec^{-2}

$\rho = 998$ kg.m^{-2}

$\theta = 0°$

Soln. : $\gamma = \frac{1}{2} r.\rho.h.g, \quad \therefore h = \frac{2\gamma}{r\rho g}$

$$\therefore h = \frac{2 \times 72.75 \times 10^{-3}}{3 \times 10^{-4} \times 998 \times 9.80665},$$

$= 0.0495$ metre $=$ **4.95 cm.**

3. 50 drops each of water and ether weigh 3.75 gm and 0.85 gm respectively. γ_{H_2O} is 72.75 dyne.cm^{-1}. What will be γ for ether?

Soln. : By the problem, $\frac{\gamma_{ether}}{\gamma_{H_2O}} = \frac{m_{ether}}{m_{H_2O}}$.

By the problem, weight of one drop of water $= \frac{3.75}{50}$ gm and that of ether is $\frac{0.85}{50}$ gm.

$$\therefore \gamma_{ether} = \frac{m_{ether}}{m_{H_2O}} \times \gamma_{H_2O} = \frac{\frac{0.85}{50} \times 72.75}{\frac{3.75}{50}} = \mathbf{16.49\ dyne.cm^{-1}}.$$

4. The number of drops of water falling from a stalagmometer is 90, whereas for organic liquid it is 310. Surface tension of water is 72.75 dyne.cm^{-1}. What is the surface tension of organic liquid, where $\rho = 0.998$ gm/cc for water and 0.75 gm/cc for organic liquid?

Soln. : $\frac{\gamma_o}{\gamma_w} = \frac{n_{H_2O}\rho_o}{n_O \cdot \rho_w}$

$$\therefore \gamma_o = \frac{n_{H_2O}\rho_o}{n_O \cdot \rho_w} \times \gamma_w = \frac{90 \times 0.75 \times 72.75}{310 \times 0.998} = \mathbf{15.87\ dyne.cm^{-1}}.$$

5. A liquid A has half the surface tension and twice the density of liquid B at a certain temperature. If in a capillary the rise is 10 cm for A, what will be the rise of B in the same tube at the same temperature? **[C.U.(H) 1972]**

Hints : $\gamma_A = \dfrac{h.g.r.\rho}{2\cos\theta}$, $\cos\theta = 1$.

$$\therefore \gamma_A = \frac{1}{2} \times 10 \times 981 \times r \times 2\rho$$

$$\gamma_B = \frac{1}{2} h \times 981 \times r \times \rho$$

$$h = 40,\ \gamma_A = \frac{1}{2}x,\ \gamma_B = x.$$

6. Calculate how far water will rise inside a capillary of internal diameter 0.25 mm; the surface tension is 72.76 dynes/cm and its density is 0.998 gm/cc in C.G.S. system. **[C.U. 1983; B.U. '96]**

Soln. : $\gamma = \dfrac{1}{2} h\rho g r,\ h = \dfrac{2\gamma}{\rho g r} = \dfrac{2 \times 2 \times 72.76}{0.998 \times 980.665 \times 0.025}$ = **11.89 cm.**

7. Calculate the energy expended in breaking 1 cc of water into droplets having mean radius 10^{-5}cm. Given : ST of water (γ_w) = 72 dynes/cm.

Soln. : Volume of 1 sphere : $\frac{4}{3}\pi(10^{-5})^3 = 4.1893 \times 10^{-15}$cc.

$\therefore$ number of spheres $= \dfrac{1}{4.1893 \times 10^{15}} = 2.3870 \times 10^{14}$.

Total area $4\pi r^2$ $= 4 \times 3.142 \times (10^{-5})^2$
$= 1.2568 \times 10^{-9}$ sq. cm.

Energy needed in breaking,

$$\begin{aligned} E &= (4\pi r^2) \times n \times \gamma \\ &= (1.2568 \times 10^{-9}) \times n \times \gamma \\ &= 1.2568 \times 10^{-9} \times 2.3870 \times 10^{14} \times 72 \text{ ergs} \\ &= 2.15 \times 10^7 \text{ ergs} \\ &= 2.15 \text{ J.} \end{aligned}$$

8. Calculate the radius of the largest drop of a liquid of surface tension (at 17 °C) 80 dynes/cm that can evaporate at 17 °C. The latent heat of evaporation is 250 cal/gm.

Soln. : Let the radius of the largest drop be r cm.

Then energy needed = $m \times$ le, where m is mass of the drop, le is latent heat per gram.

$= V \times \rho \times$ le, where V is volume of the drop, ρ is density.

$= \frac{4}{3}\pi r^3 \times \rho \times \text{le} = 4\pi r^2 \gamma$

Let d = 1 gm/cc.

$$\therefore r = \frac{3 \times 80 \text{ dynes.com}^{-1}}{250 \text{ cal.gm}^{-1} \times 4.184 \times 10^7 \text{ erg.cal}^{-1} \times 1 \text{ gm.cc}^{-1}}$$

$$= \mathbf{2.294 \times 10^{-8} \text{ cm.}}$$

9. A liquid of density 0.9 gm/cc is kept in a beaker. A clean capillary tube of diameter 0.033 cm is dipped into it. It rises to a height of 3.8 cm. Calculate the surface tension of the liquid.

Soln. : Given, $\rho = 0.9 \text{ gm/cm}^3$

$2r = 0.033$ cm

$h = 3.8$ cm.

$$\therefore \quad \gamma = \frac{1}{2} h.g.r.\rho.$$

$$= 0.5 \times 3.8 \text{ cm} \times 980.665 \text{ cm.sec}^{-2} \times \frac{0.033}{2} \text{ cm} \times 0.9 \text{ gm/cc}$$

$$= \mathbf{27.68 \text{ dynes/cm.}}$$

10. A capillary tube of radius 0.01cm is dipped in a liquid of density 1 gm/cc. Its surface tension is 74 dynes/cm. Find the height to which the liquid will rise in the capillary tube.

Hints : $h = \dfrac{2\gamma}{gr\rho}$ = **15.1 cm.**

11. A sphere of water of radius 1 mm breaks into as million drops of same size. Find the work done (γ = 72 dynes/cm).

Soln. : Volume of sphere $= V = \dfrac{4}{3}\pi r^3$

$$= \frac{4}{3} \times 3.142 \times (0.1)^3 = 4.189 \times 10^{-3} \text{ cc.}$$

Surface area $S = 4\pi r^2 = 4 \times 3.142 \times 0.1^2 = 0.12568 \text{ cm}^2$.

Energy $= 4\pi r^2\gamma = 0.12568 \text{ cm}^2 \times 72$ dynes/cm

$= 9.04896$ ergs.

4.189×10^{-3} cc contains 10^6 drops.

Volume of 1 drop $= \dfrac{4.189\times10^{-3}}{10^6} = 4.189 \times 10^{-9}$ cc.

$$\therefore \quad \frac{4}{3}\pi r^3 = 4.189 \times 10^{-9}$$

$$\text{or,} \quad r^3 = 9.9992 \times 10^{-10}$$

$$\therefore \quad r = 9.9973 \times 10^{-4} \text{ cm.}$$

Energy of 10^6 drops = area of one drop $\times \gamma \times 10^6$

$= 4 \times 3.142 \times (9.99 \times 10^{-4} \text{ cm})^2 \times 72 \text{ dynes/cm} \times 10^6 = 904.8$ ergs.

Work done = (904.8 – 9.04896) ergs = **895.75 ergs.**

12. An atomiser produces 10^5 droplets of same size from 0.5 cc of a liquid. Find the increase in surface energy, if the S.T. is 80 ergs/cm^2.

Soln. : 0.5 cc gives 10^5 droplets.

$$\therefore 0.5 = \frac{4}{3}\pi r^3 \times 10^5$$

$$r = \left(\frac{3\times0.5}{4\times3.142\times10^5}\right)^{\frac{1}{3}} = 0.010607385 \text{ cm} = 0.0106 \text{ cm.}$$

$$E = 4\pi r^2\gamma \times n$$
$$= 4 \times 3.142 \times 0.0106^2 \times 80 \times 10^5 \text{ ergs}$$
$$= 11297.12 \text{ ergs}$$
$$= \mathbf{1.1297 \times 10^4 \text{ ergs.}}$$

13. At 293 K, 10 cc of water gave 29 drops and 10 cc of ether gave 87 drops in the same stalagmometer. The density of ether is 0.7 gm/cc and γ_{H_2O} = 72 dynes/cc. Calculate the surface tension of ether.

Hints : $\gamma_1 = \gamma_{ether}$, $\gamma_2 = \gamma_{H_2O}$

$$\frac{\gamma_1}{\gamma_2} = \frac{n_2\rho_1}{n_1\rho_2} \text{ or, } \frac{\gamma_1}{\gamma_2} = \frac{29\times 0.7}{87\times 1}, \therefore \gamma_1 = \mathbf{16.8.}$$

14. A spherical soap bubble of volume $\frac{\pi}{6}$ cm^3 stands suspended in air. What is the excess pressure inside the bubble? Given the interfacial tension for soap solution air interface = 27 dynes/cc. **[C.U. 2000]**

Soln. : The volume occupied by soap bubble of radius r is $\frac{4}{3}\pi r^3$. By the problem,

$$\frac{4}{3}\pi r^3 = \frac{\pi}{6}.$$

$$\text{or, } r^3 = \frac{1}{8} \text{ cm}^3.$$

$$\therefore \quad r = 0.5 \text{ cm.}$$

If P is the excess pressure in the concave side (internal)

$$P = \frac{4\gamma}{r} = \frac{27\times 4}{0.5} = \frac{108}{0.5} = \mathbf{216 \text{ dynes/cm.}}$$

15. Find the change in surface energy when two identical mercury droplets of diameter 2 mm merged isothermally to form 1 drop (γ_{Hg} = 49 dynes/cm). **[C.U. 2000]**

Soln. : Given, γ_{Hg} = 49 dynes/cm, $d = 2r$ = 2 mm.

$$\therefore \quad r = 0.1 \text{ cm.}$$

Volume of one droplet $V = \frac{4}{3}\pi r^3 = 4.189 \times 10^{-3}$cc

For 2 droplets : $V = 8.378 \times 10^{-3}$ cc.

Total surface energy $= 4\pi r^2\gamma \times 2$

$$= 4 \times 3.142 \times 0.1^2 \times 2 \times 49$$
$$= 12.3166 \text{ ergs.}$$

Two merged : total volume = 8.378×10^{-3} cm^3.

$$\frac{4}{3}\pi r^3 = 8.378 \times 10^{-3}$$

$$r = \sqrt[3]{\frac{3\times 8.378\times 10^{-3}}{4\times 3.142}} = \sqrt[3]{1.99984\times 10^{-3}}$$

$$\therefore \quad r = 0.12538763 \text{ cm.}$$

Surface area of one large drop SA = $4 \times 3.142 \times (0.125)^2$
$= 0.196375 \text{ cm}^2$.

Total surface energy of merged drops = SA × 49
$= SE = 0.196375 \times 49 = 9.622375$ ergs.

SE of mixed drop : 9.62 ergs.

Decrease in surface energy = **2.69 ergs.**

16. At 20 °C the interfacial tension between water and benzene is 35 mN/m. If the surface tension γ = 28.85 mN/m for benzene and 72.75 mN/m for water, calculate :

(i) the work of adhesion between water and benzene, (ii) the work of cohesion for benzene and for water, (iii) spreading coefficient for benzene on water. **[B.U. 2003]**

Soln. : Work of adhesion $= (\gamma_A + \gamma_B - \gamma_{AB})$

By the problem, $\gamma_A = 72.75 \times 10^{-3}$ N/m

$\gamma_B = 28.85 \times 10^{-3}$ N/m

$\gamma_{AB} = 35 \times 10^{-3}$ N/m.

∴ work of adhesion $= (72.75 + 28.85 - 35)\ 10^{-3}$ N/m

$= 66.60 \times 10^{-3}$ N/m. or J/m^2

$=$ **66.6 mJ/m^2.**

Work of cohesion $= 2\gamma_i$

Therefore, work of cohesion for benzene

$= 2\ \gamma_B = 2 \times 28.85 \times 10^{-3}$ N/m

$= 57.7$ mJ/m^2.

For water $= 2 \times 72.75 =$ **145.5 mJ/m^2.**

Spreading coefficient for benzene on water,

$\sigma_{BA} - \gamma_A - \gamma_B - \gamma_{AB}$

$= 8.9 \times 10^{-3}$ N/m

$=$ **8.9 mJ/m^2.**

17. Show that for a given total volume, the surface area : volume ratio of a cube is about 5/4 times of a sphere. **[B.U.(H) 2002]**

Soln. : Let the total volume of cube be 1000 cc ($a^3 = 1000$; $a = 10$). Total surface area of a cube is $6a^2 = 6 \times 100 = 600$. So surface area : volume ratio in a cube is 0.6.

Let a sphere with volume 1000 cc be of radius r.

$$\therefore \quad \frac{4}{3}\pi r^3 = 1000 \text{ cc}$$

$$\therefore \quad r = \left(\frac{3 \times 1000}{4\pi}\right)^{\frac{1}{3}} = 6.2032348 \text{ cm.}$$

So surface area is $4\pi r^2 = 4 \times 3.142 \times (6.2032348)^2$

$= 483.61742 = 480 \text{ cm}^2$.

$$\frac{\text{Surface area}}{\text{Volume}} = \frac{480}{1000} = 0.48.$$

$$\text{So,} \quad 0.48 \times \frac{5}{4} = 0.6.$$

Thus, the surface area / volume ratio of a cube is $\frac{5}{4}$ times of a sphere.

18. An excess pressure of 364 Pa is required to produce a hemispherical bubble at the end of a capillary tube of 0.3 mm diameter immersed in acetone. Calculate γ.

[B.U. 2000]

Soln. : P_{excess} for the formation of hemispherical bubble = 364 Pa.

Given, $2r = 0.3$ mm.

$\therefore\ r = 0.15 \text{ mm} = 0.015\text{cm} = 0.015 \times 10^{-2}\text{m}.$

Now, $P = \frac{2\gamma}{r}$ [γ = surface tension, r = radius of bubble, P = excess pressure]

$$\therefore\ \gamma = \frac{P \times r}{2} = \frac{364 \times 0.015 \times 10^{-2}}{2} = 0.0273 \text{ N.m}^{-1}$$

$= 27.3 \times 10^{-3}$ N/m = **27.3 dynes/cm, i.e., mN/m.**

19. A spherical drop of a liquid weighing 0.04 gm is dispersed into 1500 microglobules of radius 0.02 cm by a suitable experimental device. Find the resultant increase in surface energy.

[C.U. 1998]

Density = 0.8 gm/cc

γ = 27 dynes/cm^2

Soln. : $\rho = \frac{m}{V}$, $\therefore\ V = \frac{m}{\rho} = \frac{0.04 \text{ gm}}{0.8 \text{ m / cc}} = 0{\cdot}05$ cc.

$$\therefore\ \frac{4}{3}\pi r^3 = 0.05$$

$$\text{or, } r^3 = \frac{3 \times 0.05}{4\pi} = 0.01193662.$$

$$\therefore\ r = 0.228539 = 0.228 \text{ cm.}$$

Initially surface energy,

$$E = 4\pi r^2\gamma = 4 \times 3.142 \times (0.228)^2\text{cm}^2 \times 27 \text{ dynes/cm}$$

$= 17.64004262$ ergs = 17.64 ergs.

After formation of 1500 microglobules,

$$1500 \times \frac{4}{3}\pi r^3 = 0.05.$$

Hence, r = radius of one microglobule = 0.019 cm.

$$E = 4\pi r^2\gamma.n$$

$= 4 \times 3.142 \times (0.019)^2 \times 27 \times 1500$

$= 183.7504$ ergs.

Hence, the increase in surface energy

$= 183.7504 - 17.64 =$ **166.1104 ergs.**

20. A capillary tube of radius 0.001 cm is inclined at an angle of 45° to the surface of a liquid. The liquid wets the wall. It has a density of 0.85 gm.cm^{-3} and surface tension of 36 dynes.cm^{-1}. Calculate d, the distance along the capillary to the meniscus.

Soln. : By the problem,

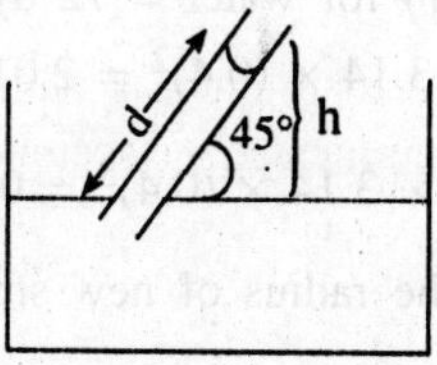

The vertical height of the meniscus is given as :

$$h = \frac{2\gamma}{\rho g r} = \frac{2 \times 36}{0.85 \times 980.665 \times 0.001} = 86.38 \text{ cm}$$

$$h = d \sin 45°$$

$$d = \frac{h}{\sin 45°} = \mathbf{122.16 \text{ cm.}}$$

21. For a water-oil interface at 25 °C and 1 atm, calculate the rise of water in a capillary tube of inside diameter 0.2 mm. The surface tension of water is 72 dynes/cm at 25 °C. The densities of water and air at 25 °C and 1 atm are 0.997 gm/ml and 0.001 gm/ml respectively. Should the rise be lower at 40 °C? Discuss briefly. **[B.U. 2002]**

Soln. : Given, $r = \frac{0.02}{2}$ cm = 0.01 cm

γ = 72 dynes/cm

ρ_{H_2O} = 0.997 gm/cc

ρ_{air} = 0.001 gm/cc

$$h = \frac{2\gamma}{\rho g r} = \frac{2 \times 72 \text{ dynes / cm}}{(0.997 - 0.001)\text{ gm / cc} \times 980.66 \text{ cm.sec}^{-2} \times 0.01 \text{ cm}}$$

$$= \frac{144}{0.996 \times 980.665 \times 0.01} \text{ cm} = 14.74 \text{ cm.}$$

The rise will be lower at 40 °C since γ decreases with increase in temperature.

22. A tube of internal diameter 4 mm is dipped into a liquid of surface tension 98 dynes/cm. Find the approximate value of the capillary rise. **[B.U. 2003]**

Hints : $h = \frac{2\gamma}{\rho g r} = \frac{2 \times 98}{980.665 \times 0.2 \times d}$.

23. At 20 °C pure water with an absolute viscosity of 0.01009 dyne.sec.cm^{-2} requires 102.2 sec to flow through a capillary of an Ostwald viscometer. At 20 °C toluene requires 68.9 sec. If the density of water and toluene be 0.998 gm/cm^3 and 0.866 gm/cm^3 respectively, calculate the viscosity of toluene. **[V.U. 1997]**

Hints : $\frac{\eta_{water}}{\eta_{toluene}} = \frac{\rho_{H_2O} t_{H_2O}}{\rho_{toluene} t_{toluene}}$

$$\therefore \eta_{toluene} = \left(\frac{0.1009 \times 0.866 \times 68.9}{0.998 \times 102.2}\right) = \mathbf{0.059 \text{ dynes.sec.cm}^{-2}}.$$

24. A spherical drop of water, 0.4 cm in radius is split into 125 identical tiny spherical droplets. Find the increase in surface energy. [γ for water = 72 dynes/cm] **[C.U. 2002]**

Soln. : Area of the drop = $4\pi r^2 = 4 \times 3.14 \times (0.4)^2 = 2.01$ cm^2.

$$\text{Volume occupied} = \frac{4}{3}\pi r^3 = \frac{4}{3} \times 3.14 \times (0.4)^3 = 0.268 \text{ cc.}$$

If it is divided into 125 drops and r be the radius of new small drops, then

$$125 \times \frac{4}{3}\pi r^3 = 0.268 \text{ cc}$$

$$\text{or,} \quad r^3 = \frac{3 \times 0.268}{125 \times 4 \times 3.14}$$

$$\therefore \quad r = 0.0799 \text{ cm} = 0.08 \text{ cm.}$$

So, total area for this drops

$= 4\pi r^2 \times 125$

$= (0.08)^2 \times 4 \times 3.14 \times 125 = 10.05$ cm^2.

Hence, increasing area = $(10.05 - 2.01)$ cm^2 = 8.04 cm^2.

So the increase in surface energy

= increasing area $\times \gamma$

= 8.04×72 = **578.88 ergs.**

25. A steel ball of density 10.0 gm/cc having diameter of 4 mm is dropped into a column of liquid. It takes 5 seconds to fall through a distance of 10 cm. Calculate the viscosity of the liquid. Explain the terms of your working formula; given, d_{liquid} = 3.8 g/cc. **[B.U. 1995]**

Soln. : We know that a falling body of mass m through gravity g produces a force = mg.

Again, viscous force is $6\pi r\eta v$,

where the terms having normal meaning.

$\therefore mg = 6\pi r\eta v.$

But m = volume $\times$ density.

The volume of a spherical ball $= \frac{4}{3}\pi r^3 = V$

$$\therefore m = V\rho = V(\rho_s - \rho_l)$$

$$= \frac{4}{3}\pi r^3(\rho_s - \rho_l).$$

$$\therefore mg = \frac{4}{3}\pi r^3(\rho_s - \rho_l) \times g$$

$$\text{or, } 6\pi r\eta v = \frac{4}{3}\pi r^3(\rho_s - \rho_l)\, g$$

$$\therefore \eta = \frac{2}{9}\frac{r^2(\rho_s - \rho_l)g}{v}.$$

Velocity is 2 cm/sec because it takes 5 sec to fall through a distance of 10 cm.

Here $r = 0.2$ cm, $v = 2$ cm/sec, $(\rho_s - \rho_l) = (10 - 3.8)$ gm/cc.

$$\therefore \eta = \frac{2}{9}.\frac{(0.2)^2 cm^2 \times (10-3.8) gm / cc \times 980.665 cm.sec^{-2}}{2 \ cm / sec}$$

$$= 27.02276889 \ gm.sec^{-1}.cm^{-1} = \mathbf{27 \ poise.}$$

26. A droplet of water ($r = 10^{-2}$cm) is falling in air at 25 °C. ($\eta = 1.8 \times 10^{-4}$ poise) What will be the velocity of fall?

Soln. : By the problem, $mg = \frac{4}{3}\pi r^3 \times g$ (assuming d = 1 gm/cc)

$$= \frac{4}{3} \times 3.142 \times (10^{-2})^3 \times 980.665$$

$$= 4.108 \times 10^{-3} \ gm.cm.sec^{-2} = 6\pi r \eta v.$$

i.e., $6 \times 3.142 \times 10^{-2} \times 1.8 \times 10^{-4} \times v = 4.108 \times 10^{-3}$.

i.e., $$v = \frac{4.108 \times 10^{-3} \ gm.cm.sec^{-2}}{6 \times 3.142 \times 10^{-2} cm \times 1.8 \times 10^{-4} \ gm.cm^{-1} \ sec^{-1}}$$

$$= 121.0599522 \ cm/sec = \mathbf{121 \ cm/sec.}$$

27. At 20 °C, pure water with an absolute viscosity of 0.01009 dyne. sec.cm^{-2} requires 102.2 sec to flow through the capillary of an Ostwald viscometer. At 20 °C, toluene requires 68.9 sec. If the densities of water and toluene be 0.998 gm/cm^3 and 0.866 gm/cm^3 respectively, calculate the viscosity of toluene. **[V.U. 1997]**

Soln. : We know that $\frac{\eta_1}{\eta_2} = \frac{\rho_1 t_1}{\rho_2 t_2}$,

where η_1 and η_2 are viscosities of liquid (1) (tolune) and liquid (2) (water) and similarly d_1 and t_1, d_2 and t_2 are respectively the densities and time taken to fall through the same length.

Given, η_2 = 0.01009 poise.

$$\therefore \eta_1 = \eta_2 \times \frac{\rho_1 t_1}{\rho_2 t_2}$$

By the problem, ρ_1 = 0.866 gm/cm^3

t_1 = 68.9 sec

ρ_2 = 0.998 gm/cm^3

t_2 = 102.2 sec

$$\therefore \eta_1 = \frac{0.01009 \times 0.866 \times 68.9}{0.998 \times 102.2} \text{ poise}$$

$$= 5.90264 \times 10^{-3} \text{ poise} = \mathbf{5.9 \times 10^{-3} \ poise.}$$

28. If there is 1% error in the value of r, the radius of the capillary, what will be the error in the viscosity coefficient value calculated by using Poiseuille equation? **[B.U. 1995]**

Soln. : We know that by Poiseuille equation,

$$\eta = \frac{\pi P r^4 t}{8 l v}, \text{ where the terms have usual meaning.}$$

Let r_1 and r_2 be correct and wrong radii respectively and η_1 and η_2 be correct and wrong viscosity coefficients respectively.

Now, $\frac{\eta_2}{\eta_1} = \left(\frac{r_2}{r_1}\right)^4$, where others have usual value.

$$\therefore \quad \frac{\eta_2}{\eta_1} = \left(\frac{r_2}{r_1}\right)^4 = \left(\frac{1.01}{1}\right)^4 = 1.04.$$

$\therefore \quad \frac{\eta_2}{\eta_1} = 1.04$. Hence if η_1 is the original value, the error is 0.04, i.e., **4%.**

29. In the absolute method of determination of viscosity coefficient η by Poiseuille formula, what should be the error in radius if error in η is to be kept with 4%. **[C.U. 1995]**

Soln. : We know from problem no. 28 that

$\frac{\eta_1}{\eta_2} = \left(\frac{r_1}{r_2}\right)^4$, where η_1, r_1 are the values in error and η_2, r_2 are correct values.

$$\therefore \quad \frac{\eta_1}{\eta_2} = 1.04 = \left(\frac{r_1}{r_2}\right)^4.$$

$$\therefore \quad \left(\frac{r_1}{r_2}\right) = (1.04)^{\frac{1}{4}} = 1.00985340 = 1.01.$$

i.e., the error is **1%.**

30. In a tall bath of oil of viscosity 54 poise and density 1.8 gm/cc, steel balls of radius 2 mm when allowed to fall down, the column of oil acquires uniform velocity of 1 cm/sec. What is the density of steel? **[C.U. 1978]**

Soln. : From problem no. 25, we have the formula—

$$v = \frac{2}{9}\frac{gr^2(\rho_s - \rho_l)}{\eta}$$

$$\text{or,} \quad 1 = \frac{2}{9} \times \frac{981 \times (0.2)^2 \times (\rho_s - 1.8)}{54}$$

$$\text{or,} \quad d_s - 1.8 = \frac{54 \times 9}{2 \times 981 \times 0.04}$$

$$\therefore \quad d_s = 7.991 = \mathbf{8\ gm/cc.}$$

31. At room temperature, the time taken by equal volumes of water and toluene to flow through a capillary tube are 102 sec and 69 sec respectively. Their densities at the same temperature are 0.989 gm/cc and 0.86 gm/cc respectively. If the viscosity of water is 0.01 P, calculate the viscosity of toluene.

Soln. : We know that $\frac{\eta_1}{\eta_2} = \frac{\rho_1 t_1}{\rho_2 t_2}$ [η_1 = viscosity of water, η_2 = viscosity of toluene]

Here, $\eta_1 = 0.01$ P, $\rho_1 = 0.989$ gm/cc, $t_1 = 102$ sec

$\eta_2 = ?$, $\rho_2 = 0.86$ gm/cc, $t_2 = 69$ sec

$$\therefore \quad \eta_2 = \frac{0.01 \times 69 \times 0.86}{0.989 \times 102} = \mathbf{5.88 \times 10^{-3}\ poise.}$$

32. In a viscometer, water and another liquid flow in 100 sec and 250 sec. Calculate the viscosity of that liquid, if its density is 0.8 gm/cc (η_w = 0.01 poise).

Soln. : We know, $\frac{\eta_1}{\eta_2} = \frac{\rho_1 t_1}{\rho_2 t_2}$.

Given, $\eta_2 = \eta_w = 0.01$ poise,

$t_2 = 100$ sec, $t_1 = 250$ sec, $\rho_1 = 0.8$ gm/cc and $\rho_2 = 1$ gm/cc, $\eta_1 = ?$

$$\therefore \eta_1 = \frac{\rho_1 t_1}{\rho_2 t_2} \times \eta_2 = \frac{0.8 \times 250 \times 0.01}{1 \times 100} = \mathbf{0.02\ poise.}$$

33. A sphere of radius 5×10^{-2} cm and $d = 1.10$ gm/cc falls at constant velocity through a liquid of density 1 gm/cc and viscosity 1 poise. What is the velocity of fall?

[C.U. 1987; B.U. 1978]

Hints : $mg = 6\pi r \eta v$

$$mg = \frac{4}{3}\pi r^3 (\rho_s - \rho_l)g = 6\pi r \eta v$$

$$\therefore \quad v = \frac{4\pi r^3(\rho_s - \rho_l)g}{3 \times 6\pi r\eta} = \frac{4r^2(\rho_s - \rho_l)g}{18\eta}$$

34. Two soap bubbles of radius 4 cm and 5 cm are joined together so as to have a common surface. Find the radius of the surface. [**Ans.** r = 20 cm]

35. Distilled water rose to a height of 4.8 cm in a certain capillary tube. The same tube filled with mercurry of density 13.55 gm/cc contained 50 mg/cm of tube length. Calculate γ for water. [**Ans.** 75.45 dynes/cm]

36. Two limbs of a vertical U-tube have internal diameters 0.05 mm and 0.3 mm respectively. It is particularly filled with a liquid of density 0.3 gm/cc. The surface tension of the liquid is 60 dynes/cm. What is the difference in the levels of the liquid in the two limbs. [**Ans.** 1.1 cm]

37. In a tall bath of oil of viscosity 54 poise and density 1.8 gm/cc, steel ball of radius 2 mm when allowed to fall down the column of oil acquires uniform velocity 1 cm/sec. What is the density of steel? [**Ans.** 8 gm/cc]

38. A ball of radius 1 cm and specific gravity 10.5 requires 50 sec to fall a distance of 1 m through a liquid of specific gravity 1.5. Calculate η. **[B.U. 2003]**

Soln. : We know that $mg = 6\pi r\eta v$

or, $\frac{4}{3}\pi r^3 (\rho_s - \rho_l)g = 6\pi r\eta v$.

By the problem,

$$\therefore \quad \frac{4}{3}\pi 1^3 (10.5 - 1.5) \times 980.665 = 6\pi r\eta v$$

$$\therefore \quad v = \frac{100\,\text{cm}}{50\,\text{sec}} = 2 \text{ cm/sec.}$$

$$\therefore \quad \frac{4}{3}(9) \times 980.665 = 12 r\eta$$

$r = 1$ cm

$$\therefore \quad \eta = \frac{4}{3} \times \frac{9 \times 980.665}{12} = \mathbf{980.665\ dynes/cm.}$$

39 For a certain liquid, log η vs $^1/_T$ plot yields a slope value of 600 K. Estimate the activation energy for flow. **[C.U. 2006]**

Soln. : $\eta = Ae^{Evis/RT}$

$$\therefore \log \eta = \log A + \frac{E_{vis}}{2.303\,RT}.$$

$$\text{By the problem, } 600 = \frac{E_{vis}}{2.303\,R}.$$

$$\therefore E_{vis} = 2.303 \times 600 \text{ K} \times 8.314 \text{ J.K}^{-1}.\text{mol}^{-1}$$

$$= \mathbf{11.488 \text{ kJ.mol}^{-1}}.$$

40. Calculate the work done in blowing a soap bubble in air of radius 10 cm. Surface tension of soap solution = 30 dynes/cm. **[C.U. 2004]**

Soln. : Work done in blowing a soap bubble

= surface energy of the bubble

= surface area × γ

= $2 \times 4\pi r^2 \times \gamma$ (since bubble has two surfaces, total surface area = $2 \times 4\pi r^2$)

= $2 \times 4 \times \pi \times 10^2$ cm^2 × 30 dynes/cm = **75398.22 ergs.**

41. The surface tension of water is 72.8 dynes.cm^{-1}. Calculate the energy required to disperse one spherical drop of radius 3.0 mm in spherical droplets of radius 3.0×10^{-3} mm. **[C.U. 2004]**

Hints : Surface energy of one spherical drop = $4 \times \pi \times (3 \times 10^{-1})^2 \gamma$

$$N \times \frac{4}{3} \times \pi\,(3 \times 10^{-3})^3 = \frac{4}{3} \times \pi \times 3^3$$

$$N = 10^9$$

Energy required = $4 \times \pi \times (3 \times 10^{-3} \times 10^{-1})^2 \times \gamma \times 10^9 - 4 \times \pi \times (3 \times 10^{-1})^2 \times \gamma$

= **82252.53 ergs.**

42. A steel ball of density 7.90 gm/cc and 4 mm diameter requires 55 sec to fall a distance of 1 metre through a liquid of density 1.10 gm/cc. Calculate the coefficient of viscosity of the liquid. **[V.U. 2004]**

Hints : $h = \frac{2}{9}\frac{gr^2(\rho_s - \rho_l)}{V}$ = 32.6 poise.

43. One end of a capillary tube 10 cm in length is connected to water supply which has a pressure of 1.8 atm and other end to a vessel open to air at 25 °C; the tube delivers 1.2 litre/minute. Calculate the diameter of the capillary, if viscosity coefficient for water at 25 °C be 0.01 poise. **[V.U. 2004]**

Soln. : According to Poiseuille equation,

$$\eta = \frac{(P_2 - P_1)\pi r^4 t}{8Vl}.$$

Here, η = 0.01 poise, P_2 = 1.8 atm, P_1 = 1 atm

V = 1.2 litre = 1200 cc, t = 60 sec, l = 10 cm

$$\therefore\ r^4 = \frac{8\eta Vl}{(P_2 - P_1)\pi t} = \frac{8 \times 0.01\,\text{gm.cm}^{-1}\,\text{sec}^{-1} \times 1200\,\text{cc} \times 10\,\text{cm}}{(1.8-1)\,\text{atm} \times 1.01325 \times 10^6\,\text{dyne.cm}^{-2}/\text{atm} \times \pi \times 60\,\text{sec}}$$

$= 6.283 \times 10^{-6}$ cm^4.

Hence, $r = 0.05$ cm.

$\therefore$ diameter of the capillary $= d = 2r =$ **0.1cm.**

44. A large glass bottle is fitted with a siphon made of capillary glass tubing. Compare the time taken to empty the bottle when it is filled with (a) water of density 10^3 kg/m^3 and (b) petroleum of density 0.8×10^3 kg/m^3. Given that viscosity coefficient of water is half that of petroleum.

Hints : $\frac{\eta_1}{\eta_2} = \frac{\rho_1 t_1}{\rho_1 t_2}$, where 1 corresponds to petroleum and

2 corresponds to water.

$\left(\frac{t_1}{t_2} = 2.5\right)$.

45. One end of a 10 cm long capillary tube is connected to water supply which has a pressure of 1.8 bar and the other open end to air of 1 bar pressure. The tube delivers 0.2l water per min. Calculate the diameter of the tube if η of water is 0.01 P. (All are at a fixed temerature of 298 K). **[V.U. 2006]**

Soln. : By Poiseuille's equation, $\eta = \frac{\pi(P_1 - P_2)r^4}{8Vl}$

By the problem,

$P_1 - P_2 = (1.8 - 1)$ bar $= 0.8$ bar $\times 10^6\ \frac{\text{dyne/cm}^2}{\text{bar}} = 0.8 \times 10^6$ dyne.cm^{-2}

$l = 10$ cm $\qquad$ r = radius of the tube

$V = 0.2$ l/min $= \frac{0.2 \times 10^3}{60}$ cc.sec^{-1}, $\eta = 0.01$ poise

$$\text{Thus, } 0.01 \text{ poise} = \frac{3142 \times 0.8 \times 10^6\,\text{dyne.cm}^{-2} \times r^4}{8 \times \frac{0.2 \times 10^3}{60}\,\text{cc.sec}^{-1} \times 10\,\text{cm}}$$

$$r^4 = \frac{0.01\,\text{gm.cm}^{-1}.\text{sec}^{-1} \times 8 \times 10\,\text{cm} \times 0.2 \times 10^3\,\text{cm}^3.\text{sec}^{-1}}{60 \times 3.142 \times 0.8 \times 10^6\,\text{gm.cm}^{-2}.\text{sec}^{-2}}, \quad \therefore r = 0.032 \text{ cm.}$$

Thus, diameter of the tube $= (0.032 \times 2)$ cm $= 0.064$ cm.

CHAPTER 4

SOLID STATE

Required Formulae

1. Bragg's equation :

$$n\lambda = 2d_{hkl} \sin \theta,$$

where n is order,

λ is wavelength of X-rays,

d_{hkl} is distance between consecutive planes,

θ is glancing angle at which X-ray is diffracted by the lattice plane

2. For cubic crystals,

$$\rho = \frac{ZM}{N_A a^3},$$

where ρ = density,

M = moleculer mass,

N_A = Avogadro number,

a = length of the edge of the cube,

Z = no. of atoms per unit cell.

3. If OA, OB, OC are intercepts of a crystalline plane on X, Y, Z respectively and a,b,c are unit distances along respective axes, then

$$\frac{a}{OA} = h \quad \frac{b}{OB} = k \quad \frac{c}{OC} = l.$$

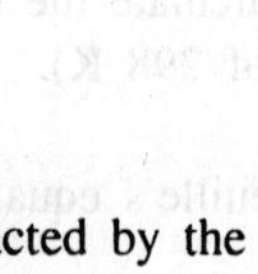

(hkl) are Miller indices.

4. For an orthorhombic lattice, $a \neq b \neq c,$

$\alpha = \beta = \gamma = 90°.$

If d_{hkl} is the distance between two hkl planes or length of the perpendicular on the hkl plane from origin, then

$$\frac{1}{d_{hkl}^2} = \left(\frac{h}{a}\right)^2 + \left(\frac{k}{b}\right)^2 + \left(\frac{l}{c}\right)^2.$$

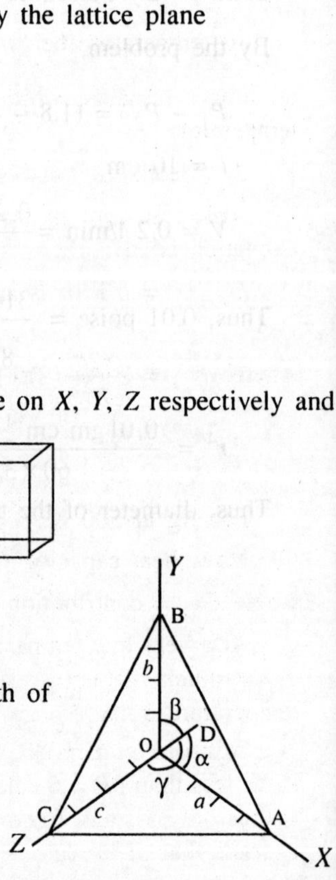

5. Rules for determining the diffracting (*hkl*) planes, responsible for appearance of X-ray diffraction patterrn, in a cubic lattices :

Simple cube	(h,k,l) any value
b.c.c	$(h + k + l)$ = even
f.c.c	(h, k, l) all odd or all even

Short Questions

1. What is Einstein characteristic temperature? Explain.

Ans. : Einstein characteristic temperature $\theta = \frac{h\nu}{k}$,

where h = Planck constant,

k = Boltzmann constant,

ν = single monochromatic frequency with which an isotropic solid vibrates

Heat capacity $C_v = 3Rf(\theta/T)$.

This has a fixed value for a fixed term.

For different substances different values of ν and θ are obtained, and these are carried out purely empirically to fit experimental results.

2. What is Debye characteristic temperature?

Ans. : Debye characteristic temperature $\theta_m = \frac{h\nu_m}{k}$, where h = Planck const., k = Boltzmann const., ν_m = maximum frequency with which an isotropic solid can vibrate.

At temperature below $\frac{\theta}{10}$, solid follows T^3 law, $\theta = 7.45TC_v^{-1/3}$. As T increases, θ no longer remains constant. θ can be calculated independently of atomic heat at low temperature.

3. Metals do not obey T^3 law at a very low temperature—comment.

Ans. : In case of metals, excitation of the electrons must be considered. Electronic excitation contributes to specific heat—a term linear in T.

$C_{v,el} = b.T$ [b is around 10^{-4} to 10^{-3} cal.mol^{-1}K^{-2}],

where $b = \frac{N_A m}{h^2}(2\pi k)^2\left(\frac{\pi}{3n}\right)^{\frac{2}{3}}$.

n = no. of electrons per unit volume.

N_A = Avogadro number.

For Ag at 300 K, b = 0.046 cal gm^{-1} (about 7% of total).

Now, heat capacity, $C_v = C_{v,el} + C_{v,l}$,

where $C_{v,l}$ = contribution to C_v due to lattice vibration.

At a very low temperature, lattice vibrations are negligible. Thus, the term containing T has much higher contribution. For Ag at (1.5-0.3) °K, T^3 law is invalid and above temperature 3 °K T^3 law is valid. For Ni at (1.1-9·22) °K T^3 law becomes invalid.

At room temperature electronic contribution to C_v is 0.03 to 0.3 cal.mol^{-1}.K^{-1}, which is far less than $3R$ = 6 cal.mol^{-1}.K^{-1}. At moderate temperature, the electronic contribution is thus only a small fraction of C_v. At a very low temperature, the electronic contribution exceeds that of the lattice vibration since $C_{v,l} = aT^3$, where a is a constant, goes to zero faster than bT, T^3 law is not valid at low temperature.

4. What is the primitive lattice corresponding to f.c.c. and b.c.c.?

Ans. : For f.c.c., the primitive lattice is rhombohedral and for b.c.c. the primitive lattice is triclinic.

5. Explain why hexagonal closed packed structure and a cubic closed packed structure for a given element would be expected to have the same density.

Ans. : The two structures have the same coordination number and hence the same packing fraction. So, they have the same density.

6. Show that in a crystal lattice there cannot be five-fold rotational axis of symmetry.

Ans. : Let *AB* be two points in a crystal lattice with length *a*. If it was a $\frac{360}{\theta}$ fold axis of symmetry, then a rotation by angle θ clockwise and anticlockwise will lead to points *C* and *D* respectively. Then the distance *CD* should be integer multiple of *a*, i.e., *ma*, where *m* is an integer.

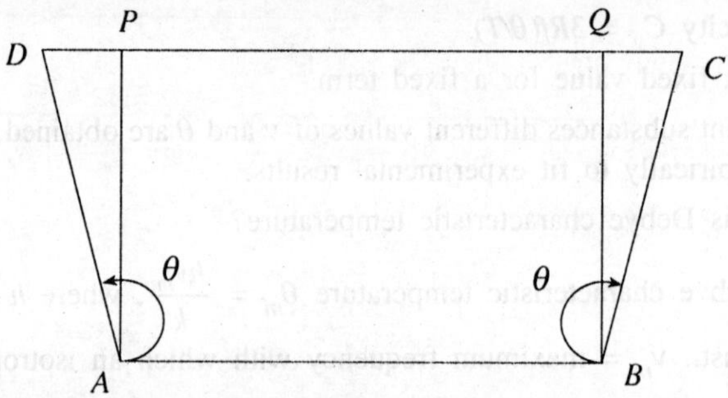

AP and *BQ* are perpendiculars drawn from *A* and *B* respectively on CD. Amongst the triangle *DPA* and *QCB*, angle $\angle DAP = \angle CBQ = \theta - 90°$. $AD = BC = a$, $AP = BQ$. Hence, the two triangles are identical. So, $DP = QC = x$ (say).

But, $\frac{DP}{AD} = \sin(\theta - 90°)$

or, $DP = a \sin(\theta - 90°)$.

Now, $CD = a + 2x = a + 2a \sin(\theta - 90°) = ma$.

$\therefore$ $ma = a + 2a \sin(\theta - 90°) = a - 2a \sin(\theta - 90°) = a - 2a \cos\theta$.

$\therefore$ $1 - 2\cos\theta = m$

or, $\cos\theta = \frac{1-m}{2}$.

$\cos\theta$ can vary from 0 to 1. Hence, *m* can be 0, 1, 2 and 3. This is shown in the following table :

m	$\cos\theta$	θ	n-fold axis
0	$\frac{1}{2}$	60	6
1	0	90	4
2	$-\frac{1}{2}$	120	3
3	-1	180	2

Thus, there cannot be five-fold axis.

7. Show that for an orthorhombic lattice the distance between two planes d_{hkl} is given by

$$\frac{1}{d_{hkl}^2} = \left(\frac{h}{a}\right)^2 + \left(\frac{k}{b}\right)^2 + \left(\frac{l}{c}\right)^2.$$

Ans. : Let us consider a plane having Miller indices *hkl*. Its equation is $hx + ky + lz + d = 0$. Let d_{hkl} be its distance from the centre. It can be shown that the distance between two planes is equal to the distance between the plane and the centre.

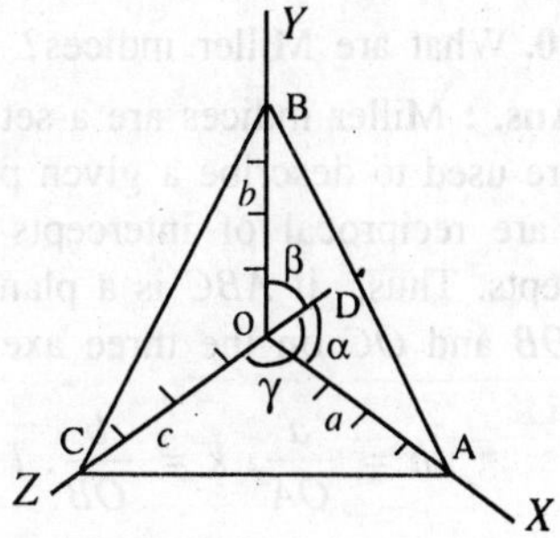

Let the axes and axial ratios *a, b, c* be as shown and the plane be *ABC*.

$$OD = d_{hkl}$$

OD is perpendicular on the plane from centre.

$$d_{hkl} = OA \cos\alpha = \left(\frac{a}{h}\right)\cos\alpha$$

$$= OB.\cos\beta = \frac{b}{k}\cos\beta.$$

$$= OC\cos\gamma = \frac{c}{l}\cos\gamma.$$

$$\cos\alpha = \frac{h}{a}\, d_{hkl}$$

$$\cos\beta = \frac{k}{b}\, d_{hkl}$$

$$\cos\gamma = \frac{l}{c} d_{hkl}$$

$\cos\alpha$, $\cos\beta$, and $\cos\gamma$ are direction cosines.

Here, $\cos^2\alpha + \cos^2\beta + \cos^2\gamma = 1$.

$$\therefore \left(\frac{h}{a}\right)^2 d_{hkl}^2 + \left(\frac{k}{b}\right)^2 d_{hkl}^2 + \left(\frac{l}{c}\right)^2 d_{hkl}^2 = 1$$

$$\text{or,} \quad \frac{1}{d_{hkl}^2} = \left(\frac{h}{a}\right)^2 + \left(\frac{k}{b}\right)^2 + \left(\frac{l}{c}\right)^2.$$

In case of a cube, $a = b = c$.

$$\therefore \frac{1}{d_{hkl}^2} = \frac{h^2 + k^2 + l^2}{a^2}.$$

8. Define a unit cell.

Ans. : The smallest collection of atoms from which the whole crystal can be built up by mentally translating the unit parallel to itself in the direction of its edges by distances equal to the length of its edges is called *unit cell of the crystal*.

It represents the symmetry of the corresponding crystal.

9. What is unit cell content?

Ans. : The unit cell content Z is the number of particles contained within the unit cell. A primitive unit cell contains one particle per unit cell.

A non-primitive unit cell is edge centred, face centred, or body centred and contains more than one point or particle per unit cell.

10. What are Miller indices?

Ans. : Miller indices are a set of integers (hkl) that are used to describe a given plane in a crystal. They are reciprocal of intercepts in terms of the intercepts. Thus, if ABC is a plane with intercepts OA, OB and OC on the three axes, then

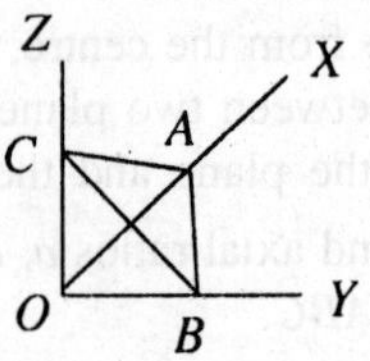

$$h = \frac{a}{OA}, k = \frac{b}{OB}, l = \frac{c}{OC}.$$

11. What is the highest order that can be observed in Bragg's reflection from a solid by X-ray?

Ans. : We know that $n\lambda = 2d \sin\theta$.

where n is the order of reflection by the lattice planes with interplanar separation d and λ is the wave length of X-ray.

$\therefore \quad n\lambda \leq 2d$

Since $(\sin\theta)_{max} = 1$,

$$\text{hence, } n \leq \frac{2d}{\lambda}.$$

So, the highest order that can be observed depends both on lattice plane separation and wave length of the X-ray and it is $\frac{2d}{\lambda}$.

12. What is Bravais lattice?

Ans. : A Bravais lattice is an array of equivalent points that reflect the symmetry of representative crystal. It is a repeating unit but devoid of any physical properties.

13. What are the close-packed planes in a (i) b.c.c. lattice; (ii) f.c.c. lattice

Ans. : (i) In b.c.c lattice, (110) planes are close packed.

(ii) In, f.c.c lattice (111) planes are close packed.

14. What is the closest distance between two neighbouring atoms in (a) f.c.c lattice; (b) b.c.c lattice.

Ans. : (a) In f.c.c lattice, the close packed direction is along face diagonal of the cubic unit cell. If the length of the unit cell is a Å, the closest distance between two atoms $= \frac{\sqrt{2}a}{2}$

$= \frac{a}{\sqrt{2}}$.

(b) In b.c.c lattice, the close packed direction is along body diagonal of the cubic unit cell. If the length of the unit cell is a Å, the closest distance between two atoms $= \frac{\sqrt{3}a}{2}$

Numerical Problems

1. A crystal plane intercepts the three crystallographic axes at the following multiples of unit distance $\frac{3}{2}$, 2 and 1. What will be the Miller indices of the plane? **[C.U. 1980]**

Soln. : Miller indices of the plane,

$$\frac{a}{\frac{3}{2}a} : \frac{b}{2b} : \frac{c}{c} = \frac{2}{3} : \frac{1}{2} : 1 = 4 : 3 : 6.$$

$\therefore$ (hkl) = **(4 3 6).**

2. Tungsten has body centred cubic lattice and each lattice point is occupied by one atom. Calculate the length of the cube edge (density of Tungsten is 19.30 gm/cc and its atomic wt. is 183.9). **[C.U. 1978]**

Soln. : We know, $\rho = \frac{MZ}{N_A . a^3}$.

Given, ρ = 19.30 gm/cc.

Z = 2

M = 183.9 gm

N_A = 6.022×10^{23}

$$19.30 = \frac{183.9 \times 2}{6.022 \times 10^{23} \times a^3}$$

$$\text{or,} \quad a^3 = \frac{183.9 \times 2}{19.30 \times 6.022 \times 10^{23}} = 3.1645 \times 10^{-23}$$

$\therefore$ a = 3.163039×10^{-8} cm = 3.163 Å

So, the length of the cube edge is 3.163 Å.

3. Aluminium crystallizes with a face-centred cubic lattice. The inter-ionic distance (shortest) in a unit cell of Aluminium is 2.86 Å. Calculate the density of Aluminium (At. wt. of Al = 27).

Soln. : We know, $\rho = \frac{MZ}{N_A . a^3}$.

By the problem, $4r = \sqrt{2}a$.

Here, M = 27 gm

Z = 4

N_A = 6.022×10^{23}

$2r$ = 2.86 Å.

Now, $2 \times 2r = \sqrt{2}a$

or, 2×2.86 Å $= \sqrt{2}a$.

$\therefore a = 2.86\sqrt{2} \times 10^{-8}$ cm.

$$\therefore \rho = \frac{27 \text{ gm} \times 4}{6.022 \times 10^{23} \times \left(2.86 \times 10^{-8} \times 1.414\right)^3 \text{cc}}$$

= 2.809 gm/cc = **2.81 gm/cc.**

4. At room temperature sodium has body centred cubic lattice with a cell of edge length 4.29 Å. At –200 °C, the density is only 4% larger but the cell edge is now 5.35 Å. What type of cubic unit cell does sodium possess at –200 °C?

Soln. : At room temperature,

$$\rho = \frac{MZ}{N_A . a^3}$$

$$= \frac{23\,\text{gm} \times 2}{6.022 \times 10^{23} \times \left(4.29 \times 10^{-8}\ \text{cm}\right)^3}$$

$= 0.967487$ gm/cc $= 0.97$ gm/cc.

At (–200 °C) the density is 0.97×1.04

$= 1.006186$ gm/cc $= 1.0062$ gm/cc.

Again, we know, $\rho = \dfrac{MZ}{N_A a^3}$.

$$\therefore \quad Z = \frac{\rho . N_A . a^3}{M} = \frac{1.0062 \times 6.022 \times 10^{23} \times \left(5.35 \times 10^{-8}\right)^3}{23} = 4.0342 = 4.$$

Hence, at –200 °C, sodium has f.c.c. lattice.

5. Caesium bromide lattice contains 8 bromide ions in the corner of a cube with a Cs^+ ion at the centre. Its density is 4.5 gm/cc. Calculate the edge length.

Soln. : We know, $\rho = \dfrac{MZ}{N_A a^3}$

Now, $N_A = 6.022 \times 10^{23}$, $\rho = 4.5$ gm/cc, $M_{Cs^+} = 133$ gm and $M_{Br^-} = 80$ gm

$$\therefore \ a^3 = \frac{MZ}{N_A \rho} = \frac{\left(M_{Cs^+} + M_{Br^-}\right)Z}{N_A \rho}$$

$$= \frac{(133\,\text{gm} + 80\,\text{gm}) \times 1}{6.022 \times 10^{23} \times 4.5\,\text{gm.cm}^{-3}} = 7.86 \times 10^{-23}\ \text{cm}^3.$$

$\therefore \ a = \mathbf{4.283 \times 10^{-8}}$ **cm.**

6. The 222 planes of Tantalum (cubic) give a reflection at 54° 7.2′, when the X-ray has $\lambda = 1.5418$ Å. At what angles would these planes give reflection, when $\lambda = 0.7107$ Å? What is the edge length? What are the spacings between 100, 110 and 111 planes?

Soln. : Here $\lambda = 1.5418 \times 10^{-8}$ cm

$\theta = 54°\ 7.2'$

Now, $\sin\theta = \sin 54°\ 7.22' = 0.81$

Again, $\lambda = 2d \sin\theta$

$$\text{or, } d = \frac{\lambda}{2\sin\theta} = \frac{1.5418 \times 10^{-8}}{2 \times 0.81} = 0.95 \times 10^{-8}\ \text{cm} = 0.95\ \text{Å}.$$

$$\text{Again, } \sin\theta = \frac{\lambda}{2d} = \frac{0.7107 \times 10^{-8}}{2 \times 0.95 \times 10^{-8}} = 0.37405.$$

$\therefore \quad \theta = 21.96577 =$ **21° 58′.**

We know, $d = \dfrac{a}{2\sqrt{3}}$ (for 222 planes)

or, $a = d \times \sqrt{3} \times 20.95 \times 10^{-8} \times \sqrt{3} \times 2$

$= 1.6454 \times 10^{-8} \times 2 =$ **3.28 Å.**

$\therefore \; d_{100} : d_{110} : d_{111}$

$$= a : \frac{a}{\sqrt{2}} : \frac{a}{\sqrt{3}}$$

$$= 3.28 : \frac{3.28}{1.414} : \frac{3.28}{1.732} = \mathbf{3.28 : 2.32 : 1.9.}$$

7. The crystal structure of magnesium oxide is similar to sodium chloride. The distance between Mg^{++} and $O^{=}$ ions in the cell is 2.10 Å. What is the density of MgO?

Soln. : Here, $a = 2d_{MgO}$ $[a = 2(r_+ + r_-)]$

$= 2 \times 2.10 \times 10^{-8}$ cm $= 4.20 \times 10^{-8}$ cm.

$$\therefore \quad \rho = \frac{MZ}{N_A a^3}$$

Here M = 40 gm

Z = 4

N_A = 6.022 × 1023

$a = 4.20 \times 10^{-8}$ cm

ρ = density.

$$\therefore \rho = \frac{40 \times 4}{6.022 \times 10^{23} \times \left(4.20 \times 10^{-8}\right)^3} = 3.586 \text{ gm/cc} = \mathbf{3.6 \text{ gm/cc.}}$$

8. KCl has a face centred cubic lattice. Density of KCl is 2 gm/cc at a certain temperature and edge length is 6.28 Å. Find the number of K^+ and Cl^- ions per gram.

Soln. : Given,

ρ = 2 gm/cc

$a = 6.28 \times 10^{-8}$ cm

Z = 4.

Volume of unit cube $= a^3 = (6.28 \times 10^{-8} \text{ cm})^3 = 2.477 \times 10^{-22}$ cc.

Hence, weight of unit cube $= 2 \times 2.477 \times 10^{-22}$ gm $= 4.954 \times 10^{-22}$ gm.

Unit cube contains $4K^+$ and $4Cl^-$ ions.

$$\therefore \text{ number of } K^+ \text{ or } Cl^- \text{ ions per gram} = \frac{4}{4.954 \times 10^{-22}} = 8.1 \times 10^{21} \text{ ions/gm.}$$

9. The element chromium exists in a body centred cubic lattice whose unit cell edge is 2.88×10^{-8} cm. The density of chromium is 7.2 gm/c.c. Find out the Avogadro number (At. wt. of Cr = 52).

Soln. : We know, $\rho = \dfrac{MZ}{N_A a^3}$

Here, ρ = 7.2 gm/cc

M = 52 gm

Z = 2

$a = 2.88 \times 10^{-8}$ cm

$\therefore \quad N_A = \mathbf{6.047 \times 10^{23}}$.

10. For LiBr the heat of formation is reported as 97.4 kcal. The heat of vaporisation of Li is 37 kcal. Ionisation energy of Li as 123.9 kcal per gram-atom. Calculate the lattice energy of LiBr. Assume that electron affinity of Br_2 is 87 kcal/gm and the heat of dissociation of bromine is 53 kcal.

Soln. : As given in the problem,

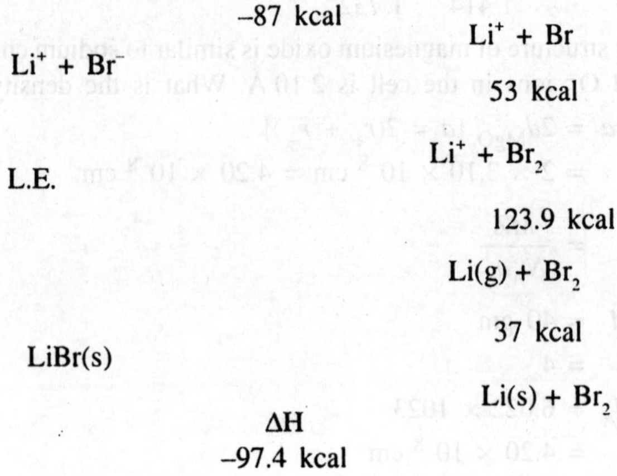

$$37 + 123.9 - 87 + 53 + \text{L.E.} = -97.4$$

$$\therefore \quad \text{L.E.} = -37 - 123.9 + 87 - 53 - 97.4 = \mathbf{-229.3\ kcal.}$$

11. A powder pattern is obtained for α-phase of a metal. Successive $\sin^2\theta$ values are in the ratio : 1, 2, 3, 4, 5, 6, 7, 8. Which type of cubic unit cell is present? Discuss the probable *hkl* value responsible for $\sin^2\theta = 7$ in the above series. The density of α-metal is 7.91 gm/cm^{-3}. The wavelength of the X-rays is 0.5 Å. Calculate *a* for unit cell. Calculate the value of θ for Bragg reflection for 110 type planes and 100 type planes. Calculate the radius of the atom. (atomic wt. of the metal is 56). The β-form is f.c.c. cube. Its $a = 3.60 \times 10^{-8}$ cm. Also calculate the radius of the metal in α- and β-forms. Calculate the ratio of their densities.

Soln. : 7 cannot be the sum of squares and it must actually be 14. The series is then 2, 4, 6, 8, 10, 12, 14, 16.

The crystal (α- form) is thus b.c.c.

$$h^2 + k^2 + l^2 = 14$$

hkl = 321, not necessarily, may be 231 or 123 in that crystal.

$$\therefore \quad a^3 = \frac{MZ}{N_A\rho} = \frac{56 \times 2}{6.022 \times 10^{23} \times 7.91}.$$

Thus, $a = 2.864839 \times 10^{-8}\ \text{cm} = \mathbf{2.86\ Å}$

$$d_{100} = \frac{a}{2} = \frac{2.86\ Å}{2} = 1.43\ Å$$

$$\sin\theta = \frac{0.5}{2.86} = 0.174825, \ \therefore\ \theta = \mathbf{10.07°}.$$

$$d_{110} = \frac{a}{\sqrt{2}} = \frac{2.86\ Å}{\sqrt{2}} = 2.022325\ Å = 2.02\ Å$$

$$\sin\theta = \frac{0.5}{2\times 2.02} = 0.123762, \quad \therefore \theta = \mathbf{7.11°}.$$

for α-form,

$$4r = \sqrt{3}a$$

$$\therefore \quad r = \frac{\sqrt{3}a}{4} = \frac{\sqrt{3}\times 2.86\times 10^{-8}}{4} = 1.238416\text{ Å} = \mathbf{1.24\text{ Å}}$$

$$\therefore \quad \rho_1 = \frac{MZ}{N_A.a^3} = \frac{112}{\left(2.86\times 10^{-8}\right)^3 N_A}$$

For the β-form,

$$\rho_2 = \frac{MZ}{N_A.a^3} = \frac{224}{N_A\times\left(3.6\times 10^{-8}\right)^3} \text{ (since } Z = 4)$$

and $4r = \sqrt{2}a$

$$\therefore \quad r = \frac{\sqrt{2}\times 3.6\times 10^{-8}}{4} = \mathbf{1.27\text{ Å}}$$

Now
$$\frac{\rho_1}{\rho_2} = \frac{112}{\left(2.86\times 10^{-8}\right)^3 N_A}\times\frac{N_A\times\left(3.6\times 10^{-8}\right)^3}{224}.$$

$$\therefore \quad \frac{\rho_1}{\rho_2} = \frac{(3.6)^3}{(2.86)^3\times 2} = \frac{46.656}{46.787} \approx \mathbf{1.}$$

12. An orthorhombic unit cell has the following parameters : a = 0.5 Å, b = 1 Å, c = 1.5 Å. What is the spacing of the (123) planes?

[C.U. 1990 1996; B.U. 1988]

Soln. : For orthorhombic unit cell, $a \neq b \neq c, \alpha = \beta = \gamma = 90°$.

The distance between two planes having Miller indices (hkl) is

$$\frac{1}{d_{hkl}^2} = \frac{h^2}{a^2}+\frac{k^2}{b^2}+\frac{l^2}{c^2}$$

or,
$$\frac{1}{d_{hkl}^2} = \frac{1^2}{(0.5)^2}+\frac{2^2}{(1)^2}+\frac{3^2}{(1.5)^2}$$

or,
$$\frac{1}{d_{hkl}^2} = \frac{1}{0.25}+\frac{4}{1}+\frac{9}{2.23} = \frac{9+9+9}{2.25} = \frac{27}{2.25} = 12$$

$$\therefore \quad d_{hkl}^2 = \frac{1}{12}$$

$$\therefore \quad d_{hkl} = 0.2886\text{ Å} = 0.289\text{Å} = 0.289\times 10^{-8}\text{ cm} = \mathbf{2890\text{ pm}}.$$

13. Find the simplest formula of a solid containing A and B atoms in a cubic arrangement in which A occupies corner and B the centre of the faces of unit cell. If side length is 5 Å, estimate the density of the solid assuming atomic weights of A and B as 60 and 90 respectively. **[C.U. 1977]**

Soln. : A atoms occupy 8 corners. Each corner is shared by 8 cubes. So, the single cube shares $\frac{1}{8}$th of each point. Thus, there is one 'A' atom ($8 \times \frac{1}{8}$A). The B atoms occupy the centre of faces of the unit cell. Thus, there are ($6 \times \frac{1}{2}$) = 3B atoms occupied by the single cube. Thus, the formula of the compound is $\mathbf{AB_3}$.

We know that density $\rho = \frac{MZ}{N_A a^3}$. Here, MZ is the full weight of the occupying atoms which here is weight of the AB_3 = (60 × 1) + (90 × 3) = 330 gm.

$$\text{Thus, density} = \frac{330}{6.022\times10^{23}\times\left(5\times10^{-8}\right)^3} = 4.3839 \text{ gm/cc} = \mathbf{4.39\ gm/cc.}$$

14(a). An element occurs in two crystalline forms α and β. The α-form has f.c.c. with a = 3.68 Å and the β-form has b.c.c. with a = 2.92 Å. Calculate the ratio of their densities. **[C.U. 1978]**

Soln. : As given in the problem—

$$\rho_\alpha = \frac{4M}{N_A a^3} = \frac{4M}{6.022\times10^{23}\times\left(3.68\times10^{-8}\right)^3}$$

and

$$\rho_\beta = \frac{2M}{N_A a^3} = \frac{2M}{6.022\times10^{23}\times\left(2.92\times10^{-8}\right)^3}.$$

$$\therefore \quad \frac{\rho_\alpha}{\rho_\beta} = \frac{4M}{2M}\times\frac{(2.92)^3}{(3.68)^3}$$

$$\text{or,} \quad \frac{\rho_\alpha}{\rho_\beta} = \frac{2\times(2.92)^3}{(3.68)^3}$$

$$\therefore \quad \frac{\rho_\alpha}{\rho_\beta} = \frac{49.794176}{49.836032} = 0.9991660125.$$

$$\therefore \quad \frac{\rho_\alpha}{\rho_\beta} \approx 1.$$

Thus, the densities are in the ratio **1 : 1.**

14(b). Same as 14(a), only difference is now a = 3.65 Å for f.c.c. and a = 2.90 Å for b.c.c. Calculate the ratio $d_\alpha : d_\beta$ as above. **[C.U. 1999]**

$$\textbf{Soln. :} \quad \frac{\rho_\alpha}{\rho_\beta} = \frac{4\times(2.90)^3}{2(3.65)^3} = 1.0031.$$

$$\therefore \quad \frac{\rho_\alpha}{\rho_\beta} \approx 1.$$

15. A metal has a b.c.c. cubic lattice and the length of a unit cell is 2.95 Å. If the density of the metal be 9.95 gm/cc, calculate the atomic weight of the metal. **[C.U. 1995]**

Hints : $\rho = \dfrac{MZ}{N_A a^3}$,

where $Z = 2$

$a = 2.95 \times 10^{-8}$ cm

$\rho = 9.95$ gm/cc

$\therefore \quad M = ?$

Ans. : $M = 76.91302$ gm = **76.9 gm.**

16. Potassium crystallizes with a b.c.c. lattice and has a density of 0.835×10^3 kg.m^{-3}. What is the length of the side of the unit cell 'a' and the distance between (200) planes and that between (110) planes. **[C.U. 1993]**

Soln. : By the problem,

$Z = 2$ for b.c.c.

At. wt. of potassium

$= 39 \times 10^{-3}$ kg/gm-atom.

$\rho = 0.835 \times 10^3$ kg.m^{-3}

$$\therefore \quad \rho = \frac{MZ}{N_A a^3}$$

$$\text{or,} \quad a^3 = \frac{39 \times 10^{-3} \times 2}{6.022 \times 10^{23} \times 0.835 \times 10^3}$$

$\therefore \quad a = 5.37306 \times 10^{-10}$ m = 537.306 pm.

$$\therefore d_{200} = \frac{a}{2} = \frac{537.306}{2} = \mathbf{268.653\ pm.}$$

$$d_{110} = \frac{a}{\sqrt{2}} = \frac{537.306}{\sqrt{2}} = \mathbf{379.93\ pm.}$$

17. A face centred lattice has all the positions occupied by atom A. The body centred octahedral hole in it is occupied by an atom B. For such a crystal calculate the void space per unit volume of unit cell. Predict the formula of the compound. **[V.U. 2005]**

Soln. : In the face centred lattice $4r_A = \sqrt{2}a$, where a is edge length of the cube and r_A is the radius of A atom.

$\therefore a = 2\sqrt{2}\, r_A$.

The body centred octahedral hole h is occupied by B atom.

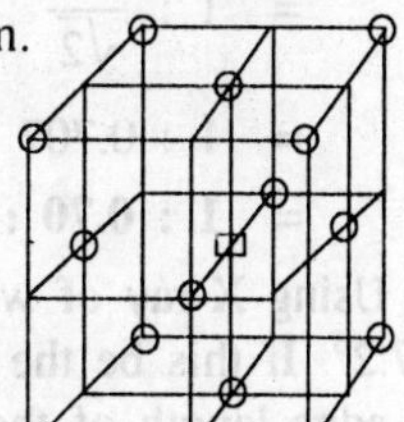

$\therefore \; 2r_B + 2r_A = a = 2\sqrt{2}r_A$.

$\therefore \; 2r_B = (2\sqrt{2} - 2)r_A$.

$\therefore \; \dfrac{r_B}{r_A} = (\sqrt{2} - 1) = 0.414$.

Volume of the cube is $a^3 = (2\sqrt{2}r_A)^3 = 16\sqrt{2}\, r_A^3$

Occupied volume by A and B :

$$4 \times \frac{4}{3}\pi r_A^3 + \frac{4}{3}\pi r_B^3 = \frac{4}{3}\pi\left(4r_A^3 + r_B^3\right).$$

Volume occupied per unit volume of the unit cell is

$$\frac{\frac{4}{3}\pi\left(4r_A^3+r_B^3\right)}{\left(2\sqrt{2}r_A\right)^3} = \frac{\frac{4}{3}\pi\left(4r_A^3+r_B^3\right)}{16\sqrt{2}r_A^3}$$

$$= \frac{\pi}{12\sqrt{2}}\left[4+\left(\frac{r_B}{r_A}\right)^3\right]$$

$$= \frac{3.14}{12\times1.414}\left(4+0.414^3\right)$$

$$= \frac{3.14\times4.070957944}{12\times1.414} = \frac{3.14\times4.071}{12\times1.414} = 0.7534.$$

So, 75.34% is occupied.

∴ void space per unit volume of unit cell = **24.66%.**

There are four A atoms and one B atom per unit cell. So the probable formula is **BA_4**.

18. Calculate the ratio of the distances among (100), (110) and (111) planes of a simple cubic crystal with a lattice side of 1.40 Å. **[C.U. 1994]**

Soln. : Let *a* be the length of the unit cell. We know that, for a simple cube the ratio of the distances are—

$$d_{100} : d_{110} : d_{111} = a : \frac{a}{\sqrt{2}} : \frac{a}{\sqrt{3}}.$$

In this case, a = 1.40 Å

Hence, $d_{100} = a = 1.40$ Å

$$d_{110} = \frac{a}{\sqrt{2}} = \frac{1.40}{\sqrt{2}}\text{Å}$$

$$d_{111} = \frac{a}{\sqrt{3}} = \frac{1.40}{\sqrt{3}}\text{Å}.$$

Thus, $d_{100} : d_{110} : d_{111}$

$$= a : \frac{a}{\sqrt{2}} : \frac{a}{\sqrt{3}}$$

$$= 1 : \frac{1}{\sqrt{2}} : \frac{1}{\sqrt{3}}$$

$$= 1 : 0.707 : 0.577$$

$$= \mathbf{1 : 0.70 : 0.58.}$$

19. Using X-ray of wavelength $\lambda = 1.79 \times 10^{-8}$ cm, a metal produces reflection at $2\theta = 47.2°$. If this be the first order reflection from the (110) plane of b.c.c. unit cells, what is edge length of the cube? **[V.U. 1994]**

Soln. : Given that, $\lambda = 1.79 \times 10^{-8}$ cm.

By Bragg's law, for 1st order reflection, $\lambda = 2d \sin\theta$.

$$\therefore \sin\theta = \frac{\lambda}{2d} = \frac{1.79\times10^{-8}}{2d}$$

For b.c.c. unit cell,

$$d_{110} = \frac{a}{\sqrt{2}}.$$

By the problem, $\theta = \frac{47.2°}{2} = 23.6°$.

So, $1.79 \times 10^{-8} = 2d \sin 23.6 = 2d \times 0.40035$.

$$\therefore\ 2d_{110} = \frac{1.79\times10^{-8}}{0.40035}$$

$$d_{110} = \frac{1.79\times10^{-8}}{2\times0.40035} = 2.23554 \times 10^{-8} = 2.24 \text{ Å}.$$

$\therefore$ Now $d_{110} = \frac{a}{\sqrt{2}}$.

$$\therefore\ a = \sqrt{2}\times d_{110}$$

$$= \sqrt{2}\times 2.24 \text{ Å} = 3.1678 \text{ Å} = \mathbf{3.17\ Å}.$$

20. The first order reflection from a crystal plane in a cubic crystal occurs at 13°41′. Find the Miller indices of the plane. Given, a = 5.63 Å, λ = 1.54 Å [$\sin^2 13° 41' = 0{\cdot}056$].

[B.U. 1992]

Hints : Use the formula,

$$n\lambda = 2d \sin \theta.$$

$$n = 1,\ \lambda = 1.54 \times 10^{-8} \text{ cm}$$

$$\therefore\ d_{hkl} = \frac{\lambda}{2\sin\theta}.$$

Again, $d_{hkl} = \frac{a}{\sqrt{h^2+k^2+l^2}} = 3.25 \times 10^{-8}$ cm.

Thus, $\sqrt{h^2+k^2+l^2} = \frac{a}{d_{hkl}} = \frac{5.63\times10^{-8}}{3.25\times10^{-8}} = 1.73.$

$$\therefore\ h^2 + k^2 + l^2 = (1.73)^2 = 3.$$

Hence, $h = k = l = 1$.

$\therefore$ Miller indices **(111).**

21. The characteristic K_α lines of Cr, Fe and Ni have wavelengths of 2.29 Å; 1.94 Å and 1.66 Å respectively. Which of these may be used to determine a lattice spacing of 1 Å?

[B.U. 1994]

Soln. : By Bragg's equation we know,

$$n\lambda = 2d\sin \theta.$$

Now, for maximum value of λ, $\sin \theta = 1$ and $n = 1$.

$\therefore\ \lambda = 2d.$

Given, d = 1 Å, $\therefore\ \lambda$ = 2 Å.

Thus, K_α **lines of Fe and Ni** can be used since the value of λ has to be lesser than or equal to 2 Å.

22. Sodium crystallizes in b.c.c. structure with a = 4.24 Å. Calculate (i) theoretical density and (ii) radius of Na atom. **[B.U. 1994]**

Hints : $\rho = \frac{MZ}{N_A a^3}$

Here, Z = 2, N_A = 6.022 × 10^{23}, a = 4.24 × 10^{-8} cm, M = 23 gm.

∴ ρ = ?

$$\rho = \frac{23 \times 2}{6.022 \times 10^{23} \times (4.24 \times 10^{-8})^3}$$

$$= 1.0021194 \text{ gm/cc} = 1 \text{ gm/cc.}$$

Again, $r = \frac{\sqrt{3}a}{4} = \frac{\sqrt{3}}{4} \times 4.24 \times 10^{-8}$ = 1.83597 Å.

∴ r = **1.84 Å.**

23. Metallic gold crystallizes in f.c.c. lattice, the length of the cubic unit cell being 4.070 Å (1) Calculate the closest distance between gold atoms. (2) How many nearest neighbour does each gold atom have at the distance in (1)? (3) Calculate the density of gold.

Soln. : (1) In f.c.c., the closest distance of an atom is to the atom at the centre of the face.

Now, for f.c.c., $\sqrt{2}a = 4r$.

∴ the closest distance between the gold atoms = $2r = \frac{a}{\sqrt{2}} = \frac{4.070}{\sqrt{2}}$ Å = **2.878** Å.

(2) Each gold atom has 12 neighbours at a distance of **2.878 Å**

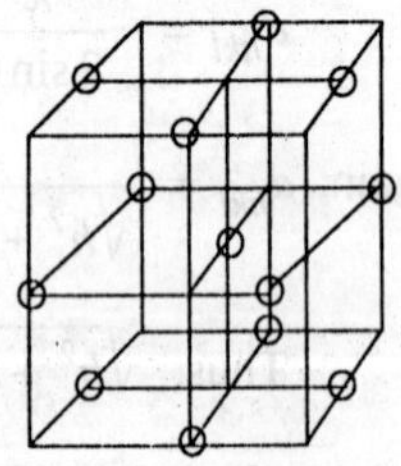

o –Au atom

(3) $\rho = \frac{MZ}{N_A a^3}$

$$= \frac{197 \times 4}{6.022 \times 10^{23} \times (4.070 \times 10^{-8})^3}$$

= 19.408 = **19.41 g/cc.**

24. The density of NaCl is 2.17 × 10^3 kg/m^3 and (100) reflection using X-ray of wavelength λ occurs at θ = 6°. Calculate λ.

Soln. : $\rho = \frac{MZ}{N_A a^3}$

ρ = 2.170 × 10^3 kg/m^3

Z = 4

M = 58.45 × 10^{-3} kg

By the problem,

$$2.170 \times 10^3 = \frac{4 \times 58.45 \times 10^{-3}}{6.022 \times 10^{23} \times a^3}.$$

$$\therefore \quad a^3 = \frac{4\times 58.45\times 10^{-3}}{2.170\times 6.022\times 10^{23}\times 10^3}$$

$$= 1.789 \times 10^{-28} m^3.$$

$$\therefore \quad a = 5.634 \times 10^{-10} \text{ m}.$$

Thus, $\lambda = 2d_{100} \sin\theta$

$$= 2\frac{a}{2}\sin\theta \text{ (since, in NaCl, } d_{100} = d_{200} = \frac{a}{2})$$

$$= a \sin\theta = 5.634 \times 10^{-10} \sin 6°$$

$$= 5.889 \times 10^{-11} \text{ m} = 58.89 \text{ pm} = \mathbf{0.589 Å}.$$

25. The first order reflection from (100) plane of KCl occurs at 5°30′ but for NaCl, it occurs at 6°, for the rays of same wavelength. The length of NaCl unit cell is 564 pm, then what is the length of KCl cell? The densities of NaCl and KCl are 1.99 gm/cc and 2.17 gm/cc respectively. Do these values support the X-ray analysis?

Soln. : $\lambda = 2d_{100} \sin\theta$.

$$\therefore \quad d_{100} = \frac{\lambda}{2\sin\theta}.$$

$$\therefore \quad \frac{d_{100}(\text{KCl})}{d_{100}(\text{NaCl})} = \frac{a_{\text{KCl}}}{a_{\text{NaCl}}} = \frac{\sin\theta\,(\text{NaCl})}{\sin\theta\,(\text{KCl})} = \frac{\sin 6°}{\sin 5°30'} = \frac{0.104528}{0.095845} = 1.09058.$$

$$\therefore \quad a_{\text{KCl}} = 1.09058\, a_{\text{NaCl}}$$

$$= 1.09058 \times 564 \text{ pm} = 615.08 \times 10^{-12} \text{ m} = 6.15 \text{ Å}.$$

Now, both NaCl & KCl have f.c.c. structures.

$$\therefore \quad \rho_{\text{NaCl}} = \frac{58.44\times Z}{N_A\cdot a_{\text{NaCl}}^3} \text{ and } \rho_{\text{KCl}} = \frac{74.45\times Z}{N_A\cdot a_{\text{NaCl}}^3}.$$

$$\therefore \quad \frac{\rho_{\text{NaCl}}}{\rho_{\text{KCl}}} = \frac{58.44}{74.45}\times\frac{a_{\text{KCl}}^3}{a_{\text{NaCl}}^3} = \frac{1.99}{2.17}.$$

$$\frac{a_{\text{KCl}}}{a_{\text{NaCl}}} = \left(\frac{1.99}{2.17}\times\frac{74.45}{58.44}\right)^{\frac{1}{3}} = (1.168282409)^{1/3} = 1.053212357.$$

$$\therefore \quad a_{\text{KCl}} = 1.0532 \times 564 = 594.0118 \text{ pm} = 5.94 \text{ Å}.$$

Thus these values **do not support the X-ray analysis.**

26. In a powdered diffraction result a KCl crystal gave lines at the following distances from central spot, when 70.8 pm X-ray was used in a camera of radius 5.74 cm : 13.2 mm, 18.42 mm, 22.81 mm, 26.21 mm, 29.42 mm, 32.2 mm, 39.65 mm. Identify the unit cell and determine the dimension.

Soln. : We know, $\sin^2\theta = \frac{\lambda^2}{4a^2}(h^2 + k^2 + l^2)$ (Sec 4.28 from Vol. 1)

$$= A\,(h^2 + k^2 + l^2) \text{ [where } A = \frac{\lambda^2}{4a^2}].$$

In a powdered diffraction, distances of the lines from central spot are related to the corresponding diffraction angles by $\theta = \frac{D}{2R}$, where R is radius of camera.

$$\theta = \frac{D}{2R};\ n\lambda = 2\,d\,\sin\theta,\ \theta = \sin^{-1}\frac{n\lambda}{2d}.$$

$$\therefore\ \theta/\text{degrees} = \frac{1}{2}\left(\frac{D}{57.4\text{ mm}}\right)\left(\frac{360}{2\pi}\right) = \frac{1}{2}D/\text{mm}.$$

The following table is drawn :

D/mm	13.2	18.42	22.81	26.21	29.42	32.2	39.65
θ/deg	6.6	9.21	11.41	13.11	14.71	16.1	19.82
100 $\sin^2\theta$	1.32	2.56	3.91	5.14	6.44	7.69	10.2

The common divisor is $\frac{1.32}{100}$. Divide by common divisor to identify $h^2 + k^2 + l^2$:

$h^2 + k^2 + l^2$	1	2	3	4	5	6	8
The corresponding indices are :	(100)	(110)	(111)	(200)	(210)	(211)	(220)

$h^2 + k^2 + l^2 \neq 7$ since, all possible $(h\,k\,l)$ planes are present, the unit cell is simple cube.

$$\frac{\lambda^2}{4a^2} = \frac{\sin^2\theta}{h^2+k^2+l^2} = \frac{1.32}{100} = 0.0132 \text{ (since, } h^2 + k^2 + l^2 = 1)$$

$$\text{or,}\quad 4a^2 = \frac{\lambda^2}{0.0132} = \frac{(70.8)^2}{0.0132}$$

$$\text{or,}\quad a^2 = \frac{(70.8)^2}{4\times 0.0132} = 94936.36$$

$$\therefore\quad a = \mathbf{308.1\ pm.}$$

27. The molar volume of KCl is known to be 1.3 times that of NaCl. If the glancing angle for first order Bragg reflection from the (100) plane of NaCl is 5.9°, calculate the same for KCl. **[B.U. 1997]**

Hints : For NaCl, Z = 4.

$\therefore$ molar vol. of NaCl $= \frac{x_1^3 \times N_A}{4}$ and that of KCl $= \frac{x_2^3 \times N_A}{4}$,

where x_1 and x_2 are side-lengths of NaCl and KCl respectively.

Given, $\frac{x_2^3 \times N_A}{4} = \frac{x_1^3 \times N_A \times 1.3}{4}$.

$\therefore\ x_2 = 1.091 x_1$

For NaCl, $d_{100} = \frac{x_1}{2}$ (first order reflection).

For KCl, $d_{100} = \frac{1.091 \times x_1}{2}$ (first order reflection).

$\therefore\ \lambda_{NaCl} = 2 \times \frac{x_1}{2} \sin 5.9°$

and $\lambda_{KCl} = 2 \times \frac{1.091}{2} x_1 \sin\theta.$

$\lambda_{NaCl} = \lambda_{KCl}$, $\therefore$ $\boldsymbol{\theta = 5.40°}$

28. CaF_2 is f.c.c. lattice with four CaF_2 groups per unit cell. The (III) reflection with X-rays of wavelength 0.1542 nm has θ = 14.18°, sin θ = 0.245. Calculate (i) the length of the unit cell edge and (ii) the density of the crystal. **[C.U. 1997]**

Hints : For f.c.c., $d_{111} = \frac{a}{\sqrt{3}}$.

$\therefore\ a = \sqrt{3}d_{111}$

Now, $\lambda = 2d\ \sin\theta$ [first order reflection]

$$\therefore\ d = \frac{\lambda}{2\sin\theta} = \frac{0.1542}{2\times\sin 0.245} = 0.3147 \text{ nm.}$$

(i) $\therefore\ a = (\sqrt{3}\times 0.3147)$ nm = $\mathbf{5.450\times 10^{-8}}$ **cm.**

(ii) $\therefore$ density $\rho = \frac{MZ}{N_A a^3}$

$$= \frac{78\times 4\,\text{gm}}{6.022\times 10^{23}\times\left(5.45\times 10^{-8}\right)^3\text{cc}}$$

$$= \frac{78\times 4\times 10}{6.022\times(5.45)^3}\ \text{gm/cc} = \mathbf{3.2005\ gm/cc.}$$

29. Ag is known to crystallise in f.c.c. form and the distance between the nearest neighbour atoms is 2.87 Å. Calculate the density of Ag. [At. wt. of Ag = 108] **[B.U. 2000]**

Hints : For f.c.c., $4r = \sqrt{2}a$.

$$\therefore\ a = \frac{4}{\sqrt{2}}\cdot r = \left(\frac{4}{\sqrt{2}}\times 1.435\right)\text{Å}$$

$a = 4.058\times 10^{-8}$ cm.

$$\therefore\ \text{density } \rho = \frac{MZ}{N_A a^3} = \frac{108\times 4}{6.022\times 10^{23}\times\left(4.058\times 10^{-8}\right)^3}$$

$$= \frac{108\times 4\times 10}{6.022\times(4.058)^3}\ \text{gm/cc} = \mathbf{10.735\ gm/cc.}$$

30. Polonium is the only metal whose lattices are primitive with a = 334·4 pm. X-rays of wavelength 153.6 pm is now used in powdered diffraction experiment. What will be its density? If it were a f.c.c. crystal with the same atomic radius, what will be its density?

Soln. : We know that

$$\sin\theta_{(hkl)} = \frac{\lambda}{2a}\sqrt{\left(h^2+k^2+l^2\right)}$$

λ = 153.6 pm

$$\therefore\ \text{density } \rho = \frac{MZ}{N_A a^3}$$

At. wt. of Polonium = 210 and for simple cube, Z = 1.

$\therefore\ \rho = \dfrac{210}{6.022\times10^{23}\times a^3}$

Given, a = 334.4 pm = 334.4 × 10^{-12} m = 3.344 × 10^{-8} cm.

$$\therefore\ \rho = \frac{210\times1}{6.022\times10^{23}\times\left(3.344\times10^{-8}\right)^3} = \textbf{9.326 gm/cc.}$$

Radius of the atom (r) $= \dfrac{a}{2}$(for simple cubic)

$= \dfrac{334.4}{2}$ pm = 167.2 pm.

In f.c.c. lattice, $4r = \sqrt{2}a$.

$$\therefore\ a = \frac{4r}{\sqrt{2}} = \frac{4\times167.2}{\sqrt{2}}\text{pm} = 472.9130 \text{ pm} = 4.73\times10^{-8}\text{ cm.}$$

$$\therefore\ \text{density} = \frac{4\times M}{N_A\cdot a^3} = \frac{4\times210}{6.022\times10^{23}\times\left(4.75\times10^{-8}\right)^3} = \textbf{13.18 gm/cc.}$$

31. α-phase of Fe crystallizes in the cubic system. In a powder pattern experiment, ratio of $\sin^2\theta$ values obtained are 1, 2, 3, 4, 5, 6, 7, 8, etc. Which type of cubic unit cell is present? What is the probable Miller indices for $\sin^2\theta = 7$?

The density of alpha iron is 7.92 gm/cc. The wavelength of X-ray used is 45 pm. Calculate the length of side of unit cell. Also calculate the value of θ for Bragg reflection from 100 type planes and radius of iron atom. (atomic weight of iron = 55.85 gm-mole^{-1})

Soln. : As $h^2 + k^2 + l^2$ can never be 7, it must actually be $14(3^2 + 2^2 + 1^2 = 14)$. So the $\sin^2\theta$ are 2, 4, 6, 8, 10, 12, 14 and 16, etc. The **unit cell is b.c.c.** Probable Miller indices are (**321**) (not necessarily in this order).

Thus, $\rho = \dfrac{MZ}{N_A a^3}$

$$\text{or, } a^3 = \frac{MZ}{N_A\rho} = \frac{55.85\times2}{\left(6.022\times10^{23}\right)\times7.92} = 2.342\times10^{-23}\text{ cm}^3.$$

$\therefore\ a = 2.861\times10^{-8}$cm = **286.1 pm.**

For (200) type planes, $d = \dfrac{a}{2}$.

For b.c.c. lattice, first order diffraction is observed from (200) planes rather than (100) planes.

$$\sin^2\theta = \frac{\lambda^2}{4a^2}\left(h^2+k^2+l^2\right)$$

$$= \frac{45^2\times4}{4\times(286.1)^2} = \frac{45^2}{286.1^2} = 0.024739408.$$

or, $\sin\theta$ = 0.15728766

$\therefore\ \theta$ = 9.04949731 = **9.05°.**

The system is b.c.c. Thus $\sqrt{3}a = 4r$.

$$r = \frac{\sqrt{3}a}{4} = \frac{\sqrt{3}\times 286.1}{4} \text{ pm} = 123.884934 \text{ pm} = \mathbf{123.9 \text{ pm.}}$$

32. MnO has a cubic unit cell with a = 447 pm. The density is 5200 kg.m^{-3}.

(a) Calculate the number of MnO units per unit cell and determine from this whether it is simple, b.c.c. or f.c.c. Find out the coordination number of Mn and oxygen.

(b) Assuming the unit to be b.c.c. of side 447 pm, calculate sin θ for reinforcement of 50.0 pm X-ray reflected from (210) planes. From which of the following three planes the first order reflection will be more intense and why? Planes : (100), (110), (211). Given at. wt. of Mn = 55.

Soln. : (a) ρ = 5200 kg.m^{-3}

a = 447 pm = 447 × 10^{-12} m

N_A = 6.022 × 10^{23}, X = ?

$$\rho = \frac{MZ}{N_A a^3}$$

or, $$Z = \frac{dN_A a^3}{M}$$

$$= \frac{5200\times 6.022\times 10^{23}\times\left(447\times 10^{-12}\right)^3}{71\times 10^{-3}} = \frac{52\times 10^{-8}\times 6.022\times(447)^3}{71}$$

$$= 3.939 = 4.$$

∴ **the unit cube is f.c.c.**

Coordination no. of M_n^{2+} = 6 and coordination number of O^{2-} = 6, since there are 6 M_n^{2+} ions surrounding each O^{2-} ion and vice versa.

(b) λ = $2d\sin\theta$ (for Ist order reflection)

$$d_{210}^2 = \frac{a^2}{h^2+k^2+l^2} = \frac{a^2}{4+1+0}$$

$$\therefore\ d_{210} = \frac{a}{\sqrt{5}} = \frac{447\text{ pm}}{2.236} = 199.91055 \text{ pm.}$$

$$\sin\theta = \frac{\lambda}{2d} = \frac{50.0\text{ pm}}{2\times 199.91055\text{ pm}} = \mathbf{0.125056.}$$

$$\therefore\ \theta = 7.1839°.$$

The first order reflection from **(211) planes is more intense** than the other two planes. In f.c.c. lattice for other two types of planes, (100) and (110), another plane is interposed halfway between two successive planes, that destructively interfere with the radiation scattered from the respective planes, thereby reducing intensity.

33. A certain solid crystallizes in a b.c.c. lattice. The first order X-ray (λ = 0.154 nm) reflection maximum from a set of (200) planes occurs at an angle of 16.6°. Calculate the dimension of the unit cell. **[M.Sc. Admission Test, 1994, I.I.T. Bombay]**

Soln. : We know that $n\lambda = 2d \sin\theta$.

Here $n = 1$, $\lambda = 0.154$ nm, $\sin\theta = 0.286$.

$$\text{So, } d_{200} = \frac{n\lambda}{2\sin\theta} = \frac{0.154 \text{ nm}}{2\times 0.286} = 0.269 \text{ nm.}$$

$$d_{200} = \frac{a}{2}$$

$$\text{So, } a = 2d_{200} = \mathbf{0.538\ nm.}$$

34. CsF crystallizes in CsCl type structure and KCl in NaCl type structure. Lattice length of KCl is 1.5 times that of CsF and molecular weight of CsF is twice that of KCl. Calculate the ratio of density of KCl to that of CsF.

Soln. : $\rho = \dfrac{MZ}{N_A a^3}$

NaCl has f.c.c. type structure and CsCl has b.c.c. type structure.

Thus, for CsF, Z = 1.

$$\rho_{CsF} = \frac{M_{CsF}}{N_A a_{CsF}^3}$$

For KCl, Z = 4, $d_{KCl} = \dfrac{4\times M_{KCl}}{N_A a_{KCl}^3}$.

$$\text{Thus, } \frac{\rho_{KCl}}{\rho_{CsF}} = 4\times\frac{M_{KCl}}{M_{CsF}}\times\left(\frac{a_{CsF}}{a_{KCl}}\right)^3$$

$$= 4\times\frac{1}{2}\times\left(\frac{1}{1.5}\right)^3 \quad (\text{Given, } M_{CsF} \doteq 2M_{KCl},\ a_{KCl} = 1.5a_{CsF})$$

$$= \mathbf{0.59.}$$

35. KCl crystallizes in NaCl type structure. The ionic radius of Na^+ is 0.5 times that of Cl^- and 0.7 times that of K^+. Calculate (a) the ratio of the side of the unit cell of NaCl to that of KCl. (b) the ratio of density of NaCl to that of KCl.

Soln. : KCl and NaCl have f.c.c. type structure. Thus, $2(r_+ + r_-) = a$ = side of unit cell.

Given, $r_{Na^+} = 0.5\ r_{Cl^-}$

$r_{Na^+} = 0.7\ r_{K^+}$.

$$\text{(a) } \frac{a_{NaCl}}{a_{KCl}} = \frac{r_{Na^+} + r_{Cl^-}}{r_{K^+} + r_{Cl^-}} = \frac{r_{Na^+} + 2r_{Na^+}}{\frac{r_{Na^+}}{0.7} + 2r_{Na^+}} = \frac{3}{3.42} = \mathbf{0.88.}$$

$$\text{(b) } \rho = \frac{MZ}{N_A a^3}$$

Here, $Z = 4$ (for both cases).

$$\text{Thus, } \frac{\rho_{NaCl}}{\rho_{KCl}} = \frac{M_{NaCl}}{M_{KCl}}\times\left(\frac{a_{KCl}}{a_{NaCl}}\right)^3 = \frac{58.5}{74.5}\times\left(\frac{3.42}{3}\right)^3 = \mathbf{1.16.}$$

36. The density of Lithium metal is 0.53 gm.cm^{-3} and the separation of the (100) planes of the metal is 350 pm. Determine whether the lattice is f.c.c. or b.c.c.

[M(m) Li = 6.941 gm.mol^{-1}]

Soln. : Given, d_{100} = 350 pm = a.

Again, $\rho = \frac{ZM}{N_A a^3}$.

Here, ρ = 0.53 gm.cm^{-3}, M = 6.941 gm.mol^{-1}

$$\therefore \quad Z = \frac{0.53 \times 6.022 \times 10^{23} \times (350 \times 10^{-10})^3}{6.941} = 1.9 = 2.$$

Hence, the **lattice is b.c.c.**

37. Polonium crystallizes in a simple cubic lattice. It has an atomic mass = 209 and density = 9.4 gm/cm^3. What is the edge length of its unit cell? What are the distances between two neighbouring (110) and two (111) planes. What are (i) the crystalline atomic radius of Po and (ii) per cent void in Po crystal? **[V.U. 1997]**

Hints : $Z = 1$, M = 209 gm, ρ = 9.4 gm/cm^3

$$\rho = \frac{ZM}{N_A a^3} = 9.4,\ d_{110} = \frac{a}{\sqrt{2}},\ d_{111} = \frac{a}{\sqrt{3}}$$

(a = 3.32 Å, d_{110} = 2.35 Å, d_{111} = 1.92 Å, r = 1.66 Å, 48%)

38. The density of CaO is 3.35 × 10^3 kg. m^{-3}. This oxide crystallizes in one of the cubic system with a = 480 pm. Calculate the number of molecules in unit cell. Which type of cubic system is it? **[V.U. 1996]**

Hints : ρ = 3.35 × 10^3 kg.m^{-3}, a = 480 × 10^{-12} m, M = 56 × 10^{-3}kg, Z = ?

[Ans. Z = 4, f.c.c.]

39. Calculate the wavelength of X-ray which shows a second order Bragg reflection angle of 14°10′ from the (100) plane of KCl. The density of KCl is 1.984 gm.cm^{-3} and there are four atoms in the unit cell. **[V.U. 2002]**

Hints : $2d_{100}\sin\theta = n\lambda$

Here, n = 2, d_{100} = a, θ = 14°10′, ρ = 1.984 g.cm^{-3}, Z = 4

$$\rho = \frac{4 \times M}{N_A a^3}$$ **[Ans.** λ = 1.54 Å]

40. The effective radius of an iron atom is 124 pm. Iron occurs in b.c.c. structure. Calculate the mass density of iron in SI system.

(Fe = 56, 1 pm = 10^{-12} m, N_A = 6.023 × 10^{23} mol^{-1}) **[V.U. 2003]**

Hints : Here, Z = 2, M = 56 × 10^{-3} kg, r = 124 pm = 124 × 10^{-12} m, $\sqrt{3}a = 4r$

[Ans. ρ = 7918.96 kg.m^{-3}]

41. A crystal having simple cubic lattice has the length of its unit cell a_o pm. One of its planes show a first order Bragg reflection at an angle of 60°. Taking the wavelength of the X-ray as a_o pm, find Miller indices for the plane.

Hints : $\lambda = a_o$, $a = a_o$, $2d\sin\theta = \lambda$, $d = \frac{a_0}{\sqrt{h^2+k^2+l^2}}$. **[Ans.** (111) planes]

42. Is it possible to obtain Bragg reflection from (100) planes of simple cubic crystals with edge 1.5 Å by using X-ray with wavelength of 3.5 Å. Explain. **[B.U. 1981]**

Soln. : For simple cubic crystal, $d_{100} = a = 1.5$ Å.

Now, $2d_{100} \sin\theta = \lambda$.

For maximum possible value of λ, $\sin\theta = 1$,

$\lambda = 2 \times 1.5 = 3.0$ Å.

Hence, **X-ray with wavelength of 3.5 Å cannot be used.**

43. X-ray powder diffraction angles of silver crystal are obtained at 19.11°, 22.18°, 32.34°, 38.84° and 40.89°. (a) Find the type of unit cell. (b) What is the radius of Ag atom, if λ of X-ray is 154.43 pm.

Soln. : θ	19.11°	22.18°	32.34°	38.84°	40.89°
$\sin^2\theta$	0.11	0.14	0.28	0.39	0.42
Dividing by common divisor 0.11	1	1.3	2.5	3.5	3.8
Multiplying by 3	3	4	8	11	12
$h^2 + k^2 + l^2$	3	4	8	11	12
The corresponding indices are	(111)	(200)	(220)	(311)	(222)

Hence, **the unit cell is f.c.c.**

Now, $2d\sin\theta = \lambda$.

$$\sin^2\theta = \frac{\lambda^2}{4d^2} = \frac{\lambda^2}{4a^2}(h^2 + k^2 + l^2).$$

$$\text{Hence, } 0.11 = \frac{154.43^2}{4a^2}(1 + 1 + 1) \text{ [Given, } \lambda = 154.43 \text{ pm]}$$

$a^2 = 162604.26$

$\therefore \quad a = 403.24$ pm.

Now, $\sqrt{2}a = 4r$.

Thus, radius of Ag atom $= r = \dfrac{\sqrt{2}}{4} \times 403.24$ m $= 142.56$ pm $=$ **1.42 Å**.

44. The first six observed Bragg diffraction angles from a powdered sample of iridium crystal are 21.97°, 25.60°, 37.66°, 45.75°, 48.43°, 59.75°. Determine the type of unit cell and assign corresponding Miller indices for every observed angle.

Soln. : θ	21.97°	25.60°	37.66°	45.75°	48.43°	59.75°
$\sin^2\theta$	0.14	0.19	0.37	0.51	0.56	0.75
Dividing by common	1	1.36	2.64	3.64	4.00	5.36
divisor 0.14 multiplying by 3	3	4	8	11	12	16
$h^2 + k^2 + l^2$	3	4	8	11	12	16
Corresponding Miller indices are	**(111)**	**(200)**	**(220)**	**(311)**	**(222)**	**(400)**

The unit cell is f.c.c.

45. X-ray power diffraction angles from the Mo crystal are at 20.26°, 29.30°, 36.82°, 43.82°, 50.70°, 58.80°, 66.30°. Find the type of unit cell and also corresponding Miller indices for the planes responsible for the above diffraction angles.

Soln. : θ	20.26°	29.30°	36.82°	43.82°	50.70°	58.80°	66.30°
$\sin^2\theta$	0.12	0.24	0.36	0.48	0.60	0.73	0.84
Dividing by 0.12 and	1	2	3	4	5	6	7
multiplying by 2,	2	4	6	8	10	12	14
since 7 cannot be sum of squares. $h^2 + k^2 + l^2$	2	4	6	8	10	12	14
Corresponding Miller indices are	**(110)**	**(200)**	**(211)**	**(220)**	**(310)**	**(222)**	**(321)**

The unit celll is b.c.c.

46. Calculate the distance between successive layers of graphite if carbon-carbon distance in the molecular layers is 1.43 Å. Its density is 2.23 gm/cc.

Soln. : Graphite has a layer structure with each layer consisting of hexagonal network of carbon atoms. Three layers of graphite constitute a unit cell.

Let carbon-carbon distance in the molecular layers be a cm and distance between successive layers be d cm.

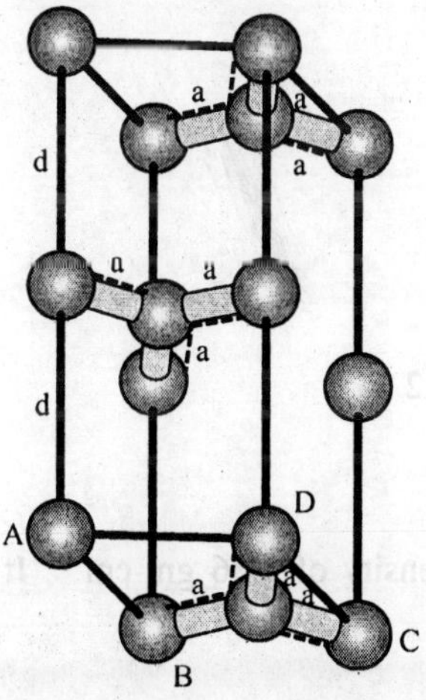

Fig. 1 : Unit Cell of graphite

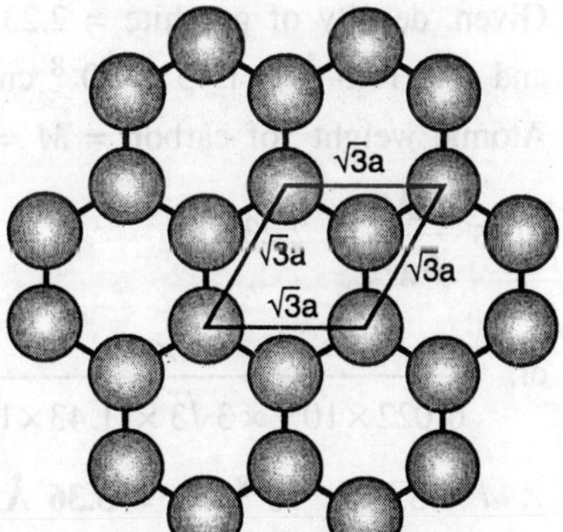

Fig. 2 : A layer in graphite

Solid lines are the boundaries of the graphite unit cell that intersects the bottom layers.

Thus, if we draw the quadrilateral constituting the bottom face of the unit cell reducing the carbon atoms to points

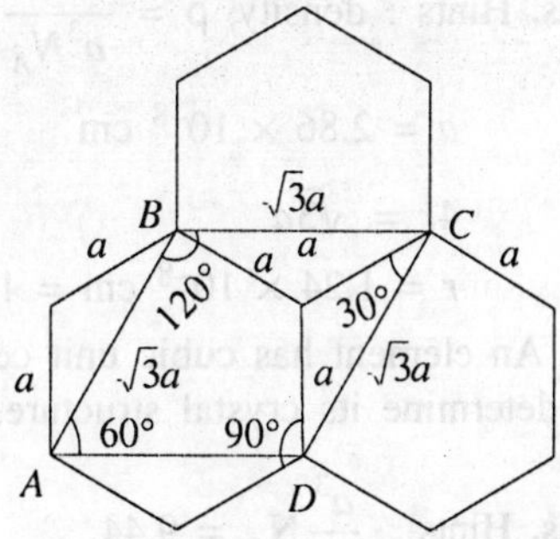

$$DB = \sqrt{3}a \text{ cm}$$

$$AO = \sqrt{3}a \cos 30° = \frac{3}{2}a \text{ cm.}$$

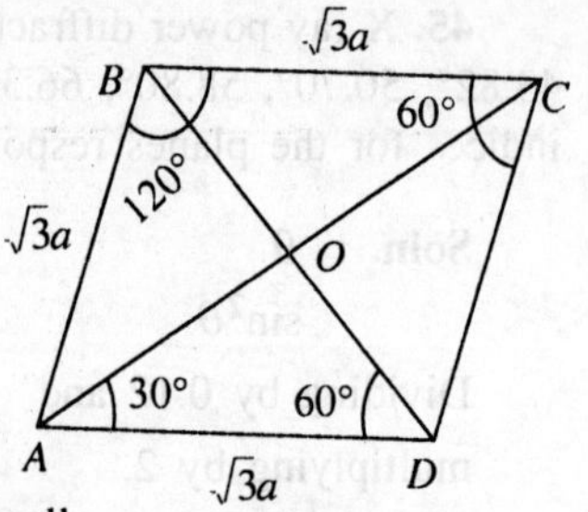

$$\text{Area of } \Delta ADB = \frac{1}{2} DB \times AO = \frac{3\sqrt{3}}{4} a^2 \text{ cm}^2$$

Thus, area of the quadrilateral (rhombus) $ABCD$

= area of the top face or bottom face of the unit cell

$$= 2 \times \frac{3\sqrt{3}}{4} a^2 = \frac{3\sqrt{3}}{2} a^2 \text{ cm}^2$$

Thus, volume of the unit cell,

$$V = \frac{3\sqrt{3}}{2} a^2 \times 2d \text{ cm}^3 = 3\sqrt{3}\, a^2 d \text{ cm}^3$$

The unit cell has 8 atoms at corners, 2 atoms at the centres of top and bottom faces, 4 atoms at the middle of vertical edges and 1 atom at the centre of the unit cell.

Thus, number of atoms in each unit cell

$$= Z = \frac{1}{8} \times 8 + 1 \times 2 + \frac{1}{4} \times 4 + 1 \times 1 = 4.$$

Given, density of graphite = 2.23 gm/cc

and a = 1.43 Å = 1.43 × 10^{-8} cm

Atomic weight of carbon = M = 12 gm

$$\text{Thus } \frac{ZM}{N_A V} = 2.23$$

$$\text{or, } \frac{4 \times 12}{6.022 \times 10^{23} \times 3\sqrt{3} \times \left(1.43 \times 10^{-8}\right)^2 \times d} = 2.23$$

$\therefore$ d = 3.36 × 10^{-8} cm = **3.36 Å**.

47. Fe below 912 °C has f.c.c. structure and density of 7.86 gm-cm^{-3}. If its atomic mass is 55.85 g, find its atomic radius.

Ans. Hints : density, $\rho = \dfrac{MZ}{a^3 N_A}$

$$a = 2.86 \times 10^{-8} \text{ cm}$$

$$4r = \sqrt{3}a$$

$$\therefore \quad r = 1.24 \times 10^{-8} \text{ cm} = 1.24 \text{ Å}$$

48. An element has cubic unit cell. If its molar volume is 9.44 cm^3-mol^{-1} and a = 3.15Å, determine its crystal structure.

Ans. Hints : $\dfrac{a^3}{Z} N_A = 9.44,$

Z = 2, So the structure is b.c.c.

CHAPTER 5

FIRST LAW OF THERMODYNAMICS, CARNOT CYCLE, CLAUSIUS-CLAPEYRON EQUATION AND THERMOCHEMISTRY

Convention Adopted

1. W = Work done upon a system (+)ve
   = Work done by a system (–)ve
   [Reverse of these is the old convention.]
2. Q = Heat absorbed by a system (+)ve
   = Heat evolved by a system (–)ve

Required Formulae

1. $đW$ and $đQ$ are inexact differential.
   dU and dH are exact differential.

$$\therefore \quad \int_1^2 dH = H_2 - H_1 = \Delta H.$$

$$\int_1^2 dU = U_2 - U_1 = \Delta U.$$

But $\int đW = W$, $\int đQ = Q$.

2. Formula of mechanical work

$đW = -P_{ex}dV$ (old convention $đW = P_{ex}\,dV$)

P_{ex} = external pressure on the system. W is (–)ve, when dV is (+)ve (expansion) and (+)ve, when dV is (–)ve (contraction).

$$W = -\int_1^2 P_{ex}dV \text{ can be integrated.}$$

If P_{ex} remains constant throughout the process,

$W = -P(V_2 - V_1) = -P\Delta V$, $(P_{ex} = P) = P(V_1 - V_2)$.

If P_{ex} remains infinitesimally close to P_{system} throughout the process (i.e., reversible process), $-\int PdV$ can be integrated if P is known as a function of V.

Reversible Isothermal change :

Ideal gas : $W = -\int_1^2 Pdv = -\int_1^2 nRT\frac{dV}{V} = -nRT\ln\frac{V_2}{V_1} = nRT\ln\frac{V_1}{V_2}$.

van der Waals gas, $W = -\int_1^2 \left(\frac{nRT}{V-nb} - \frac{an^2}{V^2}\right)dV$

$$= nRT\ \ln\frac{V_2 - nb}{V_1 - nb} - an^2\left(\frac{1}{V_2} - \frac{1}{V_1}\right).$$

$$= nRT\ \ln\frac{V_1 - nb}{V_2 - nb} + an^2\left(\frac{1}{V_1} - \frac{1}{V_2}\right).$$

3. For solid or liquid, $\alpha = \frac{1}{V}\left(\frac{\partial V}{\partial T}\right)_P$,

where α = coefficient of thermal expansion,

and $\beta = -\frac{1}{V}\left(\frac{\partial V}{\partial P}\right)_T$

where β = isothermal compressibility.

$V = f(P, T)$

$$dV = \left(\frac{\partial V}{\partial T}\right)_P dT + \left(\frac{\partial V}{\partial P}\right)_T dP$$

$$= \alpha V\ dT - \beta V\ dP.$$

$$W = -\int PdV = -\int_1^2 P(\alpha VdT - \beta VdP) = \frac{\beta V}{2}\left(P_2^2 - P_1^2\right)\ (dT = 0).$$

K = Bulk modulus $= \frac{1}{\beta}$.

4. $\left(\frac{\partial P}{\partial T}\right)_V = \frac{\alpha}{\beta}$.

5. According to first law of thermodynamics, for a system at rest in absence of external field,

$$dU = đQ + dW.$$

For the system doing only mechanical work,

$$dU = đQ - P_{ex}\ dV$$

$$đQ = dU + P_{ex}\, dV.$$

If the system has external Kinetic energy (T) and is subject to a conservative external force (e.g., gravitational force), then

$$dE = đQ + đW,$$

where $E = T + V + U$

and V = potential energy due to external field.

6. $nC_P = \left(\dfrac{đQ}{dT}\right)_P = \left(\dfrac{\partial H}{\partial T}\right)_P$,where H = enthalpy.

$nC_V = \left(\dfrac{đQ}{dT}\right)_P = \left(\dfrac{\partial U}{\partial T}\right)_V$, where U = internal energy

7. For ideal gas,

$$\left(\frac{\partial U}{\partial V}\right)_T = \left(\frac{\partial U}{\partial P}\right)_T = \left(\frac{\partial H}{\partial P}\right)_T = \left(\frac{\partial H}{\partial V}\right)_T = 0.$$

$$C_P - C_V = R.$$

8. For solid, liquid, real gas, i.e., for any system,

$$C_P - C_V = \frac{\alpha^2 T V_m}{\beta}.$$

where V_m = molar volume.

9. For exact differential dz,

$$\oint dz = 0$$

$$z = f(x,y).$$

If $dz = Mdx + Ndy$,

then $\left(\dfrac{\partial M}{\partial y}\right)_x = \left(\dfrac{\partial N}{\partial x}\right)_y$.

10. Adiabatic work :

$W = nC_v(T_2 - T_1)$, if C_v is constant and the system is an ideal gas.

For adiabatic reversible change in an ideal gas,

$$PV^{\gamma} = \text{constant}.$$

$$TV^{\gamma-1} = \text{constant}.$$

$$\frac{T}{P^{\frac{R}{C_P}}} = \text{constant}.$$

11. For free expansion, $W = 0$, since $P_{ex} = 0$.

12. Considering source of gravitational force part of the system for free fall of a body from a height h, $W = 0$.

13. For ideal monatomic gas : $C_v = \frac{3}{2}R$.

For ideal diatomic gas : $C_v = \frac{5}{2}R$.

14. Generally we assume, $U = f(T, V), H = f(T, P)$

$$dU = \left(\frac{\partial U}{\partial V}\right)_T dV + \left(\frac{\partial U}{\partial T}\right)_V dT$$

$$dH = \left(\frac{\partial H}{\partial P}\right)_T dP + \left(\frac{\partial H}{\partial T}\right)_P dT$$

For ideal gas,

$$dU = \left(\frac{\partial U}{\partial T}\right)_V dT = nC_v dT.$$

$$dH = \left(\frac{\partial H}{\partial T}\right)_P dT = nC_p dT.$$

For adiabatic work against constant P_{ext} for ideal gas,

$dU = dW.$

If C_v is constant,

$$nC_v\, dT = -P_{ext}\, dV$$

or, $nC_v (T_2 - T_1) = -P_{ex}(V_2 - V_1)$

or, $nC_v (T_1 - T_2) = P_{ext}\left(\frac{nRT_2}{P_2} - \frac{nRT_1}{P_1}\right)$.

In general, $P_{ext} = P_2$.

$$\therefore\ C_v (T_1 - T_2) = P_2\left(\frac{RT_2}{P_2} - \frac{RT_1}{P_1}\right).$$

Thus, $\frac{3}{2} R (T_1 - T_2) = R\left(T_2 - T_1\frac{P_2}{P_1}\right)$.

$$\therefore\ \frac{3}{2}(T_1 - T_2) = \left(T_2 - T_1\frac{P_2}{P_1}\right).$$

Short Questions

1. State with reasons whether the following statements are correct or incorrect :

(i) Every isolated system is closed.

Ans. : For closed system, mass is not **exchanged** with surroundings. For isolated system neither mass nor heat is exchanged with surroundings. So the statement is correct.

(ii) Every closed system is isolated.

Ans. : The statement is incorrect. Closed system exchanges heat with surroundings but isolated system neither exchanges heat nor mass with the surroundings.

(iii) Thermodynamics is dependent in the microscopic properties of matter such as work, heat.

Ans. : In thermodynamics, states are defined by a number of microscopic properties. w and q are not properties of states—these are dependent on path of change. So the statement is incorrect.

(iv) ΔH is a state function.

Ans. : A state function refers to a state. Although H is a state function, ΔH cannot be a state function since it refers to the difference in H between two states (which, however, is path independent). Thus, the above statement is not correct.

(v) C_V is independent of temperature for a perfect gas.

Ans. : C_V, the molar heat capacity, is defined as $\left(\frac{đQ}{dT}\right)_V$, where $đQ$ = heat change for one mole and $đQ = dU + P_{ext}dV$ for a system performing mechanical work only. So at constant V, $đQ = dU$. Hence, $C_V = \left(\frac{dU}{dT}\right)_V$. Now, for a perfect gas, U dpends only on temperature. Therefore, C_V will depend only on temperature for a perfect gas. Hence, the above statement is false.

(vi) A process in which the final temperature is equal to initial temperature must be an isothermal process—comment.

Ans. : It is no doubt that in an isothermal process, temperature is constant and so initial and final temperatures are same. But for an adiabatic free expansion of an ideal gas the final temperature is equal to the initial temperature. So, the above statement is not correct.

(vii) U remains constant in every isothermal process in a closed system.

Ans. : The above statement is true only for an ideal gas. Because only for an ideal gas $U = f(T)$. But, for real materials, U depends on temperature and volume. Thus, U cannot remain constant for any material in an isothermal process. So, the above statement is not correct.

(viii) $Q = 0$ for every cyclic process.

Ans. : We know that $Q = \Delta U + W$. For a cyclic process, $\oint dU = 0$ [since, U is a state function].

But, W may not be zero because it depends on path and, therefore, Q may not be zero. Thus, the statement is not true.

(ix) $\Delta U = 0$ for every cyclic process.

Ans. : U is a state function and, therefore, $\oint dU = 0$. Thus, $\Delta U = 0$ for a cyclic process and the statement is correct.

(x) When a system expands, work is done on the system.

Ans. : If the external pressure is less than internal pressure, then the system expands. Thus, work is done by the system. So, the statement is false.

(xi) If a closed system at rest in the absence of external fields undergoes an adiabatic process that has $W = 0$, then the temperature of the system must remain constant.

Ans. : We know that for an adiabatic process, $đW = -dU$.

Now, $U = f(V, T)$.

$$\because \quad dU = \left(\frac{\partial U}{\partial V}\right)_T dV + \left(\frac{\partial U}{\partial T}\right)_V dT,$$

$$\therefore \quad đW = -\left(\frac{\partial U}{\partial V}\right)_T dV - \left(\frac{\partial U}{\partial T}\right)_V dT.$$

If $W = 0$, $0 = -\int_1^2 nC_V dT - \int_1^2 \left(\frac{\partial U}{\partial V}\right)_T dV$.

If C_v is constant, then

$$0 = -nC_v (T_2 - T_1) - \int_1^2 \left(\frac{\partial U}{\partial V}\right)_T dV$$

i.e., $0 = nC_v(T_1 - T_2) - \int_1^2 \left(\frac{\partial U}{\partial V}\right)_T dV$.

Thus, if $\left(\frac{\partial U}{\partial V}\right)_T \neq 0$, T_1 is not necessarily equal to T_2 and for an ideal gas, $\left(\frac{\partial U}{\partial V}\right)_T = 0$.

The above statement is thus true for an ideal gas and is incorrect for any other material.

(xii) For a closed system doing mechanical work only, a constant pressure process that has $Q > 0$ must have $\Delta T > 0$.

Ans. : $đQ = dU + P_{ext}\, dV$

If C_v is constant, and $U = f(T, V)$, then

$$\therefore \quad đQ = nC_v dT + \left[\left(\frac{\partial U}{\partial V}\right)_T dV + P_{ext}.dV\right]$$

$$\text{or, } đQ = nC_v dT + \left[\left(\frac{\partial U}{\partial V}\right)_T + P_{ext}\right] dV.$$

$$\therefore Q = nC_v (T_2 - T_1) + \int \left[\left(\frac{\partial U}{\partial V}\right)_T + P_{ext}\right] dV.$$

Now, if $Q > 0$, ΔT may be zero, less than zero and greater than zero, because it depends on the second term in the above equation.

(xiii) P - V work is usually negligible for solids and liquids.

Ans. : For any system, $đW = P_{ext}\, dV$. For solid and liquid, P_{ext} has a very negligible effect and coefficient of compression (β) is around 10^{-6} atm^{-1}. Thus, P - V work is usually negligible for solids and liquids. The above statement is correct.

2. Show that if $Z = PV$, then Z is a state function.

Ans. : $Z = PV$

Thus, $dZ = PdV + VdP$.

Now, $\left(\frac{\partial P}{\partial P}\right)_V = 1$

$\left(\frac{\partial V}{\partial V}\right)_P = 1$

Therefore, $\oint dZ = 0$, so, Z is a state function.

3. A gas obeys the equation of state $P(V - b) = RT$, where b is a constant. For the gas,

(a) plot V vs. T at constant P.

(b) plot P vs. V at constant T.

(c) plot P vs. T at constant V.

Ans. : (a) Plot of V vs. T at constant P

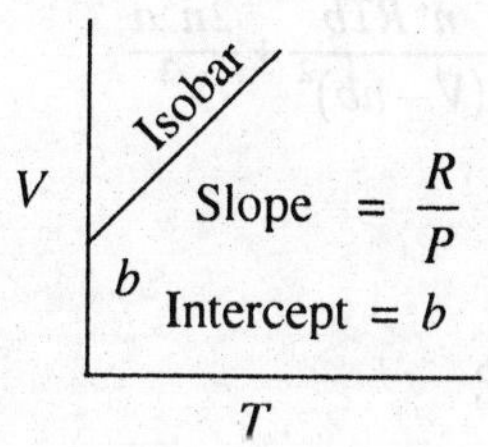

The equation is

$$(V - b) = \frac{RT}{P}.$$

$$V = \frac{R}{P}.T + b.$$

(b) Plot of P vs. V (at constant T) :

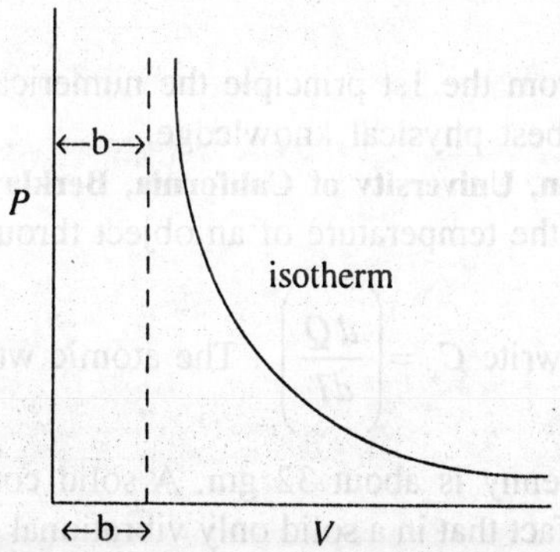

The equation is

$$P = \frac{\text{const.}}{V - b}$$

$$\therefore \quad P \propto \frac{1}{V - b}$$

$V \to \infty$ as $P \to 0$

$V \to b$ as $P \to \infty$

(c) Plot of P vs. T at constant V

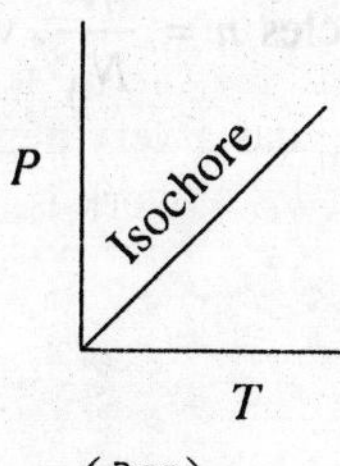

Slope $= \frac{R}{V - b}$. The equation is $P = \frac{RT}{\text{const.}}$

Intercept = 0.

4. Find $\left(\frac{\partial H}{\partial V}\right)_T$ for van der Waals and ideal gas.

Soln. : $H = U + PV$

$$dH = dU + PdV + VdP$$

$$\left(\frac{\partial H}{\partial V}\right)_T = \left(\frac{\partial U}{\partial V}\right)_T + P + V\left(\frac{\partial P}{\partial V}\right)_T$$

$$= T\left(\frac{\partial P}{\partial T}\right)_V + V\left(\frac{\partial P}{\partial V}\right)_T, \text{ since } \left(\frac{\partial U}{\partial V}\right)_T = T\left(\frac{\partial P}{\partial T}\right)_V - P.$$

For van der Waals gas, $P = \dfrac{nRT}{V-nb} - \dfrac{an^2}{V^2}$.

$$T\left(\frac{\partial P}{\partial T}\right)_V = \frac{nRT}{V-nb}$$

$$V\left(\frac{\partial P}{\partial V}\right)_T = V\left(-\frac{nRT}{(V-nb)^2} + \frac{2an^2}{V^3}\right) = -\frac{nRTV}{(V-nb)^2} + \frac{2an^2}{V^2}.$$

Hence, $\left(\dfrac{\partial H}{\partial V}\right)_T = \dfrac{nRT}{V-nb}\left(1 - \dfrac{V}{V-nb}\right) + \dfrac{2an^2}{V^2} = -\dfrac{n^2RTb}{(V-nb)^2} + \dfrac{2n^2a}{V^2}.$

For ideal gas, $P = \dfrac{nRT}{V}$.

$$T\left(\frac{\partial P}{\partial T}\right)_V = \frac{nRT}{V} = P, \quad V\left(\frac{\partial P}{\partial V}\right)_T = -\frac{nRT}{V} = -P.$$

Hence, $\left(\dfrac{\partial H}{\partial V}\right)_T = 0.$

5. Define heat capacity C_v and calculate from the 1st principle the numerical value of C_v of a Cu penny in calorie/°C using your best physical knowledge.

[PhD Entrance Examination, University of California, Berklay, USA]

Ans. : The amount of heat needed to raise the temperature of an object through 1°C is defined as its heat capacity. It is sufficient to write $C_v = \left(\dfrac{đQ}{dT}\right)_v$. The atomic wt. of Cu is 63.54 gm, i.e., roughy 64 gm.mol^{-1}. A Cu penny is about 32 gm. A solid consisting of N atoms has $U = 3NkT$. It comes out from the fact that in a solid only vibrational degrees of freedom are possible. From the equipartition principle we know that energy per vibrational degree of freedom is kT. For a system of N particles the vibrational degrees of freedom are $3N$. Thus, total energy is $3NkT$. If there are n moles of particles $n = \dfrac{N}{N_A}$, where N_A is Avogadro number. Thus,

$$U = 3NkT = 3N\frac{R}{N_A}T = 3nRT.$$

$\therefore\ C_v = 3nR.$

By the problem, $W_{Cu} = 32$gm.

Thus, $n_{CU} = 0.5$ mol

Hence, $C_v = 3 \times 0.5 \text{ mol} \times 1.987 \text{ cal.K}^{-1}\text{.mol}^{-1}$
$= 2.9805 \text{ cal.K}^{-1} = 2.9805 \text{ calorie/}^\circ\text{C}$

6. An insulated chamber is divided into two halves of same volume; the right-half is evacuated and left-half contains an ideal gas at temperature T. Now a small hole is opened between the two halves. Thus the gas will flow from left-half to right-half and the system ultimately comes to equilibrium. If no heat is exchanged with the walls, find the final temperature of the system. **[PhD Entrance Examination, Columbia University, USA]**

Soln. : The gas flows continuously from left to right into vacuum. Since opposing pressure (P_{ext}) is zero, $W = 0$. Q is also zero. Since by the first law of thermodynamics, $\Delta U = Q + W$, $\Delta U = 0$. For an ideal gas, internal energy (U) is a function of temperature only. Thus, the temperature will not change. It will remain the same after the expansion against vacuum as before.

7. What will happen if in the above case the gas is real?

Ans. : For a real gas,

$U = f(T, V)$.

$$\therefore\ dU = \left(\frac{\partial U}{\partial T}\right)_V dT + \left(\frac{\partial U}{\partial V}\right)_T dV = C_V\, dT + \left(\frac{\partial U}{\partial V}\right)_T dV.$$

$$\therefore\ \Delta U = \int_1^2 C_V dT + \int_1^2 \left(\frac{\partial U}{\partial T}\right) dV.$$

But again, $W = 0$, $Q = 0$, $\therefore\ \Delta U = 0$.

$$\text{Thus, } \int_1^2 C_V dT = -\int_1^2 \left(\frac{\partial U}{\partial V}\right)_T dV.$$

In real gas, U generally increases as volume increases. So, $\left(\frac{\partial U}{\partial V}\right)_T$ is +ve.

$$\text{Thus, } \int_1^2 C_V dT \text{ must be } -\text{ve}.$$

$$\therefore\ C_V(T_2 - T_1) = C_V(T_2 - T)\ (\because T_1 = T)$$
$$= -\text{ve}$$

i.e. $T > T_2$.

Hence, temperature will fall, i.e., decrease.

8. Two systems with heat capacities C_1 and C_2 respectively interact thermally and the temperature becomes T_F. The initial temperature of system-1 is T_1. What was the initial temperature of the system-2? The total energy of the system is constant.

[PhD Entrance Examination, Wisconsin University, USA]

Soln. : Let the initial temperature of the system-2 be T_2, ($T_2 > T$). Then by the problem,

$C_1(T_F - T_1) = C_2(T_2 - T_F)$

or, $C_2T_2 = C_1T_F - C_1T_1 + C_2T_F$

$$\therefore\ T_2 = \frac{C_1}{C_2}(T_F - T_1) + T_F$$

9. A solid object has a density d_1. Show that the pressure P and the heat capacities C_P and C_V are related by $C_P - C_V = \dfrac{3\alpha MP}{d_1}$, where M is mass of the object and α is the coefficient of linear expansion. **[PhD Entrance Examination, Wisconsin University, USA]**

Soln. : By the first law of thermodynamics, $dU = đQ + đW$.

$\therefore\ đQ = dU - đW = dU + PdV$ (since, $dW = -PdV$).

$$\therefore \left(\frac{đQ}{dT}\right)_P - \left(\frac{\partial U}{\partial T}\right)_P = P\left(\frac{\partial V}{\partial T}\right)_P$$

For a solid there is almost no difference between $\left(\frac{\partial U}{\partial T}\right)_P$ and $\left(\frac{\partial U}{\partial T}\right)_V$.

$$\therefore C_P - C_V = P\left(\frac{\partial V}{\partial T}\right)_P.$$

We know that $\alpha = \frac{1}{V}\left(\frac{\partial V}{\partial T}\right)_P$ [$\because \left(\frac{\partial V}{\partial T}\right)_P = \alpha V$ for one dimension]

Thus, $C_P - C_V = PV.\frac{1}{V}\left(\frac{\partial V}{\partial T}\right)_P = PV\alpha.$

For, three dimensions, $\left(\frac{\partial V}{\partial T}\right)_P = 3\alpha V.$

$\therefore\ C_P - C_V = 3\alpha\frac{M}{d_1}P$ (where $V = \frac{M}{d_1}$).

10. An ideal gas has been kept in a large jar of volume V_0. A metal ball of mass M which is fixed snugly in the glass tube of cross-sectional area A is fitted to the jar. The equilibrium pressure in the jar is slightly higher than atmospheric pressure because of ball's weight. If the ball is displaced slightly from equilibrium it will undergo S.H.M. If the gas state represents a quasistatic adiabatic process, find the relation between oscillation frequency and the variable of the problem. **[PhD Entance Examination, University of California, Berkeley]**

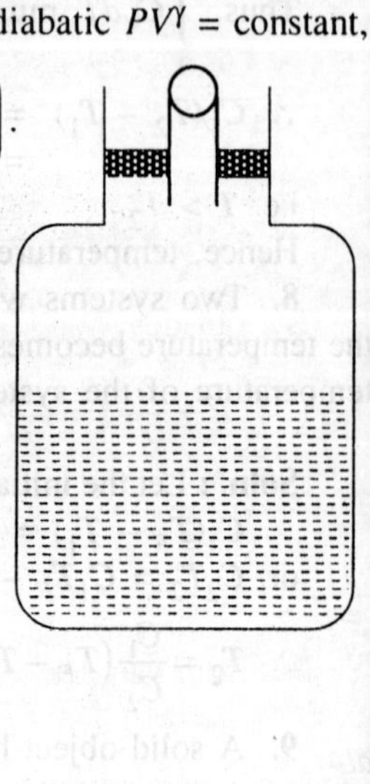

Soln. : Let the pressure in the jar be P. As the process is adiabatic PV^γ = constant, where V is the volume and γ is the ratio of specific heat $\left(\frac{C_P}{C_V}\right)$.

$$\therefore \frac{dP}{P} + \gamma\frac{dV}{V} = 0.$$

$$\therefore dP = -\frac{\gamma P dV}{V}.$$

Now $F = A.dP$

[where A is the area, dP is the change in pressure]

But a system undergoes S.H.M. So, $F = -k.l$

[where l = length $= \frac{dV}{A}$, i.e., change of volume per area]

$\therefore\ AdP = -kl.$

$$\therefore A.\left(-\gamma\frac{PdV}{V}\right) = -kl$$

$$\Rightarrow A\gamma P\frac{dV}{V} = k\frac{dV}{A}$$

$$\Rightarrow \quad \frac{A\gamma P}{V} = \frac{k}{A}.$$

$$\therefore \; k = \frac{A^2\gamma P}{V},$$

where P is the pressure in the jar and higher than atmospheric pressure.

$$P = P_0 + \frac{mg}{A}.$$

Now oscillation frequency $\nu = \frac{\omega}{2\pi} = \frac{1}{2\pi}\sqrt{\frac{k}{m}}$ $[k = 4\pi^2 m\nu^2]$

$$= \frac{1}{2\pi}\sqrt{\frac{A^2\gamma P}{mV}} \; [\text{where } \omega = \frac{\gamma}{c},\; h\gamma = hc\omega].$$

11. 1 mole of a gas with molar internal energy given by $U = CT - \frac{a}{V}$, where V is molar volume. and a and c are constant. Calculate C_P and C_V. The gas obeys van der Waals equation. **[PhD Entance Examination, Wisconson University]**

Soln. : We know that $C_V = \left(\frac{\partial U}{\partial T}\right)_V$.

$$C_P = \left(\frac{\partial H}{\partial T}\right)_P = \left(\frac{\partial U}{\partial T}\right)_P + P\left(\frac{\partial V}{\partial T}\right)_P.$$

$$\therefore \; \left(\frac{\partial U}{\partial T}\right)_P = \left(\frac{\partial U}{\partial T}\right)_V + \left(\frac{\partial U}{\partial V}\right)_T \cdot \left(\frac{\partial V}{\partial T}\right)_P$$

Now, $U = f(T, V)$

or, $dU = \left(\frac{\partial U}{\partial T}\right)_V dT + \left(\frac{\partial U}{\partial V}\right)_T dV$

$$C_P = C_V + P\left(\frac{\partial V}{\partial T}\right)_P + \left(\frac{\partial U}{\partial V}\right)_T \left(\frac{\partial V}{\partial T}\right)_P = C_V + \left[P + \left(\frac{\partial U}{\partial V}\right)_T\right]\left(\frac{\partial V}{\partial T}\right)_P$$

$$= C_V + \left(P + \frac{a}{V^2}\right)\left(\frac{\partial V}{\partial T}\right)_P.$$

By van der Waals equation,

$$\left(P + \frac{a}{V^2}\right) = \frac{RT}{V-b} \; [U = CT - \frac{a}{V},\; \left(\frac{\partial U}{\partial V}\right)_T = \frac{a}{V^2}]$$

$$\therefore \; C_P = C_V + \frac{RT}{V-b}\left(\frac{\partial V}{\partial T}\right)_P$$

But from van der Waals equation, $\left(P + \frac{a}{V^2}\right)(V - b) = RT$. Expanding and negleeting $\frac{ab}{V^2}$ in comparsion with other terms and replacing V in the denominator by $\frac{RT}{P}$ (the ideal gas value), $PV - Pb + \frac{ab}{RT} = RT.$

$$PV = RT + P\left(b - \frac{a}{RT}\right). \quad (1)$$

$$\therefore\ V = \frac{RT}{P} + \left(b - \frac{a}{RT}\right). \quad (2)$$

$$\therefore\ \left(\frac{\partial V}{\partial T}\right)_P = \frac{R}{P} + \frac{a}{RT^2}.$$

Again, from equation (1),

$$\frac{(V-b)}{T} = \frac{R}{P} - \frac{a}{RT^2}.$$

$$\therefore\ \frac{R}{P} = \frac{V-b}{T} + \frac{a}{RT^2}$$

$$\text{Now } C_P = C_V + \frac{RT}{V-b} \times \left(\frac{V-b}{T} + \frac{2a}{RT^2}\right).$$

$$\therefore C_P = C_V + R + \frac{2a}{(V-b)T}$$

$$\Rightarrow C_P = C_V + R + \frac{2a}{VT} \quad [\text{since, } V >> b]$$

$$\Rightarrow C_P = C_V + R + \frac{2aP}{RT^2}.$$

Numerical Problems

1. A mass of mercury at a pressure of 1 atm and a temperature of 0 °C is kept at a constant volume and the temperature is raised to 10 °C. Calculate the final pressure.

Given, $\alpha = 1{\cdot}81 \times 10^{-4}\ \text{deg}^{-1}$; $\beta = 3{\cdot}87 \times 10^{-6}\ \text{atm}^{-1}$.

Soln. : Given, $\alpha = 1{\cdot}81 \times 10^{-4}\text{K}^{-1}$

$\beta = 3{\cdot}87 \times 10^{-6}\ \text{atm}^{-1}$

We know that $\left(\frac{\partial P}{\partial T}\right)_V = \frac{aK^{-1}}{\beta\,\text{atm}^{-1}} = \frac{\alpha}{\beta}\ \text{atm.K}^{-1}.$

$$\therefore\ dP = \frac{\alpha}{\beta} dT.$$

$$\therefore\ P_2 - P_1 = \frac{\alpha}{\beta}(T_2 - T_1).$$

$$\therefore\ P_2 = P_1 + \frac{\alpha}{\beta}\cdot 10 \ (\text{Given, } T_2 = 10\ °\text{C}, T_1 = 0\ °\text{C})$$

$$= P_1 + \frac{1.81\times10^{-4}}{3.87\times10^{-6}}\cdot 10\ \text{atm.K}^{-1}.\text{K} = (1 + 467.70026)\ \text{atm (given, } P_1 = 1\ \text{atm})$$

$= 468.70026$ atm $=$ **468.7 atm.**

2. A metal whose volume expansivity is $5.0\times10^{-5}\ \text{deg}^{-1}$ and isothermal compressibility is $1.2\times10^{-6}\ \text{atm}^{-1}$ is at a pressure of 1 atm and a temperature of 20 °C. A thick surrounding cover of invar of negligible expansibility and compressibility is fitted to the metal very snugly.

(a) What will be the final pressure if the temperature is raised to 32 °C?

(b) If the surrounding cover can withstand a pressure of 120 atm, then what is the highest temperature to which the system can be raised?

Soln. : (a) Proceeding as in Problem 1,

$$P_2 = P_1 + \frac{5\times10^{-5}}{1.20\times10^{-6}} \times 12$$

$$= \mathbf{501\ atm} \text{ (given, } P_1 = 1 \text{ atm)}$$

(b) $P_2 = 120$ atm

$$\therefore \ \Delta P = P_2 - P_1 = 119 = \frac{5\times10^{-5}}{1.20\times10^{-6}} \cdot \Delta T$$

$$\therefore \ \Delta T = 119 \times \frac{1.2\times10^{-6}}{5\times10^{-5}} = 2.85 \text{ °C}$$

$$\therefore \ T_2 = 20 + 2.85 = \mathbf{22.85\ °C.}$$

3. A block of the same metal as in Problem 2 occupies 5 litres. As a result, temperature is raised to 32 °C, the volume is increased by 0.5 cc. Calculate the final pressure.

Soln. : $P = f(T, V)$

$$dP = \left(\frac{\partial P}{\partial T}\right)_V dT + \left(\frac{\partial P}{\partial V}\right)_T dV$$

$$\therefore \ \Delta P = \frac{\alpha}{\beta}\Delta T + \int_1^2 \left(\frac{\partial P}{\partial V}\right)_T dV$$

Given, $V_2 = 5.0005$ l, $P_1 = 1$ atm,

$V_1 = 5$ l, $\Delta T = T_2 - T_1 = (32 - 20)$ °C $= 12$ °C,

$\alpha = 5 \times 10^{-5}$ deg^{-1}, $\beta = 1.2 \times 10^{-6}$ atm^{-1}.

$$P_2 - P_1 = \Delta P = \frac{\alpha}{\beta}.\Delta T - \frac{1}{\beta}\ln\frac{V_2}{V_1} \quad \left[\text{since } \beta = -\frac{1}{V}\left(\frac{\partial V}{\partial P}\right)_T\right]$$

$$= \frac{5\times10^{-5}}{1.2\times10^{-6}} \times 12 - \frac{10^{-6}}{1.2}\ln\frac{5.0005}{5}$$

$$= 500 - 83.329 = 416.67 \text{ atm.}$$

$$P_2 = 416.67 \text{ atm} + 1 \text{ atm} = \mathbf{417.67\ atm.}$$

4. A block of copper at a pressure of 1 atm and 0 °C is kept at constant volume. What will be the final pressure if the temperature is raised by 10 °C? What will be the final pressure if the volume changes from 1000 cc to 1000.1 cc ?

Given, $\alpha = 4.8 \times 10^{-5}$deg^{-1}

Bulk modulus $= 1.3 \times 10^{12}$ dyne.cm^{-2}

Soln. : $K = \dfrac{1.3\times10^{12}}{1.01325\times10^{6}}$ atm $= 1.2830002 \times 10^{6}$ atm

$$\beta = \frac{1}{K} = 7.79423 \times 10^{-7} \text{ atm.}$$

Proceeding as Problem 1 and Problem 2,

$$P_2 = P_1 + \frac{\alpha}{\beta} \times 10$$

$= \mathbf{616.840\ atm}$ (given, $P_1 = 1$ atm, $\alpha = 4.8 \times 10^{-5}$ deg^{-1})

If V changes from 1000 cc to 1000.1 cc,

$$\Delta P = \frac{\alpha}{\beta}\Delta T - \frac{10^7}{7.79423} \ln \frac{1000.1}{1000}$$

$= 615.8402$ atm $- 128.29362$ atm $= 487.54656$ atm $= 487.5$ atm.

$\therefore$ final pressure = 1 atm + 487.5 atm = **488.5 atm.**

5. 2 mol of an ideal gas undergo isothermal and reversible expansion from V_1 to 10 V_1 litre. The initial pressure is 100 atm. Calculate the initial volume and temperature, if the work done by the system is 40000 cal.

Soln. : Given, $n = 2$ moles, $V_2 = 10V_1$, $P_1 = 100$ atm.

$W = -\ 40000$ cal, $T_1 = T_2$

The work done, $W = -nRT \ln \frac{10V_1}{V_1} = -2RT \ln 10.$

$\therefore\ -40.000$ cal $= -2$ mol $\times$ 1.987 cal.K^{-1}.mol^{-1} $\times$ T ln 10.

$$T = \frac{40.000}{2 \times 1.987 \times \ln 10} \text{ K} = 4371.35865 \text{ K} = 4371.36 \text{ K}.$$

$PV = nRT = 2\ RT.$

Thus, $$V_1 = \frac{2RT}{P_1} = \frac{2 \text{ mol} \times 0.08206\, l - \text{atm.K}^{-1}.\text{mol}^{-1} \times 4371.36 \text{K}}{100 \text{ atm}}$$

$= 7.174276032$ litres $=$ **7.17 litres**

$T_2 = T_1 =$ **437.36 K.**

6. 2 mol of an ideal gas undergo isothermal and reversible expansion at 1096 K from 2 litres to 20 litres. Calculate the work done.

Hints : $W = -nRT \ln \frac{V_2}{V_1}$. **Ans.** -41.96 kJ.

7. 2 mol of Nitrogen at STP is compressed reversibly to 10 litres. Calculate the work done upon the system if (a) the work is carried out isothermally and (b) adiabatically.

Hints : $$V_1 = \frac{2RT_1}{P_1} = \frac{2 \times 0.08206\, l - \text{atm.K}^{-1}.\text{mol}^{-1} \times 273.15 \text{K}}{1 \text{ atm}} = 44.829 \text{ litres}$$

(a) $$W_1 = -\ 2 \times R \times T \ln \frac{10}{44.829}$$

$$= -\ 2 \times 8.314 \text{ J.K}^{-1}.\text{mol}^{-1} \times 273.15 \text{ K} \times \ln \frac{10}{44.121} = 6.814 \text{ kJ}.$$

(b) $W = 2 \times C_v \times (T_2 - T_1)$

$$T_2 = T_1 \left(\frac{V_1}{V_2} \right)^{\gamma - 1}$$

$$= 273.15 \left(\frac{44.829}{10}\right)^{1.4-1} = 497.77 \text{ °K.}$$

Ans. : 9.34 kJ

8. Calculate the rise in temperature of a piece of iron (mass 10 kg) at 25 °C if it is dropped from a height of 100 metres. What is the K.E. of this piece just before it hits the ground? What is its speed? Let the gravitational acceleration constant (g to be 9.80 m/sec^2 and let the moler heat capacity of iron be C_p = 25.12 J.mol^{-1}.K^{-1}.

Soln. : Just before the mass hits the ground, all of the potential energy will be converted into kinetic energy. So,

$$\Delta U = mgh = 10 \text{ kg} \times 9.80 \text{ m.s}^{-2} \times 100 \text{ m}$$
$$= 9800 \text{ kg.m}^2\text{.s}^{-2} = 9800 \text{ J} = 9.80 \text{ kJ.}$$

$\therefore$ kinetic energy is thus 9.800 kJ. This is given by $\frac{1}{2}mv^2$,

where v is velocity; $v = \sqrt{\frac{2\,\text{K.E.}}{m}}$

$$= \sqrt{\frac{2 \times 9800 \text{ J}}{10 \text{ kg}}}$$

$$= \sqrt{2 \times 980} \text{ m.s}^{-1} = 44.27188724 \text{ m.s}^{-1}$$

$$= 44.3 \text{ m.s}^{-1}.$$

For solids, $C_p = C_v$.

$\therefore \Delta U = nC_p\, \Delta T$.

$\therefore \Delta T = \frac{\Delta U}{nC_p}$.

Given, C_p = 25.12 J.mol^{-1}.K^{-1}. M = 55.85 × 10^{-3} kg.mol^{-1}.

Hence, $\Delta T = \frac{9800 \text{ J} \times 0.05585 \text{ kg.mol}^{-1}}{10 \text{ kg} \times 25.12 \text{ J.mol}^{-1}\text{.K}^{-1}}$

$$= 2.178861465 \text{ K} = \mathbf{2.18 \text{ K}}.$$

$\therefore$ the final temperature is **27.18 °C.**

9. Calculate the rise in temperature of a piece of copper (C_p = 6.4 cal.deg^{-1}.mol^{-1}) when dropped from the top of Qutab Minar (height 270 feet) assuming that no heat is lost.

Hints : Since, $W = 0$, $Q = \Delta U = mgh = nC_v(T_2 - T_1)$. [**Ans.** 1.92 °C]

10. Water is brought to boil under a pressure of 1 atm. When an electric current of .6 ampere from a 10 V source is passed for 5 minutes through a resistance in contact with , 0.8 gm is vaporised. Calculate ΔU, ΔH and W.

Soln. : Given, I = 0.6 A, V = 10 volt, t = 5 min, m = 0.8 gm, T = 100 °C = 373.15 °K

Heat $Q = I$ ampere × V volt × t sec

= 0.6 ampere × 10 volt × 5 × 60 sec

= 1800 J = **1.8 kJ.**

Work done, $W = -P\Delta V = -P(V_g - V_l)$, $(V_g >> V_l)$.

$\therefore W = -PV_g = n(\text{mol})\ R(\text{J.K}^{-1}.\text{mol}^{-1})T\ (\text{K})$

$= -0.0444 \times 8.314 \times 373.15\ \text{J} = \mathbf{-137.8\ J.}$

$\Delta U = (1800 - 137.8)\ \text{J} = 1662.2\ \text{J} = \mathbf{1.67\ kJ.}$

The heat change has taken place at constant pressure.

$\therefore Q_p = \Delta H = \mathbf{1.8\ kJ.}$

11. **200 gm of $CaCO_3$ was heated to 727 °C when it decomposed. The operation was carried out in a container closed by a piston, which was originally resting on the sample and was retained throughout by the atmosphere. (a) How much work was done? (b) What would be the work in an open vessel? (c) What would be the work in a closed vessel?**

Hints : (a) 200 gm $CaCO_3$ = 2 moles

$2CaCO_3 = 2CaO(s) + 2CO_2(g)$

$T = 727\ °C = 1000\ K.$

CO_2 occupies a volume of 2 × 22.414 = 44.828.

$$\frac{1 \times V_2}{1000} = \frac{1 \times 44.828}{273}$$

Then the volume at 1000 K and 1 atm, $V_2 = \dfrac{44.828 \times 1000}{273}$ litre.

Work done against constant pressure of 1 atm,

$$W = -P\Delta V = -PV_g = -\frac{1 \times 44.828 \times 1000\text{l.atm}}{273} \times 101.325 \frac{1}{\text{l.atm}}$$

$= -16638.08462\ \text{J} = \mathbf{-16.64\ kJ.}$

(b) **−16.64 kJ (c) 0.**

12. **1 mole of a van der Waals gas at 300 K expands isothermally and reversibly from 20 l to 60 l ($a = 0.556\ \text{m}^6.\text{Pa}.\text{mole}^{-2}$ and $b = 0.064\ \text{dm}^3.\text{mole}^{-1}$). For van der Waals $\left(\dfrac{\partial U}{\partial V}\right)_T = \dfrac{a}{V^2}$. Calculate W, Q, ΔU and ΔH.**

Soln. : van der Waals equation of gas is $\left(P + \dfrac{an^2}{V^2}\right)(V - nb) = nRT.$

So, $P = \dfrac{nRT}{V-b} - \dfrac{an^2}{V^2}$.

$$W_{12} = -\int_1^2 PdV = -\int_1^2 \left(\frac{nRT}{V-b} - \frac{an^2}{V^2}\right)dV = -nRT\ \ln\frac{V_2 - nb}{V_1 - nb} - n^2a\left(\frac{1}{V_2} - \frac{1}{V_1}\right)$$

$$= -1\ \text{mol} \times 8.314\ \text{J.K}^{-1}.\text{mol}^{-1} 300\ \text{K}\ \ln\frac{V_2 - b}{V_1 - b} -$$

$$1\text{mol}^2 \times 0.556\text{m}^6.\text{Pa.mol}^{-2}\left(\frac{1000}{60} - \frac{1000}{20}\right)^{-3} \text{m}^{-3}$$

$$= -2494.2\,\text{J}\ \ln\frac{59.9361}{19.9361} + 18.533\ \text{m}^3.\text{Pa}$$
$$= -2494.2\,\text{J} \times \ln 3.006420 + 18.533\ \text{m}^3.\text{Pa}$$
$$= -(2745.4911 - 18.533)\,\text{J}$$
$$= -2726.96\,\text{J}.$$

$\therefore\ W_{\text{iso, rev}} = \mathbf{-\,2.727\ kJ.mol^{-1}}$.

$$\Delta U = \int_1^2 \frac{a}{V^2}dV = \frac{a}{V_1} - \frac{a}{V_2} = \mathbf{18.533\ J.mol^{-1}}$$

$$Q = \Delta U - W = 18.533\,\text{J} + 2.727\,\text{kJ}$$
$$= 18.533\ \text{J} + 2727\ \text{J} = 2745.5\ \text{J} = \mathbf{2.746\ kJ}$$

$$\Delta H = \Delta U + (P_2V_2 - P_1V_1)$$
$$= 18.533\ \text{J} + \left[\left(\frac{RT.V_2}{V_2-b} - \frac{a}{V_2}\right) - \left(\frac{RTV_1}{V_1-b} - \frac{a}{V_1}\right)\right]\text{J}$$
$$= 18.533\ \text{J} + \left[\left(\frac{RT.V_2}{V_2-b} - \frac{RTV_1}{V_1-b}\right) + \left(\frac{a}{V_1} - \frac{a}{V_2}\right)\right]\text{J}$$
$$= 18.533\,\text{J} + 8.314 \times 300\left(\frac{60}{59.936} - \frac{20}{19.936}\right)\text{J} + 0.556\left(\frac{1}{20} - \frac{1}{60}\right)\times 1000\ \text{J}$$
$$= (18.533\,\text{J} - 2494.2 \times 2.142 \times 10^{-3} + 18.5333)\,\text{J}$$
$$= (18.5333 - 5.3426 + 18.5333)\ \text{J} = 31.72\ \text{J}.$$

ΔH can also be calculated in the following way :

$$\left(\frac{\partial H}{\partial V}\right)_T = T\left(\frac{\partial P}{\partial T}\right)_V + \left(\frac{\partial P}{\partial V}\right)_T .V\ .$$

For van der Waals gas,

$$\left(\frac{\partial H}{\partial V}\right)_T = -\frac{RTb}{(V-b)^2} + \frac{2a}{V^2}.$$

Hence, $\Delta H = RTb\left(\dfrac{1}{V_2-b} - \dfrac{1}{V_2-b}\right) + 2a\left(\dfrac{1}{V_2} - \dfrac{1}{V_1}\right)$.

Putting the corresponding values same result will be obtained.

13. 16 gm of O_2 at 27 °C and 10 atm undergoes reversible adiabatic expansion to 1 atm. Calculate the final temperature, work done, ΔU and ΔH. If the same expansion takes place against a constant pressure of 1 atm, what will be the final temperature, W, ΔV and ΔH?

If the same expansion takes place against a constant pressure of 1 atm but till the volume is trebled (then pressure is raised to match the internal pressure), what will be the final temperature, W, ΔU and ΔH?

Soln. : (a) $P_2 = T_1\left(\frac{P_2}{P_1}\right)^{\frac{R}{C_P}}$. For diatomic gas, $C_v = \frac{5}{2}R$; $C_p = \frac{7}{2}R$

By the probem, $T_1 = 300$ K $\quad P_2 = 11$ atm

$P_1 = 10$ atm $\quad n = \frac{1}{2}$ mol

$$\therefore\ T_2 = 300\left(\frac{P_2}{P_1}\right)^{\frac{2}{7}} = 300 \times \left(\frac{1}{10}\right)^{\frac{2}{7}} = 155.38424 \text{ K} = 155.4 \text{ K}.$$

$$V_1 = \frac{nRT_1}{P_1} = \frac{\frac{1}{2}\text{mol} \times 0.08206\, l.\text{atm.K}^{-1}.\text{mol}^{-1} \times 300\,\text{K}}{10\,\text{atm}} = 1.2309\text{ l}.$$

$$W = nC_v\,(T_2 - T_1) = -\frac{1}{2}\times\frac{5}{2}R(300 - 155.38) = -1.502 \text{ kJ}.$$

$\Delta U = -1.502$ kJ; $\Delta H = nC_p(T_2 - T_1) = -2104.585$ J $= -2.104$ kJ.

(b) Against a constant pressure of 1 atm

$P_{ex}\,(V_2 - V_1) = nC_v\,(T_1 - T_2)$

or, $P_2\left(\frac{nRT_2}{P_2} - \frac{nRT_1}{P_1}\right) = nC_v\,(T_1 - T_2)$

or, $T_2 - 30 = 750 - \frac{5}{2}T_2$

$\frac{7}{2}T_2 = 780$, or, $T_2 = \frac{2}{7} \times 780 = 221.86$ K $= 51.15$ °C.

$\therefore\ W = nC_v(T_2 - T_1) =$ **−812.06 J**.

$\Delta U =$ **−812.06 J**

$\Delta H = nC_p(T_2 - T_1) = -1136.89$J $=$ **−1.14 kJ**.

(c) $P_{ex}\,(V_2 - V_1) = nC_V\,(T_1 - T_2)$

or, $P_{ex}\,(3V_1 - V_2) = n \times \frac{5}{2}R\,(300 - T_2)$ or, $1 \times 2V_1 = n \times \frac{5}{2} \times R(300 - T_2)$

or, $1 \times 2 \times \frac{nRT_1}{P_1} = n \times \frac{5}{2} \times R(300 - T_2)$ or, $\frac{5}{2}T_2 = 690$.

$\therefore\ T_2 = 276$ K $= 2.85$ °C

$W = nC_v(T_2 - T_1)$

$= \frac{1}{2}$mol $\times \frac{5}{2} \times 8.314$ J.K^{-1}.mol^{-1} $\times$ (276 − 300) K $=$ **− 249.42 J**.

$\Delta H = -\frac{7}{4}\times 8.314 \times 24 =$ **−349.14 J**, $\quad P_2 = \frac{nRT_2}{V_2} =$ **3.06 atm**.

14. (a) 1 mol of an ideal gas is allowed to expand against a piston which supports 0.041 MPa, the temperature being constant at 0 °C (273.15 K). The initial pressure is 1.01325 MPa and the final pressure is 0.041 MPa. Calculate W, Q, ΔU and ΔH.

(b) If the change is carried out in such a way that the confining pressure at all time is less than the gas pressure, find the value of W, Q, ΔU and ΔH.

Soln. : Given, P_1 = 1.01325 MPa = 1.01325×10^6 Pa = 10.1325 bar.

Similarly, $P_2 = P_{ex}$ = 0.041MPa = 0.41 bar.

T °C = 273.15K, R = 0.08314 bar.litre.K^{-1}.mol^{-1}.

$$V_1 = \frac{RT}{P_1} = \frac{0.08314\,l.\text{bar}.K^{-1}.\text{mol}^{-1} \times 273.15\,K}{10.1325\,\text{bar}} = 2.2412 \text{ l.mol}^{-1}.$$

$$V_2 = \frac{RT}{P_2} = \frac{0.08314\,l.\text{bar}.K^{-1}.\text{mol}^{-1} \times 273.15\,K}{0.41\,\text{bar}} = 55.389 \text{ l.mol}^{-1}.$$

(a) $W = -P_{ex}(V_2 - V_1) = -0.41\ (55.389 - 2.241) = -21.79052$ l. bar

$= \mathbf{-2.179\ kJ.}$

$-W = Q = \mathbf{2.179\ kJ.}$

$\Delta H = \Delta U = \mathbf{0.}$

(b) $W = -RT\ln\frac{V_2}{V_1} = -8.314 \times 273.15 \ln\frac{55.389}{2.2414}$

$= -7283.63$ J $= \mathbf{-7.28\ kJ.}$

$Q = -W = \mathbf{7.28\ kJ.}$

$\Delta U = \Delta H = \mathbf{0.}$

15. 104 gm of He were kept in a cylinder covered by a piston at 127 °C and 25 atm pressure. The pressure is suddenly released to 1 atm adiabatically. (a) Calculate the final temperature and work done. (b) What would be the final temperature and work done if the change was carried out reversibly and adiabatically? (c) What would be the work done if the change was reversible and isothermal? Calculate ΔU and ΔH.

Soln. :

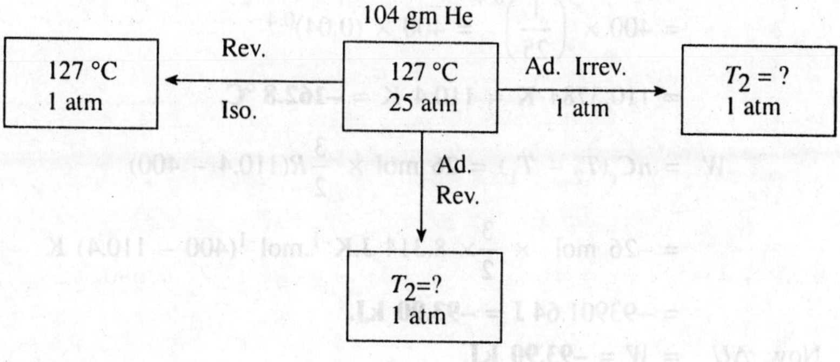

(a) P_1 = 25 atm P_2 = 1 atm

n = 26 mol T_1 = 400 K

T_2 = ?

Assuming ideal monatomic gas, $C_v = \frac{3}{2}R.$

For adiabatic irreversible expansion,

$$P_{ex}(V_2 - V_1) = nC_v(T_1 - T_2)$$

$$\text{or, } P_2\left(\frac{nRT_2}{P_2} - \frac{nRT_1}{P_1}\right) = n \times \frac{3}{2}R(T_1 - T_2)$$

$$\text{or, } nR\left(T_2 - \frac{T_1 . P_2}{P_1}\right) = nR.\frac{3}{2}(T_1 - T_2)$$

$$\text{or, } T_2 - \frac{400}{25} = \frac{3}{2}(400 - T_2) \quad \text{or, } T_2 - 16 = 600 - 1.5T_2$$

or, $2.5T_2 = 616$.

$$\therefore T_2 = \frac{616}{2.5} = 246.4\,\text{K} = -26.6\,°\text{C}.$$

$$\therefore W = nC_v(T_2 - T_1)$$

$$= 26 \text{ mol} \times \frac{3}{2}R\text{J.K}^{-1}.\text{mol}^{-1}(246.4 - 400)\text{K} = -26 \times \frac{3}{2} \times 8.314 \times 153.6 \text{ J}$$

$= -49804.186$ J $=$ **$-$49.80 kJ.**

$\therefore \Delta U =$ **$-$49.80 kJ.**

$$\Delta H = nC_p(T_2 - T_1) = -26 \times \frac{5}{2} \times 8.314 \times (400 - 246.4) \text{ J} = \mathbf{-83.01 \text{ kJ}}.$$

(b) For an adiabatic reversible process,

$$\frac{T}{P^{R/C_p}} = \text{constant.}$$

$$\therefore T_2 = T_1\left(\frac{P_2}{P_1}\right)^{R/C_p} = 400 \times \left(\frac{1}{25}\right)^{R/\frac{5}{2}R}$$

$$= 400 \times \left(\frac{1}{25}\right)^{0.4} = 400 \times (0.04)^{0.4}$$

$= 110.3784$ K ≈ 110.4 K $=$ **$-$162.8 °C.**

$$W = nC_v(T_2 - T_1) = 26 \text{ mol} \times \frac{3}{2}R(110.4 - 400)$$

$$= -26 \text{ mol} \times \frac{3}{2} \times 8.314 \text{ J.K}^{-1}.\text{mol}^{-1}(400 - 110.4) \text{ K}$$

$= -93901.64$ J $=$ **$-$93.90 kJ.**

Now, $\Delta U = W =$ **$-$93.90 kJ.**

$$\Delta H = nC_p(T_2 - T_1) = -26 \text{ mol} \times \frac{5}{2} \times 8.314 \text{ J.K}^{-1}.\text{mol}^{-1} \times 289.6 \text{ K}$$

$=$ **$-$156.50 kJ.**

(c) **Hints :** $W_{rev,iso} = -nRT \ln \frac{V_2}{V_1}$; $Q = -W$; $\Delta U = \Delta H = 0$.

16. What would be the work done, final temperature, ΔH, etc., in the above case if the pressure is suddenly released adiabatically to 5 atm and then (a) pressure is changed to 1 atm reversibly and adiabatically and (b) the pressure is released to vacuum, and after attainment of the volume corresponding to 1 atm, the pressure is made 1 atm. What is the volume?

Soln. :

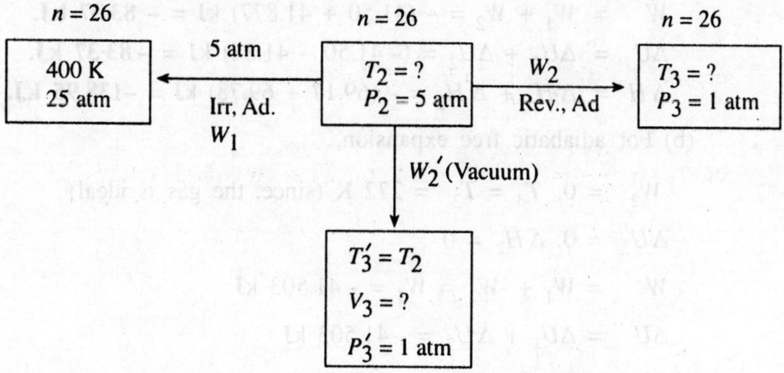

(a) Given, n = 26 mol, $C_v = \frac{3}{2}R$, T_1 = 400 K, P_1 = 25 atm.

$W = W_1 + W_2$ or, $W_1 + W_2'$

For irreversible, adiabatic expansion,

$W_1 = nC_v(T_2 - T_1) = -P_{ex}(V_2 - V_1)$

or, $n \times \frac{3}{2}R(T_2 - 400) = -5\left(\frac{nRT_2}{P_2} - \frac{nRT_1}{P_1}\right)$ (given, $P_2 = P_{ex}$ = 5 atm)

or, $\frac{3}{2}(400 - T_2) = 5\left(\frac{T_2}{5} - \frac{400}{25}\right) = T_2 - 80$

or, $600 - 1.5T_2 = T_2 - 80$

or, $2.5T_2 = 680.$

$\therefore T_2 = \frac{680}{2.5} = 272$ K

$W_1 = nC_v(T_2 - T_1) = 26 \times 1.5\text{R}(272 - 400)$

$= -41503.488 \text{ J} = -41.503 \text{ kJ}$

$\Delta U_1 = -41.503$ kJ

$\Delta H_1 = 26 \times 2.5 \times 8.314 \times (272 - 400) = -69172.48 \text{ J} = -69.172$ kJ.

For rev. adiabatic expansion : $W_2 = nC_v(T_3 - T_2)$.

$$T_3 = T_2\left(\frac{P_3}{P_2}\right)^{R/C_p} = 272\left(\frac{1}{5}\right)^{0.4} \text{ K}$$

$= \mathbf{142.88\ K.}$

$W_2 = 26 \text{ mol} \times \frac{3}{2} \times 8.314 \text{ J.K}^{-1}\text{.mol}^{-1} \times (142.88 - 272)$ K

$= -26 \times 1.5 \times 8.314 \times 129.12 \text{ J} = -41{,}866.64$ J

$= -41.87$ kJ.

$\Delta U_2 = -41.87$ kJ.

$\Delta H_2 = -26 \text{ mol} \times \frac{5}{2} \times 8.314 \text{ J.K}^{-1}.\text{mol}^{-1} \times (272 - 142.88)$ K

$= -69{,}777.7392$ J $= -69.78$ kJ.

$W = W_1 + W_2 = -(41.50 + 41.877)$ kJ $=$ **$-$ 83.37 kJ.**

$\Delta U = \Delta U_1 + \Delta U_2 = (-41.50 - 41.87)$ kJ $=$ **$-$ 83·37 kJ.**

$\Delta H = \Delta H_1 + \Delta H_2 = -(69.17 + 69.78)$ kJ $=$ **$-$138.95 kJ.**

(b) For adiabatic free expansion,

$W_2' = 0,\ T_3' = T_2 = 272$ K (since, the gas is ideal)

$\Delta U_2' = 0,\ \Delta H_2' = 0$

$W = W_1 + W_2' = W_1 = -41.503$ kJ

$\Delta U = \Delta U_1 + \Delta U_2' = -41.503$ kJ

$\Delta H = \Delta H_1 + \Delta H_2' = -69.172$ kJ

$$V_3 = \frac{nRT_3'}{P_3'} = \frac{26\,\text{mol} \times 0.08206\,\text{l.atm.K}^{-1}.\text{mol}^{-1} \times 272\,\text{K}}{1\,\text{atm}}$$

$= 580.32832$ l $=$ **580.3 l.**

17. If 1 mol of an ideal gas undergoes a reversible polytropic expansion, the relation $PV^n = C$ holds where n and C are constants with $n > 1$. (a) Calculate the work done i[f] the gas expands from V_1 litres to V_2 litres with $T_1 = 300$ K and $T_2 = 200$ K and $n = 2$. (b) If $C_v = \frac{5}{2}R$, calculate Q, ΔU and ΔH. **[B.U. 1994; C.U. 2004]**

Soln. (a) $W = -\int_1^2 P_{ex} dV = -\int_1^2 PdV$, since the process is reversible.

Now, $PV^2 = C$, $\therefore P = \frac{C}{V^2}$, $n = 1$.

$$W = -\int_1^2 \frac{C}{V^2} dV = \left[\frac{C}{V}\right]_1^2$$

$$= \left(\frac{C}{V_2} - \frac{C}{V_1}\right) = \left(\frac{P_2V_2^2}{V_2} - \frac{P_1V_1^2}{V_1}\right), \text{ since } C_1 = P_1V_1^2$$

$= (P_2V_2 - P_1V_1) = R(T_2 - T_1)$, since the gas is ideal

$= -8.314 \text{ J.K}^{-1}.\text{mol}^{-1} \times 1 \text{ mol} \times 100$ K (given, $T_1 = 300$ K, $T_2 = 200$ K)

$=$ **$-$831.4 J.**

(b) $\Delta U = nC_v(T_2 - T_1)$

$= n\frac{5}{2}R(200 - 300) = -1 \text{ mol} \times \frac{5}{2} \times 8.314 \text{ J.mol}^{-1}.\text{K}^{-1} \times 100$ K

(given, $n = 1$ mol)

$= -2078.5$ J $=$ **$-$2.078 kJ.**

$$\Delta H = nC_p(T_2 - T_1) = 1 \text{ mol} \times \frac{7}{2} \times 8.314 \text{ J.mol}^{-1}.\text{K}^{-1} \times (200 - 300)\text{K}$$

$$= -\frac{7}{2} \times 8.314 \times 100 \text{ J} = \mathbf{-2.909 \text{ kJ.}}$$

$$Q = \Delta U - W = (-2078.5 + 831.4) \text{ joule}$$

$$= \mathbf{-1.247 \text{ kJ.}}$$

18. At 25 °C, α for water is 2.07×10^{-4} deg^{-1} and its density is 0.9970 gm/cc. If the temperature of 200 gm of water is raised from 25 °C to 50 °C under a constant pressure of 2 atm, calculate W, and if C_p = 75.3J, calculate Q, ΔH and ΔU.

Soln. : $\alpha = \frac{1}{V}\left(\frac{\partial V}{\partial T}\right)_p$

$\therefore$ $dV = \alpha V dT.$

$\therefore$ $W = -\int_1^2 PdV = -\int_1^2 P\alpha VdT$

$$= -P \alpha V \int_1^2 dT = -P \alpha V(T_2 - T_1).$$

Given, $\alpha = 2.07 \times 10^{-4} \text{ deg}^{-1}$ $\quad T_2 = 323$ K $\quad m = 200$ gm

$P = 2$ atm $\quad T_1 = 298$ K $\quad d = 0.9970$ gm/cc

$$\therefore V = \frac{200 \text{ gm}}{0.9970 \text{ gm / cc}}$$

$$= 200.60 \text{ cc} = 200.60 \times 10^{-3} \text{ litre.}$$

$$\therefore W = -2 \text{ atm} \times 2.07 \times 10^{-4} \text{ K}^{-1} \times 200.60 \times 10^{-3} \text{ litre} \times 25 \text{ K}$$

$$= -2.0762 \times 10^{-3} \text{ l-atm}$$

$$= -(2.076 \times 101.325 \times 10^{-3}) \text{ J [since 1 litre-atm = 101.325 J]}$$

$$= \mathbf{-0.21035 \text{ J.}}$$

$$\Delta H = nCpdT = \frac{200}{18} \text{ mol} \times 75.3 \text{ J.mol}^{-1}.\text{K}^{-1} \times 25 \text{ K}$$

$$= \left(\frac{200}{18} \times 75.3 \times 25\right)\text{J} = \mathbf{20.917 \text{ kJ.}}$$

At constant pressure, $Q = \Delta H = 20.917$ kJ

$$\Delta U = Q + W$$

$$= (20.917 - 0.002) \text{ kJ}$$

$$= \mathbf{20.915 \text{ kJ}}$$

19. 2 mol of an ideal monatomic gas with C_v = 15 J.K^{-1} initially at 300 K and 1 atm is put through the following reversible cycle :

Path A : state 1 $\rightarrow$ 2, heated at constant volume till the temperature is 500 K.

B : state 2 $\rightarrow$ 3, expanded adiabatically till the temperature is 300 K.

C : state 3 $\rightarrow$ 1, compressed isothermally back to state 1.

Calculate Q, W, ΔU, ΔH for each path and whole cycle.

Soln. :

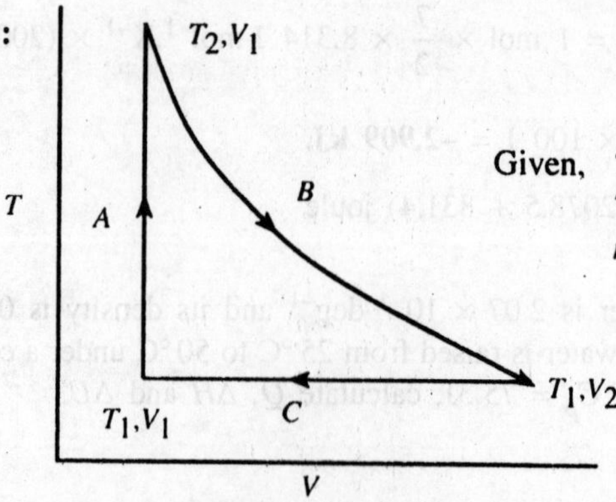

Given, $T_1 = 300$ K
$T_2 = 500$ K
$n = 2$ mol, $C_v = 15$ J.K^{-1}

Path A : $W_A = 0$ (since, $\Delta V = 0$).

$$Q_A = \Delta U_A = nC_v\,(T_2 - T_1) = 2 \times 15 \times 200 \text{ J} = 6000 \text{ J} = \mathbf{6\ kJ.}$$

$$\Delta H_A = nC_p\,(T_2 - T_1) = 2 \times 23.314 \times 200 \text{ J} = 9325.6 \text{ J} = \mathbf{9.32\ kJ.}$$

Path B : $Q_B = 0,\ W_B = \Delta U_B = 2 \times 15 \times (T_1 - T_2)$

$$= -2 \times 15 \times 200$$

$$= \mathbf{6\ kJ.}$$

$$\Delta H_B = -9.32 \text{ kJ.}$$

Path C : $\Delta U_C = \Delta H_C = \mathbf{0}$ (since the process is isothermal).

$$Q_C = -W_C = nRT_1 \ln\frac{V_1}{V_2} \quad (đW = -PdV).$$

Since path B is reversible adiabatic,

$$T_2V_1^{\gamma-1} = T_1V_2^{\gamma-1}.$$

$$\therefore \frac{V_1}{V_2} = \left(\frac{T_1}{T_2}\right)^{\frac{1}{\gamma-1}} = \left(\frac{T_1}{T_2}\right)^{\frac{C_V}{R}}.$$

$$\therefore Q_C = -W_C = nRT_1 \ln\left(\frac{T_1}{T_2}\right)^{\frac{C_V}{R}} = nC_VT_1\ln\left(\frac{T_1}{T_2}\right)$$

$$= 2 \times 15 \times 300 \ln\left(\frac{300}{500}\right) \text{J}$$

$$= -4597.4 \text{ J} = \mathbf{-4.597\ kJ.}$$

$$Q_{cy} = W_{cy} = W_A + W_B + W_C = \mathbf{-1.4\ kJ},\ \Delta U_{cy} = \Delta H_{cy} = \mathbf{0.}$$

20. 10 gm of copper initially at 0 °C and 1000 atm is compressed to 1 atm isothermally and reversibly. Calculate the work done. Given that $\beta = 0.725 \times 10^{-6}$ atm, and density of Cu = 8.93 gm/cc.

Soln. : $\beta = -\dfrac{1}{V}\left(\dfrac{\partial V}{\partial P}\right)_T$

or, $dV = -\beta V dP$.

Now, work done $W = -\int_1^2 P_{ex}dV = -\int_1^2 PdV$ [for rev. process]

$$= \int_1^2 P\beta V dP = \frac{\beta V}{2}(P_2^2 - P_1^2).$$

Given, $\beta = 0.725 \times 10^{-6}$ atm^{-1}, $P_1 = 10^3$ atm $m = 10$ gm

$P_2 = 1$ atm $d = 8.93$ gm/cc

$$\therefore\ V = \frac{10\,\text{gm}}{9.93\,\text{gm/cc}} = 1.1198 \text{ cc} = 1.1198 \times 10^{-3} \text{ litre.}$$

$$\therefore \text{ work done, } W = \frac{\beta V}{2}(P_2^2 - P_1^2)$$

$$= \frac{(0.725\times10^{-6}\,\text{atm}^{-1})\times(1.1198\times10^{-3}\,\text{litre})\times(1-10^6)\,\text{atm}^2}{2}$$

$$= \frac{(0.725\times10^{-9}\times1.1198)\times(1-10^6)\,\text{litre.atm}}{2}$$

$= -4.0592 \times 10^{-4}$ litre.atm $= -(4.0592 \times 10^{-4} \times 101.325)$ J $=$ **– 0.04113 J.**

21. An automobile tire contains air at 300 kPa at 20 °C. The valve stem is removed and air is allowed to expand adiabatically against the constant external pressure of 50 kPa until the pressure is same inside and outside. Calculate the final temperature, W, Q, ΔU and ΔH per mole. Given, $C_v = \frac{5}{2}R$. Assume air to be an ideal gas.

Hints : $P_{ex}(V_1 - V_2) = nC_v(T_2 - T_1)$

or, $50\left(\frac{nRT_1}{P_1} - \frac{nRT_2}{P_2}\right) = n\frac{5}{2}R(T_2 - 293)$

or, $\frac{7}{2}T_2 = 2.5 \times 293 + \frac{293}{6}$.

$\therefore$ $T_2 = 223.24$ K $= 223$ K $=$ **–50 °C.**

$W = \frac{5}{2}R(T_2 - T_1) =$ **–1.45 kJ.**

$\Delta U =$ **–1.45 kJ**, $\Delta H =$ **–2.04 kJ.**

22. A 32 gm sample of methane gas initially at 1 atm and 27 °C is heated to 277 °C. The empirical equation for the molar heat capacity of methane at constant pressure is $C_p = 3 + (2 \times 10^{-2})T$ cal.mol^{-1}.K^{-1}. Assuming ideal behaviour calculate Q, ΔH, ΔU for an isobaric reversible process. **[V.U. 1997]**

Soln. : Work $W = -\int_{V_1}^{V_2} P_{ex}dV = -\int_{V_1}^{V_2} PdV$

Since P is constant,

$$W = -P(V_2 - V_1) = -P\left(\frac{nRT_2}{P_2} - \frac{nRT_1}{P_1}\right), P = P_2.$$

Given, m = 32 gm, P_1 = 1 atm, T_1 = 27 °C = 300.15 K

T_2 = 277 °C = 550.15 K, P_2 = 1 atm, $C_p = 3 + (2 \times 10^{-2})T$ cal.mol^{-1}.K^{-1}

32 gm methane = $\frac{32}{16}$ or 2 mol methane.

Hence, $W = -2 \text{ mol} \times 8.314 \text{ J.K}^{-1}.\text{mol}^{-1}\left(T_2 - \frac{P_2T_1}{P_1}\right)\text{K}$

$= -2 \text{ mol} \times 8.314 \text{ J.K}^{-1}.\text{mol}^{-1} \times (550.15 - 300.15) \text{ K}$

$= -2 \text{ mol} \times 8.314 \text{ J.K}^{-1}.\text{mol}^{-1} \times 250 \text{ K} = -4157 \text{ J} =$ **– 4.16 kJ.**

$$\Delta U = \int_1^2 nC_V dT$$

$$= 2 \text{ mol} \int_1^2 (3 - 1.987) + \left(2 \times 10^{-2} T\right)\text{cal.mol}^{-1}.\text{K}^{-1} dT \text{ (since, } C_v = C_p - R)$$

$$= 2\int_1^2 1.013 dT + 4 \times 10^{-2}\int_1^2 T dT$$

$$= 2 \text{ mol} \times 1.013 \text{ cal.K}^{-1}.\text{mol}^{-1} (T_2 - T_1)\text{ K} + \frac{4\times 10^{-2}}{2} \text{cal.K}^{-2}.(T^2{}_2 - T^2{}_1)\text{K}^2$$

$$= 2 \text{ mol} \times 1.013 \text{ cal.K}^{-1}.\text{mol}^{-1} \times 250 \text{ K} + 2 \times 10^{-2}\left(550^2 - 300^2\right)\text{cal.}$$

$= 506.5$ cal + 4250 cal = 4756.5 cal = 4756.5 cal × 4.184 J/cal

= 19901.196J = **19.90 kJ.**

Similarly, $\Delta H = 2 \times 3 \times 250 \text{ cal} + 4 \times 10^{-2}\left(\frac{550^2}{2} - \frac{300^2}{2}\right)\text{cal}$

$$= 1500 \text{ cal} + \frac{4\times 10^{-2}}{2} \times 212500 \text{ cal} = (1500 + 4250) \text{ cal}$$

= 5750 cal × 4.184 J/cal = 24058 J = **24.058 kJ.**

23. One mole of an ideal monatomic gas at 25 °C and 5 atmosphere pressure is expande to a final pressure of 1 atm (a) isothermally against a constant pressure of 1 atmospher (b) adiabatically reversibly. Calculate in each case (i) the final temperature, (ii) the hea absorbed, (iii) the increase in internal energy and (iv) the increase in enthalpy of the gas **[V.U. 2000**

Soln. : Given, n = 1 mole, T_1 = 25 °C = 298.15 °K, P_1 = 5 atm, P_2 = 1 atm.

(a) For isothermal expansion, $T_2 = T_1 = T$ = **298.15 K.**

Work done by the gas, $W = -P_{ex}(V_2 - V_1)$

$$= -P_2\left(\frac{nRT}{P_2} - \frac{nRT}{P_1}\right) = -nRT\left(1 - \frac{P_2}{P_1}\right)$$

$$= -1\text{mol} \times 8.314 \text{ J.K}^{-1}.\text{mol}^{-1} \times 298.15 \text{ K} \times \left(1 - \frac{1}{5}\right)$$

$$= -1983.055 \text{ J} = -1.983 \text{ kJ}.$$

Since the gas is ideal, internal energy (U) and enthalpy (H) are functions of T only.

$\therefore \quad \Delta U = \Delta H = \mathbf{0}.$

Hence, heat absorbed, $Q = W = \mathbf{1.983 \text{ kJ}}$.

(b) For reversible adiabatic expansion,

$$W = C_v(T_2 - T_1)$$

$$\text{and } T_2 = T_1\left(\frac{P_2}{P_1}\right)^{R/C_P} \quad \left(\frac{R}{C_P} = \frac{R}{\frac{5}{2}R} = 0.4\right).$$

$$\therefore \quad T_2 = 298.15\text{K} \times \left(\frac{1}{5}\right)^{0.4} = \mathbf{156.62 \text{ K}}.$$

$$Q = 0$$

$$W = n\frac{3}{2}R(156.62 - 298) \text{ K}$$

$$= -1 \text{ mole} \times 1.5 \times 8.314 \text{ J.mol}^{-}.\text{K}^{-1} \times 141.38 \text{ K}$$

$$= -1763.15 \text{ J} = -1.76 \text{ kJ}.$$

$$\Delta U = W = \mathbf{-1.76 \text{ kJ}}.$$

$$\therefore \quad \Delta H = \frac{5}{2} \times 8.314 \times 141.38 \text{ J} = 2938.58 \text{ J} = \mathbf{2.94 \text{ kJ}}.$$

24. The pressure of a gas is represented by $\frac{20}{V}$ atm, where V is the volume of the .s. If the gas expands from 5 l to 50 l and undergoes an increase in internal energy by 0 cal, how much heat will be absorbed during the process? **[V.U. 2002]**

Soln. : Since $\frac{20}{V}$ is in atm unit, let the unit of 20 be l-atm, where V in litre.

$$W = -\int P_{ex}dV = -\int 20\frac{dV}{V} \text{ (given, } P = \frac{20}{V})$$

$$= -20 \ln\frac{V_2}{V_1}\text{(l-atm)} = -20 \ln 10 \text{ l-atm } (V_2 = 50 \text{ l}, V_1 = 5 \text{ l})$$

$$= -46.051702 \text{ l-atm} \times 101.325 \text{ J/l.atm}$$

$$= -4666.1887 \text{ J} = -4.67 \text{ kJ}.$$

$$Q = \Delta U - W = 200 \text{ cal} \times 4.184 \text{ J/cal} + 4.67 \text{ kJ}$$

$$= 836.8 \text{ J} + 4.67 \text{ kJ} = 0.837 + 4.67 \text{ kJ} = 5.507 \text{ kJ} = \mathbf{5.51 \text{ kJ}}.$$

25. One gm-mole of an ideal gas is kept at 10 atm and 300 K. The pressure is reduced to 1 atm adiabtically and reversibly, (ii) the external pressure is made 1 atm d the gas is allowed to expand till its pressure is also 1 atm, (iii) the external pressure is reduced to 1 atm and the gas is allowed to expand adiabatically till the volume is doubled and then the pressure is raised to stop the expansion. Calculate W, Q, ΔH and ΔU.

Soln. : $W_{ad} = nC_v(T_2 - T_1)$, $Q = 0$

$$\Delta U = nC_v(T_2 - T_1) = W$$

$$\Delta H = nC_p(T_2 - T_1)$$

$$C_p = \frac{5}{2}R.$$

By the problem, $n = 1$ mol, $P_1 = 10$ atm, $T_1 = 300$ K.

(i) $T_2 = T_1\left(\frac{P_2}{P_1}\right)^{R/Cp} = 300\left(\frac{1}{10}\right)^{0.4} = 300 \times 0.1^{0.4}$ (given, $P_2 = 1$ atm)

$= 119.4321512\text{ K} = \mathbf{119\ K.}$

$W = 1\text{ mol} \times 1.5\ R\text{J.K}^{-1}.\text{mol}^{-1}(119\text{ K} - 300\text{ K})$

$= -1.5 \times 8.314 \times 181\text{ J} = -2257.251\text{ J} = \mathbf{-2.26\ kJ.}$

$\Delta U = W = \mathbf{-2.26\ kJ.}$

$\Delta H = 2.5R(119 - 300)\text{ J}$

$= -2.5 \times 8.314 \times 181\text{ J} = -3762.085\text{ J} = \mathbf{-3.76\ kJ.}$

(ii) For irreversible adiabatic process, $nC_v(T_2 - T_1) = -P_{ex}(V_2 - V_1)$

$$= -P_2\left(\frac{nRT_2}{P_2} - \frac{nRT_1}{P_1}\right)$$

or, $n\frac{3}{2}R(T_2 - 300) = -nR\left(T_2 - T_1\frac{P_2}{P_1}\right)$, $n = 1$

or, $450 - \frac{3}{2}T_2 = T_2 - \left(\frac{300 \times 1}{10}\right)$

or, $\frac{5}{2}T_2 = 480.$

$\therefore T_2 = \frac{2 \times 480}{5} = \mathbf{192\ K.}$

$Q = 0$, $W = n\frac{3}{2}R(192 - 300).$

$\therefore W = -1\text{ mol} \times 1.5 \times 8.314\text{ J.K}^{-1}.\text{mol}^{-1} \times 108\text{ K}$

$= -1.5 \times 8.314 \times 108\text{ J} = -1346.868\text{ J} = \mathbf{-1.35\ kJ.}$

$\therefore \Delta U = \mathbf{-1.35\ kJ.}$

$\Delta H = 2.5 \times 8.314\ (-108)\text{ J} = -2244.78\text{J} = \mathbf{-2.24\ kJ.}$

(iii) $C_v(T_2 - T_1) = -P_2(2V_1 - V_1) = -P_2V_1 = -P_2 \times \frac{nRT_1}{P_1}.$

$\therefore \frac{3}{2}(T_2 - 300) = -1 \times 30$, or, $450 - 30 = \frac{3}{2}T_2$

or, $420 \times \frac{2}{3} = T_2$, $\therefore T_2 = \mathbf{280\ K.}$

$\therefore \quad W = -\frac{3}{2} \times 8.314 \times 20 = \mathbf{-249.42\ J.}$

$\Delta U = -249.4 = \mathbf{-249.4\ J.}$

$\Delta H = -\frac{5}{2} \times 8.314 \times 20 = \mathbf{-415.7\ J.}$

26. An ideal gas is expanded adiabatically against a constant pressure P_2 until it doubles its volume. If the initial temperature and pressure are T_1 and P_1 respectively, calculate the final temperature. **[V.U. 2005]**

Soln. : See Problem 25.

27. Calculate ΔH and ΔU for the transformation of one mole of ideal gas from 27 °C and 1 atm to 327 °C and 17 atm. Given, $C_p = 20.9 + 0.042\ J.K^{-1}.mol^{-1}$. **[V.U. 2005]**

Soln. : $C_p = 20.9 + 0.042T\ J.K^{-1}.mol^{-1}$.

$C_v = 12.586 + 0.042T\ J.K^{-1}.mol^{-1}$.

Given, $n = 1$ mol, $T_1 = 27\ °C = 300.15\ K$, $P_1 = 1$ atm,

$T_2 = 327\ °C = 600.15\ K$, $P_2 = 17$ atm.

$$\Delta U = n\int C_V dT = n\left(\int_1^2 12.184 + \int_1^2 0.042T\right) J.K^{-1}.mol^{-1} dT.$$

Since the gas is ideal, ΔU and ΔH for change of P are zero.

$$\therefore \Delta U = 1\ mol \times 12.586\ J.K^{-1}.mol^{-1} \times (600 - 300)K + 1\ mol \times \frac{0.042\ J.K^{-2}.mol^{-1}}{2} \times (600^2 - 300^2)K^2$$

$= 1\ mole \times 12.586\ J.K^{-1}.mol^{-1} \times 300\ K + 0.021 \times 900 \times 300\ J$

$= 12.586 \times 300\ J + 5670\ J$

$= 9445.8\ J = \mathbf{9.44\ kJ.}$

Similarly,

$\Delta H = 20.9(600 - 300)\ J + 0.021 \times 900 \times 300\ J$

$= 11940\ J = \mathbf{11.94\ kJ.}$

28. One mole of an ideal gas expands from 10 litres to 50 litres at 27 °C in the following two ways : (i) isothermally reversibly and (ii) against a constant pressure of 1 atm. Show by calculation in which case more heat will be absorbed during expansion. **[C.U. 1993]**

Soln. : Given, $n = 1$ mole, $V_1 = 10$ l, $V_2 = 50$ l, $T_1 = 27\ °C = 300.15$ K.

(i) For isothermal reversible expansion,

$$Q = -W = \int_1^2 nRT\frac{dV}{V} = nRT\ \text{In}\frac{V_1}{V_2}$$

$$= 1\ mol \times 8.314\ J.K^{-1}.mol^{-1} \times 300.15\ K\ \text{In}\frac{50}{10} = 4016.267\ J = \mathbf{4.016\ kJ.}$$

(ii) In the second case,

$$W = -P_{ext}(V_2 - V_1) = -1 \times (50 - 10) = -40 \text{ litre-atm}$$

$$= -40 \text{ litre-atm} \times 101.325 \text{ J/litre-atm} = -4053 \text{ J} = \mathbf{-4.053 \text{ kJ}}.$$

In case (ii) more work is done, and more heat may be absorbed. The second change cannot take place isothermally—the temperature has to change. Ideal gas can expand only up to 25 l against a pressure of 1 atm if the process is done isothermally. Since in the second case gas expanded to 50 l, temperature must have changed.

29. One mole of an ideal gas with $C_v = 3$ cal.deg^{-1}.mol^{-1} initially at S.T.P. is put through the following reversible cycle :

(i) State 1 to State 2—heated at constant volume to twice the initial temperature.

(ii) State 2 to State 3—expanded adiabatically until it is back to initial temperature.

(iii) State 3 to State 1—compressed isothermally back to state 1.

Depict the cycle in a *V-T* diagram and calculate *Q, W* and ΔU for the cycle.

[C.U. 1993; V.U. 2004]

Soln. :

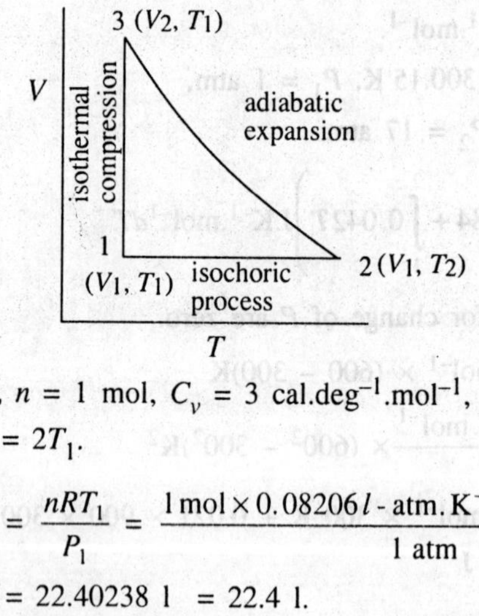

Given, $n = 1$ mol, $C_v = 3$ cal.deg^{-1}.mol^{-1}, $T_1 = 273$ K, $P_1 = 1$ atm, $T_2 = 2T_1$.

(i) $$V_1 = \frac{nRT_1}{P_1} = \frac{1 \text{ mol} \times 0.08206\, l\text{-atm.K}^{-1}.\text{mol}^{-1} \times 273 \text{ K}}{1 \text{ atm}}$$

$$= 22.40238 \text{ l} = 22.4 \text{ l}.$$

At constant volume, $W = \mathbf{0}$,

$$Q = \Delta U$$
$$= nC_v(T_2 - T_1) = nC_v(2T_1 - T_2)$$
$$= 1 \text{ mole} \times 3 \text{ cal.deg}^{-1}.\text{mol}^{-1} \times 273 \text{ K}$$
$$= 819 \text{ cal} = 819 \times 4.184 \text{ J}$$
$$= 3426.69 \text{ J} = \mathbf{3.427 \text{ kJ}}.$$

$$\Delta H = nC_p \times 273 \text{ J} = 1 \text{ mol} \times 4.987 \text{ cal.deg}^{-1}.\text{mol}^{-1} \times 273 \text{ K}$$
$$= 1361.45 \text{ cal} = 1361.445 \times 4.184 \text{ J}$$
$$= 5696.307 \text{ J} = 5.696 \text{ kJ}.$$

(ii) Adiabatic reversible change,

$$Q = \mathbf{0}$$

$$\Delta U = W = nC_v(T_1 - T_2)$$
$$= -1 \text{ mol} \times 3 \text{ cal.deg}^{-1}.\text{mol}^{-1} \times (2T_1 - T_1) \text{ K}$$

$= -3 \times 273$ cal $= -819$ cal $\times$ 4.184 J/cal

$=$ **−3.427 kJ.**

W $=$ **−3.427 kJ.**

For adiabatic reversible change,

$$T_2V_1^{\gamma-1} = T_1V_2^{\gamma-1}$$

$$2T_1V_1^{1.66-1} = T_1V_2^{1.66-1} \quad (\gamma = 1.66)$$

or, $\quad 2 \times 273 \times (22.4)^{0.66} = 273V_2^{0.66}$

$V_2^{0.66} = 15.5665972$ l

$\therefore\ V_2 = 64.02566123$ l $= 64.02$ litres.

(iii) Isothermal compression.

$$Q = -W = nRT_1 \ln\frac{V_1}{V_2}$$

$$= 1 \text{ mol} \times 8.314 \text{ J.K}^{-1}.\text{mol}^{-1} \times 273 \text{ K} \times \ln\frac{22.4}{64.02}$$

$= -2383.513549$ J $=$ **−2.384 kJ.**

$\Delta U =$ **0.**

Total ΔU $= 3.427$ kJ $- 3.427$ kJ $+ 0$

$=$ **0.**

Total W $= -(0 + 3.427$ kJ $- 2.384$ kJ$) =$ **−1.043 kJ.**

Total Q $= 3.427$ kJ $+ 0 - 2.384$ kJ

$=$ **1.043 kJ.**

30. 1 mol of an ideal gas with $C_v = 20$ kJ.mole^{-1} initially at 1 bar and 300 K is put through the following reversible cycle :

A. State 1 to State 2 by heating at constant volume to thrice the initial temperature.

B. State 2 to State 3 expanded adiabatically until back to initial temperature.

C. State 3 to State 1, compressed isothermally back to state 1. Calculate Q, W, ΔU and ΔH for all the steps and the cycle.

Draw the change in a V–T diagram (See Problems 19, 29).

31. 0.5 mol of an ideal monatomic gas initially at 5 atm pressure and 0 °C is allowed to expand against a constant external pressure of 0.5 atm. Conditions are such that the final volume is 10 times the initial volume. The final gas pressure equals the external pressure. Calculate Q, W, ΔU, ΔH, ΔS, ΔG for the process. **[C.U. 1999, 2005]**

Soln. :

Given, $n = 0.5$ mol

$P_{ex} = P_2 = 0.5$ atm

$P_1 = 5$ atm

$T_1 = 0\,°\text{C} = 273.15$ K

$$\therefore \quad V_1 = \frac{0.5\,\text{mol} \times 0.082061 - \text{atm.K}^{-1}.\text{mol}^{-1} \times 273\,\text{K}}{5\,\text{atm}}$$

$= 2.2414689$ l $= 2.24$ l.

By the problem, $V_2 = 10V_1 = 22.414689$ l.

$$\therefore \quad T_2 = \frac{P_2V_2}{nR} = \frac{0.5\,\text{atm} \times 22.4146891}{0.5\,\text{mole} \times 0.082061 - \text{atm.K}^{-1}.\text{mol}^{-1}}$$

$$= 273.15 \text{ K}.$$

Since the process is isothermal,

$$\Delta U = \mathbf{0}$$

$$\Delta H = \mathbf{0.}$$

$$Q = -W = P_2(V_2 - V_1) \quad (\because đW = P\,dV)$$

$$= P_2\left(\frac{nRT_2}{P_2} - \frac{nRT_1}{P_1}\right) = nR\left(T_2 - T_1\frac{P_2}{P_1}\right)$$

$$= 0.5 \text{ mol} \times 8.314 \text{ J.K}^{-1}.\text{mol}^{-1} \times \left(T - \frac{TP_2}{P_1}\right)$$

$$= 4.157 \text{ J.K}^{-1} \times 273.15\left(1 - \frac{0.5}{5}\right)\text{K}$$

$$= 4.157 \times 273.15 \times 0.9\,\text{J} = 1021.9361\,\text{J} = \mathbf{1.021\ kJ.}$$

See chapter 6 for the calculation of ΔS and ΔG.

$$\Delta S = nR \ln\frac{V_2}{V_1} = \frac{1}{2}\text{mol} \times 8.314 \text{ J.K}^{-1}.\text{mol}^{-1}\ln 10$$

$$= 19.14 \text{ J.K}^{-1} \times \frac{1}{2} = \mathbf{9.57\ J.K^{-1}}.$$

$$\Delta G = \frac{1}{2}RT\ln\frac{V_2}{V_1}$$

$$= -\frac{1}{2} \times 8.314 \text{ J.K}^{-1}.\text{mol}^{-1} \times 273.15 \text{ K} \ln 10 = -2614.5498 \text{ J} = \mathbf{-2.61\ kJ.}$$

32. One mol of oxygen expands adiabatically against a constant external pressure of 1 atmosphere until the pressure balances. The initial temperature and volume are 200 °C and 20 litres respectively.

(i) Explain whether the process is reversible or not.

(ii) Calculate the final temperature, work done and the change in entropy.

(iii) Will the quantaties be different if *n* moles of oxygen were taken. Give reason. Assume ideal behaviour.

(iv) Calculate ΔU and ΔH. **[C.U. 1997]**

Soln. : (i) The change is not reversible since increase or decrease of pressure cannot reverse the process.

(ii) $Q = 0$

$\Delta U = W = nC_v(T_2 - T_1)$, if C_v is constant.

But $W = -P_{ex}(V_2 - V_1)$

$P_{ex} = P_2 = 1$ atm; $T_1 = 200$ °C $= 473.15$K.

$V_1 = 20$ l,

$$\therefore P_1 = \frac{nRT_1}{V_1} = \frac{1\,\text{mol} \times 0.08206\ l\text{-atm.mol}^{-1}.\text{K}^{-1} \times 473\,\text{K}}{20\,l}$$

$$= 1.941334 \text{ atm} = 1.941 \text{ atm.}$$

$$\therefore W_{ad} = -nC_v(T_2 - T_1) = -P_{ex}(V_2 - V_1)$$

$$= -P_{ex}\left(\frac{nRT_2}{P_2} - \frac{nRT_1}{P_1}\right).$$

$$\therefore n \times \frac{5}{2}R(473.15 - T_2) = P_2\left(\frac{nRT_2}{P_2} - \frac{nRT_1}{P_1}\right)$$

$$= nR\left(T_2 - T_1\frac{1}{1.94}\right) = nR\left(T_2 - \frac{473.15}{1.94}\right)$$

$$\text{or, } \frac{5}{2} \times 473.15 + \frac{473.15}{1.94} = \frac{7}{2}T_2.$$

$$\therefore T_2 = \frac{2.5 \times 473.15 + 243.89175\,\text{K}}{\frac{7}{2}} = 407.627071 \text{ K} = \mathbf{407.63\ K.}$$

$$W = n\frac{5}{2}R(407.63 - 473.15)$$

$$= -1\text{mole} \times \frac{5}{2} \times 8.314 \text{ J.K}^{-1}.\text{mole}^{-1} \times 65.52 \text{ K}$$

$$= -1361.833 \text{ J} = \mathbf{-1.362\ kJ.}$$

$$\Delta U = \mathbf{-1.362\ kJ.}$$

$$\Delta H = -\frac{7}{2} \times 8.314 \times 65.52 \text{ J} = -1906.566 \text{ J} = \mathbf{-1.906\ kJ.}$$

$$\Delta S = nC_p \ln\frac{T_2}{T_1} - nR \ln\frac{P_2}{P_1}$$ (See second law of thermodynamics in Chapter 6)

$$= \frac{7}{2}R \ln\frac{407.63}{473.15} - R \ln\frac{1}{1.94} = \mathbf{+1.172\ J.K^{-1}.}$$

(iii) If there are n moles, then ΔU, ΔH, ΔS, etc. will change. T_2 will remain the same, since T is intensive and U, H, S are extensive. W will change since adiabatic work is extensive.

33. A cylinder is fitted with a frictionless piston and is kept in a thermostat. It contains 2 moles of an ideal gas at 27 °C and 2 atmospheric pressure. Following (i), (ii) and (iii) are three separate experiments carried out independently with the above.

(i) The piston is all on a sudden withdrawn to a position where pressure is reduced to 1 atm and equilibrium is restored.

(ii) Pressure is reduced at a single step from 2 to 1 atm.

(iii) Pressure is slowly reduced to 1 atm in such a way that the position of the piston remains unaltered if left to itself at any moment during the operation. Calculate in each case ΔU, ΔH, Q and W.

Soln. : Given, n = 2 moles, P_1 = 2 atm, T_1 = 27 °C = 300.15 K.

(i) The expansion of the gas from 2 to 1 atm is against vacuum. So W = **0**, ΔU = **0**, ΔH = **0**; Q = **0**.

Since the cylinder is kept in a thermostat the processes (i), (ii), and (iii) are isothermal.

(ii) So $W = -P_{ex}(V_2 - V_1) = -P_2(V_2 - V_1)$

$$= -P_2\left[\frac{nRT_2}{P_2} - \frac{nRT_1}{P_1}\right] = -nR\left[T_2 - T_1\frac{P_2}{P_1}\right]$$

$$= -2 \text{ mol} \times 8.314 \text{ J.K}^{-1}.\text{mol}^{-1}\ T\left(1 - \frac{1}{2}\right)\text{K}$$

$$= -2 \text{ mol} \times 8.314 \text{ J.K}^{-1}.\text{mol}^{-1} \times 300.15 \times \frac{1}{2}\text{K}$$

$$= -2495.45 \text{ J} = -\ \mathbf{2.495\ kJ.}$$

For ideal gas, $U = f(T)$ only,

Thus, $Q = -W$, $\Delta U = 0$, $\Delta H = 0$.

(iii) The work is reversible.

$$-W = Q = nRT \ln\frac{P_1}{P_2}$$

$$= 2 \text{ mol} \times 8.314 \text{ J.K}^{-1}.\text{mol}^{-1} \times 300.15 \text{ K} \ln 2$$

$$= 3459.42 \text{ J} = \mathbf{3.459\ kJ.}$$

ΔU = **0**, ΔH = **0**.

34. One mole of a monatomic ideal gas at 80 °C and 5 atmosphere pressure is expanded adiabatically against a constant external pressure 1 atmosphere in such a way that the final pressure of the gas becomes 1 atmosphere. Calculate the final temperature, ΔU, ΔH, ΔS for the process. **[C.U. 1999]**

Soln. : Given, n = 1 mol.

$$T_1 = 353 \text{ K},\ P_1 = 5 \text{ atm}.$$

$$P_{ex} = P_2 = 1 \text{ atm};\ C_v = \frac{3}{2}R,\ C_p = \frac{5}{2}R$$

$$W = -P_{ex}(V_2 - V_1) = nC_v(T_2 - T_1)$$

or, $$-P_2\left(\frac{nRT_2}{P_2} - \frac{nRT_1}{P_1}\right) = n\frac{3}{2}R(T_2 - T_1)$$

or, $$nR\left(T_2 - T_1\frac{P_2}{P_1}\right) = n \times \frac{3}{2}R(T_2 - T_1)$$

or, $$T_2 - \frac{353 \times 1}{5} = \frac{3}{2}(353 - T_2)$$

or, $$\frac{5}{2}T_2 = 1.5 \times 353 + 0.2 \times 353$$

$$= 1.7 \times 353$$

$$\therefore\ T_2 = \frac{2}{5} \times 1.7 \times 353 = 240.04 \text{ K} = \mathbf{240\ K.}$$

$$W = 1 \text{ mol} \times \frac{3}{2} \times 8.314 \text{ J.K}^{-1}\text{.mol}^{-1}(240 - 353) \text{ K}$$

$$= -1.5 \times 8.314 \times 113 \text{ J}$$

$$= -1409.223 \text{ J} = -1.410 \text{ kJ.}$$

$$Q = 0.$$

$$\Delta U = W = \mathbf{-1.41\ kJ.}$$

$$\Delta H = \frac{5}{2} \times 8.314\ (240 - 353) \text{ J}$$

$$= -2.5 \times 8.314 \times 113 \text{ J} = -2348.705\text{J} = \mathbf{-2.35\ kJ.}$$

$$\Delta S = nC_p \ln\frac{T_2}{T_1} - nR\ln\frac{P_2}{P_1}$$

$$= 1 \text{ mol} \times \frac{5}{2} \times 8.314 \text{ J.K}^{-1}\text{.mol}^{-1} \times \ln\frac{240}{353} - 1 \text{ mol} \times 8.314 \text{ J.K}^{-1}\text{.mol}^{-1} \times \ln\frac{1}{5}$$

$$= \mathbf{5.361\ J.K^{-1}.}$$

35. 2 moles of an ideal monatomic gas initially at 100 °C and 5 atmosphere pressure expands adiabatically and reversibly to 1 atm pressure. Calculate (i) the work done by the gas, (ii) final molar volume, and (iii) change in enthalpy. **[C.U. 2000]**

Hints : $T_2 = T_1\left(\frac{P_2}{P_1}\right)^{\frac{R}{C_p}} = 373\left(\frac{1}{5}\right)^{0.4} = 195.93\text{K} = 196 \text{ K}.$

$$W = nC_v\ (T_2 - T_1) = \Delta U,\ C_v = \frac{3}{2}R.$$

$$\Delta H = nC_p\ (T_2 - T_1);\ V_2 = \frac{nRT_2}{P_2}.$$

36. For a diatomic ideal gas near room temperature, what fraction of the heat supplied is available for external work if the volume of gas expands at constant pressure and at constant temperature? **[PhD Qualification Exam., Wisconsin University]**

Soln. : Let the volume increase from V_1 to V_2, temperature change from T_1 to T_2 at constant external pressure P, then

$$PV_1 = nRT_1$$

$$PV_2 = nRT_2.$$

Work done by the system, $W = P(V_2 - V_1)$.

$$W = -P(V_2 - V_1)$$

$$= -P\left(\frac{nRT_2}{P_2} - \frac{nRT_1}{P_1}\right),\ P_1 = P_2 = P$$

$$= nR(T_1 - T_2).$$

$$\Delta U = nC_v\ (T_2 - T_1).$$

Fraction of heat available for work $= \dfrac{-W}{Q} = \dfrac{-W}{\Delta U - W} = \dfrac{nR}{nC_v + nR} = \dfrac{R}{\frac{5}{2}R + R}$.

Since the gas is ideal diatomic, $C_v = \frac{5}{2}R$.

$$\therefore \frac{-W}{Q} = \frac{R}{\frac{7}{2}R} = \frac{2}{7}.$$

At constant T, $\Delta U = 0$ (since the gas is ideal).

$\therefore Q = -W$.

Hence, $\dfrac{-W}{Q} = \dfrac{Q}{Q} = 1$.

37. Will the fraction $\dfrac{W}{Q}$ remain the same for both the cases in the last problem if atomicity of the gas changes?

Soln. : W, whether done at constant pressure or at constant T, will not be affected with the change of atomicity. But heat supplied at constant P changes with the change of atomicity. Hence, $\dfrac{W}{Q}$ remains the same for the process done at constant T but changes for the process done at constant P, if atomicity changes.

38. 1 mole of an ideal monatomic gas at 298 K expands to double its volume at constant pressure. Calculate heat absorbed by the gas. **[C.U. 2006]**

Soln. : $P_1 = P_2 = P$, $V_2 = 2V_1$

Hence, $T_2 = V_2 \times \dfrac{T_1}{V_1} = 2T_1 \quad [\because V_2 = 2V_1]$

Heat absorbed by the gas $= nC_p\,(T_2 - T_1) = nC_pT_1$

$= 1 \text{ mol} \times \frac{5}{2} \times 8.314 \text{ J.K}^{-1}.\text{mol}^{-1} \times 298 \text{ K} = \mathbf{6.19\ kJ}$.

39. An ideal gas undergoes the following cycle—

(i) State 1 $\rightarrow$ State 2, heating at constant volume to raise the temperature from T_1 to T_2.

(ii) State 2 $\rightarrow$ State 3, expansion with decrease of temperature from T_2 to T_1. Work done by the gas is 1000 J while 500 J heat enters the system.

(iii) State 3 $\rightarrow$ State 1, isothermal compression, which needs 600 J work.

Calculate W, Q, ΔU for each step.

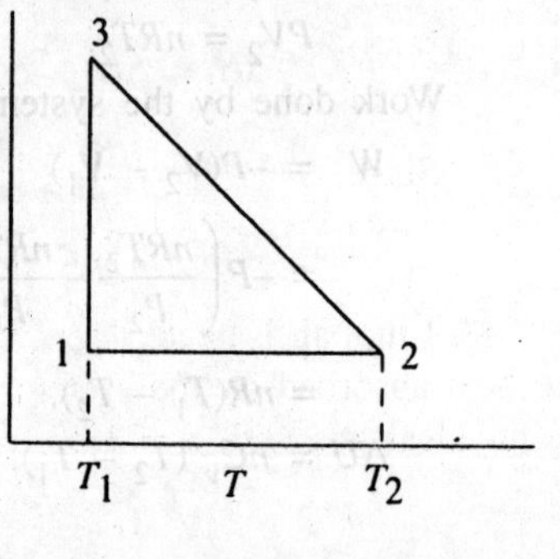

Soln. : State 1 $\rightarrow$ State 2, $W_1 = 0$, $\Delta U_1 = Q_1$.

State 2 $\rightarrow$ State 3, $W_2 = -1000$ J, $Q_2 = 500$ J,

$\Delta U_2 = Q_2 + W_2 = -500$ J.

State 3 $\rightarrow$ State 1, $\Delta U_3 = 0$, $W_3 = -Q_3 = 600$ J.

Now, ΔU cycle $= 0$

or, $\Delta U_1 + \Delta U_2 + \Delta U_3 = 0, \therefore \Delta U_1 = -\Delta U_2 = 500$ J.

Hence, $Q_1 = 500$ J.

40. 2 moles of steam at 120 °C are condensed isothermally and reversibly to water at 120 °C. Vapour pressure of water at 120 °C is 1490 mm of Hg, $l_e = 526$ cal.gm^{-1}. Calculate W, Q, ΔU and ΔH.

Soln. : Given, $n = 2$ moles

$l_e = 526$ cal.gm^{-1}

$T_1 = 120$ °C $= 393.15$ °K

$V_g = 1490$ mm of Hg

$W = -P(V_1 - V_g) = PV_g \; (V_1 << V_g)$

$= nRT = 2 \text{ mol} \times 8.314 \text{ J.K}^{-1}.\text{mol}^{-1} \times 393.15 \text{ K} = 6537.30 \text{ J} =$ **6.53 kJ.**

$Q = \Delta H = -2 \text{ moles} \times 526 \text{ cal.gm}^{-1} \times 18 \text{ gm.mol}^{-1} \times 4.184 \text{ J.cal}^{-1}$

$= -79228.22 \text{ J} = -79.23 \text{ kJ}$

$\Delta U = Q + W = (-79.23 + 6.53) \text{ kJ} =$ **−72·7 kJ.**

41. At 40°C, the dissociation pressure of the system $CuSO_4, 5H_2O \rightleftharpoons CuSO_4, 3H_2O + 2H_2O$ is 23.2 mm of Hg. At the same temperature vapour pressure of water is 55.1 mm of Hg. Calculate maximum work for the change.

$CuSO_4, 3H_2O + 2H_2O = CuSO_4 + 5H_2O$

Soln. : $P_1 = 55.1$ mm of Hg, $P_2 = 23.2$ mm of Hg, $n = 2$ moles

$$W = -nRT \ln \frac{P_1}{P_2} = -2 \text{ moles} \times 8.314 \text{ J.K}^{-1}.\text{mol}^{-1} \times 313 \text{ K} \ln \frac{55.1}{23.2}$$

$= -4501.93 \text{ J} =$ **−4·5 kJ.**

42. 1 gm-mole of methane (van der Waals gas) is compressed reversibly and isothermally from a pressure of 1 atm to 400 atm at 0°C. Calculate Q, W, ΔU and ΔH ($a = 2.264$ atm.litre^2/mole^2, $b = 0.0428$ litre/mole). For volume calculation assume ideal gas behaviour.

[C.U. 1979]

Hints :
$$W = -RT \ln \frac{V_2 - b}{V_1 - b} - a\left(\frac{1}{V_2} - \frac{1}{V_1}\right)$$

$$V_1 = \frac{RT}{P_1}, \; V_2 = \frac{RT}{P_2}.$$

Hence, $W =$ **12.8 kJ.**

Again, $$\left(\frac{\partial U}{\partial V}\right)_T = \frac{a}{V^2}.$$

$$\therefore \quad \Delta U = a\left(\frac{1}{V_1} - \frac{1}{V_2}\right) = -4.08 \text{ kJ.}$$

$$\Delta H = \Delta U + \Delta(PV) = \Delta U + P_2V_2 - P_1V_1 = \Delta U.$$

Since, $V = \dfrac{nRT}{P}$ (by the problem)

$Q = \Delta U - W =$ **−16.9 kJ.**

43. 1 mol air is heated from 291 K and 21 l to a final state of 319 K and 22 l. Calculate the heat absorbed by the system as well as change of internal energy. Air is assumed to be an ideal gas with $\gamma = 1.40$.

Soln : Initial pressure = $P_i = \dfrac{nRT_i}{V_i}$

$$= \frac{1\,\text{mol} \times 0.08206\ l.\text{atm}^{-1}\text{K}^{-1}.\text{mol}^{-1} \times 291\,\text{K} \times 1\,\text{mol}}{21 l}$$

= 1.137 atm = 1.14 atm

Final pressure = $P_f = \dfrac{nRT_f}{V_f}$

$$= \frac{1\,\text{mol} \times 0.08206\,l.\text{atm}^{-1}\text{K}^{-1}.\text{mol}^{-1} \times 319\,\text{K} \times 1\,\text{mol}}{22\,l}$$

= 1.189 atm = 1.19 atm.

The corresponding PV diagram is

P_A = 1.14 atm, V_E = 21l

P_B = 1.19 atm, V_D = 22l

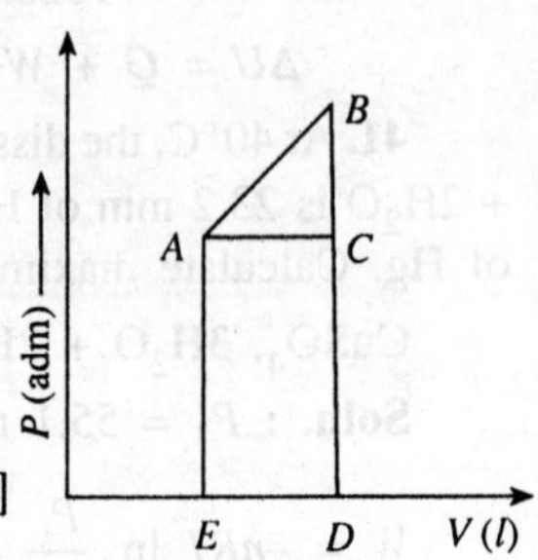

The work done is – (are a $ABDEA$)

= –[area of rectangle $ACDEA$ + area of the triangle ABC]

Thus, work done = $-[(ED \times AE) + \dfrac{1}{2}(AC \times BC)]$

$= -[(1l \times 1.14\ \text{atm}) + (\dfrac{1}{2} \times 1l \times 0.05\ \text{atm})] = -(1.14 + 0.025)\ l.\text{atm}$

$= -1.165\ l.\text{atm} = -1{,}165 \times 101.324\ \text{J} = -118.04\ \text{J}$

$\Delta U = nC_V(T_f - T_i)$

By the problem, $\gamma = 1.4 = \dfrac{7}{5} = \dfrac{\frac{7}{2}}{\frac{5}{2}} = \dfrac{\frac{7}{2}R}{\frac{5}{2}R}$.

$\gamma = \dfrac{C_P}{C_V}$

Thus, $C_V = \dfrac{5}{2}R$.

$\therefore\ \Delta U = 1\ \text{mol} \times \dfrac{5}{2}R \times (319 - 291)\ \text{K}$

$= 1\ \text{mol} \times 2.5 \times 8.314\ \text{J}\ \text{k}^{-1}.\text{mol}^{-1} \times 28\text{K}$

$= 581.98\ \text{J}$

$Q = \Delta U - W = 581.98\text{J} - (-118.04)\text{J} = 700.02\ \text{J}.$

44. A cylinder contains 1l of ideal gas which is kept at 3 atm pressure. How much work is done by the gas if it is expanded isothermally and reversibly to 1 atm. Compare this work with the work between the same two states but proceeding in a straight line in the P-V diagram.

Soln. : The processes in the P-V diagram are as shown below :

The isothermal reversible expansion work done by the gas $= W = nRT \ln \frac{V_A}{V_B}$.

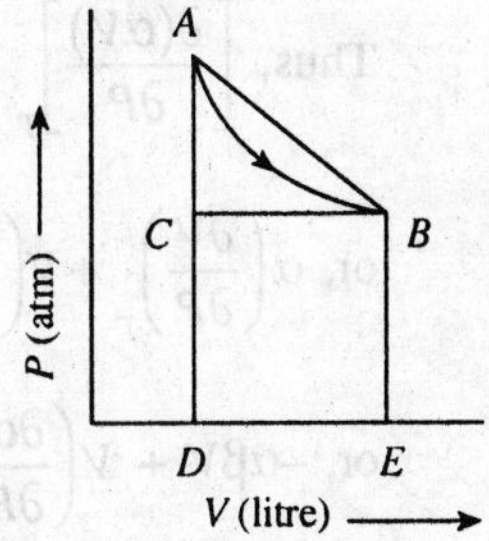

By the problem, $V_A = 1$ litre, $P_B = 1$ atm

$P_A = 300$ atm, $T_A = T_B$.

$$\therefore \; W = P_A V_A \ln \frac{P_B}{P_A} \text{ (since, } P_A V_A = nRT = P_B V_B)$$

$$= 3 \ln \frac{1}{3} \text{ litre-atm}$$

$$= -3.2958 \text{ lit. atm} \times 10^{-3} \frac{\text{m}^3}{\text{lit}} \times 1.01325 \times 10^5 \frac{\text{N}}{\text{m}^2.\text{atm}}$$

$$= -333.95 \text{ J.}$$

If A and B are joined by a straight line the work done = – (Area of $ABEDCA$)

= –(Area of triangle ABC + Area of rectangle $CBED$)

$$= -(DE \times EB) - \frac{1}{2} \times BC \times AC.$$

Now, $DE = CB = V_B - V_C = (3 - 1)$ lit $= 2$ lit

$AC = P_A - P_C = P_A - P_B = (3 - 1)$ atm $= 2$ atm

$EB = P_B = 1$ atm.

Thus, $W_{AB(\text{linear})} = (-2 \times 1) \text{ lit. atm} - \left(\frac{1}{2} \times 2 \times 2\right) \text{lit.atm} = -(2 + 2) \text{ lit.atm}$

$= -4$ lit.atm $= -4 \times 101.325$ J $= -405.3$ J.

45. The bulk modulus B of water at 0 °C is 2.032×10^4 atm. The average volume expansion coefficient $\alpha = 4.33 \times 10^{-4}$ deg^{-1} at 1 atm. Find out the same coefficient at 1000 atm. It is further given that the bulk modulus is 2.137×10^4 atm at 100 °C.

Soln. : By the problem, $\alpha_1 = 4.33 \times 10^{-4}$ deg^{-1}.

Now the coefficient of compressibility, $\beta = \frac{1}{B}$, when B is bulk modulus.

$$\therefore \; \beta_0 \text{ at } 0\,°\text{C} = \frac{1}{2.032 \times 10^4} \text{ atm}^{-1} = 4.92 \times 10^{-5} \text{ atm}^{-1}$$

$$\text{and } \beta_{100} \text{ at } 100\,°\text{C} = \frac{1}{2.137 \times 10^4} \text{ atm}^{-1} = 4.68 \times 10^{-5} \text{ atm}^{-1}.$$

$$\text{Now, } \alpha = \frac{1}{V}\left(\frac{\partial V}{\partial T}\right)_P \quad (1a)$$

$$\beta = -\frac{1}{V}\left(\frac{\partial V}{\partial P}\right)_T. \quad (1b)$$

Again, $V = f(P, T)$

or, $dV = \left(\frac{\partial V}{\partial P}\right)_T dP + \left(\frac{\partial V}{\partial T}\right)_P dT = (-\beta V)dP + (\alpha V)dT$ [from Eqns. (1a) and (1b)]

Since, V is a state function, dV is perfect differential

Thus, $\left[\frac{\partial(\alpha V)}{\partial P}\right]_T = \left[\frac{\partial(-\beta V)}{\partial T}\right]_P$

or, $\alpha\left(\frac{\partial V}{\partial P}\right)_T + V\left(\frac{\partial \alpha}{\partial P}\right)_T = -\beta\left(\frac{\partial V}{\partial T}\right)_P - V\left(\frac{\partial \beta}{\partial T}\right)_P$

or, $-\alpha\beta V + V\left(\frac{\partial \alpha}{\partial P}\right)_T = -\beta\alpha V - V\left(\frac{\partial \beta}{\partial T}\right)_P$ [from Eqns. (1a) and Eqns. (1b)]

or, $\left(\frac{\partial \alpha}{\partial P}\right)_T = -\left(\frac{\partial \beta}{\partial T}\right)_P$. (2)

Let temperature coefficient of β be x.

Hence, $\left(\frac{\partial \beta}{\partial T}\right)_P = x$

or, $d\beta = x\, dT$ or, $\beta_{100} = \beta_0 + x(T_2 - T_1)$.

Now, $T_2 = 100 + 273 = 373$ K, $T_1 = 273$ K.

$\therefore\ x = \frac{4.68\times10^{-5} - 4.92\times10^{-5}}{373-273}$ atm^{-1}.K^{-1} = -2.4×10^{-8} atm^{-1}. K^{-1}.

Let pressure coefficient of α be y.

Thus, $\left(\frac{\partial \alpha}{\partial P}\right)_T = y$.

Now, from Eqn. (2). $y = -x = 2.4 \times 10^{-8}$ atm^{-1} K^{-1}.

Thus, $\alpha_{1000} = \alpha_1 + 2.4 \times 10^{-8}$ atm^{-1}.K^{-1} $(P_2 - P_1)$.

Given that, $P_2 = 1000$ atm, $P_1 = 1$ atm.

Thus, $\alpha_{1000} = 4.33 \times 10^{-4}$K^{-1} + 2.4×10^{-8} atm^{-1}.K^{-1}(1000 − 1) atm
$= (4.33 \times 10^{-4} + 2.4 \times 10^{-8} \times 999)$K^{-1} = 4.57×10^{-4} K^{-1}

Thus, volume expansion at 1000 atm = 4.57×10^{-4} K^{-1}.

46. A gas enclosed in a cylinder with a moveable piston undergoes reversible adiabatic expansion according to the following equation P^3V^4 = constant.

(a) Find the work done by the system.

(b) System can also reach the final state (B) along any of the three paths ACB, ADB and direct linear AB as shown in the figure. Calculate W and Q along each path.

Soln. : (a) Since the process is adiabatic $Q = 0$.

Given that, P^3V^4 = constant. (1)

Thus, $P = \frac{\text{constant}}{V^{4/3}}$. (2)

Hence, $P_A V_A^{4/3} = P_B V_B^{4/3} = \text{constant} = K.$ (3)

Now, $W_{AB}^{\text{adiabatic}} = -\int_A^B PdV = -K\int_A^B \frac{dV}{V^{4/3}}$

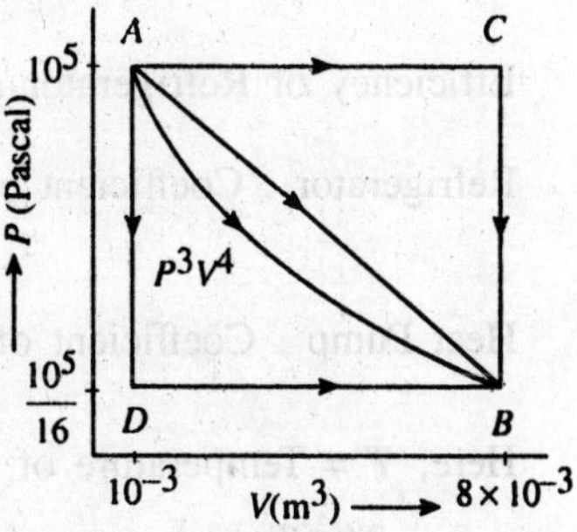

$$= -P_A V_A^{4/3}\left[\frac{V_B^{-1/3}}{-\frac{1}{3}} - \frac{V_A^{-1/3}}{-\frac{1}{3}}\right] \text{Pa.m}^3$$

$$= 3\times10^5\times\left(10^{-3}\right)^{\frac{4}{3}}\left[\left(\frac{1}{8\times10^{-3}}\right)^{\frac{1}{3}} - \left(\frac{1}{10^{-3}}\right)^{\frac{1}{3}}\right]\text{N.m} = -150\text{ J.}$$

(b) For adiabatic work, $\Delta U_{AB} = W_{AB}^{\text{Adiabatic}} = -150\text{ J}$, since $Q_{AB} = 0$.

$W_{ACB} = W_{AC} + W_{CB}$

$W_{AC} = -10^5\ (8\times10^{-3} - 10^{-3})\text{ J} = -7\times10^{-3}\times10^5\text{ J} = -700\text{ J.}$

$W_{CB} = 0$, since $\Delta V = 0$.

Hence, $W_{ACB} = -700\text{ J}$

$Q_{ACB} = \Delta U_{ACB} - W_{ACB}$

$= \Delta U_{AB} - W_{ACB}$ (since U is a state function $\Delta U_{AB} = \Delta U_{ACB}$)

$= (-150 + 700)\text{ J} = 550\text{ J}$

$W_{ADB} = W_{AD} + W_{DB}$

$$= 0 - \frac{10^5}{16}\times10^{-3}\ (8-1)\text{ J} = -7\times10^{-3}\times\frac{10^5}{16}\text{ J} = -43.75\text{ J.}$$

$Q_{ADB} = (-150 + 43.75) = -106.25\text{ J.}$

$$W_{AB}^{\text{linear}} = -\left[\frac{1}{2}\left(10^5 - \frac{10^5}{16}\right)\left(8\times10^{-3} - 10^{-3}\right) + \frac{10^5}{16}\times\left(8\times10^{-3} - 10^{-3}\right)\right]\text{J}$$

$= -371.88\text{ J.}$

$Q_{AB}^{\text{linear}} = \Delta U_{AB} - W_{AB}^{\text{linear}} = (-150 + 371.88)\text{ J} = 221.88\text{ J.}$

Carnot Cycle and Thermochemistry

Required Formulae and Information

1. ΔH_f° is standard heat of formation per mole. For elements in their standard states ΔH_f° is taken as zero at 298.15 K and one bar pressure.

2. For reaction between ideal gases, $\Delta H = \Delta U + \Delta nRT$ at constant temperature.

3. The heat change for the forward reaction is exactly identical in magnitude but opposite in sign to heat change for the reverse reaction.

4. The enthalpy change of a reaction is independent of the number of steps.

5. Efficiency of Carnot engine, $\eta = \frac{-W_{cy}}{Q} = \frac{-W_{cy}}{-W_1} = \frac{W_{cy}}{W_1} = \frac{T-T'}{T}$.

Efficiency of Refrigerator and Heat Pump Refrigerator :

Refrigerator : Coefficient of Performance, $P = \frac{Q'}{W_{cy}} = \frac{-W_3}{W_1+W_3} = \frac{T'}{T-T'}$.

Heat Pump : Coefficient of Performance $P = \frac{-Q}{W_{cy}} = \frac{W_1}{W_1+W_3} = \frac{T}{T-T'}$.

Here, T = Temperature of hot reservoir

T' = Temperature of cold reservoir

$-W_{cy}$ = work done by the engine

W_{cy} = work done upon refrigerator or heat pump

Q = heat absorbed by the engine

Q' = heat absorbed by refrigerator or heat pump

$-Q'$ = heat rejected by engine

$-Q$ = heat rejected by refrigerator or heat pump

Also, for carnot engine $= -W_{cy} = -(W_1 + W_2 + W_3 + W_4)$

$= -(W_1 + W_3) = Q - (-Q') = Q + Q'$.

W_{cy} (or $-W_{cy}$) is written also as W (or $-W$).

6. Clapeyron equation,

$$\frac{dP}{dT} = \frac{\Delta H}{T(V_2 - V_1)} \text{ for phase change } 1 \rightarrow 2,$$

where T is the phase transition temperature.

7. Clausius-Clapeyron equation, $\frac{d\ln P}{dT} = \frac{\text{Le}}{RT^2}$,

where Le = latent heat of evaporation per mole.

8. Kirchhoff equation :

$$\Delta H_2 = \Delta H_1 + \int_1^2 \Delta C_p dT.$$

9. Effect of external pressure on vapour pressure of liquid :

If P_1 and P_2 are the vapour pressures of liquid at external total pressures P_{1t} and P_{2t} respectively.

$$\ln\frac{P_2}{P_1} = \frac{V_m}{RT}(P_{2t} - P_{1t})$$

where V_m is molar volume of liquid at temperature T. If P_1 is the vapour pressure when there is no external pressure then $P_1 = P_{1t}$.

Short Questions

1. We know that heat passes from high to low temperature. How does, then, heat pass isothermally from reservoir to system in Carnot cycle at the high temperature (HTR) and from reservoir to system at low temperature.

Ans. : In case of source (HTR), it is assumed that the source is at temperature T + T while the system is at T. Heat passes from HTR to system reversibly. In case of LTR, is assumed that the source is at $T' - dT'$, so that heat passes from system at T' to LTR.

2. Two systems with heat capacities C_1 and C_2 respectively, interact isothermally and ome to a common temperature T_F. If the initial temperature of system 1 was T_1, what as the initial temperature of system 2? **[PhD Qualification Exam., Wisconsin University]**

Ans. : Let the initial temperature of system 2 be T_2. Assuming that total energy of e combined system remains unchanged, heat released by system 1 is equal to heat absorbed y system 2.

Assuming $T_2 > T_1$,

$$C_1(T_F - T_1) = C_2(T_2 - T_F)$$

$$C_1T_F - C_1T_1 + C_2T_F = C_2T_2.$$

$$\therefore \quad T_2 = \frac{C_1}{C_2}(T_F - T_1) + T_F.$$

3. A compressor designed to compress air is used instead to compress helium. It is und that the compressor overheats. Explain this effect, assuming that the compression approximately adiabatic and starting pressure is the same for both gases.

[PhD Entrance, Wisconsin University]

Soln. : Let the change be adiabatic reversible, then

$$P\left(\frac{V}{V_0}\right)^{\gamma} = P_0,\ \gamma = \frac{C_P}{C_V},$$

and P_0 are final and initial pressures respectively and V and V_0 are final and initial volumes.

Because of compression, $V_0 > V$ and $\gamma_{He} > \gamma_{air}$.

Thus, $P_{He} > P_{air}$

But, $PV = nRT$.

$\therefore \quad T_{He} > T_{air}$.

Same result can also be obtained using T-V relation instead of P-V relation in the lowing way.

$$T_0 V_0^{\gamma_{He}-1} = T_{He} V^{\gamma_{He}-1}$$

and $$T_0 V_0^{\gamma_{air}-1} = T_{air} V^{\gamma_{air}-1}$$

Hence, $$\frac{T_{He}}{T_{air}} = \left(\frac{V_0}{V}\right)^{\gamma_{He}-1}\left(\frac{V}{V_0}\right)^{\gamma_{air}-1}$$

$$= \left(\frac{V_0}{V}\right)^{\gamma_{He}-\gamma_{air}},$$

since $V_0 > V$ and $\gamma_{He} > \gamma_{air}$, $T_{He} > T_{air}$.

4. 10 litres of a gas at atmospheric pressure is compressed isothermally to a volume 1 litre and then allowed to expand adiabatically to 10 litres.

(a) Sketch the process in P-V diagram for a monatomic ideal gas.

(b) Make a similar sketch for a diatomic ideal gas.

(c) Is a net work done on or by the system?

(d) Is it greater or less for diatomic gas? **[PhD Entrance, Wisconsin University]**

Soln. : $1(10\,l, l \text{ atm}) \rightarrow 2(l\,l, P_2)$

Since $1 \rightarrow 2$ is isothermal, $P_1V_1 = P_2V_2$

$\therefore P_2 = 10$ atm.

$2 \rightarrow 3$ is adiabatic. Assuming reversiblity,

$P_2V_2^\gamma = P_3V_3^\gamma$.

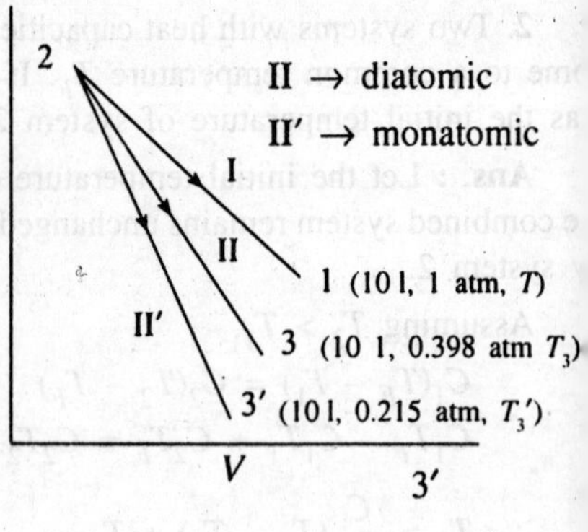

(a) For monatomic gas, $\gamma = \frac{5}{3} = 1.6$.

$$P_3' = P_2\left(\frac{V_2}{V_3}\right)^\gamma = 10\left(\frac{1}{10}\right)^{\frac{5}{3}} = 0.215 \text{ atm.}$$

(b) For a diatomic gas, $\gamma = \frac{7}{3} = 1.4$.

$$P_3 = 10\left(\frac{1}{10}\right)^{1.4} = 0.398 \text{ atm.}$$

Work done on the system in the process I is similar for both gases and is greater than the work done by the gas during processes II and II′. Hence, net work is done on the gas. It is less for diatomic gas since work done during II is greater than the work done during II′.

5. The temperature T K is the temperature of a building which is heated by a heat pump using a river at T_0 K as a source of heat. The heat pump has a perfect performance and consumes power W. The building loses heat at the rate $\alpha(T - T_0)$, where α is a constant.

(a) Show that the equilibrium temperature T_e of the building is given by :

$$T_e = T_0 + \frac{W}{2\alpha}\left[1+\sqrt{\left(1+\frac{4\alpha T_0}{W}\right)}\right].$$

(b) The heat pump is replaced by simple heater which consumes power W(constant) and converts this into heat with 100% efficiency. Show explicitly why this is less desirable than a heat pump. **[PhD Entrance Examination, Columbia University]**

Soln. We know that for a heat pump, $\frac{-Q}{W} = \frac{T}{T-T_0}$.

$$\therefore -Q = \frac{WT}{T-T_0}.$$

By the problem, $-Q = \alpha(T - T_0)$.

Let T_e be the equilibrium temperature of the building.

At equilibrium, $\frac{WT_e}{T_e-T_0} = \alpha(T_e - T_0)$

or, $WT_e = \alpha(T_e^2 - 2T_eT_0 + T_0^2)$ or, $\alpha T_e^2 - (W + 2T_0\alpha)T_e + \alpha T_0^2 = 0$.

$$\therefore T_e = \frac{W + 2T_0\alpha + \sqrt{(W+2T_0\alpha)^2 - 4\alpha^2T_0^2}}{2\alpha}$$

$$\text{or, } T_e = \frac{W}{2\alpha} + T_0 + \frac{1}{2\alpha}\sqrt{W^2 + 4T_0^2\alpha^2 + 4WT_0\alpha - 4\alpha^2 T_0^2}$$

$$= T_0 + \frac{W}{2\alpha} + \frac{1}{2\alpha}\sqrt{W^2 + 4T_0\alpha W}$$

$$\text{Thus, } T_e = T_0 + \frac{W}{2\alpha}\left[1 + \sqrt{\left(1 + \frac{4\alpha T_0}{W}\right)}\right].$$

(b) By the problem, for a heater $-Q = W$.

Now $-Q = \alpha(T - T_0)$.

At equilibrium, $W = \alpha(T_e' - T_0)$

$$T_e' = T_0 + \frac{W}{\alpha}$$

$T_e' < T_e$. Thus consuming same power heat pump keeps the room warmer than ordinary heater.

6. The temperature of a room is T_2. It loses heat to the outside at temperature T_1. The rate is $A(T_2 - T_1)$ where A is a constant. It is warmed by a heat pump operating between T_1 and T_2 ($T_2 > T_1$). The power supplied by the heat pump is $\frac{dW}{dt}$.

(a) What is the maximum rate $\frac{dQ_{max}}{dt}$ at which heat is delivered by the heat pump to the room. Evaluate the gain when T_1 is 275 K and T_2 is 300 K.

(b) Derive the expression for equilibrium temperature of the room T_2 in terms of T_1, A and $\frac{dW}{dt}$. **[PhD Entrance Examination, California University, Berkeley]**

Soln.: We know that for a heat pump,

$$-\frac{Q}{W} = \frac{T_2}{T_2 - T_1}.$$

$$\therefore\ -đQ_{max} = \frac{T_2}{T_2 - T_1}\, đW.$$

But the rate of heat lost to the outside $= -\frac{đQ}{dt} = A(T_2 - T_1)$.

Thus, at equilibrium,

$$A(T_2 - T_1) = \frac{T_2}{T_2 - T_1}\frac{dW}{dt}.$$

$$\text{Hence, } -\frac{đQ_{max}}{dt} = A(T_2 - T_1).$$

Given, $T_2 = 275\text{ K}$, $T_1 = 300\text{ K}$

Thus, $\dfrac{dQ_{max}}{dW} = \dfrac{300}{300-275} = \dfrac{300}{25} = 12.$

12 is the maximum rate at which heat pump delivers heat to the room.

(b) At equilibrium

$$A(T_2 - T_1) = \frac{T_2}{T_2 - T_1}\frac{dW}{dt}, \quad \therefore A(T_2 - T_1)^2 = T_2\frac{dW}{dt}.$$

Let $\dfrac{dW}{dt} = Y.$

$$\therefore A\left(T_2^2 - 2T_1T_2 + T_1^2\right) = T_2.Y \text{ or, } AT_2^2 - (Y + 2T_1A)T_2 + AT_1^2 = 0.$$

$$\therefore T_2 = \frac{(Y+2T_1A) \pm \sqrt{(Y+2T_1A)^2 - 4A^2T_1^2}}{2A}$$

$$= \frac{Y + 2T_1A \pm Y\sqrt{1+\dfrac{4T_1A}{Y}}}{2A}$$

$$= T_1 + \frac{Y}{2A}\left(1 \pm \sqrt{1+\frac{4T_1A}{Y}}\right)$$

Since $T_2 > T_1$ and $\sqrt{1+\dfrac{4T_1A}{Y}} > 1,$

$$T_2 = T_1 + \frac{Y}{2A}\left(1+\sqrt{1+\frac{4T_1A}{Y}}\right).$$

Now, $Y = \dfrac{dW}{dt}.$

$$\text{Thus, } T_2 = T_1 + \frac{1}{2A}\left(\frac{dW}{dt}\right) + \frac{1}{2A}\sqrt{\left(\frac{dW}{dt}\right)^2 + 4AT_1\left(\frac{dW}{dt}\right)}.$$

7. A building at temperature T °K is heated using the heat from surrounding at T_0 K by an ideal heat pump. The building loses heat at a rate of $\alpha(T - T_0)$. The heat pump consumes a power W. Find the equilibrium temperature of the building.

[PhD Entrance Examination, MIT]

Hints : See problem 5.

8. Define efficiency of a heat engine. Derive an expression for the efficiency of a reversible cyclic process (following fig.) with 1 mole of an ideal gas.

Given that the isothermal steps are $A \to B$, $C \to D$ and $E \to F$, adiabatic steps are $B \to C$, $D \to E$, $F \to A$, where V_A, V_B, V_C, V_D etc. are the volumes of the systems at the respective steps. It is also given that $\dfrac{V_A}{V_B} = \dfrac{V_C}{V_D} = \dfrac{V_F}{V_E}.$

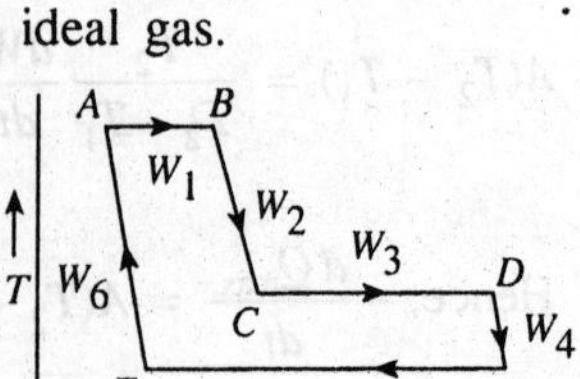

Ans. The efficiency of heat engine is given by $\eta = -\dfrac{W}{Q}$, where W is work done by the engine on the surrounding and Q is the heat absorbed at some high temperature for doing some work in a cycle.

For the heat engine working in the given reversible cycle,

$-W$ = total work done by engine $= -(W_1 + W_2 + W_3 + W_4 + W_5 + W_6)$. (1)

For an ideal gas subject to isothermal processes, $\Delta U = 0$.

Now, $\Delta U = Q + W$,

where Q = heat absorbed by the engine

W = work done upon it.

Thus, for isothermal steps.

$$W_1 = W_{A\to B} = -Q_1 = RT_1 \ln\frac{V_A}{V_B}, \quad (2)$$

$$W_3 = W_{C\to D} = -Q_2 = RT_2 \ln\frac{V_C}{V_D} \quad (3)$$

and $$W_5 = W_{E\to F} = -Q_3 = RT_3 \ln\frac{V_E}{V_F}. \quad (4)$$

Since $Q = 0$ for an adiabatic process,

$W_2 = \Delta U_2,\ W_4 = \Delta U_4,\ W_6 = \Delta U_6.$

Thus, $W_2 = C_V(T_2 - T_1)$, $\quad W_4 = C_V(T_3 - T_2)$, $\quad W_6 = C_V(T_1 - T_3)$.

Hence, $W_2 + W_4 + W_6 = 0$. (5)

Hence, $-W = -(W_1 + W_3 + W_5)$ [From eqns. (1) and (5)]

$$= -\left(RT_1 \ln\frac{V_A}{V_B} + RT_2 \ln\frac{V_C}{V_D} + RT_3 \ln\frac{V_E}{V_F}\right).$$

By the problem, $\dfrac{V_A}{V_B} = \dfrac{V_C}{V_D} = \dfrac{V_F}{V_E}$.

Thus, $W = -R(T_1 + T_2 - T_3) \ln\dfrac{V_A}{V_B}$. (6)

Now heat absorbed $Q = Q_1 + Q_2$.

Thus, $Q = -(W_1 + W_3)$ [From eqns. (2) and (3)]

$$= -\left(RT_1 \ln\frac{V_A}{V_B} + RT_2 \ln\frac{V_C}{V_D}\right) = -R(T_1 + T_2) \ln\frac{V_A}{V_B}.$$

Hence, efficiency $$\eta = \frac{W_{\text{done}}}{Q} = \frac{-W}{-(W_1 + W_3)} = \frac{-R(T_1 + T_2 - T_3)\ln\dfrac{V_A}{V_B}}{-R(T_1 + T_2)\ln\dfrac{V_A}{V_B}}$$

$$= \frac{T_1 + T_2 - T_3}{T_1 + T_2} = 1 - \frac{T_3}{T_1 + T_2}.$$

Numerical Problems

Carnot Cycle

1. A steam engine operates between 14 °C and 30 °C. What is the minimum amount of heat that must be withdrawn to obtain 100 J of work?

Soln. : We know that, $\frac{-W_{cy}}{Q} = \frac{T - T'}{T}$.

By the problem,

$-W = 100$ J, $T' = 14\,°C = 287\,°K$, $T = 30\,°C = 303\,°K$.

The heat that must be withdrawn from high temp. reservoir = Q

$$= -W \times \frac{T}{T - T'} = 100 \times \frac{303}{16} \text{ J} = 1893.75 \text{ J} = \mathbf{1.894\ kJ.}$$

2. An ideal heat engine operating between a source having a temperature 0 °C, and a sink T_2 produces 1000 cal of work per cycle rejecting 6400 cal of heat. Calculate the efficiency of the engine and the temperature of the sink. **[V.U. 1987]**

Soln. : By the problem,

$-Q' = 6400$ cal, $-W_{cy} = 1000$ cal.

$Q = -Q' - W_{cy} = 6400 + 1000 = 7400$ cal.

$T = 273.15\,°K$ (HTR).

$$\therefore \quad \frac{-W_{cy}}{Q} = \frac{T - T'}{T}$$

$$\text{or,} \quad \frac{1000}{7400} = \frac{273.15 - T'}{273.15} = 1 - \frac{T'}{273.15}$$

$$\text{or,} \quad \frac{T'}{273.15} = 1 - \frac{1000}{7400} = \frac{6400}{7400}.$$

$$T' = \frac{64}{74} \times 273.15 = \mathbf{236.2378\ K.}$$

3. Calculate the minimum amount of work in ergs required to freeze 1 gm of water at 0 °C by means of a refrigerator which operates in surrounding at 25 °C. How much heat in calorie is given to the surrounding? Given that latent heat of fusion = 80 cal/gm for ice at 0 °C. What will be change in entropy of the room?

Soln. : For a refrigerator, $\frac{Q'}{W} = \frac{T'}{T - T'}$,

where Q' is the heat absorbed from LTR, W is work done on the refrigerator and Q is the heat rejected at T (25 °C)

By the problem, $Q' = 80$ cal/gm, $T = 25\,°C = 298.15$ K, $T' = 0\,°C = 273.15$ K

$$\therefore \quad \frac{80}{W} = \frac{273.15}{298.15 - 273.15} = \frac{273.15}{25}.$$

$$\therefore W = 80 \text{ cal} \times \frac{25}{273.15} = 7.321984 \text{ cal} \times 4.184 \text{ J/cal}$$

$$= 30.63518106 \text{ J} = 3.0635 \times 10^8 \text{ ergs.}$$

$Q \;= 80 + 7.32 \text{ cal} = 87.32 \text{ cal}.$

$$\Delta S_{room} = \Delta S_{surr} = \frac{Q}{T} = \frac{87.32\,\text{cal}}{298.15\,\text{K}} = \mathbf{0.293\ cal.K^{-1}}.$$

4. An ideal operating Carnot cycle operates on a temperature difference of 100 °C and half of the heat absorbed at high temperature T is wasted as heat at low temperature T'. The cycle does 150 J of work. Find Q, Q', T and T'.

Soln. : By the problem,

$$-Q' = \frac{1}{2}Q, T - T' = \mathbf{100}.$$

$$\therefore \quad \frac{-W}{Q} = \frac{T-T'}{T} = \frac{100}{T}$$

$$\eta \;= \frac{-W}{Q} = \frac{Q-(-Q')}{Q} = \frac{1}{2}$$

$-W = 150\text{ J}.$

$$\therefore \frac{150}{Q} = \frac{1}{2}, \mathbf{Q = 300\ J.}$$

$$\text{Again, } \frac{-W}{Q} = \frac{1}{2} = \frac{100}{T}, \mathbf{T = 200\ K.}$$

$$\therefore 1 - \frac{T'}{200} = \frac{1}{2}$$

$$\frac{1}{2} = \frac{T'}{200}, \therefore \mathbf{T' = 100\ K.}$$

5. An ideal Carnot engine operates on a temperature difference of 250 °C and $\frac{1}{4}$th of heat absorbed at high temperature T is wasted as heat discharged at lower temperature T'. The cycle does 500 J of work. Calculate Q, Q', T and T'

[**Ans.** Q = 666.67 cal, Q' = −166.67 cal, T = 333.333 K, T' = 83.333 K]

6. A reversible heat engine absorbed Q heat at 800 K per cycle rejecting heat Q' at 200 K. Its work output is used to run a hoist and owing to the friction in the pulleys 10% of the work is converted into heat at 200 K. For the system engine + pulley the total heat evolved is 10000 J. Calculate Q, Q' and W.

Soln. : Let heat be absorbed = Q, heat rejected = $-Q'$.

$$\text{Thus, efficiency of the engine } \frac{-W}{Q} = \frac{T-T'}{T} = \frac{800-200}{800} = \frac{600}{800} = 0.75.$$

$\therefore -W = 0.75\ Q.$

10% $W = 0.075Q$, converted into heat.

Now, $Q - (-Q') = -W$.

$-Q' = Q - 0.75Q = 0.25Q.$

Heat released = 0.25 Q + 0.075 Q = 0.325 Q.

But by the problem, 0.325Q = 10000 J

$$Q = \frac{10.000}{0.325} = 30769.23 \text{ J} = \mathbf{30.77 \text{ kJ.}}$$

$$Q' = -0.25Q = -7692.3075 = \mathbf{-7.692 \text{ kJ.}}$$

$$W = -0.75Q = -23076.9225 = \mathbf{-23.08 \text{ kJ.}}$$

7. A reversible heat engine absorbs heat Q at 1000 K per cycle and evolves heat Q' at 350 K. Its work output (W) is used to run a machine. 20% of the work is converted into heat due to friction. For the system engine plus machines, the total heat evolved is 15000 J per cycle. Calculate Q, Q' and W.

[**Ans.** Q = 31250 J, Q' = –10937.5 J, W = –20312.5 J]

8. A heat engine operating between 800 °C and 25 °C produces work (absorbing heat Q at 800 °C) that is entirely used to run a refrigerator operating between 0 °C and 25 °C. Calculate the ratio of the heat absorbed by the pump at 0 °C (Q') to that absorbed by the engine (Q). Assuming ideal operation for both, calculate Q, Q', W and the heat rejected by the heat pump at 25 °C, if Q' heat absorbed at 0 °C leads to conversion of 100 gm of water to ice, C_p = 325 J/gm. How much heat is rejected by the engine at 25 °C?

Soln. : By the problem,

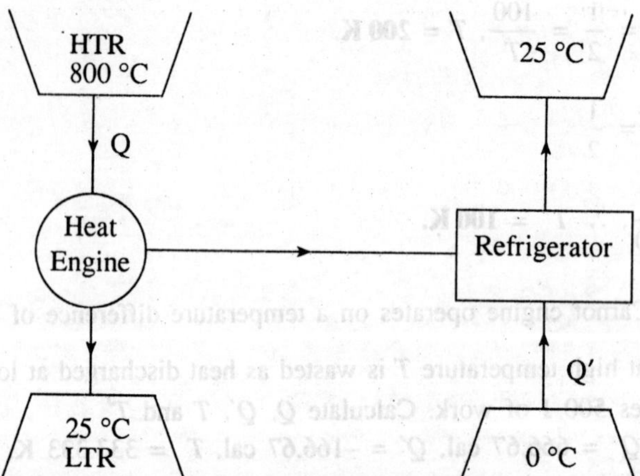

For engine, $\frac{-W}{Q} = \frac{T-T'}{T} = \frac{775}{1075}$ (25 °C = 298 K, 800 °C = 1073 K).

$$\therefore\ -W = Q\frac{775}{1073}.$$

$$\therefore\ W = -Q\frac{775}{1073}$$

For the refrigerator,

$$\frac{Q'}{W} = \frac{T'}{T-T'} = \frac{273}{25}.$$

$\therefore\ Q' = \frac{273}{25}.W = \frac{273}{25} \times \frac{775}{1073} \times Q.$ [since engine runs the refrigerator]

$\therefore \quad \frac{Q'}{Q} = 7.814404432 = 7.81$

or, $Q' = 7.81\ Q$

i.e., $Q = 0.13\ Q'$.

Thus, $\frac{Q}{Q'} = 0.13.$

By the problem, $Q' = 325 \times 100$ J = **3.25 × 10⁴ J.**

Heat rejected to high temperature reservoir

$$= Q' + W$$

$$= 3.25 \times 10^4 \text{ J} + \frac{25 \times 3.25 \times 10^4}{273} \text{ J}$$

$$= 32.5 \text{ kJ} + 2.98 \text{ kJ} = \mathbf{35.48\ kJ.}$$

$$Q = 0.13Q' = 4225 \text{ J} = \mathbf{4.2225\ kJ.}$$

$$W = 2976.19 \text{ J} = \mathbf{2.98\ kJ.}$$

9. An ideal refrigerator works between 0 °C and T °C. It freezes 2 kg of water at 0 °C per hour. At the same time, the total heat output to the room is 200 kcal.h^{-1}. Calculate T °C. Latent heat of fusion of water at 0 °C = 80 cal/gm.

Soln. : 2 kg of water freezes at 0 °C. So Q′ (heat absorbed) = 2 × 80 000 cal = 160 kcal per hour. The refrigerator works between 0 °C and T °C.

Given that Q = 200 000 cal.

Now, $\frac{Q'}{W} = \frac{T'}{T - T'}$

$-Q = Q' + W = 200\ 000$ cal

$W = -Q - Q' = (200\ 000 - 160\ 000)$ cal $= 40\ 000$ cal.

Thus, $\frac{W}{Q'} = \frac{T - T'}{T'} = \frac{T}{T'} - 1.$

$\therefore \quad \frac{40000}{160000} + 1 = \frac{1}{4} + 1 = \frac{T}{T'}.$

$\therefore T = T' \times \frac{5}{4} = 273 \times \frac{5}{4} =$ **341.25 K.**

10. An ideal refrigerator works between 0 °C and a higher temperature T °C. It freezes 3 kg of water at 0 °C per hour (l_f = 80 cal/gm) and leaves 300 kcal per hour at T °C. Calculate T.

Hints : $\frac{W}{Q'} = \frac{T - T'}{T'}$

$T' = 273$ K

$Q' = 240\ 000$ cal

$W = Q - Q' = 60\ 000$ cal

T = **341.25 K.**

11. A heat engine operating between 100 °C and 25 °C absorbs 1000 J work from high temperature reservoir. How much heat is rejected to low temperature reservoir?

Soln. : Given, Q = 1000 J, T = 100 °C = 373.15 K.

T' = 25 °C = 298.15 K

$$\text{Thus, } \eta = \frac{-W}{Q} = \frac{-W}{1000\text{J}} = \frac{T-T'}{T} = \frac{75}{373.15} = 0.201.$$

$\therefore\ -W$ = 1000 J × 0.201 = 201 J.

Hence, heat rejected = $Q' = Q - (-W)$ = (1000 − 201) J = **799 J.**

12. Two idential bodies of constant heat capacity at temperatures T_1 and T_2 respectively are used as reservoir for a heat engine. If they remain at constant pressure, show that the amount of work obtainable is $W = C_p(2T - T_1 - T_2)$, where T is the final temperature attained by both the bodies. Show that when $T = \sqrt{T_1 T_2}$, then W is maximum.

Soln. : The heat engine, working in a cycle, takes $đQ_1$ heat from high temperature reservoir (HTR) at T_1 and rejects $đQ_2$ heat to low temperature reservoir (LTR) at T_2. Eventually temperature of HTR decreases to T and that of LTR increases to T. If work obtained from the engine is W, and Q_1 is the total heat gained by the engine and Q_2' is the heat rejected by the engine, then

$$-W = Q_1 + Q_2 = C_p(T_1 - T) - C_p(T - T_2)$$
$$= C_p(T_1 + T_2 - 2T).$$

Hence, $W = C_P(2T - T_1 - T_2)$.

If the engine is a Carnot engine, then W will be maximum.

For a Carnot engine for one cycle,

$$\frac{đQ_1}{T_1} + \frac{đQ_2}{T_2} = 0.$$

Now, $đQ_1 = C_p dT_1$, and $-đQ_2 = -C_p dT_2$

$$\therefore\quad \frac{dT_1}{T_1} = -\frac{dT_2}{T_2}.$$

$$\text{Now,}\quad \int_{T_1}^{T} \frac{dT_1}{T_1} = -\int_{T_2}^{T} \frac{dT_2}{T_2}$$

$$\text{or,}\quad \ln\frac{T}{T_1} = -\ln\frac{T}{T_2} = \ln\frac{T_2}{T}.$$

$$\therefore\quad T = \sqrt{T_1 T_2}.$$

13. A reversible engine converts $\frac{1}{6}$th heat which it absorbs at heat source into work. When the temperature of the heat sink is reduced by 82 °C, its efficiency is double. Calculate the temperature of the souce and the sink.

Soln. : We know efficiency, $\eta = \frac{-W}{Q} = \frac{T-T'}{T}$.

By the problem, $-W = \frac{1}{6}Q$.

Thus, $\eta = \frac{1}{6} = \frac{T-T'}{T}$. (1a)

$\therefore\ T' = \frac{5T}{6}$, (1b)

when $T'_{new} = T' - (273.15 + 82)\text{K} = T' - 355.15\text{K}$

Then $\eta_{new} = 2\eta = \frac{1}{3}$.

Thus, $\frac{1}{3} = \frac{T-(T'-355.15\text{K})}{T}$ (2a)

or, $T = 3T - 3T' + 1065.45\text{K} = 3 \times \frac{5T}{6} - 1065.45$ K [from eqn. (1b)]

or, $0.5T = 1065.45$ K $\therefore\ T = 2130.9$ K. (2b)

From eqn. (1a), $\frac{1}{6} = \frac{2130.9\text{K} - T'}{2130.9\text{K}}$

or, $1 - \frac{T'}{2130.9\text{K}} = \frac{1}{6}$

or, $\frac{T'}{2130.9\text{K}} = \frac{5}{6}$.

$\therefore\ T' = 1775.95$ K.

Thus, temperature of the source = 2130.9 K and temperature of the sink = 1775.95 K

14. 1 mol monatomic gas is used in a Carnot engine working between 700 K and 350 K. The smallest volume occupied by the gas during the course of operation is 500 ml and it absorbs 910 J heat per stroke.

(a) Calculate W, Q, ΔU, ΔS for each step in Carnot cycle.

(b) Calculate W_{cy}, Q_{cy}, ΔU_{cy}, ΔS_{cy}

Soln. : Efficiency of a Carnot engine irrespective of nature of the working material and its amount,

$= \eta = \frac{-W}{Q} = \frac{T-T'}{T}$. (1)

Given : $T = 700$ K, $T' = 350$ K, $Q = 910$ J, $n = 1$ mol.

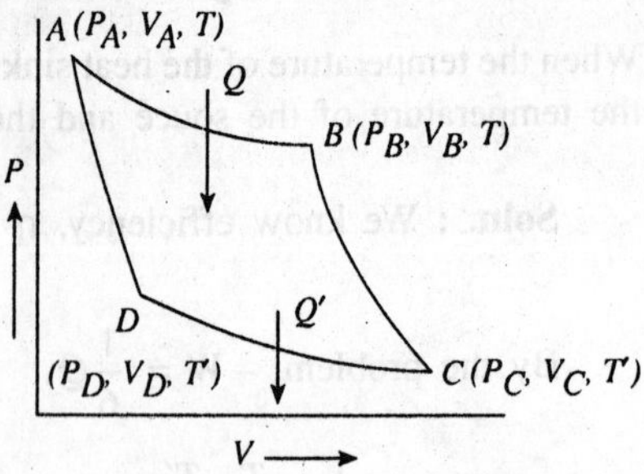

Thus, $\eta = \dfrac{700-350}{700} = 0.5.$ (2)

(a) Step A → Step B,

Assuming the gas to be ideal, $\Delta U = 0$.

Hence, $-W_1 = Q = 910$ J (3)

$\Delta S_1 = \dfrac{910}{700}$ J.k^{-1} = 1.3Jk^{-1}

(b) Step B → Step C

$Q = 0$,

$\Delta U_1 = W_2 = nC_V(T' - T)$

$= \dfrac{3}{2}R(350 - 700)\text{K}$ ($\because C_V = \dfrac{3}{2}R$ for a monatomic gas)

$= \dfrac{3}{2} \times 8.314 \times (-350)$ J = −4364.85 J = −4.364 kJ. (4)

$\Delta S = 0$ (since the process is reversible adiabatic)

Step C → Step D, $\Delta U = 0$, $W_3 = -Q'$

From eqn. (1), $-W = 0.5 \times 910$ J = 455 J. (5)

$-W = Q + Q' = 455$ J ($\because Q - (-Q') = Q + Q'$)

$Q = 910$ J

Thus, $-Q' = (910 - 455)$J = 455J (6)

Hence $W_3 = 455$ J, $\Delta S_2 = \dfrac{-455}{350}$ J.K^{-1} = −1.3 JK^{-1}

Step D → Step A

Q = 0

$\Delta U_2 = W_4 = nC_V(T - T') = 4.364$ kJ (7)

$\Delta S = 0$.

(b) From eqns. (5) and (6), $W_{cy} = -455$ J

$Q_{cy} = 455$ J, $\Delta U_{cy} = 0$, $\Delta S_{cy} = 0$.

15. An air conditioner is nothing but a refrigerator in which work done upon it and heat removed from the cooler inside which is rejected at the warmer outside.

Such an air conditioner operating on a Carnot cycle (which is reversible) taking heat from the inside at temperature T_2 and rejecting heat to the outside at temperature T_1 consumes W J/sec from the power line.

(a) The air conditioner absorbed Q_2 (J) from the inside and rejected Q_1 at the outdoor. Derive a formula for the term $\dfrac{Q_2}{W}$ as a function of T_1 and T_2 ($T_1 > T_2$).

(b) Heat leakage into the house follows the equation $Q = A(T_1 - T_2)$. Develop a formula for T_2 as a function of T_1, W and A for the continuous operation of the conditioner under constant outside temperature T_1, and T_2 is the inside temperature (uniform)

(c) The air conditioner works between room temperature 293 K and the outside temperature 303 K. The machine operates only 30% of the time. Find the highest outside temperature (H.T.R.) for which it can maintain 293 K temperature.

(d) In the winter, the cycle is reversed and the system becomes a heat pump that absorbs heat from outside and releases heat into the inside, find the lowest temperature in K for which it can maintain 293K inside. ($T_2 > T_1$)

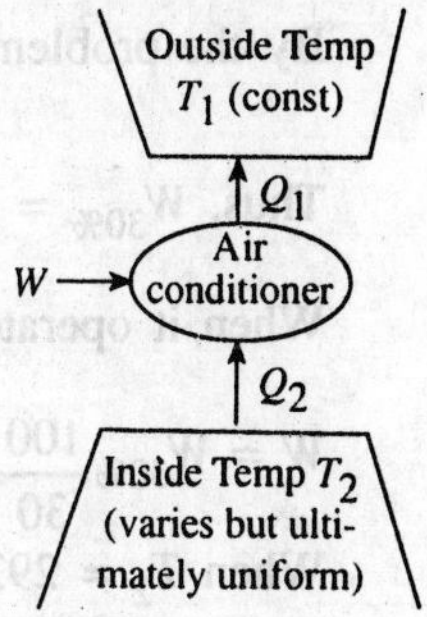

Soln. : (a) According to the problem, $Q_2 + W = -Q_1$. (1)

Since air conditioner works on a cyclic process, $\Delta S_{sys} = 0$. (2)

Since the cycle is reversible,

$$\Delta S_1 = \frac{Q_2}{T_2}, \ \Delta S_2 = -\frac{Q_1}{T_1}. \quad (3)$$

From eqn. (2) $\Delta S_1 = \Delta S_2$.

Hence, $-\dfrac{Q_1}{T_1} = \dfrac{Q_2}{T_2}$

or, $-\dfrac{Q_1}{Q_2} = \dfrac{T_1}{T_2}$ or, $\dfrac{Q_2 + W}{Q_2} = \dfrac{T_1}{T_2}$ [From eqn. (1)]

or, $\dfrac{W}{Q_2} = \dfrac{T_1}{T_2} - 1 = \dfrac{T_1 - T_2}{T_2}$ or, $\dfrac{Q_2}{W} = \dfrac{T_2}{T_1 - T_2}$.

(b) Now $Q_2 = W\left(\dfrac{T_2}{T_1 - T_2}\right) = A(T_1 - T_2)$ [at equilibrium] (4)

Hence, $T_2 W = A(T_1 - T_2)^2$

or, $T_2 W = A(T_1^2 - 2T_1T_2 + T_2^2)$

or, $AT_2^2 - (W + 2AT_1)T_2 + AT_1^2 = 0$. (5)

$$\text{Thus, } T_2 = \frac{(W + 2AT_1) \pm \sqrt{(W + 2AT_1)^2 - 4A^2T_1^2}}{2A}$$

$$= \frac{W}{2A} + \frac{2AT_1}{2A} \pm \frac{\sqrt{W^2 + 4WAT_1 + 4A^2T_1^2 - 4A^2T_1^2}}{2A}$$

$$= \frac{W}{2A} + T_1 \pm \frac{\sqrt{W^2 + 4T_1WA}}{2A}$$

$$= T_1 + \frac{1}{2}\left[\frac{W}{A} \pm \frac{1}{A}\left(\sqrt{W^2 + 4T_1WA}\right)\right]$$

$$= T_1 + \frac{1}{2}\left[\frac{W}{A} \pm \left(\sqrt{\frac{W^2}{A^2} + \frac{4T_1W}{A}}\right)\right].$$

But $T_1 > T_2$.

$$\text{Thus, } T_2 = T_1 + \frac{1}{2}\left[\frac{W}{A} - \sqrt{\left(\frac{W}{A}\right)^2 + \frac{4T_1W}{A}}\right].$$

(c) When the machine is 30% efficient then $W_{30\%} = \dfrac{A(T_1 - T_2)^2}{T_2}$

By the problem, $T_1 = 303$ K; $T_2 = 293$ K

Thus, $W_{30\%} = \dfrac{A(10^2)}{293} = \dfrac{100}{293}A.$

When it operates continuously, i.e., 100% of the time, then

$W = W_{30\%}\dfrac{100}{30} = \dfrac{100}{293}A \times \dfrac{100}{30} = 1.1377A.$

When $T_2 = 293$ K,

we solve equation for T_1 in terms of T_2, W and A.

From Equation (5),

$AT_1^2 - 2AT_1T_2 - (WT_2 - AT_2^2) = 0$

$T_1^2 - 2T_1T_2 - \left(\dfrac{WT_2}{A} - T_2^2\right) = 0$

$$T_1 = \frac{2T_2 \pm \sqrt{(2T_2)^2 + 4\left(\dfrac{WT_2}{A} - T_2^2\right)}}{2} = \frac{2T_2 \pm \sqrt{\dfrac{4WT_2}{A}}}{2}$$

$$T_1 = T_2 \pm \sqrt{\frac{WT_2}{A}}. \tag{6}$$

Since $T_1 > T_2$,

$\therefore\ T_1 = T_2 + \sqrt{W.\dfrac{T_2}{A}}.$

When $T_2 = 293$ K, we get $T_1 = T_2 + \sqrt{W.\dfrac{T_2}{A}} = 293 + \sqrt{1.1377A \times \dfrac{293}{A}}$

$= 293 + \sqrt{1.1377 \times 293}$

$= 293 + 18.257 = (293 + 18.26)\text{ K} = 311.26\text{ K} = 38.26\ °\text{C}.$

(d) Hence, $Q_1 + W = -Q_2$.

Thus, $W = -Q_2 - Q_1$.

As a heat pump the coefficient of performance,

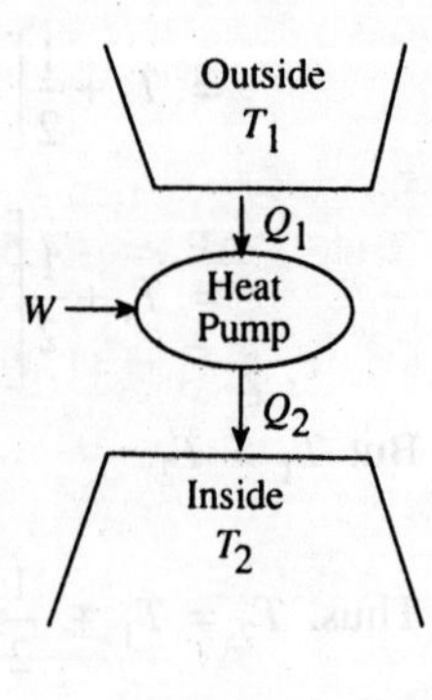

$P = \dfrac{-Q_2}{W} = \dfrac{T_2}{T_2 - T_1}.$

Solving eqn. (3) for T_1, we get

$$T_1 = \frac{2AT_2 \pm \sqrt{4A^2T_2^2 - 4A(AT_2^2 - WT_2)}}{2A}$$

$$= T_2 \pm \sqrt{T_2\frac{W}{A}}.$$

Since $T_2 > T_1$,

$$T_2 = T_1 + \sqrt{T_2 \frac{W}{A}}. \quad (7)$$

Hence, $T_1 = T_2 - \sqrt{\frac{W}{A} T_2}$

$$= 293 - \sqrt{1.1377 \times 293} = 293 - 18.26 = 274.74\text{ K} = 1.74\text{ °C}.$$

16. With an intake temperature at 300 °C and exhaust temperature of 100 °C in a Carnot cycle in operation is a steam turbine. What is maximum work the turbine can do for a given heat intake Q.

Soln. : For a Carnot cycle with heat absorbed Q at higher temperature T and heat rejected Q' at lower temperature T' the efficiency is $\eta = \frac{-W}{Q} = \frac{T - T'}{T}$.

By the probllem, $T = 300\text{ °C} = 573.15\text{ K}$

$T' = 100\text{ °C} = 373.15\text{ K}$

$-W = Q + Q'$

$(Q + Q')$ is equal to $\left(1 - \frac{T'}{T}\right)Q$.

$$-W_{max} = \left(1 - \frac{T'}{T}\right)Q = \left(1 - \frac{373.15}{573.15}\right)Q = 0.35Q.$$

It is the maximum work a turbine could do since it is working in a reversible cycle. The work will be less than $0.35Q$ if the turbine operates irreversibly.

Clausius-Clapeyron Equation

1. The pressure at triple point of water is 4.56 mm of Hg. Under this condition the volume of liquid water per gram is 1.0001 cc and that of ice is 1.0906 cc. Find out the temperature of the triple point. $l_f = 80$ cal/gm. **[C.U. 1968]**

Soln. : We know, $\Delta T = \frac{T(V_2 - V_1)}{\Delta H} \Delta P$ [Clapeyron equation]

Let the change from $l \rightarrow s$.

$$\therefore \quad \Delta T = \frac{T(V_s - V_1)}{\Delta H} \Delta P.$$

But, $\Delta P = -(4.56 - 760)$ mm of Hg $= -\frac{755.44}{760}$ atm.

$\therefore V_s - V_1 = (1.0906 - 1.0001)$ cc

$= 0.0905$ cc/gm $= 0.0905 \times 10^{-3}$ litre/gm

for $l \rightarrow s$, $l_f = -80$ cal/gm.

$$\therefore \quad \Delta T = \frac{273.15\text{ K} \times 0.0905 \times 10^{-3}\text{ litre. gm}^{-1}}{-80\text{ cal. gm}^{-1}} \times \frac{(-755.44)}{760}\text{atm} \times 24.21 \frac{\text{cal}}{\text{litre. atm}}$$

$= 0.0074$ °C .

Thus, for a decrease of pressure from 760 mm of Hg to 4.58 mm of Hg, freezing point of water increases from 0 °C by 0.0074 °C. Hence, temperature of triple point = **0.0074 °C**. This may also be calculated as follows :

$$\frac{dP}{dT} = \frac{\Delta H}{T\Delta V} = \frac{-80\,\text{cal.gm}^{-1}}{0.0905\times10^{-3}\,\text{litre.gm}^{-1}\times T} = -883977.9006\ \text{cal.l}^{-1} \times \frac{1}{T}$$

$$\text{or, } dP = -883977.9\frac{dT}{T}$$

$$\text{or, } P_2 - P_1 = -883977.9 \ln\frac{T_2}{T_1}$$

$$\text{or, } \ln\frac{T_2}{T_1} = \frac{755.44\times24.21}{760\times883977.9}.$$

$\therefore\ T_2 = 273.1574\ \text{K} = $ **0.0074 °C.**

2. The vapour pressure of water changes by 27.17 mm from 99.5 °C to 100.5 °C. The specific volume of water and steam are 1.04 cc/gm and 1674 cc/gm respectively. Calculate l_e.

Soln. : $V_l = 1.04$ cc/gm

$V_g = 1674$ cc/gm

$$\Delta T = \frac{T(V_2 - V_1)}{\Delta H}\Delta P \qquad \text{[Clapeyron equation; for } l \to g]$$

By the problem, $\Delta T = (100.5 - 99.5)°\text{C} = 1°\text{C} = 1°\text{K}$

$\Delta P = 27.17$ mm

$T = 373.15$ K.

$$\therefore \Delta H = \frac{T(V_2 - V_1)}{\Delta T}\Delta P = \frac{373.15\times(1674-1.04)\text{cc.gm}^{-1}}{1\text{K}}27.17\ \text{mm}$$

$$= \frac{373.15\times1672.96\times27.17}{1\times10^3\times760}\text{litre-atm.gm}^{-1}$$

$= 2261.3\ \text{J.gm}^{-1}$ [since 1 litre.atm = 101.325J] = 2.26 kJ.gm^{-1}

3. L_e for H_2O is 540 cal/gm at 100 °C. At what temperature will water boil at 800 mm of Hg?

Hints : Here, $L_e = 540$ cal/gm

$T_1 = 373$ K, $P_1 = 1$ atm

$$P_2 = 800\ \text{mm} = \frac{800}{760}\ \text{atm}$$

$T_2 = ?$

$$\ln\frac{P_2}{P_1} = \frac{L_e}{R}\left[\frac{T_2 - T_1}{T_1T_2}\right]$$

$$\text{or,}\quad \ln\frac{800}{760} = \frac{540\times18}{1.987}\left[\frac{T_2 - 373.15}{T_2\times373.15}\right]$$

$$T_2 = 374.62 \text{ K}$$

$$\therefore \quad T_2 = 101.62 \text{ °C.}$$

4. Given that $\frac{dP}{dT}$ is 2.72 cm of Hg per degree for water at 100 °C. Calculate the approximate heat the vaporisation in cal/gm at this temperature. Calculate ΔV.

Hints : $\frac{dP}{dT} = \frac{L_e}{T.\Delta V}$

or, $\frac{1}{P}\frac{dP}{dT} = \frac{L_e}{RT^2}$ $[V_g >> V_l]$

$$\therefore L_e = \frac{RT^2}{P}.\frac{dP}{dT}$$

$$= \frac{8.314\,\text{J.K}^{-1}.\text{mole}^{-1} \times (373.15)^2 \text{K}^2}{1\,\text{atm}} \times \frac{2.72\,\text{atm.K}^{-1}}{76} = \mathbf{41431.65\ J.mol^{-1}.}$$

$$\therefore L_e = 2301.7583 \text{ J.gm}^{-1}$$

$$= \mathbf{550.133\ cal.gm^{-1}.}$$

ΔV Calculation

Again, from Claperyon equation,

$$\frac{\Delta P}{dT} = \frac{L_e}{T.\Delta V}$$

$$\text{or,} \quad \Delta V = \frac{L_e}{T\frac{dP}{dT}} = \frac{41431.65\text{ J.mol}^{-1}}{373.15\text{ K} \times \frac{2.72}{76}\text{ atm.K}^{-1} \times 101.325\text{ J/l.atm}}$$

$$= 30.6 \text{ litres/mol} = \mathbf{1.7\ litre/gm.}$$

5. The latent heat of fusion of deuterium at 18.56 K is 47.0 cal/gm-mole. The slope $\frac{dP}{dT}$ is 40.5 kg/cm^2.deg^{-1}. Calculate the difference between molar volume of solid and liquid deuterium at this temperature. **[C.U. 1981]**

Soln. : From Clapeyron equation,

$$\frac{dP}{dT} = \frac{L_e}{T.\Delta V}$$

$$\text{or,} \quad V_l - V_s = \Delta V = \frac{L_e}{T\frac{dP}{dT}} = \frac{47\,\text{cal.mole}^{-1} \times 4.184\,\text{J/cal}}{18.56\,\text{K} \times 40.5 \times 9.80665\,\text{m.sec}^{-2} \times 10^4\,\text{kg.m}^{-2}}$$

$$= \frac{196.648\,\text{N.m.mole}^{-1}}{73714626.7\,\text{kg.m}^{-1}.\text{sec}^2} = 2.668 \times 10^{-6}\ \text{m}^3.\text{mole}^{-1} = \mathbf{2.667\ cc.mol^{-1}.}$$

6. The latent heat of evaporation of water is about 2.25 × 10^6 J/kg and the vapour density is 0.0598 kg.m^{-3}. Find the rate of change of boiling point with altitude near sea level in °C per km. The air temperature may be assumed to remain constant at 300 K.

Soln. : By Clausius-Clapeyron equation,

$$\ln\frac{P_2}{P_1} = \frac{L_e}{R}\frac{T_2 - T_1}{T_1 T_2}.$$

According to barometric formula, $P_2 = P_1 e^{-\frac{Mgh}{RT}}$.

$$\therefore \ln\frac{P_2}{P_1} = -\frac{Mgh}{RT}.$$

By the problem, $T = 300$ K, $M = 29.04$ (mol. wt of air), $h = 10^5$ cm,

$$L_e = 2.25 \times 10^6 \text{ J/kg} = \frac{2.25\times10^6\times10^7\times18}{10^3} \text{ erg/mole}$$

$$-\frac{Mgh}{RT} = \frac{2.25\times10^6\times10^7\times18}{10^3\times R}\left(\frac{1}{373.15}-\frac{1}{T_2}\right).$$

$$\text{or, } -\frac{29.04\times980.665\times10^5}{R\times300} = \frac{2.25\times10^{11}\times18}{10^3\times R}\left(\frac{1}{373.15}-\frac{1}{T_2}\right).$$

$\therefore T_2 = 369.91$°K $= 96.76$ °C.

$\therefore \Delta T = T_2 - T_1 = 96.76 - 100 = -3.24$ °C.

$\therefore$ rate of change of boiling point = **–3.24 °C per km.**

7. For $MgSO_4$, $7H_2O$ the dissociation pressure is 35.6 mm of Hg at 35 °C and 47.2 mm of Hg at 40 °C. Calculate ΔH for $MgSO_4$, $7H_2O = MgSO_4 + 7H_2O$.

Soln. : $\ln\dfrac{P_2}{P_1} = \dfrac{\Delta H}{R}\left[\dfrac{T_2 - T_1}{T_1 T_2}\right]$

$$\text{or, } \quad \ln\frac{47.2}{35.6} = \frac{\Delta H}{8.314}\left[\frac{5}{313\times308}\right]$$

$\therefore \quad \Delta H$ = **45.212 kJ.**

But if the reaction occurs producing H_2O as a gas,

$MgSO_4$, $7H_2O = MgSO_4 + 7H_2O(g)$

$K_d = P^7_{H_2O}$

$\Delta H = 7 \times 45.212$ kJ = **316.5 kJ.**

8. The vapour pressure of liquid fluorine expressed in mm of Hg is reproduced by the equation $\log_{10} P = 70.1 - \dfrac{350.6}{T}$. Calculate the molar heat of evaporation of fluorine.

Soln. : $\log_{10} P = 7.01 - \dfrac{350.6}{T}$

$\therefore \quad \log_{10} P_1 = 7.01 - \dfrac{350.6}{T_1}.$

And $\log_{10} P_2 = 7.01 - \dfrac{350.6}{T_2}.$

Putting P_1 = 100, $\log_{10}100 = 7.01 - \dfrac{350.6}{T_1}$ (i)

and P_2 = 200; $\log_{10}200 = 7.01 - \dfrac{350.6}{T_2}$. (ii)

Solving (i) and (ii), we get $T_1 = 69.98$; $T_2 = 74.45$.

Now, P_1 = 100 mm of Hg, when T_1 = 69.98 K

P_2 = 200 mm of Hg, when T_2 = 74.45 K.

From Clausius-Clapeyron equation,

$$\ln\frac{P_2}{P_1} = \frac{Le}{R}\left[\frac{T_2 - T_1}{T_1T_2}\right]$$

or, $\ln 2 = \dfrac{Le}{8.314}\times\left[\dfrac{74.45-66.98}{69.98\times74.45}\right]$, $\therefore$ Le = **6716.86 J.mole^{-1} = 6.72 kJ.mole^{-1}**.

9. The vapour pressure of a solid selenium is given by $\log P(\text{mm}) = -\dfrac{7440}{T} + 12.72$ and of liquid selenium by $\log P\,(\text{mm}) = -\dfrac{5390}{T} + 8.63$. Determine the triple point temperature of selenium. **[C.U. 2003]**

Soln. : At triple point,

$$-\frac{7440}{T_P} + 12.78 = -\frac{5390}{T_P} + 8.63.$$

$$\therefore\quad -\frac{7440}{T_P} + \frac{7440}{T_P} = 8.63 - 12.78.$$

$\therefore$ T_P = 493.97 K = **220.82 °C.**

10. The vapour pressure of *n*-butyl alcohol is given by the equation $\log P(\text{mm}) = -2443.0/T + 9.136$. Calculate the heat of vaporization per mole of the liquid at the normal boiling point, 117 °C. **[C.U. 2006]**

Soln. : According to Clausius-Clapeyron equation for l → v transition,

$$\frac{d\ln P}{dT} = \frac{\Delta H}{RT^2}.$$

$$\therefore\ \ln P = -\frac{\Delta H}{RT} + \text{constant}$$

$$\text{or, } \log P = -\frac{\Delta H}{2.303RT} + \frac{\text{const.}}{2.303}.$$

By the problem, $\dfrac{\Delta H}{2.303\times R} = 2443.$

$\therefore$ $\Delta H = 2.303 \times 2443 \times 8.314$ = 46776.47 J.mole^{-1}

= **46.78 kJ.mole^{-1}.**

11. Calculate the change in pressure required to change the freezing point of water by 1°C. At 0 °C the heat of fusion of ice is 335.5 $J.gm^{-1}$. The density of water is 0.9998 $gm.cm^{-3}$ and the density of ice is 0.9168 $gm.cm^{-3}$. **[V.U. 2003]**

Hints : $\frac{dP}{dT} = \frac{\Delta H}{T(V_w - V_{ice})}$

dT = 1°C, ΔH = 335.5 J/gm = $\frac{335.5}{101.325}$ $l.atm.gm^{-1}$

T = 273.15K, ρ_w = 0.9998 gm/cc = 0.9998 × 10^3gm/l

ρ_{ice} = 0.9168 gm/cc = 0.9168 × 10^3 gm/l

$$\therefore dP = \frac{335.5}{101.325 \times 273.15 \times \left(\frac{1}{\rho_w} - \frac{1}{\rho_{ice}}\right)} = \mathbf{-133.87\ atm.}$$

Thermochemistry

1. Given the following heats of reaction at 25 °C,

$C_2H_4(g) + 3O_2(g) = 2CO_2(g) + 2H_2O(l)$ ΔH = –337.3 kcal.

$2H_2(g) + O_2(g) = 2H_2O(l)$; ΔH = – 136.6 k cal.

$2C_2H_6(g) + 7O_2(g) = 4CO_2(g) + 6H_2O(l)$; ΔH = –745.6 kcal.

Calculate the ΔH for the reaction,

$C_2H_4(g) + H_2(g) = C_2H_6(g)$ at 25 °C.

Soln. : By the problem,

$2C_2H_4(g) + 6O_2(g) = 4CO_2(g) + 4H_2O(l)$; ΔH = –674.6 kcal

$2H_2(g) + O_2(g) = 2H_2O(l)$; ΔH = –136.6 kcal

$4CO_2(g) + 6H_2O(l) = 2C_2H_6(g) + 7O_2(g)$; ΔH = 745.6 kcal.

Adding : $2C_2H_4(g) + 2H_2(g) = 2C_2H_6(g)$; ΔH = 65.6 kcal.

$\therefore C_2H_4(g) + H_2(g) = C_2H_6(g)$, $\therefore \Delta H$ = 32.8 kcal.

Hence, the heat of hydrogenation of C_2H_4 is +32.8 kcal, i.e., 137.24 kJ.

2. Calculate the heat of formation of cane sugar from the following data :

(a) $C + O_2 = CO_2$; ΔH = –94 000 cal

(b) $2H_2 + O_2 = 2H_2O$; ΔH = –136 000 cal

(c) $C_{12}H_{22}O_{11} + 12O_2 = 12CO_2 + 11H_2O$; ΔH = –1 350 000 cal.

Soln. : By the problem,

$12C(s) + 12O_2(g) = 12CO_2(g)$; ΔH = –1128 kcal.

$\frac{11}{2} \times [2H_2(g) + O_2(g)] = 11H_2O(l)$; ΔH = –748 kcal.

$12CO_2(g) + 11H_2O(l) = C_{12}H_{22}O_{11} + 12O_2(g)$; ΔH = 1350 kcal.

Adding, $12C(s) + 11H_2(g) + \frac{11}{2}O_2(g) = C_{12}H_{22}O_{11}$; $\Delta H = -526$ kcal.

Hence, the heat of formation of cane sugar is – 526 kcal = **–2200.78 kJ.**

3. Heat of solutions of 1 mole of KCl in 200 moles of water under 1 atm pressure is 4339 cal at 21 °C and 4260 cal at 23 °C. Determine the heat of solution at 25 °C. What are the assumptions needed?

Soln. : If ΔH_2 and ΔH_1 are heat of reaction at T_2 and T_1 respectively and ΔC_P is difference between heat capacities of products and reactants, by Kirchhoff equation,

$$\Delta H_2 = \Delta H_1 + \int_1^2 \Delta C_P dT.$$

$$\therefore \quad \Delta H_2 = \Delta H_1 + \Delta C_P (T_2 - T_1).$$

Here, basic assumption is that ΔC_P is constant of temperature.

By the problem, $\Delta H_2 = 4339$ cal, $T_2 = 21\ °C = 294.15$ K

$\Delta H_1 = 4260$ cal, $T_1 = 23\ °C = 296.15$ K.

$\therefore \Delta C_P = -39.5$ cal.K^{-1}.

Hence, ΔH at 25 °C $= \Delta H_{298.15} = \Delta H_{296.15} + \Delta C_P(298.15 - 296.15)$ K

$= 4260$ cal $- 39.5$ cal. K$^{-1} \times 2 =$ **4181 cal.**

4. Both cis- and trans-azobenzene can be reduced to aniline by $SnCl_2$ and HCl. When 400 ml of ethanolic of 0.1995(M) $SnCl_2$ and 1.085(M) HCl are added to 100 ml of 0.2426(M) trans-azobenzene in a calorimeter, the temperature rise was 5.405 °C. When a similar treatment was given to 100 ml of 0.2423 (M) cis-azobenzene in the same calorimeter, the temperature was found to rise by 6.38 °C. The heat capacities of the solution were determined by electrical heating. Passage of 0.69 A at 48.0 volts for 190 sec produced a rise of 4.65 °C. Calculate the heat of conversion of cis- to trans-azobenzene.

Soln. : By the problem, if x moles of cis- or trans-azobenzene in 100 cc are added to 400 ml of ethanolic solution of 0.1995 (M) $SnCl_2$ and 1.085 (M) HCl, then in both the cases the final product is aniline. The given solutions of $SnCl_2$ and HCl are heated by 0.69 A current at 48 volts for 190 sec to lead to a temperature rise of 4.65 °C. So, water equivalent of the solution is

$\frac{Q}{\Delta T}$, where ΔT = rise in temperature

Q = heat supplied

$= iVt$

$= 0.69$ amp $\times$ 48 volts $\times$ 190 sec = 6292.8 J.

$$\therefore \text{ water equivalent} = \frac{6292.8}{4.65} = 1353.29 \text{ J.K}^{-1}.$$

Again, heat generated is (1353.29 × 5.405) when 100 cc of 0.02426 moles of trans-azobenzene is added. [100 ml 0.2426 molar = 0.02426 moles]

Hence, ΔH for conversion of trans-azobenzene to aniline $= -\frac{1353.29 \times 5.405}{0.02426}$

$= -301505.87$ J.mol^{-1}.

Thus, $\Delta H_{trans} = 301505.87$ J.mol^{-1}.

Similarly, $\Delta H_{cis} = \dfrac{1353.29 \times 6.38}{0.02423} = 356334.72 \text{ J.mol}^{-1}$.

$\therefore$ heat of conversion from cis- to trans-azobenzene

$= \Delta H_{cis \to trans}$

$= \Delta H_{trans} - \Delta H_{cis}$

$= (301505.87 - 356334.72) \text{ J.mol}^{-1}$

$= -54828.85 \text{ J.mol}^{-1}$

$= \mathbf{-54.83 \text{ kJ.mol}^{-1}}$.

5. A solution of composition $CuCl_2.\ 10H_2O$ is diluted with *a* mol of water, the hea generated is given by $\dfrac{a \times 5023}{a + 21.24}$. The heat of solution of $CuCl_2$ in 600 moles of wate is –11.08 kcal. Find the heat of solution of anhydrous $CuCl_2$ in 10 mol of water. How much heat will be evolved if a solution of $CuCl_2.600H_2O$ be infinitely diluted?

Soln. : By the problem, if $CuCl_2.10H_2O$ is converted to $CuCl_2.600H_2O$ by additio of 590 moles of water, the heat change is $Q = -\dfrac{590 \times 5023}{590 + 21.24} = -4.848$ kcal.

Again, $CuCl_2 + 600H_2O = CuCl_2.600H_2O;\ \Delta H = -11.08$ kcal; (i)

$CuCl_2.10H_2O + 590H_2O = CuCl_2.600H_2O;\ \Delta H = -4.848$ kcal. (ii)

$\therefore$ $CuCl_2.600H_2O = CuCl_2.10H_2O + 590H_2O;\ \Delta H = 4.848$ kcal. (iii)

And adding (iii) and (i), we get

$CuCl_2 + 10H_2O = CuCl_2.10H_2O;\ \Delta H. = (-11.08 + 4.848) \text{ kcal} = -6.232$ kca

$= -26.07$ kJ.

$CuCl_2.10H_2O + 590H_2O = CuCl_2.600H_2O,\ \Delta H = -4848$ cal.

But, $CuCl_2.10H_2O$ + aq. (= infinity water) = $CuCl_2.aq.\Delta H = -5023$ cal.

Since, when a = infinity, $Q = -\dfrac{a \times 5023}{a + 21.24} = -\dfrac{a \times 5023}{a} = -5023$ cal.

$\therefore$ $CuCl_2.600H_2O$ + aq. = $CuCl_2$.aq; $\Delta H = (-5023 + 4848) = -175 \text{ cal} = \mathbf{-732.2}$ **J**

6. Calculate the heat of formation of $AlCl_3$ from the following data :

			ΔH
$2Al + 6HCl\ (aq.)$	=	$2\ AlCl_3.aq + 3H_2$	–1003.2 kJ
$H_2 + Cl_2$	=	$2HCl$	–184.0 kJ
$HCl + aq$	=	$HCl.aq$	–72.40 kJ
$AlCl_3 + aq$	=	$AlCl_3.aq$	–321.5 kJ

[Vikram Univ. 1978

Soln. : By the problem—

			ΔH in kJ
$2Al + 6HCl(aq.)$	=	$2AlCl_3(aq.) + 3H_2$	–1003.2
$3H_2 + 3Cl_2$	=	$6HCl$	–(184 × 3) = –552

$6HCl + aq.$	$= 6HCl(aq)$	$-72.4 \times 6 = -434.4$
$2AlCl_3(aq.)$	$= 2AlCl_3 + aq$	$321.5 \times 2 = 643$
Adding, $2Al + 3Cl_2$	$= 2AlCl_3$	-1346.6 kJ.
$\therefore\ Al + \frac{3}{2}Cl_2$	$= AlCl_3$	$\frac{-1346.6}{2}$ kJ.

$\therefore\ \Delta H = -673.3$ **kJ.mole⁻¹.**

7. Calculate the heat of formation (ΔH) of solid Al_2Cl_6 from the following data :

$Al(s) + 3HCl(aq.) = AlCl_3(aq) + \frac{3}{2}H_2(g)$	$\Delta H = -127.0$ kcal
$H_2(g) + Cl_2(g) = 2HCl(g)$	$\Delta H = -44.0$ kcal
$HCl(g) + aq. = HCl(aq.)$	$\Delta H = -17.5$ kcal
$Al_2Cl_6(s) + aq. = 2AlCl_3(aq.)$	$\Delta H = -155.8$ kcal

[V.U. 2004]

Soln. :

$2Al(s) + 6HCl\ (aq) = 2AlCl_3(aq) + 3H_2(g)$	$\Delta H = -254.0$ kcal
$3H_2(g) + 3Cl_2(g) = 6HCl(g)$	$\Delta H = -132.0$ kcal
$6HCl(g) + aq. = 6HCl(aq)$	$\Delta H = -105.0$ kcal
$2AlCl_3(aq.) = Al_2Cl_6(s) + aq.$	$\Delta H = 155.8$ kcal
Adding, $2Al(s) + 3Cl_2(g) = Al_2Cl_6(s)$	$\Delta H = -335.2$ **kcal.**

8. 8 gm of NH_4NO_3 were dissolved in 400 gm of water taken in a calorimeter (water equivalent of the calorimeter 50). A fall in temperature of 1.3 °C was noted. Calculate the enthalpy of solution.

Soln. : 8 gm NH_4NO_3 = 0.1 gm-mole NH_4NO_3

Assuming heat capacity of water remains same after addition of NH_4NO_3, total heat capacity of calorimeter and solution = (400 + 50) cal.K⁻¹ = 450 cal.K⁻¹.

This leads to the decrease in temperature by 1.3 °C.

$\therefore$ total heat change = -450 cal.K⁻¹ × 1.3K = -585 cal.

The heat change is due to dissolution of 0.1 gm.mole NH_4NO_3.

Hence, enthalpy of solution = heat change for 1 mole

= -585×10 cal.mole⁻¹

= -5850 cal.mole⁻¹

= -24476.4 J.mole⁻¹ = **-24.48 J.mole⁻¹.**

9. A sample of sucrose $C_{12}H_{22}O_{11}$ weighing 0.1265 gm is burned in a bomb calorimeter. After the reaction is over, it is found that to produce an equal temperature increase electrically, 2082.3 J must be expended. Calculate the heat of combustion of sucrose, if the temperatrue increment is 1.743 °C. What is the heat capacity of calorimeter and its contents?

Soln. : Molecular weight of $C_{12}H_{22}O_{11}$ = 342 gm

0.1265 gm releases 2082.3 J heat

$\therefore$ 342 gm release $\frac{2082.3}{0.1265} \times 342 = 5629.62$ kJ heat.

∴ heat of combustion = **5629.62 kJ.**

Now, 2082.35 J raises the temperature by 1.743 °C.

Thus 1°C is raised by $\frac{2082.3\text{ J}}{1.743°\text{C}}$ = 1194.66 J.deg^{-1}.

Thus, heat capacity of calorimeter = **1194.66 J per degree.**

10. Calculate the enthalpy change at 1500 K for the reaction $2H_2(g) + O_2(g) = 2H_2$ for the following data.

$\Delta H_{300\,K}$ = –115.0 kcal and the values of C_P (cal.deg^{-1}.mole^{-1}) are for

$H_2(g)$ = 6.95 – 0.002 T

$O_2(g)$ = 6.10 + 0.0032 T

$H_2O(g)$ = 7.19 + 0.0024 T **[C.U.**

Soln. : $$\Delta H_{1500} = \Delta H_{300} + \int_{300}^{1500} \Delta C_P dT$$

$$\Delta C_P = 2C_{P,\,H_2O} - 2C_{P,\,H_2} - C_{P,\,O_2}$$
$$= (2 \times 7.19 - 2 \times 6.95 - 6.10) + (2 \times 0.0024 + 2 \times 0.0002 - 0.0C$$
$$= -5.62 + 0.002\text{ T.}$$

Given, ΔH_{300} = –115.0 kcal = –115000 cal.

Hence, $$\Delta H_{1500} = -115000\text{ cal} - \int_{300}^{1500} 5.62 dT\text{ cal} + \int_{300}^{1500} 0.002 T\, dT\text{ cal}$$

$$= -115000\text{ cal} - 6.52 \times 1200\text{ cal} + \frac{0.002}{2} \times 1800 \times 1200\text{ c}$$

$$= -119584\text{ cal} = \mathbf{-119.58\text{ kcal.}}$$

11. The heat of neutralization of HCN by NaOH is 2900 cal. Calculate the h ionization of 1 mole of HCN; given that 13800 cal as heat of reaction of $H^+ + OH^-$ = **[B.U.**

Soln. : By the problem,

	ΔH
$HCN = H^+ + CN^-$	X
$H^+ + OH^- = H_2O$	Y = –13800 cal
$HCN + OH^- = H_2O + CN^-$	Z = –2900 cal

By Hess's law, $X + Y = Z$

$$X = Z - Y = -2900 + 13800$$
$$= -10900\text{ cal} = \mathbf{-45.6\text{ kJ.}}$$

12. The heat of formation of $NH_3(g)$ at constant pressure is 46.1 kJ .mol 27 °C. Calculate the heat of formation of $NH_3(g)$ at constant volume assuming behaviour. How would the value differ if the gases under consideration were van der gas? **[B.U.**

Soln. : The reaction is $\frac{1}{2}N_2 + \frac{3}{2}H_2 = NH_3$

$$\Delta n = 1 - \frac{1}{2} - \frac{3}{2} = -1$$

$$\Delta H = \Delta U + RT\Delta n = \Delta U - RT$$

$$46100 = \Delta U - 8.314 \times 300$$

$$\Delta U = 48594.2 \text{ J} = 48.6 \text{ kJ}.$$

Calculation of heat of formation at constant volume from the corresponding value at tant pressure consists of the following two steps :

reactant $(P, V, T) \underset{\rightarrow}{\Delta U_1}$ product $(P, V', T) \underset{\rightarrow}{\Delta U_2}$ product (P', V, T)

$$\Delta U = \Delta U_1 + \Delta U_2 = \Delta H - P\Delta V + \Delta U_2$$

$$= \Delta H - RT\Delta n + \Delta U_2.$$

For ideal gas, $\Delta U_2 = 0$ as U is independent of volume, but for van der Waals gas it nds on V and so ΔU_2 exists. Thus, ΔU will change depending how U depends on V.

13. Molar heat capacities of gases are given (in calorie) by $C_P(N_2) = 6.5 \times 10^{-3}T$, $_2) = 6.5 + 9 \times 10^{-4}T$ and $C_P(NH_3) = 8.0 + 7\times 10^{-4}T$. Calculate ΔH (300 °K) for eaction $N_2(g) + 3H_2(g) = 2NH_3(g)$, when ΔH (0 °K) = –19 kcal. **[B.U. 1992]**

Soln. : Given, $C_P(N_2) = 6.5 \times 10^{-3}T$

$$C_P(H_2) = 6.5 + 9 \times 10^{-4}T$$

$$C_P(NH_3) = 8.0 + 7 \times 10^{-4}T$$

$$\Delta H(0\text{ °K}) = -19 \text{ kcal} = -19000 \text{ cal}.$$

The reaction is : $N_2(g) + 3H_2(g) = 2\,NH_3(g)$

$$\Delta H_{300} = \Delta H_0 + \int_0^{300} \Delta C_P dT$$

$$= -19000 + \int_0^{300} \left[2C_P(NH_3) - 3C_P(H_2) - C_P(N_2)\right]dT$$

$$= -19000 + \int_0^{300} -3.5\,dT - \int_0^{300} 7.8\times10^{-3}T\,dT$$

$$= -19000 - 3.5 \times 300 - \frac{7.8\times10^{-3}}{2} \times 300^2 = -20401 \text{ cal} = \mathbf{-20.401 \text{ kcal.}}$$

4. Calculate the latent heat of fusion of ice at –20 °C, if that at 0 °C is 1440 cal ole, the specific heat of ice being 8.7 cal.mole^{-1}. **[C.U. 1998]**

oln. : The phase transition is ice→water

$$T_1 = 0\text{ °C} = 273.15 \text{ K}, T_2 = -20\text{ °C} = 253.15 \text{ K}$$

$$\Delta H_1 = 1440 \text{ cal.mole}^{-1}, C_{P,ice} = 8.7 \text{ cal.mole}^{-1}.$$

$$\text{Thus, } \Delta H_2 = \Delta H_1 + \int_{273.15}^{253.15} (C_P, water - C_P, ice)dT$$

$$= 1440 + (18 - 8.7)(253.15 - 273.15) = \mathbf{1254 \text{ cal.mole}^{-1}}.$$

15. Calculate bond energy of O–H bond in $H_2O(g)$ from the following data :

$$H_2(g) + \frac{1}{2}O_2(g) \rightarrow H_2O(g) \qquad \Delta H^\circ = -241.814 \text{ kJ/mole}$$

$$H_2(g) \rightarrow 2H(g) \qquad \Delta H^\circ = 435.994 \text{ kJ/mole}$$

$$O_2(g) \rightarrow 2O(g) \qquad \Delta H^\circ = 498.34 \text{ kJ/mole}$$

Soln. : By the problem,

$$H_2O(g) \rightarrow H_2(g) + \frac{1}{2}O_2(g) \qquad \Delta H^\circ = 241.814 \text{ kJ/mole}$$

$$H_2(g) \rightarrow 2H(g) \qquad \Delta H^\circ = 435.994 \text{ kJ/mole}$$

$$\frac{1}{2}O_2(g) \rightarrow O(g) \qquad \Delta H^\circ = 249.17 \text{ kJ/mole}$$

Adding, $H_2O(g) \rightarrow 2H(g) + O(g) \qquad \Delta H^\circ = 926.978$ kJ/mole.

Thus, bond energy of O–H bond

$$= \frac{\Delta H^\circ}{2} = \mathbf{463.49\ kJ/mole.}$$

16. The combination of N_2 and O_2 at 27 °C occurs according to $2N_2 + O_2 = 2N_2O$. $\Delta H = 39.4$ kcal. What will be the heat of formation at 227 °C? Use the relation $C_P = a + bT$ in calculation where C_P is the molar heat capacity.

g	*a*	$b \times 10^3$
N_2	6.52	1.25
O_2	6.15	3.1
N_2O	7.0	7.1

[C.U. 1976]

Hints : $2N_2 + O_2 = 2N_2O$

$$\Delta C_p = 2C_{P,N_2O} - 2C_{P,N_2} - C_{P,O_2}$$

$$= (2a_{N_2O} - 2a_{N_2} - a_{O_2}) + (2b_{N_2O} - 2b_{N_2} - b_{O_2}) \times 10^{-3}T$$

$$= 5.19 + 8.6 \times 10^{-3}T.$$

$$\Delta H_{500} = \Delta H_{300} + \int_{300}^{500} \Delta C_P dT = 39.050 \text{ kcal.}$$

$$\therefore \Delta H_f = \frac{39.050}{2} = \mathbf{19.525\ kcal.}$$

17. At 18 °C the heat of solution of anhydrous $CuSO_4$ in large volume of water –15800 cal/mole while that of $CuSO_4.5H_2O$ is –2750 cal/mole. Find at 18 °C the heat c reaction $CuSO_4(s) + 5H_2O(l) = CuSO_4.5H_2O(s)$. [C.U. 1984

Soln. : $CuSO_4(s)$ + aq. = $CuSO_4$.aq. $\Delta H = -15800$ cal/mole (1

$CuSO_4.5H_2O(s)$ + aq. = $CuSO_4$.aq. $\Delta H = -2750$ cal/mole (2

Rearranging eqn. (2), $CuSO_4$.aq. = $CuSO_4.5H_2O(s)$ + aq. $\Delta H = 2750$ cal/mole (3

Adding eqn. (1) and (3),

$CuSO_4(s) + 5H_2O(l) = CuSO_4.5H_2O(s)$, $\Delta H = -13050$ cal/mole.

18. Enthalpies of formation of $CO_2(g)$ and $H_2O(l)$ under standard conditions are –94.8 and 68.3 kcal/mole respectively. If the enthalpy of combustion of acetaldehyde is –279.07 kcal/mole, find its enthalpy of combustion. **[Delhi University, B.Sc., 1975]**

Hints : $\Delta H°$

$$2CO_2(g) + 2H_2O(l) = CH_3CHO + \frac{5}{2}O_2 \qquad 279{\cdot}07 \text{ kcal}$$
$$2C + 2O_2 = 2CO_2(g) \qquad -94{\cdot}8 \times 2 \text{ kcal}$$
$$2H_2 + O_2 = 2H_2O(l) \qquad -68{\cdot}3 \times 2 \text{ kcal}$$
$$\therefore 2C + 2H_2 + \frac{1}{2}O_2 = CH_3CHO \qquad \mathbf{-47{\cdot}13 \text{ kcal.}}$$

19. Calculate O—O bond strength. Given that

$$H_2O(g) \to 24(g) + O(g) \qquad \Delta H° = 930 \text{ kJ/mole}$$
$$H_2O_2(g) \to 2H(g) + 2O(g) \quad \Delta H° = 1070 \text{ kJ/mole.}$$
[C.U. 2006]

Hints : O–H bond strength $= \frac{930}{2}$ kJ/mole.

O–O bond strength $= 1070 - 2 \times \frac{930}{2} = 140$ kJ/mole.

20. Calculate increase in vapour pressure of water per atmosphere rise in external ressure at 10°C (vapour pressure of water at 10 °C in 9.2 mm). **[V.U. 2002]**

Soln. : Given $P_1 = 9.2$ mm, $V_m = 18$ cc, $T = 10$ °C $= 283.15$ K

$$P_{2t} - P_{1t} = \Delta P = 1 \text{ atm}$$

$$\ln \frac{P_2}{P_1} = \frac{V_m \Delta P}{RT}$$

$$\text{or, } \ln \frac{P_2}{9.2} = \frac{1 \text{ atm} \times 18 \text{ cc.mol}^{-1} \times 10^{-3} \, l/\text{cc}}{0.8206 \, l.\text{atm.deg.mol}^{-1} \times 283.15 \text{K}} = 7.747 \times 10^{-4}$$

Hence, $P_2 = 9.207$ mm

Miscellaneous

1. Let a perfect gas be forced isobarically through an orifice into an evacuated rigid essel until its pressure in the vessel rises to the isobaric value and the process stops. Show at $T_2 = \gamma T_1$, where T_2 is the final temperature of the gas in the vessel before it loses eat, and T_1 is the temperature before entering the vessel.

Soln. : From first law, $W_{ad} = nC_P \ dT = nC_v (T_2 - T_1)$ (i)

Total work done $= P_1V_1 - P_0V_0$

$= P_1V_1$ [for expansion into evacuated vessel, $P_0 = 0$]

$= nRT_1$. (ii)

$\therefore \quad nRT_1 = nC_v (T_2 - T_1)$

or, $\quad T_1 - T_2 = -(\gamma - 1) T_1$ [since $C_p - C_v = R$ and $\frac{C_p}{C_v} = \gamma$]

or, $\quad T_1 + (\gamma - 1) T_1 = T_2$

or, $\gamma T_1 = T_2$.

$\therefore \quad T_2 = \gamma T_1$ **(Proved).**

2. A thick-walled insulated chamber contains n_i mol of He at a high pressure P_i. It is connected through a valve with a large and almost empty gas holder where the pressure is maintained at a constant value P_0, very nearly atmospheric. The valve is opened slightly and He flows slowly and adiabatically into the gas holder until the pressure on the two sides of the valve is equalised. Let n_F be the final number of moles of He left in the chamber and P_F be the final pressure. Also let U_i, U_F and U' be initial, final molar energy of He in the chamber and molar energy of helium in the gas holder, and $h' = U' + P_0V_0$ (V_0 = molar vol. of He gas in the gas holder), T_F and T' be the final temperatures of the chamber and gas holder, and assume that He is an ideal gas with constant heat capacity. Now, show that—

(1) $\dfrac{n_F}{n_i} = \dfrac{h_i' - U_i}{h_i' - U_F}$ (2) $\dfrac{n_i}{n_F} = \left(\dfrac{P_i}{P_F}\right)^{\frac{1}{\gamma}}$

(3) $T_F = T_i\left(\dfrac{P_F}{P_i}\right)^{\frac{\gamma-1}{\gamma}}$ (4) $T' = \dfrac{T_i}{\gamma} \cdot \dfrac{1 - \dfrac{P_F}{P_i}}{1 - \left(\dfrac{P_F}{P_i}\right)^{\frac{1}{\gamma}}}$.

Soln. : Let there be a chamber fitted with an adiabatic jacket which contains n_i moles of He with internal energy U_i at temperature T_i. This is joined to a holder at nearly atmospheric pressure P_0 having very large volume V. The volume V is so large that addition or removal of finite gram-moles do not alter the volume. But if n moles of He are introduced, the piston is slightly pushed and the work done is nP_0V_0, where V_0 = molar volume of He in holder.

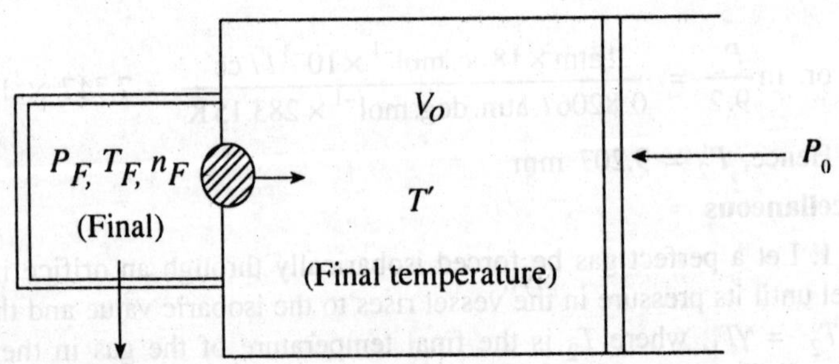

Let the adiabatic enclosure contain a valve at temperature T_i, the valve is opened and the gas molecule escapes to the holder until the pressure P_i becomes equal to P_F. Let there be n_F moles of the gas after exit of the gas from the adiabatic enclosure to the holder.

Initial number of moles $= n_i$

Temperature $= T_i$

Energy/mole $= U_i$

Pressure $= P_i$

Assuming very closely reversibility, we known that

$$T_F = T_i\left(\frac{P_F}{P_i}\right)^{\frac{\gamma-1}{\gamma}}, \text{ where } \gamma = \frac{C_P}{C_v} \textbf{ [Proved (3)].}$$

But, $n_iRT_i = P_iV_i$

$n_FRT_F = P_FV_F.$

Now, $\therefore \dfrac{n_i}{n_F} = \dfrac{T_F}{T_i}.\dfrac{P_i}{P_F}.\dfrac{V_i}{V_F}$ (since $V_i = V_F$)

$$= \left(\frac{P_F}{P_i}\right)^{\frac{\gamma-1}{\gamma}} \cdot \left(\frac{P_F}{P_i}\right)^{-1} = \left(\frac{P_F}{P_i}\right)^{\frac{\gamma-1}{\gamma}-1} = \left(\frac{P_F}{P_i}\right)^{\frac{-1}{\gamma}} = \left(\frac{P_i}{P_F}\right)^{\frac{1}{\gamma}} \textbf{ [Proved (2)].}$$

Assuming, the whole system to be isolated, the energy lost in the chamber is equal to the energy gained in the holder.

Loss of energy in the chamber $= n_iU_i - n_FU_F$

Gain of energy in the holder $= (n_i - n_F)U' + (n_i - n_F)P_0V_0$

$\therefore \; n_iU_i - n_FU_F = (n_i - n_F)U' + (n_i - n_F)\,P_0V_0.$

By interchanging,

$$n_iU_i - n_i(U' + P_0V_0) = n_FU_F - n_F(U' + P_0V_0)$$

or, $n_iU_i - n_i(U' + P_0V_0) = n_FU_F - n_F(U' + P_0V_0)$

or, $n_i(U_i - h'_i) = n_F(U_F - h'_i)$ [$\therefore h' = U' + P_0V_0$].

$$\therefore \; \frac{n_i}{n_F} = \frac{U_F - h_i'}{U_i - h_i'} = \frac{h_i' - U_F}{h_i' - U_i}$$

or, $\dfrac{n_F}{n_i} = \dfrac{h_i' - U_i}{h_i' - U_F}$ **[Proved (1)].**

But, for monatomic ideal gas, $U = \dfrac{3}{2}RT/\text{mole}$

$h = \dfrac{5}{2}RT/\text{mole}.$

$$\therefore \; \frac{n_F}{n_i} = \frac{\frac{5}{2}RT' - \frac{3}{2}RT_i}{\frac{5}{2}RT' - \frac{3}{2}RT_F} = \frac{5T' - 3T_i}{5T' - 3T_F}.$$

On cross-multiplying,

$$n_F(5T' - 3T_F) = n_i(5T' - 3T_i).$$

$$\therefore \; \frac{5}{3}T' = \frac{n_FT_F - n_iT_i}{n_F - n_i}$$

$$\gamma T' = \frac{T_i\left(n_F.\dfrac{T_F}{T_i} - n_i\right)}{n_F - n_i}.$$

$$\therefore T' = \frac{\dfrac{T_i}{\gamma}\left(\dfrac{T_F}{T_i} - \dfrac{n_i}{n_F}\right)}{1 - \dfrac{n_i}{n_F}}.$$

$$= \frac{\frac{T_i}{\gamma}\left[\left(\frac{P_F}{P_i}\right)^{\frac{\gamma-1}{\gamma}} - \left(\frac{P_i}{P_F}\right)^{\frac{1}{\gamma}}\right]}{1-\left(\frac{P_i}{P_F}\right)^{\frac{1}{\gamma}}}.$$

Dividing and multiplying by $\left(\frac{P_F}{P_i}\right)^{\frac{1}{\gamma}}$,

$$T' = \frac{\frac{T_i}{\gamma}\left[\left(\frac{P_F}{P_i}\right)^{\frac{\gamma-1}{\gamma}} - \left(\frac{P_i}{P_F}\right)^{\frac{1}{\gamma}}\right]}{1-\left(\frac{P_i}{P_F}\right)^{\frac{1}{\gamma}}} \times \frac{\left(\frac{P_F}{P_i}\right)^{\frac{1}{\gamma}}}{\left(\frac{P_F}{P_i}\right)^{\frac{1}{\gamma}}}$$

$$= \frac{\frac{T_i}{\gamma}\left[\left(\frac{P_F}{P_i}\right) - 1\right]}{\left(\frac{P_F}{P_i}\right)^{\frac{1}{\gamma}} - 1} = \frac{\frac{T_i}{\gamma}\left[1-\left(\frac{P_F}{P_i}\right)\right]}{1-\left(\frac{P_F}{P_i}\right)^{\frac{1}{\gamma}}} \quad \textbf{[Proved (4)].}$$

3. A polytropic process is one that obeys the relation $PV^n = c$, where n and c are constants. Derive a general expression for W, Q, ΔU for such a change from state I to state II carried out reversibly. Show that they provide familiar relation when $n = 0, 1, \infty, \gamma$.

Soln. : $W = -\int_1^2 PdV = -n\int_1^2 C.\frac{dV}{V^n}$

$$= -\int_1^2 CV^{-n}dV$$

$$= \frac{1}{(n-1)}\left[\frac{C}{V_2^{n-1}} - \frac{C}{V_1^{n-1}}\right]$$

$$= \frac{1}{(n-1)}\left[\frac{P_2V_2^n}{V_2^{n-1}} - \frac{P_1V_1^n}{V_1^{n-1}}\right] \qquad \left[\begin{array}{l}\because PV^n = C \\ \because P_1V_1^n = P_2V_2^n = C\end{array}\right]$$

$$= \frac{1}{(n-1)}[RT_2 - RT_1] \qquad [\because PV = RT]$$

$$= \frac{RT_1}{(n-1)}\left[\frac{T_2}{T_1} - 1\right] = \frac{RT_1}{n-1}\left[\frac{P_2V_2}{P_1V_1} - 1\right].$$

Now, $Q = \Delta U - W$

$$= C_v (T_2 - T_1) + \frac{RT_1}{n-1}\left[1 - \frac{P_2V_2}{P_1V_1}\right]$$

$$= \frac{RT_1}{n-1}\left\{\left(1 - \frac{P_2V_2}{P_1V_1}\right) + \frac{n-1}{RT_1}C_v(T_2 - T_1)\right\}$$

$$= \frac{RT_1}{n-1}\left\{\left(1 - \frac{P_2V_2}{P_1V_1}\right) + \frac{n-1}{R}C_v\left(\frac{T_2}{T_1} - 1\right)\right\}$$

But $P_2V_2 = RT_2$, $P_1V_1 = RT_1$

$$= \left[\frac{1}{n-1}(RT_1 - RT_2)\right] + C_v(T_2 - T_1) \quad = \left[\frac{RT_1}{n-1}\left(1 - \frac{T_2}{T_1}\right) - C_v(T_1 - T_2)\right]$$

$$= \frac{RT_1}{n-1}\left[\left(1 - \frac{T_2}{T_1}\right) - \frac{C_v(n-1)}{R}\left(1 - \frac{T_2}{T_1}\right)\right] = \frac{RT_1}{n-1}\left[1 - \frac{T_2}{T_1}\right]\left[1 - \frac{C_v}{R}(n-1)\right]$$

$$= \frac{RT_1}{n-1}\left[1 - \frac{T_2}{T_1}\right]\left(1 - \frac{n-1}{\gamma-1}\right) = \frac{RT_1}{n-1}\,\frac{\gamma-n}{\gamma-1}\left[1 - \frac{T_2}{T_1}\right]$$

$$= \frac{\boldsymbol{RT_1}}{\boldsymbol{n-1}}\,\frac{\boldsymbol{\gamma-n}}{\boldsymbol{\gamma-1}}\left[\boldsymbol{1 - \frac{P_2V_2}{P_1V_1}}\right].$$

$$PV^n = C$$

$$\therefore \quad \frac{\partial P}{\partial V} = -\frac{nP}{V}.$$

When $n = 0$, $\frac{\partial P}{\partial V} = 0$, isobaric process ($P$ = constant)

$n = 1$, $\frac{\partial P}{\partial V} = -\frac{P}{V}$, isothermal process ($T$ = constant)

$n = \infty$, $\frac{\partial P}{\partial V} = \infty$, isochoric process ($V$ = constant)

$n = \gamma$, $\frac{\partial P}{\partial V} = -\gamma\frac{P}{V}$, reversible adiabatic process ($Q = 0$)

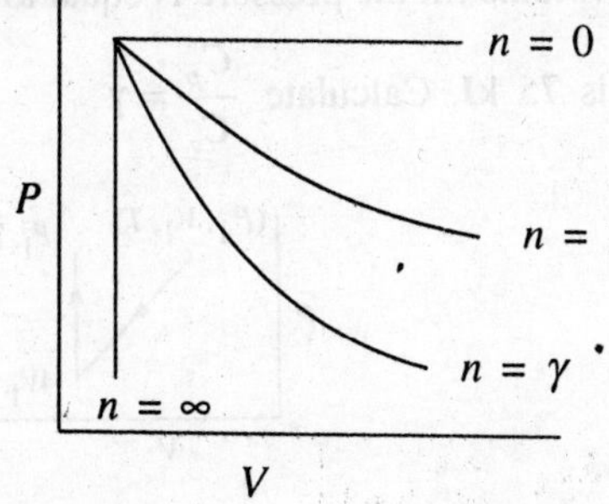

4. For reversible polytropic process described by the general relation $PV^n = C$, derive the following expression for work and heat.

$$W = \frac{RT_1}{n-1}\left\{\left(\frac{P_2}{P_1}\right)^{\frac{n-1}{n}} - 1\right\}$$

$$Q = \frac{(n-\gamma)RT_1}{(n-1)(\gamma-1)}\left\{\left(\frac{P_2}{P_1}\right)^{\frac{n-1}{n}} - 1\right\}.$$

Hints : Derive the general formula of work and heat,

$$W = \frac{RT_1}{n-1}\left[\frac{P_2V_2}{P_1V_1} - 1\right].$$

Then put $V_1 = \left(\frac{C}{P_1}\right)^{\frac{1}{n}}$ and $V_2 = \left(\frac{C}{P_2}\right)^{\frac{1}{n}}$ to obtain the above expression.

$$Q = \frac{RT_1}{(n-1)}\left(\frac{\gamma-n}{\gamma-1}\right)\left[1 - \frac{T_2}{T_1}\right] \text{ (as found in the last problem)}$$

$$= \frac{RT_1}{(n-1)}\left(\frac{\gamma-n}{\gamma-1}\right)\left[1 - \frac{P_2V_2}{P_1V_1}\right]$$

$$= \frac{RT_1}{(n-1)}\left(\frac{\gamma-n}{\gamma-1}\right)\left[1 - \left(\frac{P_2}{P_1}\right)^{\frac{n-1}{n}}\right] \quad \left[\text{putting } V_1 = \left(\frac{C}{P_1}\right)^{\frac{1}{n}},\ V_2 = \left(\frac{C}{P_2}\right)^{\frac{1}{n}}\right]$$

$$= \frac{(n-\gamma)RT_1}{(n-1)(\gamma-1)}\left\{\left(\frac{P_2}{P_1}\right)^{\frac{n-1}{n}} - 1\right\}.$$

5. 3 moles of an ideal gas at 300 K is isothermally expanded to 4 times its initial volume and then heated against constant volume till the pressure is equal to its initial value. The total heat absorbed in the process is 75 kJ. Calculate $\frac{C_p}{C_v} = \gamma$.

Soln. : Let $V_1 = V$

Given, $T = 300$ K

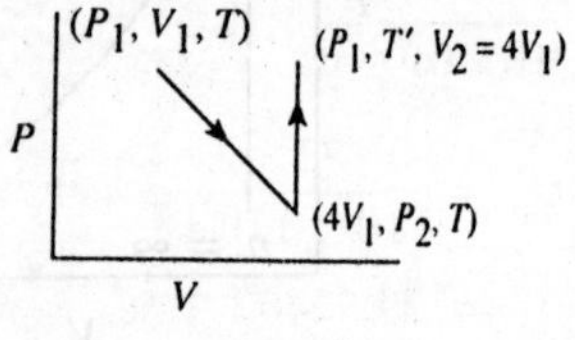

$$P_2 \times 4V = 3R \times 300$$

$$P_1 \times 4V = 3R \times T'.$$

$$\therefore \frac{P_2}{P_1} = \frac{300}{T'}.$$

But, $P_2 \times 4V = P_1 \times V.$

$\therefore \frac{P_1}{P_2} = 4.$

$\therefore \ T' = 300 \times 4 = 1200$ K.

$Q = Q_1 + Q_2$

$Q_1 = -W_1 = nRT \ln \frac{V_2}{V_1} = 3 \times 300 \times 8.314 \ln 4 = 10373.08619$ J.

By the problem, $W_2 = 0$

$Q_2 = nC_v (T' - T) = 3 \times C_v (1200 - 300) = 2700 C_v.$

$\therefore \ 2700 C_v + 10373.08 = 75000$

$\therefore \ C_v = 23.935$ J = 23.94 J.

$\therefore \ C_p = C_v + R = 32.249$ J.

$\therefore \ \gamma = \frac{C_p}{C_v} = \mathbf{1.35}.$

6. A gas at 500 K and 1 atm is passed at the rate of 100 cc/sec into an apparatus where it flows over an electrically heated wire of resistance 200 ohms. When the heating current is 0.06 A, the gas leaves the apparatus at 510 K. Calculate C_p and C_v.

Soln. : Let V_0 be the volume at NTP.

$$\frac{P_1V_1}{T_1} = \frac{P_0V_0}{T_0}$$

or, $$\frac{1 \times 100}{500} = \frac{1 \times V_0}{273}.$$

$\therefore \ V_0 = \frac{273 \times 100}{500} = 54.6$ cc = 0.0546 litre.

Heat $Q = i^2Rt = (0.06)^2 \times 200 \times 1$ J

$= 0.72 \text{ J} = \frac{0.72}{4.184}$ cal = 0.172 cal.

0.172 cal heat leads to an increasse of 10 °C for 0.0546 litre gas.

$\therefore \ \frac{0.172}{10}$ cal leads to an increase of 1 °C for 0.0546 litre.

or, $\frac{0.172}{10 \times 0.0546}$ cal leads to an increase of 1 °C for 1 litre.

or, $\frac{0.172}{10 \times 0.0546} \times 22.4$ cal leads to an increase of 1 °C for 22.4 litres.

$\therefore \ C_p = \frac{0.172 \times 22.4}{10 \times 0.0546}$ cal.deg^{-1}.mole^{-1}. = 7.0564 cal.deg^{-1}.mole^{-1}.

= **29.52402 J.deg^{-1}.mole^{-1}**

$\therefore \ C_v = (29.5240 - 8.314)$ J.deg^{-1}.mole^{-1}

= **21.21 J.deg^{-1}.mole^{-1}.**

7. A gas at 273 °C and 1 atm is passed at the rate of 80 cc/sec into an apparatus where it flows over an electrically heated wire of resistance 150 ohms. When the heating current is 0.04 A, the gas leaves the apparatus at 278 °C and 1 atm. Calculate C_p and C_v assuming the gas to be ideal.

Soln. : $\dfrac{P_1V_1}{T_1} = \dfrac{P_0V_0}{T_0}$, or, $\dfrac{1\times 80}{546} = \dfrac{1\times V_0}{273}$.

$$\therefore\ V_0 = \frac{80\times 273}{546} = 40 \text{ cc} = 0.04 \text{ litre.}$$

The volume is 40 cc at NTP.

The amount of heat is i^2Rt joule

$$= (0.04^2 \times 150 \times 1)\text{ J} = 24 \times 10^{-2}\text{ J} = 0.05736 \text{ cal.}$$

$\therefore\ \dfrac{0.05736}{5}$ cal heat requires for rise in 1°C for 0.04 litre

or, $\dfrac{0.05736}{5\times 0.04}$ cal heat requires for rise in 1°C for 1 litre

or, $\dfrac{0.05736\times 22.4}{5\times 0.04}$ cal heat requires for rise in 1°C for 22.4 litre.

$$\therefore\ C_p = \frac{0.05736\times 22.4}{5\times 0.04}\text{ cal.deg}^{-1}.\text{mole}^{-1}$$

$$= 6.424 \text{ cal.deg}^{-1}.\text{mole}^{-1} = \mathbf{26.88\ J.deg^{-1}.mole^{-1}.}$$

$$\therefore\ C_v = (6.424 - 1.987)\text{ cal.deg}^{-1}.\text{mole}^{-1}$$

$$= 4.437 \text{ cal.deg}^{-1}.\text{mole}^{-1} = \mathbf{18.564\ J.deg^{-1}.mole^{-1}.}$$

8. A bottle at 27 °C contains an ideal gas at 126.9 kPa pressure. The gas expands adiabatically after the removal of rubber stopper against the constant pressure of the atmosphere at that time, 101.9 kPa. Some gas is expelled and the stopper is quickly replaced when the pressure of the bottle equals atmospheric pressure. The gas which is cooled in the adiabatic expansion slowly warms up to 27 °C. The final pressure now becomes 111.07 kPa. Calculate γ.

Soln. : This is Clement-Desorms method for determination of γ.

Let the initial temperature and pressure be T_1 and P_1 respectively. Let the initial volume be V_1, and it expands (or contracts) to V_2 and the final volume be V_3. Since the bottle volume is fixed, $V_2 = V_3$. Number of moles expelled or entered is assumed to be negligible.

$$\therefore\ P_1V_1 = P_3V_3$$

$$V_2 = V_3$$

$$T_1 = T_3$$

Assuming the change to be reversible,

$$\frac{P_1}{P_2} = \left(\frac{V_2}{V_1}\right)^{\gamma} = \frac{(V_2)^{\gamma}}{\left(\dfrac{P_3V_3}{P_1}\right)^{\gamma}} = \left(\frac{P_1}{P_3}\right)^{\gamma}.$$

$$\therefore\ \gamma = \frac{\log \frac{P_1}{P_2}}{\log \frac{P_1}{P_3}}.$$

By the problem, $P_1 = 126.9$ kPa

$P_2 = 101.9$ kPa

$P_3 = 111.07$ kPa.

$$\gamma = \frac{\log \frac{126.9}{101.9}}{\log \frac{126.9}{111.9}} = \frac{\log 1.245338567}{\log 1.34048257} = \frac{0.0952874}{0.0546315} = \mathbf{1.74.}$$

Assuming the change to be irreversible,

$$W = -P_{ex}(V_2 - V_1) = nC_v\,(T_2 - T_1)$$

$$P_2(V_2 - V_1) = nC_v\,(T_1 - T_2)$$

$$R\left(T_2 \frac{P_2}{P_2} - T_1 \frac{P_2}{P_1}\right) = C_v\,(T_1 - T_2)$$

$$\therefore\ R\left(T_2 - T_1 \frac{P_2}{P_1}\right) = C_vT_1 - C_vT_2$$

$$\text{or, } T_2(R + C_v) = T_1\left(C_V + R\frac{P_2}{P_1}\right)$$

$$\therefore\ T_2 = \frac{C_V + R\frac{P_2}{P_1}}{C_V + R} T_1$$

$T_1 = T_3;\ V_3 = V_2$

$P_1 = 126.9$ kPa; $P_2 = 101.9$ kPa; $P_3 = 111.07$ kPa.

$$\frac{P_3}{T_1} = \frac{P_2}{T_2} = \frac{P_2}{\left(\frac{C_v + R\frac{P_2}{P_1}}{C_v + R}\right) T_1}.$$

$$\therefore\ P_3 = \frac{P_2(C_v + R)}{C_v + R\frac{P_2}{P_1}}.$$

$$\therefore\ \frac{C_v}{R} = \frac{P_2 - \frac{P_3P_2}{P_1}}{P_3 - P_2}.$$

$$\text{Thus, } \frac{C_v}{R} = \frac{101.9 - \dfrac{111.07 \times 101.9}{126.9}}{111.07 - 101.9} = \frac{101.9 - 89.1886}{9.17} = \frac{12.7114}{9.17} = 1.39.$$

$$\gamma = \frac{\dfrac{C_v}{R} + 1}{\dfrac{C_v}{R}} = \frac{2.39}{1.39} = \mathbf{1.72.}$$

9. Let the volume of the bottle in Problem 8 be 20 l. Calculate the amount of the ideal gas that is expelled if its molecular weight is 28. [**Ans.** 3.46 gm]

10. Consider the case similar to Problem 8. But now let the pressure be 151.2 kPa and the atmospheric pressure be 100.8 kPa. What will be the final pressure, if $\gamma = 1.66$? [**Ans.** 118.4 kPa]

11. The molar specific heat at constant volume of an ideal gas is $\frac{3}{2}R$. Suppose that 1 mole of such a gas is subjected to a cyclic quasistatic process which appeared as a circle in the P vs. V diagram (see fig.). Show that—

(i) The net work done by the gas in one cycle is about 3.14 J.

(ii) The internal energy difference of the gas between states, C and A is about 600 J.

(iii) The heat absorbed by the gas in going from A to C via ABC is 1107 J.

Calculate Q, ΔU and W for path CDA.

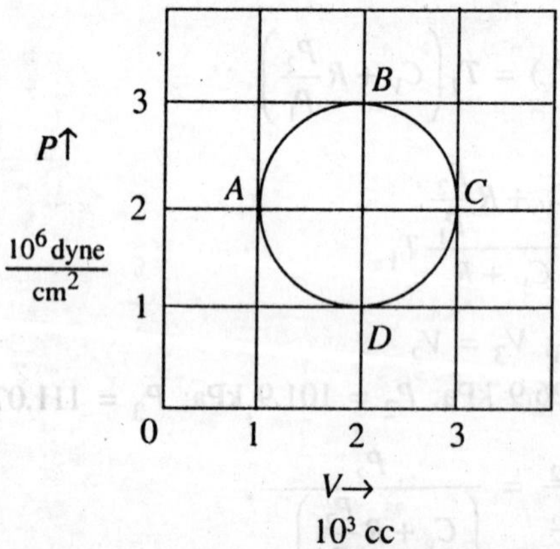

Soln. : (i) Net work done by the gas = – area of the ellipse $ABCD$ = $-\pi ab$

$$= -3.14 \times 10^6 \ \frac{\text{dyne}}{\text{cm}^2} \times 10^3 \text{ cc}$$

$$= -3.14 \times 10^6 \times 10^3 \text{ ergs} = \mathbf{-314\ J.}$$

(ii) $T_A = \dfrac{P_A V_A}{R}$, $T_C = \dfrac{P_C V_C}{R}$

Hence, $\Delta U_{CA} = U_C - U_A = C_v(T_C - T_A)$

$$= \frac{3}{2}R\left(\frac{P_C V_C}{R} - \frac{P_A V_A}{R}\right)$$

$$= \frac{3}{2}P_C(V_C - V_A) \text{ (since } P_A = P_C\text{)}$$

$$= \frac{3}{2} \times 2 \times 10^6 (3 - 1) \times 10^3 \ \frac{\text{dyne}}{\text{cm}^2} \text{cc}$$

$$= 6 \times 10^9 \text{ ergs} = \mathbf{600\ J.}$$

(iii) Work done for the path ABC = –(area under ABC)

= –(area of half ellipse + area of 4 boxes)

$$= -(\frac{1}{2} \times 314 + 4 \times 100) \text{ J}$$

(since area of each box = $10^6 \times 10^3$ ergs)

= **– 557 J.**

Thus, Q for the path $ABC = \Delta U_{CA} - W_{ABC}$ = (600 + 557) J = **1157 J.**

For return path CDA, ΔU = **– 600 J.**

W = (area of 4 boxes – area of half ellipse)

= (400 – 157) J = **243 J.**

Q = –600 – 243 = **–843 J.**

12. Show that C_v and C_P of a perfect gas are independent of both volume and pressure. Do they depend on temperature?

Soln. : $C_P = \left(\frac{\partial H}{\partial T}\right)_P$, $C_V = \left(\frac{\partial U}{\partial T}\right)_V$

$$\left(\frac{\partial C_V}{\partial V}\right)_T = \left[\frac{\partial}{\partial V}\left(\frac{\partial U}{\partial T}\right)_V\right]_T = \left[\frac{\partial}{\partial T}\left(\frac{\partial U}{\partial V}\right)_T\right]_V = 0,$$

since for ideal gas, $\left(\frac{\partial U}{\partial V}\right)_T = 0.$

Similarly, $\left(\frac{\partial C_V}{\partial P}\right)_T = \left[\frac{\partial}{\partial P}\left(\frac{\partial U}{\partial T}\right)_V\right]_T = \left[\frac{\partial}{\partial T}\left(\frac{\partial U}{\partial P}\right)_T\right]_V = 0,$

since for ideal gas, $\left(\frac{\partial U}{\partial V}\right)_T = 0.$

$$\left(\frac{\partial C_P}{\partial P}\right)_T = \left[\frac{\partial}{\partial P}\left(\frac{\partial H}{\partial T}\right)_P\right]_T = \left[\frac{\partial}{\partial T}\left(\frac{\partial H}{\partial P}\right)_T\right]_P = 0,$$

since for ideal gas, $\left(\frac{\partial H}{\partial P}\right)_T = 0.$

Similarly, $\left(\frac{\partial C_P}{\partial P}\right)_T = 0,$

Since U and H depend on temperature, C_P and C_V will depend on temperature.

13. Find the value of $\left(\frac{\partial U}{\partial V}\right)_T$ and $\left(\frac{\partial H}{\partial P}\right)_T$ for the gas obeying $PV = RT + Pb$ and hence comment which one provides better thermodynamic criteria for describing an ideal gas.

Soln. : $\left(\frac{\partial U}{\partial V}\right)_T = T\left(\frac{\partial P}{\partial T}\right)_V - P$

For the gas obeying $PV = RT + Pb$,

$$\left(\frac{\partial P}{\partial T}\right)_V = \left[\frac{\partial}{\partial T}\left(\frac{RT}{V-b}\right)\right]_V = \frac{R}{V-b}.$$

Hence, $\left(\frac{\partial U}{\partial V}\right)_T = \frac{RT}{V-b} - P = P - P = 0.$

Now, $\left(\frac{\partial H}{\partial P}\right)_T = V - T\left(\frac{\partial V}{\partial T}\right)_P.$

$\therefore \left(\frac{\partial H}{\partial P}\right)_T = \frac{R}{P}.$

Hence, $\left(\frac{\partial H}{\partial P}\right)_T = V - T\left(\frac{\partial V}{\partial T}\right)_P = \frac{RT}{P} + b - \frac{RT}{P} = b.$

Thus, $\left(\frac{\partial H}{\partial P}\right)_T$ is zero only for ideal gas but $\left(\frac{\partial U}{\partial V}\right)_T$ is zero not only for ideal gas but also for a gas having little deviation from ideal behaviour. Hence, $\left(\frac{\partial H}{\partial P}\right)_T$ **= 0 is better thermodynamic criteria for an ideal gas than** $\left(\frac{\partial U}{\partial V}\right)_T$ **= 0.**

14. What is perpetual motion machine of first kind? Does it exist? Is the orbital motion of elecctron around the nucleus an example of perpetual motion?

Ans. : A machine which creates its own energy itself for doing work is called perpetual motion machine of 1st kind.

It does not exist as it contradicts the 1st law of thermodynamics which states that energy cannot be created or destroyed.

Orbital motion of electron arround the nucleus is due to the Coulomb's force of attraction between nucleus and electron. So, it is not an example of perpetual motion.

15. Write the thermodynamic equation of state.

Ans. : $\left(\frac{\partial U}{\partial V}\right)_T = T\left(\frac{\partial P}{\partial T}\right)_V - P.$

16. Show that $\left(\frac{\partial H}{\partial V}\right)_T = 0$ for an ideal gas, given that $\left(\frac{\partial U}{\partial V}\right)_T = 0.$

Ans. : $H = U + PV$

or, $\left(\frac{\partial H}{\partial V}\right)_T = \left(\frac{\partial U}{\partial V}\right)_T + \left(\frac{\partial P}{\partial V}\right)_T V + P$

$$= \left(\frac{\partial P}{\partial V}\right)_T V + P \text{ [given, } \left(\frac{\partial U}{\partial V}\right)_T = 0]$$

For ideal gas, $PV = RT$.

$$\therefore \left(\frac{\partial P}{\partial V}\right)_T = -\frac{RT}{V^2} = -\frac{P}{V}.$$

Hence, $\left(\frac{\partial H}{\partial V}\right)_T = -P + P = 0.$

17. A perfact gas is allowed to expand adiabatically against vacuum to a finite volume and then it is compressed reversibly and adiabatically to its initial volume. Is this a cyclic process? Indicate the process in *P-V* diagram.

Ans. : State 1→ State 2, adiabatic expansion against vacuum.

Thus, $Q = 0$, $W = 0$, $\Delta U = 0$.

Since the gas is ideal, $T_1 = T_2$.

By the problem,

$$\underset{1}{(P_1, V_1, T_1)} \xrightarrow[\text{expansion against vacuum}]{\text{Adiabatic}} \underset{2}{(P_2, V_2, T_1)} \xrightarrow[\text{Rev. contraction}]{\text{Adiabatic}} \underset{3}{(P', V_1, T')}$$

Now, $P'V_1^{\gamma} = P_2V_2^{\gamma}$

and $P_1V_1 = P_2V_2$.

$$\therefore P' = P_1 \left(\frac{V_2}{V_1}\right)^{\gamma-1}.$$

Since $\gamma > 1$,

$V_2 > V_1$, hence $P' > P_1$.

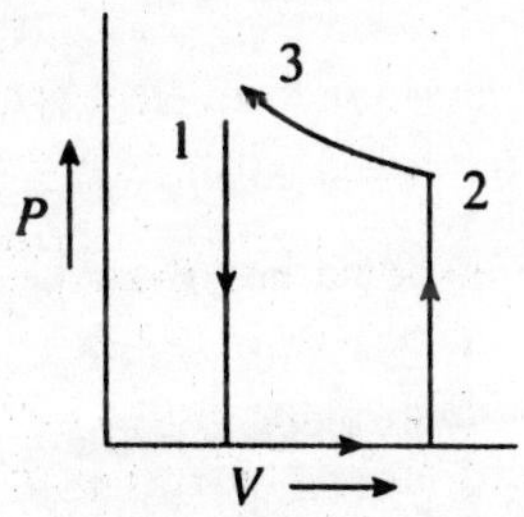

State 1→ (P_1, V_1, T_1)
State 2→ (P_2, V_2, T_1)
State 3→ (P', V_1, T')

Thus, the process is not cyclic, since state 3 is not same as state 1.

18. When will be $\Delta H = \Delta U$?

Soln. : For a reaction at constant pressure involving ideal gases if Δn (difference of total number of moles of products and reactants) is zero, then $\Delta H = \Delta U$.

$\Delta H = \Delta U + RT\Delta n$,

if $\Delta n = 0$, $\Delta H = \Delta U$.

Also for solid or liquid if any process leads to the change of temperature only, then $\Delta H \approx \Delta U$.

19. In which case will the efficiency of Carnot cycle be higher : when the temperature of the hot reservoir is increased by ΔT or when the temperature of the sink is decreased by the same amount.

Soln. : Let the temperature of the high-temperature reservoir be T and that of low-temperature reservoir be T', the efficiency, $\eta = \dfrac{T-T'}{T} = \dfrac{x}{T}$.

If T is increased by ΔT, the efficiency, $\eta_1 = \dfrac{T+\Delta T-T'}{T+\Delta T} = \dfrac{x+\Delta T}{T+\Delta T}$

and when T' is decreased by ΔT, the efficiency

$$\eta_2 = \frac{T-(T'-\Delta T)}{T} = \frac{T-T'+\Delta T}{T} = \frac{x+\Delta T}{T}.$$

Obviously, $\eta_2 > \eta_1$. Hence, the efficiency is higher when the temperature of the sink is decreased by ΔT.

For further problems and short questions, consult the book on Physical Chemistrry (Vol. I) by Dr S. Pahari.

CHAPTER 6

SECOND LAW OF THERMODYNAMICS

Required Formulae

1. Entropy change $\Delta S = \dfrac{Q_{rev}}{T}$

2. ΔS_{mix} (per mole) $= -R \sum_i X_i \ln X_i$

   X_i = mole fraction of ith component in the mixture.

3. $dU = TdS - PdV$

   $dH = TdS + VdP$

   dF (or dA) $= -PdV - SdT$

   $dG = VdP - SdT.$

4. $\Delta G = nRT \ln \dfrac{P_2}{P_1}$ (for ideal gas)

   $= V\Delta P$ (for liquid)

   $= nRT \ln \dfrac{P_2}{P_1} + n\left(b - \dfrac{a}{RT}\right)(P_2 - P_1)$ (for van der Waals gas).

5. $\Delta S_{\text{phase change}} = \dfrac{L_f}{T}$ or $\dfrac{L_e}{T}$.

6. $\left(\dfrac{\partial(\Delta G / T)}{\partial T}\right)_P = -\dfrac{\Delta H}{T^2}$

   $\Delta G = A - BT \ln T + I_c . T,$ if $\Delta H = A + BT.$

7. $\Delta S = \int nC_P \dfrac{dT}{T} - \int nR \dfrac{dP}{P}$ (for ideal gas)

   $= \int nC_V \dfrac{dT}{T} + \int nR \dfrac{dV}{V}$ (for ideal gas).

Short Questions

1. A body at constant heat capacity C_p at a temperature T_1 is put in contact with a reservoir at a higher temperature T_2. The body comes to equilibrium with the reservoir at constant pressure. Show that the entropy change of the universe is

$$nC_p\left\{\frac{T_1-T_2}{T_2}-\ln\left(1+\frac{T_1-T_2}{T_2}\right)\right\}.$$

Soln. : Let there be n moles with constant C_p.

$$\Delta S_{sy} = nC_p \ln\frac{T_2}{T_1}.$$

Reservoir remains at constant temperature T_2.

$$\Delta H_{sy} = nC_p\,(T_2 - T_1)$$

$$\Delta H_R = -nC_p(T_2 - T_1) = nC_p\,(T_1 - T_2)$$

$$\Delta S_R = \frac{\Delta H_R}{T_2} = \frac{nC_P(T_1-T_2)}{T_2}$$

$$\Delta S_{un} = \Delta S_{sy} + \Delta S_R = -nC_p \ln\frac{T_1}{T_2} + nC_p\frac{(T_1-T_2)}{T_2}$$

$$= -nC_p \ln\left(1+\frac{(T_1-T_2)}{T_2}\right) + nC_p\frac{(T_1-T_2)}{T_2}.$$

$$= nC_p\left[\frac{T_1-T_2}{T_2}-\ln\left(1+\frac{T_1-T_2}{T_2}\right)\right].$$

2. Show that

$$\left(\frac{\partial S}{\partial P}\right)_V = \frac{\beta C_V}{\alpha T} \text{ and } \left(\frac{\partial S}{\partial V}\right)_P = \frac{C_P}{aVT}.$$

Soln. : We know that, $dS = C_v\frac{dT}{T}+\left(\frac{\partial P}{\partial T}\right)_V dV$ for 1 mole of gas.

or, $dS = C_v\frac{dT}{T} + \frac{\alpha}{\beta}dV$

or, $$\left(\frac{\partial S}{\partial P}\right)_V = \frac{C_V}{T}\left(\frac{\partial T}{\partial P}\right)_V = \frac{\beta C_V}{\alpha T}\left[\text{since,}\left(\frac{\partial P}{\partial T}\right)_V = \frac{\alpha}{\beta} \text{ and} \left(\frac{\partial P}{\partial T}\right)_V = \frac{1}{\left(\frac{\partial T}{\partial P}\right)_V}\right].$$

Again, $dS = C_p\frac{dT}{T} - \left(\frac{\partial V}{\partial T}\right)_P dP$

or, $dS = C_p\frac{dT}{T} - \alpha VdP.$

$$\therefore \left(\frac{\partial S}{\partial V}\right)_P = \frac{C_P}{T}\left(\frac{\partial T}{\partial V}\right)_P = \frac{C_P}{V\alpha T} \left[\text{since}, \alpha = \frac{1}{V}\left(\frac{\partial V}{\partial T}\right)_P \text{ and} \left(\frac{\partial T}{\partial V}\right)_P = \frac{1}{\left(\frac{\partial V}{\partial T}\right)_P}\right].$$

3. The compressibility factor for a van der Waals gas is given by

$\frac{PV}{RT} = 1 + \left(b - \frac{a}{RT}\right)\frac{P}{RT}$. Show that $\left(\frac{\partial H}{\partial P}\right)_T = b - \frac{2a}{RT}$.

Soln. : By the problem,

$$PV = RT + \left(b - \frac{a}{RT}\right)P$$

or, $V = \frac{RT}{P} + \left(b - \frac{a}{RT}\right)$ or, $\left(\frac{\partial V}{\partial T}\right)_P = \frac{R}{P} + \frac{a}{RT^2}$.

$$\therefore \quad T\left(\frac{\partial V}{\partial T}\right)_P = \frac{RT}{P} + \frac{a}{RT}.$$

But $\left(\frac{\partial H}{\partial P}\right)_T = V - T\left(\frac{\partial V}{\partial T}\right)_P$.

Since, $dH = TdS + VdP$,

$$\therefore \quad \left(\frac{\partial H}{\partial P}\right)_T = T\left(\frac{\partial S}{\partial P}\right)_T + V$$

$$= V - T\left(\frac{\partial V}{\partial T}\right)_P \quad \left[\text{by Maxwell relation}, \left(\frac{\partial S}{\partial P}\right)_T = -\left(\frac{\partial V}{\partial T}\right)_P\right]$$

Hence, $\left(\frac{\partial H}{\partial P}\right)_T = \frac{RT}{P} + \left(b - \frac{a}{RT}\right) - \frac{RT}{P} - \frac{a}{RT} = b - \frac{2\alpha}{RT}$.

4. Define coefficient of compressibility at constant entropy.

Show that $\beta_S = \frac{1}{\gamma}\beta_T$.

Soln. : We know that $\beta_T = -\frac{1}{V}\left(\frac{\partial V}{\partial P}\right)_T$ and $\beta_S = -\frac{1}{V}\left(\frac{\partial V}{\partial P}\right)_S$.

$$\text{Now, } dS = C_p\frac{dT}{T} - \alpha VdP \tag{i}$$

$$= C_v\frac{dT}{T} + \frac{\alpha}{\beta_T}dV. \tag{ii}$$

$\therefore \quad \left(\frac{\partial V}{\partial T}\right)_S = -\frac{\beta_T C_V}{\alpha T}$ [putting $dS = 0$ in eqn. (i) and eqn. (ii)]

and $\left(\frac{\partial P}{\partial T}\right)_S = \frac{C_P}{\alpha TV}$.

$$\therefore \quad \left(\frac{\partial V}{\partial P}\right)_S = \left(\frac{\partial V}{\partial T}\right)_S \left(\frac{\partial T}{\partial P}\right)_S = -\frac{\beta_T C_V}{\alpha T} \cdot \frac{\alpha T V}{C_P} = -\beta_T \frac{V}{\gamma} \; [\text{since, } \frac{C_P}{C_V} = \gamma]$$

or, $$-\frac{1}{V}\left(\frac{\partial V}{\partial P}\right)_S = \beta_s = \frac{\beta_T}{\gamma}.$$

5. Show that, $C_v = \dfrac{\alpha^2 T \beta_S V}{\beta_T(\beta_T - \beta_S)}$.

Soln. : We know that, $\beta_S = -\dfrac{1}{V}\left(\dfrac{\partial V}{\partial P}\right)_S = \dfrac{\beta_T}{\gamma}$

$$\beta_T = -\frac{1}{V}\left(\frac{\partial V}{\partial P}\right)_T$$

Now, $$\frac{\alpha^2 T V \beta_S}{\beta_T(\beta_T - \beta_S)} = \frac{\alpha^2 T V \beta_T}{\gamma \beta_T \left(\beta_T - \dfrac{\beta_T}{\gamma}\right)}.$$

$$= \frac{\alpha^2 T V}{\gamma \beta_T \left(1 - \dfrac{1}{\gamma}\right)} = \frac{C_P - C_V}{\gamma - 1} \left[\text{since, } C_P - C_V = \frac{\alpha^2 T V}{\beta_T}\right]$$

$$= \frac{C_P - C_V}{\dfrac{C_P}{C_V} - 1} = C_V.$$

6. Show that $C_p = \dfrac{\alpha^2 T V}{\beta_T - \beta_S}$.

Hints : Proceed as Problem 5 and use $\beta_S = \dfrac{\beta_T}{\gamma}$.

7. Show that $\left(\dfrac{\partial F}{\partial V}\right)_S = S\left(\dfrac{\partial P}{\partial S}\right)_V - P.$

Soln. : We know that $dF = -PdV - SdT$.

$$\therefore \quad \left(\frac{\partial F}{\partial V}\right)_S = -P - S\left(\frac{\partial T}{\partial V}\right)_S.$$

According to Maxwell relation, $\left(\dfrac{\partial T}{\partial V}\right)_S = -\left(\dfrac{\partial P}{\partial S}\right)_V.$

Thus, $$\left(\frac{\partial F}{\partial V}\right)_S = S\left(\frac{\partial P}{\partial S}\right)_V - P.$$

8. For each of the following processes, state which of the following quantities ΔU, ΔH, ΔS, ΔF and ΔG are equal to zero for the system specified.

(a) A non-ideal gas is taken around a Carnot cycle.

Hints : All zeros since cyclic change of state function is zero.

(b) A non-ideal gas expanded through a throttling valve.

Hints : Joule-Thomson effect, $\Delta H = 0$.

(c) Water is vaporised at 100 °C and 1 atm.

Hints : $\Delta G = 0$.

(d) H_2 and O_2 react to form H_2O in a thermally isolated box.

Hints : All non-zeros.

(e) An ideal gas is adiabaticallly expanded through a throttling valve.

Hints : $\Delta H = 0, \Delta U = 0$.

(f) An ideal gas is allowed to expand from $P_1 \to P_2$ at constant T.

Hints : $\Delta U = \Delta H = 0$.

(g) HCl and NaOH react to form H_2O and NaCl at constant T and P.

Hints : All non-zeros.

9. At what condition is $\Delta U = \Delta G = -W$ true ?

Ans. : By first law, $đQ = dU + đW$. Under adiabatic condition $đQ = 0$,

$\therefore\ đW = -dU$, or $-W = \Delta U$. For a system performing mechanical work only, $-W = \Delta G$ at constant temperature and pressure. So if an adiabatic change takes place at constant T and P, then $\Delta G = \Delta U = -W$.

10. "Any spontaneous process is always accompanied by an increase in the entropy (ΔS) of the system." Is the statement correct? Briefly explain.

Ans. : The second law of thermodynamics predicts that ΔS of the universe is increasing. $\Delta S_{universe}$ increases for spontaneous change. This prediction is for isolated system Thus, the statement is wrong.

11. A solid body initially at a higher temperature is left in air. It is observed that the body will spontaneously cool down to room temperature. Does the free energy increase, decrease or remain constant? Does your calculation contradict the second law of thermodynamics? Give reasons for your answer.

Ans. : It can be shown that for a spontaneous process ΔG decreases at constant T and P. At constant pressure, $dG = -SdT$ and $\left(\frac{\partial \Delta G}{\partial T}\right)_P = -\Delta S$. Decrease of temperature leads to decrease in entropy, so that for a temperature decrease, G_{system} increases. A solid body at high temperature will cool down with increase in ΔG.

Here temperature is not constant. There is no prediction for ΔG regarding spontaneity when T changes. So this cooling is not against the prediction of thermodynamics.

12. Correct or justify the following observations—

(a) Below 0 °C, supercooled water changes spontaneously into ice.

(b) If the same change of state from state A to state B be effected in a system in different ways, the free energy change in each case will be same provided the processes are reversible.

Ans. : (a) For the system ice→water, as the temperature decreases, the vapour pressure of both ice and water decreases. But vapour pressure of ice decreases at a higher rate so that below 0 °C, vapour pressure of ice is lower than that of water. Here, ΔG is –ve for liquid to solid at constant T and P, since $\Delta G = V\Delta P$. So supercooled water changes spontaneously to ice.

(b) For a state change, ΔG = constant. and it does not depend on path history. For state change from $A \rightarrow B$, $\Delta G = G_B - G_A$ and it is immaterial whether the change is reversible or irreversible.

13. What is the unit of entropy? 100 gm of water is kept at 100 °C, what will happen? Will the entropy of water increase or decrease or remain constant? Is your finding against the second law of thermodynamics? Explain.

Ans. : Unit of entropy is $J.K^{-1}$. In the given case, system will spontaneously cool down. So entropy of water will decrease.

By second law of thermodynamics, entropy of universe has to increase for a spontaneous change ($\Delta S_{universe}$ is (+)ve). There is no prediction for entropy change of closed or open system regarding spontaneity. Second law predicts that for a spontaneous change ΔS_{system} is (+)ve, if it is isolated. Here system is not isolated. It loses heat to surrounding, thereby increasing latter's entropy. So it is in conformity with the second law.

14. One kg of water at 25 °C and 1 atm is subjected to a pressure of 100 atm, the temperature remaining constant. Calculate the change in free energy. Is ΔG positive? Is your finding in conformity with the 2nd law?

Ans. : For the given change let the volume be constant. If the density is assumed to be 1 gm/cc then volume is 1000 cc and $\Delta G = V\Delta P = 1 \times 99$ lit.atm = 99 lit.atm. So ΔG increases. By second law, ΔG is (–)ve for a spontaneous change at constant P and T. There is no prediction for ΔG regarding spontaneity if P or T changes. Thus it is in conformity with the second law.

15. What are the differences between J-T cooling and adiabatic cooling?

Adiabatic Cooling	***J-T Cooling***
1. Temperature decreases and ΔH decreases.	1. Temperature decreases below inversion temperature of a real system, but $\Delta H = 0$.
2. There is no inversion temperature.	2. There is an inversion temperature above which a real gas gets heated on passing from high to low pressure.
3. In adiabatic cooling, ideal gases are also cooled due to expansion from high to low pressure.	3. For ideal gas, there is no J-T cooling.

16. An adiabatic process is always isoentropic.—Comment.

Ans. : The above statement is incorrect. For adiabatic reversible process, $Q = 0$; this is reversibly taken or, released heat. So $\Delta S = \dfrac{Q_{rev}}{T} = 0$. In adiabatic irreversible process $Q = 0$, but that is irreversible heat change. There is no relation between ΔS and irreversible heat change. Actually irreversible heat change leads to increase in entropy of system and the universe.

17. If $\Delta H^\circ = 0$ for a reaction, then show $\Delta S^\circ = 0$ for the same reaction.

Ans. : We know that for reaction $\left(\dfrac{\partial \Delta H^\circ}{\partial T}\right)_P = \Delta C_P^\circ$.

$\therefore\ d\Delta H^\circ = \Delta C_P^\circ.dT.$

$\therefore \ \Delta H^\circ = \Delta C_P^\circ T + I_c$ [on integrating].

Since $\Delta H^\circ = 0$, ΔC_P° and I_c must be zero.

We know that $\left(\dfrac{\partial \Delta S^\circ}{\partial T}\right)_P = \dfrac{\Delta C_P^\circ}{T}$.

$\therefore \ \Delta S^\circ = n\Delta C_P^\circ \ln T.$

Now $\Delta C_P^\circ = 0.$

$\therefore \ \Delta S^\circ = 0.$

18. If the standard Gibbs function change of a reaction is zero, the reaction is thermodynamically impossible. State whether the statement is true or false. Justify.

Ans. : The statement is false.

The standard Gibbs energy change ΔG° is the difference between the Gibbs free energy between the unmixed products and unmixed reactants at their standard states. There is no relation between standard free energy change and spontaneity. For a spontaneous change, ΔG has to be negative at constant T and P. Again, $\Delta G^\circ = -RT \ln K_p$. If ΔG° is zero, then $K_p = 1$. A reaction with $K_p = 1$ is quite possible.

19. State briefly indicating reasons whether the following statements are correct or incorrect :

(A) Any spontaneous process is always accompanied by a decrease in free energy.

Ans. : The above statement is incorrect. The free energy, G, decreases for a spontaneous process at constant temperature (T) and pressure (P).

$$G = H - TS.$$

At constant temperature,

$$\Delta G = \Delta H - T\Delta S = \Delta H - Q_{rev}.$$

At constant P, $\Delta G = Q - Q_{rev} = 0$, for reversible change.

$= (-)$ve, for spontaneous irreversible change, since $Q_{rev} > Q$.

Thus, the above statement is true only at constant T and P.

(B) For any adiabatic process the entropy change $\Delta S = 0$, since $Q = 0$.

Ans. : The above statement is incorrect. ΔS is given by $\Delta S = \dfrac{Q_{rev}}{T}$, where Q_{rev} is the reversible heat change at temperature T and there is no relation between entropy change and irreversible heat change. In an irreversible adiabatic process, the heat involved is zero, but it is irreversible heat (Q_{irrev}) and ΔS is not equal to $\dfrac{Q_{irrev}}{T}$.

For a reversible adiabatic the heat term Q_{rev} is zero and ΔS is $\dfrac{Q_{rev}}{T} = \dfrac{0}{T} = 0$. Thus, $\Delta S \neq 0$ for any adiabatic process.

(C) When water is evaporated at 100 °C and at 1 atm, the change in free energy is zero.

Ans. : We know that, $\Delta G = 0$ at constant T and P at equilibrium. But in such a case, the equilibrium can be attained if liquid water is kept in the vessel at 100 °C and 1 atm covered by a movable piston. Evaporation means a process where the vessel is open to atmosphere, and vapour escapes. Naturally, equilibrium between two phases cannot be attained. ΔG is not equal to zero. The process is spontaneous vaporisation and ΔG will be (–)ve. However, at 1 atm and 100 °C the vapour may not escape since the external pressure is also 1 atm. Thus ΔG may, in fact, be equal to zero.

(D) Free energy increases with increase of temperature at constant pressure.

Ans. : The above statement is incorrect. We know that

$$\left(\frac{\partial G}{\partial T}\right)_P = -S$$

or, $$\Delta G = -\int S\,dT.$$

S increases with increase of temperature. Since ΔG is negative, G decreases with increase of temperature and the above statement is incorrect. The plot of G vs. T is shown in figure.

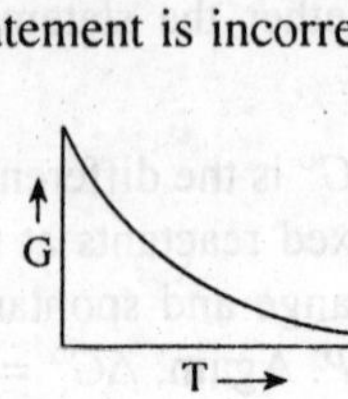

(E) Free energy increases with increase of pressure at constant temperature.

Ans. : We know that $\left(\frac{\partial G}{\partial P}\right)_T = V$. Thus, the slope of G vs. P is positive and G increases with increase of pressure. Hence, the above statement is correct. For ideal gas,

$$\Delta G = nRT \ln\left(\frac{P_2}{P_1}\right)$$

and for solid or liquid, $\Delta G = V(P_2 - P_1)$.

(F) Entropy is a state function and hence does not depend on the amount of matter in the system.

Ans. : The above statement is incorrect. We know that entropy is an extensive property. Hence, it depends on the amount of matter in the system and thus the portion 'does not' is incorrect.

Entropy is a state function, i.e., a property. But a property may be either extensive or intensive. Thus, the fact that entropy is a property does not mean that it is an extensive property. As such the word 'hence' is wrong. The correct statement is :

Entropy is a state function and it depends on the amount of matter in the system.

20. Predict the sign of ΔS for each of the following processes :

(a) C(s, graphite) $\rightarrow$ C(s, diamond)

(b) $Br_2(l) \rightarrow Br_2(g)$

(c) Hard boiling of an egg

(d) $O_2(g) \rightarrow 2O(g)$

(e) Devitrification of glass

(f) $N_2(g) + 3H_2(g) \rightarrow 2NH_3(g)$

(g) N_2 (g, 10 atm) $\rightarrow$ N_2(g, 1 atm)

(h) Desalination of sea-water

(i) $C(s) + H_2O(g) \rightarrow CO(g) + H_2(g)$

Ans. :

(a) Negative. Diamond, being a harder solid, would be expected to have more restricted atomic motions within the crystal.

(b) Positive. S is always greater for a gas than for its corresponding liquid.

(c) Positive. The fundamental process in the "boiling" of an egg is not a literal boiling, in the sense of vaporization, but a denaturation of the egg protein. A protein is a large molecule which exists in a particular configuration in the so-called native state but may occupy a large number of almost random configurations in the denatured state, resulting from rotations around the bonds. The increase in the number of possible configurations is analogous to the melting process.

(d) Positive. There is an increase in the number of gas molecules.

(e) Negative. Devitrification is the onset of crystallization in an amorphous solid, i.e., supercooled liquid.

(f) Negative. There is a decrease in the number of gas molecules.

(g) Positive. Entropy increases on expansion.

(h) Negative. Desalination is the opposite of dissolution in which solute must be removed from a solution.

(i) Positive. There is an increase in the number of gas molecules.

21. Why ΔS is not a state function?

Ans. : dS is exact differential and $\int_1^2 dS = \Delta S$. It is the change of S, the entropy, which is a state function and thus, ΔS depends only on states 1 and 2 and not on the nature of the path. However, a state function refers to a state but ΔS does not refer to any state. It refers to a state change and thus not a state function.

22. H_2 at room temperature is warmed as a result of passage from high to low pressure in the J-T cooling—explain.

Ans. : We know that J-T co-efficient $\mu_{J\text{-}T} = \left(\frac{\partial T}{\partial P}\right)_H$. For a van der Waals gas $\left(\frac{\partial T}{\partial P}\right)_H$ $= \frac{\left(\frac{2a}{RT} - b\right)}{C_P}$, where a and b are van der Waals constants and T is the temperature at which J-T experiment is carried. At inversion temperature, $T_i = \frac{2a}{Rb}$, there will be no J–T cooling or heating.

$$\therefore \quad \left(\frac{\partial T}{\partial P}\right)_H = \frac{\left(\frac{2a}{RT} - b\right)}{C_P}.$$

$$\therefore \quad dT = \frac{\left(\frac{bT_i}{T} - b\right)}{C_P}.dP$$

$$\text{or,} \quad \Delta T = \frac{b(T_i - T)}{TC_P}.\Delta P.$$

If $T > T_i$, the system will be heated, if ΔP is (–)ve. For H_2, T_i is much below room temperature and thus, it is heated in the J–T cooling.

23. Show that, $\Delta G_{system} = -T.\Delta S_{universe}$ at constant temperature and pressure.

Ans. : We know that, $G_s = H_s - T_s S_s$ where G is the Gibbs free energy of the system, S is the entropy of the system, T is temperature and H is the enthalpy of the system. For a large change at constant temperature,

$$\Delta G_s = \Delta H_s - T_s \Delta S_s \text{ ['s' indicates 'system'.]}$$

But at constant pressure,

$$\Delta H_s = Q_P$$

$$\therefore \quad \Delta G_s = Q_P - T_s \Delta S_s.$$

The system receives Q_P heat from the surrounding which loses Q_P heat at temperature T. Thus,

$$Q_{system} = -Q_{surrounding}.$$

$$\therefore \quad \Delta G_s = -Q_{surrounding} - T_s \Delta S_s.$$

But surrounding is assumed to exchange heat reversibly.

$$\therefore \quad \Delta S_{surrounding} = \frac{Q_{surrounding}}{T}$$

$$\therefore \quad \Delta G_s = -T\Delta S_{surrounding} - T\Delta S_{system}$$

$$= -T(\Delta S_{surrounding} + \Delta S_{system})$$

$$\therefore \quad \Delta G_s = -T\Delta S_{universe}. \text{ [Proved].}$$

24. Ice is melted at 0 °C and 1 atm. State which of ΔU, ΔH, $\Delta S_{universe}$, ΔA and ΔG must be zero for process.

Ans. : For this change which is an equilibrium at constant temperature and pressure, $\Delta G = 0$. As because it is a reversible process $\Delta S_{universe} = 0$ ($\Delta G_{sysem} = -T\Delta S_{universe}$, thus if $\Delta G = 0$, $\Delta S_{universe}$ must be zero).

25. What are the values of $\left(\frac{\partial S}{\partial V}\right)_P$ and $\left(\frac{\partial S}{\partial P}\right)_V$ for an ideal gas?

Ans. : From problem 2, $\left(\frac{\partial S}{\partial V}\right)_P = \frac{C_P}{\alpha VT}$, and $\left(\frac{\partial S}{\partial P}\right)_V = \frac{\beta C_V}{\alpha T}$.

$$\alpha = \frac{1}{V}\left(\frac{\partial V}{\partial T}\right)_P.$$

For ideal gas, $V = \frac{RT}{P}$. (i)

$$\text{Thus, } \alpha = \frac{1}{V}\left(\frac{\partial V}{\partial T}\right)_P = \frac{R}{PV}.$$

$$\text{Hence, } \left(\frac{\partial S}{\partial V}\right)_P = \frac{C_P PV}{RVT} = \frac{PC_P}{RT} = \frac{C_P}{V} \text{ [from (i).]}$$

$$\text{Again, } \beta = -\frac{1}{V}\left(\frac{\partial V}{\partial P}\right)_T.$$

Thus, $-\frac{1}{V}\left(\frac{\partial V}{\partial P}\right)_T = \frac{RT}{P^2V}$.

Hence, $\left(\frac{\partial S}{\partial P}\right)_V = \frac{RTC_VPV}{P^2VRT} = \frac{C_V}{P}$.

26. Find the values of $\left(\frac{\partial S}{\partial V}\right)_P$ and $\left(\frac{\partial S}{\partial P}\right)_V$ for a gas obeying $PV = RT + Pb$, where b is independent of temperature and pressure.

Ans. : Proceeding as Problem 27,

$$\left(\frac{\partial S}{\partial V}\right)_P = \frac{C_PP}{RT} \text{ and } \left(\frac{\partial S}{\partial P}\right)_V = \frac{C_V}{P}.$$

27. Calculate $\left(\frac{\partial C_P}{\partial P}\right)_T$ for ideal gas and van der Waals gas.

Ans. : $\left(\frac{\partial C_P}{\partial P}\right)_T = \left[\frac{\partial}{\partial P}\left(T\frac{\partial S}{\partial T}\right)_P\right]_T = T\left[\frac{\partial}{\partial P}\left(\frac{\partial S}{\partial T}\right)_P\right]_T$

$$= T\left[\frac{\partial}{\partial T}\left(\frac{\partial S}{\partial P}\right)_T\right]_P$$

By Maxwell relation, $\left(\frac{\partial S}{\partial P}\right)_T = -\left(\frac{\partial V}{\partial T}\right)_P$.

Thus, $\left(\frac{\partial C_P}{\partial P}\right)_T = -T\left(\frac{\partial^2 V}{\partial T^2}\right)_P$

For an ideal gas, $PV = nRT$.

Hence, $\left(\frac{\partial V}{\partial T}\right)_P = \frac{nR}{P}$.

$\therefore \left(\frac{\partial^2 V}{\partial T^2}\right)_P = 0.$

Thus, $\left(\frac{\partial C_P}{\partial P}\right)_T = 0.$

For van der Waals gas, $P = \frac{nRT}{V-nb} - \frac{an^2}{V^2}$.

It can be written as, $V = \frac{nRT}{P} + n\left(b - \frac{a}{RT}\right)$

Thus, $\left(\frac{\partial V}{\partial T}\right)_P = \frac{nR}{P} + \frac{na}{RT^2}$.

$$\therefore \quad \left(\frac{\partial^2 V}{\partial T^2}\right) = -\frac{2na}{RT^3}.$$

Hence, $\left(\frac{\partial C_P}{\partial P}\right)_T = \frac{2na}{RT^2}.$

Alternative Method : See Q. 12 of Miscellaneous in Chapter 5 for ideal gas.

$$\left(\frac{\partial H}{\partial P}\right)_T = V - T\left(\frac{\partial V}{\partial T}\right)_P \quad \text{[using Maxwell relation, } \left(\frac{\partial S}{\partial P}\right)_T = -\left(\frac{\partial V}{\partial T}\right)_P \text{].}$$

For van der Waals gas, $V = \frac{nRT}{P} + n\left(b - \frac{a}{RT}\right).$

$$\left(\frac{\partial V}{\partial T}\right)_P = \frac{nR}{P} + \frac{na}{RT^2}.$$

Thus, $\left(\frac{\partial H}{\partial P}\right)_T = V - \frac{nRT}{P} - \frac{na}{RT} = \frac{nRT}{P} + nb - \frac{na}{RT} - \frac{nRT}{P} - \frac{na}{RT}$

$$= nb - \frac{2na}{RT}.$$

Thus, $\left(\frac{\partial C_P}{\partial P}\right)_T = \left[\frac{\partial}{\partial T}\left(\frac{\partial H}{\partial P}\right)_T\right]_P = \frac{2na}{RT^2}.$

Here a and b are assumed to be temperature and pressure independent.

28. Find out the value of $\left(\frac{\partial C_V}{\partial V}\right)_T$ for ideal gas and van der Waals gas.

Ans. : $\left(\frac{\partial C_V}{\partial V}\right)_T = \left[\frac{\partial}{\partial V}\left(T\frac{\partial S}{\partial T}\right)_V\right]_T = T\left[\frac{\partial}{\partial V}\left(T\frac{\partial S}{\partial T}\right)_V\right]_T = T\left[\frac{\partial}{\partial T}\left(\frac{\partial S}{\partial V}\right)_T\right]_V.$

By Maxwell relation, $\left(\frac{\partial S}{\partial V}\right)_T = \left(\frac{\partial P}{\partial T}\right)_V.$

Thus, $\left(\frac{\partial C_V}{\partial V}\right)_T = T\left(\frac{\partial^2 P}{\partial T^2}\right)_V.$

For ideal gas, $PV = nRT,$

$$\left(\frac{\partial^2 P}{\partial T^2}\right)_V = 0.$$

Hence, $\left(\frac{\partial C_V}{\partial V}\right)_T = 0.$

For van der Waals gas, $P = \frac{nRT}{V - nb} - \frac{an^2}{V^2}.$

Again, $\left(\frac{\partial^2 P}{\partial T^2}\right)_V = 0$. (since a and b are temperature independent)

So, $\left(\frac{\partial C_V}{\partial V}\right)_T = 0$.

Alternative Method : See Q. 12 of Miscellaneous in chapter 5.

$$\left(\frac{\partial U}{\partial V}\right)_T = T\left(\frac{\partial P}{\partial T}\right)_V - V \text{ [using Maxwell relation, } \left(\frac{\partial S}{\partial V}\right)_T = \left(\frac{\partial P}{\partial T}\right)_V \text{]}$$

For van der Waals gas, $P = \frac{nRT}{V-nb} - \frac{an^2}{V^2}$

$$\left(\frac{\partial P}{\partial T}\right)_V = \frac{nR}{V-nb}$$

Thus, $\left(\frac{\partial U}{\partial V}\right)_T = \frac{nRT}{V-nb} - \frac{nRT}{V-nb} + \frac{an^2}{V^2} = \frac{an^2}{V^2}$.

Hence, $\left(\frac{\partial C_V}{\partial V}\right)_T = \left[\frac{\partial}{\partial T}\left(\frac{\partial U}{\partial V}\right)_T\right]_V = 0$ (since a and b are considered to be temperature independent)

29. Give examples of a process in each of which for the system (i) $\Delta G = 0$, $\Delta S > 0$, (ii) $\Delta G < 0$, $\Delta S < 0$, (iii) $\Delta G = 0$, $\Delta S < 0$, (iv) $\Delta G < 0$, $\Delta S > 0$.

Ans. : (i) Melting of ice to water at 0 °C and 1 atm : $\Delta G = 0$, $\Delta S > 0$.

(ii) Freezing supercooled water to ice at –5 °C and 1 atm : $\Delta G < 0$, $\Delta S < 0$.

(iii) Condensation of vapour to water at 100 °C and 1 atm : $\Delta G = 0$, $\Delta S < 0$.

(iv) Vaporization of superheated water to vapour at 105 °C and 1 atm : $\Delta G < 0$, $\Delta S > 0$.

30. The Helmholtz energy of one mole of a certain gas given by $A = -\left(\frac{a}{V}\right) - RT \ln (V - b) + f(T)$, where a and b are constants and $f(T)$ is a function of T only. Set up the expression for pressure of the gas.

Ans. : $A = \left(\frac{a}{V}\right) - RT \ln(V-b) + f(T)$

or, $\left(\frac{\partial A}{\partial V}\right)_T = \frac{a}{V^2} - \frac{RT}{V-b} = -P$.

$\therefore \quad P = \frac{RT}{V-b} - \frac{a}{V^2}$.

31. Though entropy is a fundamental state function and free energy is a derived one, the latter can be used more conveniently.—Explain.

Ans. : Both the entropy (S) and the free energy (G) indicate direction of a process. For a spontaneous process, ΔG of the system is (–)ve at constant temperature and pressure, and $\Delta G = 0$ at the same condition indicates equilibrium, while positive value of ΔS for universe indicates spontaneous process, and zero value of the same indicates reversible process. Sum of entropy change of system (ΔS_{system}) and surrounding (ΔS_{sur}) comprises entropy change for universe. Calculation of ΔS_{surr} is not always easy. So free energy is preferred to entropy as far as direction of a process is considered.

Numerical Problems

1. Consider the following cycle using 1 mole of an ideal gas initially at 25 °C and 1 atm.

(a) Isothermal expansion against zero pressure to double the volume.

(b) Isothermal reversible compression from $\frac{1}{2}$ to 1 atm.

Calculate W, ΔS for a and b in the cycle for the system and $\Delta S_{universe}$.

Soln. : By the problem, n = 1 mole, T = 298.15 K, P = 1 atm, $V = V$.

(a) For free expansion, $W = \mathbf{0}$.

For ideal gas $\Delta U = \mathbf{0}$ at constant T. $\therefore Q = 0$.

$$\text{But } \Delta S_{system} = nR \ln\frac{V_2}{V_1} = 8.314 \text{ J.K}^{-1} \ln 2 = \mathbf{5.76 \text{ J.K}^{-1}}.$$

$$\Delta S_{surr} = 0. \quad \therefore \Delta S_{universe} = \mathbf{5.76 \text{ J.K}^{-1}}.$$

$$\text{(b)} \quad \Delta S_{system} = nR \ln\frac{P_1}{P_2} = \mathbf{-5.76 \text{ J.K}^{-1}}.$$

$$W = -nR\,T\ln\frac{1}{2} = RT \ln 2 = \mathbf{1.72 \text{ kJ}}.$$

$$\mathbf{Q = -W = -1.72 \text{ kJ}}$$

$$\Delta S_{surr} = \frac{1.72}{298.5} = \mathbf{5.76 \text{ J.K}^{-1}}.$$

$$\Delta S_{universe} = 0.$$

For this cyclic change, $\Delta S_{universe} = \mathbf{5.76 \text{ J.K}^{-1}}$.

2. A quantity of ideal gas expands (i) isothermally and reversibly at 500 °K from a volume of V_1 to V_2. During the expansion the gas absorbs 400 cal of heat from the reservoir in contact with it. Find ΔS for (a) gas, (b) reservoir and (c) the whole system. (ii) If the expansion was carried out isothermally but irreversibly for the same volume change from V_1 to V_2 absorbing 200 cal of heat, what would be ΔS_g, ΔS_R and $\Delta S_{universe}$?

Soln. : (i) By the problem, T = 500 K

Q = 400 cal.

Since the change is reversible, $\Delta S_g = \frac{Q}{T} = \frac{400}{500} = 0.8 \text{ cal.K}^{-1} = \mathbf{3.35 \text{ J.K}^{-1}}$.

$\Delta S_R = -\ \mathbf{3.35 \text{ J.K}^{-1}}$. $\therefore \Delta S_{universe} = \Delta S_g + \Delta S_R = \mathbf{0}$.

(ii) $\Delta S_g = 3.35 \text{ J.K}^{-1}$ but $\Delta S_R = -\frac{200}{500} = -0.4 \text{ cal.K}^{-1} = \mathbf{-1.675 \text{ J.K}^{-1}}$.

$\therefore \Delta S_{universe} = (3.35 - 1.675) \text{ J.K}^{-1} = 1.675 \text{ J.K}^{-1} = \mathbf{1.68 \text{ J.K}^{-1}}$.

3. The molecular specific heat of CO_2 at constant pressure varies with absolute temperature according to the equation,

$$C_p = (7.0 + 0.0071\ T - 1.86 \times 10^{-6}\ T^2) \text{ cal.}$$

Calculate ΔS when 50 gm of CO_2 is brought from –10 °C to 50 °C.

Soln. : $\Delta S = \int nC_P \frac{dT}{T}$

$$= \int \frac{50}{44}\ [7.0\ \ln \frac{323.15}{263.15} + 0.0071(323.15 - 263.15) - \frac{1.86}{2}(323.15^2 - 263{\cdot}15^2)\times 10^{-6}]$$

$$= \frac{50}{44}(1.4377 + 0.426 - 0.0327155) = 2.08\ \text{cal.K}^{-1} = \mathbf{8.7 J.K^{-1}}.$$

4. For benzene, $\alpha = 1.24 \times 10^{-3}$ deg^{-1} and $\beta = 4.93 \times 10^{-5}$ atm^{-1} at 20 °C and 1 atm. Assuming α and β to be constant, find the change in molar volume which will be required to produce an entropy change of 6.75 J.deg^{-1}.

Soln. : According to the given problem—

$\alpha = 1.24 \times 10^{-3}\ K^{-1}$ constant at a constant temperature.

Similarly, $\beta = 4.93 \times 10^{-5}$ atm^{-1} constant at constant temperature.

Here, $T = 293.15$ K.

Now, $\Delta S = \frac{\alpha}{\beta}\Delta V.$

$$\therefore\ 6.75\ \text{J.deg}^{-1} = \frac{1.24\times10^{-3}\ \text{K}^{-1}}{4.93\times10^{-5}\ \text{atm}^{-1}} \times \Delta V\ \text{lit} \times 101.325\ \text{J/lit.atm.}$$

$$= 2548.5396 \times \Delta V\ \text{lit.}$$

$$\text{or,}\ \Delta V = \frac{6.75}{2548.5396} = 2.6486 \times 10^{-3}\ \text{lit} = \mathbf{2.6486\ cc.}$$

5. Calculate, ΔV in the above case if the entropy change is 0.82 cal/deg.

Hints : Same as Problem 4. **[Ans.** 1.34 cc]

6. At 30 °C, α for a certain liquid is 1.11×10^{-3} deg^{-1}. Its density is 1.03 gm/cc and 33 gm of the liquid was taken. The liquid was kept at 10 atm and the pressure was released to 1 atm. Calculate ΔS.

Soln. : According to the given problem,

$\alpha = 1.11 \times 10^{-3}\ K^{-1}$, which is constant at a constant temperature.

$\rho = 1.03$ gm/cc., $m = 33$ gm.

Hence, volume of liquid is $\frac{33}{1.03} = 32.03883495$ cc.

We know that, $\Delta S = \alpha V(P_1 - P_2)$

$$= 1.11\times10^{-3}\ \text{K}^{-1} \times \frac{32.0388}{1000}\ \text{lit} \times (10-1)\ \text{atm} \times 101.325\ \text{J/lit atm.}$$

$$= \mathbf{0.032\ J.K^{-1}}.$$

7. For an ideal gas, $C_p = 9.037 + 35.6 \times 10^{-3}T - 216.9 \times 10^{-7}T^2$ cal.mol^{-1}.deg^{-1}. Calculate ΔS when 5 moles of the gas are heated from 500 °K to 1000 °K, and the volume changes from 100 lit to 50 lit.

Soln. : According to the problem,

$C_p = 9.037 + 35.6 \times 10^{-3}\ T - 216.9 \times 10^{-7}\ T^2$ cal.mol^{-1}.deg^{-1}.

Here, n = 5 moles, T_1 = 500 K, T_2 = 1000 K, V_1 = 100 lit, V_2 = 50 lit.

$$\Delta S = n\int_{500}^{1000}\left(C_V\frac{dT}{T}\right) + nR\ln\frac{V_2}{V_1}$$

$$= 5[7.05\ \ln\frac{1000}{500} + 35.6\times10^{-3}(500) - \frac{216.9}{2}\times10^{-7}(1000^2 - 500^2)] + 5R\ \ln 2$$

Here, $C_v = C_p - R = 9.037 + 35.6\times10^{-3}T - 216.9\times10^{-7}T^2 - 1.987$

$= 7.05 + 35.6\times10^{-3}T - 216.9\times10^{-7}T^2$

$\therefore$ ΔS = **333.3 J.K^{-1}.**

8. For a certain ideal gas $C_p = 7.05 + 17.8\times10^{-3}T - 308\times10^{-7}T^2$ cal/mole deg Calculate ΔS if the state changes from 500 °K and 50 atm to 1000 °K and 120 atm and there are 10 moles of the gas.

Hints : Same as Problem 7. [**Ans.** 20.81 J.K^{-1}]

9. 3 moles of N_2 originally at 1 atm are mixed with 5 moles of H_2 at 1 atm isothermally to yield a mixture whose pressure is also 1 atm. Calculate ΔS.

Soln. : In the given problem, we have 3 moles of N_2 and 5 moles of H_2. So, mole fraction of $N_2 = \frac{3}{8}$ and that of $H_2 = \frac{5}{8}$.

We know that, $\frac{\Delta S}{\text{mole}} = -R\Sigma x_i\ \ln x_i$

$$= -R\left(\frac{3}{8}\ln\frac{3}{8} + \frac{5}{8}\ln\frac{5}{8}\right)\text{J.K}^{-1}.$$

Now, $\Delta S_{total} = -R(3\ \ln\frac{3}{8} + 5\ \ln\frac{5}{8})$ J.K^{-1}

$= 8.314\ (3\ \ln\frac{8}{3} + 5\ \ln\ \frac{8}{5})$J.K^{-1}

$= 8.314(2.94 + 2.35)$ J.K^{-1} = 8.314×5.29 J.K^{-1}

$= 43.9812$ J.K^{-1} = **44 J.K^{-1}.**

10. If two moles of an ideal gas A at 120 °C and 1 atm is mixed with 3 moles of another gas B at 70 °C and 1 atm and as a result the pressure remains at 1 atm. Calculate ΔS.

Soln. : By the problem, n_A = 2 moles, T_A = 120 °C = 393.15 K, P = 1 atm, and n_B = 3 moles, T_B = 70 °C = 343.15 K, P = 1 atm.

Here heat transfer occurs from A to B. Let the final temperature be T. Thus,

$2(120 - T) = 3(T - 70)$

$\therefore$ T = 90 °C = 363.15 K.

Let the gas A and B be mixed at 90 °C. Thus, this is an example of free mixing.

So, $\frac{\Delta S}{\text{mole}} = -R\Sigma x_i\ln x_i = -\ 8.314\ \text{J.K}^{-1}.\text{mole}^{-1}\left(\frac{2}{5}\ln\frac{2}{5} + \frac{3}{5}\ln\frac{3}{5}\right)$

$= 5.59542$ J.K^{-1}.mole^{-1}.

$\therefore$ ΔS_{total} = 5 mole × 5.59542 J.K^{-1}.mole^{-1} = 27.977 J.K^{-1}.

Now, entropy change due to temperature change is

$$\Delta S = n_A C_{P_A} \ln \frac{363.15}{393.15} + n_B C_{P_B} \ln \frac{363.15}{343.15}$$

$$= 2 \times \frac{3}{2} R \ln \frac{363.15}{393.15} + 3 \times \frac{3}{2} R \ln \frac{363.15}{343.15} = 0.1396 \text{ J.K}^{-1}.$$

Hence, total entropy change = ΔS_{mix} = (27.977 + 0.13960) J.K^{-1} = **28.12 J.K^{-1}.**

Alternative way :

$$\Delta S_A = n_A C_{P_A} \ln \frac{T}{T_A} - n_A R \ln \frac{P_A}{P} \quad (P_A = \frac{2}{5}, P = 1)$$

$$\Delta S_B = n_B C_{P_B} \ln \frac{T}{T_B} - n_B R \ln \frac{P_B}{P} \quad (P_B = \frac{3}{5}, P = 1)$$

and $\Delta S_{mix} = \Delta S_A + \Delta S_B$.

11. If two indentical monatomic perfect gases A and B with same pressure and the same number of molecules N kept in 2 bulbs of volumes V_A and V_B at temperatures T_A and T_B respectively are mixed together, calculate ΔS for mixing.

Soln. :

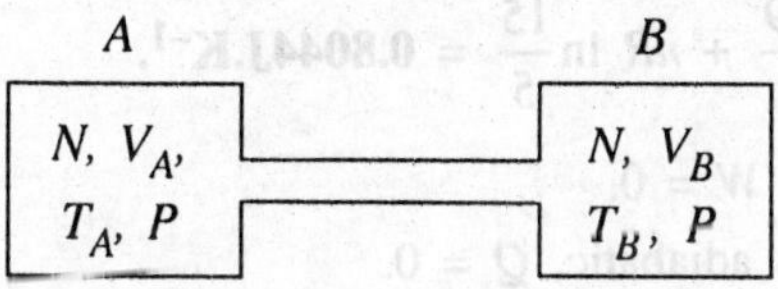

Let T' be the final temperature attained by both gases. Thus,

$$C_A(T_A - T') = C_B(T' - T_B).$$

$$\therefore T' = \frac{T_A + T_B}{2}. \text{ since } C_A = C_B.$$

Let, after mixing, P_A and P_B be partial pressures of gases A and B respectively. By the problem, $P = P_A = P_B$, since two gases are idential.

$$\Delta S_{mix} = \Delta S_A + \Delta S_B$$

$$= n_A C_{P_A} \ln \frac{T'}{T_A} + n_B C_{P_B} \ln \frac{T'}{T_B} = \frac{5}{2} N K_B \ln \frac{T'}{T_A} + \frac{5}{2} N K_B \ln \frac{T'}{T_B}$$

$$= \frac{5}{2} N K_B \ln \frac{T'^2}{T_A T_B} = \mathbf{5\ N K_B \ln \frac{(T_A + T_B)}{2\sqrt{T_A T_B}}}.$$

12. One mole of an ideal gas, originally at the volume of 5 lit at 500 °K is allowed to expand adiabatically until the final volume is 15 lit. For the gas, $C_v = \frac{3}{2} R$. Calculate final temperature and ΔS, when

(a) the expansion takes place reversibly,

(b) the expansion takes place against a constant pressure of 3 atm,

(c) the change in volume is due to free expansion.

Soln. :

5 l 500 °K n = 1 mole	⟶	15 l T_2°K : n = 1 mole

$C_v = \frac{3}{2}R$

(a) For reversible and adiabatic process, we know that—

$$T_1 V_1^{\gamma-1} = T_2 V_2^{\gamma-1}, \text{ where } \gamma = \frac{C_p}{C_v}$$

$$\text{or, } T_2 = T_1\left(\frac{V_1}{V_2}\right)^{\gamma-1} = 500\left(\frac{5}{15}\right)^{\frac{5}{3}-1} = \mathbf{240.5\ K.}$$

Since it is an adiabatic and reversible process, $\Delta S = 0$.

(b) We know that, $P_{ex}(V_2 - V_1) = nC_v(T_1 - T_2)$

$$\text{or, } 3(15-5) = \frac{3}{2}R(500 - T_2) \text{ (Here, } R = 0.0821 \text{ l-atm.K}^{-1}\text{.mole}^{-1})$$

$$\text{or, } T_2 = \mathbf{256.39.}$$

$$\therefore \Delta S = nC_v \ln\frac{256.39}{500} + nR\ln\frac{15}{5} = \mathbf{0.8044 J.K^{-1}}.$$

(c) For free expansion, $W = 0$.

Since the process is adiabatic, $Q = 0$.

$\therefore \Delta U = Q + W = 0$.

Since the gas is ideal, $\Delta T = 0$.

$\therefore T_2 = T_1 = 500$ K.

$$\text{Hence, } \Delta S = nR\ln\frac{V_2}{V_1} = 8.314 \ln 3 = \mathbf{9.13\ J.K^{-1}}.$$

13. For the fusion of ice, ΔH is given by $\Delta H = 1100 - 9T$. Calculate ΔG for the change— H_2O (l, 1 atm)→H_2O(s, 1atm) at –10 °C.

Soln. : At 273.15 K, $\Delta H = 1100 - 9T = 1100 - 9 \times 273.15$, (–)ve.
Here l→s transition is considered.

$$\text{We know that, } \left[\frac{\partial\left(\frac{\Delta G}{T}\right)}{\partial T}\right]_P = -\frac{\Delta H}{T^2}$$

$$\text{or, } d\left(\frac{\Delta G}{T}\right) = -\int\frac{\Delta H}{T^2}dT = -\int\frac{1100-9T}{T^2}dT.$$

$$\text{or, } \frac{\Delta G}{T} = \frac{1100}{T} + 9\ln T + I_c \text{ [where } I_c = \text{Integration Constant]}$$

or, $\Delta G = 1100 + 9T\ln T + I_c.T$. [Since the temperatures are different, there is a difference in these cost values.]

At 273.15 K, $\Delta G = 0$ for the above change.

$\therefore\ I_c = -54.52.$

$\therefore\ \Delta G = 1100 + 9\,T \ln T - 54.52\,T.$

Now, ΔG at -10 °C is

$\Delta G = 1100 + 9 \times 263.15 \ln 263.15 - 54.52 \times 263.15 = -48.78$ cal $=$ **−204.10 J.**

14. Find out ΔG at 1 atm per mole at −17 °C for $H_2O(l) \rightarrow H_2O(s)$. Given, $C_p(H_2O,\ l) = 18$ cal/mole, $C_p\ (H_2O,\ s) = 9$ cal/mole, $\Delta H_{273} = -1440$ cal.mole^{-1}.

Soln. : ΔC_p for the change $= -9$ cal/mole

[since, $\Delta C_p = C_p\,(H_2O,s) - C_p(H_2O,\ l) = 9 - 18$]

Now, we know that, $\left(\dfrac{\partial \Delta H}{\partial T}\right)_p = \Delta C_p.$

or, $\quad d\Delta H = \Delta C_p dT$

or, $\Delta H = \Delta C_p.T + I_c$ [where I_c = Integration Constant]

At 273.15, $\Delta H = -1440$, and $\Delta C_p = -9$.

$\therefore\ I_c = -1440 + (9 \times 273.15) = 1018.35$

$\therefore\ \Delta H = 1018.35 - 9T.$

Now, $\left[\partial\left(\dfrac{\Delta G}{T}\right)\Big/\partial T\right]_P = -\dfrac{\Delta H}{T^2}.$

Following Q. 13 we can write,

$\Delta G = 1018.35 + 9T \ln T + I_c.T.$

But at 273.15, $\Delta G = 0.$

$\therefore\ I_c = -54.22$

$\therefore\ \Delta G = 1018.35 + 9\,T \ln T - 54.22\,T.$

At −17 °C,

$\Delta G = 1018.35 + (9 \times 256.15 \ln 256.15) - (54.22 \times 256.15)$

$= -85.18$ cal.mole^{-1} $=$ **−356.39 J.mole^{-1}.**

15. 1 mole of He initially at 27 °C and 1 atm is heated to 327 °C, the final pressure of the gas being adjusted to 10 atm. Calculate ΔS and ΔG assuming ideal behaviour. Given that, $S^{\circ}_{27} = 30.20$ cal/deg.

Soln. : The process may be depicted as—

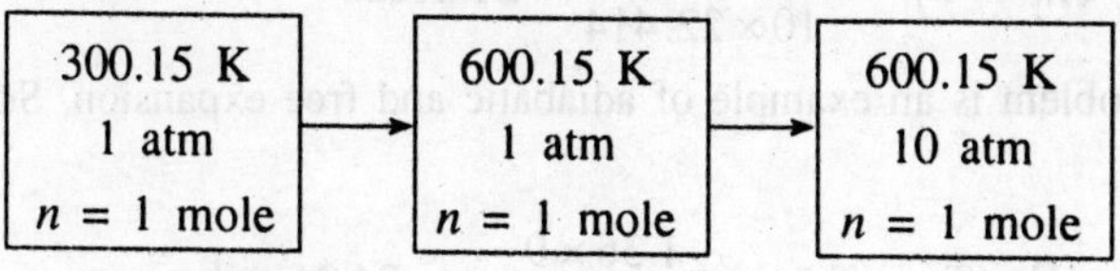

Here, $\Delta S = nC_p \ln\dfrac{600.15}{300.15} - n\,R \ln\dfrac{10}{1}.$

Putting $n = 1$ and $C_p = \dfrac{5}{2}R$, we get

$\Delta S = -1.13$ cal/deg $=$ **− 4.74 J/deg.**

Now, $\Delta G_1 = \Delta H - (T_2S_2 - T_1S_1).$

Again, $\Delta H = nC_p \, dT$ and $S_2 = S_1 + nC_P \ln\left(\frac{T_2}{T_1}\right)$.

Thus, $S_2 = S° + C_p \ln\frac{T_2}{T°}$ and $S_1 = S° + C_p \ln\frac{T_1}{T°}$.

By the problem, $T_1 = T°$ and $S_1 = S° = 30.20$ cal/deg.

Hence, $\Delta G_1 = (2.5 \times 1.987 \times 300)$ cal $- 600.15(S° + C_p \ln \frac{T_2}{T°}) + 300.15 \times S°$

$= 1490.25 - 600.15\ (30.2 + 2.5 \times 1.987 \times 0.693) + 300.15 \times 30.2$

$= -9635.45$ cal.

Again, $\Delta G_2 = +1.987 \times 600.15 \ln \frac{10}{1}$ cal $= +2745.8282$ cal.

$\therefore \quad \Delta G = (-9635.45 + 2745.8282)$ cal

$= -6889.6218$ cal $\times$ 4.184 J/cal = **−28.83 kJ.**

16. 1 mole of O_2 escapes adiabatically into an evacuated chamber of such size that its final volume is 10 times the initial volume. Given that $C_v = 4.90$ cal and 'a' for oxygen is 1.36 atm.lit^2.mole^{-2}. Calculate ΔS if the initial temperature is 0 °C (For volume calculation use ideal gas equation).

Soln. :

$n = 1$ Volume $= V_1 = V$ $P = 1$ atm	Free expansion →	$n = 1$ Volume $= V_2 = 10V$.

It is an example of free expansion.
The initial volume of the gas (O_2), V

$$= \frac{nRT}{P} = \frac{1\text{mole} \times 0.08206 \text{ l.atm.K}^{-1}.\text{mol}^{-1} \times 273.15\text{K}}{1\text{ atm.}} = 22.414 \text{ lit.}$$

Now, we know that, $\int\left(\frac{dU}{dV}\right)_T dV = \int_{V_1}^{V_2} \frac{a}{V^2} dV = a.\frac{V_2 - V_1}{V_1 V_2} = a.\frac{9V}{10V^2}$

$$= \frac{1.36 \times 9}{10 \times 22.414} \text{ lit.atm.}$$

Again, $\Delta U = nC_v\ (T_1 - T_2) - \frac{1.36 \times 9}{10 \times 22.414} \times 24.21$cal.

Now, the given problem is an example of adiabatic and free expansion. So, $Q = 0$ and $W = 0$.

Hence, $\Delta U = 0 = nC_v\ (T_1 - T_2) - \frac{1.36 \times 9}{10 \times 22.414} \times 24.21$ cal

or, $4.9(273.15 - T_2) - \frac{1.36 \times 9}{10 \times 22.414} \times 24.21 = 0$

or, $1338.435 - 4.9T_2 - 1.32208 = 0$

or, $4.9T_2 = 1337.1129$ or, $T_2 = 272.88$.

Now, $\Delta S = nC_v \ln \frac{T_2}{T_1} + nR\ln 10$

$$= 4.9 \ln \frac{272.88}{273.15} \text{ cal.K}^{-1} + 4.575 \text{ cal.K}^{-1} (\text{since, } C_v = 4.9 \text{ cal.K}^{-1}.\text{mol}^{-1}).$$

$$= -0.00484 \text{ cal.K}^{-1} + 4.575 \text{ cal.K}^{-1} = 4.570 \text{ cal/K} = \mathbf{19.12\ J.K^{-1}}.$$

17. For copper $\beta = 0.8 \times 10^{-6}$ atm^{-1}. What is the change in Gibbs free energy when 63.546 gm of copper is subjected to an extra pressure of 1000 atm? What would be ΔG if β is assumed to be zero? Given that the density of copper is 8.93 gm/cc.

Soln. : We know that co-efficient of thermal compressibility,

$$\beta = -\frac{1}{V}\left(\frac{\partial V}{\partial P}\right)_T$$

$$\therefore \quad \frac{dV}{V} = -\beta\, dP$$

Integrating we get—

$\ln V = -\beta P + I_c$ [Integration Constant].

When $P = 0$, $I_c = \ln V_0$.

$$\therefore \quad \ln \frac{V}{V_0} = -\beta P$$

i.e., $V = V_0 e^{-\beta P}$.

If β is very small, $V = V_0(1 - \beta P)$.

Again, $\Delta G = \int V\, dP = \int V_0(1 - \beta P)\, dP$.

Assuming that, V_0 is the volume at zero pressure,

$$\Delta G = \int V_0(1 - \beta P)dP.$$

Integrating, we get

$$\Delta G = \int_0^{1000} V_0(1-\beta P)dP$$

$$= V_0(P_2 - P_1) - \frac{\beta V_0}{2}\left(P_2^2 - P_1^2\right)$$

$$= (V_0 \times 1000) - \frac{\beta V_0}{2}(1000)^2 \text{ [taking } P_1 = 0]$$

$$\text{or, } \Delta G = (V_0 \times 1000) - \frac{0.8\times10^{-6}}{2} \times V_0 \times (1000)^2.$$

$$\text{Again, } V_0 = \frac{63.546}{8.93} = 7.116 \times 10^{-3} \text{ lit.}$$

$$\therefore \Delta G = 7.116 \times 10^{-3} \times 1000(\text{lit.atm}) - 4\times10^{-7} \times 7.116\times10^{-3} \times (1000)^2 \text{lit.atm}$$

$$= 7.116 \text{ lit.atm} - 0.0028464 \text{ lit.atm} = 7.113 \text{ lit. atm} = 720.725 \text{ J.}$$

If $\beta = 0$, $\Delta G = 7.116$ lit-atm

$$= \mathbf{721.0287\ J.}$$

18. For water $\beta = 4.94 \times 10^{-5}$ atm^{-1} and $d = 0.997$ gm/cc at 25 °C. Calculate ΔG, when the pressure of 18 gm of water at 1 atm is raised to 101 atm at 25 °C.

Hints : $\Delta G = V_0(P_2 - P_1) - \dfrac{\beta V_0}{2}\left(P_2^2 - P_1^2\right)$ **[Ans.** $\Delta G = 182.8$ J]

19. Calculate the entropy change of the universe as a result of each of the following processes—

(a) A zinc block of 500 gm mass and with a total heat capacity at constant pressure of 150 J/deg at 100 °C is placed in a lake at 10 °C.

(b) The same block at 10 °C is dropped from a height of 100 m into the lake.

(c) Two such blocks at 100 °C and 0 °C are joined together.

Soln. : (a) The lake is a reservoir. Its temperature remains fixed at 10 °C. The zinc block attains 10 °C. Thus it rejects 150 (100 – 10) = 150 × 90 J heat and lake receives it. Thus

$$\Delta S_{Lake} = \frac{150 \times 90}{283.15} = 47.68 \text{ J/deg.}$$

$$\Delta S_{Block} = nC_p \ln\frac{T_2}{T_1} = -150 \ln\frac{373.15}{283.15} = -41.400554 \text{ J/deg.}$$

$$\therefore \Delta S_{Universe} = 47.68 - 41.40 = \mathbf{6.28\ J/deg.}$$

(b) Q = mgh = 500 × 980.665 × 10^4 erg = 490.33 J.

This heat is rejected to the lake.

$$\therefore \Delta S_{Block} = 0,\ \Delta S_{Lake} = +\frac{490.35}{283.15} = 1.73 \text{ J/deg.}$$

$$\therefore \Delta S_{Universe} = \mathbf{+1.73\ J/deg.}$$

(c) Two such blocks are joined. They have the same mass. Hence, final temperature = 50 °C = 323.15 °K.

$$\Delta S_1 = 150 \ln\frac{323.15}{273.15} \text{ and } \Delta S_2 = 150 \ln\frac{323.15}{373.15}.$$

$$\therefore \Delta S = \Delta S_1 + \Delta S_2 = \mathbf{3.63\ J/deg.}$$

20. One mole of an ideal gas is allowed to expand isothermally at 300 K from 1 atm to 0.1 atm (i) reversibly and (ii) irreversibly. Calculate ΔG for (i) and state with reason whether ΔG for (ii) would be different.

Soln. : By the given problem state changes from I to II.

State-I		State-II
n = 1 mole	→	n = 1 mole
P_1 = 1 atm		P_2 = 0.1 atm
T_1 = 300 K		T_2 = 300 K

Same state is reached both by path I and path II. G is a state function. Thus, ΔG for both paths I and II are identical. We know that,

$$\Delta G = nRT \ln\frac{P_2}{P_1} = 1 \times 8.314 \times 300 \ln\frac{1}{10} = \mathbf{-5.74\ kJ.}$$

21. 1 gm of water enclosed in a vial is placed in an evacuated flask miantaining at 298 K. By means of a lever, the vial is broken so that the water is free to vaporise, and at equilibrium it is found that half of the water has vaporised. The vapour pressure of water at 25 °C is 24 mm of Hg and the enthalpy of vaporization is 2470 J/gm. Calculate Q, W, ΔH, ΔG and ΔS.

Soln. : The process is—

$$\frac{1}{2} \text{ gm of } H_2O \text{ (l, 25 °C, 24 mm)} \rightleftharpoons \frac{1}{2} \text{ gm of } H_2O \text{ (g, 25 °C, 24 mm)}.$$

$$\therefore \quad Q = \frac{2470}{2} = \mathbf{1235\ J.}$$

Let $Q = Q_v$, $\therefore$ $W = \mathbf{0}$ (no external work is done).

or, $\Delta U = 1235$ J.

Now, $\Delta H = Q_p = Q_v + P\Delta V = Q_v + PV_g$

$$= Q_v + nRT \ [n = \frac{1}{2 \times 18} = \frac{1}{36}]$$

$$\text{or, } \Delta H = 1235 + (\frac{1}{36} \times 8.314 \times 298) = \mathbf{1304\ J.}$$

This Q_{rev} (slow expansion against P to vaporise half the water).

$$\therefore \Delta S = \frac{1304}{298} = \mathbf{4.38\ J.K^{-1}.}$$

$$\text{Now, } \Delta G = \Delta H - T\Delta S = 1304 - 298 \times \frac{1304}{298} = \mathbf{0.}$$

22. An isolated system comprises of 1 mole of ideal gas, a heat reservoir at T_1, a machine and a source of energy for the machine. The gas at T_1 first expands freely to three times its initial volume; it is then compressed by the machine (at T_1) reversibly back to the initial volume; the gas is again allowed to expand freely to twice the initial volume. Calculate ΔS for the process.

Soln. : Step-1 + Step-2 together returns the gas to the initial state.

Thus, for the gas, $\Delta S = 0$.

ΔS of the gas for Step-3 $= R \ln 2 = 5.76$ J.K^{-1}

For the Step-1, $W_1 = 0$, $Q_1 = 0$, $\Delta U = 0$,

For the Step-2, $W_2 = RT_1 \ln \frac{V_2}{V_1} = RT_1 \ln 3 = 9.13 T_1$

Since, $\Delta U = 0$, $Q = -W = -9.13\ T_1$J

$W_3 = \Delta U = Q = 0$.

Thus, 9.13 T_1 heat would be delivered to the heat reservoir in the whole process and it gains

$\frac{9.13 T_1}{T_1} = 9.13$ J.K^{-1} entropy. Hence, $\Delta S = 9.13 + 5.76 = \mathbf{14.89\ J.K^{-1}.}$

23. One mole of an ideal monatomic gas undergoes a process whereby its volume and temperature are trebled. Calculate ΔS and ΔG if the gas is initially at 273 K and 1 atm. Its molecular wt. is 40 gm/mole.

It may be assumed that for a monatomic ideal gas of molecular wt. M,

$$S_T^\circ = \left[R \ln\left(T^{\frac{5}{2}} M^{\frac{3}{2}}\right) - 2.31\right] \text{ cal.K}^{-1}\text{mole}^{-1}.$$

Soln. : We know that,

$$\Delta S = C_v \ln\frac{T_2}{T_1} + R\ln\frac{V_2}{V_1}$$

$$= C_v \ln 3 + R\ln 3$$

$$= C_p \ln 3 = \frac{5}{2}R\ln 3 = \mathbf{22.83\ J/K.}$$

Now, $\Delta G = \Delta H - \Delta(TS)$

But $\Delta H = C_p (T_2 - T_1) = \frac{5}{2}R\ [(273 \times 3) - 273] = 11.35$ kJ.

Now, $S_{273}^\circ = 1.987 \ln\left[(273)^{\frac{5}{2}}(40)^{\frac{3}{2}}\right] - 2.31 = 36.55 \text{ cal} = 152.9 \text{ J}.$

And $S_{819}^\circ = 1.987 \ln\left[(819)^{\frac{5}{2}}(40)^{\frac{3}{2}}\right] - 2.31 = 42.01 \text{ cal} = 175.8 \text{ J}.$

$\therefore\ \Delta G = \Delta H - T_2S_2 + T_1S_1.$

Let P be constant at 1 atm.

$\therefore\ S_2 = S_2^\circ,\ S_1 = S_1^\circ.$

$\therefore\ \Delta G = \Delta H - (819 \times 175.8) + (273 \times 152.9)$

or, $\Delta G = \mathbf{-\ 90.89\ kJ.}$

S_{819}° can also be calculated as—

$$S_{819}^\circ = S_{273}^\circ + nC_p \ln\frac{819}{273}$$

$$= 152.9 + \frac{5}{2} \times 8.314 \ln\frac{819}{273} = 175.8 \text{ J/K}.$$

24. 1 mole of an ideal monatomic gas is taken from the state (24.6 l, 1 atm 300 K, S° = 81 kJ^{-1}) to the state (2 atm, 320 K). Calculate ΔU, ΔH, ΔS and ΔG for this change.

Soln. : We know that, $\Delta U = nC_v\ \Delta T = \frac{3}{2}\text{R} \times 20 = \mathbf{249.42\ J}$

and $\Delta H = nC_p\ \Delta T = \frac{5}{2}R \times 20 = \mathbf{415.7\ J.}$

Now,

State-I		State-II
V = 24.6 l	→	P = 2 atm
T = 300 K		T = 320 K
n = 1 mole		n = 1 mole

$P = 1$ atm, $\therefore\ V = \dfrac{nRT}{P} = \dfrac{0.082 \times 320}{2} = 13.12$ lit.

We know that, $\Delta S = nC_p \ln\dfrac{T_2}{T_1} - nR\ln\dfrac{P_2}{P_1}$

$$= \frac{5}{2}R\ln\frac{320}{300} - R\ln\frac{2}{1} = \mathbf{-4.42\ J.K^{-1}}.$$

Calculation of ΔG :

$$\begin{matrix} 300\text{ K} & \text{Step-I} & 320\text{ K} & \text{Step-II} & 320\text{ K} \\ 1\text{ atm} & \rightarrow & 1\text{ atm} & \rightarrow & 2\text{ atm} \end{matrix}$$

$$\Delta G_1 = \Delta H_1 - T_2S_2 + T_1S_1$$

$$= \left(\frac{5}{2}R \times 20\right) - 320\left(S^\circ + C_P \ln\frac{320}{300}\right) + (300 \times 81)$$

$$= 415.7 - 320\left(81 + \frac{5}{2}R\ln\frac{320}{300}\right) + (300 \times 81) = -1.63\text{ kJ}.$$

$$\Delta G_2 = RT\ln\frac{P_2}{P_1} = 8.314 \times 320 \ln 2 = 1.84\text{ kJ}.$$

$\therefore\ \Delta G = \Delta G_1 + \Delta G_2 = \mathbf{210.5\ J.}$

25. Calculate ΔH and ΔS when a 1 kg bar of copper at 90 °C is placed in 2 kg of water at 283 K in an insulated container maintained at 1 atm.

Heat capacity : Cu ($0.41\ J.K^{-1}.gm^{-1}$)

$H_2O(l)$ ($4.2\ J.K^{-1}.gm^{-1}$).

Soln. : Let the final temperature be T_F. So,

$$1000 \times 0.41\ (90 - T_F) = 2000 \times 4.2 \times (T_F - 10)$$

or, $T_F = 13.723042$ °C $= 286.873042$ °K

Hence, $\Delta H_{Cu} = nC_p\ (T_F - T_{Cu}) = 410\ J.K^{-1}(286.873042 - 363.15)$ K

$$= \mathbf{-31273.55278\ J.}$$

$$\Delta H_{H_2O} = 2 \times 4200\ J.K^{-1}(286.873042 - 283.15)K = 31273.5528\ J.$$

$\therefore\ \Delta H = \Delta H_{Cu} + \Delta H_{H_2O} = 0.00002\ J \approx \mathbf{0.}$

Now, $\Delta S_{Cu} = 410\ln\dfrac{286.873042}{363.15} = -96.668\ J.K^{-1}$

$$\Delta S_{H_2O} = 8400\ln\frac{286.873042}{283.15} = 109.729\ J.K^{-1}$$

$\therefore\ \Delta S = \mathbf{13.06\ J.K^{-1}}.$

26. One mole of an ideal and monatomic gas initially at 300 K and 1 atm (occupying 24.6 lit) expands isothermally and reversibly to 49.2 lit under conditions that, $W = -400$ J. Calculate ΔG and ΔS.

Soln. : Since the process is isothermal, $\Delta U = 0$. Hence, from the 1st law of thermodynamics, we get—

$$Q = -W = 400\text{ J.}$$

In the given problem, $V_1 = 24.6$ lit, $V_2 = 49.2$ lit,

$P_1 = 1$ atm, $\therefore\ P_2 = \frac{1}{2}$ atm.

$$\therefore\ \Delta S = R \ln \frac{P_1}{P_2} = R \ln 2 = \mathbf{5.76\ J.K^{-1}}.$$

$$\Delta G = -RT \ln \frac{P_1}{P_2} = -RT \ln 2 = \mathbf{-1.73\ kJ.}$$

27. One mole of a perfect monatomic gas initially at volume $V_1 = 10$ lit, pressure P_1 and temperature $T_1 = 300$ K experiences the following reversible changes :

(a) isothermal compression to one-half the volume $\left(\frac{V_1}{2}, P_2\right)$

(b) cooling at constant volume until the pressure is returned to the original value of P_1, the final temperature being T_2.

(i) Calculate P_1, P_2, T_2, Q, W, ΔU.

(ii) Are the magnitudes of ΔU, Q, W same by path C.

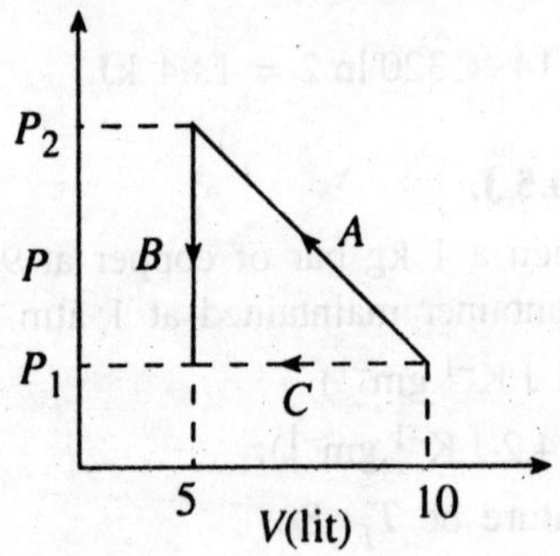

Soln. : In the given problem, $P_1 = \frac{nRT_1}{V_1}$

$$= \frac{1\,\text{mole} \times 0.08206\ \text{l.atm.K}^{-1}.\text{mole}^{-1} \times 300\,\text{K}}{10\,\text{atm}} = \mathbf{2.46\ atm.}$$

After isothermal compression, P is doubled. Hence, $P_2 = \mathbf{4.92}$ atm. To reduce this pressure to P_1 by cooling, $T_2 = \frac{T_1}{2} = \frac{300}{2} = \mathbf{150\ K.}$

$$\left[\text{since}, \frac{T_1}{P_1} = \frac{T_2}{P_2} \text{ or}, T_2 = T_1.\frac{P_2}{P_1} = \frac{T_1}{2}\right]$$

(a) For isothermal change, $\Delta U = \Delta H = 0$.

$$W = -nRT \ln \frac{V_2}{V_1} = -RT \ln \frac{1}{2} = 8.314 \times 300 \ln 2 = 1728.8477\ \text{J} = \mathbf{1.73\ kJ.}$$

$$Q = -W = -1.73\ \text{kJ}$$

$$\Delta S = nR \ln \frac{V_2}{V_1} = -R \ln 2 = -5.76\ \text{J.K}^{-1};\ \Delta G = nRT \ln \frac{V_1}{V_2} = 1.73\ \text{kJ.}$$

(b) For isochoric change, $W = \mathbf{0}$.

$\therefore \quad Q \;= \Delta U = nC_v (T_2 - T_1) = -150 \times \dfrac{3}{2} R = \mathbf{-1.87\ kJ.}$

$\Delta H \;= nC_p (T_2 - T_1) = -150 \times \dfrac{5}{2} R = -3.12\ \text{kJ}$

$\Delta S \;= nC_v \ln \dfrac{T_2}{T_1} = \dfrac{3}{2} R \ln \dfrac{1}{2} = -8.64\ \text{J.K}^{-1}$

$\Delta G \;= \Delta H - T_2 S_2 + T_1 S_1.$

Here ΔG cannot be calculated because $S°$ is not known.

N.B. : (i) $S°$ can be calculated by using the formula in Problem no. 23 and then ΔG can be calculated. Molecular wt. M is needed.

(ii) U is a state function. Hence ΔU will be same by path C. Area under C less than area under $A + B$. Thus, W will be smaller and Q will be smaller as $Q = W + \Delta U$.

28. 1 mole of H_2(g) at 373 °K is compressed adiabatically and reversibly from 1 to 10 atm. Calculate ΔU, ΔH, ΔS and ΔG assuming ideal gas behaviour.

Given, $C_p = 6.9$ cal.deg^{-1}.mole^{-1}. $S°$ at 298 = 49 cal.deg^{-1}.

Soln. : Now, $P_1 = 1$ atm, $T_1 = 373$ °K, $P_2 = 10$ atm.

$$\therefore \quad T_2 = T_1 \left(\frac{P_2}{P_1}\right)^{\frac{R}{C_p}} = 373\,(10)^{\frac{1.987}{6.9}} = 723.9\text{K}.$$

$C_v = 6.9 - 1.987 = 4.913$ cal.deg^{-1}.mole^{-1}.

$\Delta U \;= nC_v (T_2 - T_1) = 1 \times 4.913(723.9 - 373) = \mathbf{1.72\ kcal.}$

$\Delta H \;= nC_p (T_2 - T_1) = 6.9(723.9 - 373) = 2421.21 = \mathbf{2.42\ kcal.}$

$\Delta S \;= \mathbf{0}.$

ΔG can be calculated for the following change :

$n = 1$ mole		$n = 1$ mole		$n = 1$ mole
373 K, 1 atm	—I→	723.9 K, 1 atm	—II→	723.9 K, 10 atm

$$\Delta G_1 \;= \Delta H - T_2 S_2 + T_1 S_1$$

$$= 242.21 - 723.9\left(S° + C_p \ln \frac{723.9}{298}\right) + 373\left(S° + C_p \ln \frac{373}{298}\right)$$

$$= 2421.21 - 723.9\left(49 + 6.9 \ln \frac{728.9}{298}\right) + 373\left(49 + 6.9 \ln \frac{373}{298}\right)$$

$$= -\ 18628.41432\ \text{cal} = -18.628\ \text{kcal}.$$

$$\Delta G_2 = RT \ln \frac{P_2}{P_1} = 1.987 \times 723.9 \ln 10 = 3312.01376\ \text{cal}.$$

$$\therefore \quad \Delta G \;= \Delta G_1 + \Delta G_2 = (3312.01376 - 18628.41432)\ \text{cal}$$

$$= -15316.40056\ \text{cal} = -\ 15316.4\ \text{cal} = -\ \mathbf{15.32\ kcal.}$$

29. Calculate the difference in molar entropy and molar free energy (a) between liquid water at –5 °C and ice at –5 °C. (b) between liquid water and water vapour at 95 °C and 1 atm. Distinguish between ΔS_{system} and $\Delta S_{surrounding}$.

$\Delta C_{p,m}$ = 37.6 J.K^{-1}(s→l) $[C_{p,l} - C_{p,s}]$

$\Delta C_{p,m}$ = – 43.2J.K^{-1}(l→v) $[C_{p,v} - C_{p,l}]$

L_f = 6.02 kJ.mole^{-1}

L_e = 40.67 kJ.mole^{-1}

$C_{p,l}$ ⇒ specific heat at constant pressure of liquid water.

$C_{p,s}$ ⇒ specific heat at constant pressure of solid water.

$C_{p,v}$ ⇒ specific heat at constant pressure of water vapour.

Soln. : ΔS system :

(a) It is given that—

$H_2O(l)$ −5 °C —1→ $H_2O(l)$ 0 °C —2→ $H_2O(s)$ 0 °C —3→ $H_2O(s)$ −5 °C n =1

$$\Delta S = \Delta S_1 + \Delta S_2 + \Delta S_3$$

$$\text{or, } \Delta S = C_{p,l} \ln \frac{273.15}{268.15} - \left(\frac{6.02 \times 1000}{273.15}\right) + C_{p,s} \ln \frac{268.15}{273.15}$$

$$= \left(C_{p,l} \ln \frac{273.15}{268.15} - C_{p,s} \ln \frac{273.15}{268.15}\right) - 22.04$$

$$= \Delta C_{p,m} \ln \frac{273.15}{268.15} - 22.04$$

$$= 37.6 \ln \frac{273.15}{268.15} - 22.04 = 0.6946 - 22.04$$

$$= -21.3454 \text{ J.K}^{-1} = -21.34 \text{ J.K}^{-1}.$$

$$\Delta S_{surr} = -\frac{\Delta H_m(l \to s)}{T} = \frac{\Delta H_m(s \to l)}{T}.$$

Now, $\Delta H_{surr} = -C_{p,l}(T_F - T) + \Delta H_m$ (s→l) $- C_{p,s}(T - T_F)$

$= C_{p,l}(T - T_F) + \Delta H_m$ (s→l) $- C_{p,s}(T - T_F)$

$= \Delta C_{p,m}(T - T_F) + \Delta H_m$

$= (-37.6 \times 5)$ J + 6020 J = 5832 J.

$$\therefore \ \Delta S_{surr} = \frac{5832 \text{ J}}{268.15 \text{ K}} = 21.749 \text{ J.K}^{-1} = 21.75 \text{ J.K}^{-1}.$$

$\therefore$ $\Delta S_{universe}$ = (−21.34 + 21.75) J.K^{-1}.

= 0.41 J.K^{-1}.

We know that, $G = H - TS$.

$\therefore$ at constant temperature, $\Delta G = \Delta H - T\Delta S$.

$\therefore$ $\Delta G_m(l \to \text{s,T}) = \Delta H_m(l \to \text{s,T}) - T\Delta S(l \to \text{s,T})$

= −5832 − {268.15 × (−21.34)} = −5832 + 5722.321 = −109.68 J.

ΔG can be calculated in other way :

ΔH_m = 6020 J.mole^{-1}

$\Delta C_{p,m}$ = 37.6 J.K^{-1}

$$\left(\frac{\partial \Delta H}{\partial T}\right) = \Delta C_p$$

$$d(\Delta H) = \Delta C_p dT$$

$$\Delta H = \Delta C_p T + I_c \text{ [}I_c = \text{Integretion Const.]}$$

$$-6020 = (-37.6 \times 273.15) + I_c$$

$$\therefore \; I_c = 4250.44 \text{ J.mole}^{-1}$$

$$\therefore \; \Delta H = 4250.44 - 37.6\,T$$

$$\Delta G = A - BT\ln T + I_c.T \text{ [since, } \left(\frac{\partial(\Delta G/T)}{\partial T}\right)_P = -\frac{\Delta H}{T^2}\text{]}$$

$$= 4250.44 + 37.6\,T\ln T + I_c T$$

$$\therefore \; 0 = 4250.44 + 37.6 \times 273.15 \ln 273.15 + I_c \times 273.15$$

$$\text{or, } I_c = -226.49762$$

$$\therefore \; \Delta G \text{ at } 268.15$$

$$= 4250.44 + 37.6 \times 268.15 \ln 268.15 - (226.49762 \times 268.15)$$

$$= -108{\cdot}46 \text{ J.mole}^{-1}$$

(b) $H_2O(l)$ 95 °C $\xrightarrow{1}$ $H_2O(l)$ 100 °C $\xrightarrow{2}$ $H_2O(v)$ 100 °C $\xrightarrow{3}$ $H_2O(v)$ 95 °C

$$\Delta S = \Delta S_1 + \Delta S_2 + \Delta S_3$$

$$= C_{p,s} \ln \frac{373.15}{368.15} + \frac{40670}{373.15} + C_{p,v} \ln \frac{368.15}{373.15}$$

$$= 108.99102 + (C_{p,l} - C_{p,v}) \ln \frac{373.15}{368.15}$$

$$= 108.99102 - 43.2 \ln \frac{373.15}{368.15}$$

$$= 108.99102 - 0.58277 = 108.40825 \text{ J.K}^{-1}.$$

$$= 108.41 \text{ J.K}^{-1}.$$

$$\text{Now, } \Delta S_{surr} = \frac{-\Delta H_m(l \to v)}{T}$$

$$\Delta H_{surr} = -C_{p,l}(T_b - T) - \Delta H_m (l \to v, T_b) - C_{p,v}(T - T_b)$$

$$= -\Delta H_m - (C_{p,l} - C_{p,v})(T_b - T)$$

$$= -40670 - (43.2 \times 5) = -40886 \text{ J.}$$

$$\therefore \; \Delta S_{surr} = \frac{-40886}{368.15} = -111.058 \text{ J.K}^{-1}.$$

$$\therefore \; \Delta S_{universe} = -2.648 \text{ J.K}^{-1} = -2.65 \text{ J.K}^{-1}.$$

$$\Delta G = \Delta H_m (l \to v, T) - T\Delta S_m (l \to v, T)$$

$$= 40886 - (368.15 \times 108.41) = \mathbf{974.86 \text{ J.}}$$

30. An infinitesimal crystal of ice is added to 10 gm of supercooled liquid water at 10 °C in an adiabatic container. What is ΔH for the process if the system reaches equilibrium at a fixed pressure of 1 atm? (The equilibrium state will contain some ice and will, therefore, consist either of ice plus liquid at 0 °C or, of ice at or below 0 °C.) What is present at equilibrium? Calculate ΔS for the process. Given that,

$l_f = 80$ cal/gm. $C_{p,l} = 1$ cal.K^{-1}.gm^{-1}.

Soln. : 10 gm supercooled water at –10 °C is a non-equilibrium system. Addition of infinitesinal of ice will take it to equilibrium. Since there is an adiabatic jacket, heat lost for a process will be equal to heat gain. It will try to reach in equilibrium. The change may be depicted as below :

10 gm super-cooled water at –10 °C	→	10 gm water at 0 °C	→	x gm ice + $(10 - x)$ gm water at 0 °C

For moving from –10 °C to 0 °C, the system gains ΔH

$= C_p\,(T_2 - T_1) = C_p\{0 - (-10)\}$

$= C_p \times 10.$

C_p for water is 1 cal.K^{-1}.gm^{-1}.

So, $\Delta H = 10 \text{ gm} \times 1 \text{ cal.K}^{-1}.\text{gm}^{-1} \times 10\text{K} = \mathbf{100 \text{ cal.}}$

Thus, energy needed 100 cal. This must be supplied by conversion of x gm of liquid water to ice at 273.15 °K. Thus, $x = \dfrac{100\,\text{cal}}{80\,\text{cal / gm}} = 1.25$ gm.

Thus, **1.25 gm of ice and 8.75 gm of water** is present at equilibrium at 0 °C.

Now, $\Delta S = \Delta S_1 + \Delta S_2$

$$= nC_p \ln \frac{273.15}{263.15} - \frac{100}{273.15} \text{ cal.K}^{-1}$$

$$= 10 \text{ cal.K}^{-1} \ln \frac{273.15}{263.15} - \frac{100}{273.15} \text{ cal.K}^{-1}$$

$$= \left(+0.373 - \frac{100}{273.15}\right) \text{cal.K}^{-1}$$

$$= (+\ 0.373 - 0.366) \text{ cal.K}^{-1} = \mathbf{+\ 0.007 \text{ cal.K}^{-1}.}$$

31. 0.5 mole of an ideal monatomic gas initially at 5 atm pressure and 0 °C is allowed to expand against a constant external pressure of 0·5 atm. Conditions are such that the final volume is 10 times the initial volume; the final gas pressure equals the external pressure. Calculate W, Q, ΔU, ΔH, ΔS and ΔG for the process. **[C.U. 2005]**

Soln. : In the given problem,

n = 0·5 mole P = 5 atm T_1 = 273·15 °K $V = V_1$	→	n = 0·5 mole P = 0·5 atm $V_2 = 10V$ $T_2 = T$

From the given change, we get

$$\frac{P_1V_1}{T_1} = \frac{P_2V_2}{T_2}.$$

$\therefore\ T_1 = T_2$

So, this is an isothermal and irreversible change.

For such a case, ΔU = **0.**

$\therefore \quad W = -Q.$

Now, from the 1st law of thermodynamics,

$$W = -P_{ext}(V_2 - V_1)$$
$$= P_2 (V_1 - V_2)$$
$$= P_2\left[\frac{nRT}{P_1} - \frac{nRT}{P_2}\right] = nRT\left[\frac{P_2}{P_1} - 1\right].$$

$\therefore \quad W = 0.5 \times 8.314 \times 273.15\ (0.1 - 1)\ \text{J} = -1021.94\ \text{J.} = -1.02\ \text{kJ.}$

$Q = -W = 1.02$ kJ

Now, $\Delta H = nC_p\,dT = \mathbf{0}$ [for ideal gas and for isothermal process]

$$\Delta G = nRT \ln \frac{P_2}{P_1}$$

$$= 0.5\ \text{mol} \times 8.314\ \text{J.K}^{-1}.\text{mol}^{-1} \times 273.15\ \text{K}\ \ln\left(\frac{0.5}{5}\right)$$

$$= -\ 2614.5498\ \text{J} = \mathbf{-2.61\ kJ.}$$

$$\Delta S = nR \ln \frac{V_2}{V_1} = 0.5 \times 8.314 \times \ln(10) = \mathbf{9.5718\ J.K^{-1}}.$$

32. 1 mole of an ideal gas at 2 atm pressure expands irreversibly and adiabatically into vacuum to decrease its pressure to 1 atm. Calculate ΔS for the change. **[B.U. 2003]**

Soln. : $\Delta S = nR \ln \dfrac{P_1}{P_2} = 8.314 \times \ln\left(\dfrac{2}{1}\right) \text{J.K}^{-1} = \mathbf{5.763\ J.K^{-1}}.$

33. At –5 °C the equilibrium vapour pressure of ice is 3.012 mm of Hg and that of supercooled liquid water is 3.163 mm of Hg. Latent heat of fusion of ice is 5.85 kJ. mole^{-1} at –5 °C. Calculate ΔG and ΔS per mole for a transition.

$H_2O(l, -5\ °C) \rightarrow H_2O(s, -5\ °C)$

Comment on the spontaneity of this process. **[C.U. 2004]**

Soln. : $\Delta G = RT \ln\left(\dfrac{3.012}{3.163}\right)$

$$= 8.314\ \text{J.K}^{-1}.\text{mol}^{-1} \times 268.15\text{K} \times \ln\left(\frac{3.012}{3.163}\right) = \mathbf{-109.05\ J.}$$

Since equilibrium vapour pressure of ice is lower than that of supercooled water at –5 °C, freezing is not a reversible process at that temperature. Hence, for calculation of ΔS the following alternative reversible path can be conceived through which the change can be done.

$$H_2O\,(l,P_l,-5\ °C) \xrightarrow[\text{expansion}]{\text{Rev. and isothermal}} H_2O\,(l,P_s,-5\ °C) \xrightarrow[\text{at const. } T \text{ and } P]{\text{Rev. phase change}} H_2O(s,P_s,-5\ °C)$$

$$\Delta S_1 \qquad\qquad\qquad\qquad \Delta S_2$$

Assuming the vapour behaves ideally,

$$\Delta S_{system} = \Delta S_1 + \Delta S_2 = R \ln \frac{P_l}{P_s} - \frac{\Delta H_{fusion}}{T}$$

$$= 8.314 \text{ J.K}^{-1}.\text{mol}^{-1} \ln \frac{3.163}{3.012} - \frac{5.85 \times 10^3 \text{ J.mol}^{-1}}{268.15 \text{ K}}$$

$$= -21.409 \text{ J.K}^{-1}.\text{mol}^{-1}.$$

$$\Delta S_{surr} = -\frac{\Delta H(l \rightarrow s)}{T} = \frac{\Delta H_{fusion}}{T} = \frac{5850 \text{ J.mol}^{-1}}{268.15 \text{ K}} = 21.816 \text{ J.K}^{-1}.\text{mol}^{-1}.$$

Hence, $\Delta S_{universe} = \Delta S_{surr} + \Delta S_{system}$

$= (21.816 - 21.409) \text{ J.K}^{-1}.\text{mol}^{-1} = \mathbf{0.407 \text{ J.K}^{-1}.\text{mol}^{-1}}$.

Since ΔG for the process is negative at constant temperature and pressure, the process is spontaneous. Positive value of $\Delta S_{universe}$ also indicates the same thing.

34. One mole of water vapour is compressed reversibly to liquid water at 373 K. Calculate W, Q, ΔH, ΔS, ΔG and ΔA.

ΔH_{vap} at 373 K = 2258 J.gm^{-1}. **[C.U. 2002]**

Soln. : We know that, work $(W) = -P(V_2 - V_1)$.

Here, $V_2 = V_l$ and $V_1 = V_g$.

$\therefore W = -P(V_l - V_g) = PV_g$ [$\because V_g >> V_l$].

Assuming ideal behaviour, $V_g = \frac{nRT}{P} = \frac{RT}{P}$ [$n = 1$].

$\therefore \quad W = RT$

$= 8.314 \times 373 = \mathbf{3101.122 \text{ J}}$.

$\Delta H = -2258 \times 18 \text{ J} = \mathbf{-40644 \text{ J}}$.

Now, $\Delta H = \Delta U + \Delta(PV)$.

At constant pressure,

$\Delta H = \Delta U + P\Delta V$

$= \Delta U + P(V_l - V_g) = \Delta U - PV_g = \Delta U - RT$

or, $\Delta U = \Delta H + RT$.

$\therefore \quad Q = \Delta U - W = \Delta H + RT - RT = \Delta H$

or, $Q = \mathbf{-40644 \text{ J}}$.

$$\Delta S = \frac{40644 \text{ J}}{373 \text{ K}} = -108.965 \text{ J.K}^{-1}.$$

$\Delta G = \mathbf{0}$, since the process is reversible at constant temperature and pressure.

$$\Delta A = -\int_R^l P dv = -P(V_l - V_g) = PV_g = RT = \mathbf{3101.122 \text{ J}}.$$

35. Calculate the change in entropy experienced by 2 moles of an ideal gas on being heated from a pressure of 5 atm at 50 °C to a pressure of 1 atm at 100 °C. Given, $C_p = 9.88$ cal.mol^{-1}.deg^{-1}. **[V.U. 2000]**

Hints : Use formula given below,

$$\Delta S = n\left[C_P \ln\frac{T_2}{T_1} - R\ln\frac{P_2}{P_1}\right] = \mathbf{38.67\ J.K^{-1}}.$$

36. 0.5 mole water at 1 atm pressure undergoes the process :
$H_2O(l, -10\ °C) \rightarrow H_2O(s, -10\ °C)$

Compute ΔS and ΔG for the process from the following data : specific heat capacity of water and ice over the temperature range is 1.0 cal.deg^{-1}.gm^{-1} and 0.5 cal.deg^{-1}.gm^{-1} respectively, latent heat of fusion of ice is 80.0 cal.gm^{-1} at 0 °C. **[C.U. 2000]**

Soln. : The process may be depicted as below :

$H_2O(l)$, −10 °C $\xrightarrow{1}$ $H_2O(l)$, 0 °C $\xrightarrow{2}$ $H_2O(s)$, 0 °C $\xrightarrow{3}$ $H_2O(s)$, −10 °C

By the problem, n = 0.5 mole, $\therefore\ m$ = 9 gm.

P = 1 atm

$C_{p,l}$ = 1 cal.deg^{-1}.gm^{-1}

$C_{p,s}$ = 0.5 cal.deg^{-1}.gm^{-1}

l_f = 80 cal.gm^{-1}

$\Delta S = \Delta S_1 + \Delta S_2 + \Delta S_3$

$$\Delta S_1 = 9 \times 1 \times \ln\frac{273.15}{263.15} - \frac{80\times 9}{273.15} + 9 \times 0.5 \ln\frac{263.15}{273.15}$$

$= 0.335672 - 2.6359 - 0.167836$

$= -2.468064 = -2.468$ cal/K = $\mathbf{-10.326\ J.K^{-1}}$.

See Problem 14 for calculating ΔG. [$\Delta G = -85.49$ J]

37. Calculate ΔA for vaporisation of 0.1 mole of H_2O at 1 atm pressure and 100 °C. Given molar volume of $H_2O(l)$ at 100 °C is 18.8 ml.mol^{-1} and that of $H_2O(g)$ is 30.6 litre.mol^{-1}. **[C.U. 2000]**

Soln. : By the problem, V_g = 30.6 lit. mol^{-1}
= 3.06 lit for 0.1 mole.

and V_l = 0.0188 lit.mol^{-1} = 0.00188 lit for 0.1 mole.

Now $dA = -PdV - SdT$.

At constant temperature ΔA for the process given by $\Delta A = \int -PdV$.

At constant pressure,

$\Delta A = -P(V_g - V_l) = P(V_l - V_g)$

$= 1(0.00188 - 3.06)$ lit.atm

$= -3.05812$ lit. atm $= -3.05812 \times 101.325$ J

$= \mathbf{-309.864\ J}$.

38. A sample of hydrogen is confined to a cylinder fitted with a piston of 50 cm^2 cross section. It occupies 500 ml at 25 °C and 2 atm. What is the change in entropy when the piston is isothermally lifted through 10 cm? **[B.U. 1991]**

Soln. : From the given data, the initial volume of the gas (V_i) = 500 ml and final volume (V_f) = V_i + (50 × 10) ml = 500 + 500 = 1000 ml.

Now, according to the given problem—

V_i = 500 ml V_f = 1000 ml

P_i = 2 atm T_f = 298.15 °K

T_i = 298.15 °K

From the above data we can calculate the number of mole of hydrogen gas in the cylinder.

$$\text{i.e., } n = \frac{P_i V_i}{RT_i} = \frac{500 \times 2}{298.15 \times 0.082 \times 10^3} = 0.0409.$$

$$\therefore \text{ entropy change, } \Delta S = nR \ln\frac{V_f}{V_i} = 0.0409 \times 8.314 \times \ln\left(\frac{1000}{500}\right) \text{J.K}^{-1} = 0.2357 \text{J.K}^{-1}$$

39. For a certain reaction ΔG (cal/mole) = 13580 + 16.1T $\log_{10}T$ – 72.59T; find ΔS and ΔH of the reaction at 27 °C. **[C.U. 1983, '87; V.U. '97]**

Soln. : From the given problem we get

ΔG = 13580 + 16.1 $T \log_{10}T$ – 72.59 T.

$$\text{We know that, } \Delta S = -\frac{d}{dT}(\Delta G).$$

$$\therefore \Delta S = -\frac{d}{dT}(\Delta G) = -\frac{d}{dT}[13580 + 16.1\, T \log_{10}T - 72.59\, T].$$

$$= -\frac{d}{dT}(16.1\, T \log_{10}T) + \frac{d}{dT}(72.59\, T)$$

$$= -\frac{d}{dT}\left(\frac{16.1}{2.303} T \ln T\right) + 72.59 \frac{d}{dT}(T)$$

$$= -6.99088 \frac{d}{dT}(T \ln T) + 72.59 = -6.99088\, [1 + \ln T] + 72.59$$

$$= -6.99088\, [1 + \ln 300.15] + 72.59 = 25.721 \text{ cal.K}^{-1}.\text{mol}^{-1}$$

$$= \mathbf{107.62\ J.K^{-1}.mol^{-1}}.$$

Now, ΔG at 300.15 °K is—

ΔG = 13580 + (16.1 × 300.15 × $\log_{10}$300.15) – (72.59 × 300.15)

= 3763.638 cal/mol × 4.184 J/cal = **15747.06 J/mol.**

We know that,

$\Delta G = \Delta H - T\Delta S$ at constant temperature

$\therefore \Delta H = \Delta G + T\Delta S$

or, ΔH = 15747.06 J.mol^{-1} + (300.15 × 107.61) J.mol^{-1}

= 48046.2 J.mol^{-1} = **48.05 kJ.mol^{-1}.**

40. Calculate ΔH and ΔS when 1 kg bar of copper of 100 °C is placed in 2 kg of water at 0 °C in an insulated vessel maintained at 1 atm pressure. Heat capacity of H_2O(l) is 4.1 J.K^{-1}.gm^{-1} and Cu is 0.4 J.K^{-1}.gm^{-1}. Can you calculate ΔG of this change?

Soln. : Let final temperature be t °C as a result of mixing.

∴ by the given data (1000 × 0.4) (100 – t)

= 2000 × 4.1(t – 0) that is 0.4 × (100 – t) = 8.2 t

or, 40 = 8.6 t.

$\therefore \quad t = 4.6511628\ °C = 277.801628\ °K = 277.8\ °K.$

$$\therefore \ \dot{\Delta} S_{Cu} = nC_p \ln\frac{277.8}{373.15} = -118.03 \text{ J.K}^{-1}.$$

For water, $\Delta S = 2000 \times 4.1 \ln\dfrac{277.8}{373.15} = 138.42 \text{ J.K}^{-1}.$

$\therefore \ \Delta S_{total} = 20.39 \text{ J.K}^{-1}$

$\Delta H_{total} = \Delta H_{Cu} + \Delta H_{H_2O}$

$\Delta H_{Cu} = 1000 \text{ gm} \times 0.4 \text{ J.K}^{-1}.\text{gm}^{-1} \times (277.8011628 - 373.15)\text{K}$
$= -38139.53 \text{ J}.$

$\Delta H_{H_2O} = 2000 \times 4.1 \times (277.8011628 - 273.15) = 38139.53 \text{ J}.$

$\therefore \ \Delta H = 0$ and $\Delta G = \Delta H - T_2S_2 + T_1S_1$ and it (ΔG) cannot be calculated because value of at least one S is required.

41. The standard Gibbs free energy ($\Delta G°$) for a certain reaction at different temperatures (T °K) is given by

$\Delta G°(\text{cal}) = -126400 - 5.44\, T \ln T + 104.7\, T.$

Calculate the standard enthalpy change for the reaction at 2000 °K. **[C.U. 1977]**

Hints : See Problem 39. Only remember that in Problem 38, there is $\log_{10}$, but here $\ln = \log_e$.

42. Two mole of a monatomic gas initially at 4.0 bar and 47 °C undergoes reversible expansion in an insulated container. Calculate the temperature at which the pressure is reduced to 3.0 bar. **[GATE 2001]**

Soln. : We know that,

$$\left(\frac{T_2}{P_2^{\;R/C_P}}\right) = \left(\frac{T_1}{P_1^{\;R/C_P}}\right),$$

where T_1 is the temperature of state 1, P_1 and P_2 are the pressures of states 1 and 2 respectively and T_2 is the final temperature. By the problem, T_1 = 47 °C = 320.15 K, P_1 = 4 bar and P_2 = 3 bar.

$$\therefore \ T_2 = T_1\left(\frac{P_2}{P_1}\right)^{\frac{R}{C_P}} = T_1\left(\frac{P_2}{P_1}\right)^{0.4},$$

where $\dfrac{R}{C_P} = \dfrac{R}{\frac{5}{2}R} = 0.4.$

$$\therefore \ T_2 = 320.15 \times \left(\frac{3}{4}\right)^{0.4} = \mathbf{285.35\ °K.}$$

43. Establish the condition for spontaneous vaporisation of water, given ΔH = 9590 cal mole^{-1} and ΔS = 26 e.u. for the process. **[B.U. 1982]**

Soln. : The condition for any spontaneous process is that $\Delta G_{P,T} < 0$.

Now we know that $G = H - TS$.

At constant temperature $\Delta G = \Delta H - T\Delta S$.

Thus, for spontaneous vaporisation, $\Delta H - T\Delta S < 0$ [$\therefore \Delta G_{P,T} < 0$]

or, $T > \dfrac{\Delta H}{\Delta S}$

or, $T > \dfrac{9590}{26}$ °K

or, $T > 368.846$ °K or, $T > 368.85$ °K.

44. Describe a reversible process by which one mole of water at 100 °C can be converted to vapour at 0.5 atm and 100 °C. Calculate ΔH and ΔG for the process. How far these quantities would be affected if the operation were performed irreversibly? Assume water vapour to behave ideally and latent heat of vaporisation of water = 540 cal. gm^{-1}. **[C.U. 1970]**

Hints : The problem can be depicted as following :

$H_2O(l)$ 100 °C, 1 atm $\xrightarrow{1}$ $H_2O(v)$ 100 °C, 1 atm $\xrightarrow{2}$ $H_2O(v)$ 100 °C, 0.5 atm

$\Delta H = \Delta H_1 + \Delta H_2$.

Now, $\Delta H_1 = 18 \times 540$ cal.mole^{-1}

$\therefore$ $\Delta H_2 = 0$, due to isothermal change.

Similarly, $\Delta G = \Delta G_1 + \Delta G_2$.

Now, $\Delta G_1 = 0$ and $\Delta G_2 = RT\ln\dfrac{P_1}{P_2}$.

H and G are state functions and thus ΔH and ΔG would remain same if initial and final states are same irrespective of the path.

45. Calculate the change in Gibbs potential when 36 gm water initially at 100 °C and 10 atm pressure are converted to vapour at 100 °C and 0.01 atm pressure. **[C.U. 1971]**

Soln. : The process can be written as below :

$H_2O(l)$ 100 °C, 10 atm $\xrightarrow{1}$ $H_2O(l)$ 100 °C, 1 atm $\xrightarrow{2}$ $H_2O(v)$ 100 °C, 1 atm $\xrightarrow{3}$ $H_2O(v)$ 100 °C, 0.01 atm

$\Delta G = \Delta G_1 + \Delta G_2 + \Delta G_3$

Now, $\Delta G_1 = \int VdP = V(P_2 - P_1)$ lit.atm.

$= 0.036(1 - 10)$ (volume of 36 gm water = 36 cc = 0.036 l)

$= -0.324$ lit.atm $\times$ 101.325 J/lit atm $= -32.83$ J.

$\Delta G_2 = 0$, since it is an equilibrium process.

$\Delta G_3 = nRT\ln\dfrac{P_2}{P_1} = 2 \text{ mol} \times 8.314 \text{ J.deg}^{-1}.\text{mol}^{-1} \times 373.15 \text{ deg} \times \ln\dfrac{0.01\,\text{atm}}{1\,\text{atm}}$

$= -28573.875$ J

$\therefore$ ΔG = **−28606.71 J.**

46. 5 gm-moles of water initially at 27 °C are converted to a final state of vapour at 227 °C, the conversion being effected under 1 atmospheric pressure. Assuming the vapour to behave ideally, compute the total change in entropy.

[Heat capacity of water = 1 cal/gm; heat capacity of water vapour = 0.40 cal/gm; latent heat of vaporisation of water = 540 cal/gm.] **[C.U. 1972]**

Hints : The process can be depicted as following :

$H_2O(l)$ 27 °C, 1 atm $\xrightarrow{1}$ $H_2O(l)$ 100 °C, 1 atm $\xrightarrow{2}$ $H_2O(v)$ 100 °C, 1 atm $\xrightarrow{3}$ $H_2O(v)$ 227 °C, 1 atm

$$\Delta S_{total} = \Delta S_1 + \Delta S_2 + \Delta S_3$$

Now, $\Delta S_1 = m.C.\ln\frac{T_2}{T_1}$; $\Delta S_2 = \frac{m\times l_e}{T}$; $\Delta S_3 = m.C.\ln\frac{T_2}{T_1}$.

47. Calculate the entropy change when one kilogram of water at 27 °C is converted to super heated steam at 200 °C under constant atmospheric pressure. [Specific heat of liquid water = 4180 J/kg. Specific heat of steam = (1670 + 0.49 *T*) J/kg at *T* °K and latent heat of vaporisation = 23 × 10^5 J/kg.]

Soln. : The process can be written as below :

$H_2O(l)$ 27 °C, 1 atm $\xrightarrow{1}$ $H_2O(l)$ 100 °C, 1 atm $\xrightarrow{2}$ $H_2O(v)$ 100 °C, 1 atm $\xrightarrow{3}$ $H_2O(v)$ 200 °C, 1 atm

So, $\Delta S = \Delta S_1 + \Delta S_2 + \Delta S_3$.

$$\text{Now,}\quad \Delta S_1 = mC_p\ln\frac{T_2}{T_1} = 1\text{ kg}\times 4180\text{ J.kg}^{-1}\times\ln\frac{373.15}{300.15} = 909.978\text{ J.K}^{-1}.$$

$$\Delta S_2 = \frac{1\times 23\times 10^5}{373.15}\text{J.K}^{-1} = 6163.74\text{ J.K}^{-1}.$$

$$\Delta S_3 = \int_{T_1}^{T_2} mC_p\frac{dT}{T} = m\int_{T_1}^{T_2}\frac{1670+0.49T}{T}dT = 1\times\int_{T_1}^{T_2}\left(\frac{1670}{T}+0.49\right)dT$$

$$= 1670\ln\frac{T_2}{T_1} + 0.49(T_2 - T_1)$$

$$= 1670\ln\frac{473.15}{373.15} + (0.49\times 100) = 445.5\text{ J.K}^{-1}.$$

$$\therefore\quad \Delta S = \mathbf{7519.2\ J.K^{-1}}.$$

48. Calculate the change in Gibbs free energy per mole for the process :

H_2O (l, –5 °C) → H_2O (s,–5 °C)

The vapour pressure of supercooled water at –5 °C is 3.163 mm of Hg, and the vapour pressure of ice is 3.012 mm of Hg. **[C.U. 2002]**

Hints : See Problem 33 for calculation of ΔG.

49. One mole of He gas is heated from 200 °C to 400 °C at a constant pressure of 1 atm. Assuming He to be a perfect gas, calculate ΔG, ΔH and ΔS for the process. Given that the absolute entropy of He at 200 °C is 135 J.K^{-1}.mole^{-1}. **[C.U. 2003]**

Hints : See Problem 15.

$$\Delta S = nC_p\ln\frac{T_2}{T_1}$$

Here $n = 1$ and $C_p = \frac{5}{2}R$, $T_2 = 673.15\text{ K}$, $T_1 = 473.15\text{ K}$.

$$\Delta S = 7.33 \text{ J.K}^{-1}$$

$$\Delta H = \frac{5}{2}R(T_2 - T_1) = \frac{5}{2} \times 8.314 \times 200 \text{ J} = 4157 \text{ J} = 4.157 \text{ kJ}.$$

$$\Delta G = \Delta H - T_2S_2 + T_1S_1$$

$$= \Delta H - T_2(S_1 + nC_p \ln \frac{T_2}{T_1}) + T_1S_1.$$

In this problem, $S_1 = 135 \text{ J.K}^{-1}.\text{mol}^{-1}$.

50. Calculate the change in entropy when 10 gm of tin is heated from 293 K to 573 K. The melting point of tin is 505 K. Its latent heat of fusion is 14 cal.gm^{-1} and specific heat of solid tin and liquid tin are 0.055 cal.gm^{-1} and 0.064 cal.gm^{-1} respectively.

[C.U. 2005]

Soln : The process can be depicted as below :

Sn(s), 293 K —1→ Sn(s), 505 K —2→ Sn(l), 505 K —3→ Sn(l), 573 K

Here $\Delta S_{total} = \Delta S_1 + \Delta S_2 + \Delta S_3$.

Now, $\Delta S_1 = m.C_{p,s} \ln \frac{T_2}{T_1} = 10 \times 0.055 \times \ln \frac{505}{293}$ cal.K^{-1}

$$= 0.29941 \text{ cal.K}^{-1} = 1.2527 \text{ J.K}^{-1}.$$

$$\Delta S_2 = \frac{10 \times 14}{505} \text{ cal.K}^{-1} = 0.2772 \text{ cal.K}^{-1} = 1.15980 \text{ J.K}^{-1}.$$

$$\Delta S_3 = 10 \times 0.064 \times \ln \frac{573}{505} \text{ cal.K}^{-1} = 0.080849 \text{ cal.K}^{-1} = 0.3383 \text{ J.K}^{-1}$$

$$\therefore \Delta S = \mathbf{2.75 \text{ J.K}^{-1}}.$$

51. 5 moles of an ideal gas, initially at 50 atm and 300 K is expanded irreversibly where the pressure suddenly drops to 10 atm. The work involved is 4000 J. Show that the final temperature is greater than a reversible adiabatic expansion to the same pressure. If $C_v = 1.5R$, calculate the entropy change during the irreversible expansion. **[B.U. 2002]**

Soln. : Let the irreversible work be adiabatic also.

By the problem, $C_v = 1.5R$.

Now, $W = nC_v\, dT = nC_v(T_2 - T_1)$.

Here, $n = 5$ mol, $T_1 = 300$ °K, W = –4000 J

$\therefore$ –4000 J = 5 mol × 1.5 × 8.314 J.K^{-1}.mol^{-1} $(T_2 - 300)$

or, $T_2 = \mathbf{235.85}$ **°K.**

Now, under adiabatic reversible condition,

$$T_2 = T_1\left(\frac{P_2}{P_1}\right)^{\frac{R}{C_p}}.$$

By the problem, $T_1 = 300$ °K, $P_2 = 10$ atm, $P_1 = 50$ atm.

$$\therefore \quad T_2 = 300 \times \left(\frac{10}{50}\right)^{\frac{R}{C_p}} = 300 \times (0.2)^{0.4} = 157.59 \text{ °K.}$$

So, the final temperature for the given process in irreversible adiabatic process is greater than that in reversible adiabatic process.

In the irrversible process,

$T_1 = 300$ °K, $T_2 = 235.85$ °K

$P_1 = 50$ atm, $P_2 = 10$ atm.

$$\therefore \Delta S = nC_p \ln \frac{T_2}{T_1} - nR \ln \frac{P_2}{P_1} \quad [\text{Here } C_p = C_v + R = 1.5R + 5 = 2.5R]$$

$$= 5 \times 2.5 \times 8.314 \ln \frac{235.85}{300} \text{J.K}^{-1} - 5 \times 8.314 \ln \frac{10}{50} \text{J.K}^{-1}$$

$$= -25.0029 \text{ J.K}^{-1} + 66.90 \text{ J.K}^{-1} = \mathbf{41.9 \text{ J.K}^{-1}}.$$

N.B. : If irreversible work is written as $P_{ext}(V_2 - V_1)$, using $P_{ext} = P_2 = 10$ atm, the equation will be wrong, because it has not been stated that work is done against a constant pressure of 10 atm.

52. A mass of water at T is mixed isobarically and adiabatically with an equal mass of water at T'. Show that the entropy change is positive and is given by $\Delta S = 2mC_p \ln \frac{T+T'}{2\sqrt{TT'}}$.

[Here m is the mass of water.]

Soln. : The final temperature will be $\frac{T+T'}{2}$.

Since there is m gm of water in each, the entropy change for m gm from T to $\frac{T+T'}{2}$ and T' to $\frac{T+T'}{2}$ can be added.

$$\therefore \Delta S = mC_p \ln \frac{T+T'}{2T} + mC_p \ln \frac{T+T'}{2T}$$

$$= mC_p \ln \left(\frac{T+T'}{2T} \times \frac{T+T'}{2T} \right)$$

$$= mC_p \ln \frac{(T+T')^2}{4TT'} = 2mC_p \ln \frac{T+T'}{2\sqrt{TT'}} \quad \textbf{[Proved].}$$

This quantity is positive. This can be shown by the following :

$$(T + T')^2 - 4T\,T' = (T - T')^2$$

Now since, $(T - T')^2 \geq 0$,

$\therefore \quad (T + T')^2 - 4T\,T' \geq 0$

or, $(T + T')^2 \geq 4T\,T'$

or, $\dfrac{(T+T')^2}{4TT'} \geq 1$.

$\therefore \Delta S$ is positive.

53. 1 mol of H_2O at 1 atm pressure undergoes the following process H_2O (l, –10 °C) →H_2O (s,–10 °C) in a thermostate maintained at –10 °C. The heat of fusion of water is 334.720 $J.g^{-1}$ at 0 °C and specific heat capacity of water and ice over this temperature range is 4.184 $J.C^{-1}.gm^{-1}$ and 2.092 $J.C^{-1}.gm^{-1}$ respectively. Is the process spontaneous?

Soln. : See Problem 36.

$$\therefore \Delta S = 18 \times 4.184 \ln\frac{273.15}{263.15} - \frac{334.720\times 18}{273.15} + 18 \times 2.092 \ln\frac{263.15}{273.15}$$

$$= -20.653 \text{ J.K}^{-1}.$$

$$\Delta H_{263.15} = \Delta H_{273.15} + \int_{273.15}^{263.15} \Delta C_P dT$$

$$= -334.720 + (2.092 - 4.184) \times (-10) = -313.8 \text{ J.g}^{-1}.$$

$$\therefore \quad \Delta S_{surr} = -\frac{\Delta H_{263.15}(l \to s)}{263.15} = \frac{18\times 313.8}{263.15} = 21.46 \text{ J.K}^{-1}.$$

$$\therefore \quad \Delta S_{universe} = 21.46 - 20.653 = 0.807 \text{ J.K}^{-1}.$$

ΔH for the process can also be calculated in the following way :

$$\Delta H_{263.15} = 18 \times 4.184 \times (273.15 - 263.15) - 334.720 \times 18 + 18 \times 2.092(263.15 - 273.15)$$

$$= -5648.4 \text{ J}$$

$$\text{and } \Delta S_{surr} = \frac{5684.4}{263.15} = 21.46 \text{ J.K}^{-1}.$$

This process is spontaneous since for this process $\Delta S_{universe}$ is positive.

54. Calculate ΔS when 1 mol of ice at 273 K is converted to steam at 373 K at 1 atm. (l_f = 80 cal/gm, l_e = 540 cal/gm, C_p = 1 cal/gm)

Hints :

Ice 273 K, 1 atm	$\xrightarrow{\Delta S_1}$	Water 273 K, 1 atm	$\xrightarrow{\Delta S_2}$	Water 3733 K, 1 atm	$\xrightarrow{\Delta S_3}$	Vapour 373 K, 1 atm

$$\Delta S = \Delta S_1 + \Delta S_2 + \Delta S_3 = \frac{l_f \times 18}{273} + 18 \times C_p \ln\frac{373}{273} + \frac{l_e \times 18}{373}$$

$$= 36.95 \text{ cal. deg}^{-1} = 154.6 \text{ J.deg}^{-1}.$$

55. Find ΔU, ΔH, ΔS, ΔF (or ΔA), ΔG in expanding 1 litre of an ideal gas at 25 °C to 100 litres at the same temperature.

Hints : $\Delta U = \Delta H = 0$, $\Delta S = 38.3 \text{ J.K}^{-1}$, $\Delta A = \Delta G = -11.42 \text{ J.K}^{-1}$.

56. Calculate the ΔG_f° for liquid water at 25 °C. Given that ΔH° of $H_2O(l)$ is –285.8 kJ and standard molar entropies are $S^\circ_{H_2O(l)} = 69.9$ J/k, $S^\circ_{H_2(g)} = 130.6$ J/K and $S^\circ_{O_2(g)} =$ 205 J/K. **[V.U. 2007]**

Soln. : $H_2(g) + \frac{1}{2}O_2(g) = H_2O(l)$, T = 25 °C = 298.15 K.

Given, $\Delta H^\circ_{H_2O(l)} = -285.8$ kJ, $S^\circ_{H_2O(l)} = 69.9$ J/K,

$$S^\circ_{H_2(g)} = 130.6 \text{ J/K}, \; S^\circ_{O_2(g)} = 205 \text{ J/K}$$

$$\Delta G^\circ_f = \Delta H^\circ - T\Delta S^\circ$$

$$= \Delta H^\circ_{H_2O(l)} - \Delta H^\circ_{H_2(g)} - \frac{1}{2}\Delta H^\circ_{O_2(g)} - 298.15 \times \Delta S^\circ$$

$$= -285.8 \text{ kJ} - 298.15 \text{ K}\left(S^\circ_{H_2O(l)} - S^\circ_{H_2(g)} - \frac{1}{2}S^\circ_{O_2(g)}\right)$$

(since, $\Delta H^\circ_{H_2(g)} = 0$ and $\Delta H^\circ_{O_2(g)} = 0$)

$$= -285.8 \text{ kJ} - 298.15 \text{ K}(69.9 - 130.6 - \frac{1}{2} \times 205) \text{ J/K}$$

$$= -237141.92 \text{ J} = \mathbf{-237.14 \text{ kJ.}}$$

57. 2 moles of He is heated from 473 K to 673 K at a constant pressure of 1 atm. Assuming He is a perfect gas, calculate ΔG, ΔH and ΔS for the process. Given that $S^\circ_{473} = 135 \text{ J.K}^{-1}.\text{mole}^{-1}$. Is this process spontaneous?

Hints : $\Delta S_{system} = 2 \text{ mol} \times \frac{5}{2} \times 8.314 \text{ J.K}^{-1}.\text{mol}^{-1} \times \ln\frac{673}{473} = \mathbf{14.66 \text{ J.K}^{-1}}$

$$\Delta H = 2 \times \frac{5}{2} \times 8.314\ (673 - 473) = \mathbf{8314 \text{ J.}}$$

$$\Delta G = \Delta H - (T_2S_2 - T_1S_1)$$

$$= 8314 - 673\ (2 \times 135 + 2 \times \frac{5}{2} \times 8314 \ln\frac{673}{473}) + 2 \times 473 \times 135$$

$$= -55552.18 \text{ J} = \mathbf{-55.55 \text{ kJ.}}$$

$$\Delta S_{sur} = -\frac{Q_P}{T} = -\frac{\Delta H}{T} = -\frac{8314}{673} \text{ J.K}^{-1} = -12.35 \text{ J.K}^{-1}$$

Hence, $\Delta S_{uni} = 2.31 \text{ J.K}^{-1}$.

So the process is spontaneous.

58. An ideal gas expands isothermally at 27 °C, but somewhat irreversibly, producing 1500 J of work. The entropy change is 10 J.K^{-1}. Calculate the degree of irreversibility, i.e., $\frac{W_{actual}}{W_{rev}}$.

Hints : $W_{actual} = -1500$ J.

$$\Delta U = 0,\ W_{rev} = -T\Delta S = -10 \times 300 = -3000 \text{ J.}$$

$$\therefore \frac{W_{actual}}{W_{rev}} = \mathbf{0.5.}$$

59. 1 mol of water vapour at 373 K and 2 atm undergoes the following process :

H_2O(g), 373 K, 2 atm)$\rightarrow$$H_2O$ (l, 373 K, 2 atm)

Calculate ΔG, ΔH and ΔS for the process assuming the vapour to be ideal and liquid to be incompressible.

C_p for vapour = 7 cal.deg^{-1}.mol^{-1}, C_p for liquid =18 cal.deg^{-1}.mol^{-1}, L_e =10 kcal.mol^{-1}).

Hints : The process is at constant P. Thus, for the process, $\Delta H = Q_P = -10$ K.cal.mol^{-1}. The process can also be carried out in the following reversible steps.

$$\begin{array}{ccc} H_2O\ (g,\ 373\ K,\ 2\ atm) & \xrightarrow{1} & H_2O(g,\ 373\ K,\ 1\ atm) \\ & & \big\downarrow 2 \\ H_2O(l,\ 373\ K,\ 2\ atm) & \xleftarrow[3]{} & H_2O\ (l,\ 373\ K,\ 1\ atm) \end{array}$$

$$\Delta H_1 = 0,\ \Delta S_1 = R\ \ln\ \frac{P_1}{P_2} = 1.38\ \text{cal.deg}^{-1}.\text{mol}^{-1}.$$

$$\Delta G_1 = RT\ \ln\ \frac{P_2}{P_1} = -514.7\ \text{cal.mol}^{-1}.$$

$\Delta H_3 = \Delta G_3 = \Delta S_3 = 0$, since water is incompressible.

$$\Delta G_2 = 0,\ \Delta H_2 = Q_P = -10\ \text{kcal.mole}^{-1},\ \Delta S_2 = -\frac{\Delta H_2}{T} = -26.81\ \text{cal.K}^{-1}.\text{mole}^{-1}.$$

$\Delta G = -514.7$ cal.mole^{-1}, ΔH = **−10 k.cal.mole^{-1}**, ΔS = **−25.43 cal.K^{-1}.mole^{-1}.**

60. In what proportions should n-hexane and n-heptane be mixed to achieve the greatest ΔS_{mix}, assumingg ideal mixing. **[B.U. 2004]**

Soln. : ΔS_{mix} / mol $= -R\ (X_1 \ln X_1 + X_2 \ln X_2) = \Delta S$

Now, $X_1 + X_2 = 1,\ \therefore\ X_2 = 1 - X_1$.

Thus, $\Delta S = -R\ [X_1 \ln X_1 + (1 - X_1)\ \ln(1 - X_1)]$.

$$\therefore\quad \frac{d\Delta S}{dX_1} = -R\ [\ln X_1 + 1 - \ln(1 - X_1) - 1].$$

For maximum value of ΔS.

$$\frac{d\Delta S}{dX_1} = 0 = \ln X_1 - \ln\ (1 - X_1)$$

or, $X_1 = 1 - X_1,\ \therefore\ X_1 = \dfrac{1}{2}$.

$$\therefore\quad \frac{n_1}{n_1 + n_2} = \frac{1}{2}\quad \text{or, } 1 + \frac{n_2}{n_1} = 2.$$

Thus, $n_{heptane} : n_{hexane} = 1 : 1$.

Hence, n_{hexare} and n_{heptane} should be mixed in the same proportion to achieve maximum ΔS_{mix}.

61. 100 gm of water cools from 100 °C to 25 °C when kept in air. Calculate $\Delta S_{universe}$. **[V.U. 2007]**

Soln. : Given, $n_{H_2O} = \dfrac{100\ \text{gm}}{18\ \text{gm / mole}} = 5.56$ mole.

T_1 = 100 °C = 373.15 K

T_2 = 25 °C = 298.15 K

Hence, $\Delta S_{system} = nC_P \ln \dfrac{T_2}{T_1} = 5.56 \text{ mol} \times 18 \text{ cal.K}^{-1}.\text{mol}^{-1} \times \ln \dfrac{298.15}{373.15}$

$= -22.46 \text{ cal.K}^{-1}$

$Q = nC_p(T_2 - T_1)$

$= 5.56 \text{ mol} \times 18 \text{ cal.K}^{-1}.\text{mol}^{-1} \times (1298.15 - 373.15)\text{K} = -7506 \text{ cal.}$

$\therefore \Delta S_{surr} = \dfrac{7506 \text{ cal}}{298.15 \text{ K}} = 25.18 \text{ cal.K}^{-1}.$

$\therefore \Delta S_{universe} = \Delta S_{system} + \Delta S_{surr} = \mathbf{2.72 \text{ cal.K}^{-1}.}$

62. 2 mol of benzene vaporizes at 80 °C (boiling point). Calculate ΔA, ΔG, W, Q, ΔU, ΔH, ΔS if final pressure is 0.5 atm at 80 °C. Given that vapour behaves ideally and latent heat of vaporization of benezene is 422.6 J.gm^{-1}.

Hints : $\underset{2 \text{ mol}}{C_6H_6 (l)} \xrightarrow[1 \text{atm}]{80\,°\text{C}} \underset{2 \text{ mol}}{C_6H_6 (g)} \xrightarrow[0.5 \text{ atm}]{80\,°\text{C}} \underset{2 \text{ mol}}{C_6H_6 (g)}$

$\Delta H = \Delta H_1 + \Delta H_2 = 422.6 \times 78 \text{ J.mol}^{-1} \times 2 \text{ mol} + 0 = \mathbf{65.9 \text{ kJ.}}$

$W = (W_1 + W_2) = -P(V_g - V_l) - nRT \ln \dfrac{P_1}{P_2}$

$= P(V_l - V_g) + nRT \ln \dfrac{P_2}{P_1} = \mathbf{-\ 9.94 \text{ kJ.}}$

$\Delta U = \Delta U_1 + \Delta U_2 = \Delta U_1 = \Delta H_1 - P\Delta V = \Delta H_1 - nRT = \mathbf{60.03 \text{ kJ.}}$

$Q = Q_1 + Q_2 = \Delta H_1 + nRT \ln \dfrac{P_1}{P_2}$

$\Delta S = \dfrac{\Delta H_1}{T} - nR \ln \dfrac{P_2}{P_1} = \mathbf{198.13 \text{ J.K}^{-1}.}$

$\Delta G = \Delta G_1 + \Delta G_2 = 0 + nRT \ln \dfrac{P_2}{P_1}$

$\Delta A = \Delta A_1 + \Delta A_2 = -nRT + nRT \ln \dfrac{P_2}{P_1} \quad (\because \Delta A_1 = \Delta G_1 - P\Delta V).$

63. A heat engine operates through the following Carnot cycle between $T_1 = 273$ K and T_2 and produces 418 J of work per cycle. Calculate Q_1, Q_2, Q_3, Q_4 and T_2.

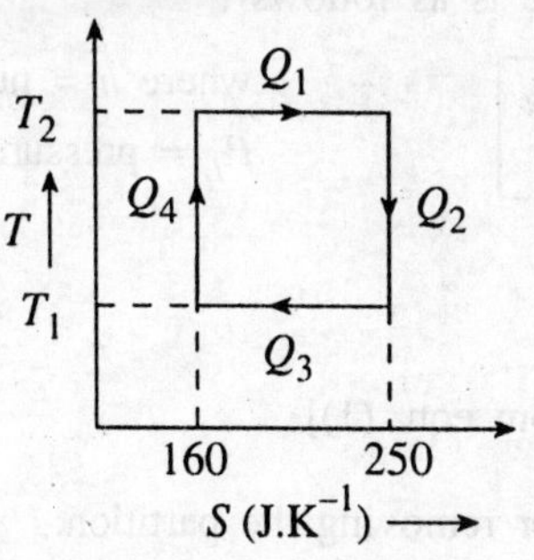

Hints :

Step 1 : Reversible Isothermal expansion, $Q_1 = \Delta S_1 \times T_2 = 90\ T_2$.

Step 2 : Reversible Isoentropic expansion, Q_2 = **0** (Rev adiabatic process).

Step 3 : Reversible Isothermal contraction,

$$Q_3 = \Delta S_3 \times T_1 = \mathbf{-24.57\ kJ.}$$

Step 4 : Reversible isoentropic (adiabatic) contraction,

$$Q_4 = 0$$
$$Q_1 + Q_2 + Q_3 + Q_4 = 418 = 90\,T_2 - 24570$$
$$T_2 = 277.6\text{ K} = \mathbf{278\ K}$$
$$\therefore\ Q_1 = 90 \times 278\text{ J} = \mathbf{25.02\ kJ.}$$

64. Calculate ΔH_m and ΔS_m for the process :

$H_2O\ (l)$ $T = 263$ K $P = 0.1$ MPa	$\longrightarrow$	$H_2O\ (s)$ $T = 263$ K $P = 0.1$ MPa

Assume that at 273 K the molar enthalpy of fusion of ice is 6006 J. mol^{-1}, the heat capacity $C_{p.m.}$(s) of ice is 38 J.K^{-1} and the heat capacity $C_{p.m.}(l)$ of liquid water is 76 J.K^{-1} mol^{-1}. Consider the heat capacities to be constants. **[JAM 2010]**

Hints :

$H_2O\ (l)$	$\longrightarrow$	$H_2O\ (l)$	$\longrightarrow$	H_2O (s)	$\longrightarrow$	H_2O (s)
$T = 263$		$T = 263$		$T = 273$		$T = 263$
$P = 0.1$ MPa		$P = 0.1$ MPa		$P = 0.1$ MPa		$P = 0.1$MPa

$$\Delta S_m = -20.58\text{ J.K}^{-1}\text{ mol}^{-1}$$
$$\Delta C_P = C_{P,m}(\text{s}) - C_{P,m}(l) = -10\text{ J.K}^{-1}$$
$$\Delta H_{273}\ (l \to \text{s}) = -\ 600\text{ J.mol}^{-1}$$
$$\Delta H_{273}\ (l \to \text{s}) = -\ \Delta H_{273}\ (l \to \text{s}) + \Delta C_P\ (268 - 273)$$
$$= -5996.38\text{ J.mol}^{-1}.$$

65. Consider a thermally insulated container divided by a partition into two compartments in such a way that left copartment is twice as big as right one. Both the compartments contain same number of moles of ideal gas at temperature T. The partition is now removed. Calculate

(a) the total change of entropy if the gases are identical;

(b) the total change of entropy if the gases are different.

Soln. : Let the volume and pressure of right compartment be V and P_R respectively. By the problem, initial state is as follows :

$2V, P_L$ T, n	V, P_R T, n

where n = number of moles of gas,
P_L = pressure of left compartment.

Thus, $P_R V = nRT$ (1)

$P_L\ 2V = nRT$

or, $P_L = \dfrac{nRT}{2V} = \dfrac{P_R}{2}$ [from eqn. (1)]. (2)

Let final pressure be P after removing the partition.

Thus $P(2V + V) = 2nRT$

or, $P = \frac{2}{3}\frac{nRT}{V} = \frac{2}{3}P_R$ [from eqn. (1)] (3a)

$= \frac{4}{3}P_L$ [from eqn. (2)]. (3b)

(a) For the right compartment,

$\Delta S_R = nR\ln\frac{P_R}{P} = nR\ln\frac{3}{2}$ [from eqn. (3a)]

For the left compartment,

$\Delta S_L = nR\ln\frac{P_L}{P} = nR\ln\frac{3}{4}$ [from eqn. (3b)]

Thus, total change in entropy, when the gases are identical,

$$\Delta S = \Delta S_R + \Delta S_L = nR\ln\frac{3}{2} + nR\ln\frac{3}{4} = nR\ln\frac{9}{8}.$$

(b) When the gases are different, partial pressure of the gas in right compartment after mixing :

$$P_R' = P.\frac{n}{n+n} = \frac{P}{2} = \frac{P_R}{3} \quad (4a)$$

and partial pressure of the gas in left compartment after mixing :

$$P_L' = \frac{P}{2} = \frac{2}{3}P_L. \quad (4b)$$

Thus, total change in entropy when the gases are different,

$$\Delta S = \Delta S_R + \Delta S_L = nR\ln\frac{P_R}{P_R'} + nR\ln\frac{P_L}{P_L'}$$

$$= nR\ln 3 + nR\ln\frac{3}{2} \text{ [from eqn. (4a) and (4b)]}$$

$$= nR\ln\frac{9}{2}.$$

66. Suppose a container of volume V is divided into two equal compartments by a thin partition with a small hole of area A in it. If it is filled with an ideal gas at temperature T such that the initial pressure on the left compartment is $P_L(0)$ and that on the right side is $P_R(0)$,

(i) find the pressure of the gas in the left compartment as a function of time.

(ii) calculate ΔS of the whole gas after the equilibrium has been reached

Soln. : Initial State :

$P_L(0)$	$P_R(0)$
T	T
$\frac{V}{2}$	$\frac{V}{2}$

If $N_L(0)$ and $N_R(0)$ are initial numbers of gas molecules in left and right compartments respectively,

$$P_L(0)\frac{V}{2} = N_L(0)\ \text{kT} \tag{1}$$

$$P_R(0)\frac{V}{2} = N_R(0)\ \text{kT}. \tag{2}$$

Let total number of molecules be N which remains constant.

Thus, $N_L(0) + N_R(0) = N.$ (3)

Let N_L and N_R be number of molecules in two respective compartments at any instant.

Hence, number of molecules effusing through hole of area A from left to right compartment in time dt

$$= dN_1 = \frac{1}{4}\frac{N_L}{\frac{V}{2}}\bar{v}Adt. \tag{4}$$

and similarly number of molecules effusing from right to left compartment in time dt

$$= dN_2 = \frac{1}{4}\frac{N_L}{\frac{V}{2}}\bar{v}Adt. \tag{5}$$

Here $\bar{v}$ = average velocity.

Net change in number of molecules in left compartment in time dt,

$$dN_L = -dN_1 + dN_2$$

$$= -\frac{1}{4}\frac{N_L}{\frac{V}{2}}\bar{v}Adt + \frac{1}{4}\frac{N_R}{\frac{V}{2}}\bar{v}Adt$$

$$= -\frac{N_L}{2V}\bar{v}Adt + \frac{(N - N_L)}{2V}\bar{v}Adt \quad \text{(since, } N_L + N_R = N\text{)}$$

$$= \frac{N_L\bar{v}Adt}{V} + \frac{N}{2V}\bar{v}Adt. \tag{6}$$

Due to effusion corresponding change in pressure in the left compartment = dP_L.

Hence, $dP_L\frac{V}{2} = dN_L\text{kT}.$

Let the final pressure in both compartments be P when effusion no longer changes the pressures of the compartments.

Hence, $PV = NKT = [N_L(0) + N_R(0)]\ \text{kT}$

$= P_L(0)\frac{V}{2} + P_R(0)\frac{V}{2}$ [from eqn. (1) and eqn. (2)]

Thus, $P = \frac{1}{2}[P_L(0) + P_R(0)].$

Hence, from eqn. (6) we obtain

$$dP_L\frac{V}{2KT} = -\frac{P_L\bar{v}Adt}{2kT} + \frac{P}{2kT}\bar{v}Adt$$

or, $$\frac{dP_L}{P-P_L} = \frac{\bar{v}A dt}{V}$$

or, $$\int_{P_L(0)}^{P_L(t)} \frac{dP_L}{P-P_L} = \frac{vA}{V}\int_0^t dt$$

or, $$-\ln \frac{P-P_L(t)}{P-P_L(0)} = \frac{\bar{v}A}{V}t.$$

$$\therefore \quad P_L(t) = P - [P - P_L(0)]\, e^{-\frac{\bar{v}At}{V}}.$$

$$= \frac{1}{2}[P_L(0)+P_R(0)] - \left[\frac{1}{2}P_L(0)+\frac{1}{2}P_R(0)-P_L(0)\right]e^{-\frac{\bar{v}At}{V}}.$$

Hence, at time t pressure at the left compartment,

$$P_L(t) = \frac{1}{2}[P_L(0)+P_R(0)] - \frac{1}{2}[P_R(0)-P_L(0)]e^{-\frac{\bar{v}At}{V}} \quad \text{[from eqn. (7)]}$$

(ii) $$\Delta S = \Delta S_L + \Delta S_R = N_L(0)k\ln\frac{P_L(0)}{P} + N_R(0)k\ln\frac{P_R(0)}{P}$$

$$= \frac{V}{2T}\left[P_L(0)\ln\frac{2P_L(0)}{P_L(0)+P_R(0)} + P_R(0)\ln\frac{2P_R(0)}{P_L(0)+P_R(0)}\right] \quad \text{[from eqn. (7)]}$$

Miscellaneous Problems

1. Show that $\left(\frac{\partial S}{\partial P}\right)_V = \frac{\beta C_v}{\alpha T}$,

$$\left(\frac{\partial S}{\partial P}\right)_P = \frac{C_p}{\alpha VT}.$$

Soln. : We know that, $đU = đQ - PdV$.

For 1 mole of gas undergoing reversible change,

$dU = TdS - PdV$

or, $$TdS = C_v dT + \left[\left(\frac{\partial U}{\partial V}\right)_T + P\right]dV$$

Again, $$\left(\frac{\partial U}{\partial V}\right)_T = T\left(\frac{\partial P}{\partial T}\right)_V - P.$$

$$\therefore\ TdS = C_v dT + T\left(\frac{\partial P}{\partial T}\right)_V dV.$$

$$\therefore\ dS = C_v\frac{dT}{T} + \left(\frac{\partial P}{\partial T}\right)_V .dV.$$

When $dS = 0$, $\left(\frac{\partial V}{\partial T}\right)_S = -\frac{C_v}{T}\frac{1}{\left(\frac{\partial P}{\partial T}\right)_V}$

$$= -\frac{\beta C_v}{\alpha T} \text{ [since } \left(\frac{\partial P}{\partial T}\right)_V = \frac{\alpha}{\beta}\text{]}.$$

By Maxwell's relation,

$$\left(\frac{\partial T}{\partial V}\right)_S = -\left(\frac{\partial P}{\partial S}\right)_V.$$

$$\therefore \left(\frac{\partial S}{\partial P}\right)_V = -\left(\frac{\partial V}{\partial T}\right)_S$$

$$\therefore \left(\frac{\partial S}{\partial P}\right)_V = \frac{\beta C_v}{\alpha T} \textbf{ [Proved].}$$

Again, $TdS = dH - VdP$

or, $TdS = C_p dT + \left[\left(\frac{\partial H}{\partial P}\right)_T - V\right]dP \left[\text{since} \left(\frac{\partial H}{\partial P}\right)_T = V - T\left(\frac{\partial V}{\partial T}\right)_P\right]$

or, $TdS = C_p dT - T\left(\frac{\partial V}{\partial T}\right)_P dP.$

or, $dS = C_p\frac{dT}{T} - \left(\frac{\partial V}{\partial T}\right)_P dP.$

$$\therefore \left(\frac{\partial P}{\partial T}\right)_S = \frac{C_p}{T}.\frac{1}{\left(\frac{\partial V}{\partial T}\right)_P} = \frac{C_p}{T\alpha V}.$$

But $\left(\frac{\partial P}{\partial T}\right)_S = \left(\frac{\partial S}{\partial V}\right)_P.$

$$\therefore \left(\frac{\partial S}{\partial V}\right)_P = \frac{C_p}{\alpha VT} \textbf{ [Proved].}$$

2. Show that for an ideal gas $\left(\frac{\partial V}{\partial S}\right)_P = \frac{RT}{PC_P}$ and $\left(\frac{\partial P}{\partial S}\right)_V = \frac{RT}{VC_V}$.

Soln. : $dS = nC_p\frac{dT}{T} - \left(\frac{\partial V}{\partial T}\right)_P dP$

or, $\left(\frac{\partial S}{\partial V}\right)_P = \frac{nC_P}{T}\left(\frac{\partial T}{\partial V}\right)_P.$

For an ideal gas, $T = \frac{PV}{nR}$, $\therefore \left(\frac{\partial T}{\partial V}\right)_P = \frac{P}{nR}.$

Thus, $\left(\frac{\partial V}{\partial S}\right)_P = \frac{T}{nC_P}\left(\frac{\partial V}{\partial T}\right)_P = \frac{RT}{PC_P}$.

$\therefore \left(\frac{\partial V}{\partial S}\right)_P = \frac{RT}{PC_P}$ **[Proved].**

Again, $dS = nC_V\frac{dT}{T} + \left(\frac{\partial P}{\partial T}\right)_V dV$

or, $\left(\frac{\partial S}{\partial P}\right)_V = \frac{nC_V}{T}\left(\frac{\partial T}{\partial P}\right)_V$.

For an ideal gas, $\left(\frac{\partial T}{\partial P}\right)_V = \frac{V}{nR}$.

Thus, $\left(\frac{\partial P}{\partial S}\right)_V = \frac{RT}{VC_V}$ **[Proved].**

3. Find the value of μ_{JT} for a gas obeying (i) $P(V-b) = RT$, (ii) $\left(P+\frac{a}{V^2}\right)V = RT$.

Comment on the results.

Ans. : $\mu_{JT} = \left(\frac{\partial T}{\partial P}\right)_H = \frac{T\left(\frac{\partial V}{\partial T}\right)_P - V}{C_V}$.

(i) $P(V-b) = RT$

or, $V = \frac{RT}{P} + b$.

$\therefore \left(\frac{\partial V}{\partial T}\right)_P = \frac{R}{P}$. Hence, $T\left(\frac{\partial V}{\partial T}\right)_P - V = \frac{RT}{P} - V = -b$.

$\therefore$ $\boldsymbol{\mu_{JT} = -b < 0}$.

Temperature will always increase on reduction of pressure since the gas has only repulsion but not attraction between its molecules. So it does not have any inversion temperature.

(ii) $\left(P+\frac{a}{V^2}\right)V = RT$

or, $V = \frac{RT}{P} - \frac{a}{RT}$ (replacing V by $\frac{RT}{P}$)

Hence, $T\left(\frac{\partial V}{\partial T}\right)_P - V = \frac{2a}{RT}$.

$\therefore \mu_{JT} = \frac{2a}{RT} > 0$.

Temperature will always decrease on reduction of pressure, since the gas has only attraction but not repulsion between its molecules. Result indicates that it does not have any inversion temperature.

4. Find out the values of the following differentials in terms of C_P, C_V, α, β and P, V, n, T :

(a) $\left(\frac{\partial U}{\partial T}\right)_V$, (b) $\left(\frac{\partial U}{\partial T}\right)_P$, (c) $\left(\frac{\partial U}{\partial V}\right)_T$, (d) $\left(\frac{\partial U}{\partial P}\right)_T$, (e) $\left(\frac{\partial U}{\partial P}\right)_V$, (f) $\left(\frac{\partial U}{\partial V}\right)_P$.

Soln. : (a) $\left(\frac{\partial U}{\partial T}\right)_V = T\left(\frac{\partial S}{\partial T}\right)_V = \boldsymbol{C_V}$.

(b) $dU = TdS - PdV$

$$\left(\frac{\partial U}{\partial T}\right)_P = T\left(\frac{\partial S}{\partial T}\right)_P - P\left(\frac{\partial V}{\partial T}\right)_P$$

$$= C_p - P\alpha V \text{ [since, } \alpha = \frac{1}{V}\left(\frac{\partial V}{\partial T}\right)_P\text{]}$$

(c) $dU = TdS - PdV$

$$\left(\frac{\partial U}{\partial V}\right)_T = T\left(\frac{\partial S}{\partial V}\right)_T - P = T\left(\frac{\partial P}{\partial T}\right)_V - P = \boldsymbol{T\frac{\alpha}{\beta} - P} \text{ [by Maxwell's relation]}$$

(d) $dU = TdS - PdV$

$$\left(\frac{\partial U}{\partial P}\right)_T = T\left(\frac{\partial S}{\partial P}\right)_T - P\left(\frac{\partial V}{\partial P}\right)_T = -P\left(\frac{\partial V}{\partial P}\right)_T - T\left(\frac{\partial V}{\partial T}\right)_P \text{ [by Maxwell's relation]}$$

$$= \boldsymbol{\beta PV - \alpha VT}, \text{ since } \beta = -\frac{1}{V}\left(\frac{\partial V}{\partial P}\right)_T.$$

(e) $dU = TdS - PdV$

$$\left(\frac{\partial U}{\partial P}\right)_V = T\left(\frac{\partial S}{\partial P}\right)_V = -T\left(\frac{\partial V}{\partial T}\right)_S \text{ [by Maxwell's relation]}$$

$$= \boldsymbol{C_V\frac{\beta}{\alpha}}\left[\because \left(\frac{\partial V}{\partial T}\right)_S = -\frac{\beta C_V}{\alpha T}\right].$$

(f) $dU = TdS - PdV$

$$\left(\frac{\partial U}{\partial V}\right)_P = T\left(\frac{\partial S}{\partial V}\right)_P - P = T\left(\frac{\partial P}{\partial T}\right)_S - P \text{ [by Maxwell's relation]}$$

$$= \boldsymbol{\frac{C_P}{\alpha V} - P}\left[\because \left(\frac{\partial P}{\partial T}\right)_S = -\frac{C_P}{\alpha VT}\right].$$

5. Find out the values of following differentials :

(i) $\left(\frac{\partial H}{\partial T}\right)_P$, (ii) $\left(\frac{\partial H}{\partial T}\right)_V$, (iii) $\left(\frac{\partial H}{\partial V}\right)_T$, (iv) $\left(\frac{\partial H}{\partial P}\right)_T$, (v) $\left(\frac{\partial H}{\partial P}\right)_V$, (vi) $\left(\frac{\partial H}{\partial V}\right)_P$.

Hints : $dH = TdS + VdP$ and then proceed as Problem 4.

(i) C_p, (ii) $\left(C_V + V\frac{\alpha}{\beta}\right)$, (iii) $\frac{1}{\beta}(\alpha T - 1)$, (iv) $V(1 - \alpha T)$, (v) $\left(V + \frac{\beta}{\alpha}.C_V\right)$, (vi) $\frac{C_P}{\alpha.V}$.

6. Draw Carnot cycle diagramatically in V vs. P, T vs. S, S vs. T, P vs. T, V vs. T, U vs. S and S vs. U with reference to P vs. V diagram. Indicate each step.

Reference plot :

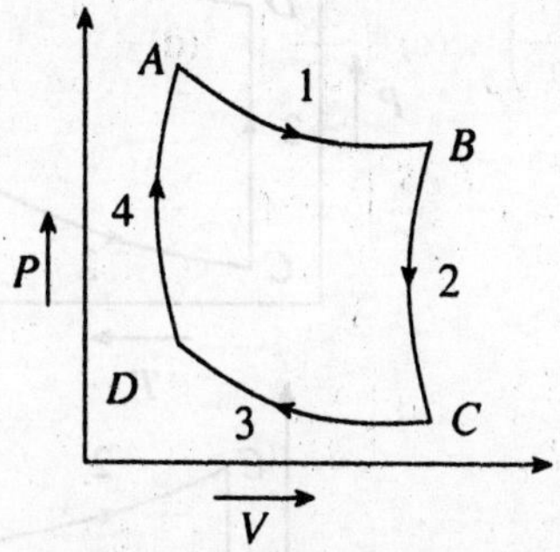

Process	*Pressure (P)*	*Volume (V)*	*Temperature (T)*	*Entropy (S)*	*Internal energy (U)*
1. Isothermal rev. expansion	Decreases	Increases	Const.	Increases	Const.
2. Adiabatic. rev. expansion	Decreases	Increases	Decreases	Const.	Decrases
3. Isothermal and rev. contraction	Increases	Decreases	Const.	Decreases	Const
4. Adiabatic rev. contraction	Increases	Decreases	Increases	Const.	Increases

Thus, the other diagrams would be as follows :

1. V vs. P

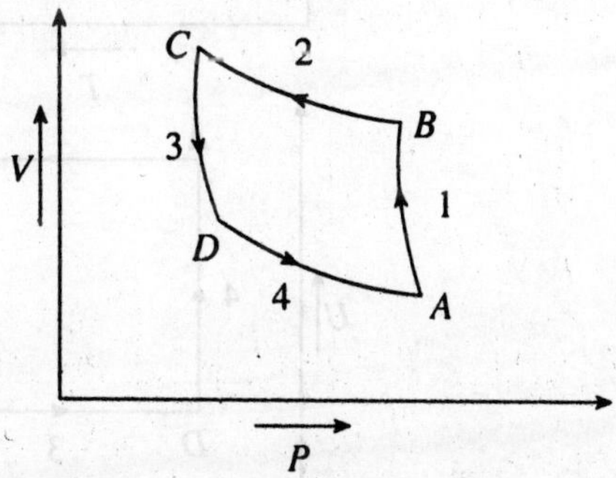

2. T vs. S

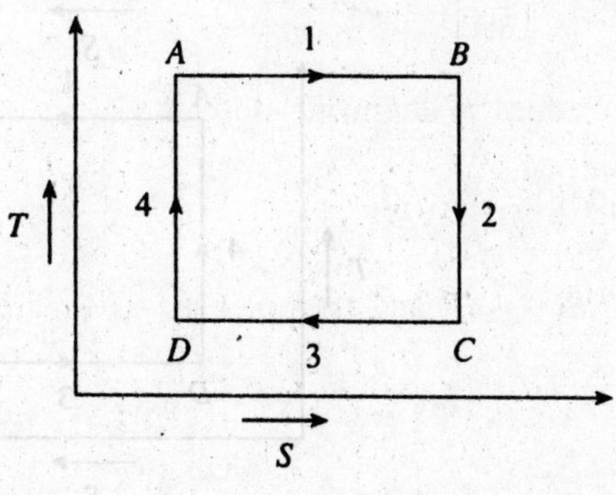

3. *P* vs. *T*

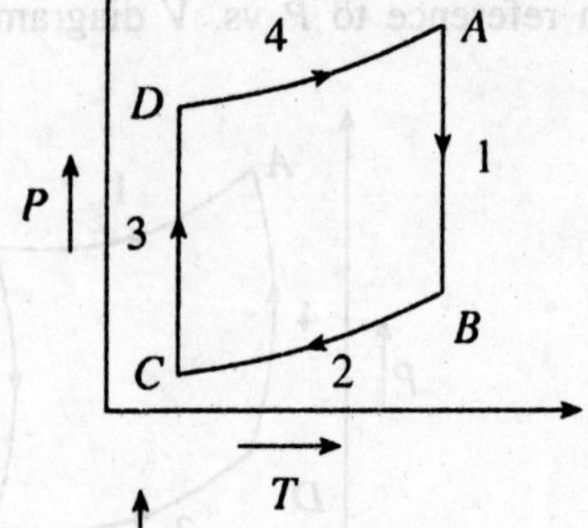

4. *V* vs. *T*

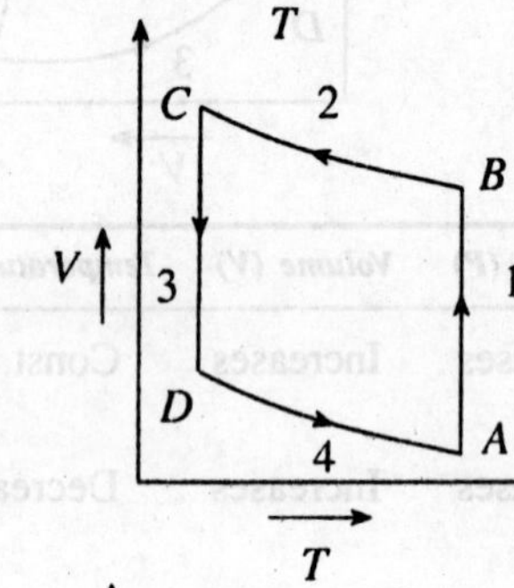

5. *S* vs. *T*

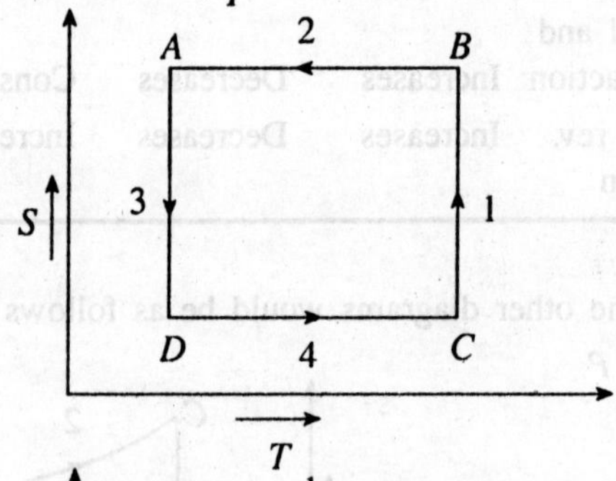

6. *U* vs. *S*

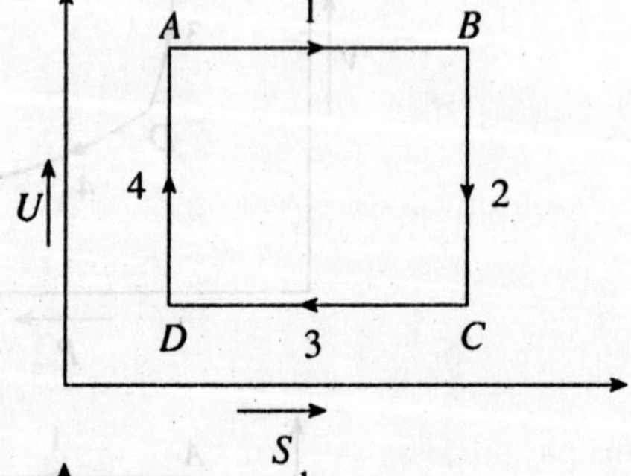

7. *S* vs. *U*

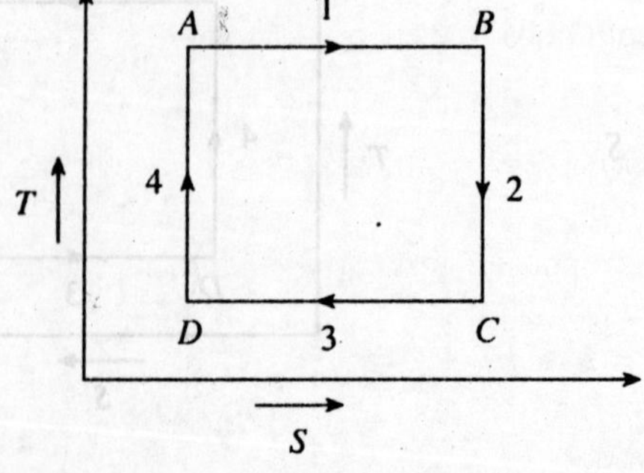

Similarities between T vs S and U vs S diagrams, and S vs T and S vs U diagrams are obvious since for ideal gas U is a function of T only.

7. Prove that for an ideal gas, $\left(\frac{\partial U}{\partial T}\right)_P = C_V$ and $\left(\frac{\partial H}{\partial T}\right)_V = C_P$.

Soln. : $dU = TdS - PdV$

or, $\left(\frac{\partial U}{\partial T}\right)_P = T\left(\frac{\partial S}{\partial T}\right)_P - P\left(\frac{\partial V}{\partial T}\right)_P = C_P - R = C_V$.

[since, for ideal gas, $PV = RT$]

or, $P\left(\frac{\partial V}{\partial T}\right)_P = P.\frac{R}{P} = R.$

Again, $dH = TdS + VdP$

or, $\left(\frac{\partial H}{\partial T}\right)_V = T\left(\frac{\partial S}{\partial T}\right)_V + V\left(\frac{\partial P}{\partial T}\right)_V = C_V + R = C_P$,

since, for ideal gas, $V\left(\frac{\partial P}{\partial T}\right)_V = V.\frac{R}{V} = R.$

It can also be proved from 1st law considering,

$$H = U + PV$$

$$\left(\frac{\partial H}{\partial T}\right)_V = \left(\frac{\partial U}{\partial T}\right)_V + V\left(\frac{\partial P}{\partial T}\right)_V = C_V + R = C_P \textbf{ [Proved].}$$

And $U = H - PV$

or, $\left(\frac{\partial U}{\partial T}\right)_P = \left(\frac{\partial H}{\partial T}\right)_P - P\left(\frac{\partial V}{\partial T}\right)_P = C_V$ **[Proved].**

8. State Trouton's rule.

Ans. : Trouton's rule state that molar entropy of vaporization of various liquids at the standard boiling point (the boiling point at 1 bar pressure) is same and it is about 88 $J.K^{-1}.mol^{-1}$.

9. Joule-Thomson process is an isoentropic process—comment.

Ans. : Joule-Thomson process is an irreversible adiabatic process. So entropy increases in this process, since for an irreversible process $dS_{system} > \frac{đQ}{T}$ and here $đQ = 0$. It is rather an isoenthalpic process.

10. Change of Gibbs free energy for a process at low temperature at a constant pressure is given by :

$\Delta G = a + bT + cT^2$, show that $b = 0$ and $\left(\frac{\partial \Delta G}{\partial T}\right)_P = \left(\frac{\partial \Delta H}{\partial T}\right)_P$ have values equal in magnitude and opposite in sign and both the values tend towards 0 as $T \to 0$.

[B.U. 2006]

Ans. : $\Delta G = a + bT + cT^2$

$$-\Delta S = \left(\frac{\partial \Delta G}{\partial T}\right)_P = b + 2Tc.$$

$\therefore\ \Delta S = -(b + 2Tc)$ and $\underset{T\to 0}{Lt}\ \Delta S = -b.$

Now, according to the third law of thermodynamics, $\underset{T\to 0}{Lt}\ \Delta S = 0.$

$\therefore\ b = 0.$

Thus, $\Delta G = a + cT^2$ and $\left(\frac{\partial \Delta G}{\partial T}\right)_P = 2Tc.$

$$\Delta H = \Delta G - T\left(\frac{\partial \Delta G}{\partial T}\right)_P = a + cT^2 - 2cT^2 = a - cT^2.$$

$$\therefore \left(\frac{\partial \Delta H}{\partial T}\right)_P = -2Tc,\ \therefore \left(\frac{\partial \Delta G}{\partial T}\right)_P = -\left(\frac{\partial \Delta H}{\partial T}\right)_P = 2Tc.$$

Thus, $\underset{T\to 0}{Lt}\left(\frac{\partial \Delta G}{\partial T}\right)_P = \underset{T\to 0}{Lt}\left(\frac{\partial \Delta H}{\partial T}\right)_P = 0.$

CHAPTER 7

COLLIGATIVE PROPERTIES OF SOLUTION

Required Formulae

1. Osmotic pressure (Π) = RTC

$R \Rightarrow$ universal gas constant.

$T \Rightarrow$ temperature

$C \Rightarrow$ concentration in molarity (M) = $\frac{n_2}{V}$

where n_2 is the number of moles of solute and V is volume in dm^3

2(a). $\frac{P^\circ - P}{P^\circ} = x_2$, where P° is the vapour pressure of pure solvent, P is the vapour pressure of solution and x_2 is the mole fraction of solute.

2(b). For ideal solution mixture of two liquids A and B, total vapour pressure

$$P = P_A + P_B = x_A \overset{\circ}{P_A} + x_B \overset{\circ}{P_B}$$

If x_A^{vap} and x_B^{vap} are mole fractions of A and B respectively in vapour state,

$P_A = x_A^{vap} P$, $P_B = x_B^{vap} P$.

3. $\Delta T_b = \frac{RT_b^{\circ 2}}{1000 l_e}$, where m is concentration in molality and is given by

$m = \frac{1000 W_2}{M_2 \times W_1}$, where $M_2 \Rightarrow$ molecular weight of solute.

W_2 gm of solute is dissolved in W_1 gm of solvent.

$\Delta T_b \Rightarrow$ elevation of boiling point

$\overset{\circ}{T_b} \Rightarrow$ normal boiling point of pure solvent.

$l_e \Rightarrow$ latent heat of evaporation per gram of solvent.

$\therefore \Delta T_b = K_b \times m$

$K_b \Rightarrow$ ebullioscopic constant.

4. $\Delta T_f = K_f \times m$,

where K_f ⇒ cryoscopic constant and $K_f = \dfrac{RT_f^{\circ 2}}{1000 l_f}$,

where T_f° ⇒ freezing point of pure solvent

l_f ⇒ latent heat of fusion per gram of solvent.

ΔT_f ⇒ depression of freezing point.

If there is association or dissociation, the quantities are multiplied by van't Hoff factor *i*.

$$i = \frac{\Pi_{obs}}{\Pi_{theo}}, \text{ etc.}$$

5. Degree of dissociation, $\alpha = \dfrac{i-1}{\nu-1}$,

where ν is the number of particles into which solute dissociates.

6. Degrees of association, $\beta = \dfrac{\nu(1-i)}{\nu-1}$,

where ν is the number of molecules of solute associating.

7. In S.I. unit,

$$\Pi = RT\frac{n_2}{V} = RT\frac{W_2 \text{ kg}}{M_2 \text{ kg.mole}^{-1} \times V \text{ m}^3}$$

$$\Delta T_{b,f} = K_{b,f}\frac{W_2 \text{ kg}}{M_2 \text{ kg.mole}^{-1} \times W_1 \text{ kg}} = \frac{RT_{b,f}^{\circ 2}}{L_{e.f}}\text{m}$$

L_e = latent heat of evaporation per kg

L_f = latent heat of fusion per kg.

Short Questions

1. What is van't Hoff factor *i*?

Ans. : When solute molecules associate or dissociate in an ideally dilute solution, the simple relations relating mole-fraction, molarity or molality of solute molecules with different colligative properties of the solution will no longer be valid, since concentration changes. So observed value deviates from calculated value. van't Hoff factor *i* is the ratio of observed colligative property to the theoretical value at the same concentration and temperature. *i* can be defined as the ratio of observed osmotic pressure, Π_{obs}, to its theoretical value Π_{theo}, at the same concentration and same temperature.

$$i = \frac{\Pi_{obs}}{\Pi_{theo}} = \frac{\Pi_{obs}}{RTC}$$

$$\therefore \Pi_{obs} = iRTC.$$

Similarly, $\left(\dfrac{P^\circ - P}{P^\circ}\right)_{obs} = i\,x_2.$

$$(\Delta T_b)_{obs} = iK_b m$$

$$(\Delta T_f)_{obs} = iK_f m.$$

2. When can the van't Hoff factor i be integral?

Ans. : The van't Hoff factor i can be integral when a solute is completely dissociated to ν particles. The colligative properties depend on number, and for a C molar solution the real concentration becomes νC (and for m molar it is νm) so that.

$$\Pi = \nu RTC$$
$$\Delta T_f = \nu K_f m$$
$$i = \nu.$$

3. "Colligative properties are intensive."—Comment.

Ans. : A property is extensive when sum of its values of constituting parts of a system gives the whole. When parts do not add to give the whole, the property is intensive. This is so about colligative properties. Let the osmotic pressure Π of a 800 cc solution of glucose of certain molarity m be A, then Π of 1 litre solution is also A. That is why colligative properties are intensive.

4. Is there any advantage in using molality instead of molarity as a concentration unit?

Ans. : Yes, there is an advantage. When the temparature of the liquid phase changes, and yet we need to compare relative concentrations at two tempartures, then the temparature independent concentration unit 'molality' may be used.

5. When can there be freezing point elevation?

Ans. : In an ideal solution the solute is insoluble in solid solvent, and ΔT_f (the freezing point depression) = $T_0 - T_f$ (T_0, freezing point for solvent and T_f, freezing point of solution) is (+)ve. Thus, there is freezing point depression. In the case when the solute is soluble in solid solvent and the solubility in the solid solvent is greater than that in liquid solvent, then there can be freezing point elevation, i.e., ΔT_f is (–)ve. It can be easily shown that $\Delta T_f = K_f(1 - K)$, where K is the ratio of mol-fraction of solute in solid solvent to mol-fraction of solute in liquid solvent. When $K > 1$, there will be freezing point elevation.

6. "Lowering of vapour pressure of a liquid by a non-volatile solute is due to the attraction of solvent molecule through solvation."—Justify or criticise.

Ans. : Lowering of vapour pressure is an entropy effect. A solvent vaporises when kept in a container because that leads to more chaos and randomness, thereby increasing entropy. In a closed vessel the molecules in the vapour phase collide with the wall giving a pressure P^o—this is the vapour pressure. Addition of non-volatile solute introduces extra randomness in liquid phase. As a result the tendency of solvent to escape to vapour decreases. Consequently in a closed container now there will be lesser amount of solvent in vapour phase and naturally some extra molecules of solvent are there in the liquid solvent. It is comparable to arresting of solvent molecules, maybe through solvation. Hence, the above statement.

7. Why is it necessary that the solid dissolved in the liquid solvent be non-volatile (in case of colligative properties)?

Ans. : When a solute is dissolved in a solvent, the chemical potential of the solvent in the liquid phase will decrease in comparison to that of vapour phase in equilibrium with it if the solute is absent in the vapour phase. Thus the solute has to be non-volatile for lowering of vapour pressure, and other related colligative properties. The chemical potential in liquid phase can, of course, be less than that in the vapour phase if the concentration of the volatile solute is less in the vapour phase in comparison to liquid phase. But it cannot be said unequivocally that addition of solute leads to the decrease in chemical potential of the solvent in the liquid phase.

8. When will the van't Hoff factor i be fractional?

Ans. : The van't Hoff factor i expresses the deviation from ideality of a real solution in their colligative properties. It can explain easily the deviation when there is association or dissociation. When there is association, complete or fractional, i must be a fraction (because, say, for complete dimerisation $i = \frac{1}{2} = 0.5$). Also i is fractional when there is partial dissociation.

9. Can we use the same Beckmann thermometer for boiling point elevation and freezing point depression measurements?

Ans. : In Beckmann thermometer, the bulb volume error has been eliminated by keeping a mercury reservoir at the top. Thus, the same Beckmann thermometer may be used for both measurements. However, the thermometer needs setting before being used so that the bulb volume is adjusted to the desired range.

10. Show that for a solution of a solute in a non-polar solvent at the particular concentration $\Delta T_b / T_b$ is independent of the nature of solvent.

Ans. : We know that $\Delta T_b = \frac{RT_b^2}{L_e} x_2 . \Delta T_b$ depends on the number of moles of solute and nature of the solvent, since T_b (boiling point) and L_e, the latent heat of evaporation per mole, depend on the nature of solvent.

$$\frac{\Delta T_b}{T_b} = \frac{RT_b}{L_e} x_2 = \frac{2}{21} x_2,$$

where $R = 2$ cal and $\frac{L_e}{T_b} = 21$ cal, by Trouton's rule.

Hence, $\frac{\Delta T_b}{T_b}$ is independent of the nature of solvent and is equal to $\frac{x_2}{10.5}$, where x_2 is mole fraction of the solute. It has been assumed that Trouton's rule is valid.

11. Which one is expected to be higher among K_f and K_b?

Ans. : $K_f = \frac{R\left(T_f^\circ\right)^2}{1000 l_f}$ and $K_b = \frac{R\left(T_b^\circ\right)^2}{1000 l_e}$

$$\therefore \quad \frac{K_f}{K_b} = \frac{\left(T_f^\circ\right)^2}{l_f} \times \frac{l_e}{\left(T_b^\circ\right)^2} = \left(\frac{T_f^\circ}{T_b^\circ}\right)^2 \times \frac{l_e}{l_f}.$$

The ratio of squares of T_f° and T_b° is generally above 0·25, whereas $\frac{l_e}{l_f}$ is rarely less than 5. So $\frac{K_f}{K_b} > 1$, i.e., $K_f > K_b$.

12. Why is camphor more suitable than water as a solvent in determination of molecular weights of organic substances by cryoscopic method?

Ans. : For determination of molecular weights of organic substances by cryoscopic method depression in freezing point (ΔT_f) of the dilute solution has to be measured. Since

$\Delta T_f = K_f m$ and m (molatity of the substance in the solvent) has to be very low, ΔT_f will be reasonably high to be detected even by ordinary thermometer, if and only if K_f is high. Camphor has appreciably high K_f (40) while that of water is only 1.86. So, camphor is preferred to water as a solvent in determination of molecular weights of organic substances.

13. Why is effervescence observed when a soda water bottle is opened?

Ans. : According to Henry's law, solubility of gas in water increases with pressure. Soda water contains CO_2 at high pressure. When bottle is opened, the pressure is released and solubility of CO_2 also decreases. So the effervescence is observed.

Numerical Problems

1. At 25 °C, 10.50 litres of N_2 at 760 mm of Hg are passed through an aqueous solution of a non-volatile solute, whereby the solution loses 0.246 gm in weight. If the total pressure above the solution is also 760 mm, what is the vapour pressure of the solution and mole fraction of the solute? Given that the vapour pressure of pure water at this temperature is 23.76 mm of Hg.

Soln. : The partial pressure of water in N_2 is the vapour pressure of the solution. Thus,

$$P_{H_2O} = x_{H_2O}.P = \frac{n_{H_2O}}{n_{H_2O} + n_{N_2}}.P$$

Now, 10.5 litres of N_2 at 25 °C contains

$$n = \frac{1 \times 10.5}{0.08205 \times 298.15} \text{ mole} = 0.42916 \text{ mole.}$$

$$\therefore P_{H_2O} = \frac{\frac{0.246}{18}}{\frac{0.246}{18} + 0.42916} \times 760 \text{ mm of Hg}$$

$$= \frac{0.013666}{0.013666 + 0.42916} \times 760 \text{ mm of Hg} = \mathbf{23.45 \text{ mm of Hg.}}$$

By Raoult's law, the mole fraction of the solute in solution is

$$x_2 = \frac{P^o - P}{P^o} = \frac{23.76 - 23.45}{23.76} = \mathbf{0.013.}$$

2. At 25 °C, 10.5 litres of pure N_2 measured at 760 mm of Hg is passed through an aqueous solution of a non-volatile solute whereby the solution loses 0.2455 gm in weight. If the total pressure above the solution is also 760 mm, what is the vapour pressure of the solution and mole fraction of the solute? P^o at 25 °C = 23.756 mm of Hg.

Hints : See Problem 1.

3. A 5% solution of cane sugar is isotonic with a 8.77% solution of urea. Calculate the molecular weight of urea (the molecular weight of cane sugar is 342)

[Agra B.Sc. 1973; Meerut T.D.C.I. 1982]

Soln. : The cane suger solution contains 50 gm of cane sugar in 1000 gm of water. Let the density of the solution be 1 gm/cc. Then

$$C = \frac{50}{342};\ \pi = \frac{50}{342}RT.$$

Similarly, for urea with molecular weight M_2,

$$\Pi = \frac{8.77}{M_2} RT.$$

Thus, by the problem— $\dfrac{50}{342} RT = \dfrac{8.77}{M_2} RT$

or, $M_2 = 59.9868 \approx$ **60.**

4. 10 gm of a substance was dissolved in 250 ml of water and the osmotic pressure was found to be 600 mm Hg at 15 °C. Find out the molecular weight of the substance.

Soln. : $\Pi = RTC$

By the problem, $\Pi = \dfrac{600}{760}$ atm.

$C = \dfrac{40}{M}$, $\left[\begin{array}{l} 10\,\text{gm in } 250\,\text{cc} \\ \therefore 40\,\text{gm in } 1000\,\text{cc} \end{array}\right]$

where M = molecular weight.

$$\therefore \frac{600}{760} = 0.08206 \times 288.15 \times \frac{40}{M}.$$

$\therefore M = 1198.0432 \approx$ **1198.**

5. How much urea must be dissolved in 10^{-2} m^3 of water to yield a solution whose osmotic pressure is 2.03×10^5 N.m^{-2} at 300 K?

Soln. : The mol. wt. of urea = 60

Now $\Pi = 2.03 \times 10^5$ N.m^{-2} $\quad M_2 = 60$ gm

$V = 10^{-2}$m^3 $\quad T = 300$ K

$W_2 = ?$ $\quad R = 8.314$ J.K^{-1}.mol^{-1}

$$\therefore \Pi = RT\frac{W_2}{M_2 V} = 8.314\ \text{J.K}^{-1}.\text{mol}^{-1} \times 300\ \text{K} \times \frac{W_2\,\text{kg}}{0.060\,\text{kg.mol}^{-1} \times 10^{-2}\,\text{m}^3}$$

$$\text{or, } W_2 = \frac{2.03 \times 10^5\,\text{N.m}^{-2} \times 0.06\,\text{kg.mole}^{-1} \times 10^{-2}\,\text{m}^3}{8.314\,\text{J.K}^{-1}.\text{mole}^{-1} \times 300\,\text{K}} = 0.04883\ \text{kg} = \mathbf{48.83\ gm.}$$

6. Consider a vertical tube with a cross-sectional area of 0.5 cm^2. The bottom of the tube is closed with a semipermeable membrane and 0.12 gm of urea is placed in the tube. The closed end of the tube is just immersed in water. What will be the height of water level in the tube at equilibrium? The final density of the solution is 1.01 gm/cc and the temperature is 27 °C. What is the osmotic pressure of the solution in S.I. unit?

Soln. : Let the height of water raised be h cm. Since the area is 0.5 cm^2, the volume is $0.5h$ cm^3. Since the weight of urea taken is 0.12 gm, $C = \dfrac{0.12 \times 10^3}{60 \times 0.5h}$ moles. lit^{-1}.

$\therefore \Pi = RTC$

$$= 0.08206\ \text{lit-atm. K}^{-1}.\text{mole}^{-1} \times 300.15\ \text{K} \times \frac{0.12 \times 10^3}{30\,h}\ \text{mole.lit}^{-1}$$

$$= \frac{98.5212}{h}\ \text{atm.}$$

Again, $\Pi = h\rho g$

$= h \text{ cm} \times 1.01 \text{ gm.cm}^{-3} \times 980.665 \text{ cm.sec}^{-2}$

$= h \times 990.47165 \text{ gm.cm}^{-1}.\text{sec}^{-2}$

$$= \frac{h \times 990.47165 \text{ dyne.cm}^{-2}}{1.01325 \times 10^6 \text{ dyne.cm}^{-2}.\text{atm}^{-1}} = h \times 9.77519 \times 10^{-4} \text{ atm.}$$

$$\therefore \quad \frac{98.5212}{h} \text{ atm} = 9.77519 \times 10^{-4} \times h \text{ atm}$$

$$\text{or, } h^2 = \frac{98.5212}{9.77519 \times 10^{-4}}$$

$\therefore h = 317.47 \text{ cm} = \mathbf{3.175 \text{ m.}}$

$$\Pi = \frac{98.5212 \text{ atm}}{317.47 \text{ cm}} \times 1.01325 \times 10^5 \frac{\text{Pa}}{\text{atm}} = \mathbf{31.44 \text{ kPa.}}$$

7. The osmotic presssure is measured between water and a solution containing 1 gm glucose (M = 180) and 1 gm sucrose (M = 342) in 1000 gm water. The temperature is maintained at 25 °C. What will be the osmotic pressure of the solution? If this pressure is measured without knowing that the solute is a mixture, what will be the resulting molecular weight? What kind of average is it?

Soln. : Total number of moles in 1000 gm of water $= \left(\frac{1}{180} + \frac{1}{342}\right)$ gm. Assuming the density of water 1 gm/cc, the molar concentration of the solution is

$$C = \left(\frac{1}{180} + \frac{1}{342}\right) \text{ moles.litre}^{-1}$$

T = 25 °C = 298.15 K

$$\therefore \quad \Pi = RTC = 0.08206 \text{ lit-atm.K}^{-1}.\text{mol}^{-1} \times 298.15 \text{ K} \times \left(\frac{1}{180} + \frac{1}{342}\right) \text{mol.lit}^{-1}$$

$= 0.20746 \text{ atm} = 0.21 \text{ atm.}$

Again, $\Pi = RTC$

$$\text{or, } 0.21 \text{ atm} = 0.08206 \text{ lit.atm.K}^{-1}.\text{mol}^{-1} \times 298.15 \text{ K} \times \frac{2}{M_2} \text{ mol.lit}^{-1}$$

[where M_2 = resulting mol. wt.]

$$\text{or, } M_2 = \frac{0.08206 \text{ lit.atm.K}^{-2}\text{mol}^{-1} \times 298.15 \text{ K} \times 2 \text{ mol.lit}^{-1}}{0.21 \text{ atm}}$$

or, M_2 = **233.011 gm.**

$$\text{The average means } \frac{\text{average weight}}{\text{mole}} = \frac{\Sigma W_i}{\Sigma n_i}.$$

8. Using van't Hoff equation, calculate the osmotic pressure of 0.865(M) sucrose solution at 20 °C and compare your result with the experimental value which is 26.64 atm. Explain the discrepancy. **[C.U.(H) 1971]**

Soln. : We know that for non-electrolytic solute, $\Pi = RTC$

or, $\Pi = 0.08206$ litre.atm.K^{-1}.mol^{-1} × 293.15 K × 0.865 mol.lit^{-1}

$= 20.808$ atm.

Now, $\pi_{expt} = 26.64$ atm and $\Pi_{theo} = 20.808$ atm.

$$\therefore i = \frac{\Pi_{expt}}{\Pi_{theo}} = \frac{26.64 \text{ atm}}{20.808 \text{ atm}} = 1.28.$$

$$\therefore \alpha = \frac{i-1}{n-1} = \frac{1.28-1}{2-1} = 0.28.$$

Since $n = 2$, sucrose breaks into glucose and fructose with $\alpha = 0.28$.

9. The normal boiling point of chloroform is 61.2 °C and its heat of vaporisation is 59.0 cal/gm. If 0.50 gm of an organic substance is dissolved in 50 gm of $CHCl_3$, the boiling point of the solution is 61.42 °C, calculate the molecular weight of the solute. **[C.U.(H) 1970]**

Soln. : Let the molecular weight of the solute be M_2. Then the molal concentration of the solution will be

$$m = \frac{1000 \times 0.5}{M_2 \times 50}.$$

$$\text{Now, } \Delta T_b = \frac{RT_b^{o^2}}{1000 l_e} \times m.$$

Here, $T_b^o = 334.35$ K

$l_e = 59$ cal/gm, $\therefore \Delta T_b = 0.22$ K.

$$\therefore \Delta T_b = \frac{1.987 \text{ cal.K}^{-1}.\text{mol}^{-1} \times (334.35)^2 \text{ K}^2}{1000 \text{ gm} \times 59 \text{ cal.gm}^{-1}} \times \frac{1000 \text{ gm}}{50 \text{ gm}} \times \left(\frac{0.5}{M_2}\right) \text{mol}$$

$$\text{or, } M_2 = \frac{1.987 \text{ cal.K}^{-1}.\text{mol}^{-1} \times (334.35)^2 \text{ K}^2 \times 1000 \text{ gm}}{1000 \text{ gm} \times 59 \text{ cal.gm}^{-1} \times 50 \text{ gm} \times 0.22 \text{ K}} \times 0.5 \text{ gm}$$

$= \mathbf{171.12}$ **gm.**

10. A and B are liquids and form ideal solution. The total vapour pressure at 50 °C of a solution containing 1 gm-mole of A and 2 gm-moles of B is 250 mm of Hg, and that of a solution containing 2 gm-moles of A and 2 gm-moles of B is 300 mm of Hg at 50 °C. Calculate the vapour pressures of the pure liquids. **[C.U.(H) 1968]**

Soln. : The total pressure P is the summation of the partial pressure of the components.

So, $P = P_A + P_B$ [P_A and P_B are partial pressure of A and B respectively]

Again, $P_i = x_i \cdot P_i^o$,

where x_i and P_i^o are the mole fraction and vapour pressure of pure i-th substance.

$\therefore P_A = x_A P_A^o$ and $P_B = x_B P_B^o$.

By the given problem—

$$\frac{1}{3} P_A^o \text{ mm of Hg} + \frac{2}{3} P_B^o \text{ mm of Hg} = 250 \text{ mm of Hg}$$

and $\frac{1}{2} P_A^o$ mm of Hg + $\frac{1}{2} P_B^o$ mm of Hg = 300 mm of Hg.

From the above two equations we get—

- P_A^o = **450 mm of Hg.**

  P_B^o = **150 mm of Hg.**

11. 122 gm of benzoic acid which completely dimerises in benzene solution are dissolved in 1000 gm of benzene at certain temperature. If the vapour pressure of pure benzene is 66.6 cm of Hg at that temperature, then what will be the vapour pressure of the solution? **[C.U.(H) 77]**

Soln. : We know that $\frac{P^o - P}{P^o} = \frac{n_2}{n_2 + n_1}$,

where P^o and P are the vapour pressures of pure solvent and solution respectively, and n_1 and n_2 are the number of mole of solvent and solute respectively.

By the given problem,

P^o = 66.6 cm of Hg

$$n_1 = \frac{1000}{78} = 12.8205 \text{ mol}$$

$$n_2 = \frac{122}{244} = 0.5 \text{ mol} \quad \left[\begin{array}{l}\text{Benzoic acid dimerises,} \\ \text{so its molcular wt.} = 2 \times 122 = 244\end{array}\right]$$

P = ?

From the above equation, we get

$$1 - \frac{P}{P^o} = \frac{n_2}{n_2 + n_1}$$

$$\text{or, } \frac{P}{P^o} = 1 - \frac{n_2}{n_2 + n_1}$$

$$\text{or, } P = \left(1 - \frac{n_2}{n_2 + n_1}\right) \times P^o$$

$$\text{or, } P = \left(1 - \frac{0.5\,\text{mol}}{0.5\,\text{mol} + 12.8205\,\text{mol}}\right) \times 66.6 \text{ cm of Hg} = \mathbf{64.1 \text{ cm of Hg.}}$$

12. Calculate the osmotic pressure of a 5% solution of cane sugar at 25 °C in SI unit.

Soln. : We know that

$\Pi = RTC$,

where T = absolute temperature

C = molar concentration.

By the given problem, concentration of solution is 5%,

i.e., 5 gm cane sugar is present in 100 cc solvent.

$$\therefore \text{ molar concentration } (M) = \frac{0.005 \times 10^4}{0.342} \text{ mol/m}^3$$

[Mol. wt. of cane sugar = 0.342 kg/mole]

$$\therefore \Pi = 8.314 \text{ J.K}^{-1}.\text{mol}^{-1} \times 298.15 \text{ K} \times \left(\frac{0.005 \times 10^4}{0.342}\right) \text{mol.m}^{-3}$$

$$= 362400.45 \text{ Pa} = \mathbf{362.4 \text{ kPa.}}$$

13. A 1% solution of $AgNO_3$ is isotonic with the 2% solution of glucose. Calculate the degree of dissociation of $AgNO_3$.

Soln. : 2% glucose $= \dfrac{20}{180}$(M) glucose solution.

[$\because$ molecular weight of glucose = 180]

Again, 1% $AgNO_3$ solution $= \dfrac{10}{169.8}$(M) $AgNO_3$ solution.

Now in case of glucose, since it is a non-electrolyte,

$\Pi = RTC_g$, where C_g is concentration of glucose solution.

Again, in case of $AgNO_3$ solution, since it is an electrolyte,

$\Pi = iRTC_{AgNO_3}$ [i = van't Hoff factor]

According to the given problem,

$$iC_{AgNO_3} = C_g$$

$$\text{or, } i = \frac{\dfrac{20}{180}\text{mole.lit}^{-1}}{\dfrac{10}{169.8}\text{mole.lit}^{-1}} = 1.8866.$$

So, degree of dissociation $(\alpha) = \dfrac{i-1}{\nu-1}$

Here, $\nu = 2$.

$$\therefore \alpha = \frac{1.8866-1}{2-1} = 0.8866.$$

$\therefore$ degree of dissociation = **0.8866.**

14. Water is used in car radiators. In winter season, ethylene glycol is added so that water may not freeze. Assuming ethylene glycol to be non-volatile, calculate the minimum amount of ethylene glycol that must be added to 6.0 kg of water to prevent it from freezing at 272.7 K. K_f for water = 1.86. **[Punjab B.Sc. 1975]**

Soln. : By the problem, in winter season the freezing point of water is 272.7 K. The freezing point of water in car radiator must be less than or equal to 272.7 K. With glycol, $(CH_2OH)_2$, (0.062 kg.mol^{-1}) as solute the freezing point of the solution has to be 272.7 K or lesser. Thus, freezing point depression is (273.15 – 272.7) = 0.45 K or greater.

$$\Delta T_f \geq 0.45\text{K}$$

$$\therefore \quad 0.45 \text{ K} \leq K_f m$$

$$\leq 1.86 \text{ kg.K.mol}^{-1} \times \frac{W_2 \text{ kg}}{0.062 \text{ kg.mol}^{-1} \times 6 \text{ kg}}$$

$$\therefore W_2 \geq \frac{0.45 \text{ K} \times 0.062 \text{ kg.mol}^{-1} \times 6 \text{ kg}}{1.86 \text{ kg.K.mol}^{-1}}$$

$$\text{or,} \quad W_2 \geq 0.09 \text{ kg.}$$

Therefore, the minimum amount of ethylene glycol to be added to 6 kg water is **0.09 kg** (or, **90 gm**).

15. How many gm of methyl alcohol should be added to 10-litre tank of water to prevent its freezing at 268 K? K_f for water is 1.86.

Hints : See Problem 14. **[Ans.** 886.022 gm]

16. A solution of $Hg(CN)_2$ containing 3 gm per litre has an osmotic pressure 231.9 mm of Hg at 298 K. What is the apparent molecular weight and degree of dissociation of $Hg(CN)_2$?

Soln. : By the problem,

$$\Pi = 231.9 \text{ mm of Hg} = \frac{231.9}{760} \text{ atm.}$$

Weight of $Hg(CN)_2$, i.e., W_2 = 3 gm per lit,

$$= \frac{3}{M_2} \text{ mol.lit}^{-1},$$

where M_2 = mol. wt. of $Hg(CN)_2$.

$$\therefore \quad \frac{231.9}{760} = RTC = 0.08206 \text{ lit.atm.K}^{-1}.\text{mol}^{-1} \times 298 \text{ K} \times \frac{3}{M_2} \text{ mol.lit}^{-1}$$

$$M_2 = \frac{0.08206 \text{ lit.atm.K}^{-1}.\text{mol}^{-1} \times 298 \text{ K} \times 3 \text{ mol.lit}^{-1}}{\frac{231.9}{760} \text{ atm}} = \mathbf{240.42 \text{ gm.}}$$

So, the apparent mol. wt. is **240.42.** It is reduced from the real value 252.6 because of the dissociation to three ions, Hg_2^{2+} and $2CN^-$. Therefore, $\Pi = iRTC$.

$$\therefore i = \frac{\Pi}{RTC} = \frac{\frac{231.9}{760} \text{ atm}}{0.08206 \text{ lit.atm.K}^{-1}.\text{mol}^{-1} \times 298 \text{ K} \times \frac{3}{252.6} \text{ mol.lit}^{-1}}$$

$$\therefore i = \frac{231.9 \times 252.6}{760 \times 0.08206 \times 298 \times 3} = 1.0506.$$

$$\text{So, } \alpha = \frac{i-1}{\nu - 1} = \frac{0.0506}{2} = 0.025$$

$\therefore$ degree of dissociation = 0.025.

17. When 2 gm of a non-volatile hydrocarbon containing 94.4% carbon is dissolved in 100 gm of benzene, the vapour pressure of benzene at 20 °C is lowered from 74.66 mm to 74.01 mm. Calculate the molecular formula of the hydrocarbon. **[C.U.(H) 1984]**

Soln. : By the given problem, the hydrocarbon contains 94.4% of C and the rest, i.e., 5.6% of H. Thus, the number of C atom may be $\frac{94.4}{12} = 7.86$ and the number of H atom may be $\frac{5.6}{1} = 5.6$ or, in the numbers of atom, C : H = 1.4 : 1.

So, the approximate formula is $(C_{1.4}H)_x$.

Now, let M be the molecular weight of the hydrocarbon.

By the problem,

$$\therefore \quad \frac{P^o - P}{P^o} = \frac{\frac{2}{M}}{\frac{2}{M} + \frac{100}{78}}$$

$$\text{or,} \quad \frac{P^o}{P^o - P} = 1 + \frac{100}{78} \times \frac{M}{2}$$

$$\text{or,} \quad \frac{100}{78} \times \frac{M}{2} = \frac{P^o}{P^o - P} - 1 = \frac{P}{P^o - P}$$

$$\therefore \quad M = \frac{P}{P^o - P} \times \frac{78 \times 2}{100}$$

$$= \frac{74.01 \text{ mm of Hg}}{0.65 \text{ mm of Hg}} \times \frac{78 \times 2}{100}$$

or, $M = 177.6$

$\therefore \quad (12 \times 1.4 + 1) \times x = 177.6$

or, $x = 10.$

So the correct formula is $(C_{1.4}H_1)_{10}$ = $\mathbf{C_{14}H_{10}}$.

18. A mixture of 1 gm of napthalene (M = 128) and 10 gm of camphor freezes at 147 °C whereas pure camphor freezes at 177.5 °C. Calculate the cryoscopic constant of camphor. **[C.U.(H) 1983]**

Soln : We know that,

$$\Delta T_f = K_f m$$

Again, $m = \dfrac{1000 \times W_2}{M_2 \times W_1}$,

where W_1 and W_2 are wt. of solvent and solute respectively.

M_2 = mol. wt. of solute.

By the given problem,

$\Delta T_f = 30.5$ °C

$W_1 = 10$ gm; $W_2 = 1$ gm; $M_2 = 128$ gm

$$\therefore 30.5\ °C = K_f \times \frac{1000 \times 1}{128 \times 10}$$

or, K_f = **39.04 K.gm.mol^{-1}**.

19. A solution containing 0.684 gm of cane sugar in 100 gm of water freezes at –0.037 °C, while a solution containing 0.585 gm of NaCl in 100 gm of water freezes at –0.342 °C. Calculate K_f for water, i and % dissociation of NaCl. **[C.U.(H) 1982]**

Soln. : We know that, for non-electrolyte,

$$\Delta T_f = K_f \times \frac{W_2}{M_2 \times W_1} \quad \text{[in S.I. unit]}$$

$$\text{or,} \quad K_f = \frac{\Delta T_f K \times M_2 \text{ kg.mol}^{-1} \times W_1 \text{kg.}}{0.684 \times 10^{-3} \text{ kg.}}$$

For water,

$$K_f = \frac{0.037\text{K} \times 0.342 \text{ kg.mol}^{-1} \times 0.1\text{kg}}{0.684 \times 10^{-3} \text{ kg}}$$

or, K_f = **1.85 kg.K.mol⁻¹.**

For NaCl solution, which is an electrolyte,

$$\Delta T_f = i.K_f \times \frac{W_2}{W_1 M_2} \text{ [in S.I. unit]}$$

$$\text{or, } i = \frac{\Delta T_f \times M_2 \times W_1}{K_f \times W_2} = \frac{0.342\text{ K} \times 0.0585 \text{ kg.mol}^{-1} \times 0.1\text{kg}}{1.85 \text{ kg.K.mol}^{-1} \times 0.585 \times 10^{-3} \text{ kg}}$$

or, i = **1.8486.**

$$\therefore \alpha = \frac{i-l}{\nu - l} = \frac{0.8486}{2-1} = 0.8486.$$

or, α = 0.85, ∴ percentage dissociation = **85%.**

20. An organic acid (molar mass = 60) associates in benzene to form a dimer when 1.65 gm of the acid is dissolved in 100 gm benzene, the boiling point of benzene is raised by 0.36 °C. Calculate the van't Hoff factor and the degree of association of the acid in benzene. K_b = 2.6. **[V.U. 2000]**

Soln. : We know that, for electrolytic solution—

$$\Delta T_b = iK_b.m$$

$$\text{or, } i = \frac{\Delta T_b}{K_b m}$$

$$\text{or, } i = \frac{0.36\text{ K}}{2.6 \text{ kg.K.mol}^{-1} \times \text{m mol.kg}^{-1}}$$

$$\text{Here } m = \frac{1.65 \times 10^{-3} \text{ kg}}{0.1\text{ kg} \times 60 \times 10^{-3} \text{ kg.mol}^{-1}}.$$

∴ i = **0.5.**

$$\text{Now, degree of association, } \beta = \frac{\nu(1-i)}{\nu - 1}$$

$$\text{or, } \beta = \frac{2 \times 0.5}{1} = 1.$$

∴ the degree of association is **1 (100% association).**

21. Benzoic acid dimerises when dissolved in benzene. The osmotic pressure of a solution of 5 gm of benzoic acid in 100 ml of benzene is 5.73 atm at 10 °C. Find van't Hoff factor and degree of association. **[B.U. 1994]**

Soln. : We know that, $\Pi_{\text{theo}} = RTC.$

$$\text{In the given problem, } C = \frac{W_2}{M_2 \times W_1} = \frac{5 \times 10^{-3} \text{ kg}}{0.122 \text{ kg.mol}^{-1} \times 0.1 \times 10^{-3} \text{ m}^3}$$

$$\therefore \Pi_{\text{theo}} = \frac{0.08206 \text{ lit.atm.K}^{-1}.\text{mol}^{-1} \times 283.15\text{ K} \times \left(5 \times 10^{-3}\right)\text{kg}}{0.122 \text{ kg.mol}^{-1} \times 0.1 \text{ lit}} = 9.523 \text{ atm.}$$

$$\therefore\ i = \frac{\Pi_{obs}}{\Pi_{theo}} = \frac{5.73\,\text{atm}}{9.523\,\text{atm}} = \mathbf{0.60173.}$$

$$\therefore \text{ degree of association } \beta = \frac{\nu(1-i)}{\nu-1} = \frac{2(1-0.6)}{2-1} = \mathbf{0.8.}$$

22. The freezing point of pure benzene is 5.44 °C and that for a solution of 2.092 gm benzaldehyde in 100 gm of benzene is 4.44 °C. Calculate the molar mass of benzaldehyde, given K_f for benzene is 5.1 kg.K.mol^{-1}. **[B.U. 1997]**

Soln. : By the problem, $\Delta T_f = 1$ K.

$$\text{Now, molal concentration } (m) = \frac{0.002092\text{ kg}}{0.1\text{ kg}\times M_2\text{ kg.mol}^{-1}} = \frac{0.02092}{M_2\text{kg.mol}^{-1}}.$$

But, $\Delta T_f = K_f m$

$$\text{or, } 1\text{K} = 5.1\text{ kg.K.mol}^{-1} \times \frac{0.02092}{M_2\text{kg.mol}^{-1}}$$

$$\text{or, } M_2 = \frac{5.1\,\text{kg.K.mol}^{-1}\times 0.02092}{1\,\text{K}}$$

$= 0.1066924$ kg.mol^{-1} $=$ **106.69 gm.**

23. Heavy water (D_2O) boils at 102 °C and its K_b value is 10% higher than that of water. How does its latent heat per mole compare with that of water? **[B.U. 1991]**

Soln. : Let K_b and K_b' be the ebullioscopic constant for water and heavy water respectively. Then by the given problem, since K_b value for D_2O is 10% higher than that of H_2O, $K_b' = 1.1K_b$.

Again, let T_b and T_b'; L_v and L_v' be the boiling temperatures and latent heats for H_2O and D_2O respectively.

Now, $T_b = 373.15$ °K; $T_b' = 375.15$ °K.

We know that,

$$K_b = \frac{RT_b^2}{1000 l_e} \text{ and } K_b' = \frac{RT_b'^2}{1000 L_e'}$$

$$\text{or, } \frac{K_b'}{K_b} = \frac{T_b'^2 L_e}{T_b^2 L_e'}$$

$$\text{or, } \frac{L_e'}{L_e} = \frac{T_b'^2}{T_b^2} \times \frac{K_b}{K_b'}$$

$$\text{or, } \frac{L_e'}{L_e} = \left(\frac{375.15}{373.15}\right)^2 \times \frac{K_b}{1.1K_b}$$

$$\text{or, } \frac{L_e'}{L_e} = 0.91886.$$

24. x gm NaCl, y gm sucrose and z gm urea are mixed and dissolved in 0.5 dm^3 of water. Suppose that no chemical reaction sets in, what should be the osmotic pressure of the solution? **[B.U. 2000]**

Soln. : We know that—

$$\Pi = RTC$$
$$R = 8.314 \text{ J.K}^{-1}.\text{mol}^{-1}$$
$$T = T \text{ °K}$$
$$C = A \text{ mol.m}^{-3}.$$

By the problem, the solution is made of x gm NaCl, i.e., $(x \times 10^{-3})$kg NaCl, y gm sucrose, i.e., $(y \times 10^{-3})$ kg sucrose and z gm urea, i.e., $(z \times 10^{-3})$ kg urea. Hence, number of moles NaCl per m^3 = $\dfrac{x \times 10^{-3}\text{ kg}}{0.0585 \text{ kg.mol}^{-1} \times 0.5 \times 10^{-3}\text{ m}^3}$.

That of sucrose is $\dfrac{y \times 10^{-3}\text{ kg}}{0.342 \text{ kg.mol}^{-1} \times 0.5 \times 10^{-3}\text{ m}^3}$.

And that of urea = $\dfrac{z \times 10^{-3}\text{ kg}}{0.060 \text{ kg.mol}^{-1} \times 0.5 \times 10^{-3}\text{ m}^3}$.

∴ total concentration to be calculated for which we assume that NaCl completely dissociates.

Its number of mole = $\dfrac{2 \times x \times 10^{-3}\text{ kg}}{0.0585 \text{ kg.mol}^{-1} \times 0.5 \times 10^{-3}\text{ m}^3}$.

$$\text{So, } C = \left(\frac{4x}{0.0585} + \frac{2y}{0.342} + \frac{2z}{0.06}\right)\text{mol.m}^{-3}.$$

$$\therefore \Pi = CRT$$

$$= \left(\frac{4x}{0.0585} + \frac{2y}{0.342} + \frac{2z}{0.06}\right)\text{mol.m}^{-3} \times 8{\cdot}314 \text{ J.K}^{-1}.\text{mol}^{-1} \times \text{T K}$$

$$= \mathbf{8.314}\, T \times \left(\frac{\mathbf{4x}}{\mathbf{0.0585}} + \frac{\mathbf{2y}}{\mathbf{0.342}} + \frac{\mathbf{2z}}{\mathbf{0.06}}\right) \textbf{ J.m}^{-3} \textbf{ or N.m}^{-2} \textbf{ or Pa}$$

25. A mixture which contains 0.55 gm of camphor and 0.045 gm of organic solute (containing 93.46% of C and 6.54% of H) freezes at 157 °C. Calcuate the molecular formula of the organic compound (Given : freezing point of camphor = 178.4 °C and K_f for camphor = 37.7 °C kg.mole^{-1}). **[C.U.(H) 1985]**

Soln. : By the problem, $\Delta T_f = 21.4$ K

$$W_1 = 0.55 \times 10^{-3}\text{kg}, \; K_f = 37.7 \text{ kg.K.mol}^{-1}$$
$$W_2 = 0.045 \times 10^{-3}\text{ kg}.$$

We know that, $\Delta T_f = K_f m$

$$\text{or, } \Delta T_f = 37.7 \text{ kg.K.mol}^{-1} \times \frac{0.045 \times 10^{-3}\text{Kg}}{M_2 \text{ kg.mol}^{-1} \times 0.55 \times 10^{-3}\text{ kg}}$$

$$\text{or, } M_2 = \frac{37.7 \text{kg.K.mol}^{-1} \times 0.045 \times 10^{-3}\text{ kg}}{0.55 \times 10^{-3}\text{ kg} \times 21.4 \text{K}}$$

or, $M_2 = 0.144$ kg.mol^{-1}.

Again, from % composition data,

$$\text{gm-atomic \% of C} = \frac{93.45}{12} = 7.78$$

$$\text{and gm-atomic \% of H} = \frac{6.54}{1} = 6.54.$$

∴ atomic ratio, C : H = 7.78 : 6.54 = 1.19 : 1.

Multiplying by 10, molecule is $C_{12}H_{10}$; molar mass = 0.154 kg.mol^{-1}.

But the error in molar mass is very high. [Taking 1.19 as 1.1, the molecular formula is $C_{11}H_{10}$ with molar mass 0.142 kg.mol^{-1}. The error % in molar mass is reasonable. So, the molecular formula may be written as $\mathbf{C_{11}H_{10}}$]

26. Calculate the freezing temperature of an aqueous solution having its boiling temperature 100.1°C. The molal ebullioscopic constant and the molal cryoscopic constant are 0.513 K.kg.mol^{-1} and 1.86 K.kg.mol^{-1} respectively. **[V.U.(H) 1987]**

Soln. : By the problem, $\Delta T_b = 0.1$ K.

$K_b = 0.513$ K.kg.mol^{-1}.

We know that, $\Delta T_b = K_b.m$

$$\text{or, } m = \frac{0.1\,\text{K}}{0.513\ \text{K.kg.mol}^{-1}} = \left(\frac{0.1}{0.513}\right)\text{mol.kg}^{-1}$$

Again, $\Delta T_f = K_f m$

$$= 1.86\ \text{K.kg.mol}^{-1} \times \frac{0.1}{0.513}\ \text{mol.kg}^{-1} = \mathbf{0.36\ K.}$$

∴ the freezing temperature is **–0.36 °C**

27. A 2.5 molal HCl solution freezes at –9.88 °C. Calculate the apparent percentage of dissociation and explain your answer ($K_f = 1.86$).

Soln. : By the problem,

$\Delta T_f = 9.88$ K

$K_f = 1.86$ K.kg.mol^{-1}

$m = 2.5$ mol.kg^{-1}.

We know that,

$\Delta T_f = iK_f m$

$$\text{or, } i = \frac{9.88\,\text{K}}{1.86\ \text{kg.K.mol}^{-1} \times 2.5\ \text{mol.kg}^{-1}} = 2.12.$$

$$\text{So, degree of dissociation, } \alpha = \frac{i-1}{\nu-1} = \frac{1.12}{2-1} = 1.12.$$

So, percentage of dissociation = **112%.**

The reason for this absurd result is that the equation $\Delta T_f = iK_f m$ is not valid for this case. HCl is soluble in solid solvent. It is volatile and the solution is reasonably concentrated.

28. (i) Benzene freezes at 5.6 °C. Its value for K_f is 5.1 Find ΔH_f.

(ii) An aqueous 20% (by wt.) solution of sucrose has density 1.08 gm.cm^{-3} at 25 °C. Express the concentration of solute in molar, molal and mole fraction units. **[B.U. 2000]**

Soln. : (i) By the problem,

$$K_f = \frac{RT_f^2}{L_e} = 5.1 \text{ kg.K.mol}^{-1} \quad [L_e = \text{latent heat per kg}]$$

$$\text{or, } L_e = \frac{8.314\,\text{J.K}^{-1}\text{mol}^{-1} \times 278.75\,\text{K}^2}{5.1\,\text{kg.K.mol}^{-1}} = 126668.78 \text{ J.kg}^{-1}$$

$\therefore$ latent heat of fusion per mole = 126668.78 J.kg^{-1} × 0.078 kg.mol^{-1}

[m. wt. of benzene = 78]

$\therefore$ ΔH_f = **9880.16 J.mol^{-1}.**

(ii) By the given problem,

ρ = 1.08 gm.cm^{-3}

M = 0.342 kg.mol^{-1}

The solution is 20% by wt.

So, 20 gm of sucrose in 80 ml of water.

Total weight of solution = 100 gm.

Density of solution = 1.08 gm.cm^{-3}.

$$\therefore C \text{ (molar concentration)} = \frac{20\,\text{gm} \times 1000\,\text{cc}}{342\,\text{gm} \times (100/1.08)\,\text{cc}} = \mathbf{0.6316 \text{ mol.lit}^{-1}}.$$

$$m \text{ (molal concentration)} = \frac{20\,\text{gm} \times 1000\,\text{gm}}{342\,\text{gm} \times 80\,\text{gm}} = \mathbf{0.7309 \text{ mol.kg}^{-1}}.$$

$$x \text{ (mole fraction)} = \frac{\frac{20}{342}}{\frac{20}{342} + \frac{80}{18}} = \mathbf{0.013}.$$

29. (i) For a certain solvent, the boiling point is 127 °C and the latent heat of evaporation is 120 cal/gm. Find its molal elevation constant K_b.

(ii) On addition of a solute the vapour pressure of a liquid reduced to 9/10th of its original value. If 2 gm of the solute (molar mass =100) is added to 100 gm of the liquid to achieve the reduction, find the molar mass of the solvent assuming ideality. **[B.U. 2003]**

Soln. : (i) By the given problem,

T_b° = 400.15 K

l_e = 120 cal/gm = 120 × 4.184 × 1000 J.kg^{-1}.

We know that,

$$K_b = \frac{RT_b^{\circ 2}}{L_e}, \quad [L_e \text{ is the latent heat per kg}]$$

$$= \frac{8.314\,\text{J.K}^{-1}.\text{mol}^{-1} \times (400.15)^2\,\text{K}^2}{120 \times 4.184 \times 1000\,\text{J.kg}^{-1}}$$

$$= \mathbf{2.6514 \text{ K.kg.mol}^{-1}}$$

(ii) By the problem,

$$P = \frac{9}{10} P^o$$

$$\therefore \quad \frac{P^o - P}{P^o} = 0.1$$

Again, $$\frac{P^o - P}{P^o} = \frac{n_2}{n_1 + n_2}$$

$$\text{or,} \quad 0.1 = \frac{\frac{2}{100}}{\frac{2}{100} + \frac{100}{M_1}}$$

or, M_1 = **555.56 gm.**

30. Two grams of benzoic acid dissolved in 25 gm of benzene (K_f = 4.90 K.kg. mol^{-1}) produces a freezing point depression of 1.62 K. Calculate the molar mass of benzoic acid. **[V.U. 2005]**

Soln. : From the given problem,

W_1 = 25 gm of benzene

W_2 = 2 gm of benzoic acid

K_f = 4.90 K.kg.mol^{-1}

ΔT_f = 1.62 K

We know that—

$$\Delta T_f = K_f m$$

$$= K_f \frac{W_2}{M_2 . W_1}$$

$$\text{or,} \quad M_2 = \frac{K_f . W_2}{\Delta T_f . W_1}$$

$$= \frac{4.9\,\text{K. kg. mol}^{-1} \times 2 \times 10^{-3}\,\text{kg}}{1.62\,\text{K} \times 25 \times 10^{-3}\,\text{kg}} = 0.241975\ \text{kg.mol}^{-1}$$

or, M_2 = **243 gm.**

Molecular weight of benzoic acid = 144 gm. Since it demerizes, M_2 gets doubled.

31. 0.01 molal aqueous solution of $K_3Fe(CN)_6$ freezes at 0.062 °C. Calculate the percentage of dissociation of the solute. Given K_f for water = 1.86 K.kg.mol^{-1}. **[V.U. 2004]**

Soln. : According to the given problem,

m = 0.01 molal

ΔT_f = 0.062 K

K_f = 1.86 kg.K.mol^{-1}.

We know that—

$$\Delta T_f = i.K_f m$$

or, $i = \dfrac{\Delta T_f}{K_f.m}$

$$= \frac{0.062\,K}{1.86\ kg.K.mol^{-1} \times 0.01\ mol.kg^{-1}} = 3.33.$$

Now, $K_3Fe(CN)_6$ dissociates producing 4 ions.

$K_3Fe(N)_6 \rightarrow 3K^+ + Fe(CN)_6^{-3}$

So, degree of dissociation $(\alpha) = \dfrac{i-1}{\nu-1}$

$$= \frac{2.33}{4-1} = \frac{2.33}{3} = 0.7433.$$

$\therefore$ percentage of dissociation is **74.33%.**

32. A 0.5% aqueous solution of potassium chloride is found to freeze at –0.24 °C. Calculate the van't Hoff factor and degree of dissociation of the solute at that concentration. $[K_f = 1.86\ K.kg.mol^{-1}]$ **[V.U. 2003]**

Soln. : According to the given problem—

the solution is 0.5%, i.e. 0.5 gm KCl in 100 cc H_2O.

So, molal concentration(m)

$$= \frac{W_2}{M_2 \times W_1}\ mole.kg^{-1}$$

$$= \frac{0.5\,kg}{100\,kg \times 0.07457\,kg.mol^{-1}} = 0.067\ mole.kg^{-1}.$$

We know that

$\Delta T_f = iK_f \times m$

or, $i = \dfrac{\Delta T_f}{K_f.m}$

$$= \frac{0.24\,K}{1.86\ K.kg.mol^{-1} \times 0.067\ mol.kg^{-1}} = \mathbf{1.925.}$$

KCl dissociates to give 2 ions.

$KCl \rightarrow K^+ + Cl^-$

$\therefore$ degree of dissociation $(\alpha) = \dfrac{i-1}{\nu-1} = \dfrac{0.925}{2-1} = \mathbf{0.925.}$

Thus, percentage of dissociation = **92.5%.**

33. Raoult's law is obeyed by a mixture of benzene and toluene. A mixture of the two containing 0.34 mole fraction of toluene boils at 88 °C under atmospheric pressure. The vapour pressures of pure benzene and toluene at this temperature are 960 mm and 380 mm respectively. Find out the composition of the vapour on its boiling. **[V.U. 2002]**

Soln. : Let x_1 and x_2 be the mole fractions of tolune and benzene respectively and P_1^o and P_2^o be their corresponding vapour pressures in pure states.

Given, $x_1 = 0.34$

$P_1^o = 380$ mm of Hg

$P_2^o = 960$ mm of Hg

$x_2 = 1 - x_1 = 0.66$

Let P = total vapour pressure of mixture

P_1 = vapour pressure of toluine in mixture.

$$x_{toluene} \text{ (in vapour state)} = \frac{P_1}{P} = \frac{x_1 P_1^o}{x_1 P_1^o + x_2 P_2^o}$$

$$= \frac{0.34 \times 380}{(0.34 \times 380) + (0.66 \times 960)}$$

$$= 0.1693759 = \mathbf{0.17.}$$

Similarly, $x_{benzene}$ (in vapour state) = $(1 - 0.17)$ = **0.83.**

34. Given the density of water to be 1 gm/ml (at a given temperature), which of the following two solutions will have lower vapour pressure?

(i) 1.0 molal aqueous solution of cane sugar.

(ii) A cane sugar solution in which the mole fraction of cane sugar is 0.1.

[V.U. 2001]

Soln. : 1 molal aqueous soln. of cane sugar means that 1 mole of cane sugar has been added to 1000 gm of water. Thus, mole fraction x_2 of cane sugar is—

$$x_2 = \frac{n_2}{n_1 + n_2},$$

where n_2 is the number of moles of solute and n_1 is the number of moles of solvent (H_2O).

$$\text{Thus, } x_2 = \frac{1}{1 + \frac{1000}{18}} = 0.01768.$$

Hence, $x_1 = (1 - 0.01768) = 0.98231 = 0.98.$

$\therefore$ vapour pressure $(P) = P^o \times 0.98.$

A cane sugar solution of mole fraction 0.1 has a mole fraction $(1 - 0.1) = 0.9$ of water. Thus, in this solution $P = 0.9\,P^o$.

Hence, the **2nd solution has lower vapour pressure.**

35. The boiling point elevation constant for benzene and water are 3 and 0.5 K.kg. mol^{-1}. When 0.585 gm of NaCl is dissolved in 100 gm of water, it causes an elevation of 0.1 °C in the boiling point. Calculate the molecular weight of NaCl. Will the elevation in boiling point of an NaCl solution (of the same molality) in benzene be higher? Assume that the solubility NaCl in benzene is 1% by weight.

[M.Sc. Adminission Test 1994, IIT Bombay]

Soln. : By the problem,

$W_2 = 0.585$ gm $= 0.585 \times 10^{-3}$ kg.

$W_1 = 100$ gm $= 0.1$ kg.

Assuming complete dissociation of NaCl, i.e., $\alpha = 1$,

we get

$$\alpha = \frac{i-1}{\nu-1}$$

or, $1 = \frac{i-1}{2-1}$; or, $i = 2$.

We know that, $\Delta T_b = i\ K_b.m = iK_b.\frac{W_2}{M_2.W_1}$.

Here $\Delta T_b = 0.1$ K and $K_b = 0.5$ K.kg.mol^{-1}.

$$\therefore\ M_2 = \frac{iK_b.W_2}{\Delta T_b.W_1}$$

$$= \frac{2\times 0.5\,\text{K.kg.mol}^{-1}\times 0.585\times 10^{-3}\,\text{kg}}{0.1\,\text{K}\times 0.1\,\text{kg}}$$

or, M_2 = **0.0585 kg. mol^{-1}.**

In benzene, the solubility of NaCl is 1% by weight, i.e., 1 kg of NaCl in 99 kg of benzene.

We know that,

$$\Delta T_b = K_b.m$$

$$= 3\ \text{K.kg.mol}^{-1}\times\frac{0.585\times 10^{-3}\,\text{kg}}{0.0585\ \text{kg mol}^{-1}\times 0.1\,\text{kg}}$$

$$= 0.3\ \text{K}.$$

So, ΔT_b will be more in benzene than in water.

36. Liquids *A* and *B* form an ideal solution. In a binary solution of A and B the mole fraction of *A* is 0.33. Calculate the composition of the vapour in equilibrium with the solution. Given : P_A^o = 75 mm of Hg and P_B^o = 22 mm of Hg.

[M.Sc. Admission Test 1993, IIT Bombay]

Soln. : By the problem

x_A = 0.33, $\therefore$ x_B = (1 − 0·33) = 0.67

P_A^o = 75 mm of Hg

P_B^o = 22 mm of Hg.

We know that, for ideal solution total vapour pressure,

$P = P_A + P_B = x_A P_A^o + x_B P_B^o$ = (0.33 × 75) + (0.67× 22) = 39.49.

Let, in the vapour phase, mole fraction of *A* be x_A' and mole fraction of *B* be x_B'

$$\therefore\ x_A' = \frac{P_A}{P} = \frac{x_A.P^o}{P} = \frac{0.33\times 75}{39.49} = \mathbf{0.62.}$$

$\therefore$ x_B' = **0.38.**

37. At 140 °C the vapour pressure of C_6H_5Cl is 939.4 mm and that of C_6H_5Br is 495.8 mm. They form an ideal solution that boils at atmospheric pressure. Find out the composition of the solution. **[M.Sc. Admission Test 1992, IIT Bombay]**

Soln. : By the problem,

$$P_1^o = 939.4 \text{ mm}$$

$$P_2^o = 495.8 \text{ mm}.$$

Atmospheric pressure (P) = 760 mm.

We know that for ideal solution,

$$P = x_1 P_1^o + x_2 P_2^o = x_1 P_1^o + (1 - x_1) P_2^o$$

or, $760 = (x_1 \times 939.4) + (1 - x_1) \times 495.8$

or, $x_1 = 0.6$

$\therefore$ $x_2 = 0.4$

$\therefore$ $x_{C_6H_5Cl} = 0.6$ and $x_{C_6H_5Br}$ **= 0.4.**

38. What is the freezing point of a 10% (by weight) solution of methanol in water? Molal freezing point depression constant for water $K_f = 1.86$ K.kg.mol^{-1}

[M.Sc. Admission Test 1993, IIT KGP]

Soln. : By the problem, the solution is 10% by weight, i.e., 10 kg CH_3OH in 90 kg of H_2O.

$$\therefore m = \frac{W_2}{M_2 . W_1} = \frac{10 \text{ kg}}{0.032 \text{ kg. mol}^{-1} \times 90 \text{ kg}} = 3.4722 \text{ mol.kg}^{-1}.$$

$$\Delta T_f = K_f m$$

$$= 1.86 \text{ K.kg.mol}^{-1} \times 3.47 \text{ mol.kg}^{-1}$$

$$= 6.454 \text{ K}$$

T °C = **−6.454 °C.**

39. What is the freezing point of a 0.01 molal solution of $K_3[Fe(CN)_6]$, which is 78% dissociated in water (K_f = 1.85)? **[M.Sc. Admission Test 1992, IIT KGP]**

Soln. : By the given problem—

$\alpha = 0.78$

$\nu = 4.$

$$\therefore \ i = 3.34 \left[\alpha = \frac{i-1}{\nu-1}\right]$$

We know that $\Delta T_f = iK_f m$

$$= 3.34 \times 1.85 \text{ K.kg.mole}^{-1} \times 0.01 \text{ mole.kg}^{-1} = 0.06.$$

$\therefore$ freezing point will be **− 0.06 °C.**

40. A solution of KI is isotonic with a 0.01 M solution of I_2 at 27 °C. When equal volume of two solutions were mixed together, the osmotic pressure dropped by 18.5% of that of the individual solutions. Calculate the percentage of conversion of I^- to I_3^-. Assume that the solutions behave ideally and the salts are completely dissociated.

[M.Sc. Admission Test 1990, IIT KGP]

Soln. : Osmotic pressure (Π) for 0.01 molar solution of I_2 at 27 °C is—

$$\Pi_{I_2} = 0.01 \times 0.08206 \times 300.15$$

$$= 0.246 \text{ atm}.$$

By the problem,

solution of KI is isotonic with 0.01M solution of I_2.

$\therefore \Pi_{KI} = 0.246$ atm at 27 °C

Let the concentration of KI solution be C mole.lit^{-1}

$\therefore \Pi = i \times RTC$

KI dissociates into K^+ and I^-.

Assuming complete dissociation, $i = 2$.

$$\therefore C = \frac{\Pi}{i \times R \times T} = \frac{0.246 \text{ atm}}{2 \times 0.08206 \text{ lit.atm.K}^{-1}.\text{mol}^{-1} \times 300.15 \text{ K}}$$

$= 0.00499$ mole.lit^{-1} = 0.005 mole.lit^{-1}.

Now since equal volume of two solutions are mixed, the concentration will be half.

i.e., $C_{KI} = \frac{0.005}{2} = 0.0025$ mol. lit^{-1}

$C_{I_2} = \frac{0.01}{2} = 0.005$ mol.lit^{-1}.

Now by the given problem, Π of resulting solution,

$\Pi_{mix} = \frac{100 - 18.5}{100} \times 0.246$ atm

$= 0.2005$ atm.

The reaction is :

$KI + I_2 = KI_3$

Let x moles of KI react. So conentration of KI and I_2 will be—

$C_{KI} = (0.0025 - x)$ mol.lit^{-1} and $C_{I_2} = (0.005 - x)$ mol.lit^{-1}

Π due to KI = $2(0.0025 - x) RT$ atm.

Π due to I_2 = $(0.005 - x) RT$ atm.

Π due to KI_3 = $2 xRT$ atm.

(since, $KI_3 \rightarrow K^+ + I_3^-$ and assuming complete dissociation $i = 2$)

So, $[2(0.0025 - x) + (0.005 - x) + 2x]\ RT = 0.2005$

or, $x = 0.00186$.

$\therefore C_{I_3^-} = 0.00186$ mole.lit^{-1}.

Before reaction, $C_{I^-} = 0.0025$ mole.lit^{-1}.

After reaction, $C_{I_3^-} = 0.00186$ mole.lit^{-1}.

$\therefore$ % of conversion = $\frac{0.00186}{0.0025} \times 100 = \mathbf{74.3\%}$.

41. The vapour pressure of a solution containing 6.69 gm of $Ca(NO_3)_2$ in 100 gm of water is 746.9 mm of Hg at 100 °C. Calculate α (M = 164). **[C.U.(H) 1974]**

Soln. : P^o at 100 °C is 760. Hence, by the problem,

$$\frac{P^o - P}{P^o} = ix_2$$

$$\text{or,}\quad \frac{760-746.9}{760} = i \times \frac{\frac{W_2}{M_2}}{\frac{W_1}{M_1}+\frac{W_2}{M_2}} = i \times \frac{\frac{6.69}{164}}{\frac{100}{18}+\frac{6.69}{164}}$$

or, $i = 2.3647$.

Thus, the degree of dissociation,

$$\alpha = \frac{2.3647-1}{\nu-1} = \frac{1.3647}{3-1} = \mathbf{0.68235.}$$

42. A solution containing 10 gm of a dibasic acid in 1000 gm of water freezes at –0.15 °C. 12 ml of 0.1(N) NaOH solution is required for the neutralisation of 10 ml of this solution. Given K_f for water is 1.86 K.kg.mol^{-1}, calculate the van't Hoff factor for the acid. **[C.U.(H) 1994]**

Soln. : By the problem, the strength of acid x is given by 12 ml × 0.1(N) = 10x,

or, x = 0.12(N).

Let the density of the solution be 1.

Then, 10 gm in 1000 cc = 0.12(N)

or, $\frac{10}{0.12}$ gm in 1000 cc ≡ 1(N) = 83.33 gm in 1000 cc.

Hence, the equivalent weight is 83.33. The acid is dibasic.

Thus, its molecular weight is 83.33 × 2 = 166.66.

Now, by the problem,

freezing point = –0.15 °C

$\therefore\ \Delta T_f = 0.15.$

$$\therefore\ 0.15\text{ K} = i \times K_f \times \frac{1000 W_2}{M_2 \times W_1}.$$

$$\text{or, } i = \frac{0.15 M_2 W_1}{K_f \times 1000 \times W_2} = \frac{0.15 \times 166.66 \times 100}{1.86 \times 1000 \times 10}$$

or, i = **1.34.**

43. A and B are two non-volatile solids. A is dissolved in a solvent to give a dilute solution which is dissociated (degree of dissociation α). In the same solvent with the same molal concentration B is dimerised (degree of association β). The elevation of boiling point in the A solution is twice than that of B solution. If α be 0.8, what should be β ? **[C.U.(H) 1993]**

Soln. : Let A dissociate to two particles. Then

$i = 1 + \alpha = 1 + 0.8 = 1.8.$

∴ by the problem,

$2 \times \Delta T_b = 1.8\ K_b.m$ and $\Delta T_b = i' K_b.m$

or, $2 = \frac{1.8}{i'}$

or, $i' = 0.9.$

We know that, $\beta = \dfrac{\nu(1-i')}{\nu-1} = \dfrac{2(1-0.9)}{2-1} = 0.2.$

[**N.B.** : If A dissocites to 3 particles, $i = 2.6$, $i' = \dfrac{2.6}{2} = 1.3.$

$\therefore$ $\beta > 1$, impossible.]

44. An aqueous solution of sucrose freezes at –0.200 °C. Calculate the normal boiling point and the molality of an aqueous solution of NaCl having the same vapour pressure. Given, K_f for H_2O = 1.86 K.kg.mole^{-1} and K_b = 0.52 kg.K.mole^{-1}.

Soln. : By the problem, $\Delta T_f = 0.200$ K.

We know that, $\Delta T_f = K_f.m.$

Hence, the molality of sucrose solution (m) is

$$m = \frac{\Delta T_f}{K_f}.$$

$$\therefore \quad m = \frac{0.200\,\text{K}}{1.86\ \text{K.kg.mol}^{-1}} = 0.1075\ \text{mole.kg}^{-1}$$

$$\therefore \Delta T_b = K_b.m$$
$$= 0.52\ \text{K.kg.mole}^{-1} \times 0.1075\ \text{mole.kg}^{-1} = 0.0559\ \text{K}.$$

$$\therefore \ T_b^o = 373.15 + 0.0559 = 373.2059\ \text{K}.$$

Assuming NaCl to be completely ionised, $i = 2$.

Since the solution of NaCl has the same v. p. as that of sucrose, it will have the same boiling point.

$\therefore \Delta T_b =$ **0.0559 K.**

$\therefore$ 0.0559 = 2 × K_b × m.

$$\therefore m = \frac{0.0559}{2\times 0.52} = 0.05375.$$

$\therefore$ molality of NaCl soln. is **0.05375.**

45. A certain mass of substance when dissolved in 100 gm of benzene lowers the freezing point by 1.28 °C. The same mass of solute dissolved in 100 gm of water lowers the freezing point of the latter by 1.40 °C. If the substance has a normal molecular weight in benzene and is completely dissociated in water, into how many ions does it dissociate in water?

Given, K_f (water) = 1.86 K.kg.mol^{-1}; K_f (benzene) = 5.12 K.kg.mol^{-1}. **[C.U. 1975]**

Hints : ΔT_f (benzene) = K_f (benzene).m

ΔT_f (H_2O) = $\nu.K_f$ (H_2O).m

ΔT_f (benzene) = 1.28

ΔT_f (H_2O) = 1.40

$\therefore$ 1.28 = 5.12 × m for benzene (i)

and 1.40 = 1.86 × m × ν for H_2O. (ii)

Dividing eqn. (ii) by eqn. (i) we get the value of ν.

$$\therefore \nu = \frac{5.12\times 1.40}{1.86\times 1.28} = 3.01 \approx \mathbf{3}.$$

46. (a) Find out osmotic pressure of a 0.001 (M) aq. K_2SO_4 solution at 27 °C.

(b) Boiling point of acetone is 56.5 °C and its latent heat of vaporisation is 6920 cal. mole^{-1}. Hence, calculate the molal boiling point elevation constant of acetone.

Hints : See previous problem. **[Ans.** (a) 0.0738 atm., (b) 1.81 kg.K.mole^{-1}]

47. A solution of HCl in water contains 18.25 gm HCl per 100 gm solution, and its density is 1.10 gm.cm^{-3}. Calculate the molarity and mole fraction of HCl.

Hints : See previous problem. **[Ans.** X_2 = 0.099, concentration = 5.5(m).]

48. Calculate the van't Hoff factor and the apparent degree of dissociation of a 0.2 molal aqueous solution of $NaNO_3$ which freezes at –0.675 °C.

[Given : K_f = 1.86 K.kg.mole^{-1}] **[C.U. 2006]**

Hints : See previous problem. **[Ans.** i = 1.814, α = 0.814]

49. The vapour pressure of A is 939.4 mm of Hg and that of B is 495.8 mm of Hg at 140 °C. Assuming that they form an ideal solution, what will be the composition of a mixture, which boils at 140 °C under 1 atm? What will be the composition of the vapour at this temperature? **[C.U. 2004]**

Hints : $760 = P_A^o x_A + P_B^o x_B$

$$x_A^{vap} = \frac{P_A^o x_A}{760} \quad \begin{bmatrix} x_A = 0.596, & x_B = 0.404 \\ x_A^{vap} = 0.737, & x_B^{vap} = 0.263 \end{bmatrix}$$

50. The total vapour pressure at 25 °C of a mixture of benzene and toluene in which the two mole fractions are equal, is 62 mm Hg. The vapour pressure of pure benzene at 25 °C is 95 mm of Hg. Calculate the mole fraction of benzene in the vapour in equilibrium with the liquid mixture (assume ideal behaviour of the mixture). **[C.U. 2001]**

Hints : See previous problem. **[Ans.** 0.77]

CHAPTER 8

CHEMICAL EQUILIBRIUM

Required Formulae

1. Let a standard reaction be

$\Sigma v_i A_i = 0.$

In this equation v_i's are negative for reactants and positive for products. Taking A_1 and A_2 as two reactants and A_3 and A_4 as two products, then

$v_3 A_3 + v_4 A_4 - v_1 A_1 - v_2 A_2 = 0$

i.e., the reaction is $v_1 A_1 + v_2 A_2 = v_3 A_3 + v_4 A_4$.

2. For this reaction if a_3 and a_4 be activities of products and a_1 and a_2 be activities of reactants and the activities are the activity at any instant except equilibrium, then reaction quotient Q_a is written as :

$$\frac{a_{A_3}^{v_3} \cdot a_{A_4}^{v_4}}{a_{A_1}^{v_1} \cdot a_{A_2}^{v_2}}. \qquad \text{(i)}$$

At equilibrium Q_a is converted to K_a, the equilibrium constant.

K_P is defined as $K_P = \dfrac{P_{A_4}^{v_4} \cdot P_{A_3}^{v_3}}{P_{A_1}^{v_1} \cdot P_{A_2}^{v_2}}$, (ii)

where the terms are equilibrium partial pressures. We may define reaction quotient Q_p.

We can define K_c also, where

$$K_C = \frac{C_{A_3}^{v_3} \cdot C_{A_4}^{v_4}}{C_{A_1}^{v_1} \cdot C_{A_2}^{v_2}} \qquad \text{(iii)}$$

K_P and K_C are related to K_a.

$K_a = K_P \cdot K_\gamma$,

where γ is the activity coefficient.

Similarly, $K_a = K_C \cdot K_\gamma$

3. Unitless K_P and K_C :

$$K_P = \frac{\left(\frac{P_{A_4}}{P^o}\right)^{\nu_4} \cdot \left(\frac{P_{A_3}}{P^o}\right)^{\nu_3}}{\left(\frac{P_{A_1}}{P^o}\right)^{\nu_1} \cdot \left(\frac{P_{A_2}}{P^o}\right)^{\nu_2}}.$$

The standard state P^o is 1 atm.

$$\text{Similarly, } K_C = \frac{\left(\frac{C_{A_4}}{C^o}\right)^{\nu_4} \cdot \left(\frac{C_{A_3}}{P^o}\right)^{\nu_3}}{\left(\frac{C_{A_1}}{C^o}\right)^{\nu_1} \cdot \left(\frac{C_{A_2}}{C^o}\right)^{\nu_2}}.$$

The standard state C^o is 1 mole.lit^{-1}.

4. Using unitless K_P and K_C,

$\Delta G = \Delta G^o + RT \ln Q_p$ and $\Delta G^o = -RT \ln K_P$.

5. K_{P_1} and K_{P_2} be equilibrium constants at temperatures T_1 and T_2, then—

$$\ln\left(\frac{K_{P_2}}{K_{P_1}}\right) = \frac{\Delta H^o}{R}\left(\frac{T_2 - T_1}{T_1 T_2}\right)$$

$$\text{and} \quad \frac{d \ln K_P}{dT} = \frac{\Delta H^o}{RT^2}.$$

6. Distribution coefficient or partition coefficient K is one type of equilibrium constant. If a solute undergoes distribution between two immiscible liquids retaining molecular weight, say water and an organic solvent, then at equilibrium

$$K = \frac{C_{or}}{C_{aq}},$$

where C_{or} is concentration of solute in gm-mole in organic solvent and C_{aq} is the corresponding concentration in aqueous layer.

Short Questions

1. Show that to a first approximation the equation of state of a gas that dimerizes to a small extent is given by

$$\frac{PV}{RT} = 1 - \frac{K_C}{V},$$

where K_C is the equilibrium constant for the formation of dimer,

$$2A \rightleftharpoons A_2;\ K_C \rightleftharpoons \frac{[A_2]}{[A]^2}.$$

Ans. : The reaction is $2A \rightleftharpoons A_2$

Let α be the degree of reaction. Then, if we take 1 mole of A initially, at equilibrium C_A is $(1-\alpha)$ moles and A_2 formed is $\alpha/2$ moles. If the total volume is V,

$$K_C = \frac{\frac{(\alpha/2)}{V}}{\{(1-\alpha)/V\}^2}.$$

Now, $1 >> \alpha_r$.

$$\therefore \quad K_C = \frac{\alpha . V}{2.1}.$$

$$\therefore \quad \frac{K_C}{V} = \frac{\alpha}{2}.$$

At equilibrium there are n moles, n being $\left(1 - \frac{\alpha}{2}\right)$. If the system is an ideal gas,

$$PV = nRT = \left(1 - \frac{\alpha}{2}\right) RT.$$

$$\therefore \quad \frac{PV}{RT} = 1 - \frac{\alpha}{2}$$

$$\text{or,} \quad \frac{PV}{RT} = 1 - \frac{K_C}{V}.$$

2. Comment on the statement : "All reactions are reversible".

Ans. : We know that all reactions are reversible and characterised by an equilibrium constant K_a, so that for the reaction,

$$\nu_1 A + \nu_2 B = \nu_3 C + \nu_4 D$$

$$K_a = \frac{a_c^{\nu_3} . a_D^{\nu_4}}{a_A^{\nu_1} . a_B^{\nu_2}},$$

where a's are activities of reactants and products.

3. Predict the nature of changes on the equilibrium for the following reactions brought about by the variation of the operational parameters mentioned against each equation within parenthesis :

Ans. : (i) $2SO_3(g) = 2SO_2(g) + O_2(g)$ (heat applied)

This is an endothermic reaction. Thus, application of heat will lead to increase in the equilibrium constant and equilibrium composition of products. This can be ascertained from the relation :

$$\ln \frac{K_2}{K_1} = \frac{\Delta H}{R} . \frac{T_2 - T_1}{T_1 T_2}.$$

Since ΔH is positive, application of heat will lead to increase of temperature and $K_2 > K_1$. Consequently, the activities, i.e., concentration of SO_2 and O_2 will be higher at higher temperature

(ii) $PCl_5(g) = PCl_3(g) + Cl_2(g)$ [pressure increased]

For this reaction, if α is the degree of dissociation, then

$$K_P = \frac{P_{PCl_3} . P_{Cl_2}}{P_{Pcl_5}} = \frac{\alpha^2 P}{1 - a^2},$$

where P_{PCl_3}, P_{Cl_2}, etc. are partial pressures of products and reactants, and P is the total pressure. Hence, if total pressure P is increased, α will decrease as K_P is constant. In other words, increase of pressure means decrease of volume, and such decrease of volume will lead to a decrease in the yield of products.

(iii) $C(s) + H_2O(g) = CO(g) + H_2(g)$ [C(s) increased at const. temperature]

For this reaction,

$$K_P = \frac{P_{CO} \cdot P_{H_2}}{P_{H_2O}} = \frac{\alpha^2 P}{1-\alpha^2},$$

where P_{CO}, P_{H_2} and P_{H_2O} are partial pressure of CO, H_2 and $H_2O(g)$ respectively, and P and α are the total pressure and degree of the reaction. Thus, if increase of C(s) at constant temperature is made at constant pressure also, then there will be no effect. However, if P increases, then the yield of CO and H_2 will decrease.

(iv) $N_2(g) + 2H_2(g) = 2NH_3(g)$

[Inert gas introduced at constant volume and constant pressure]

If we start with a moles of N_2 and b moles of H_2, then it can be shown that the K_p for the reaction,

$$K_P = \frac{4x^2(a+b-2x)^2}{(a-x)(b-3x)^3 P^2},$$

where $2x$ is the number of moles of NH_3 at equilibrium and P is the total pressure (x moles of N_2 spent and $3x$ moles of H_2 spent). The total number of moles at equilibrium is $(a + b - 2x)$ so that $PV = (a + b - 2x) RT$, where P is the total pressure, V is the total volume and T is the temperature (assuming gas to be ideal). Addition of inert gas at constant volume will lead to proportionate increase of P (at const. T) so that x, the yield, will not be affected.

But, if inert gas is added at constant P, $(a + b - 2x)$ increases and hence x, the yield must decrease to keep K_P constant.

(v) $C(s) + O_2(g) = CO_2(g)$ [system cooled]

Cooling will lead to a decrease in temperature. The above reaction is exothermic. Hence, decrease of temperature will lead to an increase in K_p and the equilibrium yield.

(vi) $N_2(g) + 3H_2(g) = 2NH_3(g)$ [The catalyst is removed after equilibrium is attained.]

The catalyst is a foreign substance and has no role in equilibrium composition. Thus. the removal of catalyst will not disturb the equilibrium composition if total pressure is constant. However, if total pressure decreases, then in the above case, the equilibrium yield of NH_3 will decrease.

4. Does the equilibrium constant of a reaction depend upon (i) the standard state chosen for the reactants and products, and (ii) the stoichiometric representation of the reaction? Justify your answer. **[C.U.(H) 2000]**

Ans. : Equilibrium constant of a reaction depends upon the standard state chosen unless $\Delta\nu$ for the reaction is zero. If the standard state is 1 gm-mole/lit, resulting K is K_C and if the standard state is 1 atm, the resulting K is K_P and they are different unless $\Delta\nu = 0$.

Since, $K_P = \Pi\left(\frac{P_i}{P^o}\right)^{\nu_i}$ and $K_P = K_C\left(\frac{C^o}{P^o}RT\right)^{\Delta\nu}$, where P^o is 1 atm pressure and C^o is 1 mol.l^{-1}.

ΔG° is an extensive property and if a reaction is written as $\frac{A}{2}+\frac{B}{2}=\frac{C}{2}+\frac{D}{2}$

and K_p is K, then $\Delta G^\circ = -RT \ln K = L$.

If the reaction is $A + B = C + D$,

the mass is doubled, $\Delta G^\circ = -2RT \ln K = -RT \ln K^2 = 2L$.

Hence, $K_P = K^2$. Thus, the equilibrium constant should depend on how we write the equation in stoichiometric representation.

5. Under what condition will equilibrium constant of a reaction not change with temperature?

Ans. : The temperature dependence of K_P is given by

$\frac{d \ln K_P}{dT} = \frac{\Delta H^\circ}{RT^2}$. Thus, if $\Delta H^\circ = 0$, i.e., the standard heat of reaction is zero and constant, then K will not change with change of temperature.

6. The relation $K_P = K_C(RT)^{\Delta \nu}$ is not compatible with unitless K_P and K_C.

Ans. : This is obvious since both R and T have units. Unitless K_P for the reaction $\nu_1 A + \nu_2 B = \nu_3 C + \nu_4 D$ is really

$$K_P = \frac{\left(\frac{P_C}{P^\circ}\right)^{\nu_3}\left(\frac{P_D}{P^\circ}\right)^{\nu_4}}{\left(\frac{P_A}{P^\circ}\right)^{\nu_1}\left(\frac{P_B}{P^\circ}\right)^{\nu_2}}.$$

$$\therefore K_P = \frac{\left(\frac{C_C RT}{P^\circ}\right)^{\nu_3}\left(\frac{C_D RT}{P^\circ}\right)^{\nu_4}}{\left(\frac{C_A RT}{P^\circ}\right)^{\nu_1}\left(\frac{C_B RT}{P^\circ}\right)^{\nu_2}} = \frac{\left(\frac{C_C}{C^\circ}\right)^{\nu_3}\left(\frac{C^\circ}{P^\circ}RT\right)^{\nu_3}\left(\frac{C_D}{C^\circ}\right)^{\nu_4}\left(\frac{C^\circ RT}{P^\circ}\right)^{\nu_4}}{\left(\frac{C_A}{C^\circ}\right)^{\nu_1}\left(\frac{C^\circ RT}{P^\circ}\right)^{\nu_1}\left(\frac{C_B}{C^\circ}\right)^{\nu_2}\left(\frac{C^\circ RT}{P^\circ}\right)^{\nu_2}}$$

$$= K_C\left(\frac{C^\circ}{P^\circ}RT\right)^{\nu_3+\nu_4-\nu_1-\nu_2} = K_C\left(\frac{C^\circ}{P^\circ}RT\right)^{\Delta\nu}.$$

If C° is 1 mole/lit, $P^\circ = 1$ atm, R is lit-atm.deg^{-1}.mole^{-1}, then $\left(\frac{C^\circ}{P^\circ}RT\right)$ is dimensionless, and this P_C, etc. in K_P should be written as $\frac{P_C}{P^\circ}$.

7. Show that $K_P = K_x\left(\frac{P}{P^\circ}\right)^{\Delta\nu}$ for $\nu_1 A + \nu_2 B \rightleftharpoons \nu_3 C + \nu_4 D$.

Ans. : $K_P = \frac{\left(\frac{P_C}{P^\circ}\right)^{\nu_3}\left(\frac{P_D}{P^\circ}\right)^{\nu_4}}{\left(\frac{P_A}{P^\circ}\right)^{\nu_1}\left(\frac{P_B}{P^\circ}\right)^{\nu_2}}$

$$= \frac{\left(\frac{X_C P}{P^\circ}\right)^{\nu_3}\left(\frac{X_D P}{P^\circ}\right)^{\nu_4}}{\left(\frac{X_A P}{P^\circ}\right)^{\nu_1}\left(\frac{X_B P}{P^\circ}\right)^{\nu_2}} = K_x\left(\frac{P}{P^\circ}\right)^{\Delta\nu}.$$

8. Obtain an expression for K_P in terms of $x(x$ = moles of N_2 reacted per mole) and the total pressure P for the reaction.

$N_2 + 3H_2 = 2NH_3$

Show that $x \to 0$ as $P \to 0$ and $x \to 1$ as $P \to \infty$.

Ans. : The reaction is $N_2 + 3H_2 = 2NH_3$. If we take 1 mole of N_2 and 3 moles of H_2, then let the equilibrium number of mole of NH_3 be $2x$, i.e., the equilibrium number of moles of $N_2 = (1 - x)$ and that of $H_2 = 3(1 - x)$.

$$\therefore K_P = \frac{4x^2 \times (4-2x)^2}{(1-x)^4 \times 27P^2}.$$

$$\therefore P^2 = \frac{4x^2(4-2x)^2}{27(1-x)^4 \cdot K_P}.$$

$$\text{Hence, } P = \frac{2x(4-2x)}{3(1-x)^2 \cdot \sqrt{3K_P}}.$$

In general, $1 >> x$, hence, $P = \dfrac{8x}{3\sqrt{3K_P}}$. Thus, $x \to 0$, as $P = 0$.

Similarly, when $x \to 1$, the denominator $\to 0$ as it contains $(1 - x)$. Thus, as $P \to \infty$, $x \to 1$.

9. N_2 and O_2 combine at a given temperature to produce NO. At equilibrium the yield of NO is x per cent by volume if $x = \sqrt{Kab} - \dfrac{K(a+b)}{4}$, where K is equilibrium constant of the reaction at the given temperature and a and b are the volume perentage of N_2 and O_2 respectively in the initial pure mixture. What should be the initial composition of the reacting mixture in order that maximum yield of NO is ensured?

Ans. : By the problem yield,

$$x = \sqrt{Kab} - \frac{K(a+b)}{4}$$

$\left(\dfrac{\partial x}{\partial a}\right)_b$ and $\left(\dfrac{\partial x}{\partial b}\right)_a$ will be equal to zero at maximum value of x.

$$\left(\frac{\partial x}{\partial a}\right)_b = \frac{\sqrt{Kb}}{2\sqrt{a}} - \frac{K}{4} = 0 \quad \text{(i)}$$

and

$$\left(\frac{\partial x}{\partial b}\right)_a = \frac{\sqrt{Ka}}{2\sqrt{b}} - \frac{K}{4} = 0. \quad \text{(ii)}$$

$\therefore$ from (i) and (ii),

$$\frac{\sqrt{Kb}}{2\sqrt{a}} = \frac{\sqrt{Ka}}{2\sqrt{b}}$$

or, $$\sqrt{\frac{b}{a}} = \sqrt{\frac{a}{b}}$$

or, $a = b$.

Hence the yield will be maximum if we take equilibrium amount of N_2 and O_2 at the beginning.

10. What would be the effect of adding solid $CaCO_3$ to the following system at equilibrium at a definite temperature, $CaCO_3(s) = CaO(s) + CO_2(g)$? **[C.U. 2006]**

Ans. : For the reaction $CaCO_3(s) + CO_2(g)$ the equilibrium constant, $K_P = P_{CO_2}$. Thus, addition of $CaCO_3$ at equilibrium at a definite temperature would not affect if pressure remains constant. However, if P changes, equilibrium will be affected.

11. Consider two equilibria,

$$A_2 = 2A \qquad \text{(i)}$$

$$AB = A + B \qquad \text{(ii)}$$

and assume that the $\Delta G°$ and, therefore, K_P are the same for both. Show that the equilibrium value of advancement for the reaction (ii) ξ_2 is greater than the corresponding value of the reaction (i), ξ_1. What is the physical reason for this result? **[V.U. 2001]**

Ans. : If ν_i^o and ν_i are number of moles of species i at the beginning and at equilibrium respectively, then

$n_i = n_i^o + |\nu_i|\xi$.

ν_i is (+)ve for products and (–)ve for reactants.

Let $n^o_{A_2}$ and n^o_{AB} be 1, and n^o_A, n^o_B be zero initially. Then by the problem,

$n_{A_2} = 1 - \xi_1$, $n_A = 2\xi_1$ for reaction (i)

and $n_{AB} = 1 - \xi_2$, $n_A = \xi_2$, $n_B = \xi_2$ for reaction (ii).

Since, K_P is same for both reactions,

$$\therefore \quad \frac{4\xi_1^2}{1-\xi_1} = \frac{\xi_2^2}{1-\xi_2}$$

or, $4\xi_1^2 = \xi_2^2$ (since, $1 >> \xi$)

$\therefore \ \xi_2 = 2\xi_1$, and thus, $\xi_2 > \xi_1$.

One molecule of A_2 gives two molecules of A while one molecule of AB gives one molecule each in both A and B. Equilibrium constants of both reactions will be same if fraction of molecules of A present at equilibrium for reaction (i) is less than the fraction of molecules of A and B present at equilibrium for reaction (ii). Hence, ξ_2 is greater than ξ_1.

12. Since $\dfrac{\Delta G^o}{RT} = -\ln K$ (the terms having their usual significance), a plot of ln K vs. $\dfrac{1}{T}$ would give a straight line passing through origin. Justify or modify the statement.

[V.U. 2006]

Ans. : Since $\Delta G°$ depends on temperature, plot of ln K vs. $\dfrac{1}{T}$ will not be a straight line passing through origin.

13. Predict the direction of the slope of $\left(\frac{\partial G}{\partial \xi}\right)_{T,P}$ and thus, direction of reaction, if (i) $Q_P > K_P$ (ii) $Q_P = K_P$ (iii) $Q_P < K_P$, where ξ = extent of reaction and Q_P is pressure quotient.

Ans. : $\left(\frac{\partial G}{\partial \xi}\right)_{T,P} = \overline{\Delta G} = \overline{\Delta G^o} + RT \ln Q_P$

$$= RT \ln\left(\frac{Q_P}{K_P}\right) \text{ (since } \overline{\Delta G^o} = -RT \ln K_P).$$

(i) If $Q_P > K_P$, slope is positive and reaction will go backward, i.e., reactants will be formed at the expense of products to reach equilibrium.

(ii) If $Q_P = K_P$, $\overline{\Delta G} = 0$ and reaction is at equilibrium.

(iii) If $Q_P < K_P$, slope is negative and reaction will go forward, i.e., product will be formed.

14. "If $\overline{\Delta G^o} = 0$ for a reaction, the reaction is thermodynamically impossible."—Justify.

Ans. : Spontaneity of a reaction is determined by $\overline{\Delta G}$ and not $\overline{\Delta G^o}$ and

$$\overline{\Delta G} = \overline{\Delta G^o} + RT \ln Q_P = -RT \ln K_P + RT \ln Q_P.$$

$\overline{\Delta G^o} = 0$ means $K_P = 1$, i.e., equilibrium concentration of reactants is same as that of products. If initially $Q_P < K_P$ so that $\overline{\Delta G} < 0$ reaction proceeds in forward direction to attain equilibrium, where $\overline{\Delta G} = 0$ at constant P and T.

15. When will the plot of $\ln K_P$ vs. $\frac{1}{T}$ give straight line? How does the plot of $\ln K_P$ vs. $\frac{1}{T}$ look for exothermic and endothermic reactions? Explain.

Ans. : By van't Hoff equation, $\frac{d \ln K_P}{d\left(\frac{1}{T}\right)} = -\frac{\Delta H^o}{R}$.

If ΔH^o is independent of temperature, then plot of $\ln K_P$ vs. $\frac{1}{T}$ will give a straight line.

For exothermic reaction, ΔH^o is –ve, thus,

$\ln K_P$

$\frac{1}{T}$

For endothermic reaction, ΔH^o is +ve, thus, $\ln K_P$

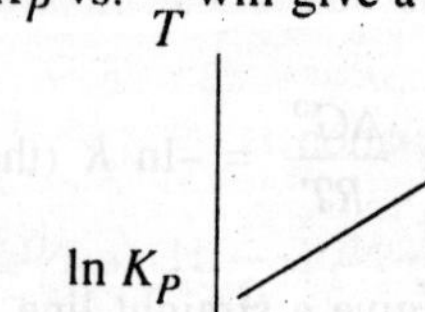

$\frac{1}{T}$

For exothermic reaction as T decreases or $\frac{1}{T}$ increases, yield increases, thereby increasing K_P. For endothermic reaction as T decreases or $\frac{1}{T}$ increases, yield decreases, thereby decreasing K_P. The shift of equilibrium in both the above cases takes place to nullify the applied change. If T decreases at equilibrium, it willl shift in such a way that more heat is generated. Hence, K_P increases for exothermic reaction and the same decreases for endothermic reaction.

Numerical Problems

1. When nitrogen and hydrogen are mixed in the proportion 1 : 3 at 50 atm and 650 °C, the equilibrium concentration of ammonia is 25% by wt. Calculate the equilibrium constant. **[C.U.(H) 1969, 1980]**

Soln. : The reaction is :

$$N_2 + 3H_2 = 2NH_3$$

Let a moles of N_2 and $3a$ moles of hydrogen be taken initially. Thus, the mass is $\{(a \times 28) + (3a \times 2)\} = 34a$ gm. This weight is constant and its 25% is $8.5a$ gm. Thus, the moles of NH_3 is $\frac{8.5}{17}a = 0.5a$. So at equilibrium,

$$\begin{array}{ccc} N_2 & + 3H_2 \rightleftharpoons & 2NH_3 \\ a(1-x) & 3a(1-x) & 2ax \end{array}$$

Now, $2ax = 0.5a$.

$\therefore x = 0.25$.

Total number of moles is $a + 3a - ax - 3ax + 2ax = (4 - 2x)a = 3.5a$.

$$\therefore P_{N_2} = \frac{0.75a}{3.5a} \times 50$$

$$P_{H_2} = \frac{2.25a}{3.5a} \times 50$$

$$P_{NH_3} = \frac{0.5a}{3.5a} \times 50.$$

(a cancels out. Thus, in problems we may forget about a and take the ratio as per equation.)

If the equation is :

$$\frac{1}{2} N_2 + \frac{3}{2} H_2 = NH_3$$

$$K_P = \frac{\left(\frac{0.5}{3.5}\right)^2 \times (50)^2}{\frac{0.75}{3.5} \times 50 \times \left\{\frac{2.25}{3.5} \times 50\right\}^3}$$

$$= \mathbf{1.4 \times 10^{-4}}.$$

2. When nitrogen and hydrogen in the molar ratio 1 : 3 are allowed to react at 100 atm and 200 °C, it was found that the conversion to ammonia is 25% by volume. Calculate the equilibrium constant for the reaction, $N_2 + 3H_2 = 2NH_3$.

Soln. : Let 1 mole of nitrogen and 3 moles of hydrogen be taken and let $2x$ mole of ammonia be formed at equilibrium.

$$\begin{array}{ccccc} N_2 & + & 3H_2 & = & 2NH_3 \\ (1-x) & & 3(1-x) & & 2x \end{array}$$

Since equal volumes contain equal number of moles, the total volume at equilibrium is $(4 - 2x)$ and volume of NH_3 is $2x$. As per problem,

$$\frac{2x}{4-2x} \times 100 = 25$$

or, $x = 0.4$.

$\therefore\ 1 - x = 0.6$ and $2x = 0.8$

or, $3(1 - x) = 1.8$.

Total number of mole = 0.6 + 1.8 + 0.8 = 3.2.

$$P_{N_2} = \frac{0.6}{3.2} \times 100$$

$$P_{H_2} = \frac{1.8}{3.2} \times 100$$

$$P_{NH_3} = \frac{0.8}{3.2} \times 100$$

$$\therefore K_P = \frac{(0.8)^2 \times (3.2)^2}{0.6 \times 1.8 \times (100)^2} = \mathbf{1.872 \times 10^{-4}}.$$

N.B. : If the conversion were 25% of original volume, then $\frac{2x}{4} \times 100 = 25$,

or, $x = 0.5$.

3. For the reaction $2A + 3B = C + 4D$, the partial pressures of A, B, C and D at equilibrium are 101.325 kPa, 10.1325 kPa, 1.01325 kPa and 202.650 kPa respectively. Calculate K_p.

Soln. : The reaction is—

$2A + 3B = C + 4D$

Let P_A, P_B, P_C and P_D be the partial pressures of A, B, C and D respectively, then—

$$K_P = \frac{\left(\frac{P_C}{P^o}\right) \cdot \left(\frac{P_D}{P^o}\right)^4}{\left(\frac{P_A}{P^o}\right)^2 \cdot \left(\frac{P_B}{P^o}\right)^3},$$

where P^o is the standard pressure = 1 atm = 101.325 kPa.

$$\therefore K_P = \frac{\left(\frac{1.01325}{101.325}\right) \cdot \left(\frac{202.650}{101.325}\right)^4}{\left(\frac{101.325}{101.325}\right)^2 \cdot \left(\frac{10.1325}{101.325}\right)^3} = 160.$$

4. A mixture of SO_2 and O_2 in the molar ratio 2 : 1 was kept over platinum catalyst in a reaction vessel at 650 °C, the total pressure at equilibrium was found to be 10 atm. If 60% SO_2 be converted to SO_3, calculate K_P for $2SO_2 + O_2 = 2SO_3$.

[B.U. 1977; C.U. 1973]

Soln. : Let 2 moles of SO_2 and 1 mole of O_2 be taken (with $2a$ and a the result will be same). Since the conversion is 60%, $x = 0.6$.

$$\begin{array}{lll} 2SO_2 \;+ & O_2 & \rightleftharpoons 2SO_3 \\ 2-2x & 1-x & \quad 2x \end{array}$$

$\therefore n_{SO_2} = 0.8$; $n_{O_2} = 0.4$; $n_{SO_3} = 1.2$.

Total number of mole $(n) = n_{SO_2} + n_{O_2} + n_{SO_3} = 2.4$.

$$\therefore P_{SO_2} = \frac{0.8}{2.4} \times 10 = \frac{10}{3}$$

$$P_{O_2} = \frac{0.4}{2.4} \times 10 = \frac{10}{6}$$

$$P_{SO_3} = \frac{1.2}{2.4} \times 10 = 5.$$

$$\therefore K_P = \frac{\left(P_{SO_3}\right)^2}{\left(P_{SO_2}\right)^2 . P_{O_2}} = \frac{(5)^2}{\left(\frac{10}{3}\right)^2 \times \frac{10}{6}} = \mathbf{1.35.}$$

5. Deduce the expression for K_P for the dissociation equilibrium

$$2H_2S(g) = 2H_2(g) + S_2(g)$$

in terms of total equilibrium pressure P and the degree of dissociation α at equilibrium starting from pure reactant (H_2S) only. Hence, calculate values of K_p, when $\alpha = 0.0118$ (1065 °C) and $\alpha = 0.026$ (1132 °C), the total pressure being 1 atm in each case. **[C.U.(H) 1982]**

Soln. : Let us consider that, there are 2 moles of H_2S and the degree of dissociation is α. Hence,

$$\begin{array}{lll} 2H_2S = & 2H_2 + & S_2 \\ 2(1-\alpha) & 2\alpha & \alpha \end{array}$$

Total number of moles = $(2 + \alpha)$.

$$\therefore P_{H_2S} = \frac{(2-2\alpha)}{2+\alpha}P$$

$$P_{H_2} = \frac{2\alpha}{2+\alpha}P \text{ and } P_{S_2} = \frac{\alpha}{2+\alpha}P.$$

$$\therefore K_P = \frac{\left(\frac{2\alpha}{2+\alpha}.P\right)^2 . \left(\frac{\alpha}{2+\alpha}.P\right)}{\left(\frac{2-2\alpha}{2+\alpha}.P\right)^2}$$

$$\text{or, } K_P = \frac{4\alpha^3}{(2-2\alpha)^2(2+\alpha)}.P.$$

When $\alpha = 0.0118$ and $P = 1$ atm,

$$K_P = \frac{4(0.0118)^3 \times 1}{4(1-0.0118)^2(2+0.0118)} = \frac{(0.0118)^3 \times 1}{(1-0.0118)^2 \times 2.0118}$$

$$= \mathbf{8.36 \times 10^{-7}}.$$

Again, when $\alpha = 0.0260$ and $P = 1$ atm,

$$K_P = \frac{4(0.026)^3 \times 1}{4(1-0.026)^3(2+0.026)}$$

$$= \mathbf{9.14 \times 10^{-6}}.$$

6. The equilibrium constant between ethyl alcohol and acetic acid to form ethyl acetate is 4.0 at 293 °K. What weight of ethyl acetate will be obtained from 50 gm of alcohol and 50 gm of acetic acid at 293 °K in a closed vessel? **[Delhi B.Sc. 1975]**

Soln. : By the problem,

$$n_{\text{alcohol}} = \frac{50}{46} = 1.087 \text{ [m. wt. of } C_2H_5OH = 46]$$

$$n_{\text{acetic acid}} = \frac{50}{60} = 0.833 \text{ [m. wt. of } CH_3COOH = 60]$$

Let x be the number of mole of ester formed.

$$\therefore \underset{(0.833 - x)}{CH_3COOH} + \underset{(1.087 - x)}{C_2H_5OH} \rightleftharpoons \underset{x}{CH_3COOC_2H_5}$$

$$\therefore 4 = \frac{x}{(1.087-x)(0.833-x)}.$$

Solving we get $x = 1.938$

and $x = 0.623$.

1.938 mole cannot be formed from 0.833 mole.

$\therefore x = 0.623$.

$\therefore$ amount of ester produced = $(0.623 \times 88) = 54.82$ gm.

[Molecular weight of $CH_3COOC_2H_5 = 88$]

7. 5 gm of PCl_5 kept in a flask of 0.4 lit capacity was completely vaporised at 523 °K. The molecular weight was found to be 158. Calculate the eqilibrium constant.

[Lucknow B.Sc. 1979]

Soln. : The reaction is $\underset{a(1-\alpha)}{PCl_5} = \underset{a\alpha}{PCl_3} + \underset{a\alpha}{Cl_2}$

where a = initial concentration of PCl_5 and

α = degree of dissociation.

We know that in this case,

$$\alpha = \frac{P - P^o}{P^o} = \frac{M_o - M}{M},$$

since pressure is inversely proportional to molecular weight at constant volume.

$\therefore\ \alpha = \dfrac{208.5-158}{158} = 0.3196.$ (molecular weight of PCl_5 = 208.5).

Initial concentration of $PCl_5 = a = \dfrac{5/208.5}{0.4} = 0.05995$ mol.l^{-1}.

At equllibrium, $C_{PCl_3} = C_{Cl_2} = 0.3196 \times 0.05995 = 0.01815$ mol.l^{-1}

$\therefore\ C_{PCl_5} = 0.05994 - 0.01915 = 0.04079$ mol.l^{-1}

$$\therefore\ K_C = \frac{C_{PCl_3}C_{Cl_2}}{C_{PCl_5}} = \frac{0.01915\times 0.01915}{0.04079} = \mathbf{9 \times 10^{-3}}.$$

8. At 100 °C and a pressure of 10^{-4} atm, CO_2 is 30% dissociated into CO and O_2. What will be the degree of dissociation if the pressure is raised to 1 atm? What will be the degree of dissociation when a mixture of 50% CO_2 and 50% O_2 is heated to 100 °C at 1 atm?

Soln. : Let 1 mole of CO_2 be taken. The reaction is :

$$CO_2 = CO + \frac{1}{2}O_2$$

$$1-\alpha \qquad \alpha \qquad \frac{1}{2}\alpha$$

$$0.7 \qquad 0.3 \qquad 0.15$$

By the problem, $\alpha = 0.3$. Hence the number of moles of CO_2, CO and O_2 at equilibrium are 0.7, 0.3 and 0.15 respectively (n = 1.15). Thus,

$$K_P = \frac{P_{CO}\cdot\left(P_{O_2}\right)^{1/2}}{P_{CO_2}} = \frac{\left(\frac{0.3}{1.15}P\right)\left(\frac{0.15}{1.15}P\right)^{1/2}}{\left(\frac{0.7}{1.15}P\right)}$$

$$= \frac{0.3\times(0.15P)^{1/2}}{0.7\times(1.15)^{1/2}}.$$

By the problem, $P = 10^{-4}$ atm.

$\therefore\ K_P = 1.5478 \times 10^{-3}$.

Let the degree of dissociation be α at 1 atm.

Hence, $\qquad CO_2 = CO + \frac{1}{2}O_2$

$$1-\alpha \qquad \alpha \qquad \frac{\alpha}{2} \qquad \text{Total} = \frac{2+\alpha}{2}$$

$$\therefore\ K_P = \frac{\left(\frac{2\alpha}{2+\alpha}\right)P.\left(\frac{\alpha}{\alpha+2}P\right)^{1/2}}{\frac{2(1-\alpha)}{\alpha+2}.P} = \frac{\alpha.\alpha^{1/2}P^{1/2}}{(2+\alpha)^{1/2}(1-\alpha)}.$$

Neglecting α in comparison to 1,

$2K_P^2 = \alpha^3$.

$\therefore\ \alpha = 0.016858537 = 0.017.$

Now, in the second case, there are 1 mole of CO_2 and 1 mole of O_2 initially. Hence, the equilibrium composition will be—

$$CO_2 = CO + \frac{1}{2}O_2.$$

$$(1-\alpha) \quad \alpha \quad 1+\frac{\alpha}{2}$$

$$\text{So, } K_P = \frac{\alpha(2+\alpha)^{1/2}}{(4+\alpha)^{1/2}(1-\alpha)}.$$

Neglecting α in comparison to 1, 2 and 4,

$$K_P = \frac{\alpha \times 2^{1/2}}{4^{1/2}} = \frac{\alpha}{\sqrt{2}}.$$

$\therefore \alpha = \sqrt{2}K_P = 2.18893 \times 10^{-3}$ = **2.2×10^{-3}.**

N.B. : Note that there is cubic power of α, and this type of problems can be solved by normal methods, only if we can neglect α in comparison to 1.

9. In the reaction $COCl_2 = CO + Cl_2$ at 550 °C, when the initial pressures of CO and Cl_2 are 250 mm of Hg and 280 mm of Hg respectively, the equilibrium pressure is found to be 380 mm of Hg. Calculate the degree of dissociation of $COCl_2$ at 1 atm. What will be the extent of dissociation when N_2 at a pressure of 0.4 atm is present, and if the total pressure is (i) 1 atm or (ii) 1.4 atm?

Soln. : The reaction is : $COCl_2 = CO + Cl_2$

$$x \quad a - x \quad b - x$$

where a = 250 mm of Hg

b = 280 mm of Hg.

Let x be the pressure of $COCl_2$ at equilibrium.

Then $P_{CO} = 250 - x$ and $P_{Cl_2} = 280 - x$

But by the problem,

$$P_{CO} + P_{Cl_2} + P_{COCl_2} = 530 - x = 380$$

$\therefore x = 150$ mm of Hg.

Hence, at equilibrium,

P_{CO} = 100 mm = 0.132 atm

P_{Cl_2} = 130 mm = 0.171 atm

P_{COCl_2} = 150 mm = 0.197 atm.

Hence, K_P for $COCl_2 \rightleftharpoons CO + Cl_2$ is

$$1-\alpha \quad \alpha \quad \alpha$$

$$K_P = \frac{P_{CO} \cdot P_{Cl_2}}{P_{COCl_2}} = \frac{0.132 \times 0.171}{0.197}$$

$$= 0.1145786 = 1.15 \times 10^{-1}.$$

$$\text{But again, } K_P = \frac{\alpha^2 P}{1-\alpha^2} = 0.115.$$

When $P = 1$, by the problem—

$$\alpha^2 = 0.115 - 0.115\alpha^2$$

or, $$\alpha^2 = \frac{0.115}{1.115}$$

$$\therefore \alpha = \sqrt{\frac{0.115}{1.115}} = 0.32115 = \mathbf{0.32.}$$

Now, N_2 is an inert gas for this reaction. When N_2 is present at 0.4 atm, the total pressure of reactants and products is 0.6 atm. Thus,

$$0.115 = \frac{\alpha_1^2 \times 0.6}{1-\alpha_1^2}.$$

$$\therefore \alpha_1 = \sqrt{\frac{0.115}{0.715}} = \mathbf{0.4.}$$

Thus, when N_2 is present at a pressure of 0.4 atm, the total pressure being 1 atm, the degree of dissociation will increase from 0.32 to 0.4. If the total pressure is 1.4 atm and the pressure of N_2 is 0.4 atm, then the pressure exterted by reactants and product is 1 atm, and hence, $\alpha = 0.32$.

10. For the dissociation of phosgene gas, the value of K_P at 100 °C is 6.7×10^{-9}, when the pressure is expressed in atmosphere. Find the fraction of phosgene dissociated at this temperature when 1 mole of phosgene is placed in a 100 lit vessel containing N_2 at a partial pressure of 1 atm. [C.U.(H) 1978]

Soln. : N_2 is an inert gas here and since volume is constant, it will have no effect on the yield. The fraction of phosgene dissociated,

$$\alpha = \sqrt{\frac{K_P}{P+K_P}}.$$

But $P >> K_P$.

$\therefore \alpha = \sqrt{\frac{K_P}{P}}$, but $P = \frac{RT}{V}$, since $a = 1$ mol, (and neglecting α in comparison to a)

$\therefore$ by the problem,

$$\alpha = \sqrt{\frac{K_P.V}{RT}} = \sqrt{\frac{6.7\times10^{-9}\times100}{0.08206\times373.15}} = \mathbf{1.48 \times 10^{-4}}.$$

11. I_2(g) is 1% dissociated to 2 I at 873 K and 25% dissociated at 1073 K. Calculate $\Delta H°$.

Soln. : The reactions is—

$$I_2 \rightleftharpoons 2I.$$

Let $2x$ be the amount of I produced.

$\therefore C_{I_2}$ at equilibrium $= (1 - x)$ and $C_I = 2x$.

$\therefore$ partial pressure of I, i.e., $P_I = \frac{2x}{1+x}P$ and $P_{I_2} = \frac{1-x}{1+x}P.$

Since P is 1 atm,

$$\therefore \quad K_P = \frac{4x^2}{1-x^2}.$$

For 1% dissociation, $x = 0.01$.

$$\therefore \quad K_{P_1} = \frac{4\times(0.01)^2}{1-(0.01)^2} = 4.0004 \times 10^{-4}.$$

For 25% dissociation,

$$K_{P_2} = \frac{4\times(0.25)^2}{1-(0.25)^2} = 0.267.$$

We know that

$$\ln\frac{K_{P_2}}{K_{P_1}} = \frac{\Delta H^\circ}{R}.\frac{T_2-T_1}{T_1T_2}.$$

$$\therefore \ln\frac{0.267}{4\times10^{-4}} = \frac{\Delta H^\circ}{8.314J}\times\frac{200}{873\times1073}$$

or, $$\Delta H^\circ = \frac{8.314\times873\times1073}{200}\times\ln\left(\frac{0.267}{4\times10^{-4}}\right) = \mathbf{253.25\ kJ.mol^{-1}}.$$

12. The vapour pressure of solid NH_4HS at 25 °C is 50 cm of Hg. Assuming complete dissociation of the vapour into NH_3 and H_2S, calculate the total pressure when NH_4HS is allowed to dissociate at 25 °C in a vessel containing ammonia at a pressure of 32 cm of Hg. [C.U. 1976]

Soln. : The reaction is $NH_4HS(s) = NH_3(g) + H_2S(g)$

Since there are two gases NH_3 and H_2S in equal amount, the mole fraction of each is $\frac{1}{2}$.

$$\therefore P_{NH_3} = 50\times\frac{1}{2} = 25 \text{ cm of Hg}$$

$$P_{H_2S} = 50\times\frac{1}{2} = 25 \text{ cm of Hg}$$

$$\therefore K_P = P_{NH_3}.P_{H_2S} = \left(\frac{25}{76}\times\frac{25}{76}\right).$$

If excess NH_3 is added, the partial pressure of each will change due to excess ammonia. Let there be x cm of Hg of NH_3 dissociated from NH_4HS.

$$\therefore P_{NH_3} = \frac{32+x}{76}$$

$$P_{H_2S} = \frac{x}{76}.$$

$$\therefore K_P = \frac{x(32+x)}{(76)^2} = \frac{(25)^2}{(76)^2}$$

or, $x^2 + 32x - 625 = 0$

$$\text{or, } x = \frac{-32+\sqrt{(32)^2+2500}}{2} = 13.68.$$

$\therefore$ total pressure $= x + x + 32 =$ **59.36 cm of Hg.**

13. Three solids A, B, C break into two gases, each as

$$A = A' + H_2S, \; B = B' + H_2S, \; C = C' + H_2S.$$

The total pressure over A is 6 cm, over B is 8 cm and over C is 24 cm of Hg, when they are kept in 3 separate vessels at 25 °C. What will be the total pressure if they are kept in the same vessel at 25 °C?

Soln. : The total pressure over A is 6 cm of Hg, when A is kept in a vessel at 25 °C.

Mole fraction of each A' and H_2S is $\frac{1}{2}$.

$$\therefore P_{A'} = \frac{1}{2} \times 6 = 3 \text{ cm of Hg and } P_{H_2S} = 3 \text{ cm of Hg.}$$

Similarly, when B and C are kept separately the pressure over B and C are 8 cm of Hg and 24 cm of Hg respectively.

$$\therefore \quad P_{B'} = \frac{1}{2} \times 8 = 4 \text{ cm of Hg}$$

$$P_{H_2S} = 4 \text{ cm of Hg}$$

$$P_{C'} = \frac{1}{2} \times 24 = 12 \text{ cm of Hg}$$

$$P_{H_2S} = 12 \text{ cm of Hg}$$

$$\therefore \quad K_A = \frac{9}{(76)^2}; \; K_B = \frac{16}{(76)^2}; \; K_C = \frac{144}{(76)^2}.$$

Let, when they are mixed together, $P_{A'} = x$ cm of Hg, $P_{B'} = y$ cm of Hg and $P_{C'} = z$ cm of Hg.

$$\therefore P_{H_2S} = (x + y + z) \text{ cm of Hg.}$$

$$\therefore K_A = \frac{x(x+y+z)}{(76)^2} = \frac{9}{(76)^2} \qquad \text{(i)}$$

$$K_B = \frac{y(x+y+z)}{(76)^2} = \frac{16}{(76)^2} \qquad \text{(ii)}$$

$$K_C = \frac{z(x+y+z)}{(76)^2} = \frac{144}{(76)^2}. \qquad \text{(iii)}$$

Adding equations (i), (ii) and (iii), we get

$$\frac{(x+y+z)^2}{(76)^2} = \frac{169}{(76)^2}$$

$$\text{or, } (x + y + z)^2 = (13)^2$$

$$\text{or, } x + y + z = 13 \text{ cm of Hg.}$$

Now, total pressure over the mixture is

$$P = P_{A'} + P_{B'} + P_{C'} + P_{H_2S}$$

$$= x + y + z + (x + y + z)$$

$$= 2(x + y + z) = 2 \times 13 = \mathbf{26 \text{ cm of Hg.}}$$

14. Two solids A and B break in the vapour phase completely to $A' + NH_3$ and $B' + NH_3$ respectively. The pressure over A is 10 cm of Hg and over B is 24 cm of Hg at 100 °C, when they are allowed to break in separate vessels. Calculate the total pressure over solids A and B, when they are kept together in a vessel at 100 °C.

Hints : See Problem 13. **[Ans.** 26 cm of Hg]

15. For the reaction $N_2O_5 \rightleftharpoons N_2O_3 + O_2$ at a certain temperture and 7 atm pressure, the yield of O_2 at equilibrium is found to be 20% of N_2O_5 present. Calculate the equilibrium constant. N_2O_5 was allowed to decompose in another vessel at 7 atm, but now CO_2 at a partial pressure of 3.5 atm was added. Find out the yield of N_2O_3.

Soln. : The reaction is :

$$\begin{array}{ccc} N_2O_5 & \rightleftharpoons N_2O_3 & + O_2 \\ 1-\alpha & \alpha & \alpha \end{array}$$

$$\therefore\ K_P = \frac{\alpha^2 P}{(1-\alpha)(1+\alpha)}.$$

Now, according to the problem,

$$\frac{\alpha}{1-\alpha} \times 100 = 20$$

$$\text{or, } 120\alpha = 20 \quad \text{or, } \alpha = \frac{1}{6}.$$

$$\therefore\ K_P = \frac{\left(\frac{1}{6}\right)^2 .7}{\left(1+\frac{1}{6}\right)\left(1-\frac{1}{6}\right)} = \mathbf{0.2.}$$

Now, CO_2 is an inert gas, here let b moles of CO_2 be added. Let the degree of dissociation now change to x. Thus, at equilibrium there is—

$(1 - x) + x + x + b = (1 + x + b)$ moles.

$$\therefore\ P_{CO_2} = \frac{b}{1+x+b}.P$$

$$\text{or, } 3.5 = \frac{b}{1+x+b}\cdot 7$$

$$\text{or, } \frac{1+x}{b} = 1, \text{ or, } (1 + x) = b.$$

$$\text{Now, } K_P = 0.2 = \frac{x^2 P}{(1-x)(1+x+b)} = \frac{x^2 P}{(1-x)(1+x+1+x)}$$

$$\text{or, } K_P = 0.2 = \frac{x^2 P}{2(1-x)(1+x)}$$

$$\text{or, } 0.4(1 - x^2) = x^2 P$$

$$\text{or, } x^2(P + 0.4) = 0.4$$

$$\text{or, } x = \sqrt{\frac{0.4}{P+0.4}}.$$

By the problem, $P = 7$ atm.

$$\therefore\ x = \sqrt{\frac{0.4}{7.4}} = 0.23.$$

16. For the reaction $H_2(g) + CO_2(g) = H_2O(g) + CO(g)$, the equilibrium constant at a particular temperature above 100 °C is found to be 1.2×10^{-2}. A mixture of 0.5 mole of H_2 and 1 mole of CO_2 was taken initially. Calculate the mole per cent of steam formed. Now a metallic oxide MO was introduced, and it was found that 20 mole per cent steam is formed. The metal oxide reacts as follows :

$$MO + H_2(g) = M + H_2O(g) \qquad \text{(i)}$$

$$\text{and } MO + CO(g) = M + CO_2(g) \qquad \text{(ii)}$$

Calculate the equilibrium constants K_1 and K_2 of the two reactions (i) and (ii).

Soln. : The reaction is :

$$H_2(g) + CO_2(g) = H_2O(g) + CO(g).$$

It is a reaction with $\Delta\nu = 0$. Hence, the total number of moles at equilibrium is 1.5. Let x moles of H_2O be formed. Then at equilibrium,

$$n_{H_2} = 0.5 - x\ ; \qquad n_{H_2O} = x$$

$$n_{CO_2} = 1 - x\ ; \qquad n_{CO} = x$$

$$\therefore\ K_P = \frac{x^2}{(0.5-x)(1-x)} = 1.2 \times 10^{-2}.$$

Solving, we get $x^2 = (0.5 - 1.5x + x^2)(1.2 \times 10^{-2})$

or, $0.988x^2 + 0.018x - 0.006 = 0.$

$$\text{or, } x = \frac{-0.018 + \sqrt{(0.018)^2 + (4 \times 0.988 \times 0.006)}}{2 \times 0.988}$$

$= 0.07$ (neglecting the negative root).

$$\therefore \quad \text{mole percent of steam formed} = \frac{0.07}{1.5} \times 100 = \mathbf{4.67}.$$

Now, the reaction (i) is :

$$MO(s) + H_2(g) = M(s) + H_2O(g).$$

Here also $\Delta\nu = 0$. Hence, the total number of moles $= 1.5$. But now the total mole per cent of steam is 20, i.e., there are 0.3 moles of steam. Thus,

$$n_{H_2} = 0.5 - 0.3 = 0.2$$

$$n_{H_2O} = 0.3$$

$$\therefore \quad K_1 = \frac{P_{H_2O}}{P_{H_2}} = \frac{0.3}{0.2} = \mathbf{1.5}.$$

The 2nd reaction (ii) is—

$$MO(s) + CO(g) = M(s) + CO_2(g)$$

$$K_2 = \frac{P_{CO_2}}{P_{CO}}.$$

But $$\frac{P_{CO} \times P_{H_2O}}{P_{H_2} \times P_{CO_2}} = 1.2 \times 10^{-2}$$

and $$\frac{P_{H_2O}}{P_{H_2}} = 1.5.$$

Thus, $$\frac{P_{CO}}{P_{CO_2}} \times 1.5 = 1.2 \times 10^{-2}.$$

$$\therefore \quad K_2 = \frac{P_{CO_2}}{P_{CO}} = \frac{1.5 \times 100}{1.2} = \mathbf{125.}$$

17. K_p for the reaction $H_2 + CO_2 = H_2O + CO$ is $\frac{9}{119}$ at 450 °C. 0.1 mole of H_2 and 0.2 mole of CO_2 are introduced in a flask at 450 °C—the equilibrium pressure is 1 atm. What will be the mole percentage of steam at equilibrium? A mixture of CoO and Co (both solid) is then introduced so that additional equilibrium :

$$CoO(s) + H_2(g) = Co(s) + H_2O(g) \quad (1)$$

$$CoO(s) + CO(g) = Co(s) + CO_2(g) \quad (2)$$

are established. The equilibrium mixture now contains 30 mole percentage of steam. Calculate K_1 and K_2.

Hints : See Problem 16. **[Ans.** 10%, 9, 119]

18. In the gaseous reaction $2A + B = A_2B$, $\Delta G° = -1200$ cal at 227 °C. What total pressure would be necessary to produce a 60% conversion of B into A_2B, when a mixture of A and B in the mole ratio 2 : 1 be used. **[C.U.(H) 1975]**

Soln. : The reaction is :

$2A + B = A_2B$ $\quad\quad$ $\Delta G° = -1200$ cal

At equilibrium, $\quad\quad$ at $T = 227$ °C + 273.15 = 500.15 K

$2A$ +	B	$\rightarrow A_2B$
$(2 - 2 \times 0.6)a$	$(1 - 0.6)a$	$0.6a$
$0.8\,a$	$0.4a$	$0.6a$

Total number of moles

$= 0.8a + 0.4a + 0.6a$
$= 1.8a$

$$P_{A_2B} = \frac{0.6a}{1.8a} P = \frac{1}{3}$$

$$P_A = \frac{0.8a}{1.8a} P = \frac{4}{9} P$$

$$P_B = \frac{0.4a}{1.8a} P = \frac{2}{9} P$$

where P = pressure required for 60% conversion of B into A_2B.

$$K_P = \frac{P_{A_2B}}{P_A^2 . P_B}$$

$$= \frac{\frac{1}{3}P}{\left(\frac{4}{9}P\right)^2 \times \frac{2}{9}P} = \frac{\frac{1}{3}\times\left(\frac{9}{4}\right)^2 \times \frac{9}{2}}{P^2}$$

$$= \frac{7.59375}{P^2}.$$

Now, $\Delta G^\circ = -RT \ln K_p$

or, $-1200 = -1.987 \times 500.15 \ln K_P$.

$$\therefore \ln K_P = \frac{1200}{1.987 \times 500.15} = 1.2075$$

$$K_P = 3.345$$

So, $\frac{7.59375}{P^2} = 3.345$

$$P^2 = \frac{7.59375}{3.345} = 2.2702$$

$P = 1.5067$ atm. $=$ **1.5 atm.**

19. What total pressure will be necessary for Problem 18, if in case, $\Delta G^\circ = +1200$ cal. **[C.U.(H) 1979]**

Soln. : The reaction is :

$2A + B = A_2B$. $\Delta G^\circ = 1200$ cal at 500.15 K

$$K_P = \frac{7.59375}{P^2}$$

Now,

$1200 = -1.987 \times 500.15 \ln K_P$.

$\therefore \ln K_P = -1.20749$

$K_P = 0.2989 \approx 0.30.$

$$\therefore P^2 = \frac{7.59375}{0.30} = 25.4056$$

$\therefore P = 5.04 \approx$ **5 atm.**

20. At 25 °C, ΔG° for the reaction $N_2O_4 = 2NO_2$ is + 1380 cal. What is the degree of dissociation (α) at 25 °C and total pressure of 10 atm? Calculate α', if the total pressure is 10 atm, but a partial pressure of CO_2 equal to 5 atm is present at equilibrium.

Soln. : The reaction is :

$N_2O_4 = 2NO_2$, $\Delta G^\circ = 1380$ cal at 25 °C.

At equilibrium, $N_2O_4 = 2NO_2$

$1 - \alpha \qquad 2\alpha$

Total number of moles present at equilibrium

$= 1 - \alpha + 2\alpha = 1 + \alpha$

$$\therefore P_{N_2O_4} = \frac{1-\alpha}{1+\alpha}P$$

$$P_{NO_2} = \frac{2\alpha}{1+\alpha}P$$

$$\therefore K_P = \frac{\left(\frac{2\alpha}{1+\alpha}P\right)^2}{\frac{1-\alpha}{1+\alpha}P} = \frac{4\alpha^2 P}{(1+\alpha)(1-\alpha)} = \frac{4\alpha^2 P}{1-\alpha^2}.$$

Now, $\Delta G^\circ = -RT \ln K_P$

or, $1380 = -1.987 \times 298.15 \ln K_P$.

$\therefore \ln K_P = -2.33$

$\therefore K_P = 0.097.$

So, $\dfrac{4\alpha^2 P}{1-\alpha^2} = 0.097$

or, $4\alpha^2 P = 0.097 - 0.097\alpha^2$

or, $40\alpha^2 + 0.097\alpha^2 = 0.097$ [$\because P = 10$ atm]

or, $40.097\alpha^2 = 0.097.$

$$\therefore \alpha = \sqrt{\frac{0.097}{40.097}} = 0.0492.$$

Now, in another condition P_{CO_2} is 5 atm, but the total pressure is 10 atm. So the pressure exerted by N_2O_4 and NO_2 is (10 atm – 5 atm) = 5 atm. Let α' be the degree of dissociation under this condition.

So now,

$$\frac{4\alpha'^2 P}{1-\alpha'^2} = 0.097, \text{ where } P = 5 \text{ atm}.$$

or, $20\alpha'^2 = 0.097 - 0.097\alpha'^2$

or $20.097\alpha'^2 = 0.097$

$\therefore \alpha' = \mathbf{0.0695.}$

21. K_P is 0.05 (pressure expressed in atm) at 20 °C for the reaction $NH_4HS(s) \rightleftharpoons NH_3(g) + H_2S(g)$. 0.06 mole solid NH_4HS is introduced into 2.4 litre flask at 20 °C.

(i) Calculate the percentage of solid that will get decomposed into NH_3 and H_2S at equilibrium.

(ii) Calculate the number of moles of NH_3 that would have to be added to the flask to reduce the decomposition of solid to 1%. (Assume all the gases to be ideal)

[C.U.(H) 2003]

Soln. : For the dissociation—

(i) $NH_4HS(s) \rightleftharpoons NH_3(g) + H_2S(g)$

$$K_P = P_{NH_3} \times P_{H_2S} = 0.05$$

If the total pressure is P atm,

$$P_{NH_3} = \frac{1}{2}P \text{ atm}$$

$$P_{H_2S} = \frac{1}{2}P \text{ atm.}$$

So, $\frac{P}{2} \times \frac{P}{2} = 0.05$

or, $P^2 = 0.05 \times 4 = 0.2$

$\therefore$ $P = 0.447$ atm.

Hence, pressure due to NH_3 or H_2S is $\frac{0.447}{2}$ atm = 0.2235 atm. Under the given condition, number of moles of NH_3 is $n = \frac{PV}{RT} = \frac{0.223 \times 2.4}{0.08206 \times 293.15} = 0.0222.$

So, the number of moles of NH_4HS that has got decomposed = 0.0222. Hence, percentage of decomposition = $\frac{0.0222}{0.06} \times 100 =$ **37%.**

(ii) Now, by adding NH_3 from outside, the decomposition of the solid has to be reduced to 1%.

Let the number of moles of H_2S be x.

So, $\frac{x \times 100}{0.06} = 1.$

$\therefore x = \frac{0.06}{100} = 0.0006$ moles.

Thus, $P_{H_2S} = \frac{nRT}{V}$

$$= \frac{0.0006 \times 0.08206 \times 293.15}{2.4} = 0.006 \text{ atm.}$$

Hence, $K_P = P_{NH_3} \times P_{H_2S}$

or, $0.05 = 0.006 \times P_{NH_3}$.

$\therefore P_{NH_3} = \frac{0.05}{0.006} = 8.333$ atm.

So, the pressure of NH_3 added from outside is (8.333 – 0.006) atm = 8.327 atm. Hence, the number of moles of NH_3 added from outside

$$= \frac{PV}{RT}$$

$$= \frac{8.327 \times 2.4}{0.08206 \times 293.15} = \mathbf{0.831}.$$

22. The equilibrium constant K_P for the reaction $H_2(g) + S(g) = H_2S(g)$ is 20.2 at 945 °C and 9.21 at 1065 °C. Calculate $\Delta H°$ assuming it to be temperature independent. **[C.U.(H) 2001]**

Soln. : By the given problem,

$K_{P_{1218 \cdot 15}} = 20.2$ at 945 °C

$K_{P_{1338 \cdot 15}} = 9.21$ at 1065 °C.

We know that, $\ln \dfrac{K_{P_2}}{K_{P_1}} = \dfrac{\Delta H^\circ}{R}\left(\dfrac{T_2 - T_1}{T_1 T_2}\right)$

or, $\ln\left(\dfrac{9.21}{20.2}\right) = \dfrac{\Delta H^\circ}{8.314\,\text{J.K}^{-1}.\text{mol}^{-1}}.\left[\dfrac{1338.15-1218.15}{1338.15\times 1218.15}\right]$

or, $\Delta H^\circ = \dfrac{8.314\times 1338.15\times 1218.15}{120}\times \ln\left(\dfrac{9.21}{20.2}\right)$

$= \mathbf{-88.7\ kJ.}$

23. For the reaction $NH_4HS(s) = NH_3(g) + H_2S(g)$, K_p is 0.06 $(atm)^2$ at 300 K. 0.08 mole of solid NH_4HS is introduced into a 2 lit flask at 300 K.

(a) Calculate the per cent of the solid that will have decomposed to NH_3 and H_2S at equilibrium.

(b) Calculate the number of moles of NH_3 that would have to be added to the flask to reduce the decomposition of solid to 2%.

(c) Having reached this state of equilibrium explain whether the addition of more $NH_4HS(s)$ would increase/decrease, if the NH_3 pressure is unchanged?

Hints : For parts (a) and (b) see Problem 21. [**Ans.** 24.9%, 0.246 moles]

For part (c) : No change. Addition of more solid does not affect its thermodynamic activity.

24. K_P has the value 10^{-9} atm^3 and 10^{-6} atm^3 at 25 °C and 50 °C respectively for the decomposition of $CuSO_4{,}3H_2O = CuSO_4 + 3H_2O(g)$.

(a) What is the minimum number of mole of H_2O vapour that must be introduced into a 3-lit flask at 25 °C in order to completely convert 0.001 mole of $CuSO_4$ to the trihydrate.

(b) Calculate ΔH° for the reaction.

Soln. : For the given reaction,

$K_P = P^3_{H_2O} = 10^{-9}$ at 25 °C.

Thus, equilibrium pressure of $H_2O = 10^{-3}$ atm.

Then the number of moles of H_2O, $n = \dfrac{PV}{RT}$

$= \dfrac{0.001\times 3}{0.08206\times 298.15} = 1.22618\times 10^{-4}$

$= \mathbf{1.23\times 10^{-4}}$ **mole.**

The total number of moles of H_2O needed

$= 3\times 0.001$ mole $+ 1.23\times 10^{-4}$ mole

$= 3.123\times 10^{-3}$.

By van't Hoff equation,

$\ln\dfrac{10^{-6}}{10^{-9}} = \dfrac{\Delta H^\circ}{R}\times\dfrac{25}{298.15\times 323.15}$

or, $\Delta H^\circ = \mathbf{221.334\ kJ.}$

25. For the reaction at 25 °C, $PCl_5(g) = PCl_3(g) + Cl_2(g)$, $\Delta H = 22$ kcal/mole. Calculate K_P at 25 °C and 600 °C, assuming ΔH to be independent of temperature. Given standard free energy of formation ΔG_f° (in kcal/mole) at 25 °C.

$Cl_2(g) = 0.0$, $PCl_3(g) = -68.42$

$PCl_5(g) = -77.6$ **[C.U.(H) 1987; V.U.(H) 1988]**

Soln. : From the given data,

$$\Delta G^\circ = -68.42 + 0 + 77.6 = 9.18 \text{ kcal.mole}^{-1}.$$

Therefore,

$$\ln K_P = -\frac{9180 \text{ cal. mole}^{-1}}{298.15 \text{ K} \times 1.987 \text{ cal. K}^{-1}.\text{mol}^{-1}} \quad [\because \ln K_P = -\frac{\Delta G^\circ}{RT}]$$

$$= -15.495657.$$

Thus, $K_{P_{298 \cdot 15}} = \mathbf{1.86 \times 10^{-7}}$.

Again, $$\ln \frac{K_{P_{873.15}}}{K_{P_{298.15}}} = \frac{\Delta H}{R} \times \left(\frac{575}{298.15 \times 873.15}\right)$$

$$= \frac{22000 \times 575}{1.987 \times 873.15 \times 298.15} = 24.455074$$

or, $$\ln \frac{K_{P_{873.15}}}{1.86 \times 10^{-7}} = 24.455074$$

or, $$K_{P_{873 \cdot 15}} = 1.86 \times 10^{-7} \times 4.17545 \times 10^{10}$$

$$= 7766.34 = \mathbf{7.766 \times 10^3}.$$

26. For the reaction $PCl_5(g) \rightleftharpoons PCl_3(g) + Cl_2(g)$, $\Delta H = 92$ kJ. mol^{-1} at 25 °C. Calculate K_P at 25 °C and 300 °C, assuming ΔH to be independent of temperature. Given standard free energy of formation ΔG_f° in kJ.mol^{-1} at 25 °C. $Cl_2(g) = 0.0$, $PCl_3(g) = -286.2$, $PCl_5(g) = -324.7$.

Hints : See Problem 25.

27. When N_2 and H_2 are mixed in 1 : 3 mole ratio, 15 gm of NH_3 is found in a mixture of 100 gm at equilibrium. The pressure is 44 atm and the temperature is 350 °C. Find an approximate value of K_P. **[B.U. 2003]**

Soln. : Let the reaction be

$$N_2 + 3H_2 = 2NH_3.$$

Then proceed as Problem 1, where $\alpha = 0.25$ but here $\alpha = 0.15$.

28. For a reaction $\Delta G^\circ = -4$ kJ.mol^{-1} at 301 K and $\Delta H^\circ = -50$ kJ. mol^{-1}. Estimate its K_P at 500 K.

Soln. : $\Delta G^\circ = -RT \ln K_P$

$\therefore \quad K_P = 4.945.$

Again, by the problem,

$$\ln \frac{K}{4.945} = \frac{-50000}{8.314} \times \frac{199}{500 \times 301}$$

Solving, $K = \mathbf{1.74 \times 10^{-3}}$.

29. At 1000 °K, the value of K_P for the reaction $2SO_2 + O_2 = 2SO_3(g)$ is 3.5, the pressure being expressed in atmospheres. Calculate the standard free energy change ($\Delta G°$) for the reaction at 1000 °K. Calculate K_C for the reaction from K_P at 1000 °K, and obtain the value of $\Delta G°$ from K_C. Clearly explain why the two values of $\Delta G°$ are different; although both represent standard free energy change for the same reaction at the same temperature. [C.U.(H) 1986]

Soln. : We know that—

$$\Delta G° = -RT \ln K_P$$
$$= -1.987 \text{ cal.K}^{-1}.\text{mol}^{-1} \times 1000 \text{ K} \ln 3.5$$
$$= -2.49 \text{ kcal.}$$

For any reaction,

$$K_P = K_C \cdot \left(\frac{C°}{P°} RT\right)^{\Delta\nu}$$

or, $$K_C = K_P \cdot \left(\frac{C°}{P°} RT\right)^{-\Delta\nu}$$

Since $\Delta\nu = -1$ for the given reaction,

$$\therefore \quad K_C = 3.5 \times 0.08206 \text{ l.atm.K}^{-1}.\text{mol}^{-1} \times 1000 \text{ K} \times \frac{1 \text{ mol.l}^{-1}}{1 \text{ atm}}$$

$$= \mathbf{287.21.}$$

Thus, $$\Delta G_C^\circ = -RT \ln K_C = -1.987 \text{ cal.K}^{-1}.\text{mol}^{-1} \times 1000 \text{ K} \times \ln 287.21$$

$$= \mathbf{-11.25 \text{ kcal.mol}^{-1}}.$$

$\Delta G°$ calculated from K_P is the free energy change when each gas is at a particular pressure of 1 atm; and $\Delta G°$ calculated from K_C is the free energy change when each gas is at a concentration 1 mole.lit^{-1}. Thus, the reference states in the two cases are different and they can be identical only when $\Delta\nu = 0$ so that $P = CRT$ becomes $P = C$ for the absence of RT terms as a whole.

30. K_P is 8×10^{-9} atm at 100 °C for the equilibrium $COCl_2(g) = CO(g) + Cl_2(g)$. $\Delta S°_{373K} = 30$ cal/deg.

(a) Calculate degree of dissociation of phosgene at 100 °C and 2 atm pressure.

(b) Calculate ΔH°_{373K} for the reaction.

(c) At what temperature phosgene will be 0.1% dissociated at 2 atm pressure? [C.U. 1998]

Soln. : (a) The reaction is :

$$COCl_2(g) = CO(g) + Cl_2(g)$$

Let us have initially 1 mole $COCl_2(g)$, and the degree of dissoiation is α. Thus, the number of moles at equilibrium are :

$$n_{COCl_2} = (1 - \alpha);\ n_{CO} = \alpha\ ;\ n_{Cl_2} = \alpha$$

Total number of moles = 1 + α.

$$P_{COCl_2} = \frac{1-\alpha}{1+\alpha}.P\ ;\ P_{CO} = \frac{\alpha}{1+\alpha}.P\ ;\ P_{Cl_2} = \frac{\alpha}{1+\alpha}.P$$

$$\therefore\ K_P = \frac{\alpha^2}{1-\alpha^2}.P$$

$$\text{or, } \alpha = \sqrt{\frac{K_P}{P}}\ [\because 1 >> \alpha, \text{ so neglecting } \alpha \text{ in comparison to } 1]$$

or, $\alpha = \mathbf{6.3 \times 10^{-5}}$.

(b) We know that

$$\Delta G° = -RT \ln K_P$$

$$\text{or, } \Delta G° = -1.987 \text{ cal.K}^{-1}.\text{mol}^{-1} \times 373.15 \text{ K} \ln(8 \times 10^{-9})$$

$$= 13.82 \text{ kcal.mol}^{-1}.$$

$$\therefore\ \Delta H^\circ_{373\,K} = \Delta G° + T\Delta S°$$

$$= 13820 + 373.15 \times 30$$

$$= 25.01 \text{ kcal.mol}^{-1} = \mathbf{104.66\ kJ.mol^{-1}}.$$

(c) By the given problem, $\alpha = 0.1\%$, i.e., $\alpha = 10^{-3}$.

$$\text{Thus, } K_P = \frac{\alpha^2 P}{1-\alpha^2} = \alpha^2 P = 2 \times 10^{-6}.$$

Now, we know that,

$$\ln\frac{K_2}{K_1} = \frac{\Delta H°}{R}\left(\frac{T_2 - T_1}{T_2 T_1}\right)$$

$$\text{or, } \ln\frac{2\times10^{-6}}{8\times10^{-9}} = \frac{25010}{1.987}\,\frac{T_2 - 373.15}{T_2\times373.15}.$$

(where T_2 = final temperature)

$\therefore\ T_2 = \mathbf{446.19\ °K}$.

31. The dissociation equilibrium constant (K_P) of $N_2O_4(g) = 2NO_2(g)$ at 25 °C and 1 atm is 0.141. Show that dissociation can be made spontaneous for the production of $NO_2(g)$ at 1 atm from $N_2O_4(g)$ at 10 atm at 25 °C. **[B.U. 1988; V.U. 2006]**

Soln. : We know that the condition of spontaneity for any reaction is that ΔG for that reaction should be negative. By the problem, the reaction is—

$$N_2O_4(g) \rightleftharpoons 2NO_2(g).$$

We know that—

$$\Delta G = \Delta G° + RT \ln \frac{P^2_{NO_2}}{P_{N_2O_4}}$$

$$\text{or, } \Delta G = -RT \ln K_P + RT \ln \frac{P^2_{NO_2}}{P_{N_2O_4}}$$

$$\text{or, } \Delta G = -1.987 \times 298.15 \ln 0.141 + 1.987 \times 298.15 \ln\frac{1}{10}$$

$$\text{or, } \Delta G = -203.55 \text{ cal.mol}^{-1}.$$

So, the reaction is spontaneous.

32. The equilibrium constant K_P for the dissociation $2H_2S(g) = 2H_2(g) + S_2(g)$ is 0.0118 at 1065 °C. Find the value of K_P at 1200 °C if heat of dissociation ΔH = 42400 cal in this temperature range. **[C.U. 1988, 1996]**

Hints : Proceed as earlier.

33. At 1000 °A the pressure of the iodine gas is found to be 0.112 atm. due to partial dissociation. Had there been no dissociation the pressure would have been 0.074 atm. Calculate K_P for the reaction, **[B.U. 1984]**

$$I_2 \rightleftharpoons 2I.$$

Soln. : By the problem, the reaction is—

$$I_2 \rightleftharpoons 2I.$$

Let there be 1 mole of I_2 initally and α be the degree of dissociation. Then at equilibrium the number of moles of $I_2 = (1 - \alpha)$ and $n_I = 2\alpha$. Total number of moles = $(1 + \alpha)$. If it is done at constant volume, then pressure ratio will be equal to the mole ratio, i.e.,

$$\frac{P_{\text{final}}}{P_{\text{initial}}} = \frac{(1+\alpha)}{1}.$$

By the problem,

$$P_{\text{final}} = 0.112$$

$$P_{\text{initial}} = 0.074.$$

$$\therefore \quad \frac{0.112}{0.074} = 1 + \alpha$$

$\therefore$ α = 0.51.

Now, we know that—

$$K_P = \frac{P_I^2}{P_{I_2}} = \frac{\left(\frac{2\alpha}{1-\alpha}P\right)^2}{\frac{1-\alpha}{1+\alpha}P} = \frac{4\alpha^2 P}{1-\alpha^2} = \frac{4\times(0.51)^2\times(0.112)}{1-(0.51)^2} = \mathbf{0.157\ atm.}$$

34. Equilibrium constants K_P for the reaction $\frac{1}{2}N_2(g) + \frac{3}{2}H_2(g) = NH_3(g)$ are 2.62×10^{-2} at 350 °C and 1.27×10^{-2} at 400 °C, pressure being expressed in atmosphere. Hence, calculate the enthalpy change, assuming it to be independent of temperature. Also calculate the standard free energy change at 400 °C. If the reaction be written as $N_2(g) + 3H_2(g) = 2NH_3(g)$, will the corresponding thermodynamic quantities will be different from the values you have calculated? **[C.U. 1972]**

Soln. : We know that—

$$\ln\frac{(K_P)_2}{(K_P)_1} = \frac{\Delta H^\circ}{R}.\frac{(T_2 - T_1)}{T_1 T_2}.$$

Here, $(K_P)_1 = 2.62 \times 10^{-2}$; T_1 = (350 + 273.15) K = 623.15 K

$(K_P)_2 = 1.27 \times 10^{-2}$; T_2 = (400 + 273.15) K = 673.15 K.

$$\ln\frac{(K_P)_2}{(K_P)_1} = \frac{\Delta H^\circ}{R}.\frac{(T_2 - T_1)}{T_1 T_2}.$$

Here, $(K_P)_1 = 2.62 \times 10^{-2}$; $T_1 = (350 + 273.15)\text{ K} = 623.15\text{ K}$

$(K_P)_2 = 1.27 \times 10^{-2}$; $T_2 = (400 + 273.15)\text{ K} = 673.15\text{ K}$.

$$\therefore \quad \Delta H^\circ = \frac{(K_P)_2}{(K_P)_1} \times \frac{R \times T_1 T_2}{T_2 - T_1}$$

$$= \ln\left(\frac{1.27 \times 10^{-2}}{2.62 \times 10^{-2}}\right) \times \frac{8.314\text{ J.K}^{-1}.\text{mol}^{-1} \times 623.15\text{ K} \times 673.15\text{ K}}{(673.15 - 623.15)\text{ K}}$$

$= -50510.00933\text{ J.mol}^{-1}$

$= \mathbf{-50.51\ kJ.mol^{-1}}$

Let the standard free energy change at 400 °C be ΔG°

$\Delta G^\circ = -RT\ln(K_P)_2$

$= -8.314 \times 673.15 \times \ln(1.27 \times 10^{-2})$

$= 24435.47856\text{ J.} = \mathbf{24.44\ kJ.}$

Equilibrium constant depends on how we write the equation. If the equation is written as $N_2(g) + 3H_2(g) = 2NH_3(g)$, K_P will be the square of the previous K_P and ΔG° will be twice the previous one and ΔH° will also be twice.

35. Starting with equimolecular amount of CO and H_2O, calculate the molar percentage of the water gas system ($H_2 + CO_2 = H_2O + CO$) at equilibrium at 600 °C. K at this temperature is 0.16.

Soln. : $H_2O + CO = H_2 + CO_2$

$(a - ax) \quad (a - ax) \quad ax \quad ax$

Let $(a - ax)$ moles of CO and H_2O be available initially.

Then after reaction,

the moles of $CO_2 = ax$

and the moles of $H_2 = ax$.

Total number of moles $= 2a(1 - x) + 2ax = 2a$.

$$\therefore \ K_P = \frac{P_{H_2} P_{CO_2}}{P_{H_2O} \cdot P_{CO}} = \frac{1}{0.16}.$$

$$\therefore \quad \frac{\frac{ax}{2a} \cdot \frac{ax}{2a}}{\frac{(a - ax)}{2a} \cdot \frac{(a - ax)}{2a}} = \frac{1}{0.16}$$

$$\text{or,} \quad \left(\frac{x}{1-x}\right)^2 = \frac{1}{0.16}$$

$$\text{or,} \quad \frac{1-x}{x} = 0.4$$

or, $1 - x = 0.4x$

or, $1.4x = 1$

$$\text{or,} \quad x = \frac{1}{1.4} = 0.7143.$$

$\therefore$ molar percentage of reactants, i.e.,

each of H_2O and CO is $= \dfrac{a(1-x)}{2a} \times 100 = \dfrac{1-0.7143}{2} \times 100$

$= 14.285 = \mathbf{14.3.}$

And molar percentage of products, i.e., each of H_2 and CO_2 is $= \dfrac{ax}{2a}$

$= \dfrac{0.7143}{2} \times 100 = \mathbf{35.7.}$

36. At high temperature CO_2 dissociates according to the equation $2CO_2(g) = 2CO(g) + O_2(g)$. At one atmosphere pressure the fraction of CO_2 dissociated is 2×10^{-5} at 1000 K. At 1400 K it is 1.27×10^{-2}. Assuming that the enthalpy change ($\Delta H°$) of the reaction is independent of temperature, calculate $\Delta G°$ and $\Delta S°$ at 1000 K. **[B.U. 1980]**

Soln. : $2CO_2(g) = 2CO(g) + O_2(g)$

$2(1-\alpha) \quad 2\alpha \quad \alpha$

Let α be the degree of dissociation.

$$\therefore K_P = \frac{P^2_{CO} \cdot P_{O_2}}{P^2_{CO_2}}$$

$$= \frac{\left(\frac{2\alpha}{2+\alpha}\right)^2 . P^2 \frac{\alpha}{2+\alpha} . P}{\left(\frac{2(1-\alpha)}{2+\alpha}\right)^2 P^2} \quad [P \text{ is the total pressure } = 1 \text{ atm}]$$

$$= \frac{(2\alpha)^2 \alpha}{[2(1-\alpha)]^2 (2+\alpha)} = \frac{\alpha^3}{2} \quad [\alpha << 1].$$

$\therefore \alpha^3 = 2 \times K_P$.

At 1000 K $\alpha = 2 \times 10^{-5}$.

$$\therefore (K_P)_{1000} = \frac{(2 \times 10^{-5})^3}{2} = 4 \times 10^{-15}.$$

$$(K_P)_{1400} = \frac{(1.27 \times 10^{-2})^3}{2} = 1.0242 \times 10^{-6}.$$

Now by van't Hoff equation,

$$\ln \frac{(K_P)_{1400}}{(K_P)_{1000}} = \frac{\Delta H°}{R}\left(\frac{T_2 - T_1}{T_1 T_2}\right).$$

$$\therefore \ln \frac{1.0242 \times 10^{-6}}{4 \times 10^{-15}} = \frac{\Delta H°}{8.314} \times \left(\frac{1400-1000}{1400 \times 1000}\right)$$

$$\text{or, } \Delta H° = \ln\left(\frac{1.0242 \times 10^{-6}}{4 \times 10^{-15}}\right) \times \frac{8.314 \times 1400 \times 1000}{400}$$

$= 563382.343$ joules $= 563.4$ kJ.

Again, $\Delta G^\circ = -RT \ln K_P$

$$(\Delta G^\circ)_{1000\text{ K}} = -8.314 \text{ J.K}^{-1}.\text{mol}^{-1} \times 1000 \text{ K} \times \ln(4 \times 10^{-15})$$

$$= 275629.7356 \text{ joules} = 275.63 \text{ kJ}$$

$\therefore\ \Delta G^\circ = \Delta H^\circ - T.\Delta S^\circ$

$$\therefore\ \Delta S^\circ = \frac{\Delta H^\circ - \Delta G^\circ}{T} \doteq \left(\frac{563.4 - 275.63}{1000}\right) \times 10^3 = \mathbf{287.77\ J.K^{-1}}.$$

37. For the gaseous reaction—

$C_2H_2 + H_2O = C_2D_2 + H_2O$; ΔH is 530 cal. At 25 °C K_P is 0.82. Calculate how much C_2D_2 will be formed if 1 mole of C_2H_2 and 2 moles of D_2O are put together at a total pressure of 1 atmosphere and at 100 °C. **[C.U.(H) 1989; V.U. 2000]**

Soln. : From van't Hoff equation,

$$\ln\frac{(K_P)_2}{(K_P)_1} = \frac{\Delta H}{R}\frac{T_2 - T_1}{T_1 T_2}.$$

Here, $(K_P)_1 = 0.82$

$T_1 = (273.15 + 25)\text{K} = 298.15 \text{ K}$

$T_2 = (273.15 + 100)\text{K} = 373.15 \text{ K}$

$\Delta\text{H} = (530 \times 4.184)$ joule.

Let $(K_P)_2$ be equilibrium constant at 100 °C.

$$\therefore\ \ln\frac{(K_P)_2}{(K_P)_1} = \frac{\Delta H}{R}\frac{T_2 - T_1}{T_1 T_2}$$

$$\text{or, } \ln\frac{(K_P)_2}{(K_P)_1} = \frac{530 \times 4.184 \times (373.15 - 298.15)}{8.314 \times 373.15 \times 298.15}.$$

$$\therefore\ \ln\frac{(K_P)_2}{(K_P)_1} = 0.1798$$

$$\text{or, } (K_P)_2 = (K_P)_1 \times e^{0.1798}$$

$$= 0.82 \times e^{0.1798} = 0.982.$$

$$\text{Now, } \quad \underset{(1-x)}{C_2H_2} + \underset{(2-x)}{D_2O} = \underset{x}{C_2D_2} + \underset{x}{H_2O}$$

Let x mole of C_2D_2 and x mole of H_2O be produced.

Then at equilibrium moles of $C_2H_2 = 1 - x$

and moles of $D_2O = 2 - x$.

Total number of moles = 3.

$$\therefore\ K_P = \frac{P_{C_2D_2} \cdot P_{H_2O}}{P_{C_2H_2} \cdot P_{D_2O}} = \frac{\frac{x}{3}P.\frac{x}{3}P}{\frac{1-x}{3}P.\frac{2-x}{3}P}$$ [Let P be the total pressure.]

$$= \frac{x^2}{(1-x)(2-x)}.$$

$\therefore \quad x^2 = K_P(2 - 3x + x^2)$

or, $x^2 - x^2.K_P + 3x\,K_P - 2K_P = 0$

or, $(1 - K_P)\,x^2 + 3x.K_P - 2K_P = 0$

or, $x = \dfrac{-3K_P \pm \sqrt{9K_P^2 + 4(1-K_P).2K_P}}{2(1-K_P)}$

$= 0.66$ [neglecting (–)ve value] (putting $K_P = 0.982$)

So, at 100 °C **0.66 moles** C_2D_2 will be produced.

38. Calculate the partial pressure of HCl gas at 27 °C above a sample of $NH_4Cl(s)$ as a result of its decomposition. Given $\Delta G°$ values of $NH_4Cl(s)$, $NH_3(g)$ and HCl(g) are –203 $kJmol^{-1}$, –16.48 $kJmol^{-1}$ and –95.3 $k.J.mol^{-1}$ respectively at 27 °C. **[V.U. 2003]**

Soln. : $NH_4Cl(s) = NH_3(g) + HCl(g)$

By the problem,

$\Delta G°_{reaction} = [(-16.48 - 95.3) - (-203)]\ kJ.mol^{-1}$

$= 91.22\ kJ.mol^{-1} = 91220\ J.mol^{-1}$.

$\therefore \quad \Delta G° = -RT \ln K_P$

or, $\ln K_P = -\dfrac{\Delta G°}{RT} = -\dfrac{91220\,J.mol^{-1}}{8.314\,J.K^{-1}.mole^{-1} \times 300.15\,K}$

or, $K_P = 1.3 \times 10^{-16}$.

But, $K_P = P_{NH_3}.P_{HCl}$

$= P^2_{HCl}$ (since, $P_{HCl} = P_{NH_3}$).

$\therefore \quad P_{HCl} = \sqrt{K_P}$

$= 1.14 \times 10^{-8}$ atm $= 8.66 \times 10^{-6}$ mm of Hg

$= \mathbf{1.16 \times 10^{-3}}$ **Pa.**

39. For the reaction $H_2(g) + I_2(g) \rightleftharpoons 2HI(g)$, $K_P = 50.0$ at 448 °C and 66.9 at 350 °C. Calculate $\Delta H°$ for the reaction. **[V.U. 2005]**

Soln. : Proceed as earlier. **[Ans. –11.1 kJ]**

40. At 2155 °C and 1 atm pressure H_2O (g) is 1.18% decomposed into H_2(g) and O_2(g) in accordance with the equation $2H_2O(g) = 2H_2(g) + O_2(g)$.

Calculate K_P for the process. **[C.U.(H) 1997]**

Hints : Here α is 0.0118.

K_P of the reaction is, $K_P = \dfrac{4\alpha^3}{(2+\alpha)(2-2\alpha)^2}$.

Putting the value of α,

we get $K_P =$ **8.36 × 10⁻⁷ atm.**

41. The equilibrium constant of the reaction,

$A_2(g) + B_2(g) \rightleftharpoons 2AB(g)$ at 100 °C is 50. If a one-litre flask containing 1 mole of A_2 is connected to a two-litre flask containing two moles of B_2, how many moles of *AB* will be formed at the said temperature? **[B.U. 2001]**

Hints : $A_2(g) + B_2(g) \rightleftharpoons 2AB(g)$

$(1-\alpha) \quad (2-\alpha) \quad 2\alpha$

Total volume = 3 litres.

$$\therefore K_C = \frac{\left(\frac{2\alpha}{3}\right)^2}{\frac{1-\alpha}{3} \cdot \frac{2-\alpha}{3}} = 50$$

$\therefore \alpha = 0.934$

$\therefore 2\alpha = (2 \times 0.934) = \mathbf{1.868.}$

42. Consider the gas phase reaction :

$A + 2B \rightarrow 4X + Y$

It was found that when 2.0 mol of A, 1.0 mol of B and 1.0 moles of Y were mixed and allowed to come to equilibrium, the resulting mixture contained 0.80 moles of X at 25 °C. Calculate the equilibrium constant of the reaction at 1.0 bar and 25 °C.

[I.I.T. Bombay 2001]

Soln. : $A + 2B \rightleftharpoons 4X + Y$

$(2-\alpha) \quad (1-2\alpha) \quad 4\alpha \quad (1+\alpha)$

Let the degree of dissociation be α.

Since 0.80 moles of X is produced,

$\therefore \quad 4\alpha = 0.8$

or, $\alpha = 0.2$.

Now, moles of $A = (2 - \alpha) = (2 - 0.2) = 1.8$

moles of $B = (1 - 2\alpha) = (1 - 0.4) = 0.6$

moles of $X = 4\ \alpha = 0.8$

moles of $Y = 1 + X = 1.2$.

Total number of moles $= 4 + 2\alpha = 4.4$.

$$\therefore P_A = \frac{1.8}{4.4} \times P \qquad P_B = \frac{0.6}{4.4} \times P$$

$$P_X = \frac{1.8}{4.4} \times P \qquad P_Y = \frac{1.2}{4.4} \times P.$$

$$\therefore K_P = \frac{P_x^4 . P_y}{P_A . P_B^2} = \frac{\left(\frac{0.8}{4.4}P\right)^4 \times \left(\frac{1.2}{4.4}P\right)}{\left(\frac{1.8}{4.4}P\right) \times \left(\frac{0.6}{4.4}P\right)^2}$$

$$= \frac{(0.8)^4 \times 1.2 \times P^2}{(1.8) \times (0.6)^2 \times (4.4)^2} = \mathbf{3.92 \times 10^{-2}}.$$

43. For the reaction, $PCl_5(g) = PCl_3(g) + Cl_2(g)$:

(A) Calculate the equilibrium constant using the following data at 25 °C :

Substance	$\Delta H°$ (kJ.mole^{-1})	$S°$(J.K^{-1}.mole^{-1})
$PCl_3(g)$	–287	312

$PCl_5(g)$	–375	365
$Cl_2(g)$	0	223

(B) Using the result from above, calculate the equilibrium constant of $2PCl_5(g) = 2PCl_3(g) + 2Cl_2(g)$. **[Joint M.Sc. Entrance 2003]**

Soln. : (A) We know that $\Delta G° = \Delta H° - T\Delta S°$.

By the problem,

$$\Delta G°_{PCl_3} = (-287000 - 298.15 \times 312) \text{ J.mole}^{-1} = -380022.8 \text{ J.mol}^{-1}$$

$$\Delta G°_{PCl_5} = (-375000 - 298.15 \times 365) \text{ J.mole}^{-1}$$

$$= -483824.75 \text{ J.mole}^{-1}$$

$$\Delta G°_{Cl_2} = (0 - 298.15 \times 223) \text{ J.mole}^{-1} = -66487.45 \text{ J.mole}^{-1}.$$

$$\therefore \quad \Delta G°_{reaction} = \Delta G°_{PCl_3} + \Delta G°_{Cl_2} - \Delta G°_{PCl_5}$$

$$\therefore \quad \Delta G°_{reaction} = 37314.5 \text{ J.mole}^{-1}.$$

Now, $\Delta G° = -RT \ln K_P$

$$\ln K_P = -\frac{\Delta G°}{RT} = -\ \mathbf{15.0533.}$$

$$\therefore \quad K_P = \mathbf{2.9 \times 10^{-7}.}$$

(B) $2PCl_5(g) = 2PCl_3(g) + 2Cl_2(g)$

$$K'_P = \frac{P^2_{PCl_3} \cdot P^2_{Cl_2}}{P^2_{PCl_5}} = (K_P)^2$$

$$= (2.9 \times 10^{-7})^2 = \mathbf{8.41 \times 10^{-14}.}$$

44. The equilibrium constant of the reaction C (graphite) + $2H_2(g) \rightleftharpoons CH_4(g)$ is 3.036×10^8 at 35 °C. The standard enthalpy of the reaction is 749 kJ.mol^{-1} in the temperature range 25 °C – 35 °C. Calculate the standard free energy change for the reaction at 25 °C.

[I.I.T. Joint Admission Test 2002]

Hints : $\ln\dfrac{(K_P)_2}{(K_P)_1} = \dfrac{\Delta H}{R}\dfrac{T_2 - T_1}{T_1 T_2}$

Here, $(K_P)_2 = 3.036 \times 10^8$

$\Delta H° = 749000 \text{ J.mol}^{-1}$

$T_2 = 35\ °C = 308.15$ K

$T_1 = 25\ °C = 298.15$ K.

From calculation, $(K_P)_1 = 1.6741 \times 10^4$.

Now, $\Delta G°_{25°C} = -RT \ln (K_P)_1$

$= -24108.04 \text{ J.mol}^{-1}$

$= \mathbf{-241.1\ kJ.mol^{-1}}$.

45. Write down the expression for the equilibrium constant of the reaction $N_2(g) + 3H_2(g) = 2NH_3(g)$, $\Delta G° = -33.4$ kJ.mol^{-1}. What is the value of K_P at 300 °K.

[I.I.T. M.Sc. Entrance Bombay 1998]

Soln. : Proceed as usual.

46. At 2000 K the standard free energy change in calories for the reaction $N_2 + O_2 \rightleftharpoons 2NO$ is given by $\Delta G^\circ = 22000 - 2.5T$. Calculate K_P at this temperature. **[C.U.(H) 2005]**

Hints : $\Delta G^\circ = (22000 - 2.5 \times 2000)$ cal $= 17000$ cal.

And $\Delta G^\circ = -RT \ln K_P$.

$\therefore \ln K_P = -\frac{\Delta G^\circ}{RT}$, $\therefore K_P = e^{\frac{-\Delta G^\circ}{RT}} = \mathbf{1.39 \times 10^{-2}}$.

47. If we like to extract at a time 51% of the iodine present in 100 ml of an aqueous solution of the same, what volume of CCl_4 is needed? Given that at the experimental temperature the distribution coefficient of I_2 between CCl_4 and H_2O is 85. **[C.U. 2006]**

Soln. : Let x gm be the weight of I_2 present in water. Let V cc be the volume of CCl_4 required to extract 51% of the I_2 at a time.

So after extraction,

I_2 present in $H_2O = (x - 0.51x)\text{gm} = 0.49x$ gm
and I_2 present in $CCl_4 = 0.51x$ gm.

Thus, concentration of I_2 in H_2O in gm-mole/litre $= C_{aq} = \dfrac{0.49x \times 1000}{100 \times M_{I_2}}$

and concentration of I_2 in CCl_4 in gm.mole/litre $= C_{org} = \dfrac{0.51x \times 1000}{V \times M_{I_2}}$

Now, $\dfrac{C_{org}}{C_{aq}} = 85.$

Hence, $\dfrac{0.51x \times 100}{V \times 0.49x} = 85.$

$\therefore V =$ **1.224 ml.**

48. The solubility of a solute is three times as high in ether as in water. Compare the amounts extracted from 100 ml of the aqueous solution by (i) 100 ml of ether in one step (ii) successive extractions each with 50 ml of ether. **[V.U. 1999]**

Soln. : Let x gm be the weight of solute in 100 ml aqueous solution. Let y gm be extracted by 100 ml of ether.

(i) So, after extraction, amount of solute in water $= (x - y)$ gm.

By the problem, $\dfrac{C_{or}}{C_{aq}} = 3.$

Now, $C_{or} = \dfrac{1000y}{100 \times M_{ether}}$ gm.mol/litre

$C_{aq} = \dfrac{1000(x-y)}{100 \times M_{ether}}$ gm.mol/litre.

$\therefore \dfrac{y}{x-y} = 3$ or, $y = \dfrac{3}{4}x = 0.75x.$

Thus, **75%** is extracted.

(ii) Let z gm be extracted by 50 ml of water.

Hence, $\dfrac{z \times 100}{50 \times (x - z)} = 3.$

$\therefore z = \dfrac{3}{5} x$ or 60% is extracted and 40% remains unextracted. Second time 50 cc will again extract 60% of the remaining unextracted solute, i.e., 24% will be extracted. So total (60 + 24)% = 84% is extracted.

Thus, extractions in several steps are much more efficient than a single step.

49. Distribution coefficient of a solute between CCl_4 and water is 25. The solute is extracted from 250 cc aqueous solution by successive extractions. Calculate the number of extractions needed for 99% recover with (i) 25 cc (ii) 50 cc CCl_4.

Hints : (i) 4 times, (ii) 3 times.

50. For a certain reaction, $\Delta G° = -4$ kJ/mole and $\Delta H° = -57$ kJ/mole at 298 K. Calculate K_P for the reaction at 500 K assuming $\Delta H°$ remains constant over the given range of temperature. **[C.U. 2006]**

Hints : $K = 1.48 \times 10^{-4}$.

51. The value of K_P for the reaction $N_2 + 3H_2 = 2NH_3$ at 400 °C is 1.6×10^{-4}. Calculate the value of the standard free energy change $\Delta G°$ with respect to K_P and K_C and interpret the value of $\Delta G°$ in two cases, if these are different. **[V.U. 2007]**

Hints : $K_P = K_C \left(\dfrac{C°}{P°} RT\right)^{-2}$, $K_C = 0.488$

$\Delta G° = -RT \ln K_P = 48.916$ kJ

$\Delta G° = -RT \ln K_C = 4.015$ kJ .

$\Delta G°$ are different in two cases as the standard states are different.

52. Calculate the percentage dissociation of N_2O_4 for the reaction $N_2O_4(g) \rightleftharpoons 2NO_2(g)$, if the equilibrium constant is 640 mm of Hg at 500 °C and the gas mixture is at 160 mm of Hg. At what pressure will the dissociation be 50%?

Hints : $N_2O_4(g) \rightleftharpoons 2NO_2(g)$

$\quad 1-\alpha \qquad\qquad 2\alpha$

where α = degree of dissociation

$$K_P = \frac{4\alpha^2 P}{(1+\alpha)(1-\alpha)} = \frac{4\alpha^2 P}{1-\alpha^2}.$$

Putting the corresponding values of K_P and P, we get $\alpha = 0.707$, or 71% N_2O_4 will dissoicate. When $\alpha = 0.5$, $P = 480$ mm of Hg.

53. At what temperature K_P and K_C of the following reaction are same?

$A(g) = B(g) + C(g)$

Soln. : $K_P = K_C \left(RT \dfrac{C°}{P°} \right)^{\Delta\nu} = K_C \left(RT \dfrac{C°}{P°} \right)$ $[\Delta\nu = 1]$

When $\quad RT \dfrac{C°}{P°} = 1$, $K_P = K_C$,

$$T = \frac{P^\circ}{C^\circ \times R} = \frac{1\,\text{atm}}{1\,\text{gm. mol. l}^{-1} \times 0.08206\,\text{l. atm. K}^{-1}.\text{mol}^{-1}} = 12.19\ \text{K.}$$

At 12.19 K, $K_P = K_C$ for the given reaction.

54. An equilibrium mixture of : $Xe\,F_2(g) + OF_2(g) = Xe\,OF_2(g) + F_2(g)$ was found to contain 0.3 mole of $Xe\,F_2$, 0.3 mole of OF_2, 0.1 mole of $Xe\,OF_2$ and 0.2 mole of F_2 in one-litre container. How many moles of OF_2 must be added to increase [$Xe\,OF_2$] to 0.3 mole?

Soln. : $K = \dfrac{0.1 \times 0.2}{0.3 \times 0.3} = \dfrac{2}{9}$.

Let x moles of OF_2 be added to increase [$XeOF_2$] to 0.3 mole.

$$\begin{array}{cccc} XeF_2 \;+ & OF_2 & \rightleftharpoons XeOF_2 \;+ & F_2 \\ 0.3 - 0.2 & 0.3 + x - 0.2 & 0.1 + 0.2 & 0.2 + 0.2 \end{array}$$

$$\frac{[0.3][0.4]}{[0.1 + x][0.1]} = \frac{2}{9}$$

$9 \times 0.12 = 2[0.01 + 0.1x] = 0.02 + 0.2x$. Thus, $x = 5.3$.

55. 98% of 2 gm NH_4Cl dissociated while heating in a one-litre flask at 300 °C. If to this flask 2.0 gm. of dry NH_3 was added, what would be the percentage of dissociation?

Hints :

$$\begin{array}{lll} NH_4Cl(s) \rightleftharpoons & NH_3(g) \quad + & HCl(g) \\ 0.037 - 0.98 \times 0.037 & 0.98 \times 0.037 & 0.98 \times 0.037 \\ 0.037 - \alpha \times 0.0377 & 0.118 + \alpha \times 0.037 & 0.037\alpha \end{array}$$

$$(0.98 \times 0.037)^2 = (0.118 + 0.037\alpha)\,0.037\alpha$$

Thus, new percentage of dissociation = **27.3%.**

56. K_P is 1×10^{-5} for the equilibrium.

$CO_2 + H_2 = CO + H_2O(g)$

At 298 °K and $\Delta S° = -10$ cal/deg, 1 mole of CO, 2 moles of H_2 and 4 moles of CO_2 are introduced in a 5-litre flask at 298 °K. Calculate (a) $\Delta G°$ and $\Delta H°$(b) equilibrium pressure (b) moles of each species present at equilibrium.

Hints : $\Delta G° = -RT \ln K_P$, $\Delta H° = \Delta G° + T\Delta S°$

$$\begin{array}{cccc} CO_2 + & H_2 = & CO + & H_2O \\ 4-x & 2-x & 1+x & x \end{array} \qquad K_P = K_x \text{ Since } \Delta\nu = 0$$

$$10^{-5} = \frac{(1+x)x}{(2-x)(4-x)}, \qquad x = 7.99 \times 10^{-5}$$

$$\begin{array}{lcccc} \text{At equilibrium,} & CO_2 \;+ & H_2 \;= & CO \;+ & H_2O \\ & 4 \text{ moles} & 2 \text{ moles} & 1 \text{ mole} & 7.99 \times 10^{-5} \text{ moles} \end{array}$$

$$\text{Equilibrium } P = \frac{nRT}{V} = \frac{7 \times 0.08206 \times 298}{5} = \mathbf{34.24\ atm.}$$

57. At 929 K the total gas pressure is 0.98 atm for the reaction.

$2\,FeSO_4(s) = Fe_2O_3(s) + SO_2(g) + SO_3(g)$

(a) Calculate K_P.

(b) What will be the equilibrium total pressure if excess $FeSO_4$ is placed in a flask at 929 K containing SO_2 at 0.4 atm

Hints : (a) $K_P = P_{SO_2} . P_{SO_3}$

$$P_{SO_2} = P_{SO_3} = \frac{0.98}{2} \text{ atm} = 0.49 \text{ atm.}$$

Thus, $K_P = 0.49^2 = \mathbf{0.2401\ atm^2}$.

(b) $(0.4 + P_{SO_3})P_{SO_3} = 0.2401$

$$P_{SO_3} = 0.329 \text{ atm}$$

$$P_{SO_2} = 0.729 \text{ atm}$$

$$P_{eq} = 1.058 \text{ atm.}$$

58. $\Delta H°$ is 30 kcal for the reaction $AgCl = Ag + \frac{1}{2}Cl_2$ at 25 °C. Standard free energy for formation of AgCl is –26 kcal at 25 °C.

(a) What is ΔG°_{298} for the above reaction?

(b) Calculate dissociation pressure of Cl_2.

(c) What is the value of $\frac{d(T \ln K_P)}{dT}$?

Hints : $\Delta G^\circ_{298} = -\Delta G^\circ_{f,AgCl} = 26$ kcal

$\Delta G° = -RT \ln K_P$, $K_P = \sqrt{P_{Cl_2}}$, $P_{Cl_2} = 7.58 \times 10^{-39}$atm

$$d(T \ln K_P) / dT = -\frac{1}{R}\frac{d\Delta G°}{dT} = \frac{\Delta S°}{R}$$

$$\Delta S° = \frac{\Delta H° - \Delta G°}{T} = 13.42 \text{ cal/deg}$$

$$\frac{\Delta S°}{R} = 6.75.$$

59. Calculate K_P for the reaction $2NOCl = 2NO + Cl_2$, if at equilibrium the total pressure is 1 atm and partial pressure of NOCl = 0.7 atm at 200 °C. Calculate $\Delta H°$, if K_P increases by 1.5% per degree around 200 °C. Also calculate the pressure at which degree of dissociation of NOCl will be 0.4 assuming $K_P = 0.1$ atm at 200 °C.

Hints : (a) $P_{NO} + P_{Cl_2} = 1 - 0.7 = 0.3$

$$P_{NO} = 0{\cdot}3 \times \frac{2}{3} = 0.2,\ P_{Cl_2} = 0.1$$

$$K_P = \frac{P^2_{NO} \cdot P_{Cl_2}}{P^2_{NOCl}}$$

$$= 8.16 \times 10^{-3}$$

(b) $\frac{d\ln K_P}{dT} = \frac{dK_P}{K_P dT} = \frac{\Delta H^\circ}{RT^2}$

$$\therefore \Delta H^\circ = \frac{RT^2\left(\frac{dK_P}{K_P}\right)}{dT} = 8.314 \text{ J.K}^{-1} \times 473.15^2 \text{ K}^2 \times 0.015 \text{ K}^{-1} = \mathbf{27.92\ kJ.}$$

When degree of dissociation of NOCl is 0.4, then

$$\therefore K_P = \frac{\alpha^3 P}{2(1-\alpha)^2\left(1+\frac{\alpha}{2}\right)} = 0.1, \quad P = \mathbf{1.35\ atm.}$$

60. N_2O_4 is 19% dissociated at a total pressure of 1 bar and at 298 K. (a) Will the addition of N_2 at constant volume affect the equilibrium? (b) If N_2 is added at a constant total pressure of 1 bar, what will be the degree of dissociation when partial pressure of N_2 is 0.6 bar?

Hints : (a) No, since, $K_P = \frac{n_{NO_2}^2}{n_{N_2O_4}}\left(\frac{P}{\sum_i n_i}\right)$, here $\Delta\nu$ is +1. Thus, addition of N_2 at constant volume increases total number of moles $\left(\sum_i n_i\right)$, but at the same time P also increases so that $\frac{RT}{V}$ remains constant. ($PV = \Sigma n_i RT$).

(b) $K_P = \frac{\alpha^2 P}{1-\alpha^2}$, here P = 0.4 bar, α = 0.292.

61. A reaction vessel of one litre containing 0.23 mol of N_2 and 0.34 of PCl_3 is heated to 523 K. Calculate K_P of the reaction $PCl_5(g) = PCl_3(g) + Cl_2(g)$, if the equilibrium total pressure is 29 bar.

Hints : $PCl_5(g) = PCl_3(g) + Cl_2(g)$

$0.34 - x \quad x \quad x$

Total number of moles = n = 0.34 + x + 0.23 = 0.57 + x

$$29 = \frac{nRT}{V}, \therefore x = 0.097 \text{ (here } R = 0.08206 \times 1.01325 \text{ bar.l.K}^{-1}\text{.mol}^{-1})$$

$$P_{N_2} = 10.00 \text{ bar}, P_{PCl_5} = 10.57 \text{ bar}, P_{Cl_2} = P_{PCl_3} = 4.22 \text{ bar.}$$

$$K_P = \frac{P_{Cl_2} \times P_{PCl_3}}{P_{PCl_5}} = \mathbf{1.68.}$$

62. Calculate K_P and ΔG° for the following reaction at 20 °C,

$CuSO_4, 4NH_3$(crystal) = $CuSO_4, 2NH_3$(crystal) + $2NH_3$(g).

The equilibrium pressure of NH_3 is 8.26 kPa.

Hints : $K_P = 6.65 \times 10^{-3}$, $\Delta G^\circ = 12.22$ kJ.mol^{-1}.

$(1.01325 \times 10^5 \text{ Pa} = 1 \text{ atm})$

CHAPTER 9

CHEMICAL KINETICS

Required Formulae

1. For an elementary reaction,
$\nu_1 A + \nu_2 B = \nu_3 C + \nu_4 D,$

the rate equation is $-\frac{1}{\nu_1}\frac{dC_A}{dt} = -\frac{1}{\nu_2}\frac{dC_B}{dt} = \frac{1}{\nu_3}\frac{dC_C}{dt} = \frac{1}{\nu_4}\frac{dC_D}{dt} = k\,C_A^{\nu_1}C_B^{\nu_2}.$

The order is $\nu_1 + \nu_2$.

2. Dimension of rate constant = $(\text{conc.})^{1-n}T^{-1}$.
Unit of k = $(\text{mole. l}^{-1})^{1-n}s^{-1}$,
n is the order of the reaction.

3. Rate equations of reactions of simple order.

1st order : $k = \frac{1}{t}\ln\frac{a}{a-x}$

a = initial concentration of reactants

x = amount spent at time t

$$t_{1/2} = \frac{0.693}{k}.$$

2nd order : $k = \frac{1}{t}\frac{x}{a(a-x)}$

$$t_{1/2} = \frac{1}{ka}, \text{ when } a = b.$$

n-th order : $k = \frac{1}{(n-1)t}\left[\frac{1}{(a-x)^{n-1}} - \frac{1}{a^{n-1}}\right]$

$$t_{1/2} = \frac{1}{(n-1)k}\left[\frac{2^{n-1}-1}{a^{n-1}}\right].$$

4. $k = \rho A e^{-Ea/RT}$

ρ = probability factor

A = frequency factor

E_a = activation energy at temperature T

$$\ln \frac{k_1}{k_2} = \frac{E_a}{R} \frac{T_1 - T_2}{T_1 T_2}.$$

5. If t_1 and t_2 are half-lives for two different initial concentrations a_1 and a_2 respectively, then

$$n = \frac{\log \frac{a_2}{a_1} + \log \frac{t_1}{t_2}}{\log \frac{a_2}{a_1}}$$

6. $\frac{t_{3/4}}{t_{1/2}} = 2^{n-1} + 1,$

where $t_{3/4}$ = three-fourth life

$t_{1/2}$ = half-life.

Theoretical Problems

1. Will the order of reaction be integral always?

Ans. : The rate equation of a reaction is an experimental quantity and may be like $-\frac{dC_A}{dt} = k.\, C_A^{n'}$ for the reaction $n_A \rightarrow$ product n' may be different from the coefficient n and is evaluated experimentally. Thus, it may be integral or fractional. It need not be integral always.

2. Can the order of a reaction be negative?

Ans. : For a reaction of the type $n_1A + n_2B \rightarrow n_3C + n_4D$, the rate equation may be of the type—

$$r = kC_A^{n'_1} C_B^{n'_2},$$

where $n'_1 + n'_2$ is the order of the reaction. n'_1 is defined as the order with respect to the reactant A and n'_2 is defined as the order with respect to B. The order is an experimental quantity and n'_1 and n'_2 may be anything positive or negative, integral or fractional. It is, however, difficult to explain negative order. The substance with negative order acts as an inhibitor. Thus, there should be more than one reactant for the exhibition of negative order by one reactant. Experimentally some reactants in heterogeneous catalytic reaction have been found to show negative order.

3. The order of a reaction may often be changed by varying the concentration of the reactants. Correct or justify.

Ans. : The order of a reaction is the sum of exponents of the concentration terms in the rate equation and is an experimental quantity. The rate equation for the reaction,

$$\nu_1 A + \nu_2 B \rightarrow \nu_3 C + \nu_4 D$$

may be $R = k C_A^m C_B^n$ under a particular set of conditions. Then the order with respect to A is m and with respect to B is n. These are constants under a particular set of conditions and is independent of rate of concentration terms.

Hence, the order of a reaction cannot be changed by varying the concentration of the reactants, and the above assertion is incorrect. However, if a reaction is such that the rate equation is changed when the concentrations of reactants are varied after a specific concentration, then the order may depend on concentration. But that is very rare.

4. **The rate of a reaction changes exponentialy with time but not the specific rate constant. Explain.**

Ans. : The rate of reaction $A \rightarrow$ product is normally given by $\frac{-dC_A}{dt} \propto C_A^n$ in simple cases, where C_A is the concentration of A at a particular time and n is an experimental quantity.

Hence, $\frac{-dC_A}{dt} = kC^n$.

k is the proportionality constant and is the specific rate constant. Hence, it is independent of both, the rate and the concentration. The integration of the above equation will give concentration C_A as a function of time, and C_A changes continuously with time usually exponentially (linearly only for zero order reaction so that in that case rate is independent of time) and since rate depends on concentration, it obviously depends on time. For first order $C_A = ae^{-kt}$, where a is initial conc. so that rate $= ake^{-kt}$, i.e., it varies exponentially with time. For other order reactions, the variation is inverse or some power inverse, etc. The concentration goes on decreasing with time never becoming zero. The variation is, thus, comparable to exponential variation for rate. Thus, although the rate varies with time, the specific rate is constant.

5. Zero order reaction must be multistep.—Explain. Does such a reaction go to completion?

Ans. : For an elementary reaction $\nu_1 A \rightarrow$ Product,

the rate equation for this reaction will be $r = k.C_A^{\nu_1}$ and the order will be ν_1. Thus, if a zero order reaction is elementary, then the reaction has to be 0A→products, which violates the law of conservation of mass. So, a zero order reaction must be multistep.

The rate equation for a zero order reaction has to be

$$-\frac{dC}{dt} = k.C_0.$$

$$\therefore \quad dC = -kdt.$$

Integrating $C = -kt + I_C$, where I_C is integration constant and C_0 is the initial concentration at $t = 0$.

$$\therefore \; C = C_0 - kt.$$

Hence, a zero order reaction will go to completion ($C = 0$), and time period is $\frac{C_0}{k}$.

6. The stoichiometric representation of a reaction is given by $A + 2B \rightarrow 3C + D$. Express the rate of the reactions in terms of change in concentration of each of the constituents and also in terms of the advancement of the reaction.

Ans. : With respect to concentration the rate equation may be written as—

$$\frac{-dC_A}{dt} = k.C_A.C_B^2$$

and $\frac{-dC_B}{dt}$, $\frac{dC_C}{dt}$ and $\frac{dC_D}{dt}$ may be defined similarly. With respect to degree of advancement

ξ, the rate may be defined as follows. Let n_A^o be the initial number of moles of A, n_B^o initial number of moles of B and n_C^o and n_D^o are initial numbers of of moles of products C and D respectively. Then, if ξ is the degree of advancement at time t, the number of moles of reactants and products at that time are :

$$n_A = n_A^o - \xi$$

$$n_B = n_B^o - 2\xi$$

$$n_C = n_C^o + 3\xi$$

$$n_D = n_D^o + \xi.$$

$$\therefore \quad -\frac{dn_A}{dt} = \frac{d\xi}{dt}$$

$$-\frac{dn_B}{dt} = 2\frac{d\xi}{dt}$$

$$\frac{dn_C}{dt} = 3\frac{d\xi}{dt}$$

$$\frac{dn_D}{dt} = \frac{d\xi}{dt}.$$

Defining $\frac{d\xi}{dt}$ as rate,

$$\frac{d\xi}{dt} = -\frac{dn_A}{dt} = -\frac{1}{2}\frac{dn_B}{dt} = \frac{1}{3}\frac{dn_C}{dt} = \frac{dn_D}{dt}.$$

If the rates are now expressed in terms of concentration C gm-moles/lit,

$$C = \frac{n}{V}.$$

$$\therefore \quad n = VC.$$

If V is constant, then

$$\frac{dn}{dt} = V.\frac{dC_A}{dt}.$$

$$\therefore \quad \frac{d\xi}{dt} = -V.\frac{dC_A}{dt} = -\frac{1}{2}V.\frac{dC_B}{dt} = \frac{1}{3}V.\frac{dC_C}{dt} = V.\frac{dC_D}{dt}.$$

Since the volume is constant, the rate is defined as $\frac{1}{V}.\frac{d\xi}{dt}$.

If the reaction is elementary, then the rate equation can be written as

$$r = \frac{1}{V}.\frac{d\xi}{dt} = \frac{dC_A}{dt} = -\frac{1}{2}\frac{dC_B}{dt} = \frac{1}{3}\frac{dC_C}{dt} = k.\,C_A C_B^2,$$

where k is the specific rate constant.

7. Arrhenius A factor always have the same unit as the rate constant. Comment.

Ans. : We know that specific rate or rate constant, $k = A.\, e^{\frac{-E_a}{RT}}$, where A is the Arrhenius A factor and, E_a is activation energy, R is the gas constant and T is the temperature in °K. Since $e^{\frac{-E_a}{RT}}$ is unitless A must have the same unit as k.

8. The specific rate constant for a reaction has the unit lit^2.mole^{-2}.sec^{-1}. What is the order of the reaction?

Ans. : For a reaction of n-th order, $\frac{-dC}{dt} = k.C^n$.

$$\therefore \quad k = \frac{dC}{C^n.dt}.$$

Therefore, the unit of k is (gm.mole/lit)$^{1-n}$.sec^{-1},

since C is generally expressed in gm-mole/lit. Thus, unit can be (gm-mole/lit)$^{-2}$.sec^{-1}, when $n = 3$. Therefore, the order is **3**.

9. For the mechanism,

$$A + B \underset{k_2}{\overset{k_1}{\rightleftharpoons}} C$$

$$C \xrightarrow{k_3} D.$$

(i) Derive the rate law assuming the steady state approximation to eliminate the concentration of C.

(ii) Assuming that $k_3 << k_2$, express the pre-exponential factor A and the overall activation energy E for the formation of D in terms of A_1, A_2 and A_3 and E_1, E_2 and E_3 for the three steps.

(iii) What is the significance of $(E_1 - E_2)$?

Ans. : (i) The rate of formation of D is $\frac{dC_D}{dt} = k_3 C_C$,

where C_D is the concentration of D and C_C is the concentration of C.

Similarly, C_A and C_B will be taken as the concentration of A and B.

Assuming steady state in C_C, the rate of formation = rate of decomposition, i.e.,

$$k_1 C_A C_B = (k_2 + k_3) C_C.$$

$$\therefore \quad C_C = \frac{k_1}{k_2 + k_3} C_A C_B.$$

$$\text{Thus, } r = \frac{k_3 k_1}{k_2 + k_3} C_A C_B.$$

(ii) If $k_3 << k_2$, then $r = \frac{k_1 k_3}{k_2}.C_A C_B$.

$$\therefore \text{ specific rate } k = \frac{k_1 k_3}{k_2}.$$

By Arrhenius equation,

$$k = A\ e^{-\frac{E}{RT}} \qquad k_1 = A_1\ e^{-\frac{E_1}{RT}}$$

$$k_2 = A_2\ e^{-\frac{E_2}{RT}} \qquad k_3 = A_3\ e^{-\frac{E_3}{RT}}.$$

where A's are frequency factors, k's are rate constant and T is the temperature in °K and E's are activation energy.

By the problem,

$$k = \frac{k_1 k_3}{k_2}.$$

$$\therefore \quad Ae^{-E/RT} = \frac{A_1 A_3}{A_2} . e^{-\frac{(E_1+E_3-E_2)}{RT}}.$$

$\therefore$ overall activation energy,

$$E = E_1 + E_3 - E_2$$

$$\text{and } A = \frac{A_1 A_3}{A_2}.$$

(iii) $E_1 - E_2$ is the difference in activation energy of reactants and products in a reversible process. Thus, it is the heat of reaction ΔH.

10. "All reactions are not necessarily characterised by an order."—Explain.

Ans. : The rate equation of a reaction is an experimental quantity and the order of a particular reactant is power of the concentration term of that reactant in the rate equation. The total order is the sum of the exponents of the concentration terms.

If the rate equation is such that there is concentration terms only in the numerator and all concentration terms have a definite and fixed power, then the reaction is characterised by a fixed order. But, if the rate equation is such that fixed power cannot be ascertained to a concentration term, then that reactant has no fixed order. If a concentration term occurs in the denominator, this may happen. Let the rate equation for the reaction,

$$\nu_1 A + \nu_2 B \rightarrow \nu_3 C + \nu_4 D,$$

be $\quad r = k\dfrac{C_A^m}{k' + C_A^x} C_B^n,$

then there is a power series in C_A, and A does not have any fixed power. Hence, the reaction does not have a fixed order. Thus, all reactions are not necessarily characterised by a fixed order.

11. What is the difference between an unstable intermediate and an activated complex?

Ans. : An unstable intermediate is an actual chemical species which perhaps may be stabilized by changing reaction conditions. The bond order of atoms is normal in these intermediates and it occurs at a minimum, albeit a small minimum, on the potential energy curve.

The activated complex is a postulated species having maximum energy in the reaction pathway of the conversion of reactants to products. Stabilization results only by a change of bond order and length, and vibration in the direction of product or reactant. The bond order of atoms of an activated complex are sometimes unusual.

12. The reaction $2NO + O_2 \rightarrow 2NO_2$ proceeds through the following steps.

$$2NO \rightarrow N_2O_2 \text{ (fast)}$$
$$N_2O_2 + O_2 \rightarrow 2NO_2 \text{ (slow)}$$

Find the order of the reaction. The overall rate constant is found to decrease with the rise in temperature. Explain.

Ans. : The rate of formation of NO_2 is $r = \frac{1}{2}\frac{dC_{NO_2}}{dt} = k_3.C_{N_2O_2}.C_{O_2}$,

where k_3 is the specific rate of the 2nd stage. Since, the first step is fast, it must be reversible with the specific rate k_1 and k_2 for the forward and backward reactions.

∴ the equation is :

$$2NO \rightleftharpoons N_2O_2$$

$$N_2O_2 + O_2 \xrightarrow{k_3} 2NO_2$$

Assuming steady state in N_2O_2,

$$k_1 C_{NO}^2 = k_2 C_{N_2O_2} + k_3 C_{N_2O_2}.C_{O_2}$$

$$\therefore \quad C_{N_2O_2} = \frac{k_1 C_{NO}^2}{k_2 + k_3.C_{O_2}}.$$

$$\text{And } r = k_3.\frac{k_1 C_{NO}^2}{k_2 + k_3.C_{O_2}}C_{O_2}.$$

If $k_2 >> k_3$,

then $r = \frac{k_1 k_3}{k_2}.C_{NO}^2.C_{O_2}.$

Hence, the order of the reaction is 3. Now, the overall specific rate,

$$k = \frac{k_1 k_3}{k_2} = K_{eq}.k_3,$$

where K_{eq} is the equilibrium constant for the 1st step.

$$\therefore \quad \ln k = \ln K_{eq} + \ln k_3.$$

$$\therefore \quad \frac{d\ln k}{dT} = \frac{d\ln K_{eq}}{dT} + \frac{d\ln k_3}{dT}.$$

$$\therefore \quad \frac{E_a}{RT^2} = \frac{\Delta H}{RT^2} + \frac{E_3}{RT^2}.$$

∴ the overall activation energy,

$$E_a = \Delta H + E_3.$$

If ΔH is highly negative, then overall E_a is –ve. Hence, rate decreases with increase in temperature.

13. Show that if A undergoes two simultaneous reactions to produce B and C according to $A \xrightarrow{k_1} B, A \xrightarrow{k_2} C$, then E_a, the observed activation energy for the disappearance of A, is given by $E_a = \frac{k_1 E_1 + k_1 E_2}{k_1 + k_2}$

Ans. : We know that

$$k_1 = A_1 . e^{\frac{-E_1}{RT}} \qquad k_2 = A_2 . e^{\frac{-E_2}{RT}},$$

where k_1 and k_2 are specific rate constants for the 1st and 2nd steps of the reaction and A_1 and A_2 are Arrhenius factors for steps 1 and 2. E_1 and E_2 are activation energies for steps 1 and 2. The specific rate for overall change for A to B and C is $k_1 + k_2$.

$\therefore\ k = k_1 + k_2.$

$$\therefore \quad \frac{d \ln k}{dT} = \frac{d \ln(k_1 + k_2)}{dT} = \frac{E_a}{RT^2},$$

where E_a is the overall activation energy for the decomposition of A.

$$\text{But } \frac{d \ln(k_1 + k_2)}{dT} = \frac{1}{k_1 + k_2} \frac{d}{dT}(k_1 + k_2) = \frac{1}{k_1 + k_2}\left(\frac{dk_1}{dT} + \frac{dk_2}{dT}\right)$$

$$= \frac{1}{k_1 + k_2}\left[\frac{d}{dT}.\left(A_1 . e^{-E_1/RT}\right) + \frac{d}{dT}.\left(A_2 . e^{-E_2/RT}\right)\right]$$

$$= \frac{1}{k_1 + k_2}\left[A_1 . \frac{E_1}{RT^2} . e^{-E_1/RT} + A_2 . \frac{E_2}{RT^2} . e^{-E_2/RT}\right]$$

$$= \frac{1}{k_1 + k_2} \frac{1}{RT^2}\left[E_1 . A_1 . e^{-E_1/RT} + E_2 . A_2 . e^{-E_2/RT}\right] = \frac{1}{k_1 + k_2} \frac{1}{RT^2}[E_1 k_1 + E_2 k_2].$$

$$\therefore \quad \frac{1}{k_1 + k_2} \frac{1}{RT^2}[E_1 k_1 + E_2 k_2] = \frac{E_a}{RT^2}.$$

$$\therefore \quad E_a = \frac{E_1 k_1 + E_2 k_2}{k_1 + k_2}.$$

14. Comment on the following :

Elementary reactions with molecularity greater than 3 generally don't occur.

Ans. : The reactants are activated as a result of molecular collisions. But the collisions generally requires high amount of energy because of internuclear and inter-electronic repulsions. For a collision two molecules are must. If the molecularity of an elementary step is 3, then three molecules must collide requiring very high energy. For a 4-body collision, energy required will be still higher and, therefore, unknown. It is far easier to have two-body collisions and that will take place. Therefore, elementary reaction of molecularity 4 or greater is unknown.

15. What is stoichiometric number ?

Ans. : Stoichiometric number of an elementary reaction is the number of times this step occurs in the mechanism for the overall reaction as given.

For example, the mechanism of decomposition of N_2O_5 may be as follows :

Step–I : $N_2O_5 \rightleftharpoons NO_2 + NO_3$

Step–II : $NO_2 + NO_3 \rightarrow NO + O_2 + NO_2$

Step–III : $NO + NO_3 \rightarrow 2NO_2$

The stoichiometric number of Steps I, II and III are 2, 1 and 1 respectively. If there is a rate limiting step in a reaction, then sometimes stoichiometric number of the rate limiting stage is defined as the stoichiometric number of the reaction.

16. "Unimolecular reactions are not always of first order."—Justify the statement using Lindemann mechanism.

Ans. : According to Lindemann mechanism even in unimolecular reactions the molecules are activated as a result of collision with a second body. Let the reactant be A and the second body be M. But there is a time lag between activation and reaction. Thus, by Lindemann theory, the mechanism of the unimolecular reaction is :

$$A + M \rightleftharpoons A^* + M$$

$$A^* \rightarrow \text{product}$$

With $k_2 << k_{-1}$ the rate

$$r = k_2 . C_A^*.$$

Assuming steady state in C_A^*,

$$k_1 C_A C_M = k_{-1} C_A^* . C_M + k_2 C_A^*.$$

$$\therefore \quad C_A^* = \frac{k_1 C_A C_M}{k_{-1} C_M + k_2}.$$

$$\text{Thus,} \quad r = \frac{k_1 k_2 C_A C_M}{k_{-1} C_M + k_2}.$$

Since $k_2 << k_{-1}$,

$$\therefore \quad r = \frac{k_1 k_2 C_A}{k_{-1}}.$$

This is the first order. But at a very low pressure the number of collision is much less. As a result at some low pressure depending on the system, k_2 will be much greater than k_{-1}. Then the rate,

$$r = k_1 \, C_A C_M .$$

If $M = A$, then $r = k_1 . C_A^2$,

i.e., order is 2.

Thus, unimolecular reactions are not always first order.

17. What is 'Bottle-neck' principle in chemical kinetics?

Ans. : The slowest step in a multistep reaction determines its rate. Steps before it are in rapid equilibrium. This slowest step serves as bottle-neck, since in the rate expression the steps after it need not be considered.

18. What is the essential condition of a chain reaction to occur?

Ans. : Chain reactions are multistep reactions where one or more intermidiates are produced. These intermidiates may be produced by collision or from radiation. The intermidiates will have to be highly reactive and may be able to reproduce the same intermidiate in later reactions so that the reaction continues through a series of steps which are termed as chain propagating steps.

19. The rate of a reaction is given by $\log k = A - \frac{B}{T} + C \log T$. Find the value of activation energy. **[B.U. 2006]**

Ans. : Given, $\log k = A - \frac{B}{T} + C \log T$

or, $\ln k = 2.303\ A - \frac{2.303B}{T} + C \ln T$

or, $\ln k - \ln T^C = 2.303\ A - 2.303\ \frac{2.303B}{T}$.

$\therefore\ k = T^C\ e^{2.303A}\ e^{-2.303B/T} = A\ e^{-E_a/RT}$

Hence, activation energy $E_a = 2.303\ BR$.

20. (a) The following initial rate data were obtained for the reaction :

$2NO(g) + O_2(g) \rightarrow 2NO_2(g)$

	Partial Pressure of		Initial rate
	NO	O_2	
Run 1	P_{NO}	P_{O_2}	v
Run 2	$2P_{NO}$	P_{O_2}	4v
Run 3	P_{NO}	$2\,P_{O_2}$	2v

(i) What is the rate law for this reaction?

(ii) One of the mechanisms proposed for this reaction is :

$$NO(g) + O_2(g) \underset{k_{-1}}{\overset{k_1}{\rightleftharpoons}} NO_3(g)$$

$$NO_3(g) + NO(g) \xrightarrow{k_2} 2NO_2(g)$$

Obtain the rate law predicted for this mechanism, assuming a steady state concentration of NO_3.

(iii) Predict the rate law for this mechanism, if the first equilibrium step is established quickly and the second step is slow. **[JAM 2009]**

Hints : (i) $r = \frac{1}{2}\frac{dC_{NO_2}}{dt} = kC_{NO}^2 C_{O_2}$.

(ii) $$\frac{dC_{NO_3}}{dt} = k_1 C_{NO} C_{O_2} - k_{-1} C_{NO_3} - 2k_2 C_{NO_3} C_{NO} = 0$$

$$\therefore\quad r = k_2 C_{NO_3} C_{NO} = \frac{k_2 k_1 C_{NO}^2 C_{O_2}}{k_{-1} + 2k_2 C_{NO}}.$$

(iii) $k_{-1} >> k_2$

$$r = k_2 \frac{k_1}{k_{-1}} C_{NO}^2 C_{O_2}$$

$$k = k_2 K_{eq}.$$

Numerical Problems

1. Determine the rate law for the reaction $A + B + C \rightarrow$ Products from the following information when the initial concentration of A is doubled, the initial rate of reaction doubles. Doubling the initial concentration of B doubles the rate twice and doubling the initial concentration of C cuts the reaction rate by half.

Soln. : Let the rate equation be $r = kC_A^{\nu_1} C_B^{\nu_2} C_C^{\nu_3}$.

Let the initial concentration of C_A be a, that of C_B be b and that of C_C be c.

Thus, $r = ka^{\nu_1} b^{\nu_2} c^{\nu_3}$.

By the problem,

if $C_A = 2a$,

$$r_1' = k(2a)^{\nu_1} b^{\nu_2} c^{\nu_3}$$

$$\frac{r_1'}{r_1} = (2)^{\nu_1} = 2^1 = 2 \text{ as } r_1' = 2r, \therefore \nu_1 = 1$$

$$r_2' = ka^{\nu_1}(2b)^{\nu_2} c^{\nu_3} = 4r$$

So, $\frac{r_2'}{r} = 2^{\nu_2} = 4 = 2^2$.

$\therefore \nu_2 = 2$.

$$r_3' = ka^{\nu_1} b^{\nu_2} (2c)^{\nu_3}$$

$$\frac{r_3'}{r} = 2^{\nu_3} = \frac{1}{2}, \quad \therefore \nu_3 = -1.$$

Thus, $r = kC_A C_B^2 C_C^{-1}$.

2. For the reaction $A + B = C + D$, it is found that doubling the concentrations of A and B, the rate increases 8-fold, but on doubling the concentration of A, the rate increases 4-fold. What is the order of reaction with respect to B?

Soln. : Let $r = kC_A^{\nu_1} C_B^{\nu_2} = ka^{\nu_1} b^{\nu_2}$.

a = initial concentration of A

b = initial concentration of B.

By the problem,

$$8r = k(2a)^{\nu_1}(2b)^{\nu_2}$$

Thus, $8 = k2^{\nu_1}2^{\nu_2} = 2^{\nu_1+\nu_2} = 2^3$.

$\therefore \nu_1 + \nu_2 = 3$.

But $4r = k(2a)^{\nu_1} b^{\nu_2}$.

$\therefore 4 = 2^{\nu_1} = 2^2$.

Thus, $\nu_1 = 2$.

Hence, $\nu_2 = 1$.

Thus, the order of reaction with respect to B = **1.**

3. Given that for the decomposition of gas of a gaseous substance at 825 K, the specific rate k is 2.89×10^{-2} litre.mole^{-1}.sec^{-1}. According to the equation : $-\frac{dC_{AB}}{dt} = kC_{AB}^2$, how many molecules of *AB* will decompose per sec per cc at atmospheric pressure? (Assume one way reaction.)

$$\textbf{Soln. :}\ C_{AB} = \frac{n}{V} = \frac{P}{RT} = \frac{1\,\text{atm}}{0.08206\,\text{l.atm.K}^{-1}.\text{mol}^{-1} \times 825\,\text{K}}$$

$$= 1.477 \times 10^{-2} \text{ mole.lit}^{-1}$$

$$\therefore -\frac{dC}{dt} = 2.89 \times 10^{-12} \times (1.477 \times 10^{-2})^2$$

$$= 6.3046 \times 10^{-16} \text{ mole.lit}^{-1}.\text{sec}^{-1}.$$

$$-\frac{dN}{dt} = 6.3046 \times 10^{-16} \times 6.022 \times 10^{23}$$

$$= 3.7966 \times 10^{8} \text{ molecules.lit}^{-1}.\text{sec}^{-1}$$

$$= 3.79 \times 10^{5} \text{ molecules.cc}^{-1}.\text{sec}^{-1}.$$

Thus, **3.79 × 10⁵** molecules will decompose at atmospheric pressure per sec per cc.

4. In the reaction $(COOH)_2 \rightarrow CO + CO_2 + H_2O$, the following results were obtained :

Time (min)	0	300	450	600	1200
Volume of $KMnO_4$ required to titrate oxalic acid (ml)	22	17.0	15.0	13.4	7.9

Show that the reaction is of 1st order, and determine the value of the velocity constant.

[Bangalore B.Sc. 1978]

Soln. : We know that $k = \frac{1}{t} \ln \frac{a}{a-x}$.

By the problem, $a = 22$.

$$\therefore\ k = \frac{1}{300} \ln \frac{22}{17} = 8.594 \times 10^{-4} \text{ min}^{-1}$$

$$k = \frac{1}{450} \ln \frac{22}{15} = 8.510 \times 10^{-4} \text{ min}^{-1}$$

$$k = \frac{1}{600} \ln \frac{22}{13.4} = 8.263 \times 10^{-4} \text{ min}^{-1}$$

$$k = \frac{1}{1200} \ln \frac{22}{7.9} = 8.534 \times 10^{-4} \text{ min}^{-1}.$$

k is approximately constant for each case. So, the reaction is of 1st order. The average value of velocity constant is $\mathbf{8.5 \times 10^{-4}\ min^{-1}}$.

5. From the following data show that the decomposition of an aq. solution of H_2O_2 is of 1st order.

t (min)	0	10	20
N (ml)	22.8	13.8	8.20

where N is the volume of $KMnO_4$ solution in ml needed to decompose a definite volume of the peroxide solution.

Hints : Choose the formula,

$$k = \frac{1}{t} \ln \frac{a}{a-x}.$$

Find out a = 22.8 ml.

Then proceed as in Problem 4.

$k = 5.1 \times 10^{-2}$ min^{-1}.

6. 1 gm-mole of ethyl acetate was hydrolysed with 1 gm-mole of NaOH when the concentration fell according to the following data. Show that the reaction is of 2nd order, and calculate the value of k.

t (min)	0	4	6	10	15	20
$(a-x)$	8.4	5.3	4.58	3.50	2.74	2.22

[Delhi University B.Sc. 1978]

Soln. : We know that $\frac{1}{a-x} - \frac{1}{a} = kt.$

$$\therefore \quad \frac{x}{a(a-x)t} = k.$$

By the problem,

a = 8.4 gm-mole/lit, (x = 0 when t = 0)

x = 3.1 gm-mole/lit, when t = 4 min

x = 3.82 gm-mole/lit, when t = 6 min

x = 4.9 gm-mole/lit, when t = 10 min

x = 5.66 gm-mole/lit, when t = 15 min

x = 6.18 gm-mole/lit, when t = 20 min

$$\therefore k = \frac{1}{4} \times \frac{3.1}{8.4 \times 5.3} = 0.0174079 \text{ lit.mole}^{-1}.\text{min}^{-1}$$

$$= \frac{1}{6} \times \frac{3.82}{8.4 \times 4.58} = 0.0165488 \text{ lit.mole}^{-1}.\text{min}^{-1}$$

$$= \frac{1}{10} \times \frac{4.9}{8.4 \times 3.5} = 0.01666 \text{ lit.mole}^{-1}.\text{min}^{-1}$$

$$= \frac{1}{15} \times \frac{5.66}{8.4 \times 2.74} = 0.016394 \text{ lit.mole}^{-1}.\text{min}^{-1}$$

$$= \frac{1}{20} \times \frac{6.18}{8.4 \times 2.22} = 0.01657 \text{ lit.mole}^{-1}.\text{min}^{-1}$$

k is approximately constant and the average value is **1.6 × 10^{-2} lit. mol^{-1}sec^{-1}**. So, the reaction is of 2nd order.

7. The rate constant (k) of a second order reaction has the value 2.4×10^{-24}, the concentration being expressed in molecules/cc and time in seconds. What will be the magnitude of k if the concentration be expressed in moles/lit and time in minute?

Soln. : Second order rate constant is 2.4×10^{-24} cc.molecule^{-1}.sec^{-1}.

Let 2.4×10^{-24} cc.molecule^{-1}.sec^{-1} = x litre.mole^{-1}.min^{-1}.

$$\therefore\ x = 2.4 \times 10^{-24}\left(\frac{cc}{litre}\right).\frac{mole}{molecule}.\frac{min}{sec}$$

$$= 2.4 \times 10^{-24}\left(\frac{cc}{1000\,cc}\right)\times 6.022 \times 10^{23}.\frac{molecule}{molecule}.\frac{60\,sec}{sec} = 0.0867.$$

∴ rate constant is **8.67 × 10^{-2} litre.mol^{-1}.min^{-1}.**

8. The activation energy for the decomposition of a gas is 80 kJ. What percentage of gas molecules possess sufficient energy to react at 500 °K?

Soln. : We know that,

$$n = n_o e^{-E_a/RT}.$$

By the problem, E_a = 80 kJ = 80000 J; T = 500 K.

$$\therefore\quad \frac{n}{n_o} = e^{-\frac{80000}{8.314\times500}}$$

$$\therefore\quad \ln\frac{n}{n_o} = -\frac{80000}{8.314\times500} = -19.25.$$

$$\therefore\quad \frac{n}{n_o} = 4.3635 \times 10^{-9}.$$

∴ the percentage of gas molecule possess sufficient energy to react at 500 °K is **4.36 × 10^{-7}**.

9. A substance decomposes according to a second order rate law. If the rate constant is 6.8 × 10^{-4} lit.mole^{-1}.sec^{-1}, calculate the half-life of the substance when the initial concentration is 0.05 mole.lit. **[C.U.(H) 1989]**

Soln. : We know that,

$$kt = \frac{1}{a-x} - \frac{1}{a}.$$

At half-life ($t_{1/2}$) $x = \frac{a}{2}$.

$$\therefore\quad t_{1/2}\ x = \frac{1}{k\times a}.$$

By the problem,

k = 6.8 × 10^{-4} lit.mole^{-1}.sec^{-1}.

a = 0.05 mole.lit.

$$\therefore\quad t_{1/2} = \frac{1}{6.8\times10^{-4}\times0.05}$$

$$= 29411.76 \text{ sec} = 8.17 \text{ hours.}$$

So, the half-life of the substance is **8.17 hours**.

10. The first order gas reaction $AB_2X_2 \rightarrow AB_2 + X_2$ has k_1 = 2.28 × 10^{-5}sec^{-1}. What per cent of sample of AB_2X_2 would be decomposed by heating at 320 °C for 2 hours. **[C.U.(H) 1989]**

Soln. : We know that, $t = \frac{1}{k}\ln\frac{a}{a-x}$.

By the problem,

$$k = 2.28 \times 10^{-5} \text{ sec}^{-1}$$

$$t = 2 \text{ hours} = 2 \times 60 \times 60 \text{ sec.}$$

$$\therefore 2 \times 60 \times 60 = \frac{1}{2.28 \times 10^{-5}} \ln \frac{a}{a-x}$$

$$\therefore \ln \frac{a}{a-x} = 2.28 \times 10^{-5} \times 2 \times 60 \times 60$$

$$\text{or, } \ln \frac{a}{a-x} = 0.16416.$$

$$\therefore \frac{a}{a-x} = 1.178402844$$

$$\text{or, } \frac{a-x}{a} = 0.848606234$$

$$\text{or, } 1 - \frac{x}{a} = 0.8486$$

$$\text{or, } \frac{x}{a} = 1 - 0.8486 = 0.1514.$$

So, the percentage is 0.1514 × 100 = **15.14.**

11. Half-life of decomposition of a compound was found to be 50 min. When the concentration to start with was halved, the half-life becomes 100 minutes. What is the order of the reaction? **[Delhi University B.Sc. 1972]**

Soln. : We know that, $n = \dfrac{\log \frac{t_1}{t_2} + \log \frac{a_2}{a_1}}{\log \frac{a_2}{a_1}}$.

By the problem,

$$a_1 = a, \; t_1 = 50 \text{ min}$$

$$a_2 = \frac{a}{2}, \; t_2 = 100 \text{ min.}$$

$$\therefore n = \frac{\log \frac{50}{100} + \log \frac{1}{2}}{\log \frac{1}{2}} = 2.$$

So, the order of the reaction is **2.**

12. A 1st order reaction has rate constant equal to 1.25×10^{-4} sec^{-1} at 298 °K and 8.5×10^{-4} sec^{-1} at 318 °K. Calculate the activation energy of the reaction.

Soln. : We know that, $\ln \dfrac{k_2}{k_1} = \dfrac{E_a}{R}\left(\dfrac{T_2 - T_1}{T_1 T_2}\right)$,

where k_1 and k_2 are two rate constant at T_1 and T_2 and E_a is the activation energy.

∴ by the problem,

$k_1 = 1.25 \times 10^{-4}\ \text{sec}^{-1}$
$k_2 = 8.5 \times 10^{-4}\ \text{sec}^{-1}$
$T_1 = 298\ °K$
$T_2 = 318\ °K.$

$$\therefore \quad \ln\frac{8.5\times10^{-4}}{1.25\times10^{-4}} = \frac{E_a}{8.314}\left(\frac{318-298}{318\times298}\right)$$

$$\text{or, } \ln\frac{8.5}{1.25} = \frac{E_a}{8.314}\times\frac{20}{318\times298}.$$

$$\therefore \quad E_a = \frac{8.314\times318\times298}{20}\ln\frac{8.5}{1.25}$$

$$= 75514.09\ \text{J} = \mathbf{75.51\ kJ.}$$

So, the activation energy of the reaction is **75.51 kJ.**

13. On raising the temperature from 27 °C to 37 °C, the rate of reaction is doubled. Calculate the acivation energy. **[C.U.(H) 1981]**

Soln. : We know that, $\ln\frac{k_2}{k_1} = \frac{E_a}{R}\frac{T_2-T_1}{T_1T_2}.$

By the problem,

$$\frac{k_2}{k_1} = 2$$

$T_2 = 37\ °C = 273.15 + 37 = 310.15\ K$
$T_1 = 27\ °C = 300.15\ K.$

$$\therefore \ln 2 = \frac{E_a}{8.314}\left(\frac{310.15-300.15}{310.15\times300.15}\right) = \frac{E_a}{8.314}\times\frac{10}{310.15\times300.15}.$$

$$\therefore E_a = \frac{8.314\times310.15\times300.15}{10}\ln 2 = 53647.02\ \text{J} = \mathbf{53.64\ kJ.}$$

So, the activation energy is **53.64 kJ.**

14. The decomposition of NH_3 catalysed by hot tungsten surface was studied at 1100 °C and the following data were obtained for the time of half decomposition with different initial pressure.

P_o mm of Hg	300	200	100
$t_{1/2}$ min	8.6	5.6	2.9

Find out the order of the reaction and the rate constant. **[C.U.(H) 1977]**

Soln. : We know that,

$$n = \frac{\log\frac{t_1}{t_2}+\log\frac{a_2}{a_1}}{\log\frac{a_2}{a_1}},$$

where t_1 and t_2 are two half-lives at pressure a_1 mm of Hg and a_2 mm of Hg.

By the problem,

$t_1 = 8.6$ min

$t_2 = 5.6$ min

$a_1 = 300$ mm of Hg

$a_2 = 200$ mm of Hg.

$$\therefore n = \frac{\log\frac{8.6}{5.6} + \log\frac{200}{300}}{\log\frac{200}{300}} = -0.058.$$

Again, using the data,

$t_1 = 5.6$ min

$t_2 = 2.9$ min

$a = 200$ mm of Hg

$a_2 = 100$ mm of Hg,

$$n = \frac{\log\frac{5.6}{2.9} + \log\frac{100}{200}}{\log\frac{100}{200}} = 0.051.$$

So, the order of the reaction $= \frac{-0.05 + 0.06}{2} = \mathbf{0.}$

For zero order reaction,

$$t_{1/2} = \frac{a}{2k}.$$

$$\therefore k = \frac{a}{2t_{1/2}}.$$

Putting $a = 300$ mm of Hg $= \frac{300}{760}$ atm and $t_{1/2} = 8.6$,

$$k = \frac{300}{2 \times 760 \times 8.6} = \mathbf{0.023\ atm.min^{-1}}.$$

15. The decomposition of a gas is a second order reaction. Using the gas at 200 mm of Hg initial pressure at a certain temperature, if 25% of the gas decomposes in 30 min, in what time 50% of the gas will decompose? Calculate the specific rate constant. **[C.U.(H) 1976]**

Soln. : We know for the 2nd order reaction,

$$k = \frac{1}{t}\frac{x}{a(a-x)}.$$

$$\therefore t = \frac{1}{k}\frac{x}{a(a-x)}.$$

$$\therefore t_{1/4} = \frac{1}{k}\frac{\frac{1}{4}a}{a \times \left(a - \frac{a}{4}\right)} = \frac{1}{k}\frac{\frac{1}{4}a}{a \times \frac{3a}{4}} = \frac{1}{3ka}.$$

Again, $t_{1/2} = \frac{1}{k}\frac{\frac{a}{2}}{a\times\left(a-\frac{a}{2}\right)} = \frac{1}{k}\frac{\frac{a}{2}}{a\times\frac{a}{2}} = \frac{1}{ka}$.

$\therefore \quad \frac{t_{1/2}}{t_{1/4}} = 3.$

$\therefore \quad t_{1/2} = 3 \times t_{1/4}.$

By the given problem, $t_{1/4} = 30$ min.

$\therefore t_{1/2} = 3 \times 30 =$ **90 minutes.**

So, the time required for 50% decomposition is 90 minutes.

$\therefore k = \frac{1}{t_{1/2}} \times \frac{1}{a},$

By the problem, $a = 200$ mm of Hg $= \frac{200}{760}$ atm.

$\therefore k = \frac{1}{90} \times \frac{1}{\frac{200}{760}} =$ **0.0422 atm.min^{-1}.**

16. At 100 °C, the gaseous reaction $A \rightarrow 3B + C$ is observed to be 1st order. Starting with pure A, it is found that at the end of 10 minutes the total pressure of the system is 130 mm of Hg and after a long time 400 mm of Hg. From these data, find (a) the initial pressure of A (b) the pressure of A at the end of 10 minutes (c) the rate constant and (d) the half-life.

Soln. : $A \rightarrow C + 3B$

Given that, after 10 minutes, pressure = 130 mm of Hg.

After a long time, pressure = 400 mm of Hg.

We may assume that after a long time, the decomposition is complete. So, one gas molecule leads to four gas molecules as shown above. Since after a long time the pressure is 400 mm of Hg with four molecules $(3B + C)$, So

(a) the initial pressure is $\frac{400}{4}$ mm of Hg = **100 mm of Hg.**

(b) Let, after 10 minutes, the pressure of A be reduced by x mm of Hg.

$$\begin{array}{lccc} & A & \rightarrow \; 3B & + \; C \\ \text{10 min,} & 100-x & 3x & x \end{array}$$

So, the total pressure, $(100 - x + 3x + x) = 100 + 3x$ mm of Hg.

By the problem, $100 + 3x = 130$

or, $\quad x = 10.$

So, the pressure of A at the end of 10 minutes is $(100 - 10)$ mm of Hg = **90 mm of Hg.**

(c) We know that for 1st order reaction,

$$k = \frac{1}{t} \ln \frac{a}{a-x}.$$

By the problem,

$t = 10$ min

$a = 100$ mm of Hg

$a - x = 90$ mm of Hg.

$$\therefore \quad k = \frac{1}{10}\ln\frac{100}{90} = \mathbf{0.0105\ min^{-1}}.$$

(d) We know,

$$t_{1/2} = \frac{0.693}{k}$$

$$\therefore \quad t_{1/2} = \frac{0.693}{0.0105} = \mathbf{66\ min.}$$

17. At 200 °C the gaseous $A \rightarrow 2B + C$ is observed to be second order. Starting with pure A it was found that at the end of 10 minutes the total pressure of the system is 176.0 mm of Hg and after a long time 270 mm of Hg. Calculate the rate constant and half-life of the substance.

Soln. : $A \rightarrow 2B + C$

We may assume that after a long time, the decomposition is complete. So, one gas molecule leads to three gas molecules as shown above.

Since after a long time the pressure is 270 mm of Hg with three molecules ($2B+C$).

So, the initial pressure is $\frac{270}{3}$ mm of Hg = 90 mm of Hg.

	A	$\rightarrow 2B$	$+ \; C$
10 min	$90 - x$	$2x$	x

So, the total pressure at 10 minutes is $(90 - x + 2x + x) = 90 + 2x$ mm of Hg.

By the problem,

$90 + 2x = 176$

$\therefore \quad x = 43$ mm of Hg.

We know that for 2nd order reaction,

$$k = \frac{1}{t}\cdot\frac{x}{a(a-x)}$$

Here $t = 10$ minutes

$x = 43$ mm of Hg

$(a - x) = (90 - 43) = 47$ mm of Hg.

$$\therefore \quad k = \frac{1}{10}\cdot\frac{43}{47\times 90} = \mathbf{1.0165 \times 10^{-3}\ mm\ of\ Hg.^{-1}min^{-1}}.$$

$$\text{Again, } t_{1/2} = \frac{1}{ka} = \frac{1}{1.0165\times 10^{-3}\times 90} = \mathbf{10.93\ min.}$$

18. A dilute solution of cane sugar was hydrolysed into glucose and fructose by addition of dil. HCl. The progress of the reaction was followed in a polarimeter tube by observing the angle of rotation at different times. The results are :

t (min)	0	10	20	30	80	∞
Angle of roation	32.4	28.8	25.5	19.6	10.3	14.1

Show that the reaction is of 1st order and calculate the half-life. **[D.U. 1982]**

Hints : $k = \frac{1}{t}\ln\frac{\theta_o - \theta_\infty}{\theta_t - \theta_\infty}$. **[Ans.** 118 min]

19. Methyl acetate is hydrolysed in approx(N) HCl at 25 °C. Aliquotes of 5 ml each are removed at intervals and titrated with 0.185(N) NaOH. Calculate the first order rate constant from the following data :

t (sec)	339	1242	2745	4546	∞
V (ml)	26.34	27.80	29.70	31.81	39.81

[Utkal 1980]

Soln. : We know that—

$$k = \frac{1}{t}\frac{V_\infty - V_0}{V_\infty - V_t}.$$

$$\therefore \quad t = \frac{1}{k}\frac{V_\infty - V_0}{V_\infty - V_t}.$$

$$t_1 = \frac{1}{k}\ln\frac{V_\infty - V_0}{V_\infty - V_{t_1}}$$

$$t_2 = \frac{1}{k}\ln\frac{V_\infty - V_0}{V_\infty - V_{t_2}}.$$

$$\therefore \quad (t_2 - t_1) = \frac{1}{k}\ln\frac{V_\infty - V_{t_1}}{V_\infty - V_{t_2}}.$$

By the problem,

t_2 = 1242 min

t_1 = 339 min

V_∞ = 39.81 ml

V_{t_1} = 26.34 ml

V_{t_2} = 27.80 ml

$$\therefore 1242 - 339 = \frac{1}{k}\ln\frac{39.81 - 26.34}{39.81 - 27.80} = \frac{0.11472535}{k}.$$

$$\therefore k = \frac{0.11472535}{903} = \mathbf{1.27 \times 10^{-4}\ min^{-1}}.$$

So, the value of rate constant is $\mathbf{1.27 \times 10^{-4}\ min^{-1}}$.

20. 1 gm-mole of ethyl acetate was hydrolysed with 1 gm-mole of NaOH when the concentration fell according to the following data. Show that the reaction is of second order and calculate k.

t (min)	0	4	6	10	15	20
$a - x$	8.04	5.3	4.58	3.50	2.74	2.22

Soln. : We know that for 2nd order reaction,

$$\frac{1}{a-x} = \frac{1}{a} + kt.$$

$$\therefore \quad \frac{1}{a-x_1} = \frac{1}{a} + kt_1 \qquad \text{(i)}$$

$$\frac{1}{a-x_2} = \frac{1}{a} + kt_2. \qquad \text{(ii)}$$

Subtracting equation (ii) from equation (i), we get

$$\frac{1}{a-x_1} - \frac{1}{a-x_2} = k\,(t_1 - t_2).$$

By the problem,

$$a - x_1 = 8.04, \text{ when } t_1 = 0$$
$$a - x_2 = 5.3, \text{ when } t_2 = 4.$$

$$\therefore \quad \frac{1}{8.04} - \frac{1}{5.3} = -4\,k$$

$$\text{or,} \quad \frac{-2.74}{8.04 \times 5.3} = -4\,k$$

$$\text{or,} \quad k = \frac{2.74}{8.04 \times 5.3 \times 4}$$

$$= \mathbf{1.607 \times 10^{-2}\ mole^{-1}.cc.min^{-1}}.$$

Again, putting $a - x_2 = 4.58$, when $t_2 = 6$, we get

$$\frac{1}{8.04} - \frac{1}{4.58} = -6k$$

or, $k = 1.567 \times 10^{-2}$ mole^{-1}.cc.min^{-1}.

Again, putting $a - x_2 = 3.50$, when $t_2 = 10$, we get

$k = 1.613 \times 10^{-2}$ mole^{-1}.cc.min^{-1}.

k is approximately constant. So, the order of the reaction is of 2nd order. The average rate constant is **1.6 × 10^{-2} mole^{-1}.cc.min^{-1}.**

21. When ethyl acetate is saponified by NaOH, the progress of the reaction can be measured by titrating the unchanged alkali against a standard acid at definite intervals. Using equal concentration of ester and alkali the following results were obtained in the experiment :

t (min)	0	5	25	55	120	∞
acid (cc)	16	10.24	4.32	2.31	1.1	0

Show that the reaction is of second order. What fraction of ethyl acetate does decompose in 30 minutes? **[C.U.(H) 1969]**

Soln. : We know that

$$k = \frac{1}{t}.\frac{x}{a(a-x)}.$$

By the problem,

when $t = 5$; $a = 16$; $(a - x) = 10.24$ and $x = (16 - 10.24) = 5.76$.

$\therefore \quad k = \frac{1}{5} \times \frac{5.76}{16 \times 10.24} = 7.03 \times 10^{-3}\ cc^{-1}.min^{-1}.$

Again, using $t = 25$; $(a - x) = 4.32$ and $x = 11.68$.

$$k = \frac{1}{25} \times \frac{11.68}{16 \times 4.32} = 6.75 \times 10^{-3}\ cc^{-1}.min^{-1}.$$

Again, using $t = 55$; $(a - x) = 2.31$; $x = 13.69$.

$$k = 6.73 \times 10^{-3}\ cc^{-1}.min^{-1}.$$

k is approximately constant and order of the reaction is second. The average rate constant is $6.83 \times 10^{-3}\ cc^{-1}.min^{-1}$.

$$k = \frac{1}{t} \frac{x}{a(a-x)}.$$

$$\therefore \quad 6.83 \times 10^{-3} = \frac{1}{30} \frac{x}{16(16-x)}$$

or, $6.83 \times 10^{-3} \times 30 = \frac{x}{16(16-x)}$

or, $3.2784 = \frac{x}{16-x}$

or, $4.2784x = 16 \times 3.2784$

or, $x = 12.26$.

So, the fraction of ethyl acetate which decomposes in 30 minutes is $\frac{12.26}{16}$ = **0.77.**

22. A certain substance A is mixed with an equal quantity of a substance B. At the end of 1 hour, 75% A reacted. How much A will be left unreacted at the end of 2 hours if the reaction is of (a) first order in A and independent of B (b) 1st order in A and first order in B and (c) zero order in A and independent of B?

[Delhi University B.Sc.(H) 1972]

Soln. : (a) $k = \frac{1}{t} \ln \frac{a}{a-x}$

$$= \frac{1}{60} \ln \frac{a}{0.25a} \text{ (by the Problem)}$$

$$= 0.0231049\ min^{-1}.$$

$\therefore$ after 2 hours,

$$\ln \frac{a}{a-x} = kt = 0.0231049 \times 120 = 2.7725887$$

or, $\frac{a}{a-x} = 16$

i.e., $a - x = \frac{1}{16} a.$

$\therefore$ A left unreacted will be $\frac{100}{16}\%$ = **6.25%.**

(b) $\frac{a}{a-x} = \frac{1}{a} + kt$

$\therefore$ by the Problem,

$$\frac{1}{0.25a} - \frac{1}{a} = 60k$$

$$\therefore \quad \frac{3}{60a} = k \text{ or, } k = \frac{1}{20a}.$$

Now, $\frac{x}{a(a-x)} = kt = \frac{1}{20a} \times 120 = \frac{6}{a}$.

or, $\frac{x}{a-x} = 6, \quad \therefore \frac{a-x}{x} = \frac{1}{6}$.

$$\therefore \quad \frac{a}{x} = \frac{7}{6} \text{ or, } \frac{x}{a} = \frac{6}{7}.$$

$$x = \frac{6}{7}a.$$

Hence, $a - x = \frac{1}{7}a = 14.285714\% =$ **14.29%.**

(c) By the problem,

$$x = kt,\ 0.75a = 60\,k.$$

$$\therefore \quad k = \frac{0.75}{60}a.$$

The reaction will be complete in time,

$$t = \frac{a}{k} = \frac{a}{\frac{0.75}{60}a} = \frac{60}{0.75} = \mathbf{80\ min.}$$

Thus, **no amount will remain** at the end of 2 hours.

23. A drug is known to be ineffective when it has decomposed to the extent of 30%. The original concentration of the sample is 500 units/ml. When analysed 20 months later the concentration was found to be 420 units/ml. Assuming that the decomposition is of first order, what will be expiration time of the sample? What is the half-life of this drug?

[C.U.(H) 1988]

Soln. : By the problem,

$$k = \frac{1}{20}\ln\frac{500}{420} = 8.71766 \times 10^{-3} \text{ month}^{-1}.$$

The expiration time t (when $x = 0.3a$)

$$= \frac{1}{k}\ \ln\frac{a}{a-0.3a}$$

$$= \frac{1}{8.71766 \times 10^{-3}} \ln \frac{1}{0.7}$$

$= 40.91405$ months $=$ **41 months.**

The half-life here is the time when it decomposes to the extent of 15%, since the drug expires after decomposition to 30%. Then

$$t_{1/2} = \frac{1}{8.71766\times10^{-3}} \ln \frac{a}{0.85a}$$

$= 18.642475$ months $=$ **19 months.**

24. The decomposition of a gas at an initial pressure of 600 mm of Hg was studied in a closed vessel at a certain temperature. The gas was found to be 50% decomposed in 30 minutes and 75% decomposed in 90 minutes. Show that the reaction is of second order, and calculate the rate constant. **[C.U.(H) 1981]**

Soln. : We know that $\frac{t_{3/4}}{t_{1/2}} = 2^{n-1} + 1.$

$$\therefore \quad \frac{90}{30} = 2^{n-1}+1$$

$$\therefore \quad n = 2.$$

Hence, $k = \frac{1}{at_{1/2}} = \frac{760}{600\times30} =$ **0.0422 atm^{-1}.min^{-1}.**

25. The thermal decomposition of acetaldehyde vapour $CH_3CHO \rightarrow CH_4 + CO$ is an irreversible process whose rate at 500 °C is given by the following two sets of data for the pressure change at constant volume and temperature :

Initial Pressure of CH_3CHO	Total Pressure after 100 sec.
400 mm of Hg	500 mm of Hg
200 mm of Hg	226 mm of Hg

Find the order with respect to CH_3CHO. Calculate the rate constant at 500 °C. The activation energy is 180 kJ. At what temperature will be rate constant be twice of what it is at 500 °C?

Soln. : Let the initial pressure of CH_3CHO be P_o. Then, the pressure of CH_3CHO after some time will be $(P_o - x)$ when the total pressure will be $(P_o + x)$.

$$\underset{P_o - x}{CH_3CHO} \rightarrow \underset{x}{CH_4} + \underset{x}{CO}$$

Thus, when $P_o = 400$ mm, after 100 sec, $x = 100$ $(P_o + x = 500)$ and so, $P_o - x = 300$, and hence,

$$r_1 = -\frac{300-400}{100} = 1 \ (a_1 = 400) \ (\text{since, } r = -\frac{dC}{dt} = \frac{dx}{dt})$$

Again, when $P_o = 200$ mm, after 100 sec, $x = 26$ and $P_o - x = 174$. And thus,

$$r_2 = -\frac{176-200}{100} = +\,0.26 \ (a_2 = 200).$$

$$\text{Now, } n = \frac{\log\frac{r_1}{r_2}}{\log\frac{a_1}{a_2}}.$$

$$\text{Therefore, } n = \frac{\ln\left(\frac{1}{0.26}\right)}{\ln\left(\frac{400}{200}\right)} = 1.9434165 = 2.$$

$$\therefore \quad k_{400} = \frac{1}{t}.\frac{x}{a(a-x)} = \frac{1}{100} \times \frac{100}{400 \times 300} = 8.33 \times 10^{-6} \text{mm}^{-1}.\text{sec}^{-1}.$$

$$k_{200} = \frac{1}{100} \times \frac{26}{200 \times 174} = 7.47 \times 10^{-6} \text{mm}^{-1}.\text{sec}^{-1}.$$

$$\therefore \quad k_{av} = \mathbf{7.9 \times 10^{-6} mm^{-1}.sec^{-1}}.$$

Again, by the problem, using Arrhenius equation,

$$\ln 2 = -\frac{180000}{8.314}\left(\frac{1}{T_2} - \frac{1}{773}\right) = \frac{180000 \times (T_2 - 733)}{8.314 \times T_2 \times 773}$$

$$\therefore \quad T_2 = 792.61578 = 793\ °\text{K} = \mathbf{520\ °C}.$$

26. A piece of wood buried by a glacier had 25.6% as much C^{14} as a recently grown piece of wood. If the amount of C^{14} in the atmosphere was the same when the old wood died as it is now, when was the wood buried? ($t_{1/2}$ for C^{14} = 5600 year).

Soln. : Let us assume that decay started when it got buried. Radioactive decay is of 1st order.

$$\therefore \quad t = \frac{1}{k} \ln\frac{N_o}{N}, \text{ but } t_{1/2} = \frac{0.693}{k}.$$

$$\therefore \quad k = \frac{0.693}{t_{1/2}}.$$

$$\therefore \quad t = \frac{t_{1/2}}{0.693} \ln \frac{1}{0.256}$$

$$= \frac{5600}{0.693} \ln \frac{1}{0.256}$$

$$= 11010.73 \text{ years} \approx \mathbf{11011 \text{ years}}.$$

Hence, the wood got burried **11011 years ago.**

27. A study by temperature jump method gave a relaxation time $t^* = 38\ \mu s$ at 20 °C for $H_2O \rightleftharpoons H^+ + OH^-$. Calculate k_f and k_b for the reaction. At 20 °C, $K_w = a_{H^+}a_{OH^-} = 1.04 \times 10^{-14}$.

Soln. : From this data,

$$x_e = \sqrt{1.04 \times 10^{-14}} = 1.0198 \times 10^{-7} \text{mole/lit}.$$

$$\text{And } K_e = \frac{C_{H^+}.C_{OH^-}}{C_{H_2O}} = \frac{1.04 \times 10^{-14}}{55.55} = 1.87218 \times 10^{-16}.$$

Now, for the reaction, $H_2O \rightleftharpoons H^+ + OH^-$,

$$t^* = \frac{1}{k_f + 2k_b x_e}.$$

$$\therefore \text{ by the problem, } 38 \times 10^{-6} = \frac{1}{(k_f + 2k_b \times 1.0198 \times 10^{-7})}.$$

But, $\frac{k_f}{k_b} = K_e$.

$\therefore \quad k_f = K_e.k_b = \mathbf{1.87218 \times 10^{-16} k_b}$.

$$\therefore \quad 38 \times 10^{-6} = \frac{1}{\left(1.87218 \times 10^{-16} \times k_b + 2 \times 1.0198 \times 10^{-7}\right)}.$$

$\therefore \quad k_b = 1.29024 \times 10^{11}$ l.mol^{-1}.sec^{-1}

$= \mathbf{1.3 \times 10^{11}}$ **lit.mole^{-1}.sec^{-1}.**

$k_f = K_e k_b = 2.41550 \times 10^{-5} = \mathbf{2.4 \times 10^{-5}}$ **sec^{-1}.**

28. For a bimolecular decomposition of HI, the rate constant is 2.8×10^{-7} lit. mole^{-1}.sec^{-1} at 556 °K. The activation energy is 44300 cal. Calculate the entropy of activation ($h = 6.6 \times 10^{-27}$ erg.sec, $R = 2$ cal/deg mole, $N_A = 6.023 \times 10^{23}$ per mole.)

[C.U.(H) 1987]

Soln. : By the eqn. 2.18.27 in Physical Chemistry, Vol. 2 by Dr S. Pahari,

$$k_2 = e^2 \frac{kT}{h}\left(\frac{RT}{P_o}\right) e^{\Delta S_o^{\ddagger}/R} e^{-E_a/RT}$$

$$= e^2 \frac{RT}{N_A h}\left(\frac{RT}{P_o}\right) e^{\Delta S_o^{\ddagger}/R} e^{-E_a/RT}.$$

Now, $\ln k_2 = \ln A - \frac{E_a}{RT}$.

$$\therefore \quad \ln A = \ln k_2 + \frac{E_a}{RT} = \ln 2.8 \times 10^{-7} + \frac{44300}{2 \times 556}.$$

$\therefore \quad A = 5.605 \times 10^{10}$.

Naturally,

$$\ln A = 2 + \ln \frac{RT}{N_A h} + \ln \frac{RT}{P_o} + e^{\Delta S_o^{\ddagger}/R}.$$

Putting $R = 8.314 \times 10^7$ erg.deg^{-1}.mole^{-1} in $\frac{RT}{N_A h}$ and

$R = 0.082061$ lit.atm.deg^{-1}.mole^{-1} in $\frac{RT}{P_o}$,

$$\ln A = 2 + \ln \frac{8.314 \times 10^7 \times 556}{6.023 \times 10^{23} \times 6.6 \times 10^{-27}} + \ln \frac{0.082061 \times 556}{1} + \frac{\Delta S_o^{\ddagger}}{R}$$

$$\text{or,} \quad 24.749 = 2 + \ln 1.162 \times 10^{13} + \ln 45.625 + \frac{\Delta S_o^{\ddagger}}{R}$$

$$\text{or,} \quad 24.749 = 35.904 + \frac{\Delta S_o^{\ddagger}}{R}$$

or, $\Delta S_o^{\ddagger} = \mathbf{-92.74}$ **J/K.**

29. The rate constant of a second order gas reaction, $H_2 + I_2 \rightleftharpoons 2HI$ is 0.0234 lit.mole^{-1}.sec^{-1} at 400 °C and the activation energy is 150 kJ/mole. Calculate $\Delta H_o^{\ddagger}$, $\Delta G_o^{\ddagger}$ and $\Delta S_o^{\ddagger}$ for this reaction.

Soln. : Since the reaction is of second order,

$$\Delta H_o^{\ddagger} = E_a - 2RT$$

$$= 150000 \text{ J.mol}^{-1} - 2 \times 8.314 \text{ J.K}^{-1}.\text{mol}^{-1} \times 673 \text{ K}$$

$$= \mathbf{138.81 \ kJ/mole.}$$

$$\text{Now, } k = \frac{RT}{N_A h}\left(\frac{RT}{P_o}\right) e^{-G_0^{\ddagger}/RT}$$

R = 8.314×10^7erg.deg^{-1}.mole^{-1}
 = 0.08206 lit.atm.deg^{-1}.mole^{-1}

T = 673.15K

N_A = 6.023×10^{23}

h = 6.626×10^{-27} erg.sec

and k = 0.0234. lit.mol^{-1}.sec^{-1}.

$$\therefore 0.0234 = 7.74674 \times 10^{14} \times e^{-G_o^{\ddagger}/RT}.$$

$$\therefore \frac{\Delta G_o^{\ddagger}}{RT} = 38.038483.$$

$$\therefore \Delta G_o^{\ddagger} = \mathbf{212.83 \ kJ.}$$

$$\Delta S_o^{\ddagger} = \frac{\Delta H_o^{\ddagger} - \Delta G_o^{\ddagger}}{T} = \frac{138.81 - 212.83}{673.15} = -0.109961 \text{ kJ.K}^{-1} = \mathbf{-110 \ J.K^{-1}}.$$

30. For a first order reaction the specific rate was determined at three different temperatures. Calculate the activation energy and the frequency factor (A). Find out $\Delta H_o^{\ddagger}$, $\Delta G_o^{\ddagger}$ and $\Delta S_o^{\ddagger}$ at 37 °C.

T °C	27 °C	37 °C	47 °C
$k_2 \times 10^2$ sec^{-1}	2.1	4.2	8

Hints : Proceed as in Problem 28. However, here $k_2 = e\dfrac{kT}{h}.e^{\frac{\Delta S_o^{\ddagger}}{R}}.e^{\frac{-E_a}{RT}}$.

Determine activation energy (E_a) using the given table :

[E_a = 53.4 kJ

A = 4.1×10^7

$\Delta H_o^{\ddagger}$ = 50.8 kJ

$\Delta G_o^{\ddagger}$ = 85.66 kJ

$\Delta S_o^{\ddagger}$ = -107.73 J.K^{-1}]

31. The effective rate constant of a 1st order reaction following Lindemann mechanism has the following values : 2.5×10^{-3} sec^{-1}, when $C = 5 \times 10^{-2}$ gm-mole.lit^{-1}, and 4.2×10^{-3} sec^{-1}, when $C = 9.8 \times 10^{-2}$ gm.mole.sec^{-1}.lit^{-1}. Find out the rate coefficient for the activation step.

Soln. : By Lindemann mechanism,

$$\frac{-dC_A}{dt} = k_{eff} C_A.$$

$$\text{Here, } k_{eff} = \frac{k_1 k_3 C_A}{k_3 + k_2 C_A}$$

$$\therefore \quad \frac{1}{k_{eff}} = \frac{1}{k_1 C_A} + \frac{k_2}{k_1 k_3}.$$

$$\therefore \quad k_{eff}^{-1} = k_1^{-1} C_A^{-1} + \frac{k_2}{k_3 k_1}.$$

Putting the value,

$$\frac{10^3}{2.5} = \frac{10^2}{5} k_1^{-1} + \frac{k_2}{k_3 k_1}$$

$$\frac{10^3}{4.2} = \frac{10^2}{9.8} k_1^{-1} + \frac{k_2}{k_3 k_1}.$$

Subtracting, we get

$$\left(\frac{10^3}{2.5} - \frac{10^3}{4.2}\right) = \left(\frac{10^2}{5} - \frac{10^2}{9.8}\right) \cdot \frac{1}{k_1}$$

$$\text{or,} \quad (400 - 238.095) = (20 - 10.204) \frac{1}{k_1}.$$

$$\therefore \quad k_1 = \frac{9.796}{161.905} = \mathbf{6.05 \times 10^{-2}\ lit.gm\text{-}mole^{-1}.sec^{-1}}$$

The k_{eff} equation comes from Lindemann mechanism as outlined in Physical Chemistry Vol.2 by Dr S. Pahari, equation no. 2.14.4. Replacing M by A, i.e., C_M by C_A,

$$r = \frac{k_3 k_1 C_A^2}{k_3 + k_2 C_A}.$$

$$\therefore \quad \frac{-dC_A}{dt} = k_{eff} C_A.$$

32. Some PH_3 is introduced into a flask at 600 °C containing some inert gas. Phosphine proceeds to decompose into P_4(g) and H_2(g) and the reaction goes to completion. The total pressure is given below as a function of time.

t (sec)	0	60	120	∞
P (mm of Hg)	260	270.5	273.14	274

The rate equation for the backward reaction is given by

$$r = k_2 P_{P4}^a . P_{H2}^b .$$

Calculate a and b. Let the initial pressure of phosphine be 20 mm of Hg and the half-life be 25 sec. Calculate k, if the decomposition is of 1st order and if it is of second order.

Soln. : The decomposition reaction may be written as

$$4PH_3 \rightarrow P_4 + 6H_2$$

If the initial pressure of phosphine is $4a$ mm of Hg and that of P_4 at some time is x mm of Hg, then at that time,

$$P_{PH_3} = 4a - 4x$$

$$P_{H_2} = 6x$$

and total pressure is $4a + 3x$. Now, there is always some inert gas. Let its pressure be b mm of Hg.

$$\therefore \quad 260 = 4a + b.$$

But at ∞, all phospine decomposes. 4 particles yield 7 particles.

$$\therefore \quad 274 = 7a + b.$$

$$\therefore \quad 3a = 14, \text{ i.e., } 4a = 4 \times \frac{14}{3} = 18.67 \text{ mm of Hg.}$$

After 60 sec,

$$4a + 3x + b = 270.5$$

$$\text{and} \quad 4a + b = 260$$

$$\therefore \quad 3x = 10.5 \text{ mm of Hg.}$$

$$\therefore \quad 4a - 4x = 18.67 - 14 = 4.67 \text{ mm of Hg.}$$

After 120 sec,

$$4a + 3x + b = 273.14$$

$$\text{and} \quad 4a + b = 260$$

$$\therefore \quad 3x = 13.14 \text{ mm of Hg}$$

and $4a - 4x = 18.67 - 17.52 = 1.15.$

Hence, assuming the reaction is of 1st order,

$$k = \frac{1}{t} \ln \frac{4a}{4a - 4x}$$

$$= \frac{1}{60} \ln \frac{18.67}{4.67} = 0.02309595 \text{ sec}^{-1}$$

$$\text{and } k' = \frac{1}{120} \ln \frac{18.67}{1.15} = 0.0232263 \text{ sec}^{-1}.$$

$$\therefore k_{average} = 0.023 \text{ sec}^{-1}.$$

Hence, $t_{1/2} = 30.13043$ sec.

Now for the above equation, the equilibrium constant

$$K = \frac{P_{P_4} \times P_{H_2}^6}{P_{PH_3}^4} = \left(\frac{k_1}{k_2}\right)^S = \left(\frac{P_{P_4}^a \times P_{P_2}^b}{P_{PH_3}}\right)^S.$$

Consequently, $S = 4$. Hence, $a = 0.025$ and $b = 1.5$. Thus, for the backward reaction, with equation written as $4PH_3 = P_4 + 6H_2$, the rate equation is :

$$r = k.\, P_{P_4}^{0.25} P_{H_2}^{1.5}.$$

If the reaction is written as

$$PH_3 \rightleftharpoons \frac{1}{4}P_4 + \frac{3}{2}H_2,$$

$$K = \frac{P_{P_4}^{\frac{1}{4}} \cdot P_{H_2}^{\frac{3}{2}}}{P_{PH_3}} = \left(\frac{k_1}{k_2}\right)^S.$$

But here $S = 1$.

$\therefore a = 0.25$ and $b = 1.5$ again. Even if we write the equation as $8PH_3 = 2P_4 + 12H_2$, $a = 0.25$, $b = 1.5$ (since $S = 8$), and if $4a = 20$ mm and $t_{1/2} = 25$, 1st order $k = \frac{0.693}{25}$ $=$ **0.02772 sec^{-1}**, 2nd order $k = \frac{760}{25 \times 20} = 1.52$ sec^{-1}.atm^{-1} $=$ **0.002 sec^{-1}.mm of Hg^{-1}**.

33. The decomposition of PH_3 at 950 K is followed by noting the change in total pressure as a function of time. The reaction is :

$$4PH_3(g) \rightarrow P_4(g) + 6H_2(g).$$

The following measurement were made on the system containing only PH_3 initially.

Time (sec)	0	50	100
P (total) (mm of Hg)	200	299	332

Show that the order of the reaction is one. Calculate the rate constant.

Soln. : By the problem, let phosphine decomposed be $4x$ and the reaction be—

$$\begin{array}{lll} 4PH_3 & \rightarrow P_4 & + 6H_2 \\ 4a - 4x & x & 6x \end{array}$$

Thus, the total pressure is $4a + 3x$.

So, $4a = 200$

$\therefore$ $4a + 3x$ at 50th second is 299.

$\therefore$ $3x = 99$.

$\therefore$ $x = 33$.

Similarly, at 100th second $4a + 3x = 332$.

$\therefore$ $3x = 132$

$\therefore$ $x = 44$.

At 50th second,

P_{PH_3} is $4a - 4x = 200 - 132 = 68$ mm of Hg.

At 100th second,

P_{PH_3} is $4a - 4x = 24$ mm to Hg

$$\therefore k = \frac{1}{t}\ln\frac{a}{a-x}$$

$$= \frac{1}{t}\ln\frac{4a}{4a-4x}.$$

$$\therefore k_{50} = \frac{1}{50}\ln\frac{200}{68} = 0.0215 \text{ sec}^{-1}.$$

$$k_{100} = \frac{1}{100} \ln \frac{200}{24}$$
$$= 0.0212 \text{ sec}^{-1}.$$

k is constant. So, the reaction is of **1st order** and $k_{average}$ = **0.0213 sec⁻¹.**

34. The decomposition of phosphine at 950 K is following by noting the change in total pressure as a function of time. The reaction is :

$$4PH_3(g) \rightarrow P_4(g) + 6H_2(g).$$

The following measurements were made on a system containing PH_3 initially.

Time (sec)	0	40	80
P (total) (mm of Hg)	100	150	166.7

Show that the order of the reaction is one. Calculate the rate constant.

Hints : See Problem 33. **[Ans.** k = 0.0275 sec^{-1}]

35. The second order rate constant for the neutralisation of 2-nitroporpane by OH^- ions in aqueous solution at T °K is given by the expression,

$$\log_{10}k = -\frac{3163.0}{T} + 11.90$$

k is expressed in litre.mole^{-1}.min^{-1}. Calculate the energy of activation, and the half-life period at 20 °C, when the initial concentrations of base and acid are each 0.008 mole. litre^{-1}. **[C.U. 1997]**

Soln. : By the problem,

$$\log_{10} k_2 = -\frac{3163}{293.15} + 11.90 \text{ at } 20\text{ °C},$$

where k_2 is specific rate constant at 20 °C.

$$\therefore \quad \log k_2 = 1.110301893$$

$$\therefore \quad k_2 = 12.89 \text{ litre.mole}^{-1}.\text{min}^{-1}$$

and $\log k_1 = -\dfrac{3163}{273.15} + 11.90$ at 0 °C,

where k_1 is specific rate constant at 0 °C.

$$\therefore \quad \log k_1 = 0.32028196.$$

$$\therefore \quad k_1 = 2.09 \text{ lit.mole}^{-1}.\text{min}^{-1}.$$

$$\therefore \quad \log \frac{k_2}{k_1} = \frac{E_a}{2.303 \times 8.314} \times \frac{20}{273.15 \times 293.15}.$$

$$\therefore \quad E_a = 60569.05 \text{ J} = \mathbf{60.569 \text{ kJ.}}$$

$$t_{1/2} \text{ at } 20\text{ °C} = \frac{1}{k.a} = \frac{1}{12.89 \text{ litre.mol}^{-1}.\text{min}^{-1} \times 0.008 \text{ mol. litre}^{-1}} = \mathbf{9.69 \text{ min.}}$$

36. In a particular reaction, the time required to complete half of the reaction was found to increase nine times when the initial concentration of the reactants was reduced to one-third. What is the order of the reaction? Mention the unit of the related rate constant. **[C.U. 1996]**

Soln. : We know that

$$n = \frac{\log\dfrac{a_2}{a_1} + \log\dfrac{t_1}{t_2}}{\log\dfrac{t_1}{t_2}}.$$

t_1 is the half-life period when initial concentration is a_1, and t_2 is the half-life period when the initial concentration is a_2.

By the problem,

$$a_2 = \frac{a_1}{3};\ t_2 = 9\,t_1.$$

$$\therefore \quad n = \frac{\log\dfrac{a_1}{3a_1} + \log\dfrac{t_1}{9t_1}}{\log\dfrac{t_1}{9t_1}} = \frac{\log\dfrac{1}{3} + \log\dfrac{1}{9}}{\log\dfrac{1}{9}} = \mathbf{1.5.}$$

$\therefore$ the unit of the rate constant is **lit$^{-1/2}$.mole$^{-1/2}$.min^{-1}.**

37. A certain reaction is 20% completed in 15 minutes at 25 °C. But for the same extent of reaction it takes only 5 minutes at 35 °C. What is the activation energy of the reaction? **[C.U. 1995]**

Soln. : The change of temperature from 25 °C to 35 °C increases the rate 3 times. Hence, by standard formula,

$$\ln 3 = \frac{E_a \times 10}{8.314 \times 308.15 \times 298.15}$$

$$\therefore \quad E_a = \mathbf{83.92\ kJ.mol^{-1}}.$$

38. A certain first order reaction is 20% complete in 15 minutes at 27 °C but for the same extent of reaction it takes 5 minutes at 37 °C. What is the activation energy of the reaction? **[C.U. 2000]**

Hints : See Problem 37. **[Ans.** 85.03 kJ.mole^{-1}**]**

39. A gas decomposes according to second order kinetics. When the initial pressure of the gas is 500 torr, 40% decomposition occurs in 30 min. Find out the time required for 75% decomposition of the gas and the value of the rate constant. **[C.U. 1999]**

Soln. : We know that the specific rate (k) for 2nd order is given by

$$k = \frac{1}{t} \cdot \frac{x}{a(a-x)}.$$

Here t is the time when the amount of decomposition is x and a is the initial amount.

By the problem,

$$k = \frac{1}{30} \cdot \frac{0.4a}{a \times (a - 0.4a)} = \frac{1}{30} \times \frac{0.4a}{a \times 0.6a}.$$

$$\therefore \quad k = \frac{1}{30} \times \frac{0.4}{0.6 \times 500}\ \text{torr}^{-1}.\text{min}^{-1} = \mathbf{4.44 \times 10^{-5}\ torr^{-1}.min^{-1}}.$$

$$\therefore t_{3/4} = \frac{1}{k}\frac{0.75a}{a \times 0.25a} = \frac{1}{k}.\frac{3}{a}$$

$$= \frac{1}{4.44 \times 10^{-5}} \times \frac{3}{500} = 135.135 \text{ min} = \mathbf{135.14 \text{ min.}}$$

40. At 1100 K the following data were obtained on the homogeneous reaction,

$2NO(g) + 2H_2(g) \rightarrow N_2(g) + 2H_2O(g)$

	[NO] (mol.dm^{-3})	[H_2] (mol.dm^{-3})	Rate, r (mol.dm^{-3}.sec^{-1})
(1)	5.0×10^{-3}	2.5×10^{-3}	3.0×10^{-5}
(2)	15.0×10^{-3}	2.5×10^{-3}	9.0×10^{-5}
(3)	15.0×10^{-3}	10.0×10^{-3}	36.0×10^{-5}

(i) Calculate the order of the reaction with respect to NO, with respect to H_2 and the overall reaction order.

(ii) Write the rate law expression for the reaction.

(iii) Calculate the rate constant for the reaction.

(iv) What will be initial rate of the reaction of $[NO] = [H_2] = 8.0 \times 10^{-3}$ mole.dm^{-3}? **[C.U. 2001]**

Soln. : (i) Let the rate equation be

$$r = k.C_{NO}^x.C_{H_2}^y.$$

By the problem, let from the first set of data,

$C_{NO} = a = 5 \times 10^{-3}$ mol.dm^{-3}. and $C_{H_2} = b = 2.5 \times 10^{-3}$ mol.dm^{-3}. and rate(r) = 3.0×10^{-5} mol.dm^{-3}.sec^{-1}.

$\therefore r = k.\ a^x.b^y.$ (i)

From the second set,

$3r = k(3a)^x.b^y$. (since, now, rate = 9.6×10^{-5} mol.dm^{-3}.sec^{-1}) (ii)

Dividing eqn. (ii) by (i),

$3 = 3^x.$

$\therefore$ $\mathbf{x = 1.}$

From the third set of data,

$12r = k.(3a)^x.(4b)^y$.[since now rate $= 36 \times 10^{-5}$ mol.dm^{-3}.sec^{-1}] (iii)

Dividing eqn. (iii) by (i),

$12 = 3^x.4^y.$

Since, $x = 1, 12 = 3.4^y$

$\therefore$ $y = 1$

(ii) Rate law is $r = k.C_{NO}.C_{H_2}$

(iii) $3 \times 10^{-5} = k \times 5 \times 10^{-3} \times 2.5 \times 10^{-3}$

$\therefore$ $k = 2.4$ dm^3.mol^{-1}.sec^{-1}.

(iv) $r = 2.4.(8 \times 10^{-3})^2 = 2.4 \times (8 \times 10^{-3})^2$

$= \mathbf{1.536 \times 10^{-4}}$ **mole.dm^{-3}.sec^{-1}.**

41. A substance decomposes according to a second-order rate law. If the constant is 6.8×10^{-4} lit.mole^{-1}.sec^{-1}, calculate the half-life of the substance if the initial concentration is 0.05 mole/lit. **[C.U. 2002]**

Soln. : We know that for 2nd order reaction,

$$t_{1/2} = \frac{1}{ka},$$

where k is the rate constant and a is the initial conc. and $t_{1/2}$ is the half-life of the reaction.

By the problem,

$$t_{1/2} = \frac{1}{6.8\times10^{-4}\times0.05} = 29411.76 \text{ sec} = 490.19 \text{ min} = \mathbf{8.17\ hours.}$$

42. The gas phase reaction, $2N_2O_5 \rightarrow 4NO_2 + O_2$ has k

$$= 2.05 \times 10^{13} \exp^{-\left(24.65 \text{ kcal.mol}^{-1}/RT\right)\text{sec}^{-1}}$$

(i) Give the values of A and E_a,

(ii) Find k at 0 °C.

(iii) Find $t_{1/2}$ at 0 °C. **[C.U. 2002]**

Soln. : (i) By the problem,

$$k = 2.05 \times 10^{13}\, e^{-\left(24.65 \text{ kcal.mol}^{-1}/RT\right)\text{sec}^{-1}}.$$

We know that by Arrhenius equation,

$$k = A.e^{\frac{-Ea}{RT}}.$$

where A is the frequency factor and E_a is the activation energy.

Comparing above two equations,

$$A = \mathbf{2.05 \times 10^{13}\ sec^{-1}.}$$

$$E_a = 24650 \text{ cal}$$

$$= \mathbf{103.136\ kJ.}$$

(ii) At 0 °C, i.e., 273.15 K the value of k is :

$$2.05 \times 10^{13}\ e^{\frac{-24650}{1.987\times273.15}}$$

$$= \mathbf{3.86748 \times 10^{-7}\ sec^{-1}.}$$

(iii) $$t_{1/2} = \frac{0.693}{k} = \frac{0.693}{3.86748\times10^{-7}} = \mathbf{1791864.47\ sec.}$$

43. A third order reaction is 50% completed in 100 sec. Calculate the time for 75% and 100% completion. **[B.U. 2000]**

Soln. : We know that for a third order reaction,

$$\frac{1}{(a-x)^2} = \frac{1}{a^2} + 2\,kt, \qquad \text{(i)}$$

where a is initial concentration and k is the rate constant and x is the amount spent in time t.

$$\therefore \; t_{1/2} = \frac{3}{2k.a^2}$$

By the problem, putting in eqn. (i), $x = 0.75a$, $k = \dfrac{3}{200a^2}$,

$$\therefore \quad \left(\frac{1}{0.25a}\right)^2 - \frac{1}{a^2} = \left(\frac{3}{100a^2}\right)t.$$

$$\text{or,} \quad \frac{1-0.25^2}{(0.25a)^2} = \frac{3}{100a^2}t \quad \text{or,} \quad \frac{1-0.25^2}{0.25^2} = \frac{3}{100}t.$$

$$\therefore \; t = \frac{15\times100}{3} = \mathbf{500\ sec.}$$

A 3rd order reaction is never complete, a can be 0, i.e., reaction is 100% complet only at $t = \infty$.

44. Consider the parallel reaction.

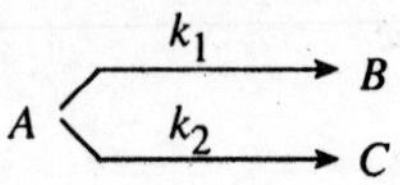

In an experiment it was observed that 60% decomposition of A takes place in 2 min and analysis of product showed that 75% of B and 25% of C are present. Calcula k_1 and k_2. **[B.U. 2000**

Soln. :

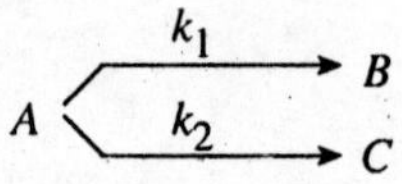

By the problem, $\dfrac{k_1}{k_2} = 3.$

$$\therefore \quad k_1 = 3k_2.$$

We know that for parallel reaction,

$$\frac{1}{t}\ln\frac{a}{a-x} = (k_1 + k_2).$$

By the problem,

$$\frac{1}{20}\ln\frac{a}{a-0.6a} = 3k_2 + k_2$$

$$\text{or,} \quad \frac{1}{20}\ln\frac{1}{0.4} = 4\,k_2.$$

$$\therefore \quad k_2 = \mathbf{0.0114\ min^{-1}}.$$

$$\therefore \quad k_1 = 3k_2 = 3 \times 0.0114 = \mathbf{0.0342\ min^{-1}}.$$

45. The first order reaction $2A \rightarrow 2B + C$ is 35% complete after 325 seconds. Ho long will it take for 70% completion? **[B.U. 199**

Soln. : We know that for first order reaction,

$$k = \frac{1}{t}\ln\frac{a}{a-x}.$$

By the problem—

$$k = \frac{1}{325}\ln\frac{a}{a-0.35a}$$

$$= \frac{1}{325}\ln\frac{1}{0.65} = 1.325 \times 10^{-3}\ \text{sec}^{-1}.$$

The time required for 70% completion is

$$t = \frac{1}{1.325\times10^{-3}}\ln\frac{a}{0.3a}$$

$$= \mathbf{908.659\ sec.}$$

46. A 1st order reaction is 20% complete in 10 min at 25 °C and 40% complete in 12 min at 40 °C. Calculate the energy of activation for the reaction. Find also the value of k at a very high termperature.

Soln. : The reaction is 20% completed in 10 min at 298.15 K. The first order rate coefficient in $k = \frac{1}{t}\ \ln\ \frac{a}{a-x}$. $x = 0.2a$. Thus $a - x = 0.8a$.

$$\therefore \quad k = \frac{1}{10}\ \ln\ \frac{1}{0.8} = 0.02231\ \text{min}^{-1}.$$

Similarly, at 313.15 °K,

$$x = 0.4a \text{ at 12 minutes}$$

$$\therefore \quad k = \frac{1}{12}\ln\frac{1}{0.6} = 0.0426\ \text{min}^{-1}$$

$$\therefore \quad \ln\ \frac{0.0426}{0.02231} = \frac{E_a}{R}\cdot\frac{15}{313.15\times298.15}.$$

$$\therefore \quad E_a = 33472.56\ \text{J} = \mathbf{33.471\ kJ.}$$

$$k = A\,e^{\frac{-E_a}{RT}}.$$

At a very high temperature **$k \to A$.**

47. At 400 K, the half-life period for the decomposition of a sample of gaseous compound initially at 55.5 kPa was 340 sec. When the pressure was 28.9 kPa the half-life period was 178 sec. Determine the order of the reaction. **[V.U. 2004]**

Soln. : We know that, the order

$$n = \frac{\log\frac{t_1}{t_2} + \log\frac{a_2}{a_1}}{\log\frac{a_2}{a_1}},$$

where t_1 is half-life when initial amount is a_1, and t_2 is half-life when the initial amount is a_2.

By the problem,

$a_1 = 55.5$ kPa $\qquad t_1 = 340$ sec

$a_2 = 28.9$ kPa $\qquad t_2 = 178$ sec

$$\therefore \quad n = \frac{\log\frac{340}{178} + \log\frac{28.9}{55.5}}{\log\frac{28.9}{55.5}}$$

$$= \frac{0.28105 - 0.28339}{-0.28339} = \frac{0.00234}{0.28339} = 8.24 \times 10^{-3} \approx \mathbf{0.}$$

48. Calculate activation energy (in SI) for a reaction whose rate coefficient, at 27 °C is doubled by 10 °C rise in temperature. (R = 8.314 J.K^{-1}.mole^{-1}) **[V.U. 2003]**

Hints : See Problem 37.

49. A solution containing equal concentration of ethyl acetate and NaOH is 25% saponified in 5 minutes. What will be the % of saponification after 10 minutes? **[V.U. 2002]**

Soln. : We know that for a 2nd order reaction with a single reactant or two reactants with same initial concentration,

$$k = \frac{1}{t}.\frac{x}{a(a-x)},$$

where a is the initial concentration and x is the amount spent at time t, and k is specific rate constant.

$$k = \frac{1}{5}\frac{0.25a}{0.75a \times a} = \frac{1}{5} \times \frac{1}{3a} = \frac{1}{15a}.$$

After 10 minutes,

$$k = \frac{1}{10}\frac{x}{a(a-x)}.$$

$$\therefore \quad \frac{1}{15a} = \frac{1}{10}\frac{x}{(a-x)a}$$

$$\Rightarrow \quad \frac{x}{a-x} = \frac{10}{15}$$

$$\Rightarrow \quad \frac{a-x}{x} = \frac{15}{10}$$

$$\Rightarrow \quad \frac{a}{x} - 1 = \frac{3}{2}$$

$$\Rightarrow \quad \frac{a}{x} = \frac{5}{2}$$

$$\Rightarrow \quad \frac{x}{a} = \frac{2}{5}.$$

So, the percentage of reactant converted is 40%.

50. Rate constants of a first order decomposition are 1.37×10^{-5} sec^{-1} and 5.15×10^{-5} sec^{-1} at 25 °C and 37 °C respectively. Calculate the activation energy and the entropy of activation for the reaction. **[V.U. 2001]**

Soln. : We know that,

$$\ln\frac{k_2}{k_1} = \frac{E_a}{R}.\left[\frac{T_2 - T_1}{T_1 T_2}\right],$$

where k_2 and k_1 are specific rates at temperature T_2 and T_1 and E_a is the activation energy, R being gas constant.

By the problem,

$k_2 = 5.15 \times 10^{-5} \text{sec}^{-1}$ at 310.15 K

$k_1 = 1.37 \times 10^{-5} \text{sec}^{-1}$ at 298.15 K.

$$\therefore \quad \ln \frac{5.15 \times 10^{-5}}{1.37 \times 10^{-5}} = \frac{E_a}{8.314} \times \frac{12}{310.15 \times 298.15}.$$

$$\therefore \quad E_a = \frac{298.15 \times 310.15 \times 8.314 \times \ln \frac{5.15 \times 10^{-5}}{1.37 \times 10^{-5}}}{12}$$

$$= 84836.815 \text{ J} = \mathbf{84.836 \text{ kJ}}.$$

For 1st order reaction,

$$\text{rate constant, } k = e \frac{RT}{N_A h} e^{\Delta S_0^{\ddagger}/R} e^{-E_a/RT}$$

putting $k = 5.15 \times 10^{-5} \text{ sec}^{-1}$ $\quad R = 8.314 \text{ J.K}^{-1}.\text{mol}^{-1}$

$T = 310.15$ K $\quad N_A = 6.022 \times 10^{23} \text{ mol}^{-1}$

$h = 6.626 \times 10^{-34}$ J.s $\quad E_a = 84836.815$ J

entropy of activation, $\Delta S_0^{\ddagger} = \mathbf{-62.11 \text{ J.K}^{-1}}$.

51. What would be $\Delta V^{\neq}$ per mole of reaction for which the rate constant increases by a factor of 1.5 for an increase of pressure from 1 to 3000 atm at 300 K.

Soln. : By equation 2.19.3, Physical Chemistry, Vol. 2 by Dr S. Pahari,

$$\Delta V^{\ddagger} = \left(\frac{\partial \Delta G^{\ddagger}}{\partial P}\right)_T$$

It can be shown from the equation that

$$\left(\frac{d \ln k_2}{dP}\right)_T = -\frac{\Delta V^{\ddagger}}{RT}.$$

$$\therefore \quad \ln \frac{k_2(P_2)}{k_1(P_1)} = -\frac{\Delta V^{\ddagger}}{RT}(P_2 - P_1),$$

where k_2 and k_1 are specific rate constants.

By the problem,

$P_1 = 1$ atm

$P_2 = 3000$ atm.

$$\therefore \quad \Delta V^{\ddagger} = -\frac{\ln 1.5 \times 0.08206 \text{ l.atm.K}^{-1}.\text{mole}^{-1} \times 300\text{K}}{2999 \text{ atm}}$$

$$= \mathbf{-3.328 \times 10^{-3} \text{ lit/mole.}}$$

52. The reaction $2NO_2 + F_2 \rightarrow 2NO_2F$ is 1st order with respect to F_2 and NO_2, where $\frac{dC_{F_2}}{dt} = kC_{F_2}C_{NO_2}$, $k = 3.40 \times 10^4 \text{ cc.mole}^{-1}.\text{sec}^{-1}$ at 25 °C. If F_2 initially at 2 mm of Hg reacts with NO_2 initially at 5 mm of Hg in rigid vessel at 25 °C, then what will be their pressures and the pressure of NO_2F after 30 sec?

The reaction is 2nd order with different initial concentrations of reactants.

Soln. : $2NO_2 + F_2 \rightarrow 2NO_2F$ Temperature, $T = 273 + 25 = 298$ °K

$b - 2x \quad a - x \quad 2x$

Thus, $\frac{1}{b-a}\ln\frac{a(b-2x)}{b(a-x)} = kt.$

or, $$\frac{3.40\times10^4 \text{ cc.mole}^{-1}.\text{sec}^{-1}\times 30\text{sec}\times10^{-3}\text{ l.cc}^{-1}}{7600\frac{\text{mm of Hg}}{\text{atm}}\times 0.08211-\text{atm.deg}^{-1}.\text{mole}^{-1}\times 298°\text{K}} = \frac{1}{b-a}\ln\frac{a(b-2x)}{b(a-x)}.$$

Now, b = 5 mm of Hg, a = 2 mm of Hg

or, $\ln\frac{2(5-2x)}{5(2-x)} = \frac{3\times3.40\times10^4\times10^{-3}\times30}{760\times0.0821\times298} = 0.16457$

or, $\frac{2(5-2x)}{5(2-x)} = 1.17888$ or, $\frac{5-2x}{2-x} = 2.9472.$

Hence, x = 0.94 mm of Hg.

Thus, after 30 sec, **pressures of NO_2, F_2 and NO_2F are 3.12 mm of Hg, 1.06 mm of Hg and 1.88 mm of Hg respectively.**

53. At 20 °C, two gases A and B_2 react as follows to give $2A + B_2 \rightarrow 2AB$(g). The rate equation is given by $r = kC_AC_{B_2}$ with $k = 3.24\times10^2$ cc.mole^{-1}.sec^{-1} at 20 °C. The initial pressure of A is 25 mm of Hg and that of B is 10 mm of Hg. What will be the pressure of A, B_2 and AB after 10 minutes?

Hints : Proceed as Problem 52. [**Ans.** 15.67, 5.34, 9.33]

54. Decomposition of N_2O_5, a first order reaction, was studied at few temperatures and the results were :

T °C	25	35	45	55	65
$k\times10^5$ sec^{-1}	1.72	6.65	24.95	75	240

Calculate frequency factor, activation energy and

$\Delta G_o^\ddagger$, $\Delta H_o^\ddagger$, $\Delta U_o^\ddagger$, $\Delta S_o^\ddagger$ at 45 °C

Hints : $\ln\frac{(k_2)_{T_2}}{(k_2)_{T_1}} = \frac{E_a}{R}\left(\frac{T_2-T_1}{T_1T_2}\right)$

For 1st order reaction,

$$k_2 = e\frac{kT}{h}\, e^{\Delta S_o^\ddagger/R}\, e^{-E_a/RT} = Ae^{-E_a/RT}.$$

Again, $\ln k_2 = \ln A - \frac{E_a}{RT}.$

$\Delta H_o^\ddagger = \Delta U_o^\ddagger = E_a - RT \quad \Delta G_o^\ddagger = \Delta H_o^\ddagger - TT\Delta S_o^\ddagger.$

$A = 2.27\times10^{13}$sec^{-1}, **E_a = 103.24 kJ, $\Delta H_o^\ddagger$ = 100.59 kJ, $\Delta U_o^\ddagger$ = 100.59 kJ, $\Delta S_o^\ddagger$ = 1.91 J.k^{-1}, $\Delta G_o^\ddagger$ = 99.98 kJ.**

CHAPTER 10

PHOTOCHEMISTRY

Required Formulae

1. $\log \frac{I_0}{I_t} = D$, where **D is extinction or absorbance or optical** density and I_0 is the intensity of the incident radiation and I_t is the intensity of transmitted radiation.

Again, $D = \varepsilon C.l$,

where ε is molar absorption or extinction coefficient or molar absorptivity, l is the path length and C is concentration.

Optical desnity (D) is additive.

$$\therefore D = \sum_{i=1}^{n} D_i.$$

2. Quantum efficiency (ϕ)

$$= \frac{\text{Number of molecules reacting}}{\text{Number of quanta absorbed}}$$

$$= \frac{\text{Number of moles reacting}}{\text{Number of einstein absorbed}}$$

$$= \frac{\text{Number of moles reacting per unit time}}{\text{Number of einstein absorbed per unit time}}$$

$$= \frac{\text{Rate}}{\text{Intensity}}.$$

If the reaction is : $v_1 A \longrightarrow P$

and I_a is the intensity in einstein of photon absorbed per unit time,

$$\phi = \frac{-\frac{1}{v_1}\frac{dC_A}{dt}}{I_a}.$$

3. The energy corresponds to one einstein of photon is :

$$E = N_A h\nu = \frac{N_A hc}{\lambda},$$

where N_A is the Avogadro number, c is the velocity of light and λ is the wavelength of the radiation.

$$\therefore\ E = \frac{6.022\times10^{23}\times6.626\times10^{-34}\text{ joule.sec}\times2.9979\times10^{8}\text{ m/see}}{\lambda\times10^{-10}\text{ m}}$$

where λ is expressed in Å.

$$E = \frac{1.1962\times10^{9}}{\lambda}\text{J} = \frac{1.1962\times10^{6}}{\lambda}\text{kJ} = \frac{2.859\times10^{5}}{\lambda}\text{k.cal.}$$

4. C is concentration of a solution in molarity C(M).

m is concentration in molality.

Theoretical Problems

1. Is the Einstein's law of photochemical equivalence always valid?

Ans. : Einstein's law of photochemical equivalence as stated for any photochemical process is often violated if there are secondary processes along with the primary process. Even for primary processes the law may be invalid for laser beam. Laser beams contain a very high density of photons, and a substance may absorb more than one photon. Sometimes as reported by E.A. Orgryzlo in 1965, a single photon can excite two molecules in contact with it. Thus, the Einstein's law is not always valid.

2. What are the factors on which the optical density of a solution depends?

Ans. : The optical density of a substance primarily depends on the wavelength of the radiation absorbed, nature of the light absorbing substance, concentration of the light absorbing substance and the length of the cell containing the solution ($D = \varepsilon.C.l.$). It also depends on the medium or solvent. Normally it does not depend on temperature.

3. Will the quantum yield be greater than one for any chain reaction?

Ans. : Generally the quantum efficiency is different from 1 if there are secondary processes. If the reaction takes place by a chain mechanism, then the primary process is the chain initiating step and the secondary processes are chain propagating (the last step being chain terminating). For a large number of chain reactions, ϕ is grater than 1. A classic example is the photochemical combination of H_2 and Cl_2 ($\phi \approx 10^6$). However, if a chain propagation step is slow (and endothermic) and requires high activation enegy, ϕ may be less than 1. A classic example is H_2 and Br_2 reaction for which $Br + H_2 \rightarrow HBr + H$ is a step requiring 75 kJ/mole as E_{act}. There is also a chain inhibiting step and ϕ is about 0.01. Therefore, for high quantum yield, the secondary processes should be fast.

4. Is it absolutely essential for light to be absorbed directly by a molecule A_2 for its photochemical decomposition?.

Ans. : The Grothuss-Draper law states that only those radiations that are absorbed by the reaction system are effective in producing a chemical change.

This law is apparently obvious since it appears that reaction by radiation cannot take place in absence of light. However, if by the system we only mean the reactant (which we normally do), then the absorption of radiation by A_2 is not a must. Let there be a substance B present which can absorb radiation, and A_2 cannot. Then B is excited to B^*, and it collides with A_2, as a result of which the energy of excitation is converted to thermal energy and A_2 reacts. Thus, it is not absolutely essential that A_2 has to absorb radiation for photochemical reaction of A_2. The above is an example of photosensitization. However, the system considered above is a mixture of A_2 and B, and the system absorbs radiation.

5. What are the exceptions to Beer's law?

Ans. : Beer's law has a number of exceptions. It is generally found that at high concentrations the substance absorbs more readily than the equation requires. The law also fails when coloured substances ionise, dissociate, or associate in solution and in highly intense light such as laser beam that contains large density of photons and induces multi-photon processes.

6. "Thermodynamically non-spontaneous reaction may also take place spontaneously in presence of light"—Explain.

Ans. : A reaction $A \longrightarrow B$ is said to be spontaneous if the equilibrium lies toward B, i.e., at equilibrium there is a very high concentration B. Normally, in such cases the free energy of the reaction defined by $\Delta G = G_B - G_A$ is negative. Thus, if the free energy of A is greater than B, then a substantial amount of A is converted spontaneously into B. Major portion of B will not be converted into A, since $G_A - G_B$ is positive (However, pure B will be converted into a mixture of A and B, since $G_f - G_i$ is always negative). However, let us assume that B can absorb a particular radiation while A cannot and let that radiation pass through the system. B then reaches an excited state of higher free energy, and now $G_A - G_B$ is negative and at equilibrium a higher proportion of A may be present. This finding is normally said to be the occurence of an apparently non-spontaneous reaction. But in reality the spontaneity is judged by $\Delta G = G_{\text{final}} - G_{\text{initial}}$, and that is always negative.

7. How can you explain the decrease in rate with temperature observed in the photochemical chlorination of benzene and some other reactions of this type?

Ans. : The effect of temperature on rate is given by the temperature coefficient of the specific rate of the reaction. For the chlorination of benzene or such other reactions, the specific rate k of the final rate equation may be expressed as $k = k'K$, where k' is a single specific rate or a product of several specific rates for the given multi-step photochemical reaction and K is the equilibrium constant involved in the formation of some of the intermediates or product. Now $\dfrac{d \ln k}{dT} = \dfrac{E_a}{RT^2}$, where E_a is the activation energy of the reaction. Hence, $E_a = E' + \Delta H°$, where $\dfrac{d \ln K}{dT} = \dfrac{\Delta H°}{RT^2}$ and $\Delta H°$ is the heat change in the reversible step. E' is the sum of the activation energy of the composite steps. Now, if $E' + \Delta H°$ is positive, E_a is positive, and the rate will increase with temperature. If $E' + \Delta H° = 0$, the rate will be independent of temperature, and if $E' + \Delta H°$ is negative, the rate will decrease with increase in temperature. Therefore, the above mentioned reaction must be such that $E' + \Delta H°$ is negative. This can easily happen if $\Delta H°$ refers to an exothermic heat change (–ve) and the magnitude is greater than $+E'$. Therefore, if there is a reversible step, as is normally assumed, then it is a highly exothermic one. (E_a can be negative also, if $\dfrac{k_1}{k_2} = K$, where k_1 and k_2 are single or product specific rates so that $E_a = E_1 - E_2$ and $E_2 > E_1$).

8. Let the energy of radiation sent through a system be slightly greater than the bond energy of A–A in the system. Will there be absorption always?

Ans. : A photochemical reaction can take place if the bond of the type *A–A* or *A–B* breaks as a result of passage of light. Thus, the electron of the system should be excited from the ground electronic state to the highest vibrational level of at least the 1st excited electronic state. Therefore, it may appear that if the energy of radiation sent to a system slightly greater than the bond energy, then absorption should take place.

However, the ground electronic state of a majority of stable systems are singlet and the corresponding 1st excited electronic state is triplet. And singlet to triplet electronic transitions are forbidden for systems like *A–A*. Naturally in most of those cases, there will be no absorption.

9. What is fluorescence?

Ans. : Fluorescence is the radiative transition between two states of same multiplicity The process may be such that frequency of transmittance radiation may be equal to or less than that of incident radiation. Generally, the frequency of the transmitted radiation is less and it takes place from the ground vibrational level of the 1st excited singlet state to the ground singlet state.

10. What is phosphorescence?

Ans. : Phosphorescence is the radiative transition between two states of diferent multiplicity. In this case the frequency of the emitted radiation is always less than the frequency of absorbed radiation.

11. Give an example of a photochemical reaction which may be used to measure the intensity of the radiation.

Ans. : A photochemical reaction of known quantum yield may be used in actinometers to measure the intensity of radiation. A common reaction is decomposition of oxalic acid sensitized by UO_2^{++}. A mixture of oxalic acid (about 0.04M) and UO_2SO_4 (0.01M) is taken. UO_2^{++} absorbs radiation in the region 250 to 440 nm and the reaction is :

$$UO_2^{++} \xrightarrow{h\nu} \left(UO_2^{++}\right)^*$$

$$\left(UO_2^{++}\right)^* + (COOH)_2 \rightarrow UO_2^{++} + CO_2 + CO + H_2O$$

The quantum efficiency is 0.57. The amount of oxalic acid decomposed is determined by the titration of standard $KMnO_4$ and hence the amount of radiation absorbed is ascertained.

12. Discuss with an example what is meant by photostationary equilibrium.

Ans. : The photostationary state is a steady state which arises if the forward or backward or both reactions are affected by the radiation. In that case, the thermal equilibrium state may be shifted to another minimum which remains the same as long as radiation is not cut off. One of the well-known examples of photostationary equilibrium is the dimerisation of Anthracene (A).

Anthracene absorbs radiation to give the dimer and hence, the equilibrium state between A and A_2 is shifted more towards A_2 in comparison to thermal conversion.

Numerical Problems

1. *k* in the expression $I = I_o\, e^{-k.lC}$ is 6.63 for light of 3130 Å for acetone vapour *C* is the concentration in moles/litre, and *l*, the cell length in cm. Calculate the percentage of light absorbed by acetone vapour at 100 mm pressure at 50°C, when a beam of light is passed through a cell 15 cm long. **[C.U. 1967]**

Soln. : By the problem,

$k = 6.63$

$$C = \frac{P}{RT} = \frac{\frac{100}{760}\text{atm}}{0.08206\,\text{lit.atm.mole}^{-1}\text{K}^{-1} \times 323.15\,\text{K}}$$

$$= \frac{100}{760 \times 0.08206 \times 323.15}\ \text{moles/lit}$$

$l = 15$ cm.

$$\therefore \quad \ln\frac{I}{I_o} = -k.lC$$

$$= -6.63 \times 15\ \text{cm} \times \frac{100}{760 \times 0.08206 \times 323.15}\ \text{moles/lit} = -0.493464.$$

$$\therefore \quad \frac{I}{I_o} = 0.6105079 = 0.61.$$

Hence, the percentage of light transmitted is 61%. Therefore, % of light absorbed is **39%.**

2. A 0.003 (M) solution of a coloured substance transmits 75% of the incident light of 500 mm, when placed in a cell of length 1 cm. Calculate the molar extinction coeficient, and hence, the optical density of a 0.001(M) solution in the same cell at the same wavelength. **[C.U. 1974]**

Soln. : By the problem,

$$\frac{I_t}{I_0} = 0.75;\ C = 0.003\ \text{M};\ \ l = 1\ \text{cm}.$$

We know that,

$$\log\frac{I_t}{I_0} = -\varepsilon.C.l.$$

$$\therefore \varepsilon = -\frac{I}{Cl}\log\frac{I_t}{I_0}$$

$$= -\frac{1}{0.003(\text{M}) \times 1\,\text{cm}} \times \log 0.75$$

$$= 41.646245\ \text{cm}^{-1}.(\text{M}^{-1}) \approx \mathbf{41.6\ cm^{-1}.(M^{-1})}.$$

$\therefore$ optical density (D)

$$= \varepsilon.C.l$$

$$= 41.6\ \text{cm}^{-1}.(\text{M}^{-1}) \times 0.001\ (\text{M}) \times 1\ \text{cm} = \mathbf{0.0416}.$$

3. A 10^{-3}(M) solution of *A* also contains some *B*, and the solution when placed in a 2 cm cell absorbs 80% of the incident light at a certain wavelength. If the extinction coefficients of *A* and *B* at this wavelength are 250 and 1000 respectively, find the concentration of *B* in the solution. **[C.U. 1967]**

Soln. : By the problem,

$$\log\frac{I_t}{I_0} = \log\frac{20}{100}.$$

$\therefore \quad D = \log \frac{I_o}{I_t} = \log 5 = 0.69897.$

But $D = D_1 + D_2$.

Again, $D_1 = \varepsilon_1 C_1 l_1$

$= 250 \text{ cm}^{-1}.(\text{M}^{-1}) \times 10^{-3} \text{ (M)} \times 2 \text{ cm}$

$= 0.5.$

$D_2 = \varepsilon_2 C_2 l_2$

$= 1000 \text{ cm}^{-1}.(\text{M}^{-1}) \times C_2 \times 2 \text{ cm}$

$= 2000\ C_2 (\text{M}^{-1}).$

$\therefore\ 2000\ C_2 (\text{M}^{-1}) + 0.5 = 0.69897.$

$\therefore \quad C_2 = \frac{0.19897}{2000} (\text{M}) = 9.9485 \times 10^{-5} (\text{M}).$

$\therefore$ the concentration of B is **9.9×10^{-5}(M).**

4. A 1×10^{-3} M solution of a dye X shows an absorbance of 0.20 at 450 mμ and an absorbance of 0.05 at 620 mμ. A 1×10^{-4} (M) solution of another dye Y shows 0.00 absorbance at 450 mμ and an absorbance of 0.42 at 620 mμ. Calculate concentration of each dye present together in a solution which exhibits an absorbance of 0.38 and and 0.71 at 450 mμ and 620 mμ respectively. The same cell is used in all measurements and its thickening is 1 cm. **[C.U. 1979]**

Soln. : By the problem,

	Dye X	Dye Y
	$C = 10^{-3}$ M	$C = 10^{-4}$ M
D^{450}	0.2	0
D^{620}	0.05	0.42
mixture of	D^{450}	D^{620}
X and Y	0.38	0.71

$\therefore \quad D_x^{450} = 0.2 = \varepsilon_x^{450} \text{C.l.}$

$= 10^{-3}.(\text{M}) \times \varepsilon_x^{450} \times 1 \text{ cm}$

$\therefore \quad \varepsilon_x^{450} = 200 \text{ cm}^{-1}.(\text{M}^{-1}).$

Again, $D_x^{620} = 0.05 = \varepsilon_x^{620} \times \text{Cl}$

$= \varepsilon_x^{620} \times 10^{-3}.(\text{M}) \times 1 \text{ cm}.$

$\therefore \quad \varepsilon_x^{620} = 50 \text{ cm}^{-1}.(\text{M}^{-1})$

$D_y^{450} = 0,\ \varepsilon_y^{450} = 0$

$D_y^{620} = 0.42 = \varepsilon_y^{620} \times \text{Cl} = \varepsilon_y^{620} \times 10^{-4} \text{ (M)} \times 1 \text{ cm}.$

$\therefore \quad \varepsilon_y^{620} = 4200 \text{ cm}^{-1}.(\text{M}^{-1}).$

Now, in the mixture at 450 mμ,

$D_{\text{mixture}}^{450} = 0.38 = \varepsilon_x^{450} \times C_x \times 1 + \varepsilon_y^{450} \times C_y \times 1$

$= 200 \text{ cm}^{-1}.(\text{M}^{-1}) \times C_x \times 1 \text{ cm} + 0 \times C_y \times 1 = 200\ C_x (\text{M}^{-1}).$

$$\therefore \quad C_X = \frac{0.38}{200} = \mathbf{1.9 \times 10^{-3}\ (M).}$$

$$\text{And } D^{620}_{\text{mixture}} = 0.71 = \varepsilon_x^{620} C_x l + \varepsilon_y^{620} C_y l$$

$$= 50\ \text{cm}^{-1}.(\text{M}^{-1}) \times 1.9 \times 10^{-3}(\text{M}) \times 1\ \text{cm} + 4200\ \text{cm}^{-1}.(\text{M}^{-1}).\ C_y \times 1\ \text{cm}$$

$$= 50 \times 1.9 \times 10^{-3} + 4200\ C_y (\text{M}^{-1})$$

$$\text{or,} \quad 0.71 = 0.095 + 4200\ C_y (\text{M}^{-1})$$

$$\text{or,} \quad 0.615 = 4200\ C_y (\text{M}^{-1})$$

$$\text{or,} \quad C_Y = \frac{0.615}{4200}\ (\text{M})$$

$$= 1.46428 \times 10^{-4}\ (\text{M})$$

$$= \mathbf{1.5 \times 10^{-4}\ (M).}$$

5. The percentage of transmittance of an aq. solution of a dye at 450 mμ and at 25 °C is 30% for a 2×10^{-3} (M) solution in a 2 cm cell. Calculate the optical density and the molar extinction coefficient. Find the conc. of the same dye in another solution where the percentage transmittancy is 20% in a 1 cm cell at the same temperature and same wavelength of light. **[C.U. 1980]**

Soln. : By the problem,

$$\frac{I_t}{I_o} = \frac{30}{100}.$$

$$\therefore \quad \frac{I_o}{I_t} = \frac{100}{30}.$$

$$\therefore D = \log \frac{I_o}{I_t} = \log \frac{100}{30} = 0.5228787 = 0.52.$$

Now, $D = \varepsilon.C.l$

$$\therefore \quad \varepsilon = \frac{D}{Cl} = \frac{0.5228787}{2 \times 10^{-3}(\text{M}) \times 2\,\text{cm}}$$

$$= 130.71961\ \text{cm}^{-1}.(\text{M}^{-1})$$

$$= 130.7\ \text{cm}^{-1}.(\text{M}^{-1}).$$

Now, $\log \dfrac{I_t}{I_o} = -\varepsilon.C.l.$

$$\therefore C = -\frac{1}{\varepsilon l} \log \frac{I_t}{I_o}$$

$$= -\frac{1}{130.7\,\text{cm}^{-1}(\text{M}^{-1}).1\text{cm}} \log \frac{20}{100}$$

$$= \mathbf{5.348 \times 10^{-3}\ (M).}$$

6. In a cell of a certain length, and at a pressure of 100 mm of Hg, gaseous acetone transmits 25.1% of the incident light of wavelength 2650 Å. Assuming Beer's law to apply, calculate the pressure at which 98% of the incident radiation will be absorbed by gaseous acetone (obeying ideal gas laws) in the same cell at the same temperature. **[C.U. 1987]**

Soln. : By the problem, $\frac{I_t}{I_0} = 10^{-\varepsilon Cl}$

or, $\log \frac{I_t}{I_0} = -\varepsilon Cl$

or, $\log 0.251 = -\varepsilon \times \left(\frac{100}{760RT}\right) \times 1.$

$\therefore \quad \varepsilon = -\frac{1}{1}\left(\frac{760RT}{100}\right) \times \log 0.251.$

Again, when 98% is absorbed, 2% is transmitted.

$\therefore \log 0.02 = -\frac{1}{l} \cdot \left(\frac{760RT}{100}\right) \times \log 0.251 \times C.l.$

or, $\log 0.02 = -\left(\frac{760RT}{100}\right) \times \log 0.251 \times \frac{P}{760\,RT}$

$\therefore$ the required pressure in mm of Hg is

$$P = -100 \times \frac{\log 0.02}{\log 0.251} = 283.00777 \text{ mm of Hg} = \mathbf{283 \text{ mm of Hg.}}$$

7. Monochromatic light is passed through a 1 mm path length cell containing 0.005 moles/dm³ solution. The light intensity is reduced to 16% of its value. Calculate the molar extinction coefficient of the sample. What would be the transmittance if the cell path is 2 mm? **[C.U. 1987]**

Soln. : By the problem, $\log \frac{I_0}{I_t} = \log \frac{100}{16} = D = \varepsilon.C.l.$

Here $C = 0.005$ moles/dm³.

$l = 1 \text{ mm} = 0.1 \text{ cm}$

$$\therefore \quad \varepsilon = \frac{D}{Cl} = \frac{\log \frac{100}{16}}{0.005 \text{ moles/dm}^3 \times 0.1 \text{ cm}} = 1591 \text{ lit/mole.cm}^{-1}.$$

If the path length is 2 mm, i.e., 0.2 cm, then

$$\therefore \quad \log \frac{I_0}{I_t} = 1591 \text{ lit.mol}^{-1}.\text{cm}^{-1} \times 0.005 \text{ mole/lit} \times 0.2 \text{ cm} = 1.591.$$

$$\therefore \quad \log \frac{100}{I_t} = 1.591.$$

$$\therefore \quad I_t = 2.56.$$

8. A 0.01 molar solution of a compound transmits 20% of Na-D line when the absorbing path length is 1.5 cm. What is the molar extinction coefficient of the substance? Solvent is assumed to be completely transparent. **[C.U. 1989]**

Hints : $D = \log \frac{100}{20}$

$$\therefore \quad \varepsilon = \frac{\log 100/20}{0.01 \times 1.5} = \mathbf{46.6\ (M^{-1}).cm^{-1}}.$$

9. An aqueous solution of a compound A of concentration 10^{-3} moles/litre absorbs 50% of incident radiation in a cell length 1 cm and another compound B of concentration 2×10^{-3} moles/litre absorbs 60% of the incident radiation at a particular wavelength. Calculate the percentage absorbed by a solution containing 10^{-3} moles/litre of A and B each in the same cell at the same wavelength. **[V.U. 1994]**

Soln. : For compound A, 50% light is absorbed.

$$\therefore \quad D = \log\frac{100}{0.50} = \varepsilon_A . Cl.$$

$$\therefore \quad \varepsilon_A = \frac{\log 1/0.50}{10^{-3}\text{moles / lit} \times 1\,\text{cm}} \text{ [by the problem].}$$

$$\therefore \quad \varepsilon_A = 301.02 \text{ lit. mole}^{-1}.\text{cm}^{-1}.$$

For compound B, 60% light is absorbed, i.e., 40% is transmitted.

$$\therefore \quad D = \log\frac{100}{40} = \varepsilon_B . Cl.$$

$$\therefore \quad \varepsilon_B = \frac{\log 100/40}{2 \times 10^{-3}\text{moles / lit} \times 1\text{ cm}} \text{ [by the problem]}$$

$$= 198.97 \text{ lit.mole}^{-1}.\text{cm}^{-1}.$$

So, for solution containing 10^{-3} moles.litre^{-1} of A and B,

$$D = (\varepsilon_A + \varepsilon_B) \times C.l$$

$$= 499.99 \text{ lit.mole}^{-1}.\text{cm}^{-1} \times 10^{-3}\text{mole.lit}^{-1} \times 1 \text{ cm}$$

$$= 0.49999.$$

$$\therefore \quad \log\frac{I_o}{I_t} = 0.49999.$$

$$\therefore \quad \log\frac{100}{I_t} = 0.49999.$$

$$\therefore \quad I_t = 31.64.$$

$\therefore$ (100 – 31.64)% = **68.36%** light is absorbed.

10. At 460 nm a blue filter transmits 75% of the light and a yellow filter transmits 40% of the light. What is the transmittance at the same wavelengths of the two filters in combination? **[V.U. 1996]**

Soln. : By the problem,

$$D \text{ of blue filter} = \log\frac{100}{75} = 0.125$$

$$D \text{ of yellow filter} = \log\frac{100}{40} = 0.398.$$

As optical density is an additive property, so optical density of combination of blue and yellow filter is (0.125 + 0.398) = 0.523.

$$\therefore \quad \log\frac{100}{I_t} = 0.523.$$

$$\therefore \quad I_t = 29.99.$$

So, the transmittance is **29.99%**.

11. A 2×10^{-4} (M) solution of a dye when placed in a cell of 1.0 cm path-length transmits 10% of the 460 nm radiation incident on it. Calculate molar extinction coeficient. If the same solution be placed in a 5 cm cell, what would be the absorbance? Will the molar absorption coefficient change if the measurement be made with (i) 5×10^{-4}(M) solution in 2 cm cell with 460 nm (ii) 2×10^{-4}(M) solution in 1 cm cell with 500 nm? **[C.U. 1997]**

Hints : See similar problem. [**Ans.** 5000 $l.mol^{-1}.cm^{-1}$, 5]

ε depends on nature of substance and wavelength. So, it will not change in case (i), but will change in case (ii).

12. Tryptophen residues in proteins absorb light with molar extinction coeficient of $\varepsilon = 54 \times 10^5$ $cm^2.mole^{-1}$. If a protein sample of 5×10^{-5} $mole.litre^{-1}$ shows absorbance $A = 0.54$ for 1 cm path-length. Calculate tryptophen residues in the sample. **[B.U. 1996]**

Hints : Calculate conc. by using Lambert Beer's law.

13. A compound absorbs U.V. radiation at wavelength 3000 Å. What would be the total energy in kilocalories absorbed by 1 mole of this compound when irradiated with this radiation? (1 cal = 4.18×10^7 erg; $h = 6.625 \times 10^{-27}$ erg-sec)

Soln. : We know that,

$$E = \frac{N_A hc}{\lambda}$$

$$= \frac{6.022 \times 10^{23} \times 6.625 \times 10^{-27} \text{ erg.sec} \times 2.9979 \times 10^{10} \text{ cm/sec}}{3000 \times 10^{-8} \text{ cm}}$$

$$= 3.9867 \times 10^{12} \text{ erg}$$

$$= 95375.6 \text{ cal} = 95.376 \text{ kcal.}$$

14. A gas, when exposed to radiation of $\lambda = 3310$ Å undergoes decomposition, and per kilocalories of light energy absorbed, 0.0230 moles of the gas is decomposed. Calculate the quantum efficiency. **[C.U. 1976]**

1 cal = 4.18×10^7 erg; $h = 6.62 \times 10^{-27}$ erg-sec

Soln. : Energy of one einstein of 3310 Å is

$$E = \frac{N_A hc}{\lambda}$$

$$= \frac{6.022 \times 10^{23} \times 6.62 \times 10^{-27} \text{ erg.sec} \times 2.9979 \times 10^{10} \text{ cm/sec}}{3310 \times 10^{-8} \text{ cm}}$$

$$= 86.379 \text{ kcal} = 1 \text{ einstein.}$$

$\therefore$ 1 kcal = 0.01157 einstein.

$\therefore$ quantum yield (ϕ)

$$= \frac{\text{No. of moles of gas decomposed}}{\text{No. of einstein absorbed}} = \frac{0.023}{0.01157}$$

$$= 1.987899 \approx \mathbf{1.99.}$$

15. The photochemical dissociation of gaseous HI to form normal H_2 and I_2 requires radiation of 4040 Å or less.

(i) Determine the molar heat of dissociation of HI.

(ii) If radiation of 2537 Å is used, how much energy will appear as kinetic energy of the atoms?

Soln. : (i) Now, $E = \dfrac{N_A hc}{\lambda}$

$$= \frac{6.022\times10^{23}\times6.625\times10^{-34}\text{ J.sec}\times2.997\times10^{8}\text{ cm / sec}}{4040\times10^{10}\text{ m}}$$

$= 296048.19$ J

$= 296.048$ kJ

$=$ **70.572 kcal.**

(ii) Now for 2537 Å,

$E_F = 471436.61$ J

$= 471.43661$ kJ

$= 112.6761$ kcal.

$\therefore\ E - E_F = (112.6761 - 70.7572)$ kcal

$=$ **41.919 kcal** energy will appear as K.E. of the atom.

16. The total amount of radiation of wavelength 4200 Å falling over a certain time on solution was calculated from radiation intensity to be 2.5 cal. In this time 1.8×10^{-5} moles of the reactant was decomposed. Calculate the quantum efficiency.

Soln. : We know that the energy of 1 einstein of photon of λ = 4200 Å,

$$E = \frac{N_A hc}{\lambda} = \frac{6.022\times10^{23}\times6.625\times10^{-34}\text{ J.sec}\times2.9979\times10^{8}\text{ m / sec}}{4200\times10^{-10}\text{ m}}$$

$= 284770.1641$ J

$= 68061.7$ cal.

$\therefore$ 1 calorie $= \dfrac{1}{68061.7}$ einstein.

$\therefore$ 2.5 calorie $= 2.5\times\dfrac{1}{68061.7}$ einstein

$= 3.67424\times10^{-5}$.

$$\therefore\ \phi = \frac{1.8\times10^{-5}}{3.67424\times10^{-5}} = 0.4898 \approx \mathbf{0.5.}$$

17. The dissociation energy of H_2 is 102.9 kcal.mole^{-1}. If H_2 is illuminated with radiation whose wavelength is 2537 Å and is thereby dissociated, how much of the radiant energy per mole will be converted to kinetic energy?

Soln. : Now, $E = \dfrac{N_A hc}{\lambda} = \dfrac{6.022\times10^{23}\times6.625\times10^{-34}\text{ J.sec}\times2.9979\times10^{8}\text{ m / sec}}{2537\times10^{-10}\text{ m}\times4.184\text{ J.cal}^{-1}}$

$= 112.676$ kcal.

Hence, the energy per mole converted to K.E. is

$(112.676 - 102.90)$ kcal $=$ **9.776 kcal.**

18. Absorption of each calorie gives 1.44×10^{-5} gm of H_2 in the photo-dissociation of HI vapour using radiation of wavelength 2070Å. Calculate the quantum yield.

Soln. : Now, energy of 1 einstein of λ = 2070 Å,

$$E = \frac{N_A hc}{\lambda}$$

$$= \frac{6.022 \times 10^{23} \times 6.625 \times 10^{-34} \text{ J.sec} \times 2.9979 \times 10^{8} \text{ m/sec}}{2070 \times 10^{-10} \text{ m} \times 4.184 \text{ J.cal}^{-1}}$$

$= 138096.21$ cal.

$\therefore$ 1 cal $= \frac{1}{138096.21}$ einstein.

Now, $2HI = H_2 + I_2$

$$\therefore \quad \phi = \frac{1.44 \times 10^{-5}}{\frac{1}{138096.21}} = \mathbf{1.99.}$$

Since, 1.44×10^{-5} gm of hydrogen is the amount in gm-atm that should result from HI per (Photon $\times$ N_A) unit amount.

19. In the photochemical combination of H_2(g) and Cl_2(g) a quantum yield of 1×10^6 is obtained with light of wavelength 4800 Å. How many moles of HCl(g) would be produced under these conditions per calorie of light energy absorbed? **[C.U. 1982]**

Soln. : By the problem,

$$E = \frac{N_A hc}{\lambda} = \frac{6.022 \times 10^{23} \times 6.625 \times 10^{-34} \text{ J.sec} \times 2.9979 \times 10^{8} \text{ m/sec}}{4800 \times 10^{-10} \text{ m} \times 4.184 \text{ J.cal}^{-1}}$$

$= 59553.99$ cal.

$\therefore$ 1 cal $= \frac{1}{59553.99}$ einstein.

Now, $H_2 + Cl_2 \xrightarrow{h\nu} 2HCl$

$$\therefore \phi = \frac{\frac{1}{2} \times \text{number of moles of HCl produced}}{\text{number of einstein absorbed}}.$$

$\therefore$ number of moles of HCl produced $= 2 \times \phi \times$ number of einstein absorbed

$$= 2 \times 10^6 \times \frac{1}{59553.99} = 33.583 = 33.6.$$

20. An uranyl oxalate actinometer is irradiated for 15 minutes with light of wavelength 4350 Å and oxalic acid equivalent to 12.0 cc. of 0.001(M) $KMnO_4$ is found to have been decomposed. The quantum efficiency of the actinometer at this wavelength is 0.58. Calculate the average intensity of the light used in (i) erg/sec and (ii) quanta/sec. **[C.U. 1983]**

Soln. : (i) The reaction is :

$2KMnO_4 + 3H_2SO_4 + 5H_2C_2O_4 = K_2SO_4 + 2MnSO_4 + 10CO_2 + 8H_2O$

Thus, 2 moles of $KMnO_4$ react with 5 moles of $H_2C_2O_4$ i.e., 1000 cc, 2 molar $KMnO_4$ = 5 moles of $H_2C_2O_4$.

Now, 1 cc 1 molar $KMnO_4 = 2.5 \times 10^{-3}$ moles of $H_2C_2O_4$.

Number of moles of oxalic acid decomposed is equivalent to 12 cc 0.001(M) $KMnO_4$

$\therefore$ number of moles of $H_2C_2O_4$ decomposed $= 2.5 \times 10^{-3} \times 12 \times 10^{-3}$ moles

$= 3 \times 10^{-5}$ moles.

$$\text{Thus,} \quad \frac{\text{number of moles of } H_2C_2O_4 \text{ decomposed}}{\text{number of einstein absorbed}} = \phi.$$

$\therefore$ number of einstein absorbed (N_E) = $\dfrac{3 \times 10^{-5}}{0.58}$ = 5.17241×10^{-5}.

(ii) Thus, number of quanta absorbed/sec

$$= \frac{5.17241 \times 10^{-5} \times 6.022 \times 10^{23}}{900} = 3.4609 \times 10^{16} = \mathbf{3.5 \times 10^{16}}.$$

Now, $E = \dfrac{N_A hc}{\lambda} = \dfrac{6.022 \times 10^{23} \times 6.625 \times 10^{-27} \text{ erg.sec} \times 2.9979 \times 10^{10} \text{ cm/sec}}{4350 \times 10^{-8} \text{ cm}}$

$= 2.7495 \times 10^{12}$ erg.

$\therefore$ energy absorbed is $\dfrac{E \times N_E}{900}$ per sec

$$= \frac{2.7495 \times 10^{12} \times 5.17241 \times 10^{-5}}{900} = \mathbf{1.58 \times 10^5 \text{ erg/sec.}}$$

21. Consider the photochemical reaction :

$$A \xrightarrow{h\nu} B + C$$

When irradiated with light of wavelength 2537 Å during a certain period, the light energy absorbed = 3.436×10^8 erg and the number of moles of B formed = 3.64×10^{-6} mole. Calculate the quantum yield. **[C.U. 1984]**

Soln. : By the problem, energy of one einstein = $\dfrac{N_A hc}{\lambda}$

$$= \frac{6.022 \times 10^{23} \times 6.625 \times 10^{-27} \text{ erg.sec} \times 2.9979 \times 10^{10} \text{ cm/sec}}{2537 \times 10^{-8} \text{ cm}}$$

$= 4.71436 \times 10^{12}$ erg.

$\therefore$ 3.436×10^8 erg = $\dfrac{3.436 \times 10^8}{4.71436 \times 10^{12}}$ einstein = 7.28837×10^{-5} einstein.

$\therefore$ $\phi = \dfrac{3.64 \times 10^{-6}}{7.28837 \times 10^{-5}} = 0.04994 \approx \mathbf{0.05}$.

22. A 100 cm^3 vessel containing H_2 and Cl_2 was irradiated with light of 400 nm. Measurements with a thermopile showed that 11×10^{-7} J of light energy was absorbed by the chlorine per second. During an irradiation of 1 minute the partial pressure of chlorine, as determined by the absorption of light and the application of Beer's law, decreased from 27.3 to 20.8 kPa (corrected to 0°). Determine the quantum yield.

Soln. : We know, $P = CRT$.

The decrease in pressure of Cl_2 is (27.3 – 20.8) kPa = 6.5 kPa

$$= \frac{6.5 \text{ kPa}}{101.325 \text{ kPa}} \text{ atm} = 0.06415 \text{ atm}.$$

Hence, the concentration of Cl_2 consumed is

$$C = \frac{0.06415 \text{ atm}}{0.08206 \text{ lit.atm.mole}^{-1}.\text{K}^{-1} \times 273.15 \text{ K}}$$

$= 2.86195 \times 10^{-3}$ moles/lit.

Since, the vessel is 100 cc, the number of moles consumed is 2.86195×10^{-4}.

The light energy absorbed in one minute was $11 \times 10^{-7} \times 60$ J = 6.6×10^{-5} J.

Now, Energy of one einstein = $\frac{N_A hc}{\lambda}$

$$= \frac{6.022 \times 10^{23} \times 6.625 \times 10^{-34} \text{ J.sec} \times 2.9979 \times 10^8 \text{ m/sec}}{4000 \times 10^{-10} \text{ m}}$$

= 299008.67 J [since 400 nm = 4000 Å].

$\therefore 6.6 \times 10^{-5}$ J = $\frac{6.6 \times 10^{-5}}{299008.67}$ einstein = $2.2072938 \times 10^{-10}$ einstein.

$$\therefore \phi = \frac{2.86195 \times 10^{-4}}{2.2072938 \times 10^{-10}} = \mathbf{1.296 \times 10^6}.$$

23. The quantum yield is 2 for the photolysis of gaseous HI to H_2 and I_2, by the light of 253.7 nm wavelength. Calculate the number of moles of HI that will be decomposed if 300 J of light of this wavelength is absorbed.

Soln. : $\phi = \frac{\text{no. of moles decomposed}}{\text{no. of einstein absorbed}}$

Now, energy of one einstein is

$$\frac{N_A hc}{\lambda} = \frac{6.022 \times 10^{23} \times 6.625 \times 10^{-34} \text{ J.sec} \times 2.9979 \times 10^8 \text{ m/sec}}{253.7 \times 10^{-9} \text{ m}}$$

= 471436.6 J = 4.71436×10^5 J.

$\therefore$ 300 J = $\frac{300}{4.71436 \times 10^5}$ einstein

= 6.3635×10^{-4} einstein.

By the problem, 2 = $\frac{\text{no. of moles decomposed}}{6.36355 \times 10^{-4}}$.

$\therefore$ number of moles decomposed

= $2 \times 6.3635 \times 10^{-4}$ = $\mathbf{1.2727 \times 10^{-3}}$.

24. The incident radiation from a monochromatic source of 2537 Å on a slit is 80 ergs/sec/sq. mm. The radiation is passed through a slit whose area is 5 sq. mm and into a cell whose face transmits 30% of the incident radiation. If the solution in the cell transmits 10% of the radiation, how much energy in calories is absorbed by the solution per hour?

Soln. : The radiation is passed through 5 sq. mm slit and hence, $80 \times 5 = 400$ ergs/sec of energy falls on the cell. In one hour, $400 \times 60 \times 60$ ergs energy falls on.

By the problem,

30% falls on the solution and 10% of it is transmitted, i.e., 3% of the original energy is transmitted. The solution, thus, absorbs (30 – 3) = 27%.

$\therefore$ light absorbed,

$\frac{27}{100} \times 400 \times 60 \times 60$ ergs = 388800 ergs

= 9.2925×10^{-3} cal = $\mathbf{9.3 \times 10^{-3}}$ **cal.** (since, 4.184×10^7 ergs = 1 calorie)

25. In an experiment to measure the quantum efficiency for a photochemical reaction, the absorbing substance was exposed to 480 nm light from a 150 W source for a period of 30 minutes. The intensity of the transmitted light was 40% of the intensity of the incident light. As a result of this, 0.35 mole of the substance decomposed. Calculate ϕ.

Soln. : Energy of 1 einstein

$$= \frac{6.022\times10^{23}\times6.625\times10^{-34}\ \text{J.sec}\times2.9979\times10^{8}\ \text{m/sec}}{480\times10^{-9}\ \text{m}} = 249173.89\ \text{J.}$$

Energy coming out from 150 W source in 30 minutes

$$= 150\frac{\text{J}}{\text{sec}} \times 30 \times 60\ \text{sec} = 270000\ \text{J.}$$

$\therefore$ number of einstein of the incident light $= \dfrac{270000}{249173.89} = 1.08358 \approx 1.08.$

$\therefore$ number of einstein absorbed $= 0.6 \times 1.08 = 0.65.$

$$\therefore \phi = \frac{0.35\ \text{mole}}{0.63\ \text{einstein}} = 0.5384615 = 0.54\ \text{mole. einstein}^{-1}.$$

26. In a photochemical reaction, the stoichiometry for which $A \rightarrow 2B+C$, the quantum efficiency with 400 nm light is 2. After exposure of 3×10^{-3} moles of A, 2.4×10^{-3} moles of B are formed. How many photons were absorbed by A? How much energy was absorbed?

Soln. : $\phi = \dfrac{\text{No. of moles reacting}}{\text{No. of einstein absorbed}}$

$\therefore$ number of photons absorbed $(N_0) = \dfrac{\text{Number of mlecules reacting}}{\phi}$.

Since 2.4×10^{-3} moles of B are formed, 1.2×10^{-3} moles of A must have reacted.

$$\therefore N_0 = \frac{1.2\times10^{-3}\times6.022\times10^{23}}{2} = \frac{3.6\times10^{20}}{6.022\times10^{23}} = 0.598 \times 10^{-3}\ \text{einstein}$$

Now, energy of one einstein

$$= \frac{6.022\times10^{23}\times6.625\times10^{-34}\ \text{J.sec}\times2.9979\times10^{8}\ \text{m/sec}}{400\times10^{-9}\ \text{m}} = 299008.67\ \text{J.}$$

$\therefore$ energy absorbed $= 0.598 \times 10^{-3} \times 299008.67$ J $=$ **178.81 J.**

27. In three seperate experiments of passage of light through uranyl oxalate actinometer, the actinometry solution consisted of (a) 5.232 gm, (b) 8.47 gm and (c) 4.852 gm of oxalic acid respectively in 50 cc of water. After exposure for 4 minutes, the remaining solution was titrated in these 3 cases seperately with $KMnO_4$, and the $KMnO_4$ solution required for complete oxidation of the remaining oxalic acid was (a) 18 cc of 0.98(M), (b) 25 cc of 1.11(M) and (c) 13 cc of 0.88(M), solution respectively. What is the average intensity of the light used in photons/sec and joules/sec in these 3 cases separately, if, in all these cases, light of wavelength 4050 Å was used. The quantum efficiency is reported as 0.55.

Soln. : (a) Number of moles of oxalic acid $= \dfrac{5.232}{90} = 0.0581333$ in 50 cc.

$$2KMnO_4 + 3H_2SO_4 + 5H_2C_2O_4 = K_2SO_4 + 2MnSO_4 + 8H_2O + 10\,CO_2$$

2000 cc 1 (M)$KMnO_4$ = 5 moles of oxalic acid

$\therefore$ 1 cc 1 (M)$KMnO_4$ $= 2.5 \times 10^{-3}$ moles of oxalic acid.

Thus,

(a) 18 cc 0.98(M) $KMnO_4$

$= 18 \times 0.98 \times 2.5 \times 10^{-3}$ moles of oxalic acid

$= 0.0441$ moles in 50 cc.

Now energy/einstein

$$= \frac{N_A hc}{\lambda} = \frac{6.022\times10^{23}\times6.625\times10^{-34}\text{ J.sec}\times2.9979\times10^{8}\text{ m / sec}}{4050\times10^{-10}\text{ m}}$$

$= 295317.21$ J

$= 295.32$ kJ.

(a) Oxalic acid decomposed $= (0.0581 - 0.0441)$ moles

$= 0.014$ moles in 240 sec

No. of moles decomposed/sec $= \frac{0.014}{240} = 5.8333 \times 10^{-5}$.

$$\therefore \phi = \frac{\text{No. of moles decomposed / sec}}{\text{Number of einstein absorbed / sec}}$$

$$\therefore \text{No. of einstein absorbed/sec} = \frac{\text{No. of moles decomposed / sec}}{\phi}$$

$$= \frac{5.8333\times10^{-5}}{0.55}$$

$= 1.0606 \times 10^{-4} = 1.0606 \times 10^{-4} \times 6.022 \times 10^{23}$ photons/sec.

$=$ **6.38693×10^{19}** photons/sec.

Energy in J/sec $= 1.0606 \times 10^{-4} \times 295317.21$ J

$= 31.321343 =$ **31.3 J/sec.**

Similarly, (b) and (c) can be calculated.

[**Ans.** (b) 1.13×10^{20}, 55.3; (c) 1.15×10^{20}, 56.6]

28. An actinometer contains 20 cc of 0.05 moles/lit oxalic acid through which light of 350 nm was passed for 2 hours. This light was absorbed by the uranyl oxalate. After exposure the solution required 34 cc $KMnO_4$ to titrate the undecomposed oxalic acid. The same volume, i.e., 20 cc oxalic acid required 40 cc of $KMnO_4$ before exposure. Calculate the energy absorbed in joule/sec, if $\phi = 0.57$.

Soln. : The solution contains $\frac{0.05}{1000} \times 20$

$= 10^{-3}$ moles of oxalic acid.

The complete reaction is :

$2KMnO_4 + 3H_2SO_4 + 5H_2C_2O_4 = K_2SO_4 + 2MnSO_4 + 8H_2O + 10\,CO_2$

$\therefore$ 1000 cc 2(M) $KMnO_4$ = 5 moles of oxalic acid.

$\therefore$ 1 cc x(M) $KMnO_4 = \frac{5}{2000} \times x$ moles of oxalic acid

$= 2.5 \times 10^{-3}\ x$ moles of oxalic acid.

Hence, 40 cc x (M) $KMnO_4 = 2.5 \times 10^{-3} \times 40x$ moles of oxalic acid.

The solution contained 10^{-3} moles of oxalic acid.

By the problem,

$$40 \times 2.5 \times 10^{-3} x = 10^{-3}$$

$$\therefore \quad x = \frac{1}{100} = 10^{-2}.$$

Hence, the strength of $KMnO_4$ is 10^{-2}(M). Again, by the problem, $H_2C_2O_4$ decomposed is equivalent to (40 – 34) = 6 cc 10^{-2}(M) $KMnO_4$.

Now, 1 cc of 1(M)$KMnO_4$ = 2.5×10^{-3} moles of oxalic acid.

∴ 6 cc of 10^{-2}(M) $KMnO_4$ = $6 \times 10^{-2} \times 2.5 \times 10^{-3} = 1.5 \times 10^{-4}$ moles of oxalic acid

∴ number of einstein absorbed

$$= \frac{1.5 \times 10^{-4}}{0.57} = 2.63157 \times 10^{-4}$$

Hence, the energy absorbed in joule/sec is :

$$2.63157 \times 10^{-4} \times \frac{6.022 \times 10^{23} \times 6.625 \times 10^{-34} \text{ J.sec} \times 2.9979 \times 10^{8} \text{ m/sec}}{350 \times 10^{-9} \text{ m} \times 2 \times 60 \times 60 \text{ sec}}$$

$= \mathbf{1.25 \times 10^{-2}}$ **J/sec.**

29. An actinometer contains 20 cc of 0.05 moles/lit oxalic acid through which light of 350 nm was passed for 2 hours. This light was absorbed by uranyl oxalate. After exposure the solution required 28 cc of $KMnO_4$ to titrate the undecomposed oxalic acid. The same volume of solution required 40 cc of $KMnO_4$ before exposure. Calculate the energy absorbed, given that quantum efficiency wavelength is 0.57.

Hints : Proceed as Problem 28. **[Ans.** 2.5×10^{-2} joule/sec]

30. At 480 nm, the quantum yield for the production of Fe^{2+} in the photolysis of $K_3Fe(C_2O_4)_3$ in 0.05 moles/lit of H_2SO_4 is 0.94. After 15 min irradiation in a cell containing 107.4 cc of solution, the solution is mixed thoroughly and 5 ml of sample is pipetted into a 25 ml volumetric flask. A quantity of 1,10 phenanthroline is added and the flask is filled to the mark with a buffer solution. A sample of this solution is placed in a 1.00 cm colorimeter cell and the transmittance is measured relative to a solution containing no iron. The value $\frac{I_t}{I_o}$ is found to be 0.548.

If the extinction coefficient is 1.15×10^3 m^2/mole for the complex, how many quanta were absorbed by the original solution? What was the absorbed intensity in J/sec and in photon/sec?

Soln. : By the problem—

$$\varepsilon = 1.15 \times 10^3 \text{ m}^2.\text{mole}^{-1}$$

$$l = 10^{-2} \text{ m}$$

$$\frac{I_t}{I_o} = 0.548.$$

We know that, $\log \frac{I_t}{I_o} = -\varepsilon.Cl.$

$$\text{or, } C = -\frac{1}{\varepsilon l} \log \frac{I_t}{I_o} = -\frac{1}{1.15 \times 10^3 \text{ m}^2.\text{mole}^{-1} \times 10^{-2} \text{ m}} \times \log 0.548$$

$$= 0.0227147 \text{ mole.m}^{-3} = 2.27 \times 10^{-5} \text{ mole.lit}^{-1}.$$

The colorimeter cell contains $\frac{2.27 \times 10^{-5}}{1000} \times 25$ moles of iron. This is present in 5 cc of original solution. Therefore, 107.4 cc of original solution contains

$$\frac{2.27 \times 10^{-5}}{1000} \times 25 \times \frac{107.5}{5} \text{ moles of iron.}$$

i.e., 1.220125×10^{-5} moles of iron as a result of exposure. Now, the reaction is :

$$2Fe^{3+} + C_2O_4^{2-} \longrightarrow 3Fe^{2+} + 2CO_2.$$

$$\text{Now, } \phi_{Fe^{+2}} = \frac{\text{no. of gm moles of } Fe^{2+} \text{ produced}}{\text{no. of einsteins absorbed } (N_E)}.$$

$$\therefore \text{ by the problems, } N_E = \frac{1.220125 \times 10^{-5}}{0.94} = 1.298 \times 10^{-5}. \; [\because \phi_{Fe^{+2}} = 0.94]$$

Thus, the no. of quanta absorbed is :

$$1.298 \times 10^{-5} \times 6.022 \times 10^{23} = 7.816556 \times 10^{18} = 7.82 \times 10^{18}.$$

Now, the energy corresponding to 1 einstein,

$$E = \frac{6.022 \times 10^{23} \times 6.625 \times 10^{-27} \times 2.9979 \times 10^{10}}{4800 \times 10^{-8} \times 10^{7}}.$$

So, the intensity absorbed in $J.sec^{-1}$ is :

$$\frac{I}{sec} = \frac{E \times N_E}{900} = \mathbf{3.59 \times 10^{-3} \; J.sec^{-1}}$$

$$\text{No. of photon/sec} = \frac{7.82 \times 10^{18}}{900} = \mathbf{8.69 \times 10^{15} \text{ photons/sec.}}$$

31. An actinometer uses a solution of $K_3[Fe(C_2O_4)_3]$ in which Fe^{3+} is reduced and the oxalate ion is oxidised. Assuming $\phi = 1.24$ at 310 nm, calculate the intensity of the incident light which produces 1.3×10^{-5} mole of Fe^{2+} in 36.5 min. The same light source is used to irradiate a sample of CH_2CO for a period of 15.2 min. If the quantum yield of C_2H_4 is 1.0 and that of CO is 2.0, determine the amount of each gas produced by the photochemical reaction. **[B.U. 1995]**

Soln. : According to the given problem,

$\phi = 1.24$; no. of mole of Fe^{2+} produced $= 1.3 \times 10^{-5}$ mole in 36.5 min.

$$\text{We know that, } \phi = \frac{\text{no. of moles of } Fe^{2+} \text{ produced}}{\text{no. of einstein absorbed } (N_E)}$$

or, no. of einsteins absorbed (N_E)

$$= \frac{\text{no. of moles of } Fe^{2+} \text{ produced}}{\phi} = \frac{1.3 \times 10^{-5}}{1.24} = 1.048 \times 10^{-5}.$$

In the given problem, time of absorption = 36.5 min = 36.5×60 sec.

So, intensity of the incident light in terms of no. of quanta absorbed/sec

$$= \frac{1.048 \times 10^{-5} \times 6.022 \times 10^{23}}{36.5 \times 60} = 2.882 \times 10^{15}/\text{sec.}$$

Now, intensity in terms of ergs/sec is :

Intensity $= n.h.\nu$,

where n = no. of quanta absorbed/sec.

h = Planck constant

ν = frequency of the incident light.

Here $$\nu = \frac{c}{\lambda} = \frac{2.9979\times10^{10}\ \text{cm / sec}}{310\times10^{-7}\ \text{cm}}$$

$$\therefore\ \text{Intensity} = \frac{2.882\times10^{15}\times6.626\times10^{-27}\times2.9979\times10^{10}}{310\times10^{-7}}$$

$= \mathbf{1.847\times10^4}$ **ergs/sec.**

$$CH_2CO + h\nu \longrightarrow \frac{1}{2}C_2H_4 + CO$$

Energy of 1 einstein

$$= \frac{6.022\times10^{23}\times6.626\times10^{-27}\ \text{erg.sec}\times2.9979\times10^{10}\ \text{cm / sec}}{310\times10^{-7}\ \text{cm}}$$

$= 3.859\times10^{12}$ ergs.

Energy absorbed by the sample in 15.2 mins

$$= 1.847\times10^4\ \text{ergs/sec}\times60\ \frac{\text{sec}}{\text{min}}\times15.2\ \text{min}$$

$= 16844640$ ergs.

$$\text{Energy absorbed in einstein} = \frac{16844640\ \text{ergs}}{3.859\times10^{12}\ \dfrac{\text{ergs}}{\text{einstein}}} = 4.36503\times10^{-6}\ \text{einstein.}$$

$$\text{Now, } \phi_{C_2H_4} = \frac{\text{no. of moles of } C_2H_4 \text{ produced}}{\text{no. of einstein absorbed}}$$

Hence, number of moles of C_2H_4 produced $= \phi_{C_2H_4}\times4.36503\times10^{-6}$

$= \mathbf{4.36503\times10^{-6}}$ $(\phi_{C_2H_4} = 1)$

$$\phi_{CO} = \frac{\text{no. of moles of CO produced}}{\text{no. of einstein absorbed}}$$

Hence, no. of moles of CO produced $= \phi_{CO}\times4.36503\times10^{-6}$

$= 2\times4.36503\times10^{-6} = \mathbf{8.73\times10^{-6}}$ **mole.**

32. An acid solution of $K_3[Fe(C_2O_4)_3]$ is used as an actinometer to measure the intensity of light incident upon a vessel. After exposure to light of wavelength 253.7 nm for 10^3 sec, the concentration of Fe^{2+} ion in a 1 lit soln. was found to be 5.0×10^{-5}(M). The fraction of light absorbed by the actinometer is 0.1. What is the intensity of the incident radiation? Given $\phi = 1.25$.

Soln. : Energy of light per einstein :

$= N_A\times h.\nu.$

$$= \frac{6.022\times10^{23}\times6.626\times10^{-34}\times2.9979\times10^{8}}{253.7\times10^{-9}}$$

$= 471.5078$ kJ.

Number of einstein per sec = 471.5078 J/sec.

$$\text{Now, } \phi = \frac{\text{no. of moles of } Fe^{2+} \text{ produced}}{\text{no. of einsteins absorbed } (E_N)}.$$

$$\therefore E_N = \frac{5\times10^{-5}}{1.25} = 4\times10^{-5}.$$

$$\text{Now, } I_o = \frac{I_a}{0.1} \text{ or, } \frac{I_a}{I_o} = 0.1.$$

Here $I_a = 4 \times 10^{-5} \times 471.5078$ J/sec = 0.01886 J/sec.

$$\therefore \quad I_o = \frac{I_a}{0.1} = \mathbf{0.1886 \ J/sec.}$$

33. A photographic film is exposed for 0.1 sec to a 250 W incident light at a distance of 3 m. If 10% of the power is emitted as visible light of 625 nm to which the film is sensitive, then estimate the number of Ag atoms that will be produced in a AgBr grain, 15 μm in diameter.

Soln. : By the problem, the power emitted as visible light sensitive to the film is

$$\frac{10}{100} \times 250 = 25 \text{ watts.}$$

Since the exposure is for 10^{-1} sec, the energy emitted is 25 × 0.1 J = 2.5 J. Now, the area receiving the energy is $4\pi r^2 = 4\pi(3)^2 = 113.112$ sq. metre.

Intercepting area of AgBr grain is πr^2 (circular exposure)

$= 3.142 \times (15 \times 10^{-6})^2 = 7.0695 \times 10^{-10}$ sq. m.

Therefore, energy received by AgBr is :

$$\frac{2.5\times7.0695\times10^{-10}}{113.112} = 1.5625 \times 10^{-11} \text{ J.}$$

Now, the energy corresponding to 1 einstein is :

$$E = \frac{N_A hc}{\lambda} = \frac{6.022\times10^{23}\times6.626\times10^{-34}\times2.9979\times10^{8}}{625\times10^{-9}} = 1.91\times10^{5} \text{ J.}$$

$$\therefore \quad 1 \text{ J} = 5.22481 \times 10^{-6} \text{ einstein.}$$

The total einstein absorbed is = $1.5625 \times 10^{-11} \times 5.22481 \times 10^{-6} = 8.16 \times 10^{-17}$ photons.

The number of Ag atoms produced = $816 \times 10^{-17} \times 6.022 \times 10^{23}$ = $\mathbf{4.91 \times 10^{7}}$.

34. The quantum efficiency for the conversion of molecule x to y is 0.5 at 300 nm. For what length of time must irradiation of 120 W source continues in order to destroy 1 mole of x?

Soln. : Energy corresponding to 1 einstein,

$$E = \frac{N_A hc}{\lambda} = \frac{6.022\times10^{23}\times6.626\times10^{-34}\times2.9979\times10^{8}}{300\times10^{-9}} = 3.9873841\times10^{5} \text{ J.}$$

Thus, a 120 W source produces—

$$= \frac{120}{3.9873841\times10^{5}} \text{ moles of photons/sec}$$

$= 3.00949 \times 10^{-4}$ moles of photons/sec.

Thus, $0.5 \times 3.00949 \times 10^{-4}$ moles are destroyed per sec. So the time required is 6645.6442 sec = **1.846 hour.**

35. A reaction responds to both red and violet light ($\lambda = 8000$ Å and 4000 Å respectively) with an equal quantum yield. Will there be more photochemical reaction per 100 cal of light in the red than in the violet or vice versa? **[C.U.(H) 1992]**

Soln. : Let ϕ_V be efficiency with violet light and ϕ_R be efficiency with red light. Let Y_R be number of moles reacting per 100 cal for red light of wavelength λ_R, and Y_v be the number of moles reacting per 100 cal of violet light of wavelength λ_v.

$$\phi_v = \frac{\text{no. of moles reacting}}{N_A \times \text{no. of violet quanta absorbed}}.$$

$$\phi_R = \frac{\text{no. of moles reacting}}{N_A \times \text{no. of red quanta absorbed}}.$$

Now, energy corresponding to 1 quanta is

$$E = \frac{hc}{\lambda} = \frac{1.9864 \times 10^{-25}\ \text{J. metre}}{\lambda}$$

$$= \frac{1.9864 \times 10^{-16}\ \text{J. metre}}{\lambda(\text{nm})}\ \text{J} = \frac{4.74763 \times 10^{-17}}{\lambda(\text{nm})}\ \text{cal}$$

$$= \frac{4.75 \times 10^{-17}}{\lambda(\text{nm})}\ \text{cal.}$$

$$\therefore \quad 1\ \text{cal} = \frac{\lambda}{4.75 \times 10^{-17}}\ \text{quanta}$$

$$\text{and } 100\ \text{cal} = \frac{\lambda}{4.75 \times 10^{-19}}\ \text{quanta.}$$

$$\therefore \quad \frac{\dfrac{Y_v}{N_A . \lambda_v}}{4.75 \times 10^{-19}} = \frac{\dfrac{Y_R}{N_A . \lambda_R}}{4.75 \times 10^{-19}}$$

$$\therefore \quad Y_R = Y_v . \frac{\lambda_R}{\lambda_v}.$$

Thus, $Y_R > Y_V$ (since, $\lambda_R > \lambda_v$).

Hence, **there will be more photochemical reactions for the red light in the given case.**

36. A 100 watts Na-vapour lamp radiates most of its energy in the yellow D-line at 589 nm. How long will it take to excite more than half the molecules of an absorbing species in a 10^{-3} mole.dm^{-3} sample if all radiant energies are absorbed by the sample? **[C.U. 1991]**

Soln. : One photon of 589 nm carries energy

$$= \frac{6.626 \times 10^{-34}\ \text{J.s} \times 2.9979 \times 10^{8}\ \text{m/sec}}{589 \times 10^{-9}\ \text{m}}\ \text{J}$$

$$= 3.3725 \times 10^{-19}\ \text{J.}$$

A 100 watts lamp emits 100 J/sec. So, no. of photons emitted per sec

$$= \frac{100}{3.375 \times 10^{-19}} = 2.963 \times 10^{20}.$$

Now, 10^{-3} mole.dm^{-3} = 6.022×10^{20} molecules.dm^{-3}. To excite more than half molecules we require more than $\frac{6.022}{2} \times 10^{20}$ photons.

i.e., > 3.011×10^{20} photons. So, time required is greater than $\frac{3.011 \times 10^{20}}{2.963 \times 10^{20}}$.

i.e., > **1.016 seconds.**

37. In the decomposition of ethylene iodide ($C_2H_4I_2$) with radiation of 424 nm, the I_2 formed after 20 min required 41.4 ml of 0.005(M) sodium thiosulphate solution. The intensity of light source was 9.15 erg.sec^{-1}. Calculate the quantum yield. **[B.U. 2000]**

Soln. : $2Na_2S_2O_3 + I_2 \longrightarrow Na_2S_4O_6 + 2NaI$

Thus, 1000 cc 1(M) NaS_2O_3 = $\frac{1}{2}$mole of I_2.

Hence, 41.4 cc 0.005 (M) $Na_2S_2O_3$ = $\frac{41.4 \times 0.005}{2 \times 1000}$ = 1.035×10^{-4} moles of I_2.

Energy absorbed by the system in 20 min from the light source of intensity of 9.15 erg. sec^{-1} = $9.15 \times 60 \times 20$ erg

= 10.980 erg.

Energy corresponding to 1 einstein of photon of λ of 424 nm

$$= \frac{6.022 \times 10^{23} \times 6.626 \times 10^{-27} \text{ erg.sec} \times 2.9979 \times 10^{10} \text{ cm / sec}}{424 \times 10^{-7} \text{ cm}}$$

$= 2.82 \times 10^{12}$ erg.

Thus, energy absorbed in einstein = $\frac{10980 \text{ erg} \times 1 \text{ einstein}}{2.82 \times 10^{12} \text{ erg}}$ = 3.8936×10^{-9} einstein.

$$\text{Hence, quantum yield} = \phi = \frac{\text{number of moles of } I_2 \text{ formed}}{\text{number of einstein absorbed}}$$

$$= \frac{1.035 \times 10^{-4}}{3.8936 \times 10^{-9}} = \mathbf{2.65 \times 10^4}.$$

38. When acetone vapour is irradiated with monochromatic light of wavelength 313 nm, it decomposes to form ethane and carbon monoxide.

$(CH_3)_2CO(g) \longrightarrow C_2H_6(g) + CO(g)$

The reaction cell used has a volume of 59 cm^3. Irradiation for 7 hours at a temperature 56.7 °C produces a change of pressure from 102.16 kPa to 104.42 kPa. If the incident intensity is 48.1×10^{-4} J.sec^{-1} and acetone vapour absorbs 91.5% of it, calculate quantum yield.

Soln. : Energy of one einstein corresponding to the radiation of λ (313 nm)

$$= \frac{6.022 \times 10^{23} \times 6.626 \times 10^{-34} \text{ J.sec} \times 2.9979 \times 10^{8} \text{ m / sec}}{313 \times 10^{-9} \text{ m}}$$

= 382177.388 J.

Now change in pressure due to decomposition of acetone

= (104.42 − 102.16) kPa = 2.26 kPa.

$$(CH_3)_2CO(g) \longrightarrow C_2H_6(g) + CO(g),\ T = (273.15 + 567)°K = 329.85°K$$

$$n_1 - x \qquad x \qquad x$$

Initial pressure, $P_{in} \propto n_1$

Final pressure, $P_{fi} \propto n_1 - x + x + x$

$\propto n_1 + x$

Thus, $x \propto P_{fi} - P_{in}$.

So, acetone decomposed in gm-mole/litre

$$= \frac{(P_{fi} - P_{in})}{RT} \text{ (since, } P = CRT)$$

$$= \frac{2.26 \times 10^3 \text{ N / m}^2}{1.01325 \times 10^5 \frac{\text{N / m}^2}{1 \text{ atm}} \times 0.0821\ l.\text{atm.deg}^{-1}.\text{mole}^{-1} \times 329.85\ °\text{K}}$$

$= 8.236300 \times 10^{-4}$ moles/litre.

The cell is of 59 cc capacity.

No. of moles of acetone decomposed

$$= \frac{8.236300 \times 10^{-4} \text{ mole / litte} \times 59 \text{ cc}}{1000 \text{ cc / litre}}$$

$= 4.8594 \times 10^{-5}$ moles.

Energy absorbed by acetone vapour in 7 hours

$$= 48.1 \times 10^{-4} \text{ J/s} \times 3600 \frac{\text{sec}}{\text{hr}} \times 7 \text{ hr} \times \frac{91.5}{100} = 110.90898 \text{ J}$$

$$\text{Thus, energy absorbed in einstein} = \frac{110.90898 \text{ J}}{3.82177 \times 10^5 \text{ J / einstein}}$$

$= 2.90203 \times 10^{-4}$ einstein.

$$\text{Hence, } \phi = \frac{4.8594 \times 10^{-5}}{2.90203 \times 10^{-4}} = \mathbf{0.167.}$$

39. The quantum yield for the photolysis of acetone $(CH_3)_2CO = C_2H_6 + CO$ with the radiation of wavelength 300 nm is 0.2. How many moles of CO are formed per second if the intensity of the 300 nm radiation absorbed is 10^{-2}J. sec^{-1}?

Soln. : Energy corresponding to 1 einstein

$$= \frac{6.022 \times 10^{23} \times 6.626 \times 10^{-34} \text{ J / sec} \times 2.9979 \times 10^8 \text{ m / sec}}{300 \times 10^{-9} \text{ m}}$$

$= 398738.4076$ J.

$$\text{Energy absorbed in einstein} = \frac{10^{-2} \text{ J}}{398738.4076 \text{ J / einstein}}$$

$= 2.50791 \times 10^{-8}$ einstein.

$$\text{Now, } \phi = \frac{\text{no. of moles of CO formed}}{\text{no. of einstein absorbed}}.$$

Thus, $0.2 = \dfrac{\text{no. of moles of CO formed}}{2.50791 \times 10^{-8} \text{ einstein}}$.

Hence, moles of CO formed = $0.2 \times 2.50791 \times 10^{-8}$ = $\mathbf{5.0158 \times 10^{-9}}$.

40. When acetone vapour is irradiated with light of wavelength 3130 Å, it decomposes to form ethane and carbon monoxide :

$$(CH_3)_2CO(g) + h\nu \longrightarrow C_2H_6(g) + CO(g)$$

Using a reaction cell of 60.3 ml capacity and a temperature of 56 °C, irradiation for 23000 sec at a rate of 85200 erg.sec^{-1} produced a change in pressure from 760 mm to 790.4 mm of Hg. Calculate (i) the number of molecules of acetone decomposed and (ii) quantum yield. **[C.U. 2006]**

Hints : Proceed as Problem 38.

[Ans. (i) 5.375×10^{19} no of acetone molecules, (ii) 0.174]

41. A solution of a coloured compound of concentration 1.0×10^{-4} (M) has 20% transmission in cell of path-length 1.0 cm at 450 nm wavelength. Calculate the molar absorption coefficient (ε) of the substance. If the path-length and the concentration are both halved, calculate the percentage transmission. Will the value of ε change if light of wavelength 550 nm is used? **[Ans.** ε = 6989.7 cm^{-1}.(M^{-1}), 67%, ε depends on wavelength]

42. The quantum yield for the photochemical decomposition $2HI \longrightarrow H_2 + I_2$ at $\lambda = 254$ nm is 1.99. 3070 J of energy is absorbed, find the number of moles of HI decomposed. **[B.U. 1998]**

Soln. : Energy of 1 einstein

$$= \frac{6.022 \times 10^{23} \times 6.626 \times 10^{-34} \text{ J.sec} \times 2.9979 \times 10^{8} \text{ m/sec}}{254 \times 10^{-9} \text{ m}} = 470950.88 \text{ J.}$$

Energy absorbed by the system = 3070 J.

It corresponds to $\dfrac{3070}{470950.88}$ einstein = 6.5187×10^{-3} einstein.

$$\phi = \frac{\text{no. of moles of HI decomposed}}{\text{no. of einstein absorbed}}$$

Hence, no. of moles of HI decomposed

$= 1.99 \times 6.5197 \times 10^{-3} = 1.29 \times 10^{-2}$ moles.

43. A dye solution of concentration 0.01 gm/cc absorbs 40% of blue light in a cell of thickness 1 cm. What should be the concentration to ensure 90% absorption in the same cell? **[B.U. 1998]**

Soln. : By the problem, $\dfrac{I_t}{I_o} = 0.6$, l = 1 cm, $C = 0.01$ gm/cc

$\log 0.6 = -\varepsilon Cl.$

Let C' be the concentration when $\dfrac{I_t}{I_o} = 0.1$.

Thus, $\log 0.1 = -\varepsilon C' l$ or, $\dfrac{C'}{C} = \dfrac{\log 0.1}{\log 0.6}$.

Hence, $C' = 0.01 \text{ gm/cc} \times \frac{\log 0.1}{\log 0.6} = 0.045 \text{ gm/cc}.$

44. The mechanism of the photolysis of HCHO vapours to H_2 and CO at 313 nm is possibly as the following :

(i) $HCHO \xrightarrow{h\nu} CO + H_2$

(ii) $HCHO \xrightarrow{h\nu} H + CHO$

(iii) $H + HCHO \xrightarrow{k_1} H_2 + CHO$

(iv) $CHO + M \xrightarrow{k_2} CO + H + M$

(v) $CHO + \text{wall} \xrightarrow{k_3} \frac{1}{2}CO + \frac{1}{2}HCHO$

If quantum yields for step (i) and step (ii) are ϕ_1 and ϕ_2 respectively, obtain the expression for quantum yield for H_2 in terms of ϕ_1 and ϕ_2.

Soln. : Let ϕ_{H_2} be quantum yield for H_2. Thus, $\phi_{H_2} = \frac{\frac{dC_{H_2}}{dt}}{I_a}$.

$$\text{Now } \frac{dH_2}{dt} = I_a\phi_1 + k_1 C_H C_{HCHO}, \quad (1)$$

$$\text{where } \phi_1 = \frac{-\frac{dC_{HCHO}}{dt}}{I_a} = \frac{\frac{dC_{H_2}}{dt}}{I_a}.$$

Again from step (ii),

$$\phi_2 = \frac{\frac{dC_H}{dt}}{I_a} = \frac{\frac{dC_{CHO}}{dt}}{I_a}$$

Applying steady state conditions to H and CHO, we obtain

$$\frac{dC_H}{dt} = 0 = I_a\phi_2 + k_2 C_{CHO} C_M - k_1 C_H C_{HCHO} \quad (2)$$

$$\frac{dC_{CHO}}{dt} \text{ and } = 0 = I_a\phi_2 + k_1 C_H C_{HCHO} - k_2 C_{CHO} C_M - k_3 C_{CHO}. \quad (3)$$

Adding Eq (2) and Eq (3), we get

$$2I_a\phi_2 - k_3 C_{CHO} = 0.$$

$$\therefore C_{CHO} = \frac{2\phi_2 I_a}{k_3}.$$

$$\text{Now } \frac{dC_{H_2}}{dt} = \phi_1 I_a + \phi_2 I_a + k_2 C_{CHO} C_M = \phi_1 I_a + \phi_2 I_a + 2\frac{k_2}{k_3}\phi_2 I_a C_M$$

$$\text{Thus, } \phi_{H_2} = \phi_1 + \phi_2 + 2\frac{k_2}{k_3}\phi_2 C_M.$$

CHAPTER 11

ELECTROLYTIC CONDUCTANCE

Required Formulae

1. $G = \frac{\kappa A}{l} = \frac{\kappa}{\text{cell constant}}$,

where G = conductance,

κ = conductivity or specific cenductance

l = distance between two parallel electrodes

A = area of the electrode.

2.(a) $\Lambda = \frac{1000\kappa}{N}$,

where Λ = equivalent conductance or equivalent conductivity

N = normality of the solution.

If N^* is concentration in gram-equivalent per cc, then

$\Lambda = \frac{\kappa}{N^*}$.

(b) $\Lambda_m = \frac{1000\kappa}{C}$.

C = molarity of the solution and $N = ZC$.

$\Lambda = \frac{\Lambda_m}{Z}$.

For an electrolyte, $A_{\nu^+}B_{\nu^-} \longrightarrow \nu^+A^{z+} + \nu^-B^{z-}$

$\Lambda = \frac{\Lambda_m}{\nu_+Z_+} = \frac{\Lambda_m}{\nu_-Z_-} = \frac{\Lambda_m}{Z}$.

3.(a) For strong electrolyte, $\Lambda° = \lambda+ + \lambda-$,

where $\Lambda°$ = equivalent conductance at infinite dilution

$\lambda+$ = equivalent conductance of positive ion

$\lambda-$ = equivalent conductance of negative ion.

(b) For an electrolyte $A_{\nu_+}B_{\nu_-}$

$$\Lambda_m = \nu_+\lambda_{m,+} + \nu_-\lambda_{m,-}$$

$\lambda_{m,+}$= molar conductance of positive ion

$\lambda_{m,-}$= molar conductance of negative ion.

4. $$v_{\pm} = u \pm \frac{\Delta\phi}{l}$$

$v_{\pm}$ = velocity of the ion

$u_{\pm}$ = mobility of the positive or negative ion

$\Delta\phi$ = potential difference between two electrodes or voltage applied

$\frac{\Delta\phi}{l}$ = potential gradient.

5.(a) $$t_{\pm} = \frac{I_{\pm}}{I_+ + I_-}$$

$t_{\pm}$ = transport number of positive or negative ion

$I_{\pm}$ = current carried by positive or negative ion.

I_+ = current carried by positive ion

I_- = current carried by negative ion.

For an electrolyte $A_{\nu_+}B_{\nu_-}$

$$t_{\pm} = \frac{Z_{\pm}C_{\pm}Fv_{\pm}A}{Z_+C_+Fv_{\pm}A + Z_-C_-Fv_-A} = \frac{v_{\pm}}{v_+ + v_-}$$

Since, $C_{\pm} = \nu_{\pm}C$ and $Z_+\nu_+ = Z_-\nu_-$

$C_{\pm}$ = conccentration of positive or negative ion in gm-mole. m^{-3}

C= concentration of electrolyte in gm-mole. m^{-3}

In a mixture where C's of different salts may vary.

$$t_B = \frac{Z_BC_Bv_B}{\Sigma Z_iC_iv_i}$$

Z_B = charge of ion B

C_B = concentration of ion B in gm.mole.m^{-3}

v_B = velocity of ion B

Z_i = change of ith ion

C_i = concentration of ith ion

v_i = velocity of ith ion

Since $v = u.\Delta\frac{\phi}{l}$, $t_B = \frac{Z_BC_BU_B}{\Sigma Z_iC_iV_i}$ and $\lambda_i = Fu_i$, thus $t_B = \frac{Z_BC_B\lambda_B}{\Sigma Z_iC_i\lambda_i}$,

where, λ_i = equivalent conductance of ith ion; $\kappa_B = Z_BC_B\lambda_B$ and $\kappa_i = Z_iC_i\lambda_i$, thus,

$$t_B = \frac{\kappa_B}{\Sigma\kappa_i}.$$

5.(b) $\lambda m,B = FZ_Bu_B$, thus $t_B = \frac{\lambda_{mB}C_B}{\Sigma\lambda_{m_i}C_i}$.

6. Transport number by moving boundary method :

$$t_i = \frac{laFN}{1000Q},$$

where l is the length through which boundary moves and a is the inner area of the tube in which ion moves. $\frac{Q}{F}$ is the charge in faraday that passes through $\frac{N}{1000}$ number of gm-equivalent per unit volume. In C.G.S., it is gm-eq.ml^{-1}.

Also, $\frac{t_i}{Z} = \frac{laFC}{Q}$

where C = concentration of electrolyte in gm.mole.m^{-3}.

7. Hittorf's Method

$t_{\pm}$ = loss or gain in gm-eq per faraday in a given chamber.

If the electrodes are inert, then loss of cation in cathode chamber is t_- gm-eq per faraday and loss of anion from the same chamber is t_- gm-eq per faraday. Thus, loss of electrolyte from cathode chamber is t_- gm-eq per faraday. Similary, loss of electrolyte from anode chamber is t_+ gm-eq per faraday.

Loss of anion from cathode chamber per faraday is t_- gm-eq. in cathode compartment.

If the electrodes are reactive, i.e., electrolysis of $CuSO_4$ between copper electrodes, of $AgNO_3$ between silver electrodes, then gain in cation, anion and salt in anode chamber is t_- gm-eq per faraday. If the cathode is reactive which will be a rare case, then the net gain of electrolyte in cathode chamber is t_+ gm-eq. For detailed analysis go through Physical Chemistry Vol. I by Dr S. Pahari (page nos. 559-561)

Short Questions

1. "During measurement of conductance a.c. is used."—Explain.

Ans. : If d.c. is used in conductance measurements and current flows through the cell, there will be electrolysis. The concentration will change as a result of this, and hence, also the conductance as the conductance depends on concentration. Moreover, there will be polarisation at each electrode due to this irreversible phenomenon. Thus, to avoid electrolysis and polarisation, a.c. is used. If a.c. is used, electrodes change sign regularly. However, use of a.c. cannot eliminate polarisation completely, but it may reduce polarisation considerably. The higher the frequency, the better is the reduction of polarisation.

2. The electrodes used in conductance measurements are platinised.—Explain.

Ans. : When a.c. is used, the polarisatioin of electrodes is greatly reduced. But unless extremely high frequency is used, polarisation is not completely eliminated. H^+ and OH^- ions, being very fast, tend to be liberated at the electrodes by successive pulses of current.

Electrodes are coated with finely divided platinum black for aiding the elimination of polarisation effects by alternating current. This is platinisation. The larger surface area of the finely divided platinum appears to catalyse the union of hydrogen and oxygen, which tend to be liberated by successive pulses of the current (eliminating bubble over voltage). What is more, platinum coated black surface absorbs gases formed effectively. The polarisation e.m.f. is thus almost completely eliminated.

3. "The relation $\Lambda = \frac{1000\kappa}{C}$ between eq. conductance and specific conductance means that Λ is inversely proportional to concentration".—Comment.

Ans. : The proportionality equations are as $y = Kx$, where K is proportionality constant, and thus, $y \propto x$. Therefore, in the above equation $\Lambda \propto \frac{1}{C}$, if $1000\ \kappa$; i.e., κ is constant.

But κ depends on concentration. Hence the above relation does not mean that $\Lambda \propto \frac{1}{C}$.

However, Λ increases as conc. (C) decreases. κ, on the other hand, decreases as C lecreases. The reason why Λ increases as C decreases is that κ decreases with decrease n concentration at a rate lower than the rate of decrease of concentration.

4. "$\frac{1}{R}$ is plotted against number of drops in a conductometric titration."—Explain.

Ans. : $\frac{1}{R}$ is the conductance G of the system. The conductance G depends on the number and type of ions for a solution at a fixed temperature and volume. If titrant is added freely, volume will change. Thus the concentration of the titrant is kept much higher han the solution to be titrated, so that the volume does not change much and titration s completed just by a few drops.

Now, since G changes regularly as per change in number and type of ions as a result of addition of titrant, the analysis of plot of G versus drops of titrant will give an idea of titrant needed for complete titration. Since the type of ions is a factor, they are so chosen hat rate of change of conductance G before and after end point are different. Thus, a tudy of G versus drop will give the end point. That is why G is plotted against drops of titrant.

5. "For weak electrolytes, Λ increases with dilution indefinitely and hence, there is o $\Lambda°$."—Comment.

Ans. : The statement is wrong. At average concentration, dilution leads to increase n the number of ions and hence, conductance increases. [An ionogen BA dissociates as

$$\underset{(1-\alpha C)}{BA} \rightleftharpoons \underset{\alpha C}{B^+} + \underset{\alpha C}{A^-}$$

where α is the degree of dissociation and C is oncentration, so that K is $\frac{\alpha^2 C}{1-\alpha}$ (approx),

ince $a = C$ as $\gamma = 1$. So, as C decreases, α increases.]

But there is electrophoretic and asymmetric effect, and hence, increase in conductance s very small. After a certain dilution, these effects start losing their importance (as the istances between ions become very large) and α also ncreases many-fold, and hence conductance (L) and quivalent conductance (Λ) increases sharply in Λ versus plot. This may appear almost vertical. If we could neasure up to $C = 0$, we would have noted that Λ merges vith the axis and that would be measured $\Lambda°$. But there a concentration below which no measurement could e done and thus increase appears as indefinite increase. is finite at theoretical $C = 0$ and is $\Lambda°$, which is the um of conductance.

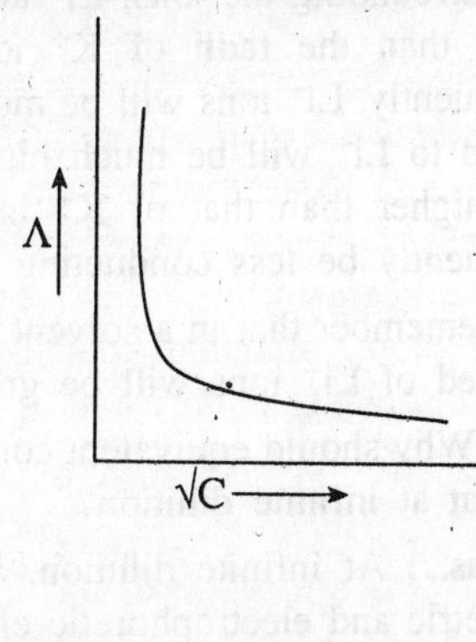

$\therefore \Lambda° = \lambda_+ + \lambda_-$.

6. What should be the nature of plot of conductance values of strong electrolytes of varying concentration measured in a particular cell against respective specific conductance values? Does this plot give any information about the cell? How essential is the information for any absolute measurement in the cell?

Ans. : We know that conductance $G = \kappa \frac{A}{l}$, where κ is specific conductance and $\frac{l}{A} = K$, the cell constant. Hence, a plot of G versus κ measured in same cell will be linear passing through origin.

The slope of the plot is $\frac{A}{l}$ or $\frac{1}{\kappa}$. So this plot gives information about the cell.

If we want to measure κ or Λ, we need to know K, the cell constant, because we measure only $G\left(\Lambda = \frac{1000\kappa}{C}\right)$. Thus the information is very essential for any absolute measurement.

7. Comment on the statement : "Debye-Hückel theory is applicable only to slightly contaminated distilled water."

Ans. : The basic assumptions in the Debye-Hückel theory are as follows :

(i) Ionophores are completely ionised at all concentrations.

(ii) Ions are point charges.

(iii) The bulk relative permitivity of the solvent controls the attraction between the ions and remains unchanged in presence of solute.

(iv) Inter-ionic attraction are due to Coulombic force only. There are no other inter-ionic force.

(v) The electrical potential energy is much smaller than its thermal energy so that Poisson equation can be linearised.

Linearisation of Poisson equation fails to allow the interaction between the ions that constitute the ion atmosphere. This can be valid only at infinite dilution. That is why it is sometimes asserted that Debye-Hückel theory has application only to slightly contaminated distilled water.

8. Why does the Li^+ ion move slower than K^+ ion in water?

Ans. : At a particular temperature and under the same set of conditions, a smaller ion should move faster than a larger ion. Li^+ ions are smaller than K^+ ions. But water is a polar solvent, and water molecules are attracted by the positive ions forming tight layer surrounding the ions. Li^+ and K^+ both are unipositive, but the radii of Li^+ ions are smaller than the radii of K^+ ions and thus the former have higher charge density. Consequently, Li^+ ions will be more solvated in water and the number of water molecules attached to Li^+ will be much higher. As a result the total apparent mass of Li^+ will be much higher than that of K^+ in water, and Li^+ ions will move slower. Li^+ ion will consequently be less conducting also.

(Remember that in a solvent where there will be no solvation or negligible solvation, the speed of Li^+ ions will be greater than those of K^+ ions.)

9. Why should equivalent conductance of an electrolyte at finite concentration be less than that at infinite dilution?

Ans. : At infinite dilution, all electrolytes are completely ionised and there is no asymmetric and electrophoretic effect. No ion pairs are formed also. Thus the equivalent conductance of an electrolyte at a particular temperature becomes the highest for any solvent (termed $\Lambda°$). At a finite concentration all these effects are present. What is more, ionogens are not even completely ionised. Consequently, Λ values will be lower than $\Lambda°$.

10. Explain why the changes in equivalent conductance of KCl and HOAC with concentration are widely different.

Ans. : KCl is a strong electrolyte or ionophore, and is completely ionised at any concentration. Therefore, the conductance at any concentration is higher than that of weak electrolytes or ionogens like HOAC which are only partially ionised (and the degree of ionisation << 1 at normal concentration.)

In ionophores, surrounding every ion, an ion atmosphere of opposite charge is present. When it moves from one place to another, the old atmosphere is destroyed and a new ion atmosphere is created. There is a time lag between the two incidents, and until the new atmosphere is created, ion is attracted or pulled backwards by old ion atmosphere. The conductance decreases with increase in concentration due to the increase in this asymmetric factor. Ion pair formation and electrophoretic effect are also responsible for the reduction of conductance. As C decreases, these effects decreases and conductance increases. This decrease or increase is linear in $\sqrt{C}$ from a very low concentration to a reasonably high concentration. The variation with concentration is small. For ionogens like HOAC, $\alpha << 1$ at normal concentration. The ions are far off and asymmetric effect, etc. are unimportant. With decrease in concentration α increases and conductance increases. But with increase in number of ions, the asymmetric effect, etc. become important. The nature of variation is different from ionophores. The variation is larger with concentration.

11. Will the conductance of a substance change if the voltage is changed?

Ans. : The voltage $\Delta\Phi$ is related to current i through the proportionality constant conductance G, i.e., $\Delta\Phi\ G = i$. Thus if $\Delta\Phi$ is changed, i will be changed and G (and hence κ and Λ) will remain unchanged. However, if Ohm's law is not valid (which may happen at extremely high voltage), then G may change on changing $\Delta\Phi$.

12. If the resistance of 0.1(N) aqueous solution of KCl is measured in different conductivity cells, will the values of conductance, specific conductance and equivalent conductance change?

Ans. : The conductance of any solution at a particular temperature, G is given by $G = \kappa.\ \frac{A}{l}$, where κ is the conductivity, A is the area of the parallel electrodes and l is the distance between the electrodes. Thus, if the cell constant $\frac{l}{A}$ of two different cells are different, the conductance will change. But the specific conductance or conductivity κ is a proportionality constant depending on the nature of the material and temperature, and thus, will not change if the cells be changed. The equivalent conductance, Λ, of a substance is related to κ and concentration, and thus, will not change on changing the cells.

13. What is the absolute mobility of ions?

Ans. : The absolute mobility of an ion, u_{ab}, is the velocity of the ion per unit applied force at infinite dilution. When ions are in a force field, the force imparts a component to velocity v_d. By the Newton's second law,

$$F = m.\ \frac{dv}{dt}, \qquad (1)$$

where m is the mass. An ion starts off after a collision. Let the mean time between collisions be τ and there be N collisions in time t. Thus,

$$\tau = \frac{t}{N}. \quad (2)$$

The average value of the velocity is the product of acceleration and average time between collisions.

$$v_d = \tau . \frac{dv}{dt} = \tau . \frac{F}{m} \quad (3)$$

$$= \frac{\tau}{m} . F = u_{ab} . F \text{ [by definition].}$$

Thus, u_{ab} is $\frac{\tau}{m}$, it is an index of how mobile an ion is.

Its unit is cm.sec^{-1}.dyne^{-1} in C.G.S. system.

14. How did you account for the abnormally high mobilities of H^+ and OH^- ions in water?

Ans. : The charge density of an ion increases with decrease in radius, and hence, hydration increases. Therefore, the effective radius should increase and the velocity should decrease. But the velocities and mobilities of H^+ and OH^- ions are very high. It has been explained to be due to Grotthuss type mechanism. In case of H^+ and OH^- movement, there is an effective motion of a proton or hydroxide ion, which involves the formation and destruction of bonds through a long chain of water molecules. The velocity and conductivity are controlled by the rate at which a water molecule can reorient in order to accept and then deliver protons and hydroxide ions from and to its neighbours.

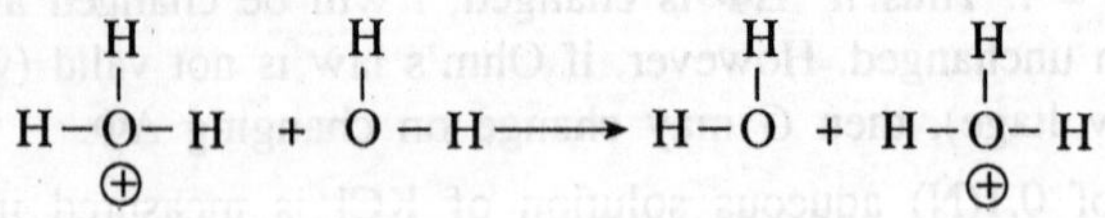

and 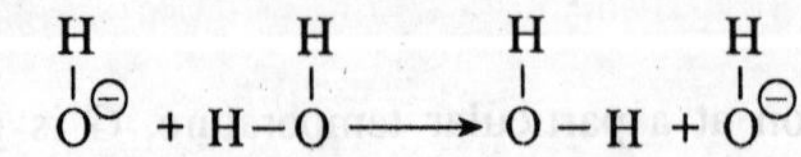

the whole process may be depicted as

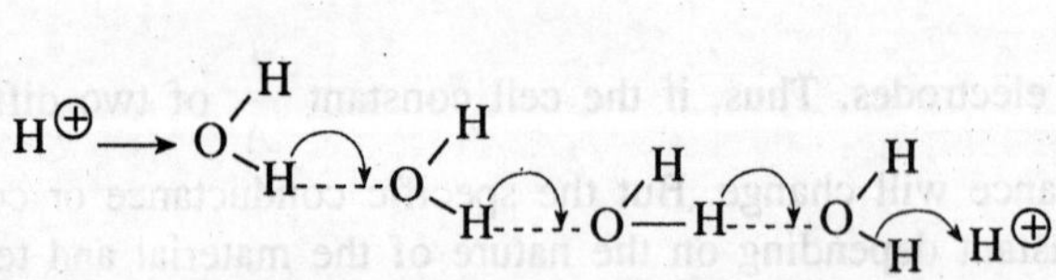

For effective motion of H^+ and OH^- are needed lengthening and shorting of bonds; reorientation for the latter,

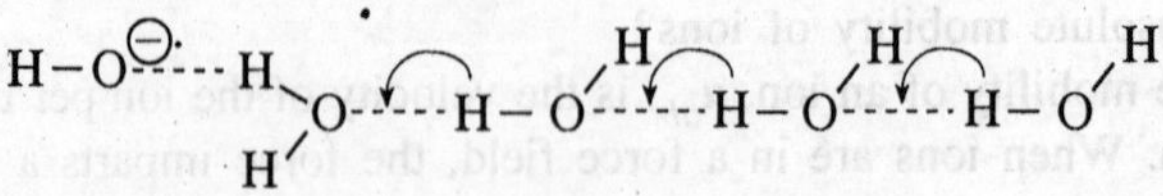

15. Conductance of H^+ ion increases as temperature increases—criticize.

Ans. : Conductance of H^+ ion increases as temperature increases. H^+ has the highest mobility of all the ions. It can be explained by Grotthuss mechanism. Mobility and hence, the conductance depend on the rate at which H_2O or ROH molcule can reorient to accept and transfer proton from and to neighbouring molecules. Hence, the process involves energy barrier. Thus, as temperature increases, the rate increases, thereby increasing the mobility and hence, conductance of H^+ ion.

16. The plot of $\frac{1}{\Lambda}$ vs ΛC for a weak electrolyte has unit slope. Express dissociation constant of the weak electrolyte in terms of α and Λ. Here all the symbols carry usual significances.

Ans. For a weak electrolyte dissociation constant $K_d = \dfrac{\alpha^2 C}{1-\alpha}$,

where C = concentration of weak electrolyte

α = degree of dissociation = $\dfrac{\Lambda}{\Lambda^\circ}$.

$$\therefore\ K_d = \frac{\left(\dfrac{\Lambda}{\Lambda^\circ}\right)^2 C}{1-\dfrac{\Lambda}{\Lambda^\circ}} \quad \text{or,} \quad \frac{1}{\Lambda}-\frac{1}{\Lambda^\circ} = \frac{\Lambda C}{\Lambda_o^2 K_d}$$

By the problem, $\Lambda_o^2 K_d = 1$

$$\text{or,} \quad \left(\frac{\Lambda}{\alpha}\right)^2 K_d = 1$$

$$\therefore \quad K_d = \left(\frac{\alpha}{\Lambda}\right)^2.$$

Problems

1. What volume of O_2 would be liberated from an aqueous solution of NaOH by a current of 5 A flowing for $2\frac{1}{2}$ hours? The temperature is 27 °C and the pressure is 1 atm.

Soln. : We know that,

$$2OH^- = H_2O + \frac{1}{2}O_2 + 2e.$$

Thus, 2 F charge liberates 11.2 litres of O_2 at NTP.

$$\text{Now,} \quad \frac{P_o V_o}{T_o} = \frac{P_1 V_1}{T_1}$$

$$\text{or,} \quad \frac{11200\times 1}{273.15} = \frac{x\times 1}{300.15}$$

$$\text{or,} \quad x = \frac{11200\times 300.15}{273.15}\ \text{cc.}$$

Hence,

$$96484 \text{ coulombs liberate } \frac{11.2\times300.15}{2\times273.15} \text{ litres of } O_2 \text{ at } 27\ °C \text{ and } 1 \text{ atm}$$

$$1 \quad " \quad " \quad \frac{11.2\times300.15}{2\times273.15\times96484} \text{ litres}$$

$$5\times2.5\times60\times60 \quad " \quad " \quad \frac{11.2\times300.15\times5\times2.5\times60\times60}{2\times273.15\times96484} \text{ litres}$$

$$= \mathbf{2.87 \text{ litres of } O_2.}$$

2. What volume of H_2 would be liberated at the anode if 2 A current flows through excess fused LiH for 2 hours. The temperature is 680 °C and pressure is 1.5 atm.

Hints : Problem $\Rightarrow$ 1 $(H^+ + e = \frac{1}{2}H_2)$ [**Ans.** 3.89 lit]

3. How long would it take a current of 2 A to reduce completely 240 cc of 0.1(M) $Fe_2(SO_4)_3$ to $FeSO_4$? How many cc of 0.1(M) $K_2Cr_2O_7$ could be reduced to $Cr_2(SO_4)_3$ by the same quantity of electricity?

Reaction : $2Fe^{3+} + 2e = 2Fe^{2+}$

Therefore,

2 faraday reduces 1000 cc of 1 molar $Fe_2(SO_4)_3$

$\frac{2\times96484}{1000}$ coulombs reduce 1 cc of 1 molar $Fe_2(SO_4)_3$

$\frac{2\times96484}{1000}\times24$ coulombs reduce 240 cc of 0.1 molar $Fe_2(SO_4)_3$

[since 240 cc 0.1(m) = 24 cc 1(m)]

Thus, $2\times t = \frac{2\times96484\times24}{1000}$ [where t is the time taken]

$$\therefore \quad t = \frac{2\times96484\times24}{1000\times2} \text{ sec}$$

$$= \frac{2315616}{1000} \text{ sec} = 38.59 \text{ minutes} = \mathbf{38 \text{ min } 35 \text{ sec.}}$$

Now, $Cr_2O_7^{=} + 14H^+ + 6e = 2Cr^{3+} + 7H_2O$.

Thus, 6F reduces 1000 cc 1 (M) $Cr_2O_7^{=}$.

Hence, the volume in this case is the $\frac{1}{3}$rd of the former one, that is **80 cc.**

4. How long would take a current of 5 A to reduce completely 100 cc 0.1(M) dichromate? How many cc of 0.1(M) $Al_2(SO_4)_3$ can be reduced by the same amount of elecricity? [**Ans.** 19.3 min, 200 cc]

5. What quantity of electricity would be required to reduce 1.5 gm of nitrobenzene to aniline? If the potential drop across the cell is 2 volts, then how much energy in joules is consumed in the process?

Soln. For the given reduction,

$$C_6H_5NO_2 + 6H^+ + 6e = C_6H_5NH_2 + 2H_2O$$

6 faradays of charge reduces 123 gm nitrobenzene.

That is, 123 gm nitrobenzene is reduced by 6 × 96484 coulombs.

∴ 1.5 gm nitrobenzene is reduced by $\dfrac{6 \times 96484 \times 1.5}{123}$ coulombs

$= 7059.8$ coulombs.

Energy consumed = 2 volt × 7059.8 C = **14119.6 joules.**

6. Calculate the total charge in coulomb carried by all Ca^{2+} ions in 500 cc 0.01(M) $CaCl_2$ solution. Assume complete ionisation.

Soln. : 1000 cc 1(M) $CaCl_2$ carries 2 faradays = (2 × 96484) coulombs.

∴ 500 cc 0.01(M) solution will carry $= \dfrac{2 \times 96484 \times 0.01 \times 500}{1000} =$ **964.84 coulombs.**

7. The molar conductance of a solution of $AlCl_3$ is found to be 130 $ohm^{-1}.cm^2.mole^{-1}$ at 298 K. What would be its equivalent conductance?

Soln. : Equivalent conductance $\Lambda = \dfrac{1000\kappa}{N}$, where N is the normality of the solution.

$\Lambda = \dfrac{1000\kappa}{ZC}$, where C is molarity and Z is valency.

∴ $\Lambda = \dfrac{\Lambda_m}{Z}$, where Λ_m is molar conductance

$\Lambda = \dfrac{130}{3}$ $ohm^{-1}.cm^2.gm\text{-}eq^{-1}$ = **43.33 $ohm^{-1}.cm^2.gm\text{-}eq^{-1}$.**

8. The resistance of an aq. solution containing 0.624 gm of $CuSO_4 . 5H_2O$ per 100 cc of solution in a conductance cell of cell constant 153.7 m^{-1} is 520 ohms at 298 K. Calculate the molar and equivalent conductivity (molar mass = 249.5).

Soln. : We know that,

$\dfrac{1}{R} = \kappa.\dfrac{A}{l} = \kappa.\dfrac{1}{l/A}$ [where $\dfrac{l}{A}$ is cell constant, R is resistance in ohm and κ is $ohm^{-1}.cm^{-1}$]

Now, 153.7 m^{-1} = 1.537 cm^{-1}.

∴ $\kappa = \dfrac{1}{R}.\dfrac{l}{A} = \dfrac{1}{520} \times 1.537$ $ohm^{-1}.cm^{-1}$

$= 2.9558 \times 10^{-3}$ $ohm^{-1}.cm^{-1}$.

6.24 gm/litre $= \dfrac{6.24}{249.5}$ molar = C molar = 0.02501 molar.

Molar conductance $\Lambda_m = \dfrac{1000\kappa}{C} = \dfrac{1000 \times 2.9558 \times 10^{-3}}{0.02501}$ $ohm^{-1}.cm^2.mole^{-1}$

= **118.2 $ohm^{-1}.cm^2.gm\text{-}mol^{-1}$.**

$\Lambda = \dfrac{118.2}{2}$ $ohm^{-1}.cm^2.gm\text{-}mol^{-1}$ = **59.1 $ohm^{-1}.cm^2.gm\text{-}eq^{-1}$.**

9. The resistance of a electrolyte solution A is 45 ohms in a given cell and that of electrolyte solution B is 100 ohms in the same cell. Equal volumes of solutions of both A and B are mixed. Calculate the resistance of this mixture, again in the same cell.

Soln. : We know that,

$$\frac{1}{R} = \kappa . \frac{A}{l}.$$

Let $\frac{A}{l} = x$ cm.

Now, $\frac{1}{R_1} = \kappa_1 . \frac{A}{l}$

[where κ_1 = conductivity of soln. A in ohm^{-1}.cm^{-1}

R_1 = resistance of soln. A in ohm]

$$\therefore \kappa_1 = \frac{1}{R_1} . \frac{1}{x} = \frac{1}{45} . \frac{1}{x} = \frac{0.0222}{x}.$$

And $\frac{1}{R_2} = \kappa_2 . \frac{A}{l}$

[κ_2 is the conductivity in ohm^{-1} of B soln., R_2 is the resistance in ohm of B soln.]

$$\text{or, } \kappa_2 = \frac{1}{R_2} . \frac{1}{x}$$

$$= \frac{1}{100} . \frac{1}{x} = \frac{0.01}{x}.$$

In the mixture,

$$\kappa_1 = \frac{0.02222}{x} \times \frac{1}{2} \text{ ohm}^{-1}.\text{cm}^{-1} \text{ and } \kappa_2 = \frac{0.01}{x} \times \frac{1}{2} \text{ ohm}^{-1}.\text{cm}^{-1}.$$

For the mixture, $\kappa = \kappa_1 + \kappa_2$

$$= \left(\frac{0.01111}{x} + \frac{0.005}{x} \right) \text{ohm}^{-1}.\text{cm}^{-1}$$

$$= \frac{0.016111}{x} \text{ ohm}^{-1}.\text{cm}^{-1}.$$

$$\therefore \frac{1}{R} = \kappa . \frac{A}{l} \text{ [where } R \text{ is the resistance of the mixture in ohm]}$$

$$= \left(\frac{0.016111}{x} . x \right) \text{ohm}^{-1} = 0.061111 \text{ ohm}^{-1}.$$

$$\therefore R = \frac{1}{0.016111} \text{ ohms} = \mathbf{62.07 \text{ ohms.}}$$

10. The resistance of an electrolyte solution A is 10 ohms and that of solution B is 50 ohms in the same cell. 1 volume of A of the same concentration is mixed with the 3 volumes of solution B of the above concentration. Calculate the approximate resistance of the mixture, again in the same cell.

Soln. : Proceeding as Problem no. 9 and assuming

$$\frac{A}{l} = x \text{ cm},$$

we get,

$$\frac{1}{R_A \text{ohm}} = \kappa_A \text{ ohm}^{-1}.\text{cm}^{-1} \times x \text{ cm}.$$

$$\therefore \quad \frac{1}{10} = \kappa_A \times x$$

$$\text{or,} \quad \kappa_A = \frac{0.1\,\text{ohm}^{-1}}{x\,\text{cm}}.$$

$$\text{Similarly,} \quad \kappa_B = \frac{0.02\,\text{ohm}^{-1}}{x\,\text{cm}}.$$

Now the given dilution will reduce κ_A to $\frac{1}{4}$th and κ_B to $\frac{3}{4}$th of the original and κ's are additive.

$\therefore$ for the mixture.

$$\kappa = \left(\frac{0.1}{x}\times\frac{1}{4}+\frac{0.02}{x}\times\frac{3}{4}\right)\text{ohm}^{-1}.\text{cm}$$

$$= \left(\frac{0.025}{x}+\frac{0.015}{x}\right)\text{ohm}^{-1}.\text{cm}.$$

$$= \frac{0.04}{x}\text{ ohm}^{-1}.\text{cm}.$$

$$\therefore \quad \text{for the mixture, } \frac{1}{R} = \left(\frac{0.04}{x}.x\right)\text{ohm}^{-1} = 0.04 \text{ ohm}^{-1}.$$

$\therefore$ R = **25 ohms.**

11. Calculate Λ and μ of a 0.1(N) solution of $Al_2(SO_4)_3$ if its κ is 0.0108 $\text{ohm}^{-1}.\text{cm}^{-1}$.

Soln. : We know that, $\Lambda = \frac{1000\kappa}{N}$.

Now, proceeding as the previous problems. [**Ans.** 108, 648]

12. The equivalent conductance of NaCl at infinite dilution is 126.4 and the transference number of Na^+ in NaCl is 0.40 at 25 °C. Calculate the ion conductance of Na^+ and Cl^-. **[C.U. 1970]**

Soln. : Let the ion conductance of Na^+ be λ_+ $\text{ohm}^{-1}.\text{cm}^2$ and the ion conductance of Cl^- be λ_- $\text{ohm}^{-1}.\text{cm}^2$. The equivalent conductance at infinite dilution is $\Lambda°$ $\text{ohm}^{-1}.\text{cm}^2$ = 126.4 $\text{ohm}^{-1}.\text{cm}^2$. Then,

$$\lambda_+ = t_+\Lambda° = (0.40 \times 126.4)\text{ ohm}^{-1}.\text{cm}^2$$

$$= \mathbf{50.56\ ohm^{-1}.cm^2.}$$

$$\lambda_- = t_-\Lambda°$$

$$= \{(1 - 0.40) \times 126.4\}\text{ ohm}^{-1}.\text{cm}^2$$

$$= \mathbf{75.84\ ohm^{-1}.cm^2.}$$

13. Calculate the ionic mobility of the cation in KCl at 25 °C, given that its transport number is 0.49 and the equivalent conductance of KCl at infinite dilution is 150.

Soln. : We know that,

$$\lambda_+ = t_+ \Lambda°.$$

That is, $Fu_+ = t_+ \Lambda°$ ohm^{-1}.cm^2.

$$\therefore \quad u_+ = \frac{t_+\Lambda°}{F} \cdot \frac{\text{ohm}^{-1}.\text{cm}^2}{\text{coulomb}}.$$

By the problem,

$$u_+ = \frac{0.49 \times 150}{96484} \frac{\text{cm}^2}{\text{volt.sec}} = 7.62 \times 10^{-4} \frac{\text{cm}^2}{\text{volt.sec}}$$

$$\left[\text{since ohm}^{-1} = \frac{\text{ampere}}{\text{volt}} \text{ and coulomb} = \text{ampere} \times \text{second}\right]$$

$$= \mathbf{7.62 \times 10^{-8}} \; \frac{\mathbf{m^2}}{\mathbf{volt.sec}}.$$

14. The equivalent conductance at infinite dilution of HCl, NaCl and NaOAC are 426.2 ohm^{-1}.cm^2, 126.5 ohm^{-1}.cm^2 and 91.0 ohm^{-1}.cm^2 respectively at 25 °C. Calculate $\Lambda°$ for CH_3COOH.

A conductance cell filled with 0.01(M) KCl has a resistance of 257.3 ohms at 25 °C. The same cell filled with 0.2(N) CH_3COOH has a resistance of 508.6 ohms. Calculate the dissociation constant of HOAC.

[κ of 0.01(M) KCl = 1.41×10^{-3} ohm^{-1}.cm^{-1}] **[V.U. 2004; C.U. 2001]**

Soln. : We know that

$$\Lambda°_{\text{HOAC}} = \Lambda°_{\text{HCl}} + \Lambda°_{\text{NaOAC}} - \Lambda°_{\text{NaCl}}$$

$$= (426.2 + 91 - 126.5) \text{ ohm}^{-1}.\text{cm}^2 = \mathbf{390.7 \; ohm^{-1}.cm^2}.$$

Again, let $\frac{A}{l} = x$ cm.

For KCl solution, $\frac{1}{R} = \kappa.\frac{A}{l} = \kappa.x.$

By the problem,

$$\frac{1}{257.3\,\text{ohm}} = 1.41 \times 10^{-3} \text{ ohm}^{-1}.\text{cm}^{-1} \times x$$

$$\therefore \quad x = \frac{1}{257.3 \text{ ohm} \times 1.41 \times 10^{-3} \text{ohm}^{-1}.\text{cm}^{-1}} = 2.7564 \text{ cm}.$$

Again, for HOAC, $\Lambda = \frac{1000\kappa}{N}$ ohm^{-1}.cm^2

$$\frac{1}{R_{\text{HOAC}}} = \kappa.x$$

$$\therefore \quad \kappa = \frac{1}{508.6} \times \frac{1}{2.7564} \text{ ohm}^{-1}.\text{cm}^{-1} = 7.133 \times 10^{-4} \text{ ohm}^{-1}.\text{cm}^{-1}.$$

$$\therefore \quad \Lambda = \frac{1000 \times \kappa}{0.2} = \frac{1000 \times 7.133 \times 10^{-4}}{0.2} \text{ ohm}^{-1}.\text{cm}^2$$

$$= 3.5665 \text{ ohm}^{-1}.\text{cm}^2.$$

Now, K_a of HOAC is $\dfrac{\alpha^2 C}{1-\alpha}$,

where α is degree of dissociation and is $\dfrac{\Lambda}{\Lambda°}$.

$$\therefore\ K_a = \frac{\Lambda^2 C}{\Lambda°(\Lambda°-\Lambda)} = \frac{(3.5665)^2 \times 0.2}{390.7(390.7-3.5665)} = \mathbf{1.68 \times 10^{-5}}.$$

15. 200 ml of 0.002(M) $BaCl_2$ is added to 300 ml of 0.003(M) Na_2SO_4. Calculate approximate κ of the resulting mixture. Determine transport number of Cl^-.

Given, $\lambda_\pm$ of $Cl^- = 76$

$Na^+ = 50$

$\frac{1}{2}Ba^{++} = 80$

$\frac{1}{2}SO_4 = 80.$

Soln. : $BaCl_2$ will react with Na_2SO_4 to give insoluble $BaSO_4$. So after mixing the molarity of the ions in solution is :

$$C_{Ba^{++}} = 0.002 \times \frac{2}{5} = 0.0008\ (M)$$

$$C_{Cl^-} = 0.004 \times \frac{2}{5} = 0.0016\ (M)$$

$$C_{Na^+} = 0.006 \times \frac{3}{5} = 0.0036\ (M)$$

$$C_{SO_4^-} = 0.003 \times \frac{3}{5} = 0.0018\ (M).$$

Ba^{2+} ions are in lesser amount.

So all the Ba^{2+} will be precipitated as $BaSO_4$. Strength of the remaining SO_4^- will be (0.0018 – 0.0008) (M) = 0.001(M).

Now, $\kappa = \dfrac{CA}{1000}$.

$\therefore\ \kappa_\pm = \dfrac{C.\lambda_+}{1000}$.

Therefore, $\kappa_{Cl^-} = \dfrac{0.0016 \times 76}{1000} = 1.216 \times 10^{-4}\ ohm^{-1}.cm^{-1}$.

Similarly, $\kappa_{SO_4^-} = \dfrac{0.001 \times 80 \times 2}{1000} = 1.6 \times 10^{-4}\ ohm^{-1}.cm^{-1}$.

$\kappa_{Na^+} = \dfrac{0.0036 \times 50}{1000} = 1.8 \times 10^{-4}\ ohm^{-1}.cm^{-1}$.

Adding, approximate κ of the soln. is $4.616 \times 10^{-4}\ ohm^{-1}.cm^{-1}$.

$$t_{Cl^-} = \frac{\kappa_{Cl^-}}{\kappa} = \frac{1.216 \times 10^{-4}}{4.616 \times 10^{-4}} = \mathbf{0.26.}$$

16. 100 ml of 0.002(M) NaCl is added to 100 ml of 0.004(M) $AgNO_3$. Calculate an approximate value for the specific conductance of the mixture. **[C.U. 1973]**

Given, $\lambda_{\pm}$ of $Na^+ = 50.1$

$Ag^+ = 61.9$

$Cl^- = 76.3$

$NO_3^- = 71.4.$

Hints : Proceed as the previous problem. Approx. κ is 2.548×10^{-4} $ohm^{-1}.cm^{-1}$

$= 2.548 \times 10^{-2}$ $ohm^{-1}.cm^{-1}$.

17. Given that $\lambda_{Ag^+} = 54$, $\lambda_{Cl^-} = 65.6$, $\lambda_{K^+} = 84.2$ and $\lambda_{NO_3^-} = 79.4$ at a particular temparature. A solution of 200 cc of 0.02(M) $AgNO_3$ is mixed with 300 cc of 0.01(M) KCl. Calculate the approximate value of κ.

Hints : Proceed as Problem 16. **[Ans.** 1.25×10^{-3}]

18. At 25 °C, the specific conductances of a saturated solution of $SrSO_4$ and that of water used are 1.482×10^{-4} $ohm^{-1}.cm^{-1}$ and 1.5×10^{-6} $ohm^{-1}.cm^{-1}$ respectively. Given the ion conductances of Sr^{2+} and SO_4^{2-} to be 59.46 $ohm^{-1}.cm^2.gm\text{-}eq^{-1}$ and 79.8 $ohm^{-1}.cm^2.gm\text{-}eq^{-1}$ respectively at this temperature. Calculate the solubility of the salt in $gm.lit^{-1}$ [M. Wt. of $SrSO_4$ = 182] **[C.U.(H) 1996]**

Soln. :

$$\Lambda^o_{SrSO_4} = (59.46 + 79.8)\ ohm^{-1}.cm^2.gm\text{-}eq^{-1}$$

$$= 139.26\ ohm^{-1}.cm^2.gm\text{-}eq^{-1}.$$

Now, $\Lambda = \dfrac{1000\kappa}{N}$ [where N is normality].

Since the salt is sparingly soluble,

$$\Lambda \approx \Lambda^o \text{ and } \kappa_{SrSO_4} = \kappa_{SrSO_4\ soln.} - \kappa_{H_2O}.$$

Thus, $\kappa = (1.482 - 0.015) \times 10^{-4}\ ohm^{-1}.cm^{-1}$
$= 1.467 \times 10^{-4}\ ohm^{-1}.cm^{-1}$.

$$\text{Hence } N = \frac{1000\kappa}{\Lambda^o} = \frac{1000 \times (1.467) \times 10^{-4}}{139.26} = 1.05342 \times 10^{-3}(N).$$

$$\text{Eq. wt. is } \frac{182}{2} = 91.$$

∴ solubility in $gm.lit^{-1}$ is $(1.05342 \times 10^{-3} \times 91) = 0.096$ **$gm.lit^{-1}$.**

19. If the eq. conductances of solutions with varying conc. of CH_3COONa, NaCl and HCl are plotted against $\sqrt{C}$, the intercept obtained are 91.0, 128 and 425 in $ohm^{-1}.cm^2.eq^{-1}$ at 25 °C respectively. If the resistance of 0.02 molar soln. of CH_3COOH in a cell with a cell constant 0.2061 cm^{-1} be 888 ohms at 25 °C, what will be the degree of dissociation of the acid at 25 °C?

Soln. : By the problem,

$$\Lambda^o_{HOAC} = (91 + 425 - 128) = 388\ ohm^{-1}.cm^2.gm\text{-}eq^{-1}$$

$$\text{Again, } \frac{1}{888\,\text{ohms}} = \kappa\ ohm^{-1}.cm^{-1} \times \left(\frac{A}{l}\right) cm$$

i.e., $\dfrac{1}{888} = \kappa \times \dfrac{1}{0.2061}$

i.e., $\kappa = \dfrac{0.2061}{888}$ ohm^{-1}.cm^{-1}

$\therefore \quad \Lambda = \dfrac{1000\kappa}{N} = \dfrac{1000 \times 0.2061}{888 \times 0.02} = 11.60473$ ohm^{-1}.cm^2.gm-eq^{-1}

$\therefore \quad \alpha = \dfrac{\Lambda}{\Lambda^\circ} = \dfrac{11.60473}{388} = 0.029909 = \mathbf{0.03.}$

20. At 25 °C in a saturated solution of $BaSO_4$, Ba^{2+} and SO_4^- ions take 151 sec and 121 sec respectively to cover a distance of 1 cm across which a potential difference of 10 volts is applied. The solubility product of $BaSO_4$ at 25 °C is 10^{-10} (considered in gm-eq.lit^{-1} unit). The conductance of the solution is 2.2×10^{-6} ohm^{-1} in a cell of which the conductance of water is 0.8×10^{-6} ohm^{-1} at 25 °C. If in the same cell conductance of 0.01(N) KCl be 1.40×10^{-3} ohm^{-1} at 25 °C, what would be its specific conductance at the same temperature? **[C.U.(H) 1994]**

Soln. : By the problem,

$$v_+ = \frac{1\,\text{cm}}{151\,\text{sec}}$$

$$v_- = \frac{1\,\text{cm}}{121\,\text{sec}}.$$

$$\therefore \quad u_+ = v_+.\frac{l}{\Delta\phi} = \frac{1}{151}(\text{cm/sec}) \times \frac{1}{10}(\text{cm/volt})$$

$$= \frac{1}{1510}\,\frac{\text{cm}^2}{\text{volt.sec}}.$$

Similarly, $u_- = \dfrac{1}{1210}$ cm^2.volt^{-1}.sec^{-1}.

$$\therefore \quad \lambda_+ = Fu_+ = 96484 \times \frac{1}{1510}(\text{A.sec}) \times \frac{\text{cm}^2}{\text{volt.sec}}$$

$$= 63.896684\ \text{ohm}^{-1}.\text{cm}^2.\text{gm.eq}^{-1}$$

$$\lambda_- = Fu_- = \frac{96484}{1210} = 79.7384\ \text{ohm}^{-1}.\text{cm}^2.\text{gm-eq}^{-1}$$

$$\therefore \ \Lambda^\circ_{BaSO_4} = \lambda_+ + \lambda_- = (63.896684 + 79.7384)$$

$$= 143.64\ \text{ohm}^{-1}.\text{cm}^2.\text{gm-eq}^{-1}.$$

Now again, solubility product

$K_s = 10^{-10}$.

$\therefore$ solubility $= 10^{-5}$ gm-eq.lit^{-1}.

And $\kappa = \dfrac{N\Lambda}{1000} = \dfrac{10^{-5} \times 143.64}{1000} = 1.4364 \times 10^{-6}$ ohm^{-1}.cm^{-1}.

[Since $BaSO_4$ is sparingly soluble, $\Lambda = \Lambda^\circ$]

Again, $G = \kappa \times \dfrac{A}{l}$

Now, $G = (2.2 - 0.8) \times 10^{-6}\ \text{ohm}^{-1} = 1.4 \times 10^{-6}\ \text{ohm}^{-1}$.

$\therefore \quad 1.4 \times 10^{-6}\ \text{ohm}^{-1} = \kappa \times \dfrac{A}{l}$

or, $\dfrac{A}{l} = \dfrac{1.4\times10^{-6}}{1.4364\times10^{-6}}\ \text{cm} = 0.97466\ \text{cm}.$

By the problem,

for KCl, $G = 1.4 \times 10^{-3}\text{ohm}^{-1}$.

$\therefore \quad 1.4 \times 10^{-3} = \kappa \times \dfrac{A}{l}$.

$\therefore \kappa = \dfrac{1.4\times10^{-3}}{0.97466}\ \text{ohm}^{-1}.\text{cm}^{-1} = \mathbf{1.4364 \times 10^{-3}\ ohm^{-1}.cm^{-1}}$.

21. The resistance of a condutivity cell was found to be 700 ohms and 800 ohms when filled with 0.01(N) KCl and 0.01(N) $AgNO_3$ solution respectively. The eq. conductance of KCl is 150 $\text{ohm}^{-1}.\text{cm}^2$. Find the eq. conductance of $AgNO_3$. **[B.U. 1994]**

Soln. : We know that,

$$\frac{1}{R_{KCl}} = \kappa_{KCl}.\frac{A}{l}$$

and
$$\frac{1}{R_{AgNO_3}} = \kappa_{AgNO_3}\frac{A}{l}.$$

$$\therefore \quad \frac{\kappa_{AgNO_3}}{\kappa_{KCl}} = \frac{R_{KCl}}{R_{AgNO_3}} = \frac{700}{800}.$$

$$\therefore \quad \frac{\Lambda_{AgNO_3}}{\Lambda_{KCl}} = \left(\frac{1000\kappa_{AgNO_3}}{N_{AgNO_3}}\right)\times\left(\frac{N_{KCl}}{1000\kappa_{KCl}}\right) = \frac{\kappa_{AgNO_3}}{\kappa_{KCl}} = \frac{700}{800}.$$

[since N for both is 0.01(N)]

$$\therefore \quad \Lambda_{AgNO_3} = \Lambda_{KCl}.\frac{700}{800} = 150.\frac{700}{800}$$

$$= \mathbf{131.24\ ohm^{-1}.cm^2.gm\text{-}eq^{-1}}.$$

22. $\kappa_{SrSO_4} = 1.5 \times 10^{-4}$ and $\kappa_{H_2O} = 1.5 \times 10^{-6}$ at 25 °C. Calculate solubility product of $SrSO_4$. Given, $\Lambda^{\circ}_{\frac{1}{2}SrCl_2} = 135.8$, $\Lambda^{\circ}_{\frac{1}{2}K_2SO_4} = 153.5$, $\Lambda^{\circ}_{KCl} = 149.85$. **[B.U. 1993]**

Soln. : Proceed as former problem. [**Ans.** 2.8×10^{-7}]

23. A 0.02(M) KCl solution having a specific conductance of $27.7\times10^{-4}\ \text{ohm}^{-1}.\text{cm}^{-1}$ gives a resistance of 175 ohms when measured in a certain conductivity cell. A 0.01(M) soln. of CH_3COONa in the same cell shows a resistance of 575 ohms. Calculate the equivalent conductance of this CH_3COONa solution.

Soln. : Proceed as usual. [**Ans.** 84.3]

24. A saturated soln. of $SrSO_4$ shows a specific conductance (κ) value 1.5×10^{-4} $ohm^{-1}.cm^{-1}$ at 25 °C. The solubility is 0.5×10^{-3} $mole.litre^{-1}$ and the equivalent conductance $\Lambda^{\circ}_{SrSO_4} = 140\ ohm^{-1}.cm^2.gm\text{-}eq^{-1}$. Calculate the approximate value of κ for water at the given temparature. **[B.U. 1997]**

Soln. : Proceed as usual. **[Ans.** 1×10^{-5}**]**

25. The conductivities of aqueous NaCl, KCl, K_2SO_4 at 25 °C at infinite dilution are 126.45, 149.84 and 306.60 $ohm^{-1}.cm^2.mole^{-1}$ respectively. If λ_{Na^+} at 25 °C is 50.11 $ohm^{-1}.cm^2.mole^{-1}$, calculate the value of $t_{SO_4^{=}}$ in Na_2SO_4 at 25 °C.

[I.I.T. M.Sc. Admission Test 1996, Kharagpur]

Soln. : $\Lambda^{\circ}_{Na_2SO_4} = 2\Lambda^{\circ}_{NaCl} + \Lambda^{\circ}_{K_2SO_4} - 2\Lambda^{\circ}_{KCl}$

$= (2 \times 126.45 + 306.60 - 2 \times 149.84)\ ohm^{-1}.cm^2.mole^{-1}$

$= 259.82\ ohm^{-1}.cm^2.mole^{-1}$.

$\lambda_{SO_4^{2-}} = \Lambda^{o}_{Na_2SO_4} - \lambda_{Na^+}$.

$= (259.82 - 2 \times 50.11)\ ohm^{-1}.cm^2.mole^{-1}$

$= 159.6\ ohm^{-1}.cm^2.mole^{-1}$.

$$\therefore \quad t_{SO_4^{2-}} = \frac{\lambda_{SO_4^{2-}}}{\Lambda^{o}_{Na_2SO_4}} = \frac{159.6}{259.82} = \mathbf{0.61.}$$

26. A conductivity cell was filled with 0.05 $mole.dm^{-3}$ NaOH and the resistance was 40 Ω. If the cell constant is 0.4 cm^{-1}, find out the molar conductivity of NaOH at that concentration. **[M.Sc. Admission Test 1992, I.I.T. Bombay]**

Soln. : We know that,

$$\text{Conductance} \quad (G) = \kappa \times \frac{A}{l}$$

$$\text{or,} \quad \frac{1}{R} = \frac{\kappa \times 1}{l/A}$$

$$\text{or,} \quad \kappa = \frac{l/A}{R}.$$

By the probem, $\kappa = \dfrac{0.4}{40} = 10^{-2}\ ohm^{-1}.cm^{-1}$.

$$\text{Now,} \quad \Lambda_{NaOH} = \frac{1000\,\kappa}{N}$$

$$= \frac{1000\kappa}{Z.C} = \frac{1000 \times 10^{-2}}{1 \times 0.05} = \mathbf{200\ ohm^{-1}.cm^2.mol^{-1}}.$$

27. Calculate the equivalent conductance of 0.001(M) H_2SO_4 at 298 K. The measured conductance of the solution was found to be $8 \times 10^{-4}\ ohm^{-1}$.

[M.Sc. Admission Test 1997, I.I.T. Bombay]

Soln. : We know that, $G = \kappa \times \dfrac{A}{l}$.

Let us assume that, $\dfrac{l}{A} = 1\ cm^{-1}$.

$\therefore\ \kappa = 8 \times 10^{-4}$ ohm^{-1}.cm^{-1}.

$$\text{Now, } \Lambda_{H_2SO_4} = \frac{1000\kappa}{N} = \frac{1000\kappa}{Z.C} = \frac{1000 \times 8 \times 10^{-4}}{2 \times 0.001}$$

$$= (4 \times 1000 \times 1000 \times 10^{-4}) \text{ ohm}^{-1}.\text{cm}^2.\text{gm-eq}^{-1}$$

$$= 400 \text{ ohm}^{-1}.\text{cm}^2.\text{gm-eq}^{-1}$$

$$= \mathbf{800\ ohm^{-1}.cm^2.mole^{-1}.}$$

28. In a moving boundary experiment wih 0.01 mole LiCl, the boundary in a tube having a cross-sectional area of 0.125 cm^2 moves through 7.3 cm in 1490 sec when a current of 1.80×10^{-3} ampere is used. Calculate t_+. **[V.U. 2005; C.U.(H) 2002]**

Soln. : By the problem, a = 0.125 cm^2, l = 7.3 cm.

Time(t) = 1490 sec.

$i = 1.80 \times 10^{-3}$ ampere

$\therefore$ charge in coulomb = $(1.80 \times 10^{-3} \times 1490)$ = 2.682.

$$\therefore\ t_+ = \frac{laFN}{1000Q} \quad (\text{for this case } N = C)$$

$$= \left(\frac{7.3 \times 0.125 \times 96484 \times 0.01}{1000 \times 2.682}\right) = 0.32826 = \mathbf{0.33.}$$

29. Determine the transference number of H$^+$ and Cl$^-$ ions from the following data obtained by moving boundary method using $CdCl_2$ as the indicator electrolyte. Concentration of HCl soln. is 0.1 mol.dm^{-3}, mass of Ag deposited in the coulometer is 0·1209 gm, movement of boundary = 7.5 cm and cross-sectional area of the tube is 1.24 cm^2. **[V.U. 2003]**

Soln. : In this problem,

charge sent is $\dfrac{0.1209}{107.8}$ faraday. Then proceeding as before (Problem no. 28) the answer is t_+ = **0.83.**

30. The transport number of Na ion in NaCl is 0.385 and the equivalent conductance at infinite dilution of NaCl is 126.5 in the usual unit at 25 °C. Estimate the distance traversed in one hour by Na ion in a very dilute solution kept in a cell with electrodes 1 cm apart, when the potential difference of 30 volts is applied between the electrodes. **[V.U. 2001]**

Soln. : Proceed as usual. [**Ans.** 54.5 cm/hour]

31. The specific conductance of pure water is 38.4×10^{-9} ohm^{-1}.cm^{-1} at 18 °C. The equivalent conductance at infinite dilution of H$^+$ and OH$^-$ are 315.2 ohm^{-1}.cm^2.gm-eq^{-1} and 173.8 ohm^{-1}.cm^2.gm-eq^{-1} respectively. Calculate the ionic product of water at 18 °C. Is this water alkaline with respect to water at 25 °C (pH = 7)? Explain with reasons. [Given : 1 litre of water at 18 °C weighs 998.50 gm] **[C.U. 2000]**

Soln. : Given $\kappa_{H_2O} = 38.4 \times 10^{-9}$ ohm^{-1}.cm^{-1}

$$\Lambda^o_{H^+} = 315.2 \text{ ohm}^{-1}.\text{cm}^2.\text{gm-eq}^{-1}$$

$$\Lambda^o_{OH^-} = 173.8 \text{ ohm}^{-1}.\text{cm}^2.\text{gm-eq}^{-1}$$

By the problem,

1 litre of water = 998.50 gm

$$\text{Now, } \Lambda = \frac{1000\kappa}{N}$$

$$\Lambda = \frac{1000\kappa}{Z.C}$$

$$\therefore \quad C = \frac{1000\kappa}{N} = \frac{1000\kappa}{\Lambda^\circ} = \frac{1000 \times 38.4 \times 10^{-9}}{(315.2+173.8)} = 7.853 \times 10^{-8}.$$

It is the conc. of H^+ and OH^- ions.

Actually, water weakly dissociates into H^+ and OH^-.

$$\text{So, } C_{H^+} = \alpha C_{H_2O} = \frac{\Lambda}{\Lambda^\circ} C_{H_2O} = \frac{1000\kappa}{C_{H_2O}\Lambda^\circ} C_{H_2O} = \frac{1000\kappa}{\Lambda^\circ} = 7.8 \times 10^{-8}.$$

$$\left(\text{Here, } C_{H_2O} = \frac{998.5}{18} = 55.47\right)$$

So, $K_w \cong C_{H^+}^2 = (7.8 \times 10^{-8})^2 = 6.084 \times 10^{-15} = 6.1 \times 10^{-15}$.

$pH = -\log_{10} C_{H+} = -\log_{10}(7.8 \times 10^{-8}) = (-\log_{10}7.8 + 8) = 7.108$.

At this temperature (18 °C) the solution will be neutral although pH = 7.108.

Since the decomposition of water is endothermic, so $(K_w)_{H_2O}$ will increase with increase in temperature and will decrease with decrease in temperature.

32. The mobility of NH_4^+ ion is 7.623×10^{-8} m^2.V^{-1}.sec^{-1}. Calculate :

(i) the ionic conductivity of the NH_4^+ ion.

(ii) the velocity of the ion if 15 volts are applied across electrodes 25 cm apart.

(iii) the transport number of the ions in $NH_4C_2H_3O_2$ solution if the mobility of $C_2H_3O_2^-$ ion is 4.239×10^{-8} m^2.V^{-1}.sec^{-1}. **[C.U.(H) 2003]**

Soln. :

(i) We know that $\lambda_+ = Fu_+$

By the problem, $\lambda_+ = (96484 \times 7.623 \times 10^{-8})$ C.m^2.V^{-1}.sec^{-1}

$= \mathbf{7.354 \times 10^{-3}}$ **ohm^{-1}.m^2.**

$$\text{(ii) Now, } v_+ = u_+ . \frac{\Delta\phi}{l} = \left(7.623 \times 10^{-8} \times \frac{15}{0.25}\right) \text{m}^2.\text{V}^{-1}.\text{sec}^{-1} \times \frac{\text{volt}}{\text{m}}$$

$= \mathbf{4.574 \times 10^{-6}}$ **m.sec^{-1}.**

$$\text{(iii) By the problem, } t_+ = \frac{u_+}{u_+ + u_-} = \left(\frac{7.623}{7.623+4.239}\right) = 0.643.$$

$\therefore t_{C_2H_3O_2^-} = (1 - 0.643) = \mathbf{0.357}$.

33. A $AgNO_3$ soln. was electrolysed in a transport cell with silver electrodes. The anode compartment contained 0.1210 gm of silver ion before electrolysis and 0.1652 gm. after electrolysis. During eletrolysis 0.0854 gm of Ag was deposited in the cathode. Calculatte t_+ and t_-. **[C.U.(H) 1973]**

Soln. : We know that,

107.8 gm silver is deposited at cathode for the passage of 1 faraday.

$\therefore$ 0.0854 gm silver is deposited for the passage of $\dfrac{0.0854}{107.8}$ faraday.

So the charge sent $= \dfrac{0.0854}{107.8}$ faraday. The gain in weight of Ag in the anode chamber for the passage of $\dfrac{0.0854}{107.8}$ faraday is, by the problem = (0.1652 − 1.1210) gm

$= 0.0442 \text{ gm} = \dfrac{0.0442}{107.8}$ gm-eq.

We know that, for reactive electrodes, the passage of 1 faraday through a soln. leads to an increase of t_- gm-eq. of cation (and anion and salt) in the anode chamber. Thus, the gain in gm-eq. of Ag in the anode chamber for the passage of 1 faraday is :

$$\frac{\frac{0.0442}{107.8}}{\frac{0.0854}{107.8}} = \frac{0.0442}{0.0854} = 0.5175644 = 0.52.$$

Thus, $t_- = 0.52$; $t_+ = (1 - 0.52) = \mathbf{0.48}$.

34. A $CuSO_4$ soln. was electrolysed in a transport cell between copper electrodes. The anode compartment contained 0.1325 gm of copper ions before electrolysis and 0.2921 gm afterward. During electrolysis, 0.069 gm. copper was deposited at the cathode. Calculate the transport number of Cu^{+2} and $SO_4^{=}$ ions.

Soln. : Proceed as Problem no. 33. **[Ans. :** $t_+ = 0.35$, $t_- = 0.65$**]**

35. A $AgNO_3$ solution was electrolysed between silver electrodes in a transport cell. The anode compartment contained 0.224 gm of $AgNO_3$ before electrolysis and 0.2921 gm $AgNO_3$ afterwards. The same amount of charge deposits 0.0251 gm Cu in a copper coulometer connected in series. Calculate t_+.

Soln. : [Proceed as Problem no. 34]

Eq. wt. of copper $\dfrac{63.6}{2}$.

Charge sent is $\dfrac{0.0251 \times 2}{63.6}$ and in this reactive anode case no. of gm-eq Ag increased in the anode compartment is $\dfrac{(0.2921 - 0.224)}{169.8}$.

$\therefore t_+ = \mathbf{0.49}$.

36. A $AgNO_3$ solution containing 0.00739 gm of $AgNO_3$ per gm of water is electrolysed between silver electrodes. At the end of the experiment 0.078 gm of Ag is found to have been deposited on the cathode. The anode soln., after electrolysis, is found to contain 23.14 gm of water and 0.236 gm of $AgNO_3$. Calculate t_+ and t_-. **[C.U.(H) 1982]**

Soln. : By the problem, the charge sent is $\dfrac{0.078}{107.8}$ faraday.

The anode soln. contained (23.14 × 0.00739) gm = 0.1710046 = 0.171 gm $AgNO_3$ before electrolysis and 0.236 gm $AgNO_3$ afterwards. Hence, the increase in wt. of $AgNO_3$ is (0.236 – 0.171) gm = 0.065 gm. That is $\frac{0.065}{169.8}$ gm-equivalent [eq. wt. of $AgNO_3$ = 169.8] of $AgNO_3$ increase takes place in the anode chamber for the passage of $\frac{0.078}{107.8}$ faraday. We know that, for the passage of 1 faraday there is an increase of t_- gm-eq. in the anode chamber.

$$\text{Thus, } t_- = \frac{0.065/169.8}{0.078/107.8} = \frac{0.065}{169.8} \times \frac{107.8}{0.078} = 0.05290537 = \mathbf{0.53.}$$

$$\therefore \quad t_+ = (1 - t_-) = (1 - 0.53) = \mathbf{0.47.}$$

37. A $CuSO_4$ soln. containing 0.0821 gm of $CuSO_4$ per gram of water is electrolysed between copper electrodes. At the end of the experiment, 0.0692 gm of Ag is found to have been deposited in a silver coulometer connected in series. The anode is found to contain 50.44 gm of water and 4.1745 gm of $CuSO_4$. Calculate the transport number of copper ion.

Soln. : Proceed as Problem no. 36. **[Ans.** $t_+ = 0.35$**]**

38. In a Hittorf cell with silver electrodes is placed a 0.01 molal $AgNO_3$ solution and a silver coulometer is connected in series. Electrolysis is continued for a time. Analysis, after electoysis, shows

(i) 32.10 mg of Ag is deposited in the coulometer.

(ii) 20.09 gm of soln. of the anode chamber of the Hittorf cell contains 39.66 mg of Ag. Calculate t_+. **[C.U.(H) 1977]**

Soln. : By the problem,

$$\text{charge sent} = \frac{0.0321}{107.8} \text{ faraday.}$$

The solution contains $AgNO_3$ apart from water. Now, the 20.09 gm soln. from anode chamber contains 0.03966 gm Ag, i.e., $\frac{0.03966}{107.8} \times 169.8 = 0.06247$ gm $AgNO_3$ = 0.0625 gm $AgNO_3$

[since 107.8 gm Ag ≡ 169.8 gm $AgNO_3$]

Thus, the amount of water in the anode chamber is (20.09 – 0.0625) gm = 20.0275 gm. Now, the solution was 0.01 molal. That is, 1000 gm water contained (0.01 × 169.8) gm $AgNO_3$.

Therefore, 20.0275 gm water contained $\frac{0.01 \times 169.8 \times 20.0275}{1000} = 0.03401$ gm before electrolysis. Hence, the increase in wt. of $AgNO_3$ in the anode chamber is (0.0625 – 0.03401) gm = 0.02849 gm.

For reactive anode, there is an increase of t_- gm.eq. of salt, i.e., $AgNO_3$ in anode chamber for the passage of 1 faraday.

Here, for the passage of $\frac{0.0321}{107.8}$ F, the increase in gm-eq. of $AgNO_3$ in the anode chamber is $\frac{0.02849}{169.8}$ gm-eq.

$$\therefore t_- = \frac{0.02849/169.8}{0.0321/107.8} = \frac{0.02849}{169.8} \times \frac{107.8}{0.0321} = 0.563467 = \mathbf{0.56.}$$

$\therefore t_+ = (1 - 0.56) = \mathbf{0.44.}$

39. In a Hittorf method experiment, an exactly 0.2000 molar $CuSO_4$ soln. was electrolysed between copper electrodes. After the experiment 18.2170 gm of the cathode soln. was found to contain 0.2207 gm of Cu. Also 0.0202 gm of Ag was deposited during the experiment in a silver coulometer connected in series. Calculate t_+. **[C.U.(H) 1985]**

Soln. : Proceed as Problem no. 37. **[Ans.** $t_+ = 0.33$]

In short : Charge sent = $\frac{0.0202}{107.8}$ faradays

Cathode soln. 18.2170 gm contains 0.2207 gm Cu, i.e., 0.5538 gm $CuSO_4$. So weight of water is (18.2170 – 0.5538) gm = 17.6632 gm.

Initial $CuSO_4$ is $(0.2 \times 159.6) \times \frac{17.6632}{1000}$ gm = 0.5638 gm.

Therefore, loss in·gm-eq. = $\frac{(0.5638 - 0.5538)\,\text{gm}}{159.6\,\text{gm}/2}$.

Thus, $t_- = 0.67$, $t_+ = \mathbf{0.33.}$

40. A soln. of $CuSO_4$ of 0.02(M) was electrolysed between copper electrodes in a Hittorf cell and a silver coulometer is connected in series. Analysis, after electrolysis, showed that—

(i) 44.2 mg of Ag was deposited in the silver coulometer.

(ii) 30.078 gm of the soln. in the cathode chamber contains 0.02982 gm Cu.

Calculate the transport number of Cu^{+2} ion.

Soln. : Proceed as Problem no. 38. **[Ans.** $t_+ = 0.36$]

41. A solution containing 3.02% by wt. of $CuSO_4$ was electrolysed using a Pt-cathode and Cu-anode. After electrolysis, the cathode solution was found to weigh 54.7 gm and to contain 0.409 gm copper. The increase in weight of the cathode was 0.408 gm. Calculate the transport number of Cu^{2+} and SO_4^{2-} ions (Atomic weight of Cu = 63.6).

Soln. : 1F deposit $\frac{63.6}{2}$ gm = 31.8 gm of Cu.

$\therefore$ charge sent = $\frac{0.408}{31.8}$F (by the problem)

The cathode solution contains $CuSO_4$ apart from water. Now,

63.6 gm of Cu $\equiv$ 159.6 $CuSO_4$

$$0.409 \text{ gm of Cu} \equiv \frac{0.409 \times 159.6}{63.6} \text{ gm of } CuSO_4$$

$= 1.02635$ gm $CuSO_4$.

Hence, the weight of water = (54.7 – 1.02635)g = 53.674 gm.

Initial percent of $CuSO_4$ is 3.02.

Thus, 100 gm soln. contains 3.02 gm $CuSO_4$. Hence, 3.02 gm $CuSO_4$ is present in (100 – 3.02) gm

= 96.98 gm of water.

$\therefore$ 96.98 gm water contains 3.02 gm of $CuSO_4$

$$53.647 \text{ gm of water contains } \frac{3.02 \times 53.647}{96.98} \text{gm} = 1.671 \text{ gm of } CuSO_4.$$

Hence, the loss in weight of $CuSO_4$ for the passage of $\frac{0.408}{31.8}$ F

$$= (1.671 - 1.026) \text{ gm} = 0.645 \text{ gm}.$$

Now, migration of anion out of cathode $= t_-$ gm-eq. per faraday. Hence, the loss of salt $= t_-$ gm-eq. per faraday.

By the problem,

for the passage of $\frac{0.408}{31.8}$F the loss of salt $= \frac{0.645}{79.8}$ gm-eq.

[since eq. wt. of $CuSO_4 = \frac{159.6}{2} = 79.8$]

$\therefore$ for the passage of 1F the loss of salt is

$$= \frac{0.645/79.8}{0.408/31.8} = \frac{0.645}{79.8} \times \frac{31.8}{0.409} = 0.6299756 = 0.63.$$

Thus, $t_+ = 1 - t_- = (1 - 0.63) = \mathbf{0.37}$.

42. 4.29 molal $FeCl_3$ soln was electrolysed between Pt-electrodes. After electrolysis, 50 gm soln. from the cathode is found to be 3.1 molal in $FeCl_3$ and 1.4 molal in $FeCl_2$. Calculate the transport number of Fe^{3+}.

Soln. : This problem can be solved by two methods :

Method 1. The reaction at the cathode is : $Fe^{3+} + e = Fe^{2+}$. Thus, production of 1 mole of Fe^{2+} ion means the passage of 1 faraday charge. Let there be x gm of water in the cathode chamber. Now,

1 molal in 1000 gm = 1 mole = passage of 1 F. 1.4 molal in x gm $= \frac{1.4 \times x}{1000}$ mole $=$ passage of $\frac{1.4 \times x}{1000}$F. Thus the charge sent is $\frac{1.4 \times x}{1000}$F. Now initially there was 4.29 molal $FeCl_3$. Thus, in 1000 gm, there were (4.29×3) moles of chloride ion (assuming complete ionisation). Hence, in x gm, there were $\frac{12.87x}{1000}$ moles. After electrolysis, therefore, there are $\frac{3.1 \times 3}{1000} x$ gm (for $FeCl_3$) and $\frac{1.4 \times 2}{1000} x$ gm (for $FeCl_2$) in x gm, i.e.,

$$\left(\frac{3.1 \times 3x}{1000} + \frac{1.4 \times 2x}{1000}\right) \text{moles } Cl^- = \frac{12.1x}{1000} \text{ moles } Cl^-.$$

For the passage of 1 faraday the cathode chamber will lose t_- gm-equivalent of an ion. Now, loss of anion from cathode chamber is $\left(\frac{12.87x}{1000} - \frac{12.1x}{1000}\right) = \frac{0.77x}{1000}$ moles of Cl^- ion, i.e., $\frac{0.77x}{1000}$ gm-eq for the passage of $\frac{1.4x}{1000}$F.

$\therefore\ t_- = \dfrac{0.77x/1000}{1.4x/1000} = \dfrac{0.77}{1.4} = \mathbf{0.55.}$

Thus $t_+ = (1 - t_-) = (1 - 0.55) = \mathbf{0.45.}$

Method 2. Considering cation, for the passage of 1 faraday, 1 gm-eq. of Fe^{+3} will be lost, but t_+ gm-eq will move in. Now, the loss of Fe^{3+} should have been $\dfrac{1.4x}{1000}$ gm-eq, as 1F reduced 1 mole of Fe^{3+}. Thus, there should be $\left(\dfrac{4.29-1.4}{1000}\right)x = \dfrac{2.89x}{1000}$ mole Fe^{3+}. But there is $\dfrac{3.1x}{1000}$ mole of $FeCl_3$ (3.1 molal). This increase, i.e., $\left(\dfrac{3.1-2.89}{1000}\right)x = \dfrac{0.21}{1000}x$ mole is due to migration of t_+ gm-eq of Fe^{3+} to the cathode chamber. Each Fe^{3+} ion carries 3 units of charge. Hence, the increase in gm-eq. is $\dfrac{0.21x\times 3}{1000}$ gm-eq.

$$\text{Hence, } t_+ = \frac{0.21x\times\dfrac{3}{1000}}{\dfrac{1.4x}{1000}} = \frac{0.21\times 3}{1.4} = \mathbf{0.45.}$$

43. A 3.33 molal soln. of $FeCl_3$ is electrolysed between Pt-electrodes. After electrolysis, the soln. from cathode chamber had a molality of 2.6 in $FeCl_3$ and 0.86 in $FeCl_2$. Calculate t_+.

Soln. : See the previous problem to solve it. **[Ans.** 0·45]

44. A 2 molal solution of $FeCl_3$ is electrolysed between Pt-electrodes. After electrolysis, the cathode compartment soln. weighing 20 gm is found to be 1.15 molal in $FeCl_3$ and 1 molal in $FeCl_2$. Calculate tthe transport number of Fe^{3+} and Cl^-. **[C.U.(H) 1999]**

Soln. : See Problem no. 42. **[Ans.** t_+ = 0.45]

45. An aq. soln. of $CuSO_4$ was electrolysed between copper electrodes. After the passage of 1.35×10^{-3}F charge, the 25.01 gm soln. from cathode chamber had a molality of 0.033. The molality was initially 0.050. Calculate t_+.

Soln. : By the problem, the charge sent is 1.35×10^{-3} F.

Now, 25.01 gm soln. from the cathode chamber had a molality of 0.033. Let there be x gm water.

Thus, 1000 gm water contains (0.033×159.6) gm of $CuSO_4$

x gm of water contains $\dfrac{0.033\times 159.6x}{1000}$ gm of $CuSO_4$

$= 0.0052668\ x$ gm of $CuSO_4$.

$\therefore$ by the problem,

$x + 0.0052668 \times x = 25.01$

or, $\quad x = \dfrac{25.01}{1.0052668} = 24.87896745 = 24.88$ gm.

Thus, there be 24.88 gm of water in the cathode chamber Now there is a loss of (0.050 – 0.033) = 0.017 molality in $CuSO_4$. Hence, the loss in weight of $CuSO_4$

$$= \frac{0.017 \times 24.88}{1000} \times 159.6 = 0.0675044 \text{ gm.}$$

But eq. wt. of $CuSO_4 = \frac{159.6}{2} = 79.8$.

Hence, the loss in number in gm-eq is $\frac{0.0675044}{79.8}$. But the loss in wt. of salt in the cathode chamber for the passage of 1F is t_- gm-eq. The above loss is for the passage of 1.35×10^{-3} F.

$$\text{Hence, } t_- = \frac{0.0675044/79.8}{1.35 \times 10^{-3}} = \frac{0.0675044}{79.8 \times 1.35 \times 10^{-3}} = 0.626607 = 0.63.$$

$\therefore t_+ = (1 - t_-) = (1 - 0.63) =$ **0.37.**

46. An aq. soln. of $NiSO_4$ was electrolysed between Ni-electrodes. After the passage of 1.42×10^{-3} F charge, 25.26 gm soln. from the cathode chamber was found to have a molality of 0.034. The molality of the soln. was initially 0.051. Calculate t_+.

[M. wt. of $NiSO_4$ = 154.7]

Soln. : Proceed as former problem. [**Ans.** 0.40]

47. A current of 10 mA was passed through a $AgNO_3$ soln. in a Hittorf cell with Ag–electrodes for 80 minutes. After electrolysis, the cathode soln. weighed 40.28 gm and was titrated with 86 cc of 0.0200 molar KSCN. The anode soln. weighed 40.40 gm and required 112 cc of 0.0200 molar KSCN. Calculate the transport number of Ag^+.

Soln. : Here, charge sent = $(10 \times 10^{-3} \text{ A}) \times (80 \times 60 \text{ sec})$

= 48 coulombs

$$= \frac{48}{96484} \text{F} \ [= x\text{F (let)}].$$

Let the weight of water in cathode be a gm and weight in anode be b gm. Now, passage of x F will lead to increase in anode chamber by xt_- gm-eq. and decrease in the cathode chamber by xt_- gm-eq.

The anode soln. required 112 cc 0.02(M) of KSCN = 112 cc 0.02(M) of $AgNO_3$

Now, 1000 cc 1(M) of $AgNO_3$ = 169.8 gm of $AgNO_3$

$$112 \text{ cc } 0.02\text{(M) of } AgNO_3 = \frac{169.8 \times 112 \times 0.02}{1000} = 0.380352$$

= 0.002249 gm-eq.

In anode the soln is 40.40 gm.

In the cathode, 1000 cc 1(M)$AgNO_3$ = 169.8 gm of $AgNO_3$.

$$\text{By the problem, 86 cc 0.02(M) } AgNO_3 = \frac{169.8 \times 86 \times 0.02}{1000} = 0.292 \text{ gm}$$

= 0.00172 gm-eq.

The cathode contains 40.28 gm soln.

So, water in anode = (40.40 – 0.380352)gm = 40.02 gm

and water in cathode = (40.28 – 0.292)gm = 39.99 gm.

Let there be a' gm-eq. $AgNO_3$ initially in cathode and b' gm-eq. $AgNO_3$ in anode.

$\therefore \; a' - x\,t_- = 0.00224$

$b' + x\,t_- = 0.00172.$

$\therefore$ adding, $a' + b' = (0.00396).$

$$\therefore \; a' = \frac{39.99}{(39.99+40.02)} \times 0.00396 = 0.00198$$

$$\text{and } b' = \frac{40.02}{(40.02+39.99)} \times 0.00396 = 0.0019807.$$

[Since, b' and a' are in the ratio of water in each chamber, i.e.,

$$\frac{b'}{a'} = \frac{40.02}{39.99}$$

$$\text{or,} \quad b' = \frac{40.02}{39.99}\, a'.]$$

In cathode, loss in weight = (0.00198 − 0.00172) = 0.00026 gm-eq.

In anode, gain in weight = (0.00224 − 0.00198) = 0.00026 gm-eq.

$$\text{Hence,} \quad t_- = \frac{0.00026 \times 96484}{48} = \mathbf{0.52}$$

$$t_+ = (1 - 0.52) = \mathbf{0.48.}$$

48. A solution containing 0.4312 gm of LiCl per 100 gm of water was electrolysed using Ag anode. After the passage of current (resulting in the deposition of 0.7394 gm of Ag in a voltmeter) 128.61 gm soln. were found to contain 0.4606 gm of LiCl. Calculate t_+.

Soln. : Proceed as usual. **[Ans.** t_+ = 0.33]

49. A 0.14941 wt% aq. KCl soln. at 25°C was electrolysed in a transport cell using two Ag-AgCl electrodes. After the experiment, it was found that 160.25 mg of Ag had been deposited in a coulometer connected in series with the cell, and the cathode compartment contained 120.99 gm of soln., that was 0.1940% KCl by wt. Calculate t_+.

Soln. : Proceed as usual. [0.48]

50. A $BaCl_2$ solution is placed between two parallel electrodes 8 cm apart and is subjected to a potential difference of 4 volts by connecting the electrodes to an external battery. What will be the electrical force on a Barium ion if the interionic attraction and dipolar effect of solvent molecules be disregarded. [D = 80] **[C.U.(H) 1976]**

Soln. : By formula,

$$\text{Force} = \frac{2 \times 4}{8} \times 1.6 \times 10^{-12} \text{ dynes}$$

$$= \mathbf{1.6 \times 10^{-12}} \textbf{ dynes.}$$

51. In a particular experiment on KCl the apparatus consisted of a tube of diameter 4.176 mm and it contained an aqueous soln. of KCl at a concentration of 0.021 mole/litre. A steady current of 1.82 mA was passed and the cationic boundary advanced 3.18 mm in 100 seconds. Find transport number of K^+ and Cl^-. **[C.U.(H) 1995, 2005]**

Soln. : Proceed as Problem no. 28. **[Ans.** t_+ = 0.48; t_- = 0.52]

52. If the equivalent conductances of solutions with varying concentration of NaOAC, NaCl and HCl are plotted against $\sqrt{C}$, the intercepts obtained are 91.0, 128 and 425 respectively in $ohm^{-1}.cm^2.eq^{-1}$ unit at 25 °C. If the resistance of 0.02 molar solution of HOAC in a cell with cell constant 0.2061 cm^{-1} be 888 ohms at 25 °C, what is the degree of dissociation of the acid at 25 °C. (C = concentration) **[C.U.(H) 1995]**

Soln. : Proceed as Problem no. 19. [**Ans.** 0.029]

53. The transference number of the ions of 1.0(N) KCl were determined by moving boundary method using a solution of 0.80(N) $BaCl_2$ as the following solution. Using a current of 0.0142 A the time required for the boundary sweep through a volume of 0.1205 cc. was 1675 sec. What are the transport numbers of K^+ and Cl^- ions? **[V.U. 1999]**

Soln. : Proceed as Problem no. 29. [**Ans.** $t_+ = 0.49$]

54. In between two platinum electrodes 8.0 cm apart and with a potential diference of 4.0 volts was placed in a very dilute aqueous solution of silver nitrate. The Ag^+ and NO_3^- ions are found to move with velocities 3.2×10^{-4} cm/sec and 3.7×10^{-4} cm/sec respectively. Find the equivalent conductance of $AgNO_3$ solution and the transport number of anion. **[V.U. 1998]**

Hints : $v_{\pm} = u_{\pm}.\dfrac{\Delta\phi}{l}$

$$\Lambda^\circ = \lambda^\circ_{Ag^+} + \lambda^\circ_{NO_3^-} = F(u_+ + u_-)$$

$$t_- = \frac{u_-}{u_+ + u_-}.$$

[**Ans.** 133.15; 0.53]

55. The resistance of a conductivity cell filled with 0.02(N) KCl at 18°C is 17.60 ohms. The resistance becomes 91.8 ohms when it is filled with 0.1(N) CH_3COOH at the same temperature. If specific conductance of 0.02(N) KCl is 2.399×10^{-3} $ohm^{-1}.cm^{-1}$ and ion conductance at infinite dilution of H^+ and CH_3COO^- ions are 315 $ohm^{-1}.cm^2.eq^{-1}$ and 35 $ohm^{-1}.cm^2.eq^{-1}$, then calculate the degree of dissociation of CH_3COOH. **[V.U. 2007]**

Hints : $\dfrac{1}{R} = \kappa\,\dfrac{A}{l}$

cell constant = 0.0422 cm^{-1}.

$$\Lambda_{CH_3COOH} = \frac{1000\,\kappa_{CH_3COOH}}{C} = \frac{1000 \times 4.59 \times 10^{-4}}{0.1} = 4.59\ ohm^{-1}.cm^2.eq^{-1}.$$

$$\therefore \text{degree of dissociation} = \alpha = \frac{\Lambda}{\Lambda^\circ} = \frac{4.59}{315+35} = \mathbf{0.013}.$$

56. At 18 °C, the mobility at infinite dilution of the ammonium ion and chlorate ion are 6.6×10^{-4} $cm^2.V^{-1}.sec^{-1}$ and 5.7×10^{-4} $cm^2.V^{-1}.sec^{-1}$ respectively. Calculate Λ°_m of ammonium chlorate and transport number of two ions. **[V.U. 2007]**

Soln. : $$\lambda^\circ_{NH_4^+} = Fu_+ = 96484 \times 6.6 \times 10^{-4} \text{ coulomb} \times \frac{cm^2}{V.sec}$$

$$= 63.68\ ohm^{-1}.cm^2.mole^{-1}.$$

$$\lambda^\circ_{ClO_3^-} = Fu_- = 96484 \times 5.7 \times 10^{-4}$$

$$= 54.99\ ohm^{-1}.cm^2.mole^{-1}.$$

$$\Lambda^{\circ}_m = \lambda^{\circ}_{NH_4^+} + \lambda^{\circ}_{ClO_3^-} = \mathbf{118.67\ ohm^{-1}.cm^2.mole^{-1}}$$

$$t_{NH_4^+} = \frac{u}{u_+ + u_-} = \frac{6.6}{6.6+5.7} = \mathbf{0.54}$$

$$t_{ClO_3^-} = 1 - 0.54 = \mathbf{0.46.}$$

57. The ionic conductance of Li^+ and K^+ are 38.7 mhos.cm^2/gm-ion and 73.5 mhos. cm^2/gm-ion respectively. How long would it take for an ion to move from one electrode to another (2 cm apart) of a conductivity cell where a potential diference of 10 volts is applied between the electrodes. **[B.U. 1991]**

Soln. : Ionic mobility of $Li^+ = u_{Li^+} = \frac{38.7}{96484} = 4.01 \times 10^{-4} cm^2.volt^{-1}.sec^{-1}$.

$$u_{K^+} = \frac{73.5}{96484} = 7.62 \times 10^{-4}\ cm^2.volt^{-1}.sec^{-1}.$$

$$v_{Li^+} = u_+ \frac{\Delta\phi}{l} = 4.01 \times 10^{-4} \times \frac{10}{2} = 2.01 \times 10^{-3}\ cm/sec.$$

$$v_{K^+} = 7.62 \times 10^{-4} \times \frac{10}{2} = 3.81 \times 10^{-3}\ cm/sec.$$

So Li^+ will take $\frac{2}{2.01\times10^{-3}}$ = **997.51 sec** and K^+ will take $\frac{2}{3.81\times10^{-3}}$ = **524.93 sec** to cover 2 cm.

58. The specific conductance of pure water at a certain temperature T was found to be 0.6×10^{-7} $ohm^{-1}.cm^{-1}$. Find the pH of water at this temperature if $\lambda^{\circ}_{H^+} = 350$ $ohm^{-1}.cm^2$ and $\lambda^{\circ}_{OH^-} = 200\ ohm^{-1}.cm^2$. **[B.U. 1998]**

Soln. : $C_{H^+} = \frac{1000\kappa}{\lambda^{\circ}_{H^+} + \lambda^{\circ}_{OH^-}} = 1.091 \times 10^{-7}$

$$pH = -\log C_{H^+} = \mathbf{6.96.}$$

59. The equivalent conductance of a 0.01(N) $CaCl_2$ is 130.36 $ohm^{-1}.cm^2.gm\text{-}eq^{-1}$. What will be the molar conductance in S.I. unit? **[B.U. 2000]**

Soln. : 0.01(N) $CaCl_2$ = 0.005(N) $CaCl_2$

Thus 1(N) $CaCl_2$ = (0.005/0.01)(M) $CaCl_2$

$$\Lambda_{CaCl_2} = 130.36\ ohm^{-1}.cm^2.gm\text{-}eq^{-1}$$

$$\Lambda_m \text{ (in SI)} = 130.36 \text{ siemen } (10^{-2}m)^2 \times \left(\frac{0.005}{0.01} \text{mole}\right)^{-1}$$

$$= \mathbf{0.026072\ siemen.metre^2.mole^{-1}.}$$

60. A conductivity cell of cell constant $1cm^{-1}$ shows a resistance of 6667 ohms when filled with 0.001(M) KCl solution at 25 °C. The same cell records a resistance of 2353 ohms when filled with 0.001(M) HCl solution at 25°C.

(i) Calculate equivalent conductance values for KCl and HCl solutions.

(ii) Calculate ion conduction of H^+ assuming that K^+ and Cl^- have same mobility; consider the solutions to be infinitely dilute.

(iii) How far will H^+ move in 10 seconds when a potential difference of 2 volts is applied between two eletrodes placed 2 cm apart ? **[B.U. 2000]**

Soln. : $\frac{1}{R} = \kappa \frac{1}{\text{cell constant}}$.

$\therefore \kappa = \frac{\text{cell constant}}{R}$.

Thus, $\kappa_{KCl} = \frac{1}{6667} = 1.49 \times 10^{-4}$ mho.cm^{-1}.

Hence, $\Lambda_{KCl} = \frac{1000\kappa}{C} = \frac{1000 \times 1.49 \times 10^{-4}}{0.001}$ = **149.99 mho.cm^2.gm-eq^{-1}**

$\Lambda_{HCl} = \frac{1000}{0.001} \times \frac{1}{2353}$ = 424.99 **mho.cm^2.gm-eq^{-1}.**

By the given problem, $\Lambda_{HCl} = \Lambda_{KCl}^{\circ} = \lambda^{\circ}_{K^+} + \lambda^{\circ}_{Cl^-}$

and $\lambda^{\circ}_{K^+} = \lambda^{\circ}_{Cl^-}$.

Hence, $\lambda^{\circ}_{Cl^-} = \frac{149.99}{2} = 74.99$ mho.cm^2.gm-eq^{-1}.

Thus, $\lambda^{\circ}_{H^+} = \Lambda_{HCl} - \lambda'_{Cl^-}$ = **350 mho.cm^2.gm-eq^{-1}.**

$u_{H^+} = \frac{350}{96484} = 3.63 \times 10^{-3}$ cm^2.volt^{-1}.sec^{-1} ($\lambda + = Fu_+$)

$v_{H^+} = u_H + \frac{\Delta\phi}{l} = 3.63 \times 10^{-3} \times \frac{2}{2} = 3.63 \times 10^{-3}$ cm.sec^{-1}.

So H^+ will move in 10 sec, $3.63 \times 10^{-3} \times 10$ = **3.63×10^{-2} cm.**

61. The equivalent conductance of a weak monobasic acid at infinite dilution is 388.5 mho.cm^2.eq^{-1} at 25 °C. Find the specifiic conductance of 0.1 molar solution the degree of dissociation of which is 6%.

Soln. : Degree of dissociation $\alpha = \frac{\Lambda}{\Lambda^{\circ}}$.

Here, $\alpha = \frac{6}{100} = 0.06$, $\Lambda^{\circ} = 388.5$ mho.cm^2.eq^{-1}.

Thus, $\Lambda = 23.31$ mho.cm^2.eq^{-1}.

Hence, $\kappa = \Lambda \frac{C}{1000} = 23.31 \times \frac{0.1}{1000}$ = **2.33×10^{-3} mho.cm^{-1}.**

62. Calculate the velocity of Li^+ in 0.1 (M) LiCl when a current of 1A is passed through a column of solution with cross-sectional area 4×10^{-4} m^2. Given that equivalent conductances of Li^+ and Cl^- are 4×10^{-3} and 7.5×10^{-3} respectively in mho.m^2.eq^{-1}.

Hints. : $v_{Li^+} = u_{Li^+} \frac{\Delta\phi}{l}$

Now, $u_{Li^+} = \dfrac{\lambda_{Li^+}}{F}$

and $\Delta\phi = i\,R = i\dfrac{l}{A\kappa}$.

Hence, $v_{Li^+} = \dfrac{\lambda_{Li^+}}{F} \times i\dfrac{l}{A\kappa \times l}$.

Again, $\Lambda = \dfrac{10^{-3}\kappa}{C}$, where, κ in mho.m^{-1} and C in gm-mole per litre.

Thus, $v_{Li^+} = \dfrac{\lambda_{Li^+}}{F} \times \dfrac{i \times 10^{-3}}{ACA} = \dfrac{4\times10^{-3}}{96484} \times \dfrac{1\times10^{-3}}{4\times10^{-4}\times0.1\times\left(\lambda_{Li^+}+\lambda_{Cl^-}\right)}$

$= \mathbf{9.013 \times 10^{-5}}$ **m.sec^{-1}.**

63. Equal volumes of 0.1 (M) HCl and 0.1 (M) NaOH are mixed so that specific conductivity of 0.1 (M) solution of NaOH decreases to 0.006 from 0.022. On addition of a further volume of HCl solution equal to that of the first portion added, specific conductivity rises to 0.0170. Calculate :

(a) Λ for NaOH, (b) Λ for NaCl, (c) Λ for HCl, (d) Λ for H^+ and OH^-.

Soln. : (a) $\Lambda_{NaOH} = \dfrac{1000 \times 0.022}{0.1} = $ **220 mho.cm^2.eq^{-1}.**

(b) $C_{NaCl} = 0.05(M)$, $\Lambda_{NaCl} = \dfrac{1000 \times 0.006}{0.05} = $ **120 mho.cm^2.eq^{-1}.**

(c) $\kappa_{NaCl} + \kappa_{HCl} = \dfrac{\Lambda_{NaCl}C_{NaCl}}{1000} + \dfrac{\Lambda_{HCl}C_{HCl}}{1000}$

$0.0170 = \dfrac{120 \times 0.033}{1000} + \dfrac{\Lambda_{HCl} \times 0.033}{1000}$

Since, $V \times 0.1 = 3V \times C_{HCl}$

$\therefore\ C_{HCl} = 0.033$

and $2V \times 0.05 = 3V \times C_{NaCl}$

$\therefore\ C_{NaCl} = 0.033$

$\Lambda_{HCl} = $ **395.15 mho.cm^2.gm-eq^{-1}.**

$\Lambda_{NaOH} + \Lambda_{HCl} - \Lambda_{NaCl}$

$= \lambda_{H^+} + \lambda_{OH^-} = 220 + 395.15 - 120 = 495.15$ mho.cm^2.gm-eq^{-1}.

64. A solution of 0.01(M) $LiNO_3$ and 0.02 (M) KCl is placed in a conductivity cell whose electrodes are 4 cm apart. A potential difference of 8 V is applied between the electrodes and the resistance of the cell is 200 ohms. Calculate t_{Li^+} and current carried by Li^+. Given, $\lambda_{Li^+} = 38.7$, $\lambda_{NO_3^-} = 71.4$, $\lambda_{K^+} = 73.5$, $\lambda_{Cl^-} = 76.3$.

Soln. : Current carried by $Li^+ = I_{Li^+} = C_+ F v^+ A$,

where C_+ = conc. of Li^+ in gm-eq per cc

v_+ = velocity of Li^+

A = cross-sectional area of electrode.

Now, $v_+ = u_+ \dfrac{\Delta\phi}{l}$,

u_+ = mobility of Li^+

$\Delta\phi$ = potential difference

l = distance between the electrodes.

Again, $\Delta\phi = \dfrac{Il}{\kappa A}$.

Hence, $I_{Li^+} = C_+ F u_+ \dfrac{I}{\kappa} = C_+ \lambda_+ \dfrac{I}{\kappa}$.

By problem, $\lambda_{Li^+} = 38.7$, $\lambda_{NO_3^-} = 71.4$, $\lambda_{K^+} = 73.5$, $\lambda_{Cl^-} = 76.3$ mho.cm^{-1}.

Thus, $\kappa = \dfrac{\Lambda C_{LiNO_3}}{1000} + \dfrac{\Lambda C_{KCl}}{1000}$

$= \dfrac{0.01(38.7+71.4)}{1000} + \dfrac{0.02(73.5+76.3)}{1000} = \dfrac{4.1}{1000}$ mho.cm^{-1}

$I = \dfrac{\Delta\phi}{R} = \dfrac{8}{200}$ A.

Hence, $I_{Li^+} = \dfrac{0.01}{1000} \times 38.7 \times \dfrac{8 \times 1000}{200 \times 4.1} = 3.78 \times 10^{-3}$ A.

$t_{Li^+} = \dfrac{I_{Li^+}}{I} = \dfrac{C_+\lambda_+}{\kappa} = \dfrac{\kappa_+}{\kappa} = \mathbf{0.943.}$

65. The resistance of a conductivity cell is 520 ohms when filled with 0.1(M) CH_3COOH and drops to 122 ohms when enough solid NaCl is added to make the solution 0.01 (M) in NaCl as well. Calculate the cell constant and concentration of H^+ in the solution. Given,

$\Lambda_{NaCl} = 126$, $\Lambda_{CH_3COOH} = 386$.

Soln. : $\kappa_{NaCl} = \dfrac{\Lambda C}{1000} = \dfrac{126 \times 0.01}{1000} = 1.26 \times 10^{-3}$ohm^{-1}.cm^{-1}.

By problem, $R_{CH_3COOH} = 520$ ohm

$R_{CH_3COOH + NaCl} = 122$ ohm.

Thus, $\dfrac{1}{520} = \kappa_{CH_3COOH} \dfrac{1}{K}$ (where K = cell constant)

and $\dfrac{1}{122} = \kappa_{mix} \dfrac{1}{K}$.

Hence, $\dfrac{520}{122} = \dfrac{\kappa_{mix}}{\kappa_{CH_3COOH}} = \dfrac{\kappa_{CH_3COOH} + \kappa_{NaCl}}{\kappa_{CH_3COOH}}$.

$\therefore \kappa_{CH_3COOH} = 3.86 \times 10^{-4}$ohm^{-1}.cm^{-1}.

Hence, cell constant = $K = 3.86 \times 10^{-4} \times 520 = \mathbf{0.2008\ cm^{-1}}$.

Now, $\Lambda_{CH_3COOH} = 386 = \dfrac{1000\kappa_{CH_3COOH}}{C}$

Hence, $C = \dfrac{1000 \times 3.86 \times 10^{-4}}{386} = 1.00 \times 10^{-3}$(M).

Thus, concentration of H^+ = **1.00×10^{-3}(M).**

66. The resistance of a conductivity cell is 220000 ohms with water, 100 ohms with 0.02 (M) KCl, 102000 ohms with water saturated with AgCl at 25 °C. Neglecting the variation of Λ with concentration, calculate :

(a) cell constant

(b) specific conductivity of the saturated solution of AgCl and

(c) solubility product of AgCl at 25 °C.

Given, $\Lambda_{AgCl} = 126.8$, $\Lambda_{KCl} = 138.3$ at 25 °C.

Hints : $220000 = \dfrac{1}{\kappa_{H_2O}} K$, where K = cell constant

$$100 = \frac{1}{\kappa_{KCl}} K$$

$$102000 = \frac{1}{\kappa_{AgCl+H_2O}}$$

$$\kappa_{KCl} = \frac{138.3 \times 0.02}{1000} = 2.77 \times 10^{-3} \text{ ohm}^{-1}.\text{cm}^{-1}$$

Cell constant K = **2.77×10^{-1} cm^{-1}.**

$\kappa_{H_2O} = 1.26 \times 10^{-6}$

$\kappa_{AgCl + H_2O} = 2.72 \times 10^{-6}$, $\kappa_{AgCl} = 1.46 \times 10^{-6}$

$$C_{Ag^+} = C_{Cl^-} = \frac{1000 \times 1.46 \times 10^{-6}}{126.8} = 1.15 \times 10^{-5}.$$

Hence, solubility product = **1.33×10^{-10}.**

67. Resistance of a conductivity cell is 7000 ohms with 0.1 (M) KCl, 2600 ohms with 0.1 (M) KCl and 0.2 (M) NaCl. Given that for 0.2 (M) NaCl, $t_{Na^+} = 0.4$, $\lambda_{Cl^-} = 75$. Calculate Λ_{KCl}.

Hints : $\Lambda_{NaCl} = \dfrac{\lambda_{Cl^-}}{1 - t_{Na^+}}$

$\dfrac{1}{R} = \kappa \times$ cell constant.

$$\kappa_1 = \kappa_{KCl} = \frac{C_{KCl}\Lambda_{KCl}}{1000}, \ \kappa_2 = \kappa_{KCl} + \kappa_{NaCl} = \frac{C_{KCl}\Lambda_{KCl}}{1000} + \frac{C_{NaCl}\Lambda_{NaCl}}{1000}$$

and $\dfrac{\kappa_1}{\kappa_2} = \dfrac{R_2}{R_1}$.

[**Ans.** $\Lambda_{KCl} = 147$ ohm^{-1}.cm^2.eq^{-1}]

68. The molar conductances at infinite dilution for $BaCl_2$, KCl, K_2SO_4 and Cl^- are 280, 150, 300 and 76 $\Omega^{-1}m^2mol^{-1}$ respectively. Calculate the transport number of Ba^{2+} in $BaSO_4$ solution at infinite dilution. **[JAM 2011]**

Hints : $BaCl_2 = Ba^{+2} + 2Cl^-$

$$\lambda_{m,BaCl_2} = \lambda_{m,Ba^{+2}} + 2\lambda_{m,Cl^-}$$

$KCl = K^+ + Cl^-$

$$\lambda_{m,KCl} = \lambda_{m,K^+} + \lambda_{m,Cl^-}$$

$$K_2SO_4 = 2K^+ + SO_4^{-2}$$

$$\lambda_{m,K_2SO_4} = 2\lambda_{m,K^+} + \lambda_{m,SO_4^{-2}}$$

$$BaSO_4 = Ba^{+2} + SO_4^{-2}$$

$$t_{Ba^{+2}} = \frac{\lambda_{m,Ba^{+2}}}{\lambda_{m,Ba^{+2}} + \lambda_{m,SO_4^{-2}}} = \frac{128}{280} = 0.46.$$

69. In a Hittorf experiment to determine the transference numbers in KCl solution, the following data were obtained. Mass of both the cathode solution and middle portion 120.99 gm; per cent KCl in cathode portion 0.19398, per cent KCl in the middle portion 0.14948. Calculate t_+ if 0.16034 gm of Ag was deposited in silver coulometer connected in series. Given that molecular weight of KCl = 74.6 and that of Ag is 107.8.

Soln. : The charge sent $= \dfrac{0.16034}{107.8}$ F.

Again, let there be x gm of water in the middle chamber.

By the data, in the middle chamber KCl present per gm of water

$$= \frac{0.14948}{100 - 0.14948} = 1.49714 \times 10^{-3}.$$

Hence, x gm water contains $1.49714 \times 10^{-3}x$ gm KCl.

So mass of middle chamber $= (x + 1.49714 \times 10^{-3}x)$ gm

By the problem,

$$x(1 + 1.49714 \times 10^{-3}) = 120.99.$$

$$\therefore \quad x = 120.8091.$$

So total amount of KCl in middle chamber = (120.99 – 120.8091) gm = 0.1809 gm.

The cathode chamber also contains 120.8091 gm of water.

Thus, after electrolysis amount of KCl present in cathode chamber =

$$\frac{0.19398}{100 - 0.19398} \times 120.8091 \text{ gm} = 0.2348 \text{ gm}.$$

Hence, increase of KCl in cathode chamber = (0.2348 – 0.1809) gm.

= 0.0539 gm

$$= \frac{0.0539}{74.6} \text{ gm mole} = 7.225 \times 10^{-4} \text{ gm.mole}.$$

This increase in KCl is due to passage of $\frac{0.16034}{107.8}$ F = 1.4874×10^{-3} F.

Hence, $t_+ = \frac{7.225 \times 10^{-4}}{1.4874 \times 10^{-3}} = 0.48.$

Thus, $t_- = 1 - t_+ = 0.52.$

CHAPTER 12

IONIC EQUILIBRIUM

Required Formulae

1. Ionic product of water (K_w) is

$K_w = a_{H^+} \, a_{OH^-}$,

where a_{H^+} is the activity of H^+ ion and a_{OH^-} is the activity of OH^- ion.

2. pH = $-\log a_{H^+}$

For dilute solutions $a_{H^+} = \frac{C_{H^+}}{C^o_{H^+}}$. Then pH = $-\log \frac{C_{H^+}}{C^o_{H^+}}$,

where $C^o_{H^+} = 1$ (M).

However, to save both space and time we will skip the denominator. Thus, pH = $-\log C_{H^+}$.

3. Ionic strength (μ) = $\frac{1}{2} \sum_i C_i Z_i^2$,

where C_i is the concentration of ith ion in moles.litre^{-1} and Z_i is the valency of ion.

4. Mean ionic activity coefficient ($\gamma_\pm$) is $\log \gamma_\pm = -AZ_+ Z_- \sqrt{\mu} + B\mu$.

$A = 0.51$ for water at 25 °C. Normally B is much smaller than A and neglected.

5. pH of weak acid = $\frac{1}{2} pK_a - \frac{1}{2} \log C$,

where pK_a is $-\log_{10} K_a$, where K_a is the dissociation constant of weak acid and C is the concentration of the acid in molarity.

6. If K_b be the dissociation constant of weak base, pH of the weak base

$= 14 - \left(\frac{1}{2} pK_b - \frac{1}{2} \log C\right)$, where K_w has been taken as 10^{-14}.

7. pH of a buffer of weak acid and its salt = $pK_a + \log \frac{C_{salt}}{C_{acid}}$.

8. pH of a buffer of weak base and its salt $= 14 - \left(pK_b + \log\frac{C_{salt}}{C_{base}}\right)$.

9. pH of a salt of weak acid and strong base $= 7 + \frac{1}{2}pK_a + \frac{1}{2}\log C$.

10. pH of a weak base and strong acid $= 7 - \frac{1}{2}pK_b - \frac{1}{2}\log C$.

11. pH of a salt of weak acid and weak base $= 7 + \frac{1}{2}pK_a - \frac{1}{2}pK_b$.

In all these cases [from 9 to 11] 7 is really $\frac{1}{2}pK_w$. It is 7, because K_w is taken as 10^{-14}.

12. Solubility product of a salt of $M_XA_Y = XM^{a+} + YA^{b-}$, so that $Xa = Yb$.

$K_s = a^X_{M^{a+}} . a^Y_{A^{b-}}$.

For a dilute solution, activity is replaced by concentration.

13. The activity of electrolyte M_XA_Y as a whole,

$$a_2 = a_+^X a_-^Y,$$

where
$$a_+ = C_+\gamma_+$$
$$a_- = C_- \gamma_-.$$

14. Mean ionic activity,

$$a_\pm = \sqrt[\nu]{a_2} = \sqrt[\nu]{C_+^X C_-^Y \gamma_+^X \gamma_-^Y} = (C_+^X C_-^Y)^{\frac{1}{x+y}} . (\gamma_+^X \gamma_-^Y)^{\frac{1}{x+y}},$$

where $X + Y = \nu$ = total number of ions resulting from 1 molecule of the electrolyte.

Now, $C_+ = XC$ and $C_- = YC$,

where C = molarity of the electrolyte and mean ionic activity coefficient of the electrolyte

$$= \gamma_\pm = \left[(\gamma_+)^X (\gamma_-)^Y\right]^{\frac{1}{X+Y}}.$$

Thus, $a_\pm = \left[(X^X Y^Y)\right]^{\frac{1}{X+Y}} C\gamma_\pm$.

Short Questions

1. Explain—"The neutralization of all bases by acids do not necessarily occur at pH 7."

Ans. : Neutralization is the reaction between an acid and a base. The product is a salt. At the equivalence point, no excess acid or base is present and it is thus expected that the pH may be 7. However, due to hydrolysis, excess acid or base may be produced and the pH may be different from 7. For example, in case of titration of weak acid by a strong base, the salt contains the conjugate strong base which accepts proton from water producing basic solution. Similarly, in case of titration of weak base and strong acid, the pH at the end point will be less than 7. In case of titration of weak acid and weak base, both the cation and anion are strong conjugate base and acid respectively, and thus they will interact with water. The pH depends on degree of interaction. However, if we consider the neutralisation of a strong base and strong acid, then the salt contains cations and anions, which are weak conjugates, and there will be no interaction and pH remains 7. Thus, as obvious, neutralisation of all acids and bases cannot take place at pH 7.

2. What will happen if HCl (g) is passed through a saturated solution of NaCl?

Ans. : In the saturated solution of NaCl, $a_{Na^+}\, a_{Cl^-} = K_s$ has been reached. If HCl (g) be passed, a_{Cl^-} will be greater. To keep solubility product (K_s) constant a_{Na^+} in solution will be reduced and thus NaCl will precipitate.

3. Do you expect the solubility of CaF_2 in aqueous KNO_3 solution to be same as lower than or higher than its solubility in pure water at the same temprature? Give reasons.

Ans. : The solubility of CaF_2 will increase in a solution of KNO_3. The solubility of any substance is normally not affected by the presence of another solute. However, if the substance is an electrolyte as is CaF_2, then another salt with uncommon ion generally increases the solubility of sparingly soluble salt. This is salt effect.

It is well-known from Debye Hückel theory that the mean ionic activity coefficient of a salt decreases with increase in ionic strength $\left(\log\gamma_{\pm} = -AZ_+Z_-\sqrt{\mu}\right)$. If we define the product of conc. terms as K_c, the product of activity coefficient as K_γ and the product of ionic activity term as K_s, then $K_s = K_c.K_\gamma$. Since K_s is a constant and K_γ decreases with increase in ionic strength, K_c must increase to keep K_s constant, i.e., the conc. of the ions will increase. That is why the addition of a salt with uncommon ion will increase the solubility of salts like CaF_2.

4. The second dissociation constant of an acid is lower in magnitude than the first dissociation constant.—Explain.

Ans. : In the first dissociation of a polybasic acid H_nA, H^+ and $H_{n-1}A^{-1}$ results with a certain K_a value. In the second dissociation $H_{n-1}A^{-1}$ ionises to $H_{n-2}A^{-2}$ and H^+; a positive charge is to be separated from a negative ion. This is much difficult and $C_{H_{n-2}A^{-2}}$ will be in smaller amount. Thus, K_2 defined by $K_2 = \dfrac{a_{H^+}.a_{H_{n-2}A^{-2}}}{a_{H_{n-1}A^{-1}}}$ will be much lower than K_1 defined as $K_1 = \dfrac{a_{H^+}.a_{H_{n-1}A^{-1}}}{a_{H_nA}}$.

5. Ionic product and dissociation constant of water are not identical—explain.

Ans. : Ionic product of water (K_w) of any aq. solution is defined as

$K_w = \left(a_{H^+}.a_{OH^-}\right)$. But the process is $H_2O \rightleftharpoons H^+ + OH^-$

Hence, the dissociation constant is $K_d = \dfrac{a_{H^+}a_{OH^-}}{a_{H_2O}}$.

In dilute solutions, a_{H^+}, etc. are very small. Thus, $K_d = \dfrac{C_{H^+}.C_{OH^-}}{C_{H_2O}}$.

$C_{H_2O} = \dfrac{1000}{18} = 55.56$, taking density of water as 1.

Hence, the dissociation constant $K_d = \dfrac{C_{H^+}.C_{OH^-}}{55.56} = \dfrac{K_w}{55.56}$.

Hence, K_d and K_w are different.

6. Phenolphthalein is not a suitable indicator for the titration of NH_3 with HCl—explain.

Ans. : An indicator changes colour roughly in the pH range (pK_{In}–1) to (pK_{In} + 1), where $pK_{In} = -\log K_{In}$, K_{In} is the indicator constant. As phenolphthalein pK_{In} is around 10 (exactly 9.6), so the colour change may take place in between 8.6 to 10.6. But in the titration of HCl, a strong acid by NH_3, a weak base, the pH at the end point is less than 7. In titrations we allowed one *drop* error. Within this error range the pH changes generally between 3 and 6. Thus, phenolphthalein is not a suitable indicator for titration of HCl by NH_3.

7. Methyl orange is a suitable indicator for the titration of strong acid and weak base. Explain.

Ans. : In the titration of strong acid by weak base the salt produced will not be at pH = 7 due to hydrolysis. Allowing for one drop error, at the end point of the titration of 0.1 (N) acid by 0.1 (N) base will lead to a change of pH generally from 3 to 6 or 7 depending on K_b of weak base. Indicators with pK_{In} below 7 are suitable since an indicator changes colour roughly between pH = pK_{In} – 1 to pK_{In} + 1. Methyl orange has pK_a = 3.7 (around 4) and changes colour between 3.1 to 4.4 (expected 2.7 to 4.7). Thus, indicators like methyl orange, methyl yellow, bromcresol green, etc are suitable indicators for the titration of strong acid by weak base.

8. You are supplied with two weak acids *A* and *B* of pka values 4.8 and 5.8 respectively. State, with reason, which one you would prefer to prepare a buffer solution of pH 5.1.

Soln. : We know, $pH = pK_a + \log\frac{C_{salt}}{C_{acid}}$.

For Acid *A*, $pK_a = 4.8$.
So at pH = 5.1,

$$\log\frac{C_{salt}}{C_{acid}} = pH - pK_a = 0.3.$$

For Acid, B, $pK_a = 5.8$.
Thus, at pH = 5.1,

$$\log\frac{C_{salt}}{C_{acid}} = -0.7.$$

A buffer acts well within pH = $pK_a \pm 1$ with maximum capacity at pH = pK_a, where $\log\frac{C_{salt}}{C_{acid}} = 0$. The given pH (5.1) falls within the range between (pK_a + 1) to (pK_a – 1) for both the weak acids *A* and *B*. However, for acid *A*, pK_a is more close to 5.1 than that of acid *B*. Thus, *A* is preferred over *B* to prepare the buffer solution of pH = 5.1.

Problems

1. Calculate the pH of (a) 0.1 (N) solution of CH_3COOH, (b) 0.01 (N) NH_4OH, (c) 0.01 (N) NH_4Cl, (d) 0.001 (N) CH_3COONa, (e) 0.02 (N) CH_3COONH_4. Given that K_a for $CH_3COOH = 2 \times 10^{-5}$ and K_b for $NH_4OH = 2 \times 10^{-5}$.

Soln. : (a) pH of weak acid $= \frac{1}{2}pK_a - \frac{1}{2}\log C$.

$$\therefore \quad pH = \left[\frac{1}{2}\left(-\log 2\times10^{-5}\right)-\frac{1}{2}\log 0.1\right] = \frac{4.699}{2} + 0.5$$

$= 2.3495 + 0.5 = 2.8495 \approx$ **2.8.**

(b) pH of weak base

$$= 14 - \left(\frac{1}{2}pK_b - \frac{1}{2}\log C\right)$$

$$= 14 - \left[\frac{1}{2}\left(-\log 2\times10^{-5}\right)-\frac{1}{2}\log 0.01\right]$$

$= 14 - 2.349 + 1 =$ **12.66.**

(c) pH of a salt of weak base and strong acid

$= 7 - \frac{1}{2}pK_b - \frac{1}{2}\log C =$ **5.65.**

(d) pH of a salt of weak acid and strong base

$= 7 + \frac{1}{2}\ pK_a + \frac{1}{2}\log C =$ **7.85.**

(e) pH of a salt of weak acid and weak base

$= 7 + \frac{1}{2}pK_a - \frac{1}{2}pK_b =$ **7.**

2. Calculate the pH of the solution obtained by mixing 10 ml of 0.2 (N) KOH with 30 ml of 0.1 (N) CH_3COOH. $K_a = 2 \times 10^{-5}$. **[C.U. 1986]**

Soln. : 10 ml 0.2 (N) KOH will react with 20 ml 0.1 (N) CH_3COOH to give 20 ml 0.1 (N) CH_3COOK

But the total volume due to mixing = (10 + 30) ml = 40 ml.

$$\therefore \quad C_{CH_3COOK} = \frac{20\times0.1}{40} = 0.5.$$

$$C_{CH_3COOH} = \frac{(30-20)\times0.1}{40} = 0.025.$$

$$\text{Now, } pH = pK_a + \log\frac{C_{salt}}{C_{acid}} = -\log\ (2\times10^{-5}) + \log\ \frac{0.05}{0.025}$$

$= 4.699 + 0.3010 =$ **5.**

3. In the titration of a 0.1(M) weak diabasic acid ($k_1 = 10^{-6}$ and $k_2 = 10^{-10}$) with a strong base, calculate the pH (i) at the start and (ii) at the first equivalence point. **[V.U. 1999]**

Soln. : (i) Let the weak dibasic acid H_2A dissociate as

$$\underset{(0.1\ -\ x)}{H_2A} \rightleftharpoons \underset{x+y}{H^+} + \underset{x}{HA^-}$$

$$\text{and} \quad \underset{x-y}{HA^-} \rightleftharpoons \underset{x+y}{H^+} + \underset{y}{A^{-2}}$$

H_2A dissociates to give x(M) HA^- while HA^- dissociates to give y(M) A^{-2}, so that concentration of $H^+ = (x + y)$ (M).

Given initial concentration of H_2A = 0.1 (M).

Now, $K_1 = \dfrac{C_{H^+}C_{HA^-}}{C_{H_2A}}$ and $K_2 = \dfrac{C_{H^+}C_{A^{-2}}}{C_{HA^-}}$.

Given, $K_1 = 10^{-6}$ and $K_2 = 10^{-10}$.

Thus, $$10^{-6} = \frac{x(x+y)}{0.1-x} \quad (1)$$

$$10^{-10} = \frac{(x+y)y}{x-y}. \quad (2)$$

Neglecting y in comparison to x, we get $y = 10^{-10}$ [from Eq. (2)]

and $x^2 + 10^{-6}x - 10^{-7} = 0$. [from Eq. (1)]

Thus, $x = 3.157 \times 10^{-4}$, -3.167×10^{-4}.

Negative value of x is neglected since it is physically unacceptable.

Thus, concentration of H^+ at the start

$C_{H^+} = (x + y)$ (M)

$= (3.157 \times 10^{-4} + 10^{-10})$ (M) $= 3.157 \times 10^{-4}$ (M).

Hence, at the start pH $= -\log C_{H^+} = -\log(3.157 \times 10^{-4}) = 3.5$.

(ii) At the first equivalence point in the titration with strong base BOH, H_2A will be half-neutralized. Thus, BOH + $H_2A \longrightarrow$ BHA + H_2O.

BHA is an acid salt. Thus, pH $= \frac{1}{2}(pk_1 + pk_2) = \frac{1}{2}(-\log 10^{-6} - \log 10^{-10}) = 8$.

4. To 50 ml of a 0.1 (M) KH_2PO_4 solution is added to (i) 10 ml and (ii) 50 ml of a 0.1 (M) NaOH solution. Calculate the pH values of the resulting buffer solution in each case. (Given for phosphoric acid $k_1 = 7.5 \times 10^{-3}$, $k_2 = 6.2 \times 10^{-8}$ and $k_3 = 4.8 \times 10^{-13}$ at 25 °C.) **[V.U. 1999]**

Soln. : (i) 10 ml 0.1 (M) NaOH reacts with 10 cc 0.1(M) KH_2PO_4 to give 10 ml 0.1 (M) $KNaHPO_4$.

$$NaOH + KH_2PO_4 \longrightarrow KNaHPO_4 + H_2O$$

Total volume of solution = 60 ml

Hence, $C_{KNaHPO_4} = \dfrac{10 \times 0.1}{60}$ (M) $= \dfrac{0.1}{6}$ (M).

KH_2PO_4 remaining = (50 − 10) ml or 40 ml 0.1 (M).

Hence, $C_{KH_2PO_4} = \dfrac{40 \times 0.1}{60}$ (M) $= \dfrac{0.2}{3}$ (M).

$$pH = pK_a + \log\frac{C_{KNaHPO_4}}{C_{KH_2PO_4}} = -\log k_2 + \log\frac{0.1 \times 3}{6 \times 0.2}.$$

Given $k_2 = 6.2 \times 10^{-8}$, Hence, pH = 6.605.

(ii) 50 ml 0.1 (M) NaOH will react with 50 ml 0.1 (M) KH_2PO_4 to produce 50 ml 0.1 (M) $KNaHPO_4$.

$$NaOH + KH_2PO_4 \longrightarrow KNaHPO_4 + H_2O$$

$KNaHPO_4$ is an acid salt.

We know, at n^{th} stage neutralization for a polybasic acid, pH $= \frac{1}{2}(pK_n + pK_{n+1})$.

It is at 2nd stage neutralization.

Hence, pH $= \frac{1}{2}(pK_2 + pK_3)$

$= \frac{1}{2}[-\log(6.2 \times 10^{-8}) - \log(4.8 \times 10^{-13})] = 9.76.$

5. Find the pH of buffer solution consisting of 0.2(M) NH_3 and 0.1 (M) NH_4NO_3. For NH_3, $K_b = 1.6 \times 10^{-5}$ at 25 °C. What is the % of change in pH after addition of 0.01 mole NaOH to 1 litre of this buffer solution? How much pH of water would change when same amount of NaOH is added to 1 litre of water?

Soln. : (i) Given : $C_{salt} = C_{NH_4NO_3} = 0.1$ (M)

$C_{base} = C_{NH_4OH} = 0.2$ (M)

and $K_b = 1.6 \times 10^{-5}$

Now, pOH $= pK_b + \log \frac{C_{salt}}{C_{base}}$

Hence, pOH $= -\log(1.6 \times 10^{-5}) + \log \frac{0.1}{0.2} = 4.49.$

Thus, pH = 14 – 4.49 = **9.51**.

(ii) When 0.01 mole of NaOH is added to 1l of this buffer solution, OH^- (from NaOH) will react with NH_4^+ to given NH_4OH.

Thus, $NaOH + NH_4NO_3 = NH_4OH + NaNO_3$

0.01(M) 0.01(M) 0.01(M)

So concentration of NH_4NO_3 = (0.1 – 0.01) (M) = 0.09 (M).

∴ concentration of NH_4OH = (0.2 + 0.01) (M) = 0.21 (M).

Hence, pOH $= -\log(1.6 \times 10^{-5}) + \log \frac{0.09}{0.21} = 4.43.$

Thus, pH = 9.57.

Hence, % change in pH $= \frac{9.57 - 9.51}{9.51} \times 100 =$ **0.63%**.

(iii) pH of water = 7

pH of 0.01 (M) NaOH = 14 – pOH = 14 + $\log C_{OH^-}$ = 12.

Hence, % change in pH $= \frac{12-7}{7} \times 100 =$ **71.43%**.

6. Calculate the pH of a solution obtained by mixing 50.00 ml of 0.20 M weak acid HA ($K_a = 10^{-5}$) and 50.00 ml of 0.20 M NaOH at room temperature. **[JAM 2006]**

Soln. : $HA + NaOH \longrightarrow NaA + H_2O$

50 ml of 0.20 (M) HA reacts with 50 ml of 0.2 (M) NaA.

Total volume = 100 ml.

Thus, strength of NaA = $\frac{50 \times 0.2}{100}$ (M) = 0.1(M)

Again, NaA is salt of weak acid and strong base. So it will undergo hydrolysis. Thus,

$$pH = 7 + \frac{1}{2}pK_a + \frac{1}{2}\log C.$$

Given that, $K_a = 10^{-5}$. Hence,

$$pH = 7 + \frac{1}{2}(-\log 10^{-5}) + \frac{1}{2}\log (0.1) = \mathbf{9}.$$

7. A solution at 25 °C contains 0.01 molar propionic acid ($K_a = 1.39 \times 10^{-5}$) and 0.03 molar HCl. Calculate α for propionic acid and pH of the mixture.

Soln. : Propionic acid ionises as

$$\underset{0.01\,(1-\alpha)}{H_2O + CH_3CH_2CH_2COOH} \rightleftharpoons \underset{0.01\,\alpha}{H_3O^+} + \underset{0.01\,\alpha}{CH_3CH_2CH_2COO^-}$$

HCl ionises completely to give 0.03 (N) H_3O^+.

$\therefore \quad C_{H_3O^+} = (0.03 + 0.01\,\alpha)$

$$\therefore \quad K_a = \frac{C_{H_3O^+}.C_{CH_3CH_2CH_2COO^-}}{C_{CH_3CH_2CH_2COOH}} = \frac{(0.03+0.01\alpha)0.01\alpha}{0.01(1-\alpha)}$$

Neglecting α in comparison to 1,

$$K_a = \frac{(0.03+0.01\alpha)0.01\alpha}{0.01} = 1.39 \times 10^{-5}$$

or, $0.01\,\alpha^2 + 0{\cdot}03\,\alpha = 1.39 \times 10^{-5}$

or, $0.01\,\alpha^2 + 0.03\,\alpha - 1.39 \times 10^{-5} = 0$

$$\text{or, } \alpha = \frac{-0.03 \pm \sqrt{(0.03)^2 + 4 \times 0.01 \times 1.39 \times 10^{-5}}}{2.01}.$$

Neglecting (–)ve sign, since α cannot be negative, $\alpha = \frac{0.0009265}{2} = 0.00046325$ $= \mathbf{4.63 \times 10^{-4}}$.

Now, $C_{H_3O^+} = (0.03 + 0.01\alpha) = 0.03 + 0.01 \times 4.63 \times 10^{-4} = 0.0300046.$

$\therefore \quad pH = -\log C_{H_3O^+} = 1.5228117 \approx \mathbf{1.5}.$

8. Calculate the pH of 0.001 (M) H_2SO_4 (given, $K_2 = 1.01 \times 10^{-2}$)

Soln. : We know that, H_2SO_4 is a strong acid.

$$\underset{0.001}{H_2SO_4} + H_2O \rightleftharpoons \underset{0.001}{HSO_4^-} + \underset{0.001}{H_3O^+}$$

$$HSO_4^- + H_2O \rightleftharpoons H_3O^+ + SO_4^=$$
$$(0.001-x) \qquad (0.001+x) \qquad x$$

Now, $C_{HSO_4^-} = (0.001 - x)$

$C_{H_3O^+} = (0.001 + x)$

$C_{SO_4^=} = x.$

$$\therefore \quad K_2 = \frac{C_{H_3O^+} \cdot C_{SO_4^{2-}}}{C_{HSO_4^-}}.$$

$$\text{or, } 1.01 \times 10^{-2} = \frac{(0.001+x)x}{(0.001-x)}$$

or, $1.01 \times 10^{-2} \times 0.001 - 1.01 \times 10^{-2} x = 0.001x + x^2$

or, $x^2 + 0.001x + 0.0101\ x - 1.01 \times 10^{-2} \times 0.001 = 0$

or, $x^2 + 0.0111x - 0.0101 \times 0.001 = 0$

$$\text{or, } x = \frac{-0.0111 \pm \sqrt{(0.0111)^2 + 4 \times 1 \times 0.0101 \times 0.001}}{2}$$

Neglecting (–)ve sign, $x = 8.4551 \times 10^{-4}$.

$\therefore\ C_{H_3O^+} = 1.84551 \times 10^{-3}$

pH $= -\log C_{H_3O^+} = 2.7338845 \approx$ **2.73.**

9. Prepare a 100 cc buffer solution of pH = 9. You are supplied with 0.1 (N) NH_4OH and 0.1 (N) HCl. Given that K_b for $NH_4OH = 2 \times 10^{-5}$. **[C.U.(H) 1972]**

Soln. : The resulting solution will be a buffer of weak base and its salt, whose pH is given by

$$pK_w - \left[pK_b + \log \frac{C_{salt}}{C_{base}} \right].$$

By the problem,

$$9 = 14 - \left[5 - \log 2 + \log \frac{C_{salt}}{C_{base}} \right]$$

$$\text{or, } \frac{C_{salt}}{C_{base}} = 2.$$

Let us consider that x cc 0.1 (N) HCl is mixed with $(100 - x)$ cc 0.1 (N) NH_4OH. Since x cc 0.1 (N) HCl will consume x cc 0.1 (N) NH_4OH, the resulting mixture will contain $(100 - 2x)$ cc 0.1 (N) NH_4OH.

$$\therefore \quad C_{salt} = \frac{x \times 0.1}{100}$$

$$\text{and } C_{base} = \frac{(100-2x) \times 0.1}{100}.$$

$$\therefore \quad \frac{C_{salt}}{C_{base}} = \frac{x}{100-2x}.$$

Therefore, $\dfrac{x}{100-2x} = 2$

or, $x = 200 - 4x$ or, $5x = 200$

or, $x = \dfrac{200}{5} = 40$ cc.

Thus **40 cc of 0.1 (N) HCl should be mixed with** (100 – 40) cc 0.1 (N), i.e., **60 cc of 0.1 (N) NH_4OH.**

10. If the final volume is 1 litre, how many moles of HCl will have to be added to 500 cc of 0.1 (M) Na_2CO_3 to adjust the pH to 10.0? (Given : for H_2CO_3, $K_1 = 4.52 \times 10^{-7}$, $K_2 = 4.69 \times 10^{-11}$)

Soln. : If the neutralisation is not complete (as apparent from pH = 10), we have a buffer of Na_2CO_3 and $NaHCO_3$. ($NaHCO_3$ is an acid.)

$$pH = pK_a + \log \frac{C_{salt}}{C_{acid}}$$

or, $10 = -\log(4.69 \times 10^{-11}) + \log \dfrac{C_s}{C_a}$

or, $\log \dfrac{C_s}{C_a} = (10 + \log 4.69 - 11)$

or, $\dfrac{C_s}{C_a} = 0.469.$

Now, 500 cc x (M) HCl will produce 500 cc x (M) $NaHCO_3$ (in total volume of 1 litre).

$\therefore C_{NaHCO_3} = \dfrac{500x}{1000} = \dfrac{x}{2}$ (M).

Now, 500 cc x (M) HCl will react with 500 cc x (M) Na_2CO_3.

Hence, there will remain 500 cc $(0.1 - x)$ (M) Na_2CO_3.

$$C_{Na_2CO_3} = \frac{500(0.1-x)}{1000} = \frac{0.1-x}{2} \text{ (M).}$$

By the problem,

$$\frac{(0.1-x)/2}{x/2} = 0.469$$

or, $\dfrac{0.1-x}{x} = 0.469$ or, $0.1 - x = 0.469x$

or, $1.469x = 0.1$

or, $x = \dfrac{0.1}{1.469} = 0.068$ (M).

Thus, 500 cc 0.068 molar HCl, i.e.,

$\dfrac{0.068 \times 500}{1000} = \dfrac{0.068}{2}$ mole = **0.034 mole of HCl is to be added.**

11. The degree of dissociation of pure water at 18 °C is found to be 1.8×10^{-9}. Find ionic product of water and its dissociation constant at 18 °C.

Soln. : Concentration of water $= \frac{1000}{18} = 55.56$ moles/litre.

Thus,	H_2O	$\rightleftharpoons$	H^+	+	OH^-
Initially :	55.56		0		0
At eq. :	55.56 (1–α)		55.56α		55.56α

α = the degree of dissociation $= 1.8 \times 10^{-9}$.

$$K_w = C_{H^+}.C_{OH^-} = \left(55.56 \times 1.8 \times 10^{-9}\right)^2 = 1 \times 10^{-14} \text{ (mole/litre)}^2.$$

Further, $K_d = \dfrac{C_{H^+}.C_{OH^-}}{C_{H_2O}}$ [K_d = dissociation constant]

$$= \frac{K_w}{C_{H_2O}} = \frac{1 \times 10^{-14}}{55.56} = \mathbf{1.8 \times 10^{-16}} \textbf{ mole/litre.}$$

12. For the reaction $Ag(CN)_2^- \rightleftharpoons Ag^+ + 2CN^-$, the equlibrium constant at 25 °C is 4×10^{-19}. Calculate C_{Ag^+} in a solution, which was originally 0.1 (M) in KCN and 0.03 (M) in $AgNO_3$.

Soln. : $Ag^+ + 2CN^{-1} \rightleftharpoons Ag(CN)_2^-$

For this reaction, $K = \dfrac{1}{4 \times 10^{-19}} = 2.5 \times 10^{18}$. The initial conc. of Ag^+ is 0.03 (M) and CN^- is 0.1 (M). Since K is so high, the constituent in lesser amount will be completely consumed. Now, 0.1 (M) CN^- will react with 0.05 (M) Ag^+. C_{Ag^+} is in lesser amount. So C_{CN^-} will almost be (0.1 – 0.06) = 0.04 (M) and $Ag(CN)_2^-$ will almost be 0.03 (M). But some Ag^+ will remain. Let the $C_{Ag^+} = x$.

$$\therefore C_{CN^-} = (0.04 + 2x)$$

and $C_{Ag(CN)_2^-} = (0.03 - x)$.

$$\therefore \frac{1}{K} = \frac{C_{Ag^+}.C^2_{CN^-}}{C_{Ag(CN)_2^-}}$$

$$\text{or, } \frac{1}{2.5 \times 10^{18}} = \frac{x.(0.04 + 2x)^2}{(0.03 - x)}.$$

Neglecting x in comparison to 0.04 and 0.03,

$$\frac{1}{2.5 \times 10^{18}} = \frac{(0.04)^2.x}{0.03}$$

$$\text{or, } x = \frac{0.03}{2.5 \times 10^{18} \times (0.04)^2} = \mathbf{7.5 \times 10^{-18}} \textbf{ (M).}$$

13. K_a for acetic acid (HOAc) and benzoic acid (HOBz) are 1.86×10^{-5} and 6.6×10^{-5} respectively. Calculate the pH of a mixture of 2 gm of HOAc and 2 gm of HOBz dissolved in 100 cc of water.

Soln. : By the problem, there is initially $\frac{2}{60}$ mole of HOAc in 100 cc (60 is the m.wt. of HOAc).

$\therefore$ initial molarity $= \frac{2 \times 10}{60} = \frac{1}{3} = a.$

Similarly, initial molarity of HOBz is $= \frac{20}{122} = \frac{10}{61} = b.$

Now the ionisation process of HOAc is

$$\begin{array}{lll} HOAc \rightleftharpoons & H^+ & + \; OAc^- \\ (a - x) & (x + y) & x \\ HOBz \rightleftharpoons & H^+ & + \; OBz^- \\ (b - y) & (x + y) & y \end{array}$$

Acetic acid breaks into H^+ and OAc^- with $C_{H^+} = x$ and benzoic acid breaks into H^+ and OBz^- with $C_{H^+} = y$. H^+ is common to both. So C_{H^+} is $(x + y)$.

Now, by the problem,

for acetic acid,

$$\frac{(x+y).x}{(a-x)} = 1.86 \times 10^{-5}.$$

Similarly, $\frac{(x+y).y}{(b-y)} = 6.6 \times 10^{-5}.$

Neglecting x and y in comparision to a and b respectively,

$(x + y)x = 1.86 \times 10^{-5}\, a$

and $(x + y)y = 6.6 \times 10^{-5}\, b.$

$\therefore (x + y)x = 1.86 \times 10^{-5} \times \frac{1}{3} = 6.2 \times 10^{-6}$ (i)

and $(x + y)y = 6.6 \times 10^{-5} \times \frac{10}{61} = 1.08 \times 10^{-5}.$ (ii)

Adding (i) and (ii),

$(x + y)^2 = 1.7 \times 10^{-5}$

or, $(x + y) = \sqrt{1.7 \times 10^{-5}} = (4.123 \times 10^{-3}).$

$C_{H^+} = 4.123 \times 10^{-3}.$

$\therefore$ pH $= - \log C_{H^+} = - \log (4.123 \times 10^{-3}) =$ **2.38.**

14. 300 cc of a solution of HOAc (m. wt. = 60) containing 10 gm/l ($K_a = 1.86 \times 10^{-5}$) is mixed with 200 cc of solution of phenyl acetic acid (m.wt. = 136) containing 5 gm/l ($K_a = 5.42 \times 10^{-5}$). Calculate the pH of the mixture.

Soln. : Initial molarity of acetic acid $= \frac{10}{60} = 0.1667.$

Initial molarity of phenyl acetic acid = $\frac{5}{136}$ = 0.0368.

After mixing, molarity of HOAc = $\frac{300}{500} \times 0.1667 = 0.1002$ (M)

and molarity of phenyl acetic acid = $\frac{200}{500} \times 0.0368 = 0.0147$ (M).

Let $C_{A_1^-} = C_{AcO^-} = x,$

$C_{A_2^-} = C_{PHAcO^-} = y,$

so that for both the acids $C_{H^+} = (x + y)$. Thus, at equilibrium the compositions are as shown :

$$HA_1 \rightleftharpoons H^+ + A_1^-$$
$$(0.1002 - x) \quad (x + y) \quad x$$

$$HA_2 \rightleftharpoons H^+ + A_2^-$$
$$(0.0147 - y) \quad (x + y) \quad y$$

Neglecting x in comparison to 0.1 and y in comparison to 0.015,

$$1.86 \times 10^{-5} = \frac{(x+y)x}{0.1002}.$$

$\therefore (x + y)x = 1.86 \times 10^{-6}.$ (A)

And $(x + y)y = 5.42 \times 10^{-5} \times 0.015 = 0.7978 \times 10^{-6}$ (B)

Adding (A) and (B),

$(x + y)^2 = 2.673 \times 10^{-6}.$

$\therefore \quad (x + y) = 1.63 \times 10^{-3}.$

$\therefore \quad pH = -\log C_{H^+} = -\log (1.63 \times 10^{-3})$ = **2.79.**

15. Calculate the ionic strength of 0.04 (M) solution of $Al_2(SO_4)_3$.

Soln. : $Al_2(SO_4)_3$ completely ionises as Al^{3+} and SO_4^{2-}.

$$Al_2(SO_4)_3 \longrightarrow 2Al^{3+} + 3SO_4^{2-}$$

$C_{Al^{3+}} = (2 \times 0.04)$ (M)

$C_{SO_4^{2-}} = (3 \times 0.04)$ (M)

$Z_{Al^{3+}} = +3$

$Z_{SO_4^{2-}} = -2.$

$\therefore$ ionic strength (μ) = $\frac{1}{2}\sum_i C_i Z_i^2 = \frac{1}{2}\left[(2\times0.04)\times(3)^2 + (3\times0.04)\times2^2\right]$

$= [(0.04 \times 9) + (3 \times 0.04 \times 2)]$ = **0.6.**

16. Using Debye-Hückel limiting law calculate the mean activity co-efficients of 0.001 (M) aq. solutions of $K_3[Fe(CN)_6]$ and $K_4[Fe(CN)_6]$. Debye-Hückel constant is 0.51.

[M.Sc. Admission Test 1995, I.I.T. Bombay]

Soln. : Mean ionic activity coefficient $\log (\gamma_\pm) = -AZ_+Z_-\sqrt{\mu} + B\mu$.

B is much smaller than A and is neglected.

Now, to calculate $\gamma_\pm$ we have to calculate at first the value of μ.

$$\mu = \frac{1}{2}\sum_i C_i Z_i^2$$

For $K_3[Fe(CN)_6]$, $K_3[Fe(CN)_6] \longrightarrow 3K^+ + Fe(CN)_6^{3-}$

$C_{K^+} = (0.001 \times 3)$

$C_{Fe(CN)_6^{3-}} = 0.001$

$\therefore \mu_{K_3[Fe(CN)_6]} = \frac{1}{2}\left[(3\times 0.001)\times 1^2 + (0.001)\times 3^2\right] = \frac{1}{2}[0.003 + 0.009] = 0.006.$

$\therefore \log(\gamma_\pm)_{K_3[Fe(CN)_6]} = \left(-0.51\times 1\times 3\times\sqrt{0.006}\right).$

$\gamma_\pm = 0.761178844 \approx 0.76$

For $K_4[Fe(CN)_6]$,

$K_4[Fe(CN)_6] \longrightarrow 4K^+ + Fe(CN)_6^{4-}$.

$\therefore C_{K^+} = (0.001 \times 4)$

$C_{Fe(CN)_6^{4-}} = 0.001.$

$\therefore \mu_{K_4[Fe(CN)_6]} = \frac{1}{2}\left[(0.001\times 4)\times 1^2 + (0.001)\times 4^2\right] = \frac{1}{2}[0.004 + 0.016] = 0.01.$

$\therefore \log(\gamma_\pm)_{K_4[Fe(CN)_6]} = -0.51 \times 1 \times 4 \times \sqrt{0.01}.$

$\therefore \gamma_\pm = 0.625173 \approx$ **0.6.**

17. The solubility product of CaF_2 at 27 °C is 3.55×10^{-11}. Using Debye-Hückel limiting law calculate the solubility of CaF_2 at 27 °C in moles/litre.

[M.Sc. Admission Test 1990, I.I.T. Kharagpur]

Soln. : $CaF_2(s) \rightleftharpoons Ca^{2+} + 2F^-$

$\therefore K_S = a_{Ca^{2+}} \cdot a_{F^-}^2$

$a_{Ca^{+2}} = \gamma_+ C_+ = \gamma_+ C$

$a_{F^-} = \gamma_- C_- = \gamma_- 2C$

$K_S = (\gamma_+ C)(\gamma_- 2C)^2 = 4C^3 \ \gamma_\pm^3.$

Since, mean ionic activity coefficient $\gamma_\pm = \sqrt[3]{\gamma_+\gamma_-^2}$.

From Debye-Huckel limiting law, $\log \gamma_\pm = -AZ_+ Z_- \sqrt{\mu}$.

Now, ionic strength (μ) of CaF_2 is very small and is taken as zero, i.e., $\mu \approx 0$. Therefore, its activity = concentration.

$\therefore \log(\gamma_\pm)_{CaF_2} = 0.$

$\therefore (\gamma_\pm) = 1, \ \therefore K_s = 4C^3$

or, $3.55 \times 10^{-11} = 4C^3$

or, $C^3 = \dfrac{3.55\times10^{-11}}{4} = 8.875 \times 10^{-12}$

or, $C = \mathbf{2.07 \times 10^{-4}}$ **mole/litre.**

18. Calculate the ionic strength of a solution obtained by mixing equal volume of 0.01 (M) NaCl and 0.02 (M) $AlCl_3$ solution. **[C.U.(H) 1989]**

Soln. : Due to mixing of the two solutions of equal volume, the concentration of each solution will be the half of the given concentration.

Therefore, $C_{NaCl} = \dfrac{0.01}{2}(M) = 0.005\ (M)$

$C_{AlCl_3} = \dfrac{0.02}{2}(M) = 0.01\ (M)$

$\therefore\quad C_{Na^+} = 0.005\ (M)$

$C_{Al^{3+}} = 0.01\ (M)$

$C_{Cl^-} = [0.005 + 3 \times 0.01]\ (M) = [0.005 + 0.03]\ (M) = 0.035\ (M).$

$\therefore\quad \mu = \frac{1}{2}\left[(0.005)\times1^2 + (0.01)\times3^2 + (0.035)\times1^2\right] = \mathbf{0.065.}$

19. Calculate the mean ionic activity coefficient at 25 °C of a 0.01 (M) solution of $K_3[Fe(CN)_6]$. Given Debye-Hückel constant A for water at 25 °C = 0.51.

The solubility of CaF_2 in water at 18 °C is 2.04×10^{-4} moles/litre. Calculate the solubility in 0.01 (M) NaF solution. **[C.U.(H) 1988]**

Soln. : For the 1st part see Problem no. 12.

2nd part : Since CaF_2 is sparingly soluble in water, its ionic strength is negligible. Therefore, its activity = concentration.

$CaF_2 \rightleftharpoons Ca^{2+} + 2F^-$

$\therefore\quad K_s = C.(2C)^2$

[K_s is the solubility product and C is the solubility in moles/litre]

$\therefore\quad 4C^3 = K_s$ or, $K_s = 4 \times (2.04 \times 10^{-4})^3$.

Now, let the solubility of CaF_2 in presence of NaF is C'(M).

$\therefore$ then $C_{Ca^{2+}} = C'$ (M)

and $C_{F^-} = (2C' + 0.01)\ (M) \approx 0.01\ (M)$.

[since 0.01 is much higher in comparison to $2C'$]

Now, $K_S = a_{Ca^{2+}}.a^2_{F^-} = \left(\gamma + C^{2+}_{Ca}\right)\left(\gamma - C_{F^-}\right)^2 = \gamma^3_{\pm}.C'.(0.01)^2.$

Now, $\log \gamma_{\pm} = -AZ_+Z_-\sqrt{\mu}$.

For 0.01 (M) NaF solution $\mu = \frac{1}{2}\sum_i C_i Z_i^2$

$= \frac{1}{2}\left[0.01(1)^2 + 0.01\times(1)^2\right] = 0.01.$

$\therefore \log \gamma_{\pm} = -0.51 \times 2 \times 1 \times \sqrt{0.01}$

$\therefore \gamma_{\pm} = 0.79.$

Now, we have calculated before, $K_S = \gamma_{\pm}^3.C'.(0.01)^2$.

$\therefore$ putting the values of K_S and $\gamma_{\pm}$, we have

$4 \times (2.04 \times 10^{-4})^3 = (0.79)^3 \times C' \times (0.01)^2$.

or, $C' = \dfrac{4\times(2.04\times10^{-4})^3}{(0.79)^3\times(0.01)^2}$ = **6.89 × 10⁻⁷ mole/litre.**

20. Calculate the ionic strength of a solution containing 0.01 (N) NaCl and 0.02 (N) Na_2SO_4.

Soln. : See solution to Problem no. 14.

21. Find out the solubility of AgCl in a solution where $C_{Cl^-} = 10^{-1}$ (N). To a saturated solution of AgCl in water, 10^{-8} (N) Cl^- is added. Find out the amount of AgCl precipitated in gm/litre. The solubility product (K_s) of AgCl is 1×10^{-10}.

Soln. : $K_S = C_{Ag^+}.C_{Cl^-}$

Since $C_{Cl^-} = 10^{-1}$ (N),

$\therefore C_{Ag^+} = \dfrac{K_S}{C_{Cl^-}} = \dfrac{10^{-10}}{10^{-1}} = 10^{-9}$.

Therefore, 10^{-9} moles/litre AgCl will dissolve. The solubility in gm/litre $= 10^{-9} \times 143.5 = 1.435 \times 10^{-7}$ gm. [m. wt. of AgCl = 143.5]

Now, if the solution is initially saturated in water, then there are 10^{-5} gm-mole/litre of AgCl, i.e., 1.435×10^{-3} gm/litre of AgCl.

Now $C_{Ag^+}.C_{Cl^-} = 10^{-10}$

Let C_{Ag^+} coming from AgCl in solution be x. So, if 10^{-8} (N) C_{Cl^-} is added,

$$x\,(x + 10^{-8}) = 10^{-10}$$

or, $x^2 + 10^{-8}x - 10^{-10} = 0$

or, $x = \dfrac{-10^{-8}+\sqrt{10^{-16}+4.10^{-10}}}{2} = \dfrac{-10^{-8}+\sqrt{4\times10^{-10}}}{2}$ [$\because 10^{-16} << 4 \times 10^{-10}$]

$= 0.9995 \times 10^{-5}$.

$\therefore$ AgCl precipitated $= (10^{-5} - 0.9995 \times 10^{-5})$ gm-mole/litre

$= 0.0005 \times 10^{-5}$ gm-mole/litre

$= 0.0005 \times 10^{-5} \times 143.5$ gm/liter

= **7.175 × 10⁻⁷ gm/litre.**

22. Calculate the solubility product of Ag_2CrO_4. Given that its solubility is 7×10^{-5} gm-mole/litre. **[V.U. 2004]**

Soln. : $Ag_2CrO_4(g) \rightleftharpoons 2Ag^+ + CrO_4^{2-}$

$C_{CrO_4^{2-}} = 7 \times 10^{-5}$

$C_{Ag^+} = 2 \times 7 \times 10^{-5}$

$\therefore (K_S)_{Ag_2CrO_4} = C^2_{Ag^+}.C_{CrO_4^{2-}} = \left\{(2\times7\times10^{-5})^2\times7\times10^{-5}\right\}$ = **1.372 × 10⁻¹².**

23. Calculate $S^{=}_{conc.}$ in a saturated solution of H_2S which is approximately 0.1 (M). $K_1 = 9.1 \times 10^{-8}$ and $K_2 = 1.2 \times 10^{-15}$.

Soln. : $H_2S \rightleftharpoons H^+ + HS^-$

$$\therefore K_1 = \frac{C_{H^+}.C_{HS^-}}{C_{H_2S}} = 9.1 \times 10^{-8}.$$

And $HS^- \rightleftharpoons H^+ + S^{2-}$.

$$\therefore K_2 = \frac{C_{H^+}.C_{S}^{2-}}{C_{HS^-}} = 1.2 \times 10^{-15}.$$

Since K_2 is very small, it may be assumed that $C_{S^{2-}}$ is very small. Thus, we may safely assume that $C_{H^+} = C_{HS^-}$. Since K_1 is also very small, $C_{H_2S} = 0.1$ (M).

$$\therefore \quad C_{H^+} = C_{HS^-} = \sqrt{K_1 \times 0.1} = \sqrt{9.1 \times 10^{-8} \times 0.1} = 9.5 \times 10^{-5}.$$

And $C_{S^{2-}} = 1.2 \times 10^{-15}$.

$$\text{Again, } K_1K_2 = \frac{\left(C_{H^+}\right)^2 \times C_{S^{2-}}}{C_{H_2S}}$$

$$C_{S^2} = \frac{K_1K_2.C_{H_2S}}{\left(C_{H^+}\right)^2} = \frac{1.1 \times 10^{-23}}{\left(C_{H^+}\right)^2} = 1.2 \times 10^{-15}.$$

Problem ⇒ Calculate $C_{S^{2-}}$ when $C_H{}^+ = 0.3$ and 10^{-7}.

[**Ans.** 1.2×10^{-22}; $1{\cdot}1 \times 10^{-9}$]

24. The solubility product of CaF_2 at 25 °C is 1.6×10^{-10}. Should precipitation occur when 50 ml of 5×10^{-2} (M) $Ca(NO_3)_2$ is mixed with 50 ml of 4.0×10^{-4} (M) NaF solution? If precipitation occurs, how much CaF_2 will be precipitated? **[C.U.(H) 1986]**

Soln. : By the problem, after mixing the concentration of Ca^{2+}

$$= \frac{50}{100} \times 5 \times 10^{-2} \text{ (M)} = 2.5 \times 10^{-2} \text{ (M)}$$

$$\text{and } C_{F^-} = \frac{50}{100} \times 4 \times 10^{-4} \text{ (M)} = 2 \times 10^{-4} \text{ (M)}.$$

$$\text{Now, } C_{Ca^{2+}}.C^2_{F^-} = \left(2.5 \times 10^{-2}\right) \times \left(2 \times 10^{-4}\right)^2 = 10 \times 10^{-10} = 10^{-9}.$$

$$\text{But } (K_S)_{CaF_2} = 1.6 \times 10^{-10}.$$

Since the product of concentration terms is greater than K_s, there will be precipitation. Let x moles/litre Ca^{2+} ion be precipitated. Then $2x$ moles/litre F^- will be precipitated.

$$K_S = C_{Ca^{2+}}.C^2_{F^-}$$

$$\text{or, } 1.6 \times 10^{-10} = (2.5 \times 10^{-2} - x)(2 \times 10^{-4} - 2x)^2.$$

The result will be a equation containing x^3 which is difficult to solve. But we know that $2.5 \times 10^{-2} >> x$.

$\therefore\ 2.5 \times 10^{-2}\ (2 \times 10^{-4} - 2x)^2 = 1.6 \times 10^{-10}.$

$\therefore\ 0.1\ x^2 - 2 \times 10^{-5}\ x + 8.4 \times 10^{-10} = 0$

or, $x = 1.4 \times 10^{-4}$ or, 6×10^{-5}.

If x is 1.4×10^{-4}, then $2x = (2 \times 1.4 \times 10^{-4})$ which is greater than 2×10^{-4}. This is not possible.

Therefore, $x = 6 \times 10^{-5}$ molar.

But the volume of the solution is 100 cc, therefore, 6×10^{-6} moles of CaF_2 will be precipitated, i.e., $78 \times 6 \times 10^{-6}$ = **4.68×10^{-4} gm** of CaF_2.

25. The solubility product of PbI_2 is 7.47×10^{-9} at 15 °C and 1.39×10^{-8} at 25 °C. Calculate (i) the molar heat of solution of PbI_2 and (ii) the solubility in moles/litre at 75 °C. **[C.U.(H) 2005]**

Soln. : By van't Hoff equation,

$$\ln \frac{(K_S)_2}{(K_S)_1} = \frac{\Delta H}{R} \frac{T_2 - T_1}{T_1 T_2},$$

where $(K_S)_2$ is the solubility product at T_2 K, $(K_S)_1$ is the solubility product at T_1 K and ΔH is a constant heat of reaction. For the given case,

$$PbI_2 \rightleftharpoons Pb^{2+} + 2I^-$$

By the problem,

$T_1 = 288.15,\ (K_s)_1 = 7.47 \times 10^{-9}$

$T_2 = 298.15,\ (K_s)_2 = 1.39 \times 10^{-8}$

$$\therefore \ln \frac{13.9}{7.47} = \frac{\Delta H}{R}\left(\frac{10}{288.15 \times 298.15}\right).$$

$$\therefore \Delta H = \frac{8.314 \times 288.15 \times 298.15}{10} \ln \frac{13.9}{7.47}$$

$= 44355.8345$ joules $= 44.36$ kJ.

Thus $K_s = K$ at 75 °C, i.e., 348.15 K

$$\ln \frac{K}{1.39 \times 10^{-8}} = \frac{44355.834}{8.314} \times \frac{50}{348.15 \times 298.15} = 2.5698603$$

$$\text{or, } \frac{K}{1.39 \times 10^{-8}} = 13.064$$

or, $K = 1.81589 \times 10^{-7} = 1.82 \times 10^{-7}$.

If the solubility is x mole/litre, then

$x.(2x)^2 = 1.82 \times 10^{-7}$

or, $4x^3 = 1.82 \times 10^{-7}$

or, $x = 3.57 \times 10^{-3}$ = **0.00357 mole/litre.**

26. The specific conductance of a saturared solution of AgCl is 1.55×10^{-6} $ohm^{-1}.cm^{-1}$. The mobilities of Cl^- and Ag^+ ions are 5.6×10^{-4} cm per sec and 6.8×10^{-4} cm per sec respectively under unit potential gradient. Calculate the solubility product of AgCl. **[C.U. 1983]**

Soln. : $\Lambda^{o}_{AgCl} = \lambda^+_{Ag^+} + \lambda^-_{Cl^-} = Fu_+ + Fu = (u_+ + u_-)F$

$= (5.6 \times 10^{-4} + 6.8 \times 10^{-4}) \times 96484 = 119.64.$

Now, $\Lambda = \dfrac{1000\kappa}{N}$ [N = normality]

$= \dfrac{100\kappa}{ZC}$ [Z is the valancy = 1]

$\therefore \quad C = \dfrac{1000\kappa}{Z.\Lambda}$.

Since the solution is extremely dilute, Λ is almost equal to Λ°, i.e., $\Lambda \approx \Lambda^\circ$

$= \dfrac{1000 \times 1.55 \times 10^{-6}}{119.64} = 1.29555 \times 10^{-5}$.

$\therefore K_s = C_{Ag^+}.C_{Cl^-} = C^2$
$= (1.29555 \times 10^{-5})^2 = \mathbf{1.68 \times 10^{-10}}$.

27. The solubility product of $PbSO_4$ is 1.31×10^{-8}. Calculate the number of moles of $PbSO_4$ that can be dissolved in 5 litres of 1×10^{-3} (M) Na_2SO_4. Show whether $PbSO_4$ will be precipitated if 20 cc of 2×10^{-4} (M) $Pb(NO_3)_2$ is mixed with 80 cc of 1×10^{-4} (M) Na_2SO_4. **[C.U.(H) 1972]**

Soln. :

$$PbSO_{4(s)} \rightleftharpoons \underset{x}{Pb^{2+}} + \underset{x}{SO_4^{-2}}$$

$$Na_2SO_4 \longrightarrow 2Na^+ + SO_4^=$$

1st Part :

$x(x + 10^{-3}) = 1.31 \times 10^{-8}$

Solving, $x = 1.293 \times 10^{-5}$.

Thus in 5 litres moles of $PbSO_4$ is

$= (5 \times 1.293 \times 10^{-5}) = \mathbf{6.465 \times 10^{-5}}$.

2nd Part : Initial conc. of Pb^{2+}

$= 2 \times \dfrac{20}{100} \times 10^{-4}$ (M) $= 2 \times 0.2 \times 10^{-4}$ (M).

Initial conc. of $SO_4^= = 1 \times \dfrac{80}{100} \times 10^{-4}$ (M) $= 0.8 \times 10^{-4}$ (M).

$\therefore \; C_{Pb^{2+}}.C_{SO_4^=} = (2 \times 0.2 \times 10^{-4}) \times (0.8 \times 10^{-4})$
$= 0.32 \times 10^{-8}$.

This is less than (K_s) $PbSO_4$. So there will be **nc precipitation of $PbSO_4$.**

28. If 20 cc of 2×10^{-3} (M) $Pb(NO_3)_2$ is mixed with 80 cc of 1×10^{-4} (M) Na_2SO_4, will there be precipitation? If so, then calculate the amount of precipitation in gm.

[Given that $(K_s)_{PbSO_4} = 1.31 \times 10^{-8}$]

Soln. : The conc. of Pb^{2+} is

$= \dfrac{20}{100} \times 2 \times 10^{-3}$ (M) $= 0.2 \times 2 \times 10^{-3}$ (M) $= 4 \times 10^{-4}$ (M)

and conc. of SO_4^{2-} is $= \dfrac{80}{100} \times 1 \times 10^{-4}$ (M) $= 0.8 \times 10^{-4}$ (M).

The product of these two conc. terms is greater than 1.31×10^{-8}. Hence, there will be precipitation.

Let x mole/litre will be precipitated.

Then $(8 \times 10^{-5} - x)(4 \times 10^{-4} - x) = 1.31 \times 10^{-8}$.

Solving, $x = 4.35 \times 10^{-5}$ (M) [neglecting (+) sign]

The solution is 100 cc.

Thus number of moles precipitated $= 4.35 \times 10^{-6}$ mole

$= (4.35 \times 10^{-6} \times 302)$ gm $=$ **0.0013 gm.**

[M. wt. of $PbSO_4 = 302$]

29. At 25 °C, the solubility product of $AgBrO_3$ is 5.77×10^{-5}. Using the Debye-Hückel limiting law, calculate its solubility in (a) pure H_2O (b) 0.01 (M) $KBrO_3$.

Soln. : (a) We know that,

$$K_s = a_{Ag^+} \cdot a_{BrO_3^-} = 5.77 \times 10^{-5}.$$

The amount of a_{Ag^+} and $a_{BrO_3^-}$ in solutions are so insignificant that the ionic strength (μ) of solution is almost zero, so that $\gamma_\pm = 1$.

$$\therefore \quad K_s = C_{Ag^+} \cdot C_{BrO_3^-} = x^2.$$

$$\therefore \quad x = \sqrt{5.77 \times 10^{-5}} = \mathbf{7.596 \times 10^{-3}\ (M).}$$

(b) In a 0.01 (M) solution of $KBrO_3$ the ionic strength is

$$\mu = \frac{1}{2}\sum C_i Z_i^2 = \frac{1}{2}[0.01\ (1)^2 + 0.01 \times (1)^2] = \frac{1}{2}[2 \times 0.01] = 0.01.$$

$$\therefore \log \gamma_\pm = -AZ_+Z_-\sqrt{\mu} + B\mu$$

B is very small in comparison to A.

So, $\log \gamma_\pm = -0.51 \times 1 \times 1\sqrt{0.01} = -0.051$

$$\therefore \quad \gamma_\pm = 0.889.$$

$$\text{Now, } K_S = a_{Ag^+} \cdot a_{BrO_3^-} = \left(C_{Ag^+} \cdot \gamma_+\right)\left(C_{BrO_3^-} \cdot \gamma_-\right)$$

$$= (x.\gamma_+)\{(x + 0.01)\gamma_-\} = x(x + 0.01)\gamma_\pm^2 \text{ (since } \gamma_+\gamma_- = \gamma_\pm^2).$$

$$\therefore \quad x\,(x + 0.01)\,(0.889)^2 = 5.77 \times 10^{-5}$$

or, $x^2 + 0.01x - 7.30 \times 10^{-5} = 0$

$$\therefore \quad x = \frac{-0.01 \pm \sqrt{(0.01)^2 + 4 \times 7.30 \times 10^{-5}}}{2} = \frac{-0.01 \pm \sqrt{3.920 \times 10^{-4}}}{2}.$$

$$= \frac{-0.01 \pm 0.01980}{2}.$$

Neglecting (–)ve sign,

$$x = \frac{-0.01 + 0.01980}{2} = \mathbf{0.0049\ (M).}$$

30. A soln. is 0.1 (M) in Cl^- and 0.001 (M) in $CrO_4^=$. If solid $AgNO_3$ is gradually added to the solutions, which will precipitate first, AgCl or Ag_2CrO_4? Assume that the addition causes no change in volume. Given that solubility products of AgCl and Ag_2CrO_4 are respectively 1.7×10^{-10} and 1.9×10^{-12}. What will be the conc. of Cl^- when Ag_2CrO_4 begins to precipitate? What percentage of Cl^- originally present will remain in solution at this point?

Soln. : C_{Ag^+} required to precipitate Ag_2CrO_4 is

$$a^2_{Ag^+}\, a_{CrO_4^=} = (K_S)_{Ag_2CrO_4}$$

$$\text{or,} \quad a^2_{Ag^+} = \frac{1.9 \times 10^{-12}}{a_{CrO_4^=}}.$$

[Assuming, $a_{Ag^+} = C_{Ag^+}$ and $a_{CrO_4^{2-}} = C_{CrO_4^=}$]

$$C^2_{Ag^+} = \frac{1.9 \times 10^{-12}}{C_{CrO_4^=}} = \frac{1.9 \times 10^{-12}}{0.001} \quad [\text{Initial conc. of } C_{CrO_4^{2-}} = 0.001(\text{M})]$$

$$= 1.9 \times 10^{-9}$$

$\therefore \quad C_{Ag^+} = 4.36 \times 10^{-5}$.

Now, since initial conc. of Cl^- is 0.1(M),

C_{Ag^+} required to precipitate AgCl is calculated as follows :

$\therefore \quad (K_S)_{AgCl} = C_{Ag^+} \cdot C_{Cl^-}$ [asuming activity = conc.]

$$\therefore \quad C_{Ag^+} = \frac{(K_S)_{AgCl}}{C_{Cl^-}} = \frac{1.7 \times 10^{-10}}{0.1} = 1.7 \times 10^{-9}.$$

Since conc. of Ag^+ required for precipitation of Ag_2CrO_4 is greater than conc. of Ag^+ required for precipitation of AgCl, therefore, if solid $AgNO_3$ is gradually added to the solution, AgCl will precipitate first.

When Ag_2CrO_4 starts precipitating, the conc. of Ag^+ in soln. is 4.35×10^{-5} (M).

Thus C_{Cl^-} at this stage is

$$1.7 \times 10^{-10} = 4.35 \times 10^{-5} \times C_{Cl^-}$$

$$\text{or,} \quad C_{Cl^-} = \frac{1.7 \times 10^{-10}}{4.35 \times 10^{-5}} (\text{M}) = \mathbf{3.9 \times 10^{-6}\ (M)}.$$

Initially there is 0.1 (M) of C_{Cl^-}.

$$\text{Hence, percentage of } Cl^- \text{ remaining} = \frac{3.9 \times 10^{-6}}{0.1} \times 100 = \mathbf{0.0039}.$$

31. A solution containing 0.01 (M) $ZnCl_2$ and $MnCl_2$ is saturated with H_2S. The solubility products of ZnS and MnS are 1×10^{-22} and 5.6×10^{-14} respectively. What is the pH at which MnS will form precipitate? What will be the conc. of Zn^{2+} remaining at the pH? (For H_2S, $\beta = K_1K_2 = 1.1 \times 10^{-21}$)

Soln. : The solution contains 0.01 (M) $ZnCl_2$ and 0.01(M) $MnCl_2$.

Now the conc. of $S^=$ required for precipitation of MnS is calculated as follows :

$$C_{Mn^{2+}} \cdot C_{S^{2-}} = (K_S)_{MnS}$$

or, $C_{S^{2-}} = \frac{(K_S)_{MnS}}{C_{Mn^{2+}}} = \frac{5.6\times10^{-16}}{0.01}$ (M) = 5.6×10^{-14} (M).

Now, H_2S dissociates as $H_2S \rightleftharpoons H^+ + HS^-$.

$$K_1 = \frac{C_{H^+}.C_{HS^-}}{C_{H_2S}}$$

and $HS^- = H^+ + S^=$

$$K_2 = \frac{C_{H^+}.C_{HS^-}}{C_{HS^-}}$$

$$\therefore \quad K_1K_2 = \frac{C^2_{H^+}.C_{S^=}}{C_{H_2S}}.$$

Now when MnS will precipitate, the conc. of S^{2-} will be $= 5.6 \times 10^{-14}$.

At this conc. C_{H^+} will be

$$\beta = K_1K_2 = \frac{C^2_{H^+}.C_{S^{2-}}}{C_{H_2S}} = 1.1 \times 10^{-21}$$

or, $C^2_{H^+} = \frac{1.1\times10^{-21}\times0.1}{5.6\times10^{-14}}$ [C_{H_2S} = 0.1 for saturated soln.]

$= 0.0196 \times 10^{-7} = 0.196 \times 10^{-8}$.

$C_{H^+} = 0.443 \times 10^{-4}$ (M)

$\therefore$ pH $= -\log C_{H^+} = -\log (0.443 \times 10^{-4})$ = **4.35.**

Now, at this pH $C_{S^{2-}} = 5.6 \times 10^{-14}$.

Let the conc. of $Zn^{2+} = C_{Zn^{2+}}$.

$\therefore \quad C_{Zn^{2+}}.C_{S^{-2}} = (K_S)_{ZnS}$

or, $C_{Zn^{2+}} = \frac{(K_s)_{ZnS}}{C_{S^{2-}}} = \frac{1\times10^{-22}}{5.6\times10^{-14}}$ = **1.79×10^{-9} (M).**

32. 10 cc of 0.1 (M) NaOH soln. are added to 50 cc of 1.1 (M) HOAc. What will be the pH of the resulting mixture? Given that K_a for HOAc = 1.75×10^{-5}.

[C.U. 1977]

Soln. : 10 cc 0.1 (M) NaOH = 10 × 0.1 cc 1 (M) = 1 cc 1 (M).

And 50 cc 1.1 (M) HOAc = (50 × 1.1) cc 1 (M) = 55 cc 1 (M).

Now, 1 cc 1 (M) NaOH will consume 1 cc 1 (M) HOAc and produce 1 cc 1 (M) salt.

Thus, HOAc remaining will be = (55 – 1) cc 1 (M) = 54 cc 1 (M).

Total volume of the mixture = 60 cc.

Now conc. of salt = $\frac{1}{60} \times 1$ (M)

and conc. of acid = $\frac{54}{60} \times 1$ (M).

Now, $pH = pK_a + \log \frac{C_{salt}}{C_{acid}} = -\log (1.75 \times 10^{-5}) + \log \frac{1}{54}$

$= 4.757 - 1.7324 = 3.0246 = \mathbf{3.0.}$

33. It is desired to prepare a buffer solution of pH 4.5 using HOAc and NaOAc. Given that K_a for HOAc = 1.8×10^{-5}. Calculate the weight of CH_3COONa that must be added to 1 litre of 1 (N) HOAc.

Soln. : We know that pH of a buffer of weak acid and its salt is

$$pK_a + \log \frac{C_{salt}}{C_{acid}}$$

or, $4.5 = -\log (1.8 \times 10^{-5}) + \log \frac{C_{salt}}{C_{acid}}$.

$\therefore \log \frac{C_{salt}}{C_{acid}} = 4.5 + \log (1.8 \times 10^{-5})$.

$\therefore \frac{C_{salt}}{C_{acid}} = 0.5692.$

Now, let x (N) CH_3COONa will have to be mixed with 1 (N) CH_3COOH.

Therefore, $\frac{x}{1} = 0.5692.$ or, $x = 0.5692$ (N) $= 0.5692 \times 82 =$ **46.67 gm.**

[$\because$ m. wt of CH_3COONa = 82]

34. A definite volume of an aq. $\frac{N}{20}$ HOAc ($K_a = 1.8 \times 10^{-5}$) is titrated with a strong NaOH soln. It is found that 75 equal sized drops of NaOH, added from a burette, effect the complete neutralisation. Find the pH values when the acid soln. is neutralised to the extent of 20%, 40% and 90% respectively. Neglect volume change due to addition of alkali.

[C.U.(H) 1979]

Soln. : Addition of alkali to a weak acid, produces a buffer of weak acid and its salt.

$$pH = pK_a + \log \frac{C_{salt}}{C_{acid}}$$

When 20% is neutralised, then 20% of salt and 80% of acid remain in solution.

$\therefore pH = -\log (1.8 \times 10^{-5}) + \log \frac{C_S}{C_A} = 4.745 + \log \frac{20}{80} = \mathbf{4.145.} \approx \mathbf{4.1.}$

For 40% neutralisation process,

$$pH = 4.745 + \log \frac{40}{60} = 4.57 \approx \mathbf{4.6.}$$

For 90% neutralisation process,

$$pH = 4.745 + \log \frac{90}{10} = 5.6992 = \mathbf{5.7.}$$

35. A buffer solution is made from 0.06 mole of HOAc and 0.04 mole of NaOAc per litre. What is the pH of the solution? Calculate the changed pH resulting from the addition of (i) 0.015 mole HCl (ii) 0.03 mole of NaOH per litre of the solution.

[$K_a = 1.8 \times 10^{-5}$]

Soln. : The pH of the solution will be

$$pH = pK_a + \log \frac{C_S}{C_A}$$

$$= -\log (1.8 \times 10^{-5}) + \log \frac{0.04}{0.06} = 4.568 \approx 4.6.$$

(i) 0.015 mole acid will react with AcO^- and produce (0.04 – 0.015) mole = 0.025 mole salt and (0.06 + 0.015) = 0.075 mole acid.

$$\therefore \; pH = pK_a + \log \frac{C_S}{C_A} = 4.745 + \log \frac{40}{60} = 4.745 - 0.4771$$

$$= 4.2679 \approx \mathbf{4.3}.$$

(ii) When 0.03 mole NaOH is added to the solution, then it produces (0.04 + 0.03) mole = 0.07 mole salt and (0.06 – 0.03) = 0.03 mole acid.

$$\therefore \; pH = pK_a + \log \frac{C_S}{C_A} = 4.745 + \log \frac{0.07}{0.03} = 4.745 + 0.37 = 5.112 = \mathbf{5.1}.$$

36. Calculate the concentration of H^+, $H_2AsO_4^-$, $HAsO_4^=$ and $AsO_4^=$ in 0.5(M) solution of H_3AsO_4. Given, $K_1 = 2.5 \times 10^{-4}$; $K_2 = 5.6 \times 10^{-8}$; $K_3 = 3 \times 10^{-12}$.

Soln. : We know that,

$$H_2O + H_3AsO_4 = H_3O^+ + H_2AsO_4^- .$$

$$K_1 = \frac{C_{H_3O^+} . C_{H_2AsO_4^-}}{C_{H_3AsO_4}}$$

$$H_2O + H_2AsO_4^- = H_3O^+ + HAsO_4^=$$

$$K_2 = \frac{C_{H_3O^+} . C_{HAsO_4^=}}{C_{H_2AsO_4^-}}$$

$$H_2O + HAsO_4^= = H_3O^+ + AsO_4^{-3}$$

$$K_3 = \frac{C_{H_3O^+} . C_{AsO_4^{3-}}}{C_{HAsO_4^=}}$$

Since K_2 and K_3 are much smaller than K_1,

$$\therefore \; C_{H_3O^+} = C_{H_2AsO_4^-} = x \text{ (say)}.$$

$$\therefore \; K_1 = \frac{x \times x}{(0.5 - x)} = 2.5 \times 10^{-4}$$

or, $x^2 = 2.5 \times 10^{-4} \; (0.5 - x)$

or, $x^2 + 2.5 \times 10^{-4} \; x - 2.5 \times 0.5 \times 10^{-4} = 0.$

Solving, $x = 0.011$ [neglecting (–)ve term]

$$\therefore \; C_{H_3O^+} = C_{H_2AsO_4^-} = 0.011 \text{ (M)}.$$

$$\text{Now, } K_2 = \frac{C_{H_3O^+} \cdot C_{HAsO_4^{=}}}{C_{H_2AsO_4^-}}$$

$$= C_{HASO_4^{2-}} \quad [\text{since } C_{H_3O^+} = C_{H_2AsO_4^-}]$$

$$\therefore C_{HAsO_4^{2-}} = 5.6 \times 10^{-8} \text{ (M)}.$$

$$\text{Again, } K_3 = \frac{C_{H_3O^+} \cdot C_{AsO_4^{3-}}}{C_{HAsO_4^{=}}} = \frac{C_{H_3O^+} \cdot C_{AsO_4^{3-}}}{K_2}.$$

$$\therefore K_2K_3 = C_{H_3O^+} \cdot C_{AsO_4^{3-}}$$

$$\text{or, } C_{AsO_4^{3-}} = \frac{K_2K_3}{C_{H_3O^+}} = 1.527 \times 10^{-17} \text{ (M)} = 1.53 \times 10^{-7} \text{(M)}.$$

37. Calculate H_3O^+, $H_2PO_4^-$, HPO_4^{2-} and PO_4^{3-} conc. in 0.1 (M) H_3PO_4 at 25 °C. [Given : $K_1 = 7.52 \times 10^{-3}$, $K_2 = 6.23 \times 10^{-8}$ and $K_3 = 4.8 \times 10^{-13}$]

Soln. : Proceed as Problem 32.

[$C_{H_3O^+} = C_{H_2PO_4^-} = 0.0239$ (M); $C_{H_3PO_4} = 0.0761$ (M); $C_{HPO_4^{-2}} = 6.23 \times 10^{-8}$ (M);

$C_{PO_4^{-3}} = 1.25 \times 10^{-18}$ (M)]

38. You are supplied with 0.12 (N) strong acid HX and 0.15 (N) weak base BOH of $K_b = 2 \times 10^{-5}$. Prepare 110 cc of a buffer of pH 9.

Soln. : We know that, the pH of a weak base and its salt is

$$14 - \left[pK_b + \log\frac{C_{\text{salt}}}{C_{\text{base}}}\right].$$

$$\therefore \quad 9 = 14 - \left[5 - \log 2 + \log\frac{C_s}{C_B}\right]$$

$$\text{or, } \log 2 = \log \frac{C_S}{C_B}$$

$$\text{or, } \frac{C_S}{C_B} = 2.$$

Let us take x cc of HX and $(110 - x)$ cc BOH.

Now, x cc of 0.12 (N) HX will react with x cc of 0.12 (N) BOH to give x cc of 0.12 (N) salt.

Since the final volume = 110 cc,

$$\therefore C_{\text{salt}} = \frac{x \times 0.12}{110}.$$

Now, let y cc of 0.15 (N) BOH react with x cc of 0.12 (N) HX.

$$\therefore \quad y = \frac{x \times 0.12}{0.15}.$$

Thus, the remaining BOH is $\left(110 - x - \frac{x \times 0.12}{0.15}\right)$.

So the strength of the remaining BOH

$$= \frac{\left(110 - \frac{0.27x}{0.15}\right)0.15}{110} = \frac{16.5 - 0.27x}{110}.$$

Now, $\frac{C_{salt}}{C_{base}} = 2$.

$$\therefore \quad \frac{\frac{x \times 0.12}{110}}{\frac{16.5 - 0.27x}{110}} = 2 \quad \text{or,} \quad \frac{x \times 0.12}{16.5 - 0.27x} = 2$$

or, $0.12x = 16.5 \times 2 - 0.27 \times 2x$ or, $0.66x = 16.5 \times 2$

or, $x = \frac{16.5 \times 2}{0.66} = 50.$

Hence, 50 cc of 0.12 (N) HX is to be mixed with (110 – 50) cc of 0.15 (N) BOH or, 60 cc of 0.15 (N) BOH.

39. Prepare a 1000 cc buffer soln. of pH 5 with 0.5 (N) HA ($K_a = 2 \times 10^{-5}$) and 0.2 (N) NaOH.

Hints : pH = $pK_a + \log \frac{C_S}{C_A}$.

$$\frac{C_S}{C_A} = 2$$

Let us take x cc 0.2 (N) OH^-.

$$\therefore \quad C_{salt} = \frac{x \times 0.2}{1000}.$$

Now, x cc 0.2 (N) NaOH will consume y cc 0.5 (N) HA.

$$\therefore y = \frac{x \times 0.2}{0.5}.$$

$$\text{So } C_{acid} = \frac{\left(1000 - x - \frac{x \times 0.2}{0.5}\right)0.5}{1000} = \frac{500 - 0.7x}{1000}.$$

$$\therefore \quad \frac{x \times 0.2}{500 - 0.7x} = 2.$$

$\therefore x = 625$ cc base.

$\therefore (1000 - x) = 375$ cc acid.

40. You have two stock solution :

HOAc ($K_a = 1.86 \times 10^{-5}$) and $PhCH_2COOH$ ($K_a = 5.42 \times 10^{-5}$) containing 10 gm/litre and 5 gm/litre of the acids respectively.

By mixing how many litres of $PhCH_2COOH$ with 1 litre of HOAC, you can prepare a solution of $C_{H_3O^+} = 1.5 \times 10^{-3}$?

Soln. : Let K litre of $PhCH_2COOH$ (HA_2) containing $\frac{5}{136} = 0.0368$ mole/litre is added to 1 litre of HOAc (HA_1) containing $\frac{10}{60} = 0.1667$ mole/litre.

They will both ionise to H^+ and A_1^- or A_2^-.

Let the equation composition be shown as below :

$$HA_1 \rightleftharpoons H^+ + A_1^-$$

$$\frac{(0.1667-x)}{1+K} \qquad 1.5 \times 10^{-3} \qquad \frac{x}{1+K}$$

$$HA_2 \rightleftharpoons H^+ + A_2^-$$

$$\frac{(0.0368-y)K}{1+K} \qquad 1.5 \times 10^{-3} \qquad \frac{yK}{1+K}$$

Hence, $$\frac{1.5\times10^{-3}\times\dfrac{x}{1+K}}{\dfrac{0.1667-x}{1+K}} = 1.86 \times 10^{-5}$$

or, $$\frac{x}{0.1667-x} = \frac{1.86\times10^{-5}}{1.5\times10^{-3}}.$$

Neglecting x in comparison to 0.1667,

$$\frac{x}{0.1667} = 1.24 \times 10^{-2}.$$

$\therefore \quad x = 1.24 \times 10^{-2} \times 0.1667; \quad \therefore x = 2.07 \times 10^{-3}.$

Again, $$\frac{1.5\times10^{-3}\times\dfrac{yK}{1+K}}{\dfrac{(0.0368-y)K}{1+K}} = 5.42 \times 10^{-5}.$$

Neglecting y in comparison to 0.0368,

$$\frac{1.5\times10^{-3}y}{0.0368} = 5.42 \times 10^{-5}.$$

$$\therefore y = \frac{5.42\times10^{-5}\times0.0368}{1.5\times10^{-3}} = 1.33 \times 10^{-3}.$$

By the problem,

$$C_{H^+} = \frac{\left(2.07\times10^{-3}+1.33\times10^{-3}K\right)}{1+K} = 1.5 \times 10^{-3}, \therefore K = \mathbf{3.4\ litres.}$$

41. 60 cc of 0.015 molar $BaCl_2$ solution is mixed with 40 cc of 0.0125 molar KCl solution.Calculate the ionic strength of the solution. **[B.U. 1987]**

Soln. : Proceed as shown earlier. **[Ans.** 0.032]

42. At 25 °C and at a dilution of 32 litres the equivalent conductance of a salt of weak base and strong acid is 98.5 ohm^{-1}.cm^2, while that of a strong acid at the same dilution is 383 ohm^{-1}.cm^2. On adding successive quantities of the free base to the solution of the salt, the equivalent conductance could be finally reduced to 89.7 ohm^{-1}.cm^2. Calculate the degree of hydrolysis, and hydrolysis constant. **[C.U.(H) 1983]**

Soln. : If α be the degree of hydrolysis of the salt and Λ_C be the equivalent conductance of the unhydrolysed salt, then

$$\alpha = \frac{\Lambda - \Lambda_C}{\Lambda_{HA} - \Lambda_C} \; [\Lambda = \text{eq. conductance of the solution}].$$

$$\therefore \quad \Lambda = (1-\alpha)\Lambda_C + \alpha\Lambda_{HA}$$

Λ_{HA} = eq. conductance of strong acid.

Now, Λ = 98.5 ohm^{-1}.cm^2

Λ_C = 89.7 ohm^{-1}.cm^2

Λ_{HA} = 383 ohm^{-1}.cm^2.

$$\text{So, } \alpha = \frac{98.5 - 89.7}{383 - 89.7} = 0.03.$$

Concentration of the solution is $C = \frac{1}{32}$ (N).

$$\text{Hydrolysis constant } (K_h) = \frac{\alpha^2 C}{1-\alpha} = \frac{(0.03)^2 \times \frac{1}{32}}{(1-0.03)} = \mathbf{2.899 \times 10^{-5}}.$$

43. Calculate the pH of a 0.1(M) K_3PO_4 solution. The third dissociation constant of ortho-phosphoric acid is 1.3×10^{-12}. The hydrolysis proceeds only in the first step. Assume $K_w = 1 \times 10^{-14}$. Do not neglect α.

Soln. : The hydrolysis reaction is :

$$PO_4^{3-} + H_2O = HPO_4^{2-} + OH^-$$

Initial :	0.1	0	0
Equilibrium :	0.1 (1 − α)	0.1α	0.1α

$$K_h = \frac{K_w}{K_a} = \frac{1\times10^{-14}}{1.3\times10^{-12}} = 7.692 \times 10^{-3}.$$

$$\text{But } K_h = \frac{\alpha^2 C}{1-\alpha}.$$

$$\therefore \frac{\alpha^2 C}{1-\alpha} = 7.692 \times 10^{-3} = 7.7 \times 10^{-3}.$$

$$\therefore \alpha^2 C + 7.7 \times 10^{-3}\,\alpha - 7.7 \times 10^{-3} = 0.$$

$$\therefore \alpha = \frac{-7.7\times10^{-3} + \sqrt{(7.7\times10^{-3})^2 + 4\times7.7\times10^{-3}\times0.1}}{2\times0.1} \quad (\text{since, } C = 0.1\text{M})$$

$$= \frac{-7.7\times10^{-3} + \sqrt{5.929\times10^{-5} + 3.08\times10^{-3}}}{2\times0.1}$$

$$= \frac{-7.7\times10^{-3} + 0.056029}{2\times0.1} = \frac{0.04833}{0.2} = 0.2416.$$

$\therefore \quad C_{OH^-} = \alpha \times C = (0.2416 \times 0.1) = 0.02416$

$\quad pOH = 1.6168 = 1.62$

$\therefore \quad pH = (14 - 1.62) = \mathbf{12.38.}$

If we neglect α, we can use the formula : $pH = 7 + \frac{1}{2}pK_3 + \frac{1}{2}\log C = \mathbf{12.44.}$

44. Will $Fe(OH)_3$ (s) precipitate from buffer solution prepared by mixing 0.5 (M) HOAc and 0.15 (M) NaOAc at 25 °C, if the solution contains 0.25 (M) Fe^{3+}?

K_{SP} of $Fe(OH)_3 = 4 \times 10^{-38}$, K_a of HOAc $= 1.74 \times 10^{-5}$.

Soln. : For weak acid and its salt the pH of the solution is :

$$pK_a + \log \frac{C_{salt}}{C_{acid}}$$

$$\text{or, } -\log C_{H^+} = -\log K_a + \log \frac{C_S}{C_A}$$

$$\text{or, } C_{H^+} = K_a \cdot \frac{C_A}{C_S} = 1.74 \times 10^{-5} \times \frac{0.5}{0.15} = 5.8 \times 10^{-5}.$$

$$\therefore C_{OH^-} = \frac{K_w}{C_{H^+}} = \frac{10^{-14}}{5.8 \times 10^{-5}} = 1.7 \times 10^{-10}.$$

Now, $C_{Fe^{3+}} \cdot C^3_{OH^-} = 0.25 \times (1.7 \times 10^{-10})^3 = 1.23 \times 10^{-30}$.

As the ionic product of Fe $(OH)_3$, i.e., 1.23×10^{-30} **is greater than the solubility** product, i.e., 4×10^{-38} So $\mathbf{Fe(OH)_3}$ **will precipitate.**

45. The passage of CO_2 gas through a saturated solution of $FeCO_3$ leads to the following reaction at 30 °C :

$$FeCO_3 + H_2CO_3 \longrightarrow Fe^{2+} + 2\,HCO_3$$

Given that the solubility of $FeCO_3$ is 6.6×10^{-6} moles/litre and the solubility of CO_2 is 0.665 volume/litre. Calculate the pH of the solution and equilibrium constant of the above reaction. (For H_2CO_3, $K_1 = 3.7 \times 10^{-7}$ and $K_2 = 5 \times 10^{-11}$)

Soln. : 22.414 vol at 273.15 °K ≡ 1 mole

i.e., $\frac{22.414 \times 303.15}{273.15}$ vol. at 30 °C ≡ 24.87 vol. ≡ 1 mole

$$\therefore 1 \text{ vol.} = \frac{1}{24.87} \text{ mole}$$

i.e., 0.665 vol. $= \frac{0.665}{24.87}$ mole $= 0.0267$ mole/litre

$$H_2CO_3 + H_2O \rightleftharpoons H_3O^+ + HCO_3^-$$

$$(1 - \alpha_1)C \qquad \alpha_1 C \qquad \alpha_1 C$$

$$\therefore K_1 = \frac{\alpha_1^2 C}{1 - \alpha_1}.$$

Since α_1 is small in comparison to 1,

$$\therefore K_1 = \alpha_1^2 C.$$

$$\therefore\ \alpha_1 = \sqrt{\frac{K_1}{C}} = \sqrt{\frac{3.7\times10^{-7}}{0.0267}} = 3.72 \times 10^{-3}.$$

$$\therefore\ C_{HCO_3^-} = \alpha_1 C = (3.72 \times 10^{-3} \times 0.0267) = 9.93 \times 10^{-5}\ \text{mole.litre}^{-1}.$$

For the reaction, $FeCO_3 + H_2CO_3 \rightleftharpoons Fe^{2+} + 2\,HCO_3^-$

$$\therefore\ K = \frac{C_{Fe^{2+}}.C^2_{HCO_3^-}}{C_{H_2CO_3}} = \frac{(6.6\times10^{-6})(9.93\times10^{-5})^2}{0.0267}$$

$[C_{H_2CO_3} = 0.0267$ mole, since $\alpha_1 << 1]$

$\therefore\ K = \mathbf{2.44 \times 10^{-12}}$**.**

Now, $C_{H_3O^+} \approx C_{HCO_3^-}$.

Since $C_{H_3O^+}$ resulting from 2nd dissociation is negligible in comparison to the first dissociation constant,

$\therefore$ pH $= -\log C_{H_3O^+} = 4.002 = \mathbf{4}$.

46. The dissociation constant of NH_4OH is 1.8×10^{-5}. The solubility product of $Mg(OH)_2$ is 1.22×10^{-11}. How many grams of solid NH_4Cl must be added to a mixture of 50 cc (N) NH_4OH solution and 50 cc (N) $MgCl_2$ solution so that ppt. of $Mg(OH)_2$ just disappears? It is assumed that the volume of the solution is not changed by dissolving solid NH_4Cl and the dissociation of the salt is complete. **[B.U. 2001]**

Hints : Conc. of OH^- required so that ppt. of $Mg(OH)_2$ just disappears $= 6.99 \times 10^{-6}$(M).

Let x gm of NH_4Cl be added so that dissociation of NH_4OH is suppressed.

$$NH_4OH \rightleftharpoons NH_4^+ + OH^-$$

$$1.8 \times 10^{-5} = \frac{C_{NH_4^+} \times 6.99\times10^{-6}}{0.5}$$

$$C_{NH_4^+} = 1.288(\text{M}).$$

Thus, $x = \frac{1.288}{1000} \times 100 \times 53.5 = \mathbf{6.89\ gm.}$

47. The value of equivalent conductance (Λ) of a 0.01(M) solution of a weak acid (HA) is 60 at 298 K while the value of $\Lambda°$ is 360 at the same temperature. Calculate pK_a of the acid considering the activities. **[B.U. 2000]**

Soln. : A weak acid HA dissociates into H^+ and A^-.

$$\begin{array}{ccccc} HA & \rightleftharpoons & H^+ & + & A^- \\ (1-\alpha)C & & \alpha C & & \alpha C \end{array}$$

$$K_c = \frac{C_{H^+}C_{A^-}}{C_{HA}} = \frac{\alpha^2 C}{1-\alpha}.$$

Dissociation constant, $K_a = \frac{a_{H^+}a_{A^-}}{a_{HA}} = \frac{C_{H^+}C_{A^-}}{C_{HA}}.\frac{\gamma_{H^+}\gamma_{A^-}}{\gamma_{HA}} = K_c\gamma_\pm^2$ (since $\gamma_{HA} = 1$)

Hence, $pK_a = -\log K_a = -\log K_c - 2\log\gamma_\pm$.

Now, $\alpha = \dfrac{\Lambda}{\Lambda^\circ}$.

Here, $\Lambda = 60$ mho.cm^2

$\Lambda^\circ = 360$ mho.cm^2

$\therefore \quad \alpha = 0.167$ and

$$K_c = \frac{0.167^2 \times 0.01}{1-0.167} = 3.35 \times 10^{-4}.$$

Again, $\log\gamma_\pm = -AZ_+Z_-\sqrt{\mu}$

$$\mu = \frac{1}{2}\sum_i C_i Z_i^2 = \frac{1}{2}\left[C_{H^+}.1^2 + C_{A^-}.1^2\right] = \frac{1}{2}[\alpha C + \alpha C]$$

$$= \alpha C = 0.167 \times 0.01 = 0.00167.$$

$\therefore \quad \log\gamma_\pm = -0.51 \times 1 \times 1 \times \sqrt{0.00167} = -0.021$

Thus, $pK_a = -\log(3.35 \times 10^{-4}) - 2 \times (-0.021) = \mathbf{3.52}$.

48. Calculate the ionic strength of a solution obtained by mixing aqueous solution of 20 ml of 0.05 (M) $MgCl_2$. 30 ml of 0.05 (M) Na_2SO_4, 25 ml of 0.04 (M) $AlCl_3$ and 25 ml of 0.01 (M) glucose at a given temperature.

Soln. : Total valume = 20 + 30 + 25 + 25 = 100 ml.

Thus, $C_{MgCl_2} = 0.01$ (M), $C_{Na_2SO_4} = 0.015$ (M), $C_{AlCl_3} = 0.01$ (M).

Glucose is non-electrolyte. So it is not considered.

$MgCl_2 = Mg^{+2} + 2Cl^-$

$Na_2SO_4 = 2Na^+ + SO_4^{-2}$

$AlCl_3 = Al^{+3} + 3Cl^-$

$$\mu = \frac{1}{2}C_i Z_i^2$$

$$= \frac{1}{2}\left[C_{Mg^{+2}}Z^2_{Mg^{+2}} + C_{Na^+}Z^2_{Na^+} + C_{Al^{+3}}Z^2_{Al^{+3}} + C_{SO_4^{-2}}Z^2_{SO_4^{-2}} + C_{Cl^-}Z^2_{Cl^{-2}}\right]$$

$$= \frac{1}{2}[0.01 \times 2^2 + 2 \times 0.015 \times 1^2 + 0.01 \times 3^2 + 0.015 \times 2^2 + (2 \times 0.01 + 3 \times 0.01) \times 1^2].$$

$= \mathbf{0.135}$.

49. Solubility of $MgCO_3$ at a particular temperature is 10^{-3} moles/litre. Calculate is solubility in 0.1 (M) KNO_3 solution. **[B.U. 1994]**

Soln. : $MgCO_3 \rightleftharpoons Mg^{+2} + CO_3^{-2}$

For $MgCO_3$ solubility product, K_{SP}

$$= a_{Mg^{+2}}\, a_{CO_3^{-2}}$$

$$= C_{Mg^{+2}}\, C_{CO_3^{-2}}\gamma_\pm^2.$$

Since $MgCO_3$ is sparingly soluble in water,

$\gamma_{\pm} = 1$ and thus $a_{Mg^{+2}} = C_{Mg^{+2}}$ and $a_{CO_3^{-2}} = C_{CO_3^{-2}}$.

Here, $C_{Mg^{+2}} = C_{CO_3^{-2}} = C = 10^{-3}$(M).

Thus, $K_{SP} = (10^{-3})^2 = 10^{-6}$.

Let the solubility of $MgCO_3$ be C' in presence of 0.1 (M) KNO_3.

Now, $K_{SP} = C'^2\ (\gamma_{\pm})^2 = 10^{-6}$

$$\log \gamma_{\pm} = -AZ_+Z_-\sqrt{\mu}.$$

Now, $\mu = \frac{1}{2}[0.1 \times 1^2 + 0.1 \times 1^2] = 0.1$

Thus, $\log \gamma_{\pm} = -0.51 \times 2 \times 2 \times \sqrt{0.1}$

$\therefore \quad \gamma_{\pm} = 0.226.$

Hence, $10^{-6} = C'^2 \times (0.226)^2$

$\therefore \quad C' = 10^{-3}/0.226 =$ **4.42 × 10⁻³ moles/litre.**

50. Solubility of a sparingly soluble salt (MA) at 25 °C is 10^{-4}(M). Calculate its solubility in 0.1(M) MCl solution. [consider the effect of ionic strength. The Debye-Hückel constant A at 25 °C = 0.51] **[B.U. 2000]**

Hints : $K_{SP} = 10^{-8} = \gamma_{\pm}^2 C'0.1$

$\log \gamma_{\pm} = -0.51 \times 1 \times 1 \times \sqrt{0.1}$

$\gamma_{\pm} = 0.6898.$

Hence, $C' =$ **2.102 × 10⁻⁷ mole/litre.**

51. Solubility of AgCl is 10^{-5} mole/litre. Calculate the solubility in (a) water (b) 0.01 (M) NaCl, (c) 0.01 (M) $NaNO_3$ (d) 0.001 (M) $Ca(NO_3)_2$. Arrange the solubility of AgCl in decreasing order.

Soln. : (a) In water, $\gamma_{\pm} = 1$.

Thus, $C_{Ag^+} = C_{Cl^-} =$ **10⁻⁵**.

(b) Let the solubility be C_1 mole/litre in 0.01 (M) NaCl.

Thus, $C_{Ag^+} = C_1$(M), $C_{Cl^-} = (0.01 + C_1) \approx 0.01$(M).

Ionic strength, $\mu = \frac{1}{2}[0.01.1^2 + 0.01 \cdot 1^2] = 0.01.$

$\log \gamma_{\pm} = -A.Z+Z-\sqrt{\mu} = -0.51 \times 1 \times 1 \times \sqrt{0.01} = -0.051.$

$\therefore \quad \gamma_{\pm} = 0.89.$

So, $10^{-10} = C_1 \times 0.01 \times (0.89)^2$, $\therefore$ $C_1 =$ **1.26 × 10⁻⁸ mole/litre.**

(c) Let the solubility of AgCl be C_2 mole/litre in 0.01(M) $NaNO_3$.

$\mu = 0.01$ and $\gamma_{\pm} = 0.89$

Now, $C_{Ag^+} = C_2$(M) and $C_{Cl^-} = C_2$(M).

Hence, $10^{-10} = C_2^2 \times (0.89)^2$.

$\therefore$ $C_2 =$ **1.12 × 10⁻⁵ mole/litre.**

(d) Let the solubility of AgCl be C_3 mole/litre in 0.01(M) $Ca(NO_3)_2$.

Hence, $C_{Ag^+} = C_{Cl^-} = C_3$(M).

$$\mu = \frac{1}{2}[0.01 \times 2^2 + 2 \times 0.01 \times 12] = 0.03.$$

$$\log \gamma_\pm = -0.51 \times 2 \times 1 \times \sqrt{0.03}$$
$$\therefore \gamma_\pm = 0.66.$$

Hence, $10^{-10} = C_3^2 \times (0.66)^2$, $\therefore$ C_3 = **1.52×10^{-5} mole/litre.**

Thus, solubility of AgCl in **$Ca(NO_3)_2$ > $NaNO_3$ > H_2O > NaCl.**

52. To one litre of a solution of 0.1 (M) in acetic acid and 0.1 (M) in CH_3COONa. 1 cc of 10 (N) HCl is added. What will be the change in pH? The change in volume may be neglected. $K_a = 1.8 \times 10^{-5}$ at 25 °C. **[B.U. 1979]**

Soln. : The pH of the solution before addition of HCl $= pK_a + \log \frac{C_S}{C_A}$.

Here, $C_S = 0.1$(M), $C_A = 0.1$(M)

Hence, $\text{pH} = -\log(1.8 \times 10^{-5}) + \log \frac{0.1}{0.1} = 4{\cdot}74.$

1cc of 10(N) HCl, i.e., 10^{-2} mole HCl is added. It will react with 10^{-2} mole CH_3COONa to produce CH_3COOH.

Thus, $C_S = 10^{-1} - 10^{-2} = 0.09$ mole
$C_A = 10^{-1} + 10^{-2} = 0.11$ mole

Hence, $\text{pH} = pK_a + \log \frac{0.09}{0.11}$
$= 4.74 - 0.09 = 4.65.$

Thus, pH will decrese by **0.09 unit.**

53. Using Debye-Hückel limiting law calculate the activity of a 2 – 1 electrolyte at a molarity of 0.01 aqueous solution at 15 °C. [$A = 0.50$ at 15 °C]

Soln. : 0.01 molar 2 –1 electrolyte (MA_2) dissociate as $MA_2 = M^{+2} + 2A^-$.

Hence, $C_{M^{+2}} = 0.01$(M) $C_{A^-} = 0.02$ molar.

So, ionic strength, $\mu = \frac{1}{2}[0.01 \times 2^2 + 0.02 \times 1^2] = 0.03.$

$\log \gamma_\pm = -0.50 \times 2 \times 1 \times \sqrt{0.03}$.

$\therefore$ mean ionic activity coefficient of MA_2 = **0.67.**

Thus, activity of the electrolyte $MA_2 = a_2 = a_{M^{+2}} a_{A^-}^2$

$$= \left(C_{M^{+2}} \gamma_{M^{+2}}\right)\left(C_{A^-} \gamma_{A^-}\right)^2$$
$$= 0.01 \times 0.02^2 \times \gamma_\pm^3$$
$$= 0.01 \times 0.02^2 \times 0.67^3$$
$$= \mathbf{1.2 \times 10^{-6}}.$$

CHAPTER 13

EMF

Required Formulae

1. Work done in carrying positive charge from point a to point b,

$$W = q_i\ (\phi_b - \phi_a) = q_i \Delta\phi\ (\phi_b > \phi_a),$$

where ϕ_b = electric potential at b; ϕ_a = electric potential at a.

If Z_i = valency of an ion,

$$W = Z_i e \Delta\phi.$$

2. Work done on potentiometer for the passage of electrons from anode to cathode in a galvanic cell.

$$W = -nFE,$$

E = emf of the cell.

n = number of moles of electron required to write a balanced cell reaction.

3. $\Delta G = W = -nFE,$

where ΔG is the free energy change in joule. This equation is strictly true for reversible cell potential.

4. $$\Delta H = nF\left\{T\left(\frac{\partial E}{\partial T}\right)_P - E\right\},$$

where ΔH = enthalpy change for a cell reaction

and $\left(\frac{\partial E}{\partial T}\right)_P$ = temperature coefficient.

5. $E = \frac{RT}{nF}\ln\frac{K_a}{Q_n}$, where K_a = equilibrium constant for a reaction taking place in a cell,

Q_a = reaction quotient, R = 8.314 J.mole^{-1}.K^{-1}.

6. $E_{M^{Z+}|M} = E^0_{M^{Z+}|M} + \frac{0.05916}{Z} \log a_{M^{Z+}}$ at 25 °C,

where $E_{M^{Z+}|M}$ = electrode potential in volt.

$E^0_{M^{Z+}|M}$ = standard electrode potential in volt.

$a_{M^{Z+}}$ = activity of M^{Z+}.

7. $E = \frac{2RT}{F} \ln \frac{C_2}{C_1}$, where C_2 and C_1 are concentration of the species in a dilute solution.

E = emf of the concentration cell without transport.

8. $E = t_- \frac{2RT}{F} \ln \frac{C_2}{C_1}$, where t_- = transport number of negative ion.

E = emf of the concentration cell with transport.

9. $E_{IJP} = (t_- - t_+) \frac{RT}{F} \ln \frac{(a_\pm)_2}{(a_\pm)_1}$,

where t_- and t_+ are the transport number of –ve and +ve ions respectively. $(a_\pm)_1$ and $(a_\pm)_2$ are the mean ionic activities of solutions 1 and 2 in LHE and RHE respectively.

10. $\overline{\mu}_{M^{Z+}} = \mu_{M^{Z+}} + Z\phi F$,

where Z = valency of the ion

ϕ = electric potential

F = Faraday

$\mu_{M^{Z+}}$ = chemical potential of the ion

$\overline{\mu}_{M^{Z+}}$ = electrochemical potential of the ion.

Short Questions

1. Define Electrochemical potential.

Ans. : The difference in chemical potential of an ion with and without the electric potential is $ZF\phi$, where ϕ is the electric potential and ZF is the charge per mole on the ion. The chemical potential of an ion in the presence of an electric potential $ZF\phi$ is the electrochemical potential $\overline{\mu}$. It follows that $\overline{\mu}_i = \mu_i + Z_iF\phi$, where $\overline{\mu}_i$ = electrochemical potential, μ_i = chemical potential, Z_i = valency of ith ion.

When there is no charge ($Z_i = 0$), the chemical and electrochemical potential are equal.

2. What is the meaning of the statement?—The standard electrode potential of the Fe^{3+}/Fe^{2+} is 0·77 volt.

Ans. : It means that the potential of the electrode Fe^{3+}/Fe^{2+} where the reaction is $Fe^{3+} + e \rightleftharpoons Fe^{2+}$ at unit activities of the ions is 0.77 volt with reference to that of standard hydrogen electrode potential taken as zero. This is the standard emf of the cell—

Pt, H_2 (g, 1 atm) | H^+ || Fe^{3+} | Fe^{2+}, Pt, i.e., it is true emf of the cell.

$$\text{Pt, } H_2 \text{ (g, 1 atm)} \left| \begin{matrix} H^+ \\ a_{H^+} = 1 \end{matrix} \right\| \begin{matrix} Fe^{3+} \\ \left(a_{Fe^{3+1}}\right) = 1 \end{matrix} \left| \begin{matrix} Fe^{2+}, Pt \\ a_{Fe^{+2}} = 1 \end{matrix} \right.$$

3. Does the emf of a cell depend on temperature?

Ans. : The emf of a cell should depend on temperature.

4. Is the emf of a cell measured potential of a cell?

Ans. : No, it is not the measured potential of the cell. It is the measured potential when the electrodes are joined by a wire of infinite resistance, i.e., when no current is flowing and no net reaction is taking place within the cell.

5. Does positive $E°$ refer to feasibility of a reaction under the given set of condition?

Ans. : The positive $E°$ only indicates that the given reaction at unit activities has an equilibrium constant K greater than unity. Thus, when equilibrium is reached, the product may have a predominance in the given case. But, it tells nothing about rate, nothing about whether it will actually proceed towards right under the given set of condition or even at unit activities.

6. The standard potentials of $Cu^{++}|$ Cu and $Zn^{++}|$ Zn systems are reported with different signs. Why? Which one is positive according to your convention?

Ans. : The standard potentials $E°$ are reported with (+)ve or (–)ve signs with respect to that of standard hydrogen electrode, potential of which is taken as zero. The hydrogen electrode is taken as either left-hand electrode or right-hand electrode in two different conventions. Thus, reported values of $E°$ refer to a reference cell. If the emf of the cell with $Cu^{++}|Cu$ is positive and giving current at unit activities, then emf of the cell with $Zn^{++}|Zn$ is negative and is not giving current and vice versa, and that is why they are reported with different signs in our conventions which is the international convention where the hydrogen electrode is L.H.E. Thus, $E°$ of $Cu^{++}|Cu$ system is the standard emf of the cell,

Pt, H_2 (g, 1 atm)$|H^+||Cu^{++}|$Cu,

where the reaction is $H_2 + Cu^{++} = Cu + 2H^+$ per two faraday. The reaction has equilibrium constant $K > 1$ at unit activities and $E°$ is +ve.

$E°$ of $Zn^{++}|Zn$ system refers to the cell Pt, H_2 (g, 1 atm) $|H^+||Zn^{++}|$Zn. Here, equilibrium constant $K < 1$ at unit activities and $E°$ is negative.

7. Why is the decomposition potential of most acid solutions 1.7 volts?

Ans. : The decomposition voltage of an electrolyte is the minimum voltage that must be applied to a particular solution of an electrolyte in order that current of appreciable strength shall be able to pass. The decomposition voltage of aqueous solution of most acids and bases were found to be around 1.7 volt. This leads to the conclusion that the same electrolytic processes are occuring in each case, that is, the decomposition of water to hydrogen and oxygen. The theoretical voltage for this process is 1.23 volt. The extra value is due to polarisation of electrodes.

8. Decomposition voltage of dilute salt solutions are greater than decomposition voltage of common acids.—Explain.

Ans. : The electrolysis of dilute solution of salt like Na_2SO_4, K_2SO_4, etc in electrolytic cell also lead to electrolysis of water. But in such cases, the decomposition voltage is greater han the voltage needed for electrolysis of dilute acids (i.e., 1.7 volt). This is beleived o be due to the fact that acid is created following the removal of OH^- at the anode, and lkali is created at the cathode following the removal of H^+ ions. Thus, in those cases, he ions are not discharged from the same solutions. H^+ ions are discharged from a more lkaline and OH^- ions are discharged from a more acidic solution resulting in an increase n overvoltage and decomposition voltage.

9. "Passage of electricity through dilute solutic of H_2SO_4, H_3PO_4, etc. leads to the ecomposition of water, but passage of electricity unrough dilute solution of HCl leads o the decomposition of HCl at a lower voltage."—Explain.

Ans. : This is due to overvoltage. The theoretical decomposition voltage of water s 1.23 volt. Due to overvoltage, from dilute acid solution like H_2SO_4, H_3PO_4, etc., the ecomposition starts at 1.7 volts. Halogen acids like HCl decompose at a lesser voltage. Because of negligible overvoltage, HCl decomposes at around 1.3 volts. (discharge potential f Cl^- = 1.3 volt) and it decomposes to H_2 and Cl_2. However, if the voltage rises to 1.7 olt, water begins to get electrolysed from dilute HCl solutions.

10. "Passage of electricity through dilute solution at both H_2SO_4 and Na_2SO_4 leads the electrolysis of water, but in the second case, higher voltage is needed."—Why?

Ans. : [**Hints :** See Question no. 8.]

11. Explain why—

KCl or NH_4NO_3 is preferred to make salt bridge.

Ans. : The salt bridge is used for the purpose of eliminating liquid junction potential. JP). The expression of LJP is

$$MX(a_\pm)_1 \vdots MX(a_\pm)_2$$

$$E_{LJP} = (t_- - t_+)\ \frac{RT}{F} \ln \frac{(a_\pm)_2}{(a_\pm)_1},$$

here t_+ and t_- are transport numbers of cations and anions. $(a_\pm)$'s are mean ionic activities nd all other terms have usual meaning.

Now, velocities at K^+ and Cl^- or NH_4^+ and NO_3^- are nearly same. i.e., $t_+ \approx t_-$. So JP is almost eliminated. That is why KCl or NH_4NO_3 is preferred to make salt bridge.

12. The standard potential of hydrogen electrode is taken as zero at all temperatures.—xplain.

Ans. : The standard potential $E°$ and emf E of cell depend on temperature. $E^o_{H^+|H_2}$ taken as zero at 25 °C. It may not be zero at other temperature, but then we can only etermine the temperature coefficient of emf of cells. The temperature coefficient of electrode annot be found out. For determining temperature coefficient of electrodes, $E°$ of $H^+|H_2$ ectrode may be taken as zero at all temperatures. Then the determined temperature coefficient a cell with $H^+|H_2$ as L.H.E. will be the temperature coefficient of R.H.E.

13. Distinguish between single electrode potential and standard electrode potential.

Ans. : Single electrode potential is the electrode potential of an electrode whereas andard electrode potential $E°$ is the emf of a cell with $H^+|H_2$ as L.H.E. at unit activities all reactants and products.

14. Can you use quinhydrone electrode above pH = 8? Explain.

Ans. : Quinhydrone electrode is an electrode where the reaction is $Q + 2H^+ + 2e = QH_2$, where Q is quinone 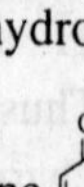 and H_2Q is 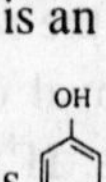. In alkaline medium, above or at pH = 8, quinone may be attacked by OH^- to give different products of aldol type and the equilibrium is disturbed. So, it cannot be used at or above pH = 8.

15. Explain : The density of H_2SO_4 in a lead storage battery increases as it is charged.

Ans. : The reaction in the acid storage cell is :

$$PbO_2 + Pb + 2H_2SO_4 = 2PbSO_4 + 2H_2O.$$

The reaction from right to left is the reaction which takes place during charging. Thus, H_2SO_4 is produced and H_2O is consumed. That is why the density of H_2SO_4 increases.

16. Cl^-, AgCl(s) | Ag electrode can be treated as Ag^+ | Ag electrode. Explain.

Ans. : The reaction taking place in the electrode Cl^-, AgCl(s) | Ag is :

$$AgCl = Ag^+ + Cl^-$$

$$Ag^+ + e = Ag$$

Adding, $AgCl + e = Ag + Cl^-$

But AgCl as shown dissociates completely to Ag^+ whichever amount of AgCl goes in solution. Therefore, Cl^-, AgCl | Ag can be treated as Ag^+ | Ag electrode, with the restriction that $a_{Ag^+} = \frac{K_s}{a_{Cl^-}}$. K_s is the solubility product of the cell.

17. For an irreversible cell $nFE° \neq RT \ln K_{eq}$—Explain

Ans. : The electrical work in a cell for the passage of charge nF is $W_{el} = nFE°$. But this is equal to $-\Delta G°$ for a reversible cell only. Thus, for reversible cell, $nFE° = RT \ln K$. This is not true for irreversible cell, because the Nernst equation is not applicable for these cells.

THE CONSTRUCTION OF CELLS :

The cell construction for any reaction involves the following steps :

(i) Writing the reaction ionically, if possible.

(ii) Selection of redox pairs or choosing the case where reduction (consumption) of an ion or a substance has taken place. That system will constitute the R.H.E.

(iii) Choosing the case where oxidation has taken place. This system will be L.H.E. If some reactant or product of R.H.E. is absent in the final equations, then the L.H.E. should contain that reactant or product. In some cases, we may have to choose L.H.E. first. Then the above applies for R.H.E.

(iv) Then the probable cell has to be written, and the cell reaction has to be checked. If it fits (usually it will fit), then the constructed cell is the required cell.

Remember that solution of any salt contains the ions. However, weak electrolytes and insoluble salts cannot be written ionically. Also remember that production of positive ion and consumption of negative ion are comparable to oxidation for our electrode construction process. Similarly, consumption of positive ions and production of negative ions are comparable to reduction. Any electrode should be reversible with respect to at least one ion. For example, let us consider the reaction $H_2 + \frac{1}{2}O_2 = H_2O$.

In this reaction, O_2 is consumed. Therefore, there must be one electrode containing O_2, i.e., $OH^-|O_2$, Pt. where the reaction is $\frac{1}{2}O_2 + H_2O + 2e = 2OH^-$. Thus, it is the R.H.E. But the final equation does not contain OH^-. So it must be consumed in L.H.E. With H_2 remaining, the electrode at R.H.E. should be $OH^-|H_2$. As L.H.E. the electrode is Pt, H_2(g, 1 atm) | OH^-. Hence, the probable cell :

Pt, H_2 (g, 1 atm) | OH^- | O_2 (g, 1 atm), Pt

18. Construct cell where the following reactions take place :

(i) $2KMnO_4 + 3H_2SO_4 + 5H_2O_2 = K_2SO_4 + 2MnSO_4 + 8H_2O + 5O_2$

Ans. : Writing ionically :

$$2K^+ + 2MnO_4^- + 6H^+ + 3SO_4^{2-} + 5H_2O_2$$

$$= 2K^+ + SO_4^{2-} + 2Mn^{2+} + 2SO_4^{2-} + 8H_2O + 5O_2$$

$$\therefore \quad 2MnO_4^- + 6H^+ + 5H_2O_2 = 2Mn^{2+} + 8H_2O + 5O_2$$

In this reaction MnO_4^- is reduced to Mn^{2+}. Thus, $MnO_4^-|Mn^{2+}$ should be R.H.E. and remaining O_2 and H_2O_2 should be a part of L.H.E. The electrode is O_2 (g, 1 atm) | H_2O_2, H^+ as L.H.E. Therefore, the cell is :

Pt, O_2(g, 1 atm) | H_2O_2, H^+ || H^+, MnO_4^- | Mn^{2+}, Pt

The reaction in this cell is as follows :

Reaction at R.H.E. : $[MnO_4^- + 8H^+ + 5e = Mn^{2+} + 4H_2O] \times 2$

Reaction at L.H.E. : $O_2 + 2H^+ + 2e = H_2O_2$ as R.H.E.

Reaction at L.H.E. : $[H_2O_2 = O_2 + 2H^+ + 2e] \times 5$

Adding : $2MnO_4^- + 6H^+ + 5H_2O_2 = 2Mn^{2+} + 8H_2O + 5O_2$

(ii) $\frac{1}{2}H_2 + \frac{1}{2}Cl_2 = HCl$

Ans. : Writing ionically : $\frac{1}{2}H_2 + \frac{1}{2}Cl_2 = H^+ + Cl^-$.

Hence, the cell is :

Pt, H_2 (g, 1 atm) | H^+, Cl^- | Cl_2 (g, 1 atm), Pt.

The reaction will fit. Therefore, the written cell is the required cell.

(iii) $HCl + NaOH = NaCl + H_2O$

Ans. : Writing ionically : $H^+ + Cl^- + Na^+ + OH^- = Na^+ + Cl^- + H_2O$.

Thus, the net reaction is $H^+ + OH^- = H_2O$.

In this reaction, H^+ is consumed, therefore, H^+ | H_2 electrode will be R.H.E., where the reaction is $H^+ + e = \frac{1}{2}H_2$.

Final equation does not contain H_2. It must be consumed at L.H.E.. With OH^- remaining the L.H.E. as R.H.E. is :

$OH^- | H_2$ (g, 1 atm). Hence, as L.H.E. it is H_2(g, 1 atm) | OH^-. The reaction at L.H.E is $\frac{1}{2}H_2 + OH^- = H_2O + e$. The cell is, thus,

Pt, H_2(g, 1 atm) | OH^- | | H^+ | H_2 (g, 1 atm), Pt.

(iv) $Zn + PbO = ZnO + Pb$.

In this reaction PbO is reduced to Pb. Thus, R.H.E. should be OH^-, PbO(s) | Pb(Hg), Pt.

Naturally Zn is oxidised to ZnO. Hence, L.H.E. is Zn | ZnO (s), OH^-. Thus, the probable cell is :

Zn | ZnO(s), OH^-, PbO (s) | Pb(Hg), Pt.

The reaction taking place in the cell at the R.H.E. :

$$PbO + H_2O = Pb^{2+} + 2OH^-$$
$$Pb^{2+} + 2e = Pb$$

Adding, $PbO + H_2O + 2e = Pb + 2OH^-$

Obviously the reaction at L.H.E. would be :

$$Zn + 2OH^- = ZnO + H_2O + 2e$$

Adding : $Zn + PbO = ZnO + Pb$

(v) $2Ag + Hg_2Cl_2 = 2Hg + 2AgCl$

The cell is Ag | AgCl, Cl^-, Hg_2Cl_2 | Hg, Pt.

Proceeding in the line of (iv) the cell may be constructed as shown above. Both the electrodes are reversible with respect to Cl^-.

19. Construct cells where the following reactions take place.

(i) $P_4 + 3NaOH + 3H_2O = 3NaH_2PO_2 + PH_3$

(ii) $3Fe^{2+} = 2Fe^{3+} + Fe$

(iii) $CaCl_2 + Na_2CO_3 = CaCO_3 + 2NaCl$

(iv) $CuSO_4\ (C_1) = CuSO_4\ (C_2)$

(v) $K_2Cr_2O_7 + 7H_2SO_4 + 6KI = 4K_2SO_4 + Cr_2(SO_4)_3 + 7H_2O + 3I_2$

(vi) $Ag^+ + Br^- = AgBr$

(vii) $AgI = Ag^- + I^-$

(viii) $ZnCl_2 = Zn + Cl_2$

(ix) $PbO + H_2O = Pb^{2+} + 2OH^-$

(x) $\frac{1}{2}Cl_2 + Br^- = Cl^- + \frac{1}{2}Br_2$

(xi) $PbCl_2$ (s) + 2AgI (s) = 2AgCl (s) + PbI_2 (s)

Hints : Proceed as the usual method, choose redox pair. Only in the last cell the reaction probably does not take place in water solution. However, we can construct cell– both the electrodes are 3rd kind.

In this reaction PbI_2 is produced. Therefore, PbI_2 should be a part of R.H.E. with Ag, AgI present. The electrode is :

$$Pb^{2+}, PbI_2(s), AgI(s) \mid Ag,$$

where the reaction is :

$$2AgI = 2Ag^+ + 2I^-$$

$$2Ag^+ + 2e = Ag$$

$$Pb^{2+} + 2I^- = PbI_2$$

Adding : $2AgI + 2e + Pb^{2+} = PbI_2 + 2Ag$

The L.H.E. is $Ag \mid AgCl\ (s), PbCl_2(s), Pb^{2+}$.

Now the reaction at L.H.E. as R.H.E. :

$$2AgCl = 2Ag^+ + Cl^-$$

$$2Ag^+ + 2e = 2Ag$$

$$Pb^{2+} + 2Cl^- = PbCl_2$$

Adding : $2AgCl + Pb^{2+} + 2e = 2Ag + PbCl_2$

Now at the L.H.E. the reaction is $2Ag + PbCl_2 = 2AgCl + 2e + Pb^{2+}$

Adding R.H.E. and L.H.E. : $PbCl_2 + 2AgI = 2AgCl + PbI_2$

Hence, the probable cell is :

$Ag \mid AgCl\ (s), PbCl_2(s), Pb^{2+}, PbI_2(s), AgI(s) \mid Ag.$

20. Show that, $\dfrac{\partial(E^\circ/T)}{\partial(1/T)} = -\dfrac{\Delta H^\circ}{nF}$.

Soln. : $\dfrac{\partial(E^\circ/T)}{\partial(1/T)} = \dfrac{\partial(-\Delta G^\circ/T)}{nF\partial(1/T)}$ (since, $\Delta G^\circ = -nFE^\circ$)

$$= \frac{1}{nF}\left[-\Delta G^\circ - \frac{1}{T}\frac{\partial \Delta G^\circ}{\partial(1/T)}\right]$$

$$= \frac{1}{nF}\left[-\Delta G^\circ + T\frac{\partial \Delta G^\circ}{\partial T}\right] \text{ (since, } \partial\left(\frac{1}{T}\right) = -\frac{1}{T^2}\,dT)$$

$$= -\frac{\Delta H^\circ}{nF} \text{ (since, } \Delta G^\circ = \Delta H^\circ - T\Delta S = \Delta H^\circ + T\frac{\partial \Delta G^\circ}{\partial T}).$$

Numerical Problems

1. Given that $E^\circ_{Sn^{4+}|Sn^{+2}} = 0.15$ V and $E^\circ_{Sn^{+2}|Sn} = -0.136$ V. Calculate $\mu^\circ_{Sn^{4+}}$, $\mu^\circ_{Sn^{+2}}$ and $E^\circ_{Sn^{4+}|Sn}$.

Soln. : $\mu^\circ_{Sn^{+2}} = Z.F.E^\circ_{Sn^{+2}|Sn}$. [Z = no. of charge, F = Faraday]

$= 2 \times 96484 \times (-0.136)$ J.mole^{-1} = **–26.2 kJ.mole^{-1}.**

The reactions are

$Sn^{4+} + 2e \rightleftharpoons Sn^{2+}\ \Delta G^\circ = -2 \times F \times 0.15$ V

$Sn^{+2} + 2e \rightleftharpoons Sn\ \Delta G^\circ = -2 \times F \times (-0.136)$ V

Adding $Sn^{4+} + 4e \rightleftharpoons Sn \Delta G^\circ = -2 \times F \times 0.014$ V.

$\therefore\ -2 \times F \times 0.014 = -4 \times F \times E^\circ$.

$$\therefore \left(E^\circ_{Sn^{4+}|Sn}\right) = \frac{0.014}{2} = 0.007 \text{ V.}$$

$\therefore\ \mu^\circ_{Sn^{4+}} = 4 \times 96484 \times 0.007$ J.mole^{-1} = **2.70 kJ.mol^{-1}.**

2. Given that $E^\circ_{Sn^{+2}|Sn} = -0.14$V and $E^\circ_{Pb^{++}|Pb} = -0.13$V. Calculate the molarity of Sn^{++} ion when Sn is added to a 0.02 M solution of $Pb(NO_3)_2$ at 25 °C.

Soln. : The reaction is $Sn + Pb^{++} = Sn^{++} + Pb$. This reaction occurring in the cell

Pt Sn | Sn^{++} || Pb^{++} | Pb(Hg), Pt

$E^\circ_{Cell} = -0.13 - (-0.14) = -0.13 + 0.14 = 0.01$ volt.

$\therefore\ \Delta G^\circ = -RT \ln K = -nFE^\circ$.

$$\therefore \ln K = \frac{nF}{RT}.E^\circ = \frac{2 \times 96484 \times 0.01}{8.314 \times 298.15} = 0.778467456.$$

$\therefore\ K = 2.178 \approx 2.18.$

Let the molarity of $Sn^{++} = x$ at equilibrium, then that of $Pb^{++} = 0.02 - x$.

The equilibrium constant of the reaction in the cell is $\frac{C_{Sn^{++}}}{C_{Pb^{++}}}$

or, $\frac{x}{0.02 - x} = 2.18$

or, $3.18\, x = 0.0436$ or, $x = 0.0137$

i.e., $C_{Sn^{++}}$ = **0.0137 (M).**

3. What is the equilibrium concentration of Fe^{++} when a piece of iron is dipped in a 0.02 (M) $CdSO_4$ solution at 25 °C.

Given, $E^\circ_{Fe^{++}|Fe} = -0.441$V, $E^\circ_{Cd^{++}|Cd} = -0.403$V.

Soln. : Proceed as Problem 2.

$C_{Fe^{++}}$ = **0.019 (M).**

4. Given that $E^\circ_{MnO_4^-|Mn^{++}} = 1.51$ V and $E^\circ_{MnO_2|Mn^{++}} = 1.23$ V in acid solution. Calculate $E^\circ_{MnO_4^-|MnO_2}$ in acid solution.

Soln. : $MnO_4^- + 8H^+ + 5e \rightleftharpoons Mn^{++} + 4H_2O$. $\Delta G^\circ = -5F \times (1.51)$

$Mn^{++} + 2H_2O \rightleftharpoons MnO_2 + 4H^+ + 2e$, $\Delta G^\circ = -2F\ (-1.23)$

Adding, $MnO_4^- + 4H^+ + 3e \rightleftharpoons MnO_2 + 2H_2O$, $\Delta G° = -3FE°$

$\therefore (-7.55 \times F) + (2.46 \times F) = -3F \times E°$

or, $-7.55 + 2.46 = -3 \times E°$

$\therefore E° = \dfrac{-5.09}{-3}$ volt = **1.696 volt.**

5. Calculate ΔG at 25 °C and 1 atm for reaction $\frac{1}{2}H_2(g) + AgCl = Ag + H^+ + Cl^-$. Given that E of the cell Pt, $H_2(g)$ | HCl, AgCl | Ag is 0.35252 V at 25 °C.

Soln. : Here, $E = 0.35252$ V,

$F = 96484$ coulombs

For the reaction, $\frac{1}{2}H_2(g) + AgCl = Ag + H^+ + Cl^-$, $n = 1$.

$\therefore \Delta G = -nFE$

$= -1 \times 96484$ coloumbs $\times 0.35252$ volt $= -34012.53$ J = **−34.01 kJ.**

6. For the cell H_2 (1 atm) | HBr (a = 1), AgBr | Ag, E is given as

$E = 0.07131 - 4.99 \times 10^{-4} (t - 25) - 3.45 \times 10^{-6} (t - 25)^2$, where t is temperature in °C. Calculate ΔG, ΔS and ΔH in joules for the reaction taking place in the cell at 25 °C.

Soln. : $E = 0.07131 - 4.99 \times 10^{-4} (t - 25) - 3.45 \times 10^{-6} (t - 25)^2$

Thus, $\left(\dfrac{\partial E}{\partial T}\right)_P = -4.99 \times 10^{-4} \quad 3.45 \times 10^{-6} \times 2t + 2 \times 25 \times 3.45 \times 10^{-6}$

Hence, at 25 °C $\left(\dfrac{\partial E}{\partial T}\right)_P = -4.99 \times 10^{-4} - 3.45 \times 10^{-6} \times 2 \times 25 + 2 \times 25 \times 3.45 \times 10^{-6}$

$= -4.99 \times 10^{-4}$ volt.K^{-1}.

$E = 0.07131$ volt at 25 °C

The reaction in the cell for the passage of 2 faraday :

$$2\,AgBr + 2e = 2Ag + 2Br^-$$
$$H_2 = 2H^+ + 2e$$

$$2AgBr + H_2 = 2Ag + 2H^+ + 2Br^-$$

Here $n = 2$.

Hence, $\Delta G = -nFE = -2 \times 96484 \times 0.01731$ J = −13760.55 J = −13.761 kJ.

$\Delta S = nF\left(\dfrac{\partial E}{\partial T}\right)_P = 96.29$ J.

Hence, $\Delta H = nF\left\{T\left(\dfrac{\partial E}{\partial F}\right)_P - E\right\}$

$= 2 \times 96484$ coulomb $\{298.15 \times (-4.99 \times 10^{-4}) - 0.07131\}$ volt

$= -42469.72$ J = **− 42.47 kJ.**

7. For a particular cell, E at 20 °C, 25 °C and 30 °C are 0.0663 V, 0.06839 V and 0.07045 V respectively. Calculate ΔG, ΔS and ΔH for the reaction taking place in the cell at 25 °C.

Soln. : $\left(\frac{\partial E}{\partial T}\right)_P = \frac{E_2 - E_1}{T_2 - T_1} = \frac{0.06839 - 0.0663}{5} = 4.18 \times 10^{-4}$ volt.K^{-1}.

Again, $\left(\frac{\partial E}{\partial T}\right)_P = \frac{0.07035 - 0.06839}{5} = 3.92 \times 10^{-4}$ volt.K^{-1}.

$\therefore$ taking average, $\left(\frac{\partial E}{\partial T}\right)_P = 4.05 \times 10^{-4}$ volt.K^{-1}.

$\therefore T\left(\frac{\partial E}{\partial T}\right)_P = 298.15 \times 4.05 \times 10^{-4}$ volt at 25 °C.

Now, $\Delta H = nF\left\{T\left(\frac{\partial E}{\partial T}\right)_P - E\right\} = 1 \times 96484\ (298.15 \times 4.05 \times 10^{-4} - 0.06839)$

$= 96484\ (0.12075 - 0.06839)$ [taking n = 1 for this cell]

$=$ **5.051 kJ.**

$\Delta G = -nFE = -1 \times 96486$ coulombs $\times$ 0.06839 volt [taking n = 1 for this cell]

$= -6.598$ kJ $\approx$ **– 6.6 kJ.**

$\Delta S = nF\left(\frac{\partial E}{\partial T}\right)_P = 1 \times 96484$ coulombs $\times 4.05 \times 10^{-4}$ volt. K^{-1} [n = 1]

= **39.076 J.K^{-1}.mol^{-1}.**

8. The emf of the cell Cd | $CdCl_2$ (1m), AgCl (s) | Ag is 0.675 volt at 25 °C. The temperature coefficient of the cell is – 6.5 × 10^{-4} volt/K. Find the changes in heat content and entropy for the electrochemical reaction that occurs in the cell when one faraday of electricity is drawn from it. **[C.U. 1975]**

Hints : Proceed as Problem 7.

In the cell, for the passage of 1 faraday the reaction is

$AgCl + e = Ag + Cl^-$

$\frac{1}{2}Cd = \frac{1}{2}Cd^{+2} + e$

$AgCl + \frac{1}{2}Cd = Ag + \frac{1}{2}Cd^{+2} + Cl^-$

ΔH = – 83.83 kJ/mol, ΔS = – 62.72 J.K^{-1}.mol^{-1}.

9. At 0 °C, a calorimetric determination of ΔH for the reaction $Zn + 2AgCl = ZnCl_2 + 2Ag$ yields the value –52.50 kcal. The emf of the cell is 1.015 volts. Calculate the temperature coefficient of the cell. **[C.U. 1979]**

Soln. : $\Delta H = -52.50$ kcal $= -219.66 \times 10^3$ J.

For the cell reaction n = 2; T = 273.15 K, E = 1.015 volt.

We know, $\Delta H = nF\left\{T\left(\frac{\partial E}{\partial T}\right)_P - E\right\}$, where $\left(\frac{\partial E}{\partial T}\right)_P$ is temperature coefficient.

$\therefore -219.66 \times 10^3 = 2 \times 96484\ \left\{273.15\left(\frac{\partial E}{\partial T}\right)_P - 1.015\right\}$

or, $\left(\frac{\partial E}{\partial T}\right)_P$ = **– 4.514 × 10^{-4} volt.K^{-1}.**

10. The Pb|$PbCl_2$ (fused)|Cl_2 (g), 1 atm shows an emf :

$$E = 1.2467 - 6.50 \times 10^{-4} (t - 55) \text{ V}; t \text{ is expressed in °C.}$$

What are the values of enthalpy and entropy of formation of $PbCl_2$ at 500 °C?

Soln. : Proceed as problem (6). Here $n = 2$

$\Delta H = -281.73$ kJ/mole

$\Delta S = -125.43$ J.K^{-1}.mole^{-1}

11. The emf of the cell (V) glass electrode | buffer solution || calomel electrode was found to be 0.0232 volt at 25 °C when the buffer has a pH = 2.5. The emf increased to 0.112 volt when another buffer solution is used. What is the pH of the latter buffer? **[C.U. 1979]**

Soln. : By the problem, $E = E_{cal} - (E^\circ_G - 0.059 \text{ pH})$

$\therefore\ 0.0232 = (E_{cal} - E^\circ_G) + 0.059\ (2.5)$

or, $0.0232 = x + 0.1475$ (where $E_{cal} - E^\circ_G = x$).

$\therefore\ x = 0.0232 - 0.1475 = -0.1243.$

Under new condition, $E = 0.112$ volt,

$\therefore\ 0.112 = -0.1243 + 0.059$ pH

or, $\text{pH} = \dfrac{0.112 + 0.1243}{0.059} = 4.005 \approx$ **4.**

12. The emf of the cell obtained by coupling through a salt bridge with quinhydrone electrode ($E^\circ = 0.6994$ V) dipping into a buffer solution and a saturated calomel electrode ($E = 0.2415$ V) is 0.0042 V. What is the pH of the buffer? **[C.U. 1982]**

Soln. : $E_{Cell} = E_{Quinhydrone} - E_{Calomel}$

$= (0.6994 - 0.059 \text{ pH}) - 0.2415 = 0.4579 - 0.059 \text{ pH}$

Given, $E_{Cell} = 0.0042$

$\therefore\ 0.0042 = 0.4579 - 0.059$ pH

$\therefore$ pH $= 7.689 =$ **7.7.**

13. Calculate the emf of a cell at 25 °C which comprises a saturated calomel electrode coupled through a salt bridge, with a platinum black electrode dipping into a solution *A* and H_2 gas at 1 atm is bubbled over the platinum surface. The solution *A* was prepared by mixing 10 ml $\left(\dfrac{N}{10}\right)$ NaOH solution with 90 ml $\left(\dfrac{N}{100}\right)$ HCl solution. (Oxidation potential of saturated calomel electrode = -0.242 V at 25 °C) **[C.U. 1976; V.U. 1989]**

Soln. : $E_{Cell} = E_{Cal} - E_{H^+/H_2} = 0.242 - 0.059 \log a_{H^+} = 0.242 + 0.059$ pH.

Now, the solution *A* is prepared by mixing 10 ml $\left(\dfrac{N}{10}\right)$ NaOH with 90 ml $\left(\dfrac{N}{100}\right)$ HCl.

[90 cc of $\left(\dfrac{N}{100}\right)$ (M) is 0.9 cc of 1(M)

10 cc 0.1(M) ≡ 1cc 1(M)]

$\therefore\ C_{OH^-} = 0.1$cc 1(M) in 100 cc.

Thus, C_{OH^-} in NaOH $= \frac{0.1}{100}$(M) $= 10^{-3}$ (M).

$\therefore$ pOH = 3.

$\therefore$ pH = 14 − 3 = 11 at 25 °C.

$\therefore$ $E_{Cell} = (0.059 \times 11) + 0.242 =$ **0.891 V.**

14. Write down the cell reaction and calculate the emf, the free energy change and the equilibrium constant of the reaction taking place at 25 °C of the following cells : **[C.U. 1984]**

(i) H_2 (1 atm) | HBr ($a_\pm = 0.2$), Hg_2Br_2 (s) | Hg

Given, $E^o_{Hg|Hg_2Br_2(s).Br^-} = -0.1385$ V.

Soln. : Reaction for the passage of two faradays :

Reaction at R.H.E. : $Hg_2Br_2 + 2e \rightleftharpoons 2Hg + 2Br^-$

Reaction at L.H.E. : $H_2 \rightleftharpoons 2H^+ + 2e$

$\therefore$ cell reaction : $Hg_2Br_2 + H_2 \rightleftharpoons 2Hg + 2H^+ + 2Br^-$

$$\therefore E = \left(E_2^0 - E_1^0\right) - \frac{0.059}{2}\log a^2_{H^+}.a^2_{Br^-}.$$

Given, $E^0_{Hg|Hg_2Br_2.Br^-} = -0.1385$ V.

$\therefore E = (0.1385) - 0.059 \log a_{H^+}.a_{Br^-}$

$= 0.1385 - 0.059 \log a_\pm^2$ [$a_\pm$ = mean ionic activity]

$\therefore E = 0.1385 - 0.059 \times 2 \times \log 0.2$

$= 0.1385 + 0.0821 =$ **0.220 V.**

$\therefore \Delta G = -nFE = -2 \times 96484 \times 0.220$ J $=$ **− 42.64 kJ.**

Now, $RT \ln K = nFE^o$

$$\text{or, } \ln K = \frac{nFE^o}{RT} = \frac{2 \times 96484 \times 0.1385}{8.314 \times 298.15}$$

or, $\ln K = 10.78177$

$\therefore K =$ **4.81×10^4.**

(ii) Zn | Zn^{++} ($a = 0.01$) || $Fe^{3+}{}_{(a=0.1)}$ | $Fe^{++}{}_{(a=0.001)}$ | Pt

Given, $E^o_{Zn|Zn^{++}} = 0.761$ V

$E^o_{Fe^{2+}|Fe^{3+}} = -0.771$ V.

Soln. : The cell reaction is $Zn + 2Fe^{3+} = Zn^{++} + 2Fe^{2+}$.

$$\therefore E = (0.771 + 0.761) - \frac{0.059}{2}\log\frac{\left(a_{Fe^{2+}}\right)^2 . a_{Zn^{++}}}{\left(a_{Fe^{3+}}\right)^2 . a_{Zn}}$$

$$\text{or, } E = 1.532 - \frac{0.059}{2}\log\frac{(0.001)^2 \times (0.01)}{(0.1)^2 \times (1)}$$

or, $E = 1.532 - \frac{0.059}{2} \log 10^{-6}$

$\therefore$ $E = 1.532 + 0.059 \times 3$

$\therefore$ $E =$ **1.702 V.**

$\therefore$ $\Delta G = -nFE = -(2 \times 96484 \times 1.702)$ J = **−328.43 kJ.**

$\therefore$ $\Delta G^\circ = -nFE^\circ = -RT \ln K$

or, $\ln K = \frac{2 \times 96484 \times 1.532}{8.314 \times 298.15}$

$\therefore$ $K = 6.229 \times 10^{51}$.

15. The potential of the cell, Zn (s) | $ZnCl_2$ (0.01021 M), AgCl | Ag was found to be 1.1566 at 25 °C. Given that $E^\circ_{Zn^{++}|Zn} = -0.7618$, $E^\circ_{Cl^-,AgCl|Ag} = 0.224$. Calculate $\gamma_\pm$.

Soln. : The cell reaction is Zn + 2AgCl $\rightleftharpoons$ $Zn^{++} + 2Cl^- + 2Ag$

Now, $E^\circ_{Cl^-,AgCl|Ag} - E^\circ_{Zn^{++}|Zn}$

$= 0.224 - (-0.7618)$V

$= (0.224 + 0.7618)$V $= 0.9858$ V.

$$\therefore E = \left(E^\circ_{Cl^-.AgCl|Ag} - E^\circ_{Zn^{+2}|Zn}\right) - \frac{0.059}{2} \log a_{Zn^{+2}} . a^2_{Cl^-}$$

Since, for the salt M_xA_y, $a_+^x a_-^y = x^x y^y C^{x+y} \gamma_\pm^{x+y}$

$$a_{Zn^{+2}} a^2_{Cl^-} = 1^1 2^2 C^3 \gamma_\pm^{\ 3} = 4C^3 \gamma_\pm^{\ 3}.$$

Hence, $1.1566 = 0.9858 - \frac{0.059}{2} \log\left(4.C_\pm^3.\gamma_\pm^3\right)$

or, $1.1566 = 0.9858 - 0.0591 \times \log(2) - \frac{3 \times 0.059}{2} \log C\gamma_\pm$

or, $1.1566 = 0.9858 - 0.01776 + 0.176238 - 1.5 \times 0.059 \log \gamma_\pm$.

$\therefore$ $\log \gamma_\pm = -\frac{0.01232}{1.5 \times 0.059}$.

or, $\gamma_\pm$ = **0.726.**

16. What will be the emf of the cell at 25 °C,

H_2 (1 atm) | 0.5 (N) formic acid || 1(N) acetic acid | H_2 (1 atm),

if the diffusion potential be neglected? The dissociation constants of formic and acetic acids are 17.7×10^{-5} and 1.8×10^{-5} respectively. Which will act as the positive electrode?

[C.U. 1975; V.U. 1988]

Hints : For the passage of one faraday 1 gm atom of H^+ ions passes from 1(N) AcOH solution to 0.5 (N) formic acid solution, emf of the cell is $-\frac{RT}{F} \ln \frac{a_{H^+(0.5NHCOOH)}}{a_{H^+(1NAcOH)}}$.

But C_{H^+} of a weak acid with acidity constant K_a is $\sqrt{C.K_a}$, where C = concentration of the acid, assuming $C_{H^+} \approx a_{H^+}$, since activity coefficients are unknown.

$$E = -0.059 \log \frac{\left(8.85\times10^{-5}\right)^{\frac{1}{2}}}{\left(1.8\times10^{-5}\right)^{\frac{1}{2}}} = -\frac{0.059}{2}\log\frac{8.85}{1.80} = -\mathbf{0.02\ volt.}$$

Thus, **R.H.E. should be –ve electrode and L.H.E. should be +ve electrode.**

17. For the reaction Fe = Fe^{3+} + 3e; $E° = 0.036$ V and $E^{o}_{Fe^{3+}|Fe^{2+}} = 0.771$ V. Calculate $E^{o}_{Fe^{2+}|Fe}$. Is the reaction Fe + $2Fe^{3+}$ = $3Fe^{2+}$ spontaneous in the forward or the reverse direction?

[Delhi University (Hons.) 1977]

Soln. : Given

	$E°$	$\Delta G°$
$Fe^{3+} + 3e \rightleftharpoons Fe;$	-0.036 V	$-3 \times F \times (-0.036)$
$Fe^{2+} \rightleftharpoons Fe^{3+} + e;$	-0.771 V	$-1 \times F \times (-0.771)$
Adding, $Fe^{2+} + 2e \rightleftharpoons Fe;$		$-2 \times F \times E°$

$$\therefore\ E° = -\frac{(3\times0.036+1.77)}{2} = -0.439\text{ V} \approx -0.44\text{ V}$$

The cell in which the reaction,

$Fe + 2Fe^{3+} \rightleftharpoons 3Fe^{2+}$, takes place is Fe | Fe^{2+} || Fe^{3+} | Fe^{2+}, Pt.

The $E°$ of the cell is $E°$ = (0.771 + 0.44) volt = **1.21 V.**

So, the reaction is spontaneous **in forward direction.**

18. Calculate the pH of the solution for which a quinhydrone electrode $E° = 0.6944$ V coupled with a saturated calomel electrode $E = 0.2415$ V through a salt bridge shows no measurable emf.

[C.U.(H) 1984; V.U. 1988]

Hints : Proceed as before. By the problem, 0.4529 V = 0.059 pH.

$\therefore$ pH = **7.76.**

19. The solubility product of $Fe(OH)_3$ at 25 °C is $10^{-38.1}$ and $E^{o}_{Fe^{3+}|Fe} = -0.036$ V. Calculate the standard emf for OH^-, $Fe(OH)_3$ | Fe electrode.

Soln. : We know that $E°$ (2nd kind with an insoluble salt MS)

$$E^{o}_{cell} = \frac{RT}{nF}\ln K_s$$

$$= \frac{0.059}{n}\ln K_s, \text{ where the temperature is 25 °C.}$$

$$\therefore\ E^{\circ}_{OH^-,Fe(OH)_3|Fe} = E^{\circ}_{Fe^{3+}|Fe} + \frac{0.059}{3}\log K_s$$

$$= -0.036 + \frac{0.059}{3}\log 10^{-38.1} = -0.036 - \frac{0.059\times38.1}{3} = \mathbf{-0.785\ V.}$$

20. What is the equilibrium concentration of Ni^{2+} when a piece of iron is dipped in a 0.005 (M) solution of $NiSO_4$ at 25 °C. Given that $E^{\circ}_{Fe^{2+}|Fe} = -0.441$ volt and

$E^{\circ}_{Ni^{++}|Ni} = -0.24$ V.

Soln. : The reaction is $Fe + Ni^{++} = Fe^{++} + Ni$ which occurs in the cell

$Fe \mid Fe^{++} \mid\mid Ni^{++} \mid Ni$

$$E° = -0.24 + 0.441 = 0.201 \text{ V}.$$

$$\therefore \Delta G° = -nFE° = -2 \times 96484 \times 0.201 = -RT \ln K.$$

$$\therefore K = 6.2444 \times 10^6.$$

$\because$ K is so high, almost all of Ni^{++} will react to give Fe^{2+}. Let coefficient of Ni^{++} remaining be x. Thus, $C_{Fe^{2+}} = 0.005 - x$. Now, assuming activity = concentration,

$$6.2444 \times 10^6 = \frac{C_{Fe^{2+}}}{C_{Ni^{++}}} = \frac{0.005 - x}{x}.$$

Neglecting x in comparison to 0.005,

$$\text{concentration } x = \frac{0.005}{6.2444 \times 10^6} = 8.0071 \times 10^{-10}.$$

Thus, the equlibrium concentration of Ni^{++} will be **8.01×10^{-10} molar.**

21. An aqueous solution containing 0.01 (M) Fe $(ClO_4)_3$, 0.01 (M) $Fe(ClO_4)_2$ and 0.01 (M) $HClO_4$ was titrated with a concentrated NaOH solution at room temperature (30 °C), so that changes in volume were negligible. Calculate the redox potential of $Fe^{3+} \mid Fe^{2+}$ system at pH values 2.2, 4.2, 6.0, 8.2 and 10 assuming new species formed during titration were $Fe(OH)_3$ and $Fe(OH)_2$ only. [$E°$ for system $Fe^{3+} \mid Fe^{+2} = +0.77$ V. Solubility product of $Fe(OH)_3 = 10^{-37.1}$ and that of $Fe(OH)_2 = 10^{-18.4}$. **[C.U. 1973]**

Soln. : The redox potential of $Fe^{3+} \mid Fe^{2+}$ system by the problem is :

$$E = 0.77 + \frac{2.3026RT}{F} \log \frac{a_{Fe^{3+}}}{a_{Fe^{2+}}}$$

$$= 0.77 + 0.06014 \log \frac{a_{Fe^{3+}}}{a_{Fe^{2+}}} \text{ at } 30 \text{ °C}$$

$$= 0.77 + 0.06 \log \frac{C_{Fe^{3+}}}{C_{Fe^{2+}}}.$$

Assuming activity to be equal to concentration.

When $Fe(OH)_3$ appears, $C_{Fe^{3+}} \times C^3_{OH^-} = K_s$.

$$\therefore C_{Fe^{3+}} = \frac{10^{-37.1}}{C^3_{OH^-}}.$$

And also when $Fe(OH)_2$ appears, $C_{Fe^{2+}} = \dfrac{10^{-18.4}}{C^2_{OH^-}}$.

Now, at the given pH values (pH increasing as a result of addition of NaOH); assuming $K_w = 10^{-14}$,

pH	C_{H^+}	$C_{OH^-} = \dfrac{Kw}{C_{H^+}}$	$C_{Fe^{3+}} \times C^3_{OH^-}$	$C_{Fe^{2+}} \times C^2_{OH^-}$	Result
2.2	$10^{-2.2}$	$10^{-11.8}$	$10^{-2} \times 10^{-35.4} = 10^{-37.4}$	$10^{-2} \times 10^{-23.6} = 10^{-25.6}$	No ppt.
4.2	$10^{-4.2}$	$10^{-9.8}$	$10^{-31.4}$	$10^{-21.6}$	$Fe(OH)_3$ ppted.
6.0	10^{-6}	10^{-8}	10^{-26}	10^{-18}	Both ppted.

Thus, at pH = 2.2,

$$E = 0.77 + 0.06 \log \frac{0.01}{0.01} = \mathbf{0.77\ V.}$$

At pH 4.2, $E = 0.77 + 0.06 \log \dfrac{10^{-37.1}/10^{-29.4}}{0.01} = 0.77 + 0.06 \log 10^{-5.7}$

$$= 0.77 - 0.342 = \mathbf{0.428\ V.}$$

At pH = 6, 8.2 and 10.

$$E = 0.77 + 0.06 \log \frac{C_{Fe(OH)_3}/C^3_{OH^-}}{C_{Fe(OH)_2}/C^2_{OH^-}}$$

$$= 0.77 + 0.06 \log \frac{10^{-37 \cdot 1}}{10^{-18 \cdot 4}} - 0.06 \log C_{OH^-}$$

$$= -0.352 - 0.06 \log C_{OH^-}$$

$\therefore$ at pH = 6,

$E = -0.352 - 0.06 \log 10^{-8} = \mathbf{0.128\ V.}$

At pH = 8.2,

$E = -0.352 - 0.06 \log 10^{-5.8}$

$= -0.352 + 0.348 = \mathbf{-0.004\ V.}$

At pH = 10,

$E = -0.352 - 0.06 \log 10^{-4} = \mathbf{-0.112\ V.}$

Thus, although there is no H^+ or OH^- in the $Fe^{3+} \mid Fe^{2+}$ system needed for balancing, E depends on pH because of precipitation of hydroxides.

22. In 1(M) $HClO_4$ solution, the formal potentials for the different stages in the reduction of MnO_4^- are given in the following emf diagram :

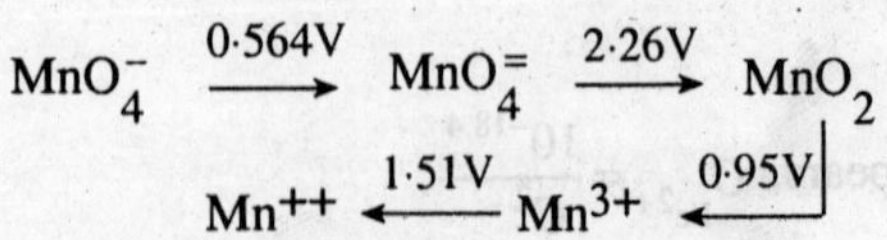

Calculate the formal potential for the reaction

$MnO_4^- \longrightarrow Mn^{++}$ and $MnO_4^{=} \longrightarrow Mn^{++}$ **[C.U. 1979]**

Soln. : We know, ΔG is additive. Thus,

	E_{formal} (V)	ΔG
$MnO_4^- + e = MnO_4^=$	0.564	$-0.564\ F$(1)
$MnO_4^= + 4H^+ + 2e = MnO_2 + 2H_2O$	2.26	$-4.52\ F$(2)
$MnO_2 + 4H^+ + e = Mn^{3+} + 2H_2O$	0.95	$-0.95\ F$(3)
$Mn^{3+} + e = Mn^{++}$	1.51	$-1.51\ F$(4)

Adding, $MnO_4^- + 8H^+ + 5e = Mn^{++} + 4H_2O$

$$\Delta G = -(0.564 + 4.52 + 0.95 + 1.51)F = -5\ FE$$

$\therefore \quad E = 1.5088$ V.

Adding (2) to (4),

$$MnO_4^= + 8H^+ + 5e = Mn^{++} + 4H_2O$$
$$\Delta G = -(4.52 + 0.95 + 1.51)\ F = -5FE.$$

$\therefore \quad E =$ **1.754 V.**

23. The emf of Weston standard cell is 1.01530 V at 20 °C and 1.01807 V at 25 °C. Calculate ΔG, ΔH and ΔS for the cell reaction at 25 °C. **[C.U. 1982]**

Soln. : $$\left(\frac{\partial E}{\partial T}\right)_P = \frac{1.01807 - 1.01530}{5}$$

$$= 5.54 \times 10^{-4} \text{ volt/deg.}$$

$$\Delta S = nF\left(\frac{\partial E}{\partial T}\right)_P$$

Here, $n = 2$

$\therefore \quad \Delta S = 2 \times 96484 \times 5.54 \times 10^{-4}$ J/K = **106.9 J/K.**

$\Delta G = -nFE = -2 \times 96484 \times 1.01807$

$= $ **−196.45 kJ.**

$$\Delta H = nF\left\{T\left(\frac{\partial E}{\partial T}\right)_P - E\right\} = 2 \times 96484\ (298.15 \times 5.54 \times 10^{-4} - 1.01807)$$

$=$ **−164.58 kJ.**

24. The potential of the cell Cd | $CdI_2\,(a_2)$, AgI(s) | Ag is 0.2860 V at 25 °C. Calculate the mean ionic activity and activity of the electrolyte.

Given, $E^o_{I^-,AgI|Ag} = -0.1522$ V and $E^o_{Cd^{++}|Cd} = -0.403$ V.

Soln. : $E^o_{Cell} = -0.1522 + 0.403 = 0.2508$ V.

The reaction is $Cd + 2AgI \rightleftharpoons 2Ag + Cd^{++} + 2I^-$

Thus at 25 °C, $E = E^o - \frac{0.059}{2} \log a_{Cd^{++}} \cdot a^2_{I^-}$

or, $0.2860 = 0.2508 - \frac{0.059}{2} \log a_2$

$\therefore \log a_2 = -\dfrac{(0.2860-0.2508)\times 2}{0.059} = -1.1932203.$

$\therefore a_2 = 0.0640884 = 0.0641.$

Now $a_2 = a_{\pm}^3$.

$\therefore a_{\pm} = 0.400184 = 0.40.$

25. The emf measurements of the cell H_2 (1 atm) | HCl (m), AgCl(s) | Ag were made using two HCl solution of diff. molalities with the following results :

m	= 0.0032	0.1238
E (volts)	= 0.5205	0.3420

Assuming infinite dilution for the first solution, find $E°$ of the cell and hence calculate the mean activity coefficient of HCl in the second solution. **[C.U. 1973]**

Soln. : For the passage of 1 faraday, the reaction taking place in the cell is

$$\frac{1}{2}H_2\,(g) + AgCl\,(s) = Ag(s) + H^+ + Cl^-$$

$\therefore E = E° - 0.059 \log a_{H^+}.a_{Cl^-}$ at 25 °C

$= E° - 2 \times 0.059 \log a_{\pm}.$

By the problem, $E° = E + 0.118 \log m$ for 1st solution

$= 0.5205 + 0.118 \log 0.0032$

$= 0.2261$ V.

Thus, for the second solution,

$E = 0.2261 - 0.118 \log m\,\gamma_{\pm}$

or, $0.3420 = 0.2261 - 0.118 \log 0.1238 - 0.118 \log \gamma_{\pm}$.

$\therefore \gamma_{\pm} = 0.841561 \approx$ **0.84.**

26. A hydrogen electrode and a calomel half-cell are used to determine the pH of a solution on a mountain top when the pressure is 500 m of Hg. The pressure was wrongly assumed to be 760 mm of Hg and the pH was found to be 4; what is the correct pH?

Soln. : The cell is H_2 (P) | solution || KCl, Hg_2Cl_2 (s) | Hg

$E = E_2 - E_1$

$$= E_2 - \left(E_1^o + 0.059 \log \frac{a_{H^+}}{P_{H_2}^{1/2}}\right) \text{ at 25 °C}$$

$$= E_{Calomel} + 0.059\text{ pH} + \frac{0.059}{2}\log P_{H_2}.$$

Thus, $\dfrac{E - E_2}{0.059} = 4$ by the problem assuming $P_{H_2} = 1$ atm.

Thus, $\dfrac{E - E_2}{0.059} = \text{pH} + \dfrac{1}{2}\log P_{H_2}.$

$\therefore 4 = \text{pH} + \dfrac{1}{2}\log \dfrac{500}{760}$

or, $pH = 4 - \frac{1}{2}\log\frac{500}{760} = 4.09092$.

$\therefore$ pH $\approx$ **4.09.**

The above is true even at any other temperature.

27. A silver electrode is immersed in 100 cc of 0.1 (M) KCl. What will be the potential of the electrode at 25 °C after addition of (a) 30 cc, (b) 45 cc, (c) 50 cc, (d) 55 cc and (e) 70 cc of 0.2 (M) $AgNO_3$ solution? Given that at 25 °C, $E^o_{Ag^+|Ag} = 0.80$ V and the solubility product of AgCl is 2.3×10^{-10}.

Soln. : Let us consider a cell :

$$Ag \mid Ag^+ \parallel Cl^-, AgCl\,(s) \mid Ag.$$

In this cell the reaction is :

$$AgCl \rightleftharpoons Ag^+ + Cl^-.$$

Therefore, $E = E^o - \frac{RT}{F}\ln a_{Ag^+}.a_{Cl^-}$.

At equilibrium, $E = 0$ and $a_{Ag^+}.a_{Cl^-} = K_s$.

K_s = solubility product of AgCl.

$$\therefore\ E^o = \frac{RT}{F}\ln K_s$$

$$E^o_{Cell} = E^o_{Cl^-,AgCl|Ag} - E^o_{Ag^+|Ag}$$

$$\therefore\ E^o_{Cl^-,AgCl|Ag} = E^o_{Ag^+|Ag} + 0.059\log 2.3 \times 10^{-10} = 0.2313 \text{ V}.$$

Stock solution is 100 cc 0.1 (M) KCl = 10 cc of 1(M) KCl. $\approx$ 50 cc of 0.2(M) KCl.

Different amounts of 0.2 (M) $AgNO_3$ solution are added. After addition of a trace of 0.2 (M) $AgNO_3$, the electrode becomes Cl^-, AgCl | Ag ($AgNO_3$ reacts with KCl giving AgCl), where the reaction is

$$AgCl + e \rightleftharpoons Ag + Cl^-.$$

$$\therefore\ E = E^o + 0.059\log\frac{1}{a_{Cl^-}}$$

$$= 0.2313 - 0.059\log a_{Cl^-}.$$

Here, activity coefficient, as it is not given, is considered as unity.

After addition of 50 cc of 0.2 (M) $AgNO_3$ the electrode becomes Ag^+ | Ag, where the reaction is $Ag^+ + e = Ag$.

Here, $E = 0.80 + 0.059\log C_{Ag^+}$

cc of 0.2(M) $AgNO_3$ added	Total vol.	C_{Cl^-} or C_{Ag^+}	E in volt
30 cc	130 cc	$C_{Cl^-} = \frac{4}{130}$(M)	**0.32 volt**
45 cc	145 cc	$C_{Cl^-} = \frac{1}{145}$ (M)	**0.36 volt**

50 cc	150 cc	$C_{Cl^-} = \sqrt{2.3\times10^{-10}}$ $= 1.516 \times 10^{-5}$ (M)	**0.5 volt**
55 cc	155 cc	$C_{Ag^+} = \frac{1}{155}$ (M)	**0.67 volt**
70 cc	170 cc	$C_{Ag^+} = \frac{4}{170}$ (M)	**0.704 volt**

28. Calculate the potential difference between a standard hydrogen electrode and half-cells comprising platinum foil coated with platinum black in contact with H_2 at a pressure of 1 atm into the following solution at 25 °C.

(i) 10^{-3} (M) $Ba(OH)_2$

(ii) 0.1 (M) acetic acid

(iii) a mixture of 20 cc of 0.25 (M) HOAc and 80 c of 0.1 (M) NaOH.

Soln. : Proceed as Problem 27.

(i) $E = E^o_{H^+/H_2} - E_{OH^-/H_2}$

$= -\left(E^o_{OH^-/H_2} - 0.059 \log a_{OH^-}\right)$

$= -0.059 \log K_w + 0.059 \log \frac{K_w}{a_{H^+}}$

$= -0.059 \log a_{H^+} = 0.059$ pH volt.

(ii) $E = E^o_{H^+/H_2} - E_{H^+/H_2}$

$= -0.059 \log a_{H^+}(0.1M\ CH_3COOH)$

$= -0.059 \log \sqrt{K_a C_{CH_3COOH}}$ volt.

(iii) $E = E^o_{H^+/H_2} - E_{OH^-/H_2}$

$= -0.059 \log a_{H^+} = 0.059$ pH volt (pOH = 1.52)

[Ans. (i) 0.67 volt, (ii) 0.17 volt, (iii) 0.74 V]

29. The emf of the cell Cd | $Cd(NO_3)_2$ (0.05M), KCN(2M) || $Cd(NO_3)_2$(0.19M)| Cd is found to be 0.578 volt.

Calculate the instability constant of the complex $Cd(CN)_4$.

Soln. : It is a concentration cell without transport for the passage of 2 faradays.

Cd^{++} from 0.19 (M) passes to Cd^{++} (0.05) molar which is taken as x, because it has taken part in complex formation $Cd(CN)_4^{=}$.

$$\therefore E = \frac{0.059}{2} \log \frac{C_{Cd^{++}}(0.19M)}{x}.$$

$$\therefore \log \frac{0.19}{x} = \frac{0.578\times2}{0.059}$$

or, $\log \dfrac{0.19}{x} = 19.5932.$

$\therefore \quad \dfrac{0.19}{x} = 3.9194 \times 10^{19}$

$\therefore \quad x = 4.847 \times 10^{-21}.$

The instability constant of $Cd(CN)_4^{=}$ is the equilibrium constant K of the reaction

$$Cd(CN)_4^{=} \rightleftharpoons Cd^{++} + 4CN^{-}.$$

Since, x is very small, almost all of Cd^{++} has been consumed.

Thus, $C_{Cd^{++}}$ remaining $= x$ and $C_{Cd(CN)_4^{=}}$ formed $= (0.05 - x)$

$\therefore \quad C_{CN^-} = 2 - [4 \times \{(0.05) - x\}] = 1.8 + 4x.$

$\therefore \quad K = \dfrac{C_{Cd^{++}} . C^4_{CN^-}}{C_{Cd(CN)_4^{=}}}.$

$\therefore \quad K = \dfrac{x(1.8+4x)^4}{(0.05-x)}.$

Neglecting x in comparison to 0.05 and 1.8.

$\therefore \quad K = \dfrac{4\cdot847\times10^{-21}\times(1\cdot8)^4}{0\cdot05} = \mathbf{1.0176 \times 10^{-18}}.$

30. The solubility of iodine in water at 25 °C is 1.33×10^{-3} moles/litre. What is the conc. of I^- in a saturated solution of I_2 in water at 25 °C, when

$C_{I^-} = 0.500$ (M)

Given, $E^o_{I^-|I_2} = 0.5355$ V

$E^o_{I^-|I_3^-} = 0.5365$ V at 25 °C.

Soln. : In a saturated solution of I_2 in I^-, the reaction is $\underset{0.5}{I_2 + I^-} \rightleftharpoons \underset{x}{I_3^-}$

The reaction takes place in the cell

$$I_3^- \mid I^- \mid I_2(s)$$

Reaction at R.H.E., I_2 (s) + $2e$ = $2I^-$

at L.H.E., $3I^- = I_3^- + 2e$

$I_2 \text{ (s)} + I^- = I_3^-$

$\therefore \quad K = \dfrac{a_{I_3^-}}{a_{I_2} \times a_{I^-}} = \dfrac{a_{I_3^-}}{a_{I^-}} \dfrac{x}{0.5} \quad [\because a_{I_2}(s) = 1]$

$\therefore \quad E_{cell} = E^o_{cell} - \dfrac{RT}{2F} \ln \dfrac{a_{I_3^-}}{a_{I^-}}.$ ...(1)

At equilibrium, $E_{cell} = 0$.

$$\therefore 0 = E^\circ - \frac{RT}{2F}\ln K$$

$$\text{or, } 0 = E^\circ - \frac{0.05916}{2}\log K \qquad (2)$$

$$\text{or, } E^\circ = \frac{0.05916}{2}\log K.$$

$\therefore \log K = -0.033806.$

$\therefore K = 0.92510996.$

$\therefore x = 0.5 \times K = 0.4626$ mole.

If we choose a different cell,

$$I_3^- \mid I^- \mid I_2 \text{ (aq. solution)},$$

then we have to know E° of $I^- | I_2$ (aq.). The reaction in the cell

$$I_2\text{ (aq)} + 2e = 2I^-$$

$$3I^- = I_3^- + 2e$$

At equilibrium, $E^\circ_{I^-|I_2(aq)} - E^\circ_{I^-|I_3^-} = \frac{0.05916}{2}\log K.$

$$I_2\text{ (aq)} + I^- = I_3^-$$

$$0.5 \qquad x$$

Let us consider the cell

$$I_2 \text{ (aq. solution)} \left| I^- \right| I_2 \text{ (s)}.$$

The reaction is

$$\frac{1}{2}I_2\text{ (s)} + e = I^-$$

$$I^- = \frac{1}{2}I_2\text{ (aq.)} + e$$

$$\frac{1}{2}I_2\text{ (s)} = \frac{1}{2}I_2\text{ (aq.)}$$

$$\Delta G = \Delta G^\circ + RT\ln\frac{a^{1/2}_{I_2(aq)}}{a_{I_2(s)}}$$

$$E = E^\circ_{I^-|I_2(s)} - E^\circ_{I^-|I_2(aq)} - \frac{0.05916}{2}\log\frac{a_{I_2(aq)}}{a_{I_2(s)}}.$$

At equilibrium,

$$0 = E^\circ_{I^-|I_2(s)} - E^\circ_{I^-|I_2(aq)} - \frac{0.05916}{2}\log 0.00133.$$

$$\therefore E^\circ_{I^-|I_2(aq)} = 0.5355 - \frac{0.05916}{2}\log 0.00133$$

$$= 0.62057 \text{ volt} = 0.6206 \text{ volt}.$$

Putting the value of eq. (1),

$$\log K = \frac{(0.6206-0.5365)\times 2}{0.05916} = \frac{0.084\times 2}{0.05916} = 2.84313.$$

$\therefore \quad K = 696.84 \approx 697.$

Now, $\quad K = \dfrac{a_{I_3^-}}{a_{I^-}a_{I_2(aq)}} = \dfrac{x}{0\cdot 00133\times 0\cdot 5} = 696.84.$

$\therefore \quad x = 0.5 \times 696.84 \times 0.00133 \approx 0.463$ (M).

If it is assumed that solid I_2 is removed from saturated solution,

$$696.85 = \frac{x}{(0.00133-x)\times 0.5}$$

$\therefore \quad x = \dfrac{696.85\times 0.00133\times 0.5}{696.85\times 0.5+1} = \dfrac{0.463}{349.425} = \mathbf{0.001325\ (M).}$

31. 25 ml of 0.2 (M) $CuSO_4$ is mixed with 25 ml of 1.0 (M) KI. Calculate the concentration of Cu^{++} ion in the resulting solution. (Given, $E^\circ_{Cu^{++}|Cu^+} = 0.17$ V, $E^\circ_{\frac{1}{2}I_2|I^-} = 0.53$ V) and solublity product K_s of CuI = 5×10^{-12}. **[C.U. 1986]**

Soln. : The reaction is $Cu^{++} + 2I^- = CuI + \frac{1}{2}I_2.$ (1)

This reaction takes place in the cell $I_2|I^-, Cu^{++}|CuI$ (s) with electrode reaction

at R.H.E., $Cu^{++} + I^- + e = CuI$

at L.H.E., $I^- = \frac{1}{2}I_2 + e.$

Now, $E_{I^-,Cu^{++}|CuI} = E^\circ_{I^-,Cu^{++}|CuI} + 0.059 \log a_{Cu^{++}}.a_{I^-}$ at 25 °C. (2)

Again, considering the R.H.E as $Cu^{++}|Cu^+$ electrode (CuI is sparingly soluble).

$$E_{I^-,Cu^{++}|CuI} = E^\circ_{Cu^{++}|Cu^+} + 0.059 \log \frac{a_{Cu^{++}}}{a_{Cu^{+1}}} = 0.17 + 0.059 \log \frac{a_{Cu^{++}}}{K_s / a_{I^-}}.$$

By the problem,

$E_{I^-,Cu^{++}|CuI} = (0.17 - 0.059 \log K_s) + 0.059 \log a_{Cu^{++}}.a_{I^-}.$ (3)

Equating eqn. (2) and eqn. (3),

$$E^\circ_{I^-,CuI|Cu^{++}} = 0.17 - 0.059 \log 5 \times 10^{-12} = 0.8367 \text{ volt.}$$

Thus, $E^\circ_{cell} = 0.8367 - 0.53 = 0.3067.$

But, $\Delta G^\circ = -FE^\circ = -RT \ln K$ so that

$$\log K = \frac{0.3067}{0.059} \text{ at } 25°C = 5.199.$$

$\therefore \quad K = 1.58 \times 10^5.$

Since, K for reaction (1) is so high, it may be assumed that Cu^{++}, the ion with lower concentration, is completely consumed. The concentration of I^- will be around $(0.5 - 2 \times 0.1) = 0.3$ (M).

If x be the concentration of Cu^{++} remaining, then by equation (1), the concentration of I^- remaining at equilibrium is $0.3 + 2x$.

$$\therefore\ K = \frac{a_{CuI} \times a_{I_2}^{1/2}}{a_{Cu^{++}} \times a_{I^-}^2} = \frac{1}{C_{Cu^{++}} . C_{I^-}^2}\ [a \approx C].$$

$$\text{Thus, } 1.58 \times 10^5 = \frac{1}{x(0.3+2x)^2}.$$

Neglecting $2x$ in comparison to 0.3,

$$x = \frac{1}{1.58 \times 10^5 \times (0.3)^2} = \mathbf{7.03 \times 10^{-5}\ (M).}$$

32. The solublity of iodine in water at 25 °C is 1.33×10^{-3} moles | litre. Calculate the equilibrium constant of the reaction $I^- + I_2 = I_3^-$ if $E^\circ_{I_3^-/I^-} = 0.534$ V and $E^\circ_{I^-/I_2} = 0.535$ V.

Soln. : The reaction is really $I_2\ (aq) + I^- = I_3^-$.

So, we have to determine E° of $I^- \mid I_2$ (aq.).

Consider the cell I_2 (aq.) | I^- | I_2 (s).

$$\text{The reaction is } \tfrac{1}{2} I_2\ (s) + e \rightleftharpoons I^-$$

$$I^- \rightleftharpoons \tfrac{1}{2} I_2\ (aq.) + e$$

$$\tfrac{1}{2} I_2\ (s) = \tfrac{1}{2} I_2\ (aq.)$$

$$\therefore\ E = E^\circ - \frac{0.05916}{2} \log \frac{a_{I_2(aq.)}}{a_{I_2(s)}}.$$

At equilibrium,

$$0 = E^\circ_{I^-|I_2(s)} - E^\circ_{I^-|I_2(aq)} - \frac{0.05916}{2} \log 0.00133.$$

$$\therefore\ E^\circ_{I^-|I_2(aq)} = 0.535 - \frac{0.05916}{2} \log 0.00133 = 0.62 \text{ volt.}$$

The reaction takes place in the cell $I_3^-|I^-|I_2(aq.)$

$E^\circ = (0.62 - 0.534)$ volt $= 0.086$ volt.

$\therefore\ RT \ln K = nFE^\circ = 2 \times 96484 \times 0.086.$

$$\therefore\ \ln K = \frac{2 \times 96484 \times 0.086}{8.314 \times 298.15}.$$

$\therefore\ \ln K = 6.69482,$

$\therefore\ K = \mathbf{808.2.}$

33. Given that, I_2 (aq.) + $2e$ = $2I^-$.

$E° = 0.6197$ V and $I_3^- + 2e = 3I^-$, $E° = 0.5355$ V.

Calculate K for the reaction I_2 (aq) + I^- = I_3^-.

Soln. : Proceed as Problem no. 32. **[Ans. K = 708]**

34. Calculate the emf of the cell :

Pt, Pb(Hg) | $PbSO_4$(s),$CuSO_4$ (a_1); $CuSO_4$ (a_2), $PbSO_4$(s), | Pb(Hg), Pt

where the molarity of the solution

$a_1 = 0.2$ (M), $\gamma_\pm = 0.11$ and $a_2 = 0.02$ (M), $\gamma_\pm = 0.32$.

$(t_+)_{Cu^{2+}} = 0.37$, which is the negative electrode.

Soln. : R.H.E. : The reaction is—

$$PbSO_4 + 2e = Pb + SO_4^{2-}\ (a_2)$$

Net change for the passage of one faraday :

Gain of SO_4^{2-} : $\frac{1}{2}$ gm-mole

Loss of SO_4^{2-} : $\frac{t_-}{2}$ gm-mole.

Net gain of $SO_4^{2-}\left(\frac{1}{2}-\frac{t_-}{2}\right) = \frac{t_+}{2}$ gm-mole.

R.H.E. : Net gain of Cu^{2+} = $\frac{t_+}{2}$ gm-mole of $CuSO_4$ (a_2)

L.H.E. : The reaction is Pb + SO_4^{2-} = $PbSO_4$ + $2e$

Loss of SO_4^{2-} : $\frac{1}{2}$ gm-mole

Gain of SO_4^{2-} : $\frac{t_-}{2}$ gm-mole

Net loss of SO_4^{2-} : $\frac{t_+}{2}$ gm-mole

Net loss of Cu^{2+} : $\frac{t_+}{2}$ gm-mole of $CuSO_4(a_1)$.

So for the passage of one faraday, $\frac{t_+}{2}$ gm-mole of $CuSO_4$ passes from a_1 to a_2.

$$\therefore\ \Delta G = \Delta G_{Cu^{2+}} + \Delta G_{SO_4^{2-}}$$

$$= \int_1^2 \frac{t_+}{2} RTd\ln a_{Cu^{2+}} + \int_1^2 \frac{t_+}{2} RTd\ln a_{SO_4^{2-}}$$

$$= \frac{t_+}{2} RT \ln \frac{\left(a_{Cu^{+2}}\right)_2}{\left(a_{Cu^{+2}}\right)_1} + \frac{t_+}{2} RT \ln \frac{\left(a_{SO_4^{-2}}\right)_2}{\left(a_{SO_4^{-2}}\right)_1}$$

$$= \frac{t_+}{2} RT \ln \frac{a_2}{a_1} + \frac{t_+}{2} RT \ln \frac{a_2}{a_1}$$

$$= t_+ RT \ln \frac{a_2}{a_1}.$$

$$a_1 = C_1\gamma_\pm, \qquad a_2 = C_2\gamma_\pm$$

$$\Delta G = t_+ RT \ln \frac{a_2}{a_1}$$

$$= t_+ RT \ln \frac{C_2\gamma_\pm}{C_1\gamma_\pm} = t_+ RT \ln \frac{C_2 \times 0.32}{C_1 \times 0.11}$$

$$= t_+ RT \ln \frac{0.02 \times 0.32}{0.2 \times 0.11}.$$

$$E = -\frac{\Delta G}{nF} = -\frac{RT}{nF} t_+ \ln \frac{0.02 \times 0.32}{0.2 \times 0.11}$$

$= 0.0117$ V (where $n = 1$, $t_+ = 0.37$)

$=$ **0.012 V.**

Since the emf is positive, the L.H.E. is negative electrode.

35. ΔG° for the reaction H_2 (g, 1 atm) + $\frac{1}{2}O_2$ (g, 1 atm) = H_2O (l) is –237.2 kJ at 298 K. Calculate the $E^\circ_{OH^-|O_2(g)}$. The cell potential producing the reaction,

H_2O (l) = H^+ ($a_{H^+} = 1$) + OH^- ($a_{OH^-} = 1$), is – 0.828 V. (i)

Soln. : The reaction $H_2 + \frac{1}{2}O_2 = H_2O$ (l) takes place in the cell.

Pt, H_2 (g, 1 atm) | OH^- | O_2 (g, 1 atm), Pt. Let the standard potential be E°. Then E° is :

$$E^\circ = \frac{237200}{2 \times 96484} = 1.2292193 \text{ V}.$$

E° of the cell is :

$$E^\circ = E^\circ_{OH^-|O_2} - E^\circ_{OH^-|H_2}$$

$\therefore$ 1.229 (V) $= E^\circ_{OH^-|O_2} - E^\circ_{OH^-|H_2}$. (ii)

The cell producing the reaction, H_2O (l) = $H^+ + OH^-$, is $H_2 | H^+ \| OH^- | H_2$.
But the standard cell potential of the cell $H_2 | H^+ \| OH^- | H_2$ is = – 0.828 V.

$E^\circ_{OH^-|H_2} = -0.828$ (since $E^\circ_{H^+|H_2} = 0$)

Putting in (ii), $1.229 = E^\circ_{OH^-|O_2} - (-0.828 \text{ V}) = E^\circ_{OH^-|O_2} + 0.828$ V.

$\therefore$ $E^\circ_{OH^-|O_2} = 1.229 - 0.828 =$ **0.401 V.**

36. Evaluate the emf of the cell at 298 K neglecting the LJP

Pt, H_2 (g, 1 atm)|CH_3COOH(1M) || HCOOH(0.5M)| O_2 (g,1 atm), Pt

Given, $E^o_{OH^-|O_2} = 0.401$ V

$K_w = 1.008 \times 10^{-14}$

$(K_a)_{CH_3COOH} = 1.85 \times 10^{-5}$

$(K_a)_{HCOOH} = 17.7 \times 10^{-5}$.

Soln. : The emf of the cell $E = E_2 - E_1$, where E_2 is the potential of R.H.E. and E_1 is the potential of L.H.E. as R.H.E.

Reaction at $OH^- | O_2$ electrode.

$$\frac{1}{2}O_2 + H_2O + 2e = 2OH^-$$

$$E_{OH^-|O_2} = E^o_{OH^-|O_2} - \frac{RT}{nF}\ln a^2_{OH^-} = E^o_{OH^-|O_2} - 0.059 \log \frac{K_w}{a_{H^+}}$$

Hence, $E^o_{H^+|O_2} = E^o_{OH^-|O_2} - 0.059 \log K_w$ (since, $a_{OH^-} = \frac{K_w}{a_{H^+}}$)

As $H^+ | O_2$ electrode reaction at R.H.E.,

$$\frac{1}{2}O_2 + 2H^+ + 2e = H_2O$$

$$E_2 = E^o_{H^+|O_2} + \frac{0.059}{2}\log a^2_{H^+} \text{ at 25 °C and 1 atm}$$

$$= E^o_{H^+|O_2} + 0.059 \log a_{H^+}$$

$$= E^o_{OH^-|O_2} - 0.059 \log K_w + 0.059 \log a_{H^+}.$$

Since $\gamma_\pm$ is not known, the concentration in molarity is used for activity.

$\therefore E_2 = 0.401 - 0.059 \log K_w + 0.059 \log (C_{H^+})$

$= 0.401 - (-0.82579) + 0.059 \log C_{H^+}$

$= 1.22679 + 0.059 \log \sqrt{C_{HCOOH}(K_a)_{HCOOH}}$

$C_{HCOOH}\,(K_a)_{HCOOH} = 0.5(M) \times 17.7 \times 10^{-5} = 8.85 \times 10^{-5}(M)$

$= 1.22679 + 0.059 \log \sqrt{8.85 \times 10^{-5}}$

$= 1.22679 - 0.119565 = 1.107$ (V).

For L.H.E. the reaction is H^+ (CH_3COOH) + e = $\frac{1}{2}H_2$ (1 atm).

$E_1 = E^o + 0.059 \log a_{H^+}$

$= 0 + 0.059 \log C_{H^+}$.

For 1 mole CH_3COOH,

$$C_{H^+} = \sqrt{C_{CH_3COOH}(K_a)_{CH_3COOH}} = \sqrt{1\times1.85\times10^{-5}}$$

$$E_1 = 0.059 \log C_{H^+} = -0.1396.$$

$$\therefore\ E = E_2 - E_1 = 1.107 + 0.1396 = \mathbf{1.25\ V.}$$

37. During discharge of a lead storage battery the density of H_2SO_4 falls from 1.295 gm/cc to 1.140 gm/cc. H_2SO_4 of density 1.295 gm.cc^{-1} is 39.2% H_2SO_4 by weight and H_2SO_4 of density 1.140 gm/cc is 20.8% H_2SO_4 by weight. The battery holds 3.6 litres of acid and the volume remains practically constant during discharge. Calculate the number of ampere-hour for which the battery must be used.

Soln. : The net reaction of discharge for the passage of one faraday is

$Pb(s) + PbO_2(s) + 2H_2SO_4 = 2PbSO_4(s) + 2H_2O\ (l).$

As per reaction the equivalent weight = 98 (molecular weight).

Number of equivalent of H_2SO_4 present initially,

$$3600 \text{ ml} \times 1.295 \text{ gm/cc} \times \frac{39.2}{100}\times\frac{1}{98} = 18.648 \text{ equivalent.}$$

Number of equivalent present after discharging,

$$3600 \times 1.14 \times \frac{20.8}{100}\times\frac{1}{98} = 8.7105.$$

Number of equivalent loss = 9.9375 eq.

Quantity of charges = 9.9375 × 96484 = 958809.75 coulombs

= 958809.75 ampere-sec.

So the required ampere-hour for which battery must be used

$$= \frac{958809.75}{3600} = \mathbf{266.34\ ampere\text{-}hour.}$$

38. A current of 4.2 micro-ampere is passed through the solution of $AgNO_3$ for 30 minutes using Pt electrode. A uniforn single large Ag atom deposited carries 45% of cathode surface. What is the total surface area of cathode? It is known that one Ag atom covers 5.404×10^{-16} cm^2.

Soln. : Total charges pass = $4.2 \times 10^{-6} \times 30 \times 60 = 7.56 \times 10^{-3}$ C = 0.0756 C.

$$\text{Weight of Ag deposited at cathode} = \frac{108\times0.0756}{96484} = 8.4629 \times 10^{-5}.$$

$$\text{Total number of Ag atoms} = \frac{8.46\times10^{-5}}{108} \times 6.022 \times 10^{23} = 4.718 \times 10^{17}.$$

So surface area at cathode covers, $4.718 \times 10^{17} \times 5.404 \times 10^{-16}$

$= 254.9607$ cm^2

$$\therefore \text{ total surface of the cathode} = \frac{254.982\times100}{45} = \mathbf{566.58\ cm^2.}$$

39. For the reaction :

$$H_2\text{ (g)} + O_2\text{ (g)} \longrightarrow H_2O\text{ (l) at } 25\text{ °C}$$

$$\Delta G° = -237.2\text{ kJ} \quad \text{and } \Delta H° = -285.9\text{ kJ.}$$

(i) Devise a cell in which above reaction would take place.

(ii) What is $E°$ of the cell?

(iii) At what rate does $E°$ change with T around 25 °C?

[M.Sc. Admission Test, I.I.T. Kharagpur 1996]

Soln. : The cell is : Pt, H_2 (g, 1 atm) | NaOH | O_2 (g, 1 atm), Pt

$$\Delta G° = -237.2\text{ kJ} = -237200\text{ J.}$$

$$\therefore \Delta H° = -285.9\text{ kJ} = -285900\text{ J.}$$

$$E° = -\frac{\Delta G°}{nF} = -\frac{237200}{2\times 96484} = 1.2292\text{ V} = \mathbf{1.23\text{ V.}}$$

$$\Delta H° = nF\left\{T\left(\frac{\partial E°}{\partial T}\right)_P - E°\right\}$$

$$-285900 = 2 \times 96484 \left\{298.15\left(\frac{\partial E°}{\partial T}\right)_P - 1.229\right\}$$

$$\therefore \quad 2 \times 96484 \times 298.15\left(\frac{\partial E°}{\partial T}\right)_P = -285900 + 1.229 \times 2 \times 96484$$

$$\therefore \quad \left(\frac{\partial E°}{\partial T}\right)_P = \mathbf{-8.472 \times 10^{-4}}.$$

40. Devise a cell where the net reaction is $Zn + Cu^{2+} \rightleftharpoons Zn^{2+} + Cu$.

Standard reduction potential ($E°$) values for $Zn^{2+}|Zn$ and $Cu^{2+}|Cu$ electrodes are -0.76 V and 0.34 V respectively. Calculate the equilibrium constant of the reaction. **[V.U. 2005]**

Soln. : The cell is Zn | Zn^{2+} || Cu^{2+} |Cu.

For reaction balancing two electrons are needed, that is $n = 2$.

$\therefore -2 \times F \times E° = -RT \ln K.$

$E° = \{0.34 - (-0.76)\}$ volt $= 1.10$ volt.

Then proceed as earlier.

Equilibrium constant $K = \mathbf{1.546 \times 10^{37}}$.

41. Calculate the free energy change and the equilibrium constant of the reaction that takes place in the following cell at 27 °C :

$$Cu \mid Cu^{++}\ (a = 0.01) \mid\mid Fe^{2+}\ (a = 0.001),\ Fe^{3+}\ (a = 0.01) \mid Pt$$

Given, $E°_{Cu^{2+}|Cu} = 0.34$ volt, $E°_{Fe^{3+}|Fe^{2+}} = 0.77$ volt. **[V.U. 2004]**

Soln. : Proceed as earlier. For the cell, $E° = 0.43$ volt. So, $K = \mathbf{2.7589 \times 10^{14}}$ and $\Delta G = \mathbf{-105.94\text{ kJ.}}$

42. Write cell reactions for the following electrochemical cell :

$$Fe, Fe^{2+}\ (0.1\ M)\ ||\ Cd^{2+}\ (0.001\ M), Cd$$

assuming that activities equal concentration. Also (i) calculate the emf of the cell and determine the direction of the spontaneous reaction, and (ii) calculate the equilibrium constant of the cell reaction. Given, $E^{o}_{Cd^{2+}|Cd} = -0.40$ V and $E^{o}_{Fe^{2+}|Fe} = -0.44$V.

[V.U. 2003]

Soln. : The reaction for the given cell,

$$Cd^{+2} + 2e = Cd$$

$$Fe = Fe^{+2} + 2e$$

$$Cd^{+2} + Fe = Cd + Fe^{+2}$$

$$-nFE^{o} = -RT \ln K$$

$$\therefore \ln K = \frac{2 \times 96484 \times 0.04}{8.314 \times 298.15} \quad (\text{since, } n = 2)$$

$$K = \mathbf{22.51.}$$

$$\text{At } 25\ °C,\ E = 0.04\ V - \frac{0.059}{2} \log \frac{a_{Fe^{2}}}{a_{Cd^{2+}}}$$

$$= 0.04 - \frac{0.059}{2} \log \frac{0.1}{0.001}$$

$$= 0.04 - 0.0295 \log 100$$

$$= 0.04 - 0.059 = \mathbf{-0.019\ volt.}$$

Hence, for the given cell, L.H.E. is not the negative electrode. Current is passing from R.H.E. to L.H.E., i.e., R.H.E. is negative electrode and should be written as L.H.E.

43. The measure emfs are 0.06839 V at 25 °C and 0.07048 V at 30 °C for the following cell Ag, AgBr (s) | KBr solution | Hg_2Br_2(s) | Hg. Calculate, after stating the reaction involved the values of free energy change and entropy change of the process at 25 °C.

[M.Sc. Admission Test 1999, I.I.T. Khargapur]

Soln. : By the problem,

$$\left(\frac{\partial E}{\partial T}\right)_P = \frac{E_{30°C} - E_{25°C}}{5}$$

$$= 4.18 \times 10^{-4}\ \text{volt/K.}$$

The reaction takes place in the cell,

$$Hg_2Br_2 + 2Ag = 2Hg + 2AgBr$$

$$\Delta G^{o} = -nFE^{o} = \mathbf{-13.2\ kJ.}$$

$$\Delta S^{o} = 2 \times F \times \left(\frac{\partial E}{\partial T}\right)_P = \mathbf{80.66\ J.K^{-1}.}$$

44. The emf of a galvanic cell :

H_2 | aq. solution (pH = x) || IN calomel is 0.5164 V at 25 °C, when the barometric pressure is 754.1 mm of Hg. Given that electrode potential of the calomel electrode is 0.28 volt and the vapour pressure of water is 23.8 mm Hg, calculate the value of x, i.e., pH of the aqueous solution. **[M.Sc. Admission Test 1993, I.I.T. Bombay]**

Soln. : For the L.H.E. the reaction as R.H.E. is $H^+ + e = \frac{1}{2}H_2$.

$$E = E^\circ + \frac{RT}{F}\ln\frac{a_{H^+}}{\sqrt{(P_{H_2})}}.$$

$$\therefore\ E = 0.059\log a_{H^+} - \frac{1}{2}0.059\log P_{H_2},\ \text{since } E^\circ = 0,$$

$$= -0.059\ \text{pH} - \frac{1}{2}0.059\log P_{H_2}.$$

By the problem,

$$P_{H_2} = \frac{754.1-23.8}{760}\ \text{atm.} = 0.96\ \text{atm.}$$

$$E = E_{cal} - E_{H^+/H_2}.$$

$$\therefore\ 0.5164 = 0.28 + 0.059\ \text{pH} + \frac{1}{2}0.059\log 0.96.$$

$$\therefore\ \text{pH} = 4.015 = \mathbf{4.02}.$$

45. Write the cell reaction and calculate the emf of the cell

Cu|Cu^{2+} (4M) || Ag^+ (0.1 M) | Ag at 25 °C from the following data :

$E^\circ_{Cu^{2+}|Cu} = 0.34$ V, $E^\circ_{Ag^+|Ag} = 0.80$ V. **[M.Sc. Admission Test 1993, I.I.T. Bombay]**

Soln. : Proceed as earlier.

$E_{cell} = 0.38$ volt

[Use concentration for activities.]

46. Calculate the solubility product of $PbCl_2$ at 25 °C.

E° for $PbCl_2$(s) | Pb(s) = – 0.268 V and E° for Pb^{2+} | Pb (s) = – 0.126 V.

[M.Sc. Admission Test 1994, I.I.T. Bombay]

Soln. : We know that for the reaction $PbCl_2 = Pb^{+2} + 2Cl^-$, the cell is

Pt, Pb (Hg) | Pb^{2+} || Cl^-, $PbCl_2$(s) | Pb(Hg), Pt.

For the given cell,

$$E^\circ = E^\circ_{Cl^-,PbCl_2|Pb} - E^\circ_{Pb^{2+}|Pb} = -0.268 - (-0.126)\ \text{V} = -0{\cdot}142\ \text{V}.$$

$$\therefore\ -0.142 = \frac{0.059}{2}\log K_s$$

$$\therefore\ \log K_s = \frac{-0.142\times 2}{0.059}$$

$$\therefore\ \log K_s = -4.813,\quad \therefore\ K_s = 1.53\times 10^{-5}.$$

47. The potential (E) of the cell Pt | H_2 (g, 1 atm) | NaOH (aq. 0.1 mol/kg), NaCl (aq. 0.01125 mol/kg | AgCl (s) | Ag at 25 °C was found to be 1.04864 V. Calculate pH and the standard free energy change of autoprotolysis of water at 25 °C. Assume activity coefficients to be equal to unity. Given, $E^{\circ}_{AgCl|Ag} = 0.2223$ V.

[M.Sc. Admission Test 1996, I.I.T. Bombay]

Soln. : The cell given is :

Pt, H_2(g, 1 atm) | NaOH (0.1 molal), NaCl (0.01125 molal), AgCl | Ag.

$$E_{Cell} = E_{R.H.E.} - E_{L.H.E.}$$

$$= \left(E^{\circ}_{Cl^-.AgCl|Ag} - \frac{RT}{F}\ln a_{Cl^-}\right) - \left(E^{\circ}_{H^+|H_2} + 0.059\log a_{H^+}\right)$$

$$= (0.2223 - 0.059 \log 0.01125) - (0 - 0.059 \text{ pH}).$$

$\therefore$ $E_{cell} = 0.3373 + 0.059$ pH

$\therefore$ $1.04864 = 0.3373 + 0.059$ pH

$\therefore$ pH $=$ **12.057.**

$\therefore$ $C_{H^+} = 8.77 \times 10^{-13}$, $\therefore$ $K_w = C_{H^+}.C_{OH^-} = 8.77 \times 10^{-14}$.

$\Delta G^{\circ} = -RT \ln K_w = -8.314 \times 298.15 \ln (8.77 \times 10^{-14})$

$= 74525$ joule $=$ **74.52 kJ.**

48. The standard potentials of Cu^{2+} | Cu and Cu^+ | Cu electrodes are 0.34 volt and 0.52 volt respectively. Calculate the standard potential for Cu^{2+} electrode.

[M.Sc. Admission Test 1991, I.I.T. Kharagpur]

Soln. : Given,

	E°	ΔG°
(i) $Cu^{2+} + 2e = Cu$	0.34 V	$-2F \times 0.34$
(ii) $Cu^+ + e = Cu$	0.52 V	$-F \times 0.52$
(iii) $Cu^{2+} + e = Cu^+$	E°	$-FE^{\circ}_{cell}$

Reaction (iii) is obtained by subtracting (2) from (1).

[The cell is Cu | Cu^+ || Cu^{+2} | Cu^+]

$\therefore$ $\Delta G^{\circ}_3 = \Delta G^{\circ}_1 - \Delta G^{\circ}_2$

$-FE^{\circ} = -0.16\ F$

$\therefore$ $E^{\circ} =$ **+ 0.16 volt.**

49. The equilibrium constant for the reaction $2Cu^+ = Cu^{2+} + Cu(s)$ at 25 °C is 1.646×10^6. (i) Construct a cell in which the reaction could occur, (ii) Calculate the standard emf for the cell at 25 °C. (iii) Evaluate the standard oxidation potential of the Cu | Cu^+ electrode at 25 °C | The standard potential of Cu^+ | Cu^{+2} electrode = – 0.153 V at 25 °C]

[C.U. 2004]

Soln. : The reaction, $2Cu^+ = Cu^{2+} + Cu$ takes place in the cell.

Pt, Cu^+ | Cu^{2+} || Cu^+ | Cu.

ΔG° for the cell is $-2 \times F \times E^{\circ} = -RT \ln K$.

$$\therefore\ E^{\circ} = \frac{RT}{F}\ln(1.646 \times 10^6) = \frac{0.059}{1}\log K = 0.366 \text{ V}.$$

$$E^{o}_{Cell} = E^{o}_{Cu^{+}|Cu} - E^{o}_{Cu^{2+}|Cu^{+}}$$

$$0.366 = E^{o}_{Cu^{+}|Cu} - (-0.153)$$

$$\therefore \; E^{o}_{Cu^{+}|Cu} = (0.366 - 0.153)V = \mathbf{0.213\ V.}$$

Therefore, standard oxidation potential of Cu | Cu^{+} = – 0.213 V.

Oxidation potential is negative of reduction potential.

50. $\text{Pt, } H_2 \text{ (g, 1 atm)} \left| \begin{array}{c} HA_2(pK_a = 6) \\ C_1 \text{ at } 25°C \end{array} \right\| \begin{array}{c} HA_1(pK_a = 4) \\ C_1 \text{ at } 25°C \end{array} \left| H_2(\text{g,1 atm}) \right| \text{Pt}$

Calculate the emf of the above cell at 25 °C. What are the assumptions which must be made in order to solve the problem. **[C.U. 2005]**

Soln. : The reaction taking place in the cell for the passage of IF :

At the R.H.E., H^{+} $(pK_{a_1} = 4) + e = H_2$

At the L.H.E., $H_2 = H^{+}$ $(pK_{a_2} = 6) + e$

Adding, H^{+} $(pK_{a_1} = 4) \longrightarrow H^{+}$ $(pK_{a_2} = 6)$

$$\Delta G = RT \ln \frac{a_{H^{+}}(pK_{a_2} = 6)}{a_{H^{+}}(pK_{a_1} = 4)}$$

$$\therefore \; E = \ln \frac{a_{H^{+}}(pK_{a_1} = 4)}{a_{H^{+}}(pK_{a_2} = 6)}.$$

Assuming that $\gamma_{\pm} = 1$, $a_{H^{+}} = C_{H^{+}}$,

$$E = \frac{RT}{F} \ln. \frac{C_{H^{+}}(pK_{a_1} = 4)}{C_{H^{+}}(pK_{a_2} = 6)}$$

At 25 °C,

$$E = 0.059 \log \frac{C_{H^{+}}(pK_{a_1} = 4)}{C_{H^{+}}(pK_{a_2} = 6)}$$

$$= 0.059 \log C_{H^{+}}(pK_{a_1} = 4) - 0.059 \log C_{H^{+}}(pK_{a_2} = 6)$$

$$= -0.059 \text{ pH } (HA_1) + 0.059 \text{ pH } (HA_2).$$

Any weak acid *HA* dissociates as

$HA \rightleftharpoons H^{+} + A^{-}$

At equilibrium, $\quad HA \rightleftharpoons H^{+} + A^{-}$

$\qquad\qquad\qquad C(1-\alpha) \quad \alpha C \quad \alpha C$

If the initial concentration is *C* and α is the degree of dissociation, then

$$K_a = \frac{C_{H^{+}} C_{A^{-}}}{C_{HA}} = \frac{\alpha^2 C}{(1-\alpha)}.$$

Neglecting α in comparison to 1,

$$K_a = \alpha^2 C.$$

$$\therefore\ C_{H^+} = \alpha C = \sqrt{C.K_a}\,.$$

$$\therefore\ pH = \frac{1}{2}pK_a - \frac{1}{2}\log C.$$

Neglecting α in comparison to 1,

$$K_a = \alpha^2 C.$$

$$\therefore\ C_{H^+} = \alpha C = \sqrt{C.K_a}\,.$$

$$\therefore\ pH = \frac{1}{2}pK_a - \frac{1}{2}\log C.$$

∴ by the problem,

$$E = -0.059\left(\frac{1}{2}\times 4 - \frac{1}{2}\log C_1\right) + 0.059\left(\frac{1}{2}\times 6 - \frac{1}{2}\log C_1\right) = .059(3-2) = \mathbf{0.059V.}$$

The assumptions made up :

(1) There is no LJP.

(2) Degree of dissociation of weak acid is negligible in comparison to 1.

(3) The addition or removal of a finite no. of gm mole does not alter the concentration.

(4) Activity coefficient is unity.

51. Calculate the emf, the free energy change and the equilibrium constant of the reaction involved for the following cell at 25 °C.

Cu | Cu^{+2} (a = 0.01) || Fe^{+2} (a = 0.001), Fe^{+3} (a = 0.01) | Pt.

Given, $E^o_{Cu^{++},Cu}$ = 0.34 V, $E^o_{Fe^{+3},Fe^{+2}}$ = 0.77V. **[V.U. 2007]**

Soln. : Same as Problem 41.

[**Ans.** E = 0.548 V, ΔG = −105.75 kJ, $K = 3.45 \times 10^{14}$]

52. For the following standard cell, Cd(Hg) | $CdSO_4$, H_2O(s) Hg_2SO_4(s) | Hg, the emf is given as a function of temperature t in °C by the following expression :

$$E = 1.01845 - (4.05 \times 10^{-5})\ (t - 25) - (9.5 \times 10^{-7})\ \ (t - 20)^2.$$

Write down the cell reaction at each electrode. Calculate ΔG, ΔH and ΔS for cell reaction at 25 °C.

Soln. : Reaction at R.H.E. : $Hg_2SO_4(s) + 2e = Hg + SO_4^{-2}$.

Reaction at L.H.E. : $Cd = Cd^{+2} + 2e$.

Hence, reaction in the cell is $Hg_2SO_4(s) + Cd = Hg + Cd^{+2} + SO_4^{-2}$.

Proceed as Problem 6, $\left(\frac{\partial E}{\partial T}\right)_P = -5 \times 10^{-5} V.K^{-1}$ at 25 °C.

E = 1.01843 V at 25 °C

[**Ans.** ΔG = −196.52 kJ, ΔH = −199.40 kJ, ΔS = −9.65 $J.K^{-1}$]

53. Calculate the emf of the following cell at 298 K :

H_2(Pt) 1 atm | HCl (a = 0.1) | H_2(Pt) 0.1 atm. **[B.U. 2004]**

Soln. : $H^+ \ (a = 0.1) + e = \frac{1}{2} H_2 \ (0.1 \text{ atm})$

$$\frac{1}{2}H_2 \ (1 \text{ atm}) \quad = H^+ \ (a = 0.1) + e$$

Adding, $\frac{1}{2}H_2 \ (1 \text{ atm}) \quad = \frac{1}{2}H_2 \ (0.1 \text{ atm})$

This is a concentration cell.

$$\text{Thus, } E = -\frac{RT}{F}\ln\frac{P_{H_2}^{1/2}(0.1\,\text{atm})}{P_{H_2}^{1/2}(1\,\text{atm})} \ (\text{since, } E^\circ_{H^+/H_2} = 0)$$

$= \mathbf{0.0295 \ V.}$

54. The emf of the following cell is 1.71 volts at 25 °C, Zn | Zn^{++}(a = 0.01) || Fe^{++} (a = 0.01), Fe^{+3} (a = 0.01) | Pt.

Calculate the equilibrium constant of the reaction :

Zn + 2Fe^{+3} = Zn^{++} + 2Fe^{+2} at 25 °C. **[B.U. 1978]**

Soln. : The reaction for the cell :

Zn | Zn^{++} (a = 0.01) || Fe^{++} (a = 0.001), Fe^{+3} (a = 0.01) | Pt is

Zn + 2Fe^{+3} = Zn^{++} + 2Fe^{+2}.

$$\text{Hence, } E = E^\circ - \frac{RT}{2F}\ \ln\ \frac{a_{Zn^{++}}\left(a_{Fe^{+2}}\right)^2}{\left(a_{Fe^{+3}}\right)^2 a_{Zn}} \quad (n = 2).$$

$$\text{or, } 1.71 = E^\circ - \frac{0.059}{2}\ \log\ \frac{0.01\times(0.001)^2}{(0.01)^2} \quad (a_{Zn} = 1).$$

$E^\circ \quad = 1.592$ V.

Proceed as Problem 14. **[Ans.** $K = 6.652 \times 10^{53}$**]**

55. A silver electrode is dipped into a 0.1 molal KCl solution saturated with AgCl. The emf of the cell formed by combining this electrode with 0.1 (N) calomel electrode is found to be 0.0494 volt at 25 °C. Calculate the solubility product of AgCl at 25 °C. (Potential of 0.1(N) calomel electrode for oxidation is – 0.3338 volt, the standard oxidation potential of the silver electrode is – 0.799 volt at 25 °C. The activity of chloride ion in the solution is 0.077) **[B.U. 1979]**

Soln. : For the cell Ag | AgCl, KCl || 0.1(N) calomel,

E = 0.0494 volt at 25 °C.

E_{red} for 0.1 (N) calomel electrode = 0.3338.

$E^\circ_{Ag^+|Ag} \quad = 0.799.$

Now, $E_{cell} \quad = E_{caiomel}$ (1)

AgCl $\quad = Ag^+ + Cl^-$ (2)

$Ag^+ + e = Ag$ (3)

Adding (1) and (2), $AgCl + e = Ag + Cl^-$

$$E^o_{Cl^-,AgCl|Ag} = E^o_{Cl^-,AgCl|Ag} - 0.059 \log a_{Cl^-}$$

$$= E^o_{Ag^+|Ag} + 0.059 \log a_{Ag^+} \text{ (disregarding eqn. 2)}$$

$$= E^o_{Ag^+|Ag} + 0.059 \log \frac{K_s}{a_{Cl^-}} \text{ (}K_s = \text{solubility product of AgCl).}$$

Hence, $E^o_{Cl^-,AgCl|Ag} = E^o_{Ag^+|Ag} + 0.059 \log K_s$.

Thus, from equation (1),

$$0.0494 = 0.3338 - (E^o_{Ag^+|Ag} + 0.059 \log K_s - 0.059 \log a_{Cl^-})$$

$$= 0.3338 - 0.799 - 0.059 \log K_s + 0.059 \log 0.077.$$

Hence, K_s = **1.46 × 10⁻¹⁰**.

56. A solution of Fe^{+2} is titrated potentiometrically using Ce^{+4} solution. Calculate the emf of the redox electrode thus formed when (i) 50% of Fe^{+2} is titrated, (ii) 90% of Fe^{+2} is titrated and (iii) 110% titration is done.

Given, $Fe^{+2} \rightarrow Fe^{+3} + e$ $\quad E^\circ = -0.77$ volts

and $Fe^{+2} + Ce^{+4} \rightarrow Fe^{+3} + Ce^{+3}$ $\quad K = 10^{14}$ **[B.U. 1986]**

Soln. : Given, $E^o_{Fe^{+3}|Fe^{+2}} = 0.77$ volts.

For 50% and 90% titration the electrode is $Fe^{+3}|Fe^{+2}$.

(i) At 50% titration $E = E^o_{Fe^{+3}|Fe^{+2}} - \frac{RT}{F} \ln \frac{a_{Fe^{+2}}}{a_{Fe^{+3}}}$

$$= E^o_{Fe^{+3}|Fe^{+2}} \text{ (since } a_{Fe^{+2}} = a_{Fe^{+3}})$$

= **0.77 V.**

(ii) At 90% titration, $E = 0.77 - 0.059 \log$ = **0.82 V.**

(iii) At 100% titration, equilibrium is attained.

At equilibrium, $E = 0 = E^\circ - 0.059 \log K$.

or, $E^o_{Ce^{+4}/Ce^{+3}} = E^o_{Ce^{+3}/Ce^{+2}} + 0.059 \log K$ (the cell is $Fe^{+2} \mid Fe^{+3} \parallel Ce^{+4} \mid Ce^{+3}$)

= 1.596 volt.

At 110% titration,

$$E_{Ce^{+4}|Ce^{+3}} = E^\circ_{Ce^{+4}/Ce^{+3}} - 0.059 \log \frac{a_{Ce^{+3}}}{a_{Ce^{+4}}}$$

$$= 1.596 - 0.059 \log \frac{100}{10}$$

= **1.537 volt.**

57. Two hydrogen electrodes at one atmosphere pressure each are inserted into 0.01 (N) NaOH and 0.01 (N) NH_4OH solution respectively. They are connected internally by a salt bridge. If the teperature be 25 °C, what will be the emf of the cell? [Given, dissociation constant of $NH_4OH = 1.85 \times 10^{-5}$ and $R = 8.31$ volt.coulomb/deg]. Which electrode does act as positive terminal of the cell? **[B.U. 1989]**

Soln. : Let the cell be :

H_2 (1 atm) | NaOH (0.01N) || NH_4OH (0.01N) | H_2 (1 atm)

Reaction occuring in the cell,

OH^-(0.01 N NaOH) = OH^- (0.01 N NH_4OH).

$$E_{cell} = \frac{RT}{F} \ln \frac{a_{OH^-(0.01\,N\,NaOH)}}{a_{OH^-(0.01\,N\,NH_4OH)}}$$

$$= 0.059 \log \frac{0.01}{\sqrt{1.85\times10^{-5}\times0.01}} = 0.081 \text{ V.}$$

Hence, NH_4OH | H_2 acts as R.H.E. and the other as L.H.E.

58. A saturated calomel electrode is coupled through a salt bridge with a quinhydrone electrode dipping in 0.1 M NH_4Cl. The observed emf at 25 °C is 0.152 volts. Find the dissociation constant of NH_4OH. Oxidation potential of saturated calomel = – 0.242 V and standard oxidation potential of quinhydrone electrode is = – 0.699 volt at 25 °C.

[C.U. 1977]

Hints : For the cell.

Pt, Hg | Hg_2Cl_2(s), KCl || solution Q | QH_2, Pt.

E_{cell} = 0.699 – 0.059 pH – 0.242.

Since NH_4Cl is a salt of strong acid (HCl) and weak base (NH_4OH),

pH = 7 – pK_b – log$C$ **[Ans. : K_b = 2.19 × 10^{-5}]**

59. Calculate the junction potential at 25 °C of HCl (0.1 N) : HCl (0.01 N), the transference number of the cation being 0.83. Assume concentration to be identical with activities. **[C.U. 1978]**

Soln. : Liquid junction potential $= E_{LJP} = (t_- - t_+)\dfrac{RT}{F}\ln\dfrac{(a_\pm)_2}{(a_\pm)_1}$

for the process $\left|\mathrm{HCl}(a_\pm)_1 \overset{t_+\mathrm{H}^+}{\underset{t_-\mathrm{Cl}^-}{\rightleftarrows}} \mathrm{HCl}(a_\pm)_2\right|$

t_+ = 0.83, t_- = 1 – 0.83 = 0.17

E_{LJP} = (0.17 – 0.83) × 0.059 log $\dfrac{0.01}{0.1}$ = **0.039 volt.**

60. For the following cell, write down the cell reaction, and calculate emf; free energy change; and equilibrium constant for the cell reaction.

H_2 (g, 1 atm) | H_2SO_4 (m = 0.05, $\gamma_\pm$ = 0.340), Hg_2SO_4(s) | Hg.

Given, $E^\circ_{Hg^+|Hg_2SO_4(s),SO_4^{-2}(OX)} = -0.6141$ V. **[C.U. 1985]**

Soln. : Reaction at R.H.E., $Hg_2SO_4(s) + 2e = 2Hg + SO_4^{-2}$

Reaction at R.H.E., $H_2(g,\ 1\ atm) = 2H^+ + 2e$

Cell reaction, $Hg_2SO_4(s) + H_2\ (1\ atm) = 2Hg + 2H^+ + SO_4^{-2}$

$$E_{cell} = E^\circ_{cell} - \frac{RT}{2F}\log (a_{H^+})^2\left(a_{SO_4^{-2}}\right)$$

$$= E^\circ_{SO_4^{-2},Hg_2SO_4(s)|Hg} - E^\circ_{H^+|H_2} - \frac{0.059}{2}\log (a_{H^+})^2\left(a_{SO_4^{-2}}\right)$$

$$= 0.6141 - \frac{0.059}{2}\log 2^2.1^1.C^3.\gamma_\pm^3\ \ (E^\circ_{H^+|H_2} = 0)$$

$$= 0.6141 - \frac{0.059}{2}\log (4 \times 0.05^3 \times 0.34^3) = \mathbf{0.7529\ V.}$$

$$\Delta G = -nFE = -2 \times 96484 \times 0.7529 = \mathbf{-145.29\ kJ.}$$

$$\ln K = \frac{nF}{RT}E^\circ$$

$$\log K = \frac{2}{0.059} \times 0.6141.$$

$$\therefore\ K = \mathbf{6.56 \times 10^{20}.}$$

61. The thermodynamic dissociation constant of a weak acid (HA) is 1.0×10^{-5} at 25 °C. Find the standard potential (E°) for the half-cell reaction :

$HA(aq.) + e \longrightarrow A^-(aq.) + H_2(g).$ **[B.U. 1999]**

Hints : $E_{cell} = E_{A^-,HA\,|\,H_2} - 0.059\log a_{A^-} = E^\circ_{H^+/H_2} + 0.059\log a_{H^+},\ a_{H^+} = \dfrac{K_a}{a_{A^-}}$

[Ans. $E^\circ = -0.295$ V**]**

62. A glass electrode is used in conjunction with a calomel electrode. The following data points are obtained :

pH	emf (V)
4.01	0.1246
6.99	0.3012

Calculate the value of F, the Faraday constant. **[B.U. 1998]**

Soln. : For glass electrode in conjunction with calomel electrode,

$$E = E_{cal} - E_g$$

$$= E_{cal} - E^\circ_g - \frac{RT}{F}\ln a_{H^+}$$

$$= x - \frac{RT}{F} \times 2.303 \log a_{H^+} \left(E_{cal} - E_g^o = x\right)$$

$$= x + \frac{8.314 \times 298.15 \times 2.303}{F} pH = x + \frac{5708.72}{F} pH.$$

So, $0.1246 = x + \frac{5708.72}{F} \times 4.01$

and $0.3012 = x + \frac{5708.72}{F} \times 6.99.$

Hence, F = **96331 coulombs.**

63. Calculate the emf of the cell :

Pt | H_2 (1 atm) | HA_1(0.1M) || HA_2(0.1 M) | H_2(1 atm) | Pt, where HA_1 and HA_2 are two weak acids of dissociation constants 10^{-4} and 10^{-5} (M) respectively. **[B.U. 1994]**

Hints : $E = -0.059 \log \frac{a_{H^+(HA_1)}}{a_{H^+(HA_2)}}$. **[Ans.** $E = -0.0295$ V**]**

64. The standard electrode potentials for

$Zn^{+2} + 2e = Zn$ $\quad E^o_{298} = -0.76$ V

and $Co^{+2} + 2e = Co$ $\quad E^o_{298} = -0.28$ V

(i) Which of the two metals is more noble?

(ii) If a solution containing 1 gm-mole/litre of both $ZnSO_4$ and $CoSO_4$ is electrolysed with Pt-electrodes at 25 °C, what should be the approximate concentration of cobalt ion when deposition of Zn starts. **[C.U. 1994]**

Soln. : (i) Since, $E^o_{Zn^{+2}|Zn} < E^o_{Co^{+2}|Co}$, Co is more noble as Co^{+2} is more readily reduced to Co than Zn^{+2} to Zn.

Thus, Zn will be more easily oxidized to Zn^{+2} than Co to Co^{+2}.

(ii) Co^{+2} will deposit first since its reduction potential is greater than Zn^{+2}. But reduction potential of Co^{+2} will decrease due to loss of its concentration in solution and when its reduction potential at certain concentration of Co^{+2} equals that of Zn^{+2}, then Zn^{+2} will start deposition. Let the concentration of Co^{+2} be x(M), when

$$E_{Co^{+2}/Co} = E_{Zn^{+2}/Zn}$$

or, $E_{Co^{+2}/Co} + \frac{0.059}{2} \log x = E_{Zn^{+2}/Zn} + \frac{0.059}{2} \log 1$ at 25°C.

or, $-0.28 + \frac{0.059}{2} \log x = -0.76.$

x = **5.36 × 10^{-17} mole / litre.**

65. Standard electrode potential for the Ag^+ | Ag electrode is 0.7991 V at 25 °C and solubility product for AgI is 8.2×10^{-17}. What is the standard electrode potential for I^-, AgI | Ag?

Soln. : The required cell for measuring solubility product of AgI is :

$$Ag \mid Ag^+ \parallel I^-, AgI \mid Ag.$$

The reaction at R.H.E. : $AgI \rightleftharpoons Ag^+ + I^-$ (1)

$Ag^+ + e = Ag$ (2)

Adding (1) and (2), $AgI = Ag + I^-$

Hence, $E^o_{I^-,AgI|Ag} = E^o_{I^-,AgI|Ag} - 0.059 \log a_{I^-}$ (at 25 °C).

$= E^o_{Ag^+|Ag} + 0.059 \log a_{Ag^+}$ (disregarding equation 1)

$= E^o_{Ag^+|Ag} + 0.059 \log \frac{K_s}{a_{I^-}}$ (K_s = solubility product = $a_{Ag^+} a_{I^-}$).

Thus, $E^o_{I^-,AgI|Ag} = E^o_{Ag^+|Ag} + 0.059 \log K_s$

$= 0.7991 + 0.059 \log (8.2 \times 10^{-17}) = -$ **0.149 V.**

66. If $(a_{\pm})_1$ and $(a_{\pm})_2$ are 9.05×10^{-3} and 1.75×10^{-2} respectively and if mobilities of hydrogen and chloride ions in a solution of hydrochloric acid are 33.71×10^{-4} cm^2.V^{-1}.sec^2 and 6.84×10^{-4} cm^2.V^{-1}.sec^2 respectively, calculate the liquid junction potential. **[V.U. 1991]**

Soln. : Let the cell be

$$\begin{array}{c} \xrightarrow{t_+H^+} \\ \left| HCl(a_{\pm})_1 \vdots HCl(a_{\pm})_2 \right| \\ \xleftarrow[t_-Cl^-]{} \end{array}$$

$$E_{LJP} = (t_- - t_+) \times 0.059 \log \frac{(a_{\pm 2})}{(a_{\pm 1})}$$

$$t_- = \frac{u_-}{u_+ + u_-} = \frac{6.84 \times 10^{-4}}{33.71 \times 10^{-4} + 6.84 \times 10^{-4}} = 0.17$$

$t_+ = 0.83$

Hence, $E_{LJP} = (0.17 - 0.83) \times 0.059 \log \frac{1.75 \times 10^{-2}}{9.05 \times 10^{-3}} = -$ **0.011 V.**

CHAPTER 14

ADSORPTION, CATALYSIS, COLLOIDS

Required Formulae

(a) Adsorption :

(i) Langmuir equation : $\theta = \dfrac{K_A P}{1+K_A P}$, where K_A = adsorption equilibrium constant

θ = fraction of surface covered.

(ii) If the substances A and B are adsorbed and the partial pressures P_A, P_B are at equilibrium, then

$$\theta_A = \frac{K_A P_A}{1+K_A P_A + K_B P_B}, \qquad \theta_B = \frac{K_B P_B}{1+K_A P_A + K_B P_B}.$$

(iii) Freundlich equation : $\dfrac{x}{m} = \theta = KP^{1/n}$, where K and n are constant, and

where x = amount of adsorbate adsorbed

m = amount of adsorbent

P = equilibrium pressure.

(iv) $\dfrac{d \ln K_p}{dT} = \dfrac{\Delta H}{RT^2}$.

But, by Langmuir equation $KP = \dfrac{\theta}{1-\theta}$.

$\therefore$ $\ln K + \ln P = \ln \dfrac{\theta}{1-\theta}$.

At constant coverage, $\dfrac{d \ln K}{dT} + \dfrac{d \ln P}{dT} = 0$.

$\therefore$ $\dfrac{d \ln P}{dT} = -\dfrac{\Delta H}{RT^2}$.

i.e., $\ln \dfrac{P_2}{P_1} = \dfrac{\Delta H}{R}\left[\dfrac{T_1 - T_2}{T_1 T_2}\right]$, where the terms have usual meaning.

(b) Colloids :

(i) $\zeta = \dfrac{4\pi\eta\nu}{D.\Delta\phi / l} \times 300^2$ in volt, where η = coefficient of viscosity, $\dfrac{\Delta\phi}{l}$ = potential gradient and ν = velocity in cm/sec, D = dielectric constant.

(ii) $\zeta = \dfrac{4\pi\eta V\kappa}{D.I.} \times 300^2$ in volt, where D = dielectric constant, I = current in ampere,

V = volume moving per unit area per second

ζ = zeta potential in volt and κ = specific conductance.

In these equations, $\dfrac{\Delta\phi}{l}$ = potential gradient in volt/cm.

(iii) $\nu = \dfrac{D.\Delta\phi / l.\zeta}{4\pi\eta \times 300^2}$ in cm/sec. [all terms having usual significance]

(iv) $N_A = \dfrac{RT}{\frac{4}{3}\pi r^3(\rho - \rho_0)g(h_2 - h_1)} \times \ln \dfrac{n_1}{n_2}$,

where N_A = Avogadro number

$(\rho - \rho_0)$ = difference in densities between sol solution (ρ) and pure solvent (ρ_o)

$(h_2 - h_1)$ = the difference in heights where the no. of particle's reading has been taken.

n_1 = no. of particles at height h_1

n_2 = no. of particles at height h_2

T = temperature in K.

r = radius of the sol particles

R = ideal gas constant in ergs.

(c) Catalysis :

(i) $J = k_0 + k_{c_1}C_{c_1} + k_{c_2}C_{c_2}$,

where k_0 = specific rate in absence of catalyst

k_{c_1} = catalytic coefficient of catalyst C_1.

C_{c_1} = concentration of catalyst C_1

k_{c_2} and C_{c_2} are respectively catalytic coefficient and catalyst concentration of catalyst 2.

(ii) $J = J_0 + k_{HB}C_{HB} + k_{B^-}C_{B^-}$,

where J = first order rate constant or total catalytic coefficient being equal to $-\dfrac{1}{C}.\dfrac{dC}{dt}$,

C = concentration of the reactant, t = time.

(iii) $J = k_{H^+}.K_a^{\frac{1}{2}}$,

where k_{H^+} = catalytic coefficient of H^+

K_a = acidity constant of the weak acid.

(iv) In enzyme catalysis,

$$\text{Turnover number} = \frac{\text{Maximum rate}}{\text{Initial enzyme concentration}} = \frac{R}{C_E^0}.$$

Short Questions

(a) Adsorption :

(i) What happens in common cases of absorption?

Ans. : In absorption the foreign substances undergo uniform penetration into the field of force that exists between atoms, ions or molecules, in the bulk of the absorbent. In some cases, this will lead to the formation of homogenous mixture, that is solution. In many other cases this may lead to compound formation.

(ii) "Physisorption must be exothermic."—Explain.

Ans. : Physisorption is a weak adsorption taking place in the surface of a solid where van der Waals type forces are operative. The process is spontaneous, and thus $\Delta G = \Delta H - T\Delta S$ must be negative. Since the translation freedom of the adsorbate is reduced on its adsorption, ΔS is negative. Thus, ΔH has to be negative to make ΔG negative. Therefore, physisorption has to be exothermic.

(iii) "Adsorption has to be exothermic."—Comment.

Ans. : Adsorption is a spontaneous process with ΔG being (–)ve. For gas with three degrees of freedom adsorption means the loss of one degree of freedom (as adsorbed gas molecule is mobile in 2 dimensions). For immobile type of chemisorption, translational degrees of freedom are reduced to zero. Thus, ΔS is negative and ΔH has to be negative for ΔG to be negative.

In this approach, the adsorbent has been treated as completely inert. But, there are cases when it may swell making ΔS to be (+)ve. What is more, in case of dissociative chemisorption, ΔS may be positive. For these cases, at low temperature $T\Delta S$ may be sufficiently negative so that ΔH need not be negative to make ΔG negative. It can be +ve. Adsorption of H_2 on glass at room temperature has been found to be endothermic.

(iv) Why do electrolytes increase the surface tension of a liquid?

Ans. : Electrolytes attract solvent molecules very strongly. The solvent-solute interaction is stronger than solvent-solvent interaction. In the bulk, the amount of solvent available is larger and solvent molecules may surround the solute molecules from all sides. This increased opportunity in bulk for attraction leads to depletion of ions from the surface. As a consequence, the surface solvent molecule apparently feels an inner side pull and surface tension (γ) increases. This is also given by Gibbs equation. $\Gamma_2 = -\frac{C}{RT}\cdot\frac{d\gamma}{dc}$. Since Γ_2, surface excess, is negative, so $\frac{d\gamma}{dc}$ is (+)ve. i.e., increase in concentration leads to increase in surface tension.

(v) Show that for a gas obeying Langmuir equation a plot of $\ln \frac{\theta}{P}$ vs. θ will be linear with a slope of -1. What will be the slope of $\ln \frac{V}{P}$ vs. V plot?

Ans. : By Langmuir equation $\theta = \frac{K_A.P}{1+K_A.P}$

or, $\frac{\theta}{1-\theta} = K_A$ or, $\frac{\theta}{P} = K_A\ (1-\theta)$.

$\therefore\ \ln \frac{\theta}{P} = \ln K_A + \ln (1-\theta)$.

When θ is small, $\ln (1-\theta) = -\theta$.

$$\text{So } \ln \frac{\theta}{P} = \ln K_A - \theta. \qquad (1)$$

Hence, a plot of $\ln \frac{\theta}{P}$ vs. θ will be linear with -1 as slope.

Again, $\theta = \frac{V}{V_m}$.

Putting the value of θ in eqn. (1),

$$\ln \frac{V}{PV_m} = \ln K_A - \frac{V}{V_m}$$

$$\ln \frac{V}{P} = \ln (K_A.V_m) - \frac{V}{V_m}.$$

Hence, a plot of $\ln \frac{V}{P}$ vs. V will be linear with $-\frac{1}{V_m}$ as slope.

(vi) Write down B.E.T. equation. Explain all the terms.

Ans. : B.E.T. equation is : $\frac{F}{V(1-F)} = \frac{1}{V_m C} + \frac{C-1}{V_m.C}.F$

$F = \frac{P}{P_0}$, where P is pressure at which the adsorption is started.

and P_0 = saturated vapour pressure of the adsorbate as a liquid.

$$C = e^{-(E_1-E_L)/RT},$$

where E_1 = heat of chemisorption in the first layer

E_L = heat of liquefaction of adsorbate gas.

V_m = monolayer volume (generally reported at NTP).

(vii) Write the B.E.T. equation. Under what condition it is converted to Langmuir equation?

Ans. : B.E.T. equation is : $\frac{F}{V(1-F)} = \frac{1}{V_m C} + \frac{C-1}{V_m.C}.F$

[where all the terms have usual meanings]

Putting $F = \frac{P}{P_0}$, and multiplying by P_0.

$$\frac{P_0.P}{V(P_0 - P)} = \frac{P_0}{V_m.C} + \frac{C-1}{V_m.C}P$$

$$c = e^{-(E_1 - E_L)/RT}.$$

Let $C >> 1$.

Now, equilibrium between gaseous and liquid adsorbate is :

$$g \rightleftharpoons 1. \quad K = \frac{1}{P_0}.$$

In B.E.T. method, adsorption is comparable with liquefaction.

Let $P_0 >> P$.

$$\therefore \quad \frac{P}{V\left(1 - \frac{P}{P_0}\right)} = \frac{1}{V_m\left(\frac{C}{P_0}\right)} + \frac{P}{V_m}.$$

$\therefore \quad \frac{P}{V} = \frac{1}{V_m}.\frac{1}{K} + \frac{P}{V_m}$ which is the Langmuir equation with $K = K_a$.

(viii) Explain the following :

(a) The decomposition of NH_3 on tungsten is of zero order.

(b) The decomposition of N_2O on gold is of first order.

(c) The recombination of H atoms on gold is of second order.

(d) The decomposition rate of NH_3 on Pt depends on $P_{(NH_3)}/P_{(H_2)}$.

(e) The decomposition rate of NH_3 on Mo is strongly retarded by N_2, but does not approach zero as the surface becomes saturated with nitrogen.

Use Langmuir-Hinshelwood model and be as brief as possible.

Soln. : (a) Ammonia is very strongly adsorbed on tungsten surface so that rate

$R = k_2.\theta$, where k_2 = rate constant, θ = fraction of surface covered.

$\therefore \quad R = k_2.\frac{K_AP}{(1+K_AP)}$, where K_A = adsorption equilibrium constant, P = pressure of the adsorbate. For very strong adsorption $K_A.P >> 1$, so that

$$R = k_2 \frac{K_AP}{K_AP}. \quad \therefore R = k_2.P^\circ.$$

Hence this is of zero order.

(b) N_2O is weakly adsorbed on gold surface, so that

$R = k_2.\theta$

$= \quad k_2.\frac{K_AP}{(1+K_AP)}$ [all terms have usual significance]

In case of N_2O, $1 >> K_AP$.

$\therefore \quad R = k_2.K_AP,$

which is of the first order.

(c) H atoms are weakly adsorbed on gold. And the rate is proportional to the collision of two adsorbed H atoms.

$$\therefore \text{ Rate } = k_2\theta^2 = k_2 \left(\frac{K_H P_H}{1+K_H.P_H}\right)^2, 1 >> K_H.P_H.$$

$$\therefore \text{ Rate } = k_2.\ (K_H)^2 (P_H)^2.$$

Thus, the rate is of second order. Here, K_H is adsorption equilibrrium constant of H atom, P_H is the partial pressure of H atom.

(d) Product H_2 of NH_3 decomposition is strongly adsorbed on active centres, where NH_3 is weakly adsorbed. Let K_A be the adsorption equilibrium constant of hydrogen maintained at a pressure P_A. K_B and P_B are those of NH_3. Then Rate = $k_2.\theta_A.\theta_B$.

$$\therefore \text{ Rate } = k_2\left(\frac{K_A.P_A}{1+K_AP_A+K_BP_B}\right)\left(\frac{K_B.P_B}{1+K_AP_A+K_BP_B}\right) = k_2 \frac{(K_A.P_A)(K_B.P_B)}{(1+K_AP_A+K_BP_B)^2}.$$

Now, $K_AP_A >> 1$ and $K_AP_A >> K_BP_B$.

$$\therefore \text{ Rate } = k_2.\frac{(K_A.P_A)(K_B.P_B)}{(K_AP_A)^2} = k_2.\frac{K_A^{-1}.K_B.P_B}{P_A}.$$

As already indicated, $P_B = P_{NH_3}$ and $P_A = P_{H_2}$.

$$\therefore \text{ Rate } = k.\frac{P_{NH_3}}{P_{H_2}},$$ where k is a constant and is equal to $k_2 K_A^{-1} K_B$.

(e) There are obviously at least two types of active centres on Mo surface. In one type of active centre, NH_3 is weakly adsorbed and N_2 is strongly adsorbed so that rate

$$= K_A.\frac{P_{NH_3}}{P_{N_2}}.$$

So, as decomposition proceeds, P_{N_2} increases and the rate decreases. But obviously, there is at least another type of active centre of low concentration on which only NH_3 is adsorbed. So, rate of decomposition of NH_3 does not become zero.

[For details see the chapter on Adsorption of Physical Chemistry, Vol. II by Dr S. Pahari.]

(b) Colloids :

(i) "Deltas are formed at a place where rivers pour water into sea."—Comment.

Ans. : River water contains mud which is clay in colloidal form. These are stable due to charge. In the sea, there are salts. So, when rivers pour water into sea, clays are mixed with salts. Their charges are neutralised. From colloids, these become suspension and get precipitated and continuous deposition through years leads to the formation of a delta.

(ii) "Colloid solutions are thermodynamically unstable."—Comment.

Ans. : Colloids are particles being in between crystalloids and suspensions in size. Colloid particles are so large (about 100–1000 Å in diameter) that they have a distinguishing surface. So, if a large number of such particles fuse, the total surface area and hence, free energy will decrease. Therefore, colloids are thermodynamically unstable showing a tendency to coalesce. The reason of their stability is electrokinetic—their motion and adsorbed charges on them prevent collision of two such particles.

(iii) What is the absolute primary role of double layer?

Ans. : The primary role of double layer is to confer kinetic stability to the thermodynamically unstable colloids.

(iv) If a very small amount of $AlCl_3$ is added to gold sol, the gold flocculates, but if a large quantity of $AlCl_3$ is added, there is no flocculation.—Why?

Ans. : Gold sols are lyophobic and negatively charged. Addition of small amount of $AlCl_3$ thus leads to reduction of ζ-potential and hence, flocculation of the sol. However, if a large amount is added, greater amount of Al^{3+} is adsorbed making gold sol positive and kinetically stable.

(v) "Presence of H_2S is essential in As_2S_3 sol, though H_2S ionises and should be precipitated as As_2S_3."—Explain.

Ans. : As_2S_3 sol can exist as a sol, only if it adsorbs some ion. As_2S_3 sol owes its stability mainly to adsorbed $S^=$ ion. Because of adsorption of charge, small particles cannot collide directly to yield a precipitate. H_2S ionises weakly to H^+ and $S^=$ ions. The concentrations of $S^=$ is extremely small. But the adsorption of this amount of $S^=$ confers some ζ-potential to the suspensoid and As_2S_3 soln. is stabilised. And, thus, although passage of H_2S to As^{3+} solution. precipitates As_2S_3; some extra H_2S must be present for existence of As_2S_3 sol.

(vi) Why do colloids precipitate when they are heated?

Ans. : Colliding colloidal particles can break through the double layer and coalesce, only if the collision is sufficiently energetic to disrupt the layers of ions and solvating molecules. Of course, this is also possible, if as a result of brisk motion, the surface accumulation of charge is stirred away. This may happen at high temperature, when the velocity (K.E. = $\frac{1}{2}mv^2$) is expected to be high, and that is why sols precipitate when they are heated:

(vii) "Uncharged colloids exist and are important, both in aqueous and nonaqueous media."—Explain their stability.

Ans. : A colloidal sol is a heterogenous system and the units tend to coalesce which is a process that reduces negative free energy and hence, these coagulate. The adsorbed charges confer kinetic stability. However, solvation also is another important factor which may prevent the collisions of particles. These may not break through their solvent layers. The solvation may be a result of dipole-dipole attraction. We generally speak of hydrogen bonds with surface in aqueous media and van der Waals attraction in non-aqueous media for explaining the reason of solvation.

(viii) "Lyophobic colloids flocculate only when the charges on the collidal particles are completely neutralised by addition of electrolytes."—Comment.

Ans. : The above assertion is not completely correct. Colloids are kinetically stable due to ζ-potential resulting from the adsorption of charge by the particles. The repulsion of the charge is a factor why the particles cannot collide and hence, cannot coalesce to form precipitate. When electrolytes are added, the charge is neutralised. Completely neutralised lyophobic colloids are unstable and must form precipitate. However, colliding colloidal particles can break through the double layer if the collision is sufficiently energetic to disrupt the layers of ions and solvating molecules. Thus, at every temperature, there will be a minimum value of ζ-potential below which this is possible. This has been found to be about 0.02–0.03 volts. Thus, for coagulations, complete neutralisation is not needed (partial neutralisation is sufficient).

(ix) Acids and Bases are particularly effective as peptising agent.—Why?

Hints : H^+ and OH^- ions are especially strongly adsorbed by uncharged surface.

(x) Organic salts are more powerful than corresponding inorganic salts.—Why?

Hints : Organic ions are more strongly adsorbed on charged surface in comparison to inorganic salts.

(xi) On what factors does the course of coagulation depend?

Ans. : Mainly two factors—Brownian motion of the particles and interaction between the particles when they are close enough.

(xii) We speak of zeta-potential in case of colloids but not in case of salt solutions, although both colloidal particles and ions in salts are charged.—Why?

Ans. : Salt solutions result from dissolution of salts in a solvent which is generally water. This process is spontaneous with negative free energy. However, the solution is homogenous and there is no surface of separation between ions and solvent. In case of colloidal sol, the solution is heterogenous for the large size of colloidal particle which are stable kinetically due to the adsorbed charge on them so that they are repelled from one another. Thus, there is a surface with charge separation, and naturally a potential difference is set up. A part of the potential is zeta-potential and we speak of ζ-potential in discussing this.

(xiii) Both emulsifier and peptising agents stabilize colloids but their actions are different.—Explain.

Ans. : Colloids are stabilized by peptising agents because colloid particles adsorb preferably one of the ions of peptising agents at their surface. As a result, ζ-potential originates and colloids become stable kinetically.

In case of emulsion, the two parts of the emulsifier are adsorbed at the two exposed liquid surfaces and the interfacial tension is reduced. The nature of the emulsifier determines the type of emulsion. The film curves toward the phase on the surface with higher tension. Univalent cation soap-like sodium oleate is more wetted by water, the surface tension is lower on water side. Calcium or aluminium oleates are less soluble in water—these are more easily wetted by oil, and water surface is curved giving water/oil emulsion. Thus, no doubt there is adsorption and double layer, but the emulsifier acts by reducing surface tension of a particular surface due to greater wetting (greater preferential adsorption).

(c) Catalysis :

(i) A catalyst remains unchanged at the end of reaction.—Explain.

Ans. : A catalyst is both the reactant and the product of a reaction. It takes part in a reaction to form a stable or unstable intermediate *I*. This may be depicted with reactant

R, product P and intermediate I as follows :

$$R + C \rightleftharpoons I \longrightarrow P + C.$$

Thus, the catalyst is regenerated in the final stage. So, the above statement is true. Although, it is chemically unchanged but physical form may be changed.

(ii) A catalyst cannot affect the position of equilibrium of a reaction.—Explain.

Ans. : Let us consider a reversible reaction,

$$A \underset{k_{-1}}{\overset{k_1}{\rightleftharpoons}} B,$$

where k_1 and k_{-1} are the specific rates for the forward and backward reactions respectively. In the given case, the steps are elementary so that the equilibrium constants,

$$K = \frac{k_1}{k_{-1}}, \therefore K = \frac{A_1 . e^{-E_1/RT}}{A_2 . e^{-E_2/RT}}$$

$$\therefore \text{ uncatalysed } K = \frac{A_1}{A_2} . e^{-\frac{(E_1 - E_2)}{RT}}.$$

A catalyst reduces the activation energy by x, so that the activation energy now is $(E_1 - x)$. Similarly, activation energy for the backward reaction is $(E_2 - x)$. There is no reason for the change of frequency factor A. Therefore, equilibrium constant of the catalysed reaction K^C is :

$$K^C = \frac{A_1^c}{A_2^c} . e^{-\left[\frac{(E_1 - x) - (E_2 - x)}{RT}\right]}$$

$$= \frac{A_1}{A_2} . e^{-\frac{(E_1 - E_2)}{RT}} = K.$$

There is no reason why A^C and A will be different. So, $K^C = K$; the equilibrium position is not changed.

(iii) "A catalyst eliminates undesirable side products."—Comment.

Ans. : A catalyst is specific and reduces the activation energy of a specific path by taking part in the formation of the activated complex of the path. Let a substance A decompose to product B and D in two different paths.

$$A \longrightarrow B$$
$$A \longrightarrow D$$

If the activation energies are comparable, then B and D will occur together. Let a catalyst C_1 reduce activation energy of $A \longrightarrow B$ pathway.

Thus, in presence of C_1, B will be formed first, and if collected, B will be the sole product. Thus, the catalyst C_1 eliminates the undesirable side product, D. Similarly, another catalyst C_2 may eliminate B and give solely D.

(iv) Negative catalysts are not catalysts.—Explain.

Ans. : The so-called negative catalysts are substances that retard the rates of reaction. These substances can do so by increasing activation energy (and /or by decreasing the frequency factor). But, catalysts are defined as substances that increase the rate of reactions. Catalysts generally reduce the activation energy by taking part in the activated complex formation. They may do so also by increasing frequency factor. Thus, negative catalysts are not catalysts. They are nowadays termed as inhibitors.

(v) It is not essential for a catalyst to decrease activation energy. Comment.

Ans. : A catalyst increases rate by increasing specific rate k. But $k = A.e^{-E/RT}$, where A = frequency factor, E = activation energy.

Hence, a catalyst may increase k by decreasing E. But it also may increase k by increasing A, and without decreasing E. It may even increase E and yet increase rate if A is increased sufficiently.

(vi) If a catalyst does not take part in intermediate formation, then there is limit of high concentration up to which the rate will depend on the catalyst concentration.

Ans. : Let a catalyst C catalyses a reaction A to P by just activating A to A^* and not taking part in the formation of A^*.

That is,

$$A + C \underset{k_{-1}}{\overset{k_1}{\rightleftharpoons}} A^* + C.$$

Now, A^* decomposes to give P.

$$A^* \xrightarrow{k_2} P.$$

Thus, $r = k_2.C_A{}^*$.

Assuming steady state in A*,

$$k_1 C_A.C_C = k_{-1}.C_A{}^* C_C + k_2.C_A{}^*,$$

so that $C_A{}^* = \dfrac{k_1 C_A.C_C}{k_{-1}C_C + k_2}$.

So, $r = \dfrac{k_2 k_1 C_A.C_C}{k_{-1}C_C + k_2}$.

So, at high C_C, r is independent of catalyst concentration.

(vii) For the reaction $R + C \longrightarrow P$, rate $= kC_C.C_R$, when R is reactant, and C is catalyst. So when $C_C = 0$, the rate is zero. Comment. **[V.U. 2002]**

Ans. : For a catalytic reaction, $R \longrightarrow P$,

we define, a first order rate constant.

$$J = -\frac{1}{C_R}\frac{dC_R}{dt} = k_0 + k_C.C_C,$$

where C_R = constant of the reactant

k_0 = first order rate constant without catalyst

k_C = catalytic coefficient

C_C = catalyst concentration.

But, $k_C.C_C$ is so much greater than k_0 that k_0 is neglected. But when C_C is zero $k_C C_C$ is zero. Rate constant becomes

$J = k_0$. Hence the rate is not zero.

(viii) If a catalyst catalyses the forward reaction, then it must catalyse the backward reaction.—Explain.

Ans. : All reactions are reversible. Let us consider a reaction, $A \rightleftharpoons B$.

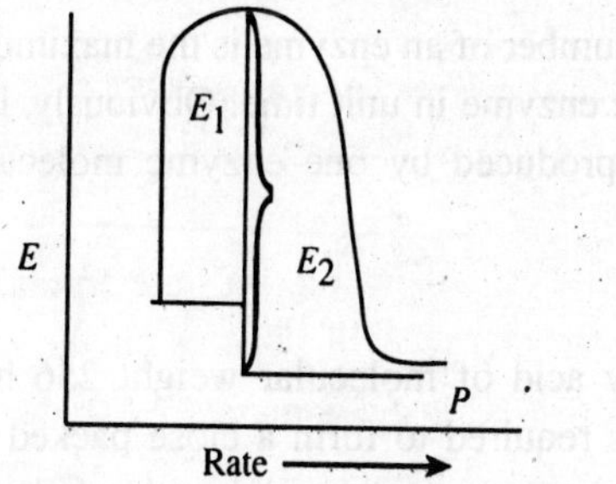

The energy vs. reaction coordinate diagram

E_1 is the activation energy of the forward path and E_2 is the activation energy of the backward path. If k_1 is the specific rate constant of the forward reaction and k_2 is the same for backward reaction, then

$$K = \frac{k_1}{k_2} = \frac{A_1 e^{-E_1/RT}}{A_2 e^{-E_2/RT}} = \frac{A_1}{A_2} e^{-\frac{(E_1 - E_2)}{RT}}$$, where A_1 and A_2 are frequency factors.

Let the catalyst reduce the activation energy by x. Then activation energy in presence of catalyst $= E_{C_1} = E_1 - x$.

and equilibrium constant K^c in presence of catalyst becomes $K^c = \frac{A_1.e^{-\frac{(E_1 - x)}{RT}}}{A_2.e^{-\frac{(E')}{RT}}}$.

Then, K^c may become equal to K only if E' is $E_2 - x$.

Thus, if a catalyst catalyses the forward reaction, it must catalyse the backward reaction the same way.

(ix) What are enzymes?

Ans. : Enzymes occuring in some natural juices, plants, animals, etc. possess incredible capacity of bringing about many complex chemical reactions. Enzymes are large protein molecules, namely—diastate, urease, pepsin, zymase, etc. They have much high molecular weights, usually between 10^4-10^6. Enzymes disperse in water forming colloidal solution and are sometimes referred to as microheterogenous catalysts.

(x) What is Michaelis constant?

Ans. : Michaelis constant is a characteristic of enzyme catalysed reaction. Let an enzyme E react with reactant S to give product P. The reaction has been suggested to follow the mechanism—

$$E + S \underset{k_{-1}}{\overset{k_1}{\rightleftharpoons}} ES \xrightarrow{k_2} E + P.$$

Here, k_1 is the specific rate constant of *ES* forming reaction.

k_{-1} and k_2 are specific rate constants of *ES* decomposition.

Michaelis constant K_M is defined as $K_M = \frac{k_{-1} + k_2}{k_1}$.

Its unit is moles/litre.

(xi) What is turnover number?

Ans. : The turnover number of an enzyme is the maximum number of moles of produ produced by 1 mole of the enzyme in unit time. Obviously, it is also the maximum numbe of molecules of product produced by one enzyme molecule in unit time.

Numerical Problems

(a) Adsorption :

1. A long chain fatty acid of molecular weight 256 has a density of 0.82 gm/cm If 0.102 mg of the acid is required to form a close packed monolayer film over 500 cm of water surface, find the cross section and length of the acid molecule. **[C.U. 198**

Soln. : Let *a* be the cross section and *h* be the length. By the problem, $\frac{0.102 \times 10^{-}}{256}$ moles occupy 500 sq. cm.

Thus, $\frac{0.102 \times 10^{-3}}{256} \times 6.022 \times 10^{23} \times a = 500.$

$\therefore\ a = 2.08386 \times 10^{-15}\ \text{cm}^2 = \mathbf{20.8\ sq.\ Å}.$

Again, $v \times \rho = m$

or, $500 \times h \times 0.82 = 0.102 \times 10^{-3}$

$\therefore\ h = 2.4878 \times 10^{-7} = \mathbf{24.9\ Å}.$

2. 5.2×10^{-5} gm of stearic acid (M = 284) when placed on water a surface film cou occupy an area of 225 sq. cm as a close packed monolayer. Find cross-sectional area stearic acid molecule. **[C.U. 197**

Soln. : If *a* be the cross-sectional area of the acid molecule, then by the proble

$$\frac{5.2 \times 10^{-5} \times 6.022 \times 10^{23}}{284} \times a = 225.$$

$$\therefore\ a = \frac{225 \times 284}{6.022 \times 5.2} \times 10^{-18} = 2.04059 \times 10^{-15}\ \text{cm}^2 = \mathbf{20.4059\ sq.\ Å}.$$

3. In a laboratory experiment 60 cc of a solution originally 0.21 (M) in AcOH w permitted to come to equilibrium with a 10 gm sample of charcoal. After equilibriu 20 cc of AcOH required 30 cc of 0.11 (M) NaOH for neutralisation. Calculate the equilibriu concentration of the acid and the weight of CH_3COOH adsorbed per gm.

Soln. : Equilibrium concentration $= \frac{30 \times 0.11}{20} = \mathbf{0.165\ (M)}.$

No. of moles of AcOH initially $= \frac{60 \times 0.21}{1000} = 0.0126.$

No. of moles of AcOH at equilibrium $= \frac{60 \times 0.165}{1000} = 0.0099.$

So, no. of moles of AcOH adsorbed $= (0.0126 - 0.0099) = 0.0027$ moles

$= (0.0027 \times 60)$ gm $= 0.162$ gm.

$\therefore$ wt. of acetic acid adsorbed per gm $= \frac{0.162}{10} =$ **0.0162 gm.**

4. For the adsorption of a substance from aqueous solution by charcoal at 25 °C, the Freundlich constants are $n = 3.0$ and $K = 0.50$, the concentration of solution being expressed in gm/litre and the amount being adsorbed in gm/gm. What weight of substance will be adsorbed by 2 gm of charcoal from 1 litre of a solution containing originally 2 gm of the substance? **[C.U.(H) 1981]**

Soln. : We know, $\frac{x}{m} = K.C^{\frac{1}{n}}$.

By the problem, $C = (2 - x)$ gm, $K = 0.5$, $n = 3$, $m = 2$ gm.

$\therefore \quad \frac{x}{2} = 0.5 \ (2-x)^{\frac{1}{3}}$

or, $x = (2-x)^{\frac{1}{3}}$

or, $x^3 = 2 - x$ or, $x^3 + x - 2 = 0$

or, $(x - 1)(x^2 + x + 2) = 0$, $\therefore x = 1$.

i.e., 1 gm of the substance will be adsorbed.

5. The adsorption of a dye from solution on charcoal is governed by the Freundlich isotherm in which $n = 2$ and $K = 6.8$. The concentration of the dye is expressed in millimoles/cc and the amount adsorbed in millimoles/gm. If 10 gm of the charcoal be shaken with 100 cc of a 0.1 (M) solution of the dye, then what will be the equilibrium concentration of the solution.?

Soln. : We know that 0.1 (M) $= 0.1 \times 10^{-3}$ moles in 1 cc $= 0.1$ millimoles/cc.

Thus, C before adsorption is 0.1 millimoles/cc. Let x millimoles/cc be adsorbed. Hence, C at equilibrium is $0.1 - x$. But 10 gm of charcoal adsorbs $100x$ millimoles from 100 cc.

So, for $\frac{X}{m} = K.C^{\frac{1}{n}}$,

$X = 100x$; $K = 6.8$, $m = 10$, $C = 0.1 - x$, $n = \frac{1}{2}$.

Thus, $\frac{100x}{10} = 6.8\sqrt{(0.1-x)}$

i.e., $10x = 6.8\sqrt{(0.1-x)}$

i.e., $100x^2 = 46.24 \ (0.1 - x)$

i.e., $100x^2 + 46.24x - 4.624 = 0$.

$\therefore \quad x = 0.0845$ millimoles/cc $= 0.0845$ (M).

$\therefore$ the equilibrium concentration $= (0.1 - 0.0845) =$ **0.0155 (M).**

6. The adsorption of a dye from solution on charcoal is governed by the Freundlich isotherm, where $n = 2$ and $K = 6.8$. The concentration of the dye is expressed in millimoles/cc and the amount adsorbed in millimoles/gm. If 10 gm of the charcoal is shaken with (a) 1 litre, (b) 10 cc at 0.1 (M) of the dye, then what will be the equilibrium concentration of the solution? **[V.U. 1995]**

Hints : Proceed as Problem 5.

(a) The equation is $\frac{1000x}{10} = 6.8\sqrt{(0.1-x)}$.

[Ans. $0.1 - x = 0.081$ (M)**]**

(b) The equation is $\frac{10x}{10} = 6.8\sqrt{(0.1-x)}$

[Ans. $0.1 - x = 0.00022$ (M)**]**

7. Calculate the surface area per gm of a finely powdered solid for which the volume of N_2(g) adsorbed per gm at STP to form a monolayer was found to be 3.01 cc. Given area of N_2 molecule is 16.2 sq. Å.

Soln. : At NTP, 22400 cc contains 6.022×10^{23} N_2 molecules.

$$\therefore \quad 3.01 \text{ cc contains } \left(\frac{6.022\times10^{23}}{22400}\times3.01\right) \text{ molecules.}$$

Hence, the surface area is $\frac{6.022\times10^{23}\times3.01}{22400} \times 16.2 \times 10^{-16}$ sq. cm

$= 131091.41$ sq. cm $=$ **13.11 sq. m.**

8. For the adsorption of a substance from aq. solution by a solid at 20 °C, the Freundlich constants are $n = 2$ and $K = 0.5$, the concentration in solution being expressed in gm/litre and the amount being expressed in gm/gm. What weight of the substance will be adsorbed by 2 gm of the solid from one litre of solution containing initially 3 gm of the substance?

Hints : Proceed as Problem 4.

Here the equation is :

$$\frac{x}{2} = (0.5)\,(3-x)^{\frac{1}{2}}.$$

$\therefore \quad x =$ **1.3 gm.**

9. The adsorption of a gas is described by Langmuir isotherm with $K_a = 0.85$ kPa^{-1}. Find the pressure at which the surface coverage is (a) 15%, (b) 95%.

Soln. : (a) 15% coverage means, $\theta = 0.15$.

$\therefore$ by Langmuir equation, $0.15 = \frac{K_a.P}{1+K_a.P}$

$$\text{or, } 0.15 = \frac{0.85kP_a^{-1}}{1+0.85kP_a^{-1}}$$

$$\text{or, } \frac{1}{0.15} = \frac{1}{0.85P} + 1$$

$$\text{or, } \frac{0.85}{0.15} = \frac{1}{0.85P}$$

$$\text{or, } 0.85P = \frac{0.15}{0.85}$$

$$\therefore\ P = \frac{0.15\times 1}{(0.85)^2} = 0.2076 \text{ kPa} = \mathbf{0.21\ kPa.}$$

(b) 95% coverage means, $\theta = 0.95$.

Therefore, proceeding as in (a),

$$\frac{0.05}{0.95} = \frac{1}{0.85P}.$$

$$\therefore\ P = \frac{0.95}{0.05\times 0.85} = \mathbf{22.35\ kPa.}$$

10. A certain solid sample adsorbs 0.44 mg of CO when the pressure of the gas is 26.0 kPa and the temperature is 300 K. The amount adsorbed when the pressure is 3 kPa and temperature is 300 K is 0.19 mg. The Langmuir isotherm describes the above process. Calculate the fraction of surface covered.

Soln. : We know, $\theta = \frac{KP}{1+KP}$.

$\theta_1 = \frac{KP_1}{1+KP_1}$, where K = adsorption equilibrium constant.

$\theta_2 = \frac{KP_2}{1+KP_2}$. Again, $\frac{\theta_1}{\theta_2} = \frac{w_1}{w_2}$

$$\therefore\ \frac{w_1}{w_2} = \frac{KP_1}{1+KP_1}\cdot\frac{1+KP_2}{KP_2} = \frac{P_1}{P_2}\cdot\frac{1+KP_2}{1+KP_1}.$$

By the problem, $w_1 = 0.44$ mg, $w_2 = 0.19$ mg, $P_1 = 26$ kPa, $P_2 = 3$ kPa.

$$\therefore\ \frac{0.44}{0.19} = \frac{26}{3}\cdot\frac{1+3K}{1+26K}$$

or, $K = 0.184$.

Now, fraction of surface covered, $\theta_1 = \frac{0.184\times 26}{1+0.184\times 26} = \mathbf{0.83.}$

$$\theta_2 = \frac{0.184\times 3}{1+0.184\times 3} = \mathbf{0.36.}$$

11. A certain solid sample adsorbs 0.52 mg of hydrogen when the pressure of the gas is 34 kPa and the temperature is 35 °C. The amount adsorbed when pressure is 5 kPa at the same temperature is 0.21 mg. The Langmuir isotherm describes the above process. Calculate the fraction of surface covered.

Soln. : Proceed as Problem 10. **[Ans.** $\theta_1 = 0.75$, $\theta_2 = 0.30$]

12. A solid with surface area independent of temperature adsorbs 2 mg of a gas, when in contact with the gas at a pressure of 10 kPa and a temperature of 283 °K. The system follows Langmuir equation. The amount of heat absorbed when 2.00 millimole of the gas is desorbed in 80 J. What is the equilibrium pressure for the adsorption of 2 mg of the gas at a temperature of 323 K?

Soln. : By the problem, $P_1 = 10$ kPa, $T_1 = 283$ K, $T_2 = 323$ K, $P_2 = ?$

Heat absorbed for 2 millimole = 80 J

∴ " " " 1 millimole = 40 J

∴ " " " 1 mole = 40×10^3 = 40 kJ.

∴ $\Delta H_a = -40$ kJ, where ΔH_a = heat of adsorption/mole.

We know that, $\frac{d \ln P}{dT} = -\frac{d \ln K_a}{dT} = -\frac{\Delta H_a}{RT^2}$,

when θ is temperature independent.

$$\therefore \quad \ln\frac{P_2}{P_1} = \frac{\Delta H_a}{R}\left[\frac{T_1 - T_2}{T_1 T_2}\right]$$

$$\text{or,} \quad \ln\frac{P_2}{10} = -\frac{40\times10^3}{8.314}\left[\frac{-40}{323\times283}\right]$$

$$\text{or,} \quad \ln\frac{P_2}{10} = 2.105334.$$

∴ P_2 = 10 × 8.2098 kPa = 82.0984 kPa = **82.1 kPa.**

13. A solid in contact with a gas at a pressure of 10 kPa at 300 K adsorbs 2 mg of the gas. The system follows Langmuir equation. The heat released for the adsorption of 1 millimole is 15 J. What is the equilibrium pressure for the adsorption of 2 mg of the gas at 37 °C. Assume that the surface area is independent of temperature.

Soln. : Proceed as Problem 12.

$$\ln\frac{P_2}{P_1} = \frac{\Delta H}{R}\left[\frac{T_1 - T_2}{T_1 T_2}\right],$$ **[Ans. P_2 = 12.14 kPa]**

14. The chemisorption of H_2 on manganese is activated. It proceeds at a rate 35% faster at 1000 K than at 600 K. Calculate the activation energy of chemisorption.

Soln. : We know that $\ln\frac{k_2}{k_1} = \frac{E_a}{R}\left[\frac{T_1 - T_2}{T_1 T_2}\right]$,

where k_1 and k_2 are specific rate constants at temperature T_1 and T_2 respectively.

By the problem,

k_2 = 135, k_1 = 100, T_2 = 1000 K, T_1 = 600 K.

$$\therefore \quad \ln\frac{135}{100} = \frac{E_a}{8.314}\times\left(\frac{400}{600\times1000}\right)$$

or, E_a = **3.74 kJ.**

15. Argon was absorbed on charcoal to the extent of 0.835 gm/cc at a pressure of 2.4 atm and a temperature of 190 K. But at 250 K the same amount can be adsorbed only when the pressure is 16 atm. Calculate the heat of adsorption.

Soln. : We know that $\ln\frac{P_2}{P_1} = \frac{\Delta H}{R}\left[\frac{T_1 - T_2}{T_1 T_2}\right]$.

By the problem, P_2 = 16 atm, P_1 = 2.4 atm., T_1 = 190 K, T_2 = 250 K.

$$\therefore \Delta H = -\frac{190\times 250\times 8.314 \ln \frac{16}{2.4}}{60}$$

$\therefore \Delta H =$ **−12.5 kJ.**

16. Adsorption is studied on a clean surface. Let a solid having density of surface sites 1.3×10^{16} cm^{-2} be placed in a high vacuum chamber at a pressure of 1.00×10^{-12} mm of Hg at 298.15 K. Let there be air of average molecular weight 29 be present. How long will it take to occupy 1% of surface sites, assuming that every striking molecule is adsorbed.

Soln. : Number of molecules striking the surface at 298.15 K and 1.00×10^{-12} mm of Hg

$$= \frac{PN_A}{\sqrt{2\pi MRT}} \text{ m}^{-2}.\text{sec}^{-1}$$

$$= \frac{\left(\frac{1.00\times 10^{-12}}{760}\times 1.01325\times 10^{5}\right)\text{N/m}^2 \times 6.022\times 10^{23}\text{ mole}^{-1}}{\sqrt{2\times \pi \times 29\times 10^{-3}\text{ kg.mole}^{-1}\times 8.314\text{ J.K}^{-1}.\text{mole}^{-1}\times 298.15\text{ K}}}$$

$= 3.778 \times 10^{12}$ m^{-2}.sec^{-1} $\times 10^{-3} = 3.778 \times 10^{8}$ cm^{-2}.sec^{-1}.

In one second the fraction of surface sites occupied by the molecules

$$= \frac{3.778\times 10^{8}}{1.3\times 10^{16}} = 2.9 \times 10^{-8}.$$

Hence, time taken by the molecules to occupy 1% of the surface sites

$$= \frac{3.778\times 10^{8}}{1.3\times 10^{16}} = \mathbf{3.45 \times 10^{5}\ sec.}$$

17. From the following data for the adsorption of N_2 gas on a solid surface, calculate the constant involved in Langmuir isotherm; $P = \infty$, $V_1 = 180$ cc, $P = 3.5$ atm, $V_2 = 100$ cc/gm [V = volume of gas adsorbed]. **[B.U. 1996]**

Soln. : By Langmuir equation, $\theta = \frac{V}{V_m} = \frac{KP}{1+KP}$.

Now, $P_1 = \infty$, $V_1 = 180$ cc/gm; $P_2 = 3.5$ atm, $V_2 = 100$ cc/gm.

$$\text{Thus, } \frac{180}{100} = \frac{KP_1}{1+KP_1}\cdot\frac{1+KP_2}{KP_2}$$

$$= \frac{1+KP_2}{KP_2} \text{ (since, } P_1 = \infty) = \frac{1+3.5K}{3.5K}.$$

K = **0.36 atm^{-1}.**

18. Adsorption of N_2(g) on a piece of solid graphite was studied at 97 K. The volumes of adsorbate are reported at different gas pressures :

P/atm	3.49	10	16.70	25.70	29.55
V/cc	3.28	4.57	4.98	5.27	5.36

Assuming that the Langmuir equation is obeyed, determine K_a and V_m, the monolayer volume.

Soln. : By Langmuir equation, $\frac{V}{V_m} = \frac{K_a P}{1+K_a P}$.

Choosing any two readings,

$$\frac{3.28}{4.57} = \frac{K_a \times 3.49}{1+K_a 3.49} \cdot \frac{1+K_a .10}{10K_a}, \; K_a = 0.37 \text{ atm}^{-1}$$

$$\frac{4.98}{5.27} = \frac{16.70\,K_a}{1+16.70\,K_a} \cdot \frac{1+25.70\,K_a}{25.70\,K_a}, \; K_a = 0.32 \text{ atm}^{-1}$$

$$\frac{5.27}{5.36} = \frac{25.70\,K_a}{1+25.70\,K_a} \cdot \frac{1+29.55\,K_a}{29.55\,K_a}, \; K_a = 0.26 \text{ atm}^{-1}.$$

Thus, $K_a = \frac{0.37+0.32+0.26}{3}$ = **0.32 atm⁻¹.**

Choosing any one reading,

$$\frac{5.36}{V_m} = \frac{0.32 \times 29.55}{1+0.32 \times 29.55}, \; \therefore \; V_m = \mathbf{5.927 \text{ cc.}}$$

19. At 27 °C and at a pressure of 1 atm, only 0.25 fraction of the total surface of an iron catalyst is covered by molecular nitrogen. What is the standard free energy change for adsorption pressure at 27 °C? **[V.U. 1998]**

Soln. : By Langmuir isotherm, $0.25 = \frac{1 \times K_a}{1+K_a}$ (since, P = 1 atm)

$K_a = 0.33 \text{ atm}^{-1}$

$\Delta G° = -RT \ln K_a = -\,8.314 \times (273.15 + 27) \times \ln 0.33 = \mathbf{2.767 \text{ kJ.}}$

20. A film containing 4.22×10^{-5} gm fatty acid of molar mass 228 gm spread on water was compressed to a monolayer at 25 °C occupying an area of 245 sq. cm. The density of fatty acid is 0.812 gm/cc. Calculate (a) the area of cross section and (b) the length of the molecule (c). Also calculate the surface tension if surface tension for water at 25 °C is 0.072 Nm^{-1}

Soln. : Proceed as Problem 1.

(a) a = area of cross section = 21.98 sq. Å

(b) h = length of the molecule = 21.21 Å

(c) Now, σ = area occupied by one mole of fatty acid

$= 6.022 \times 10^{23} \times 21.98$ sq. Å

$= 6.022 \times 10^{23} \times 21.98 \times 10^{-20}$ sq. m (10^{-10} m = 1 Å)

Now, $\pi\sigma = RT$ (π = surface pressure)

Hence, $\pi = \frac{8.314 \times (273.15+25)}{6.0222 \times 10^{23} \times 21.98 \times 10^{-20}} \text{ N.m}^{-1} = 0.0187 \text{ N.m}^{-1}$.

$\gamma_0 - \gamma = \pi$.

Thus, surface tension of the solution, $\gamma = \gamma_o - \pi = 0.072 - 0.0187 = \mathbf{0.053 \text{ N.m}^{-1}}$.

21. Find ΔH_{ads} for $N_2(g)$ at 1.03 bar given that 155 cm³ is adsorbed by 1 gm charcoal at 88 K and 15 cm³ at 273 K. **[V.U. 2000]**

Soln. : Let k_1 and k_2 be desorption rate constants at T_1 and T_2 respectively. Then, using Arrhenius equation,

$$\ln\frac{k_2}{k_1} = \frac{E_d}{R}\times\frac{T_2-T_1}{T_1T_2},$$

where E_d = activation energy for desorption.

Now, rate of desorption is inversely proportional to volumes adsorbed. So, by the problem,

$$\ln\frac{155}{15} = \frac{E_d}{8.314}\times\frac{273-88}{273\times88}.$$

$\therefore$ E_d = 2521.39438 J.

Now, assuming physical adsorption to be non-activated,

since $E_a - E_d = \Delta H_{ads}$, $E_d = -\Delta H_{ads}$ (when $E_a = 0$).

Thus, ΔH_{ads} = –2521.39 J = **–2.5214 kJ.**

(b) Colloids :

1. A quartz particle of diameter 1 × 10^{-14} cm in aqueous suspension at 25 °C (η_w = 0.8903 CP) migrates with a velocity of 3 × 10^{-3} cm/sec under an applied potential gradient of 10 V.cm^{-1}. Calculate the zeta-potential. Given, the dielectric constant of water is 78.30. **[C.U. 1986]**

Soln. : $\zeta = \dfrac{4\pi\eta v}{D.\left(\dfrac{\Delta\phi}{l}\right)} \times 300^2$ volt

Here, η = 8.903 × 10^{-3} poise

D = 78.30, v = 3 × 10^{-3} cm/sec, $\dfrac{\Delta\phi}{l}$ = 10 V.cm^{-1}.

$$\therefore\ \zeta = \frac{4\times3.142\times8.903\times10^{-3}\times3\times10^{-3}}{78.30\times10}\times300^2 = 0.0385837\ \text{V} = \mathbf{0.039\ V.}$$

2. The volume of a liquid transported through porous diaphragm is 1.8 cc/per sec. If the viscosity of the liquid is 8 × 10^{-3} poise, and the specific conductance of the solution is found to be 1.61 × 10^{-3} ohm^{-1}.cm^{-1}, calculate the ζ-potential in volts when the flowing current was found to be 6 amperes and dielectric constants of the liquid is 80.

Soln. : We know that, $\zeta = \dfrac{4\pi\eta V\kappa}{DI}\times 9\times10^4$ volt.

Here, η = 8 × 10^{-3} poise, V = 1.8 cc, κ = 1.61 × 10^{-3} ohm^{-1}.cm^{-1}, I = 6 A.

$$\therefore\ \zeta = \frac{4\times3.142\times8\times10^{-3}\times1.8\times1.61\times10^{-3}\times9\times10^{4}}{80\times6}$$

= 0.0546331 = **0.055 volts.**

3. Calculate the electrophoretic velocity of colloidal particles with zeta-potential of 0.102 volt in a medium with dielectric constant 80 and viscosity coefficient of 7.8 × 10^{-3} poise at 25 °C, if the applied potential gradient is 4.8 volt/cm.

Soln. : The velocity in cm/sec is given by $v = \dfrac{\zeta DE}{4\pi\eta\times300^2}$

Here, $\zeta = 0.102$, $\eta = 7.8 \times 10^{-3}$, $D = 80$, $E = 4.8$.

$$\therefore \quad v = \frac{0.102 \times 80 \times 4.8}{4 \times 3.142 \times 7.8 \times 10^{-3} \times 300^2} = \mathbf{4.44 \times 10^{-3}\ cm/sec.}$$

4. In equilibrium distribution of gold sol, in the gravitation field, the following data were recorded :

Radius of the particle = 629 Å.

$\Delta h = h_2 - h_1 = 44.6 \times 10^{-4}$ cm.

Density of sol = 19.31 gm/cc, density of pure solvent = 0.89 gm/cc, number of particles, $n_1 = 327.6$, $n_2 = 40.34$. $T = 293$ K. Calculate Avogadro number.

Soln. : We know, $N_0 = \dfrac{RT}{\frac{4}{3}\pi r^3(\rho - \rho_0)g\Delta h} \ln \dfrac{n_1}{n_2}$.

$$\therefore N_A = \frac{3 \times 8.314 \times 10^7 \times 293}{4 \times 3.142 \times \left(629 \times 10^{-8}\right)^3 (19.31 - 0.89) \times 981 \times 44.6 \times 10^{-4}} \ln \frac{327.6}{40.34}$$

$= 6.07235 \times 10^{23} \approx \mathbf{6.07 \times 10^{23}}$.

5. The volume of liquid transported through fine capillaries is 10 cc per sec. If the viscosity of the liquid is 0.8 CP, the specific conductance of the solution is 2×10^{-3} ohm^{-1}.cm^{-1} and the current flowing is 10 amperes, then calculate potential if the dielectric constant is 80.

Soln. : We know, $\zeta = \dfrac{4\pi\eta V\kappa}{DI} \times 9 \times 10^4$

Here, $\eta = 0.8$ CP $= 8.0 \times 10^{-3}$ poise.

$V = 10$ cc, $\kappa = 2 \times 10^{-3}$ ohm^{-1}.cm^{-1}, $D = 80.0$, $I = 10$ amperes.

Putting in equation, $\zeta = 0.226224$ V $=$ **0.23 V.**

6. If in the above case, $\kappa = 1.43 \times 10^{-3}$, $I = 6$ amperes, what will be ζ?

Soln. : Proceed as in Problem 5. **[Ans.** $\zeta = 0.27$ V]

7. A quartz particle of diameter 1×10^{-4} cm in aq. suspension at 25 °C migrates at a velocity of 3×10^{-8} cm/sec under potential gradient of 10 V/cm. Calculate zeta-potential at the quartz water interface (η of water = 0.89 CP, dielectric constant of water = 78.3). **[V.U. 1994]**

Soln. : Proceed as Problem 1. **[Ans.** $\zeta = 3.86 \times 10^{-7}$ V]

8. Calculate the electrophoretic velocity of colloidal particles with zeta potential of 0.04 V in a medium with dielectric constant 78 and viscosity coefficient 8.9×10^{-3} poise at 25 °C, if the applied potential gradient is 10.4 V/cm.

Soln. : We know, $\zeta = \dfrac{4\pi\eta v}{D.\dfrac{\Delta\phi}{l}} \times 300^2$.

$$\therefore \quad v = \frac{D.\dfrac{\Delta\phi}{l}.\zeta}{4\pi\eta \times 300^2}.$$

Here, $D = 78$, $\frac{\Delta\phi}{l} = 10.4$ V/cm, $\zeta = 0.04$ V, $\eta = 8.9 \times 10^{-3}$ poise

Putting the values,

$$v = \frac{78 \times 10.4 \times 0.04}{4 \times 3.142 \times 8.9 \times 10^{-3} \times (300)^2} \text{ cm/sec} = \mathbf{3.2 \times 10^{-3} \text{ cm/sec.}}$$

9. The flocculation value of $ZnCl_2$ for As_2S_3 sol is 0.68 (millimoles of electrolyte needed per litre of sol). How many ml of 0.24(M) $ZnCl_2$ will be sufficient to start the coagulation of 300 ml of sol?

Soln. : Let x ml of 0.24 (M) $ZnCl_2$ be needed.

$\therefore$ total volume = $(300 + x)$ ml.

By the problem, 0.24 (M) means 0.24 moles in 1 litre sol = 0.24 millimole/cc.

Thus, $0.24x$ millimole is needed.

Now, 0.68 millimole is needed per litre.

$\therefore$ $\frac{0.68}{1000}$ millimole is needed per cc.

$\therefore$ $\frac{0.68}{1000}(300 + x)$ is needed for $(300 + x)$ cc.

$$\therefore \frac{0.68 \times (300 + x)}{1000} = 0.24x$$

or, $0.68(300 + x) = 240x$ or, $204 = 239.32x$.

$$\therefore x = \frac{204}{239.32} = 0.8524 \quad \therefore x = \mathbf{0.85 \text{ ml.}}$$

10. The diameter of the micelle of a surfactant is 4 nm and the area of the head group of the constituent surfactant molecule is 0.20 nm². Calculate the number of surfactant molecules in the micelle. **[V.U. 1999]**

Soln. : Diameter of the micelle = 4 nm.

$\therefore$ radius = 2 nm.

$\therefore$ total surface area covered = $4\pi(2)^2$ (nm)^2 = 50.24 (nm)^2.

Now, total number of surfactant molecules (assuming close packing)

$$= \frac{50.24}{0.2} \approx \mathbf{251.}$$

11. 100 cc of a silver sol contains 3×10^{10} particles each of diameter 100 nm. If the solution requires 2.7×10^{-4} gm of Al^{3+} to attain iso-electric point. Calculate the charge density of silver particles. **[V.U. 1998]**

Soln. : Diameter of each particle = 100 nm = 100×10^{-9} m = 10^{-7} m = 10^{-5} cm.

$$\therefore \text{ surface area covered by each particle} = 4 \times 3.142 \times \left(\frac{10^{-5}}{2}\right)^2 \text{ cm}^2$$

$$= \frac{4 \times 3.142 \times 10^{-25}}{4} \text{ cm}^2.$$

Since, the sol contains 3×10^{10} particles, so total surface area covered

$$= \left(3\times10^{10}\times3.142\times10^{-25}\right)\text{cm}^2 = \left(3\times3.142\times10^{-15}\right)\text{cm}^2.$$

Now, charge on 1 mole (27 gm) of Al^{3+} = 3 × 96500 coulombs.

∴ charge on 2.7×10^{-4} gm Al^{3+}

$$= \frac{3\times96500\times2.7\times10^{-4}}{27} \text{ coulombs} = 2.895 \text{ coulombs.}$$

This charge neutralises the charge on sol particles of total surface area

$$= \left(3\times3.14\times10^{-15}\right)\text{cm}^2.$$

$$\therefore \text{ charge density} = \frac{2.895}{3\times3.14\times10^{-15}} = \mathbf{3.07 \times 10^{14}\ coulombs/cm^2.}$$

12. 2 gm of a water insoluble substance of density 0.89 gm/cm^3 is dispersed in 2 litre of water, leading to the formation of a colloidal sol containing 10^{14} particles of spherical shape per mm^3. Determine the radius of the particle.

Soln. : 1 mm^3 of the sol contains 10^{14} particles.

∴ 1 cm^3 of the sol solution contains 10^{17} particles.

Now, 2 litres of the sol solution contain $2 \times 10^3 \times 10^{17}$ particles

$= 2 \times 10^{20}$ particles.

Given mass = 2 gm.

Density = 0.89 gm/cm^3.

$$\therefore \text{ volume} = \frac{2}{0.89} = 2.2471 \text{ cm}^3.$$

$$\text{So, volume of each particle} = \frac{2.2471}{2\times10^{20}} = 1.12359 \times 10^{-20} \text{ cc.}$$

Hence, taking radius r (in cm),

$$\frac{4}{3}\pi r^3 = 1.12359 \times 10^{-20} \text{ cc.}$$

$$\therefore r = \left(2.68237\times10^{-21}\right)^{\frac{1}{3}} = 2.98052 \times 10^{-7} \text{ cm}$$

$$= 1.38944 \times 10^{-7} \text{ cm} = \mathbf{13.89\text{Å}}.$$

13. In a coagulation experiment 5 ml of AgI is mixed with distilled water and 0.1 (M) solution of NaCl so that total volume is 10 ml. It has been found that all tubes containing more than 4.9 ml of NaCl coagulates within 5 minutes. What will be the flocculation value of NaCl for AgI?

Soln. : Flocculation value is expressed in millimole/litre of electrolyte.

Given experiment, a tube of 10 ml is taken containing 5 ml of AgI sol and x cc of NaCl solution, so that the total solution is 0.1 (M) NaCl.

Flocculation takes place when NaCl added is 4.9 ml.

0.1 (M) solution means 0.1 mole per litre, i.e.,

0.1×10^{-3} mole/cc, i.e., 0.1 millimole/cc.

So, in the given coagulation 4.9 ml of NaCl is added so that in the total volume of 10 cc, there is 4.9 ml NaCl. Thus, there is $4.9 \times 0.1 = 0.49$ millimole in 10 cc, i.e., 0.49×100 millimole/litre.

Thus, the flocculation value is $0.49 \times 100 = 49$ millimole/litre.

(c) Catalysis :

1. The hydrolysis of an ester is catalysed by acid HA, A^- ions and H_3O^+. The rate constant of the reaction in the following 3 solutions are given in the table :

	Conc. of **HA**	*Conc. of* A^-	***Rate constant***
Soln. I	0.2 (M)	0.2 (M)	6.18×10^{-3}
Soln. II	0.2 (M)	0.1 (M)	5.95×10^{-3}
Soln. III	0.2 (M)	0.05 (M)	8.48×10^{-3}

Find the catalytic coefficent of H_3O^+, A^- and HA. (Given, pK_a of the acid 4.75)

Soln. : We know that first order rate constant $J = k_o + k_1C_{HA} + k_2C_{A^-} + k_3C_{H^+}$, where k_0 = non-catalytic rate constant.

k_1, k_2 and k_3 are respectively the catalytic coefficients of HA, A^- and H^+.

k_0 is so small that it is neglected. C_{H^+} values calculated for these three buffers are the following :

$$\text{Solution (1)} \longrightarrow \text{pH} = \text{pK}_a + \log\frac{C_{A^-}}{C_{HA}} = 4.75 + \log\frac{0.2}{0.2} = 4.75.$$

$$\therefore\ C_{H^+} = 1.78 \times 10^{-5}\ (M).$$

$$\text{Solution (2)} \longrightarrow \text{pH} = 4.75 + \log\frac{0.1}{0.2} = 4.75 - 0.3010 = 4.45.$$

$$\therefore\ C_{H^+} = 3.55 \times 10^{-5}\ (M).$$

$$\text{Solution (3)} \longrightarrow \text{pH} = 4.75 + \log\frac{0.05}{0.2} = 4.75 - 0.602 = 4.15.$$

$$\therefore\ C_{H^+} = 7.08 \times 10^{-5}\ (M).$$

Therefore, by the problem,

$$0.2k_1 + 0.2k_2 + 1.78 \times 10^{-5}\ k_3 = 6.18 \times 10^{-3} \qquad \text{(i)}$$

Again, $$0.2k_1 + 0.1k_2 + 3.55 \times 10^{-5}\ k_3 = 5.95 \times 10^{-3} \qquad \text{(ii)}$$

$$0.2k_1 + 0.05k_2 + 7.08 \times 10^{-5}\ k_3 = 8.48 \times 10^{-3}. \qquad \text{(iii)}$$

Subtracting equation (ii) from equation (i),

$$0.1\ k_2 - 1.77 \times 10^{-5}\ k_3 = 0.23 \times 10^{-3}. \qquad \text{(iv)}$$

Subtracting equation (iii) from equation (ii),

$$0.05\ k_2 - 3.53 \times 10^{-5}\ k_3 = -2.53 \times 10^{-3}. \qquad \text{(v)}$$

Multiplying (v) by 2,

$$0.1\ k_2 - 7.06 \times 10^{-5}\ k_3 = -5.06 \times 10^{-3}. \qquad \text{(vi)}$$

Subtracting equation (iv) from equation (vi),

$$k_3 = \frac{5.29 \times 10^{-3}}{5.29 \times 10^{-5}} = \mathbf{10^2}.$$

Putting k_3 in (vi),

$$0.1\ k_2 = 7.06 \times 10^{-3} - 5.06 \times 10^{-3}$$

or, $0.1\ k_2 = 2 \times 10^{-3}$, $\therefore\ k_2 = \mathbf{2 \times 10^{-2}}$.

Putting k_2 in equation (1),

$$0.2\ k_1 + 0.2\ (2 \times 10^{-2}) + 1.78 \times 10^{-5}\ (10^2) = 6.18 \times 10^{-3}$$

or, $0.2\ k_1 = 4 \times 10^{-4}$

or, $k_1 = \mathbf{2 \times 10^{-3}}$.

2. A reaction is catalysed by acetic acid, acetate ions and proton. The rate constant of the reaction at 25 °C in two solutions, one containing 0.2 mole acetic acid and acetate ions and the other containing 0.2 mole acetic acid and 0.1 mole acetate ions are 4.4×10^{-3} sec^{-1} and 2.4×10^{-3} sec^{-1} respectively. Find the catalytic coefficients of acetate ions and acetic acid respectively. (assuming $C_{H_3O^+}$ to be low enough to be ignored). **[V.U. 1995]**

Soln. : Proceed as in Problem no. 1.

However, here the two equations are :

$$k_{HA} \times 0.2 + k_{A^-} \times 0.2 = 4.4 \times 10^{-3} \quad (1)$$

$$k_A \times 0.2 + k_{A^-} \times 0.1 = 2.4 \times 10^{-3} \quad (2)$$

$\therefore\ k_{A^-} = 2 \times 10^{-2}\ sec^{-1}.\ (gm/l)^{-1}$

$k_{HA} = 2 \times 10^{-3}\ sec^{-1}.\ (gm/l)^{-1}$.

3. The hydrolysis of a substance is specifically catalysed by hydrogen ions, the rate constants being given by $k = 4.7 \times 10^{-2}\ C_{H_3O^+}\ sec^{-1}$. When the substance was dissolved in 10^{-3} (M) solution of an acid HA, the rate constant was found to be $3.2 \times 10^{-5}\ sec^{-1}$. Calculate the dissociation constant of the acid HA.

Soln. : For the acid HA,

$$HA + H_2O \rightleftharpoons H_3O^+ + A^-$$

$$(1-\alpha)C \qquad \alpha C \qquad \alpha C$$

where α = degree of dissociation, C = concentration in molarity.

$$\therefore\ K_a = \frac{\alpha^2 C}{1-\alpha},\ \text{if } 1 >> \alpha,\ \alpha = \sqrt{\frac{K_a}{C}}.$$

$$\therefore\ C_{H_3O^+} = \sqrt{K_a.C}.$$

Thus, by the problem,

$$3.2 \times 10^{-5} = 4.7 \times 10^{-2} \sqrt{K_a.C}$$

$$= 4.7 \times 10^{-2} \sqrt{K_a \times 10^{-3}}$$

$$\therefore\ K_a \times 10^{-3} = 4.6355 \times 10^{-7}$$

or, $K_a = 4.6355 \times 10^{-4} \approx \mathbf{4.6 \times 10^{-4}}$.

4. H_3O^+ acts as a homogenous catalyst in the reaction of inversion of sucrose. At pH = 3 the reaction in aqueous solution proceeds with constant half-life of 50 minutes. What value of half-life would you expect at pH = 4? **[C.U. 1997]**

Soln. : The first order rate constant $J = k_{H^+}.C_{H^+}$,

where k_{H^+} = catalytic coefficient of H^+ ion

C_{H^+} = concentraton of H^+ ion.

$$\therefore \quad k_{H^+} = \frac{J}{C_{H^+}}.$$

$$\text{Now,} \quad J = \frac{0.693}{t_{1/2}}.$$

In the first case, $J = \frac{0.693}{50}$ at pH = 3 and $C_{H^+} = 10^{-3}$.

$$k_{H^+} = \frac{0.693}{50 \times 10^{-3}} = 13.86.$$

In the second case, when pH = 4, $C_{H^+} = 10^{-4}$.

So, in the second case, $J = \frac{0.693}{50 \times 10^{-3}} \times 10^{-4} = 1.386 \times 10^{-3}$.

$$\therefore J = \frac{0.693}{500}.$$

$$\text{But } J = \frac{0.693}{t_{1/2}}$$

$$\text{or, } t_{1/2} = \frac{0.693}{J} = \frac{0.693}{1.386 \times 10^{-3}} = \mathbf{500\ min.}$$

5. The hydrolysis of a substance is catalysed by H_3O^+ and the rate is given by $r = 5.58 \times 10^{-2} C_{H_3O^+} sec^{-1}$.

When the substance is dissolved in 10^{-2} (M) solution of an acid HA, the rate constant was found to be 2.4×10^{-5}. Calculate the dissociation constant of acid.

Soln. : Considering the acid HA, the dissociation equation is :

$$HA \rightleftharpoons H^+ + A^-$$

$$\therefore \quad (1-\alpha)C \qquad \alpha C \quad \alpha C.$$

α = degree of dissociation, C = molar concentration.

$$\therefore \quad K_a = \frac{\alpha^2 C}{1-\alpha} \approx \alpha^2 C \ (\because 1 >> \alpha)$$

$$\therefore \quad \alpha = \sqrt{\frac{K_a}{C}}.$$

$$\therefore \quad C_{H_3O^+} = \alpha.C = \sqrt{K_a.C}.$$

$$\therefore \quad r = 5.58 \times 10^{-2} \sqrt{K_a.C}.$$

When $C = 10^{-2}$ (M), the rate constant is 2.4×10^{-5}.

Hence, $2.4 \times 10^{-5} = 5.58 \times 10^{-2} \sqrt{10^{-2} \times K_a}$

or, $\left(2.4\times10^{-5}\right)^2 = \left(5.58\times10^{-2}\right)^2 \times 10^{-2} \times K_a$

$\therefore \quad K_a = \mathbf{1.85 \times 10^{-5}}$.

6. The first order constant for a catalyst by a weak acid HA is 8.4×10^{-5} sec^{-1}. For this reaction $k_{H^+} = 1.4 \times 10^{-2}$ sec^{-1}. Calculate the acidity constant K_a by using $J = k_{H^+} \cdot (K_a)^{\frac{1}{2}}$.

Soln. : Given, J = first order rate constant = 8.4×10^{-5} sec^{-1}.

$k_{H^+} = 1.4 \times 10^{-2}$ sec^{-1}.

Now, $8.4 \times 10^{-5} = 1.4 \times 10^{-2} \times (K_a)^{\frac{1}{2}}$

or, $\left(8.4\times10^{-5}\right)^2 = \left(1.4\times10^{-2}\right)^2 . K_a$

$$\therefore K_a = \frac{\left(8.4\times10^{-5}\right)^2}{\left(1.4\times10^{-2}\right)^2} = \mathbf{3.6 \times 10^{-5}}.$$

7. A reaction is catalysed by acetic acid, acetate ions and H_3O^+. The rate constants of the reaction at 25 °C in two solutions, one containing 0.2 mole of acetic acid and acetate and another containing 0.2 mole of acetic acid and 0.1 mole of acetate, are 4.4×10^{-3} sec^{-1} and 2.4×10^{-3} sec^{-1} respectively. Find the catalytic coefficients of acetic acid and acetate at the same temperature. [Assume H_3O^+ concentration to be low enough to be ignored]. **[V.U. 1994]**

Soln. : Given,

	Conc. of CH_3COOH	Conc. of CH_3COO^-	Rate constant
Solution I	0.2(M)	0.2(M)	$4.4 \times 10^{-3} sec^{-1}$
Solution II	0.2(M)	0.1(M)	$2.4 \times 10^{-3} sec^{-1}$

We know, $J = k_{HA} C_{HA} + k_{A^-} C_{A^-}$ (neglecting $C_{H_3O^+}$)

Solution I : $4.4 \times 10^{-3} = k_{HA}\ 0.2 + 0.2\ k_{A^-}$ (1)

Solution II : $2.4 \times 10^{-3} = 0.2\ k_{HA} + 0.1\ k_{A^-}$ (2)

Solving equations (1) and (2), we get

$k_{A^-} = \mathbf{0.02}$

$k_{HA} = \mathbf{0.002}$.

8. An enzyme following Michaelis-Menton kinetics was found to have highest activity at 37 °C and pH 7.0. If maximum velocity V_{max} for the enzyme was 2.4×10^{-4} mol. $l^{-1}.s^{-1}$ with an initial ezyme concentration [E] = 2.4 nmol, calculate turnover frequency. **[JAM 2007]**

Soln. : Turnover frequency $= \dfrac{2.4\times10^{-4}}{2.4\times10^{-9}} = \mathbf{10^5}$ **moles/sec.**

CHAPTER 15

QUANTUM MECHANICS

Required Formulae

1. Black body radiation :

Stefan-Boltzmann law :

$$R_T = \sigma T^4,$$

where R_T = total energy radiated per unit time per unit area of black body

T = absolute temperature

σ = Stefan-Boltzmann constant = 5.67×10^{-8} W/m²K⁴.

If ν_{max} is the frequency at which R_T (ν) is maximum at temperature T, then according to Wien's displacement law,

$$\nu_{max} \propto T.$$

2. Photoelectric effect :

$$\frac{1}{2} mv^2 = h\nu - W = h\nu - h\nu_0,$$

where m is the mass of election, v is the velocity of the ejected electron, ν is the frequency of radiation striking the metal surface, W is the work-function of the metal, ν_0 is the threshold or cut-off frequency and h is Planck constant.

$$eV_0 = \frac{1}{2} mv^2,$$

where V_0 is the stopping potential, e is the charge on the electron and m and v have their usual meanings.

3. If a particle of mass m moves with velocity v, its de Brogle wavelength $\lambda = \dfrac{h}{mv} = \dfrac{h}{p}$, where p is the momentum of the particle.

4. If an electron of charge e and mass m is accelerated through a potential difference of V, then its kinetic energy $\frac{1}{2} mv^2 = eV$, where v is the velocity of electron.

$$\frac{p^2}{2m} = eV \quad \text{or,} \quad p = \sqrt{2meV}.$$

Hence, associated wavelength $\lambda = \frac{h}{\sqrt{2meV}}$.

5. Heisenberg uncertainty principle :

If Δx is the uncertainty in the position x, Δp_x is the uncertainty in the momentum p_x

$$\Delta x\ \Delta p_x \geq \frac{\hbar}{2}, \text{ where } \hbar = \frac{h}{2\pi}.$$

- If ΔE is the uncertainty in the energy E and Δt is the uncertainty in the time

$$\Delta E\ \Delta t \geq \frac{\hbar}{2}.$$

6. Probability of finding a particle described by wave function ψ between a and b

$$P = \int_a^b \psi^* \psi dx.$$

7. If N is a normalization constant and $N\psi' = \psi$, then

$$\int_{-\infty}^{+\infty} \psi^* \psi dx = 1 \text{ or, } N^2 \int_{-\infty}^{+\infty} \psi^* \psi dx = 1.$$

8. Particle in a box :

Energy of the particle in one dimensional box, $E_n = \frac{n^2h^2}{8ma^2}$ (n = 1, 2, 3, ..., etc.)

where E_n is the energy of the particle of mass m at nth level and a is the length of the box.

Excitation energy, $\Delta E = E_{n+1} - E_n = \frac{(n+1)^2h^2}{8ma^2} - \frac{n^2h^2}{8ma^2} = \frac{(2n+1)h^2}{8ma^2}$.

Energy at the particle of mass m in a cube of volume a^3,

$$E_n = \frac{(n_x^2 + n_y^2 + n_z^2)h^2}{8ma^2} = \frac{n^2h^2}{8ma^2}$$

n_x = 1, 2, 3, ..., etc.
n_y = 1, 2, 3, ..., etc.
n_z = 1, 2, 3, ..., etc.

Wave function of a particle in one dimensional box of length a at nth energy state

$$\psi_n = \sqrt{\frac{2}{a}} \sin \frac{n\pi x}{a}$$

Wave function associated with a particle in a cube of volume a^3,

$$\psi_n = \left(\sqrt{\frac{2}{a}}\right)^3 \sin \frac{n_x \pi x}{a} \sin \frac{n_y \pi y}{a} \sin \frac{n_z \pi z}{a}.$$

9. For a particle of mass m possessing energy E tunnelling through a potential barrier of width a and height V with $E < V$,

T = transmission coefficient

$$= \frac{1}{\cosh^2 \beta a + \frac{1}{4}\left(\frac{\alpha}{\beta} - \frac{\beta}{\alpha}\right)^2 \sinh^2 \beta a}$$

where $\beta = \frac{1}{h}\sqrt{2m(V-E)}$

$$\alpha = \sqrt{\frac{2mE}{h}}.$$

$$\cosh \beta a = \frac{1}{2}(e^{\beta a} + e^{-\beta a})$$

$$\sinh \beta a = \frac{1}{2}(e^{\beta a} + e^{-\beta a}).$$

10. Energy of a harmonic oscillator at nth state, $E_n = \left(n+\frac{1}{2}\right)h\nu$, $n = 0, 1, 2, \ldots$etc.,

where ν is the frequency of oscillation. $\nu = \frac{1}{2\pi}\sqrt{\frac{k}{\mu}}$, where k is the force constant and is reduced mass.

11. If ψ_1 and ψ_2 are orthogonal, $\int_{-\infty}^{+\infty} \psi_1^* \psi_2 dx = 0.$

12. $\int_0^{\infty} x^n e^{-qx} dx = \frac{n!}{q^{n+1}}, \; n > -1.$

hort Questions

1. Photoelectric work function and ionization potential of a metal are not the same—omment. **[B.U.(H) 1999]**

Ans. : Work function, a characteristic energy of a metal, is defined as the minimum nergy an electron must possess to leave the metal surface overcoming the attractive forces at normally bind the electron to the metal.

On the other hand, ionization potential is the energy required to remove an electron om an isolated gaseous atom.

Hence, it is evident from the definitions that photoelectric work function and ionization otential of a metal cannot be the same.

2. In a photoelectric experiment when the emitter metal is irradiated with light of vo different frequencies the following observations were made :

(i) electrons could not be emitted by using a very intense beam of light having wavelength λ_1.

(ii) electrons could be easily emitted using a weakly intense beam of weavelength λ_2.

How do you explain the phenomena? Which wavelength is shorter. **[B.U.(H) 1996]**

Ans. : Light of frequency ν consists of particles, called photons, each with energy of magnitude $h\nu$, where h is Planck constant. When a single photon strikes the metal urface, its entire energy is absorbed by an electron. Now, to escape the metal surface vercoming the attractive forces electron requires a minimum energy. So, if the impinging equency is not sufficient, ejection of electron would never occur, no matter how intense e light is.

Energy of the photon constituting light is $h\nu$ or $h\frac{c}{\lambda}$, where λ is the wavelength of light. So, lower the value of λ, the greater will be the energy of photon. Now, the weakly intense light of wavelength λ_2 has more energy than the other with wavelength λ_1. Thus, the weakly intense light has shorter wavelength.

3. Write notes on :

(i) The Stefan-Boltzmann law.

(ii) The Wien's displacement law indicating the quantities involved. **[C.U.(H) 2002]**

Ans. : (i) Stefan-Boltzmann law states that the total energy radiated (R_T) per unit area per unit time from the surface of a black body at a particular temperature T increases with the increasing temperature.

$$R_T = \sigma T^4$$

σ = Stefan-Boltzmann constant.

(ii) The frequency at which maximum radiation per unit time per unit area of the surface of the black body occurs at particular temperature T shifts towards higher frequency as T increases.

$$\nu_{max} \propto T$$

ν_{max} = frequency at which radiation power is most intense.

4. Make use of de Broglie relationship to calculate the total energy of H atom in the Bohr model, where radius is assumed to be equal to a_0. **[B.U.(H) 1996]**

Ans. : De Broglie relation relates wavelength (λ) of matter wave associated with the motion of a material particle to its momentum p.

According to de Broglie relationship, $\lambda = \frac{h}{p}$, where h = Planck constant.

In the Bohr model, electron of charge $-e$ and mass m is moving with velocity v around nucleus of charge $+e$ in a circular orbit of radius a_0. Circumference of allowed orbits must accomodate integral multiple of de Broglie wavelength associated with the moving electron.

Hence, $n\lambda = 2\pi a_0$, $n = 1, 2, 3, \ldots$, etc.

$$\text{or, } \lambda = \frac{2\pi a_0}{n}. \quad (1)$$

$$\text{Now, } p = \frac{h}{\lambda}.$$

$$\text{Thus, } p = \frac{nh}{2\pi a_0} = n\frac{\hbar}{a_0}, \text{ where, } \hbar = \frac{h}{2\pi}. \quad (2)$$

According to Bohr, there exists a mechanical equilibrium between centrifugal force and centripetal force in the orbit.

$$\text{Centrifugal force} = \frac{mv^2}{a_0}.$$

Centrifugal force = Coulomb force between electron and nucleus

$$= \frac{e^2}{a_0^2}$$

At equilibrium, $\dfrac{mv^2}{a_0} = \dfrac{e^2}{a_0^2}$.

$$\therefore \quad v^2 = \frac{e^2}{ma_0}. \tag{3}$$

$$\text{Again, } p = mv = n\frac{\hbar}{a_0}, \ \therefore \ v^2 = \frac{n^2\hbar^2}{a_0^2 m^2}. \tag{4}$$

$$\text{Thus,} \quad \frac{e^2}{ma_0} = \frac{n^2\hbar^2}{a_0^2 m^2}, \ \therefore \ a_0 = \frac{n^2\hbar^2}{e^2 m}. \tag{5}$$

Total energy of an electron in hydrogen atom = kinetic energy + potential energy

$$= \frac{p^2}{2m} - \frac{e^2}{a_0}$$

$$= \frac{n^2\hbar^2}{2a_0^2 m} - \frac{e^2}{a_0} = -\frac{me^4}{2n^2\hbar^2} \quad \text{[putting the value of } a_0 \text{ from equation (5)]}$$

5. How does the de Broglie hypothesis lead to one important postulate made by Bohr in his theory of hydrogen atom? **[B.U.(H) 1997]**

Ans. : De Broglie hypothesis states that every moving particle has wave associated with it that governs its motion.

Thus, according to de Broglie hypothesis, $\lambda = \dfrac{h}{p}$

where λ is the de Broglie wavelength and p is the momentum of the particle. Now, according to Bohr model of hydrogen atom, an electron of mass m moves in a circular orbit of radius r with velocity v.

Electron has a wave associated with it. Now, the circumference of a stable orbit must contain integral number of the wavelength λ.

Hence, $n\lambda = 2\pi r$, $\quad n = 1, 2, 3, \ldots$, etc.

$$\therefore \quad \lambda = \frac{2\pi r}{n}. \tag{1}$$

Again, by de Broglie relation, $p = \dfrac{h}{\lambda}$

or, $\quad p = \dfrac{nh}{2\pi r}$ [putting the value of λ from (1)]

$$\therefore \quad mvr = n\hbar, \tag{2}$$

where $\hbar = \dfrac{h}{2\pi}$, $\ p = mv$.

Equation (2) predicts that angular momentum (mvr) of electron is allowed to have only certain discrete values and this is the important postulate made by Bohr in his theory of hydrogen atom regarding the quantization of angular momentum.

6. Write action integrals for radial and angular motions in atom. **[B.U.(H) 2001]**

Ans. : $\oint L d\phi = n_\phi h$, where L = angular momentum, ϕ = polar coordinate.

$n_\phi = 1, 2, 3, \ldots$, etc.

$\oint P_r dr = n_r h$,

where p_r = radial momentum, r = polar coordinate, h = Planck constant.

$n_r = 0, 1, 2, \ldots$, etc.

7. Write down the dimensions (in M, L, T) for force, angular momentum and action.

[B.U.(H) 1996]

Ans. : Dimension of force = MLT^{-2} ($\because$ force = mass × acceleration)

Dimension of action = ML^2T^{-1} ($\because$ action = energy × time).

Dimension of angular momentum = ML^2T^{-1} ($\because$ angular momentum = mass × velocity × distance).

8. State Wilson-Sommerfeld quantization rule.

Ans. : Every physical system in which coordinates are periodic functions of time obeys a quantum condition. This condition is $\oint p_q dq = nh$ (n = 0, 1, 2, ...), where p_q is the momentum associated with the coordinate q, n is a quantum number that takes on only integral values, and $\oint$ means that the integration extends over one period of q.

The above relation is known as Wilson-Sommerfeld quantization rule.

9. Explain Bohr correspondence principle, taking the case of H atom.

[B.U.(H) 2002]

Ans. : Correspondence principle states that quantum theory approaches classical theory asymptotically in the limit of large quantum number.

In the Bohr model of hydrogen atom, an electron of mass m and charge $-e$ is rotating around nucleus of charge $+e$ in a circular orbit of radius r.

According to Bohr theory of H atom for a stationary orbit,

$$\frac{mv^2}{r} = \frac{e^2}{r^2}, \tag{1}$$

where $\frac{mv^2}{r}$ = centrifugal force

$\frac{e^2}{r^2}$ = centripetal force

and $mvr = n\hbar$ (n = 1, 2, ...) (2)

where mvr = angular momentum

n = quantum number

$\hbar = \frac{h}{2\pi}$, h = Planck constant.

From (1), we get $r = \frac{e^2}{mv^2}$. (3)

Putting the value of r in equation (2), we get

$$v = \frac{e^2}{n\hbar}. \tag{4}$$

Hence, $r = \frac{n^2\hbar^2}{me^2}$. (5)

Now, if ν be the frequency of radiation that an electron would emit when it jumps from $(n + 1)$th orbit to nth orbit, then

$$h\nu = E_{n+1} - E_n$$

$$= \frac{me^4}{2\hbar^2}\left[\frac{1}{n^2} - \frac{1}{(n+1)^2}\right] \text{ [since energy of Bohr orbit} = E_n = -\frac{e^2}{2r} = -\frac{me^4}{2n^2\hbar^2}\text{]}$$

$$\text{or, } \nu = \frac{1}{2\pi}\cdot\frac{me^4}{2h^3}\left[\frac{(n+1)^2 - n^2}{n^2(n+1)^2}\right]$$

$$\text{or, } \nu = \frac{me^4(2n+1)}{2.2\pi\hbar^3 n^4}.$$

$$\therefore \nu = \frac{me^4}{2\pi\hbar^3 n^3} \quad \text{[since } n >> 1,\ n+1 \approx n,\ 2n+1 \approx 2n\text{]}. \qquad (6)$$

In classical physics, frequency (ν_{cl}) of radiation radiated by an rotating electron,

$$\nu_{cl} = \frac{v}{2\pi r}$$

$$= \frac{1}{2\pi}\ \frac{e^2}{nh}\ \frac{me^2}{n^2h^2} \text{ [putting the values from eqns. (4) and (5)]}$$

$$= \frac{me^4}{2\pi n^3 h^3} = \nu \text{ [from eqn. (6)]}.$$

10. Apply Wilson-Sommerfeld quantization rule to obtain an energy expression for a particle in one dimensional box.

Ans. : The particle is moving back and forth within a one-dimensional box of dimension a. So, the distance covered by the particle in one period is $2a$.

Now, by Wilson-Sommerfeld quantization rule,

$\oint p_x dx = nh$, where p_x is linear momentum along x direction, n is the quantum number that takes on only integral value, h is Planck constant, and $\oint$ means that the integration is taken over one period of motion.

Thus, $\oint p_x dx = p_x \oint dx$ (since p_x is constant as no force is acting on the particle)

$= p.\ 2a = nh.$

$$\therefore p = \frac{nh}{2a}.$$

Energy of the particle $= E = \dfrac{p^2}{2m} = \dfrac{n^2h^2}{8ma^2}$, where m = mass of the particle.

11. Prove that $\dfrac{d}{dx}$ is not Hermitian in nature. **[V.U.(H) 1997]**

Soln. : An operator $\hat{A}$ is said Hermitian if it obeys the following relation :

$$\int_{-\infty}^{+\infty} \psi_1^* \hat{A} \psi_2 dx\, dx = \int_{-\infty}^{+\infty} \psi_2 \hat{A}^* \psi_1^* dx$$

Now, $\int_{-\infty}^{+\infty} \psi_1^* \frac{d}{dx} \psi_2 dx = \left(\psi_1^* \psi_2\right)_{-\infty}^{+\infty} - \int_{-\infty}^{+\infty} \psi_2 \frac{d\psi_1^*}{dx} dx$

$= -\int_{-\infty}^{+\infty} \psi_2 \frac{d}{dx} \psi_1^* dx$ (since ψ_1 and ψ_2 become zero at infinity).

$\therefore \quad \int_{-\infty}^{+\infty} \psi_1^* \frac{d}{dx} \psi_2 dx = -\int_{-\infty}^{+\infty} \psi_2 \left(\frac{d}{dx}\right)^* \psi_1^* dx \quad [\because \frac{d}{dx} = \left(\frac{d}{dx}\right)^*]$.

Hence, $\frac{d}{dx}$ is not Hermitian, rather it is an anti-Hermitian operator.

12. Prove that Hermitian operator gives us real eigenvalues **[V.U.(H) 1997]**

Soln. : Let $\hat{A}$ be a Hermitian operator and ψ is its eigenfunction with eigenvalue a, such that,

$$\hat{A}|\psi\rangle = a|\psi\rangle \tag{1}$$

$$(A|\psi\rangle)^* = \langle A\psi| = a^*\langle\psi|. \tag{2}$$

Since $\hat{A}$ is Hermitian, a^* is complex conjugate of a.

Now, $\langle\psi|\hat{A}|\psi\rangle = a\langle\psi|\psi\rangle$ (3)

Again, $\langle\psi|\hat{A}\psi\rangle = \langle\hat{A}\psi|\psi\rangle$ (applying turn-over rule)

$= a^*\langle\psi|\psi\rangle$. (4)

Now, subtracting (4) from (3), we get

$$(a - a^*)\ \langle\psi|\psi\rangle = 0$$

or, $a - a^* = 0$ (since, $\langle\psi|\psi\rangle \neq 0$)

$\therefore$ $a = a^*$.

Thus, a is real.

13. Explain the terms : linear operator, eigenvalue and stationary state. **[V.U.(H) 2001]**

Soln. : Linear operator : An operator $\hat{A}$ is called linear if it possesses the following properties :

(i) for any arbitrary functions $f(x)$ and $g(x)$,

$$\hat{A}(f(x) + g(x)) = \hat{A}f(x) + \hat{A}g(x) \text{ and}$$

(ii) $\hat{A}\, c\, f(x) = c\,\hat{A}f(x)$, where c is an arbitrary constant. e.g., $\frac{d}{dx}$

Eigenvalue : If the action of an operator on a function gives back the same function multiplied by a constant, then the constant is called eignevalue. e.g., $\hat{H}\psi = E\psi$.

E is the eigenvalue of the operator $\hat{H}$.

Stationary state : A state which does not evolve with time is called stationary state. For stationary state, the probability density $\psi^*\psi$, where ψ represents the state, does not change with time.

14. Does the function $e^{-iEt/\hbar}\psi(x)$ represent a stationary state?

Soln. : Given, $\psi(x, t) = \psi(x)\ e^{-iEt/\hbar}$.

$$\therefore \quad \psi^*(x, t)\ \psi\ (x, t) = \psi^*(x)\, e^{-iEt/\hbar}\psi(x)\ \ e^{-iEt/\hbar}$$

$$= \psi^*(x)\ \psi(x).$$

Thus, the probability density remains constant in time. Hence, the given function represents a stationary state.

15. Show that if two operators commute, they have simultaneous non-degenerate eigenfunctions. **[V.U.(H) 1994]**

Soln. : Let $|\psi\rangle$ be an eigenfunction of the operator $\hat{A}$ with an eigenvalue a.

Then $\hat{A}|\psi\rangle = a|\psi\rangle$. (1)

Let $\hat{B}$ be another operator which commutes with $\hat{A}$.

Hence, $\left[\hat{A}, \hat{B}\right]|\psi\rangle = 0$

or, $\hat{A}\hat{B}|\psi\rangle - \hat{B}\hat{A}|\psi\rangle = 0.$

$\therefore\ \hat{A}\left(\hat{B}|\psi\rangle\right) = \hat{B}\left(\hat{A}|\psi\rangle\right).$ (2)

But, $\hat{B}\hat{A}|\psi\rangle = a\hat{B}|\psi\rangle$ [from equation (1)] (3)

Comparing equations (2) and (3), we get

$A\left(\hat{B}|\psi\rangle\right) = a\left(\hat{B}|\psi\rangle\right).$ (4)

Thus, $\hat{B}|\psi\rangle$ is an eigenfunction of $\hat{A}$ with the eigenvalue a.

But, $|\psi\rangle$ is the only eigenfunction of $\hat{A}$ with the eigenvalue a, since $|\psi\rangle$ is non-degenerate eigenfunction of $\hat{A}$.

Hence, $\hat{B}|\psi\rangle$ cannot be a new eigenfunction of $\hat{A}$ with the same value a. It must be the function $|\psi\rangle$ with some constant.

Hence, $\hat{B}|\psi\rangle = b|\psi\rangle.$

Thus, $\hat{A}$ and $\hat{B}$ have same set of non-degenerate eigenfunctions.

16. Prove that if two eigenfunctions of a Hermitian operator have different eigenvalues, then the wave functions are orthogonal. **[V.U.(H) 2000]**

Soln. : Let $|\psi_1\rangle$ and $|\psi_2\rangle$ be two eigenfunctions of a Hermitian operator $\hat{A}$ with eigenvalues a_1 and a_2 respectively.

Thus, $\quad \hat{A}|\psi_1\rangle = a_1|\psi_1\rangle$ (1)

$\hat{A}|\psi_2\rangle = a_2|\psi_2\rangle$ (2)

Hence, $\left\langle \psi_1 \middle| \hat{A} \middle| \psi_2 \right\rangle = a_2 \left\langle \psi_1 | \psi_2 \right\rangle$ [by equaion (2)]. (3)

Again, $\left\langle \psi_1 \middle| \hat{A} \middle| \psi_2 \right\rangle = \left\langle \hat{A}\psi_1 \middle| \psi_2 \right\rangle$ [by turn-over rule]

$= a_1 \left\langle \psi_1 | \psi_2 \right\rangle$ [from equation (1)] (4)

Now, subtracting equation (3) from equation (4), we get

$$a_1 \left\langle \psi_1 | \psi_2 \right\rangle - a_2 \left\langle \psi_1 | \psi_2 \right\rangle = 0$$

or, $(a_1 - a_2) \left\langle \psi_1 | \psi_2 \right\rangle = 0.$

Since, $a_1 \neq a_2, \left\langle \psi_1 | \psi_2 \right\rangle = 0$, i.e., $\int \psi_1^* \psi_2 d\tau = 0$

Hence, $|\psi_1\rangle$ and $|\psi_2\rangle$ are orthogonal.

17. If two operators α and β are Hermitian, then find out the condition when αβ will be Hermitian. **[V.U.(H) 1995]**

Soln. : Since α and β are Hermitian,

$$\langle \psi_1|\alpha|\psi_2\rangle = \langle \alpha\psi_1|\psi_2\rangle \quad (1)$$

$$\langle \psi_1|\beta|\psi_2\rangle = \langle \beta\psi_1|\psi_2\rangle. \quad (2)$$

Now, $\langle \psi_1|\alpha\beta|\psi_2\rangle = \langle \alpha\psi_1|\beta|\psi_2\rangle$ [by turn-over rule]

$= \langle \beta\alpha\psi_1|\psi_2\rangle$ [by turn-over rule].

If $\alpha\beta = \beta\alpha$, i.e., $[\alpha, \beta] = 0$, then

$$\langle \psi_1|\alpha\beta|\psi_2\rangle = \langle \alpha\beta\psi_1|\psi_2\rangle.$$

Thus, if α and β commutes, αβ will be Hermitian.

18: Show that momentum operator is Hermitian.

Soln. : Momentum operator $= \hat{p}_x = -i\hbar\dfrac{\partial}{\partial x}$.

Now, $\displaystyle\int_{-\infty}^{+\infty} \psi_1^* \left(-i\hbar\frac{\partial}{\partial x}\right)\psi_2 dx = -i\hbar\int_{-\infty}^{+\infty} \psi_1^* \frac{\partial\psi_2}{\partial x} dx$

$$= -i\hbar\left[\left(\psi_1^*\psi_2\right)_{-\infty}^{+\infty} - \int_{-\infty}^{+\infty} \psi_2 \frac{\partial\psi_1^*}{\partial x} dx\right].$$

Since, ψ_1 and ψ_2 vanish at infinity,

$$\therefore \int_{-\infty}^{+\infty} \psi_1^* \left(-i\hbar\frac{\partial}{\partial x}\right)\psi_2 dx = i\hbar\int_{-\infty}^{+\infty} \psi_2 \frac{\partial\psi_1^*}{\partial x} dx$$

$$= \int_{-\infty}^{+\infty} \psi_2 \left(-i\hbar\frac{\partial}{\partial x}\right)^* \psi_1^* dx \quad [\because (-i)^* = i].$$

19. If ψ_1 and ψ_2 are eigenfunctions of linear operator $\hat{A}$ with the same eigenvalues a, show that any linear combination of ψ_1 and ψ_2 will also be an eigenfunction of $\hat{A}$ with the same eigenvalue. **[C.U.(H) 2000]**

Soln. : Given, $\hat{A}\psi_1 = a\psi_1$

$$\hat{A}\psi_2 = a\psi_2.$$

Let $\psi = C_1\psi_1 + C_2\psi_2$, where C_1 and C_2 are combining coefficients.

Hence,

$$\hat{A}\psi = \hat{A}(C_1\psi_1 + C_2\psi_2) = \hat{A}C_1\psi_1 + \hat{A}C_2\psi_2 \quad [\because \hat{A} \text{ is linear operator}]$$

$$= C_1\hat{A}\psi_1 + C_2\hat{A}\psi_2 \quad [\because \hat{A} \text{ is linear operator}]$$

$$= aC_1\psi_1 + aC_2\psi_2 = a(C_1\psi_1 + C_2\psi_2) = a\psi.$$

20. Under what condition is $\left(\hat{A}+\hat{B}\right)^2 = \hat{A}^2 + 2\hat{A}\hat{B} + \hat{B}^2$?

Soln. : $\left(\hat{A}+\hat{B}\right)^2 = \left(\hat{A}+\hat{B}\right)\left(\hat{A}+\hat{B}\right)$

$$= \hat{A}^2 + \hat{A}\hat{B} + \hat{B}\hat{A} + \hat{B}^2$$

$$= \hat{A}^2 + \hat{A}\hat{B} + \hat{A}\hat{B} + \hat{B}^2 \quad [\text{if } \hat{A}\hat{B} = \hat{B}\hat{A}, \text{ i.e., } [\hat{A},\hat{B}] = 0]$$

$$= \hat{A}^2 + 2\hat{A}\hat{B} + \hat{B}^2.$$

Hence, if $\hat{A}$ and $\hat{B}$ commute,

$$\left(\hat{A}+\hat{B}\right)^2 = \hat{A}^2 + 2\hat{A}\hat{B} + \hat{B}^2.$$

21. Evaluate $[\hat{x}^n, \hat{p}_x]$. **[C.U.(H) 2005]**

Soln. : $[\hat{x}^n, \hat{p}_x]\psi = \hat{x}^n\hat{p}_x\psi - \hat{p}_x\hat{x}^n\psi$

$$= -i\hbar x^n \frac{\partial\psi}{\partial x} + i\hbar\frac{\partial}{\partial x}\left(x^n\psi\right) \quad [\because \hat{p}_x = -i\hbar\frac{\partial}{\partial x}]$$

$$= -i\hbar x^n \frac{\partial\psi}{\partial x} + i\hbar x^n \frac{\partial\psi}{\partial x} + i\hbar n x^{n-1}\psi$$

$$= i\hbar n x^{n-1}\psi.$$

$$\therefore \quad [\hat{x}^n, \hat{p}_x] = i\hbar n x^{n-1}.$$

22. Given an arbitrary operator $\hat{A}$, show that $A^\dagger A$ is always Hermitian.

Soln. : $\langle\psi_1|A^\dagger A|\psi_2\rangle = \langle A\psi_1|A|\psi_2\rangle$ [by turn-over rule]

$$= \langle A^\dagger A\psi_1|\psi_2\rangle \text{ [by turn-over rule]}$$

Again, $\langle\psi_1|A^\dagger A|\psi_2\rangle = \left\langle\left(A^\dagger A\right)^\dagger\psi_1|\psi_2\right\rangle$.

Thus, $\left(A^{\dagger}A\right)^{\dagger} = A^{\dagger}A$.

Hence, $A^{\dagger}A$ is always Hermitian.

23. If two operators $\hat{A}$ and $\hat{B}$ are Hermitian, prove that $\hat{A}\hat{B}+\hat{B}\hat{A}$ is always Hermitian.

Soln. : $\left\langle \psi_1 \middle| \hat{A}\hat{B}+\hat{B}\hat{A} \middle| \psi_2 \right\rangle = \left\langle \psi_1 \middle| \hat{A}\hat{B} \middle| \psi_2 \right\rangle + \left\langle \psi_1 \middle| \hat{B}\hat{A} \middle| \psi_2 \right\rangle$

$= \left\langle \hat{B}^{\dagger}\hat{A}^{\dagger}\psi_1 | \psi_2 \right\rangle + \left\langle \hat{A}^{\dagger}\hat{B}^{\dagger}\psi_1 | \psi_2 \right\rangle$ [by turn-over rule]

$= \left\langle \hat{B}\hat{A}\psi_1 | \psi_2 \right\rangle + \left\langle \hat{A}\hat{B}\psi_1 | \psi_2 \right\rangle$ [$\because$ $\hat{A}$ and $\hat{B}$ are Hermitian]

$= \left\langle \left(\hat{B}\hat{A}+\hat{A}\hat{B}\right)\psi_1 | \psi_2 \right\rangle$

$= \left\langle \left(\hat{A}\hat{B}+\hat{B}\hat{A}\right)\psi_1 | \psi_2 \right\rangle$.

Again, $\left\langle \psi_1 \middle| \hat{A}\hat{B}+\hat{B}\hat{A} \middle| \psi_2 \right\rangle = \left\langle \left(\hat{A}\hat{B}+\hat{B}\hat{A}\right)^{\dagger} \psi_1 | \psi_2 \right\rangle$ [by turn-over rule].

Hence, $\left(\hat{A}\hat{B}+\hat{B}\hat{A}\right)^{\dagger} = \hat{A}\hat{B}+\hat{B}\hat{A}$.

Thus, $\hat{A}\hat{B}+\hat{B}\hat{A}$ is always Hermitian.

24. Determine, citing reasons, whether each of the following functions is acceptable or not as a wave function over the indicated intervals :

(i) e^{-x} $(0, \infty)$; (ii) e^{-x} $(-\infty, \infty)$; (iii) $\sin^{-1}x$ $(-1, 1)$; (iv) $e^{-|x|}$ $(-\infty, \infty)$

[C.U.(H) 2002]

Ans. : (i) e^{-x} is acceptable as it is finite (i.e., vanishes at infinity), continuous, single-valued and its first derivative or slope $\left(\frac{de^{-x}}{dx}\right)$ is also continuous within the indicated interval.

(ii) e^{-x} is not acceptable within the given interval as it diverges as x tends to $-\infty$.

(iii) $\sin^{-1} x$ is not acceptable, since it is a multivalued function.

(iv) $e^{-|x|}$ is not acceptable as its first derivative $\left(\frac{de^{-|x|}}{dx}\right)$ is not continuous at $x = 0$.

25. Test whether the following functions are eigenfunctions of the operator $\left(\frac{d^2}{dx^2}\right)$. Write also the corresponding eigenvalues :

(i) $\sin x$; (ii) $\sin x \cos x$; (iii) $A \sin mx + B \cos nx$. **[B.U.(H) 2004]**

Soln. : (i) $\frac{d^2}{dx^2}\sin x = \frac{d}{dx}\cos x = -\sin x$, eigenvalue $= -1$.

(ii) $\frac{d^2}{dx^2}\sin x\cos x = \frac{1}{2}\frac{d^2}{dx^2}\sin 2x = \frac{d}{dx}\cos 2x = -2\sin 2x$, eigenvalue = –2.

(iii) $\frac{d^2}{dx^2}(A\sin mx + B\cos nx) = \frac{d}{dx}(mA\cos mx - nB\sin nx)$

$= -m^2A\sin mx - n^2B\cos nx.$

$(A\sin mx + B\cos nx)$ is not eigenfunction of $\frac{d^2}{dx^2}$.

26. An operator $(\hat{A})$ is defined as : $\hat{A} = x\frac{\partial}{\partial x}$. Find $\hat{A}^2$. **[B.U.(H) 2003]**

Soln. : Given, $\hat{A} = x\frac{\partial}{\partial x}$. Let ψ be any function.

Thus, $\hat{A}^2\psi = \hat{A}.\hat{A}\psi = x.\frac{\partial}{\partial x}\left(x.\frac{\partial}{\partial x}\right)\psi$

$= \left(x\frac{\partial}{\partial x} + x^2\frac{\partial^2}{\partial x^2}\right)\psi.$

$\therefore \quad \hat{A}^2 = x\frac{\partial}{\partial x} + x^2\frac{\partial^2}{\partial x^2}.$

27. Expand the operator $\left(\frac{d}{dx}+x\right)^2$. **[B.U.(H) 2001]**

Ans. : Let ψ be any function.

$\left(\frac{d}{dx}+x\right)^2\psi = \left(\frac{d}{dx}+x\right)\left(\frac{d}{dx}+x\right)$

$= \left(\frac{d^2}{dx^2}+x\frac{d}{dx}+1+x^2\right)\psi.$

$\therefore \quad \left(\frac{d}{dx}+x\right)^2 = \frac{d^2}{dx^2}+x\frac{d}{dx}+x^2+1.$

28. Find the dimension of $h\frac{\partial}{\partial x}$. **[B.U.(H) 2003]**

Ans. : $h\frac{\partial}{\partial x}$ has a dimension of momentum, i.e., MLT^{-1}.

29. Show that $\hat{H}$ is Hermitian. **[V.U.(H) 1996]**

Soln. : $\hat{H} = -\frac{h^2}{2m}\frac{d^2}{dx^2} + V(x)$

Since $\frac{\hbar^2}{2m}$ and V are real numbers, they are automatically Hermitian.

Now, $\int_{-\infty}^{+\infty}\psi_1^*\frac{d^2\psi_2}{dx^2}dx$

$= \int_{-\infty}^{+\infty}\psi_1^*\frac{d\alpha}{dx}dx$ [putting $\frac{d\psi_2}{dx} = \alpha$]

$$= \left(\psi_1^* \alpha\right)_{-\infty}^{+\infty} - \int_{-\infty}^{+\infty} \frac{d\psi_1^*}{dx} \alpha dx$$

$$= -\int_{-\infty}^{+\infty} \frac{d\psi_1^*}{dx} \alpha dx \quad \text{[since } \psi_1 \text{ and } \frac{d\psi_2}{dx} \alpha \text{ vanish at infinity]}$$

$$= -\int_{-\infty}^{+\infty} \frac{d\psi_1^*}{dx} \frac{d\psi_2}{dx} dx \text{ [since } \alpha = \frac{d\psi_2}{dx}]$$

$$= -\left[\left(\psi_2 \frac{d\psi_1^*}{dx}\right)_{-\infty}^{+\infty} - \int_{-\infty}^{+\infty} \frac{d^2\psi_1^*}{dx^2} \psi_2 dx\right]$$

$$= \int_{-\infty}^{+\infty} \psi_2 \frac{d^2\psi_1^*}{dx^2} dx.$$

Hence, $\frac{d^2}{dx^2}$ is Hermitian. Thus, $\hat{H}$ is Hermitian.

30. What is meant by precise value in quantum mechanics?

Ans. : If a wave function associated with a system is eigenfunction of an operator which corresponds to a dynamical property in classical mechanics, then the wave function will always give some value of the property whenever is acted on by the corresponding operator. So there cannot remain any uncertainty in value and thus, it is actually a precise value. Free particle is associated with the wave function e^{ikx} which is eigenfunction of the operator $(-i\hbar)\frac{d}{dx}$ representing momentum of the particle. Thus,

$$-i\hbar \frac{d}{dx}(e^{ikx}) = \hbar k (e^{ikx}).$$

So the free particle has precise value of momentum and the value is $\hbar k$.

31. Calculate degeneracy of the level having an energy of 14 $(h^2/8ma^2)$ for a particle of mass m confined in a cubic box of dimension a. **[V.U.(H) 2002]**

Soln. : Energy of the particle in a cubic box,

$$E_n = E_{n_x} + E_{n_y} + E_{n_z}$$

$$= \frac{\left(n_x^2 + n_y^2 + n_z^2\right)h^2}{8ma^2} = \frac{n^2 h^2}{8ma^2}, \quad (1)$$

where m = mass of the particle, a = edge of the cube.

Given, $n^2 = 14$

or, $n_x^2 + n_y^2 + n_z^2 = 14.$

Now $1^2 + 2^2 + 3^2 = 14.$ (2)

Numbers on the L.H.S. of the equation (2) can be arranged in the following 6 different ways among themselves, i.e.,

n_x	n_y	n_z
1	2	3
1	3	2
2	1	3
2	3	1
3	1	2
3	2	1

So, the given energy level is six-fold degenerate.

32. Can the zero point energy of a particle in a box be zero? Answer with reasons. **[V.U.(H) 1995]**

or,

Why is the value of $n = 0$ of the quantum number not permitted for a particle in a one-dimensional box. **[C.U.(H) 2001]**

Ans. : Energy of a particle of mass m confined in a one-dimensional box of length a is $E = \frac{p^2}{2m} = \frac{n^2h^2}{8ma^2}$, where p is the momentum of the particle. When $n = 0$, E is zero. It means that momentum of the particle is also zero. Hence, there is no uncertainty in its momentum, that is, $\Delta p = 0$. Now, to maintain Heisenberg uncertainty principle which states $\Delta x\,\Delta p \geq \frac{\hbar}{2}$, the position of the particle has to be absolutely uncertain. But the particle is confined within a box of finite length a. So, the uncertainty in the position (Δx) of the particle cannot exceed the length of the box. Thus, the zero point energy of the particle in a box cannot be zero, since otherwise it would violate uncertainty principle.

33. For a particle in a box problem, show that the energy levels are continuous, if either the mass of the particle or the length of the box becomes large. **[V.U.(H) 2001]**

Soln. : The energy of the nth energy level of the particle confined in a box

$$= E_n = \frac{n^2h^2}{8ml^2},$$

where h = Planck constant, m = mass of the particle, l = length of the box.

The energy difference between two successive energy levels $= E_{n+1} - E_n$

$$= \frac{(n+1)h^2}{8ml^2} - \frac{n^2h^2}{8ml^2} = \frac{(2n+1)h^2}{8ml^2}.$$

Separation between two successive energy levels decreases if either the mass of the particle or the length of the box becomes large.

34. Show that the error in the de Broglie wavelength (λ) is related to the error in velocity (v) by the relation $d\lambda = -\frac{\lambda}{v}\,dv$. **[B.U.(H) 2003]**

Soln. : According to de Broglie formula, $\lambda = \frac{h}{p}$, (1)

where λ is the wavelength of the particle with momentum p.

Now, from (1), we get $\lambda = \frac{h}{mv}$, [$\because p = mv$]

where m = mass of the particle, v = velocity of the particle.

$$\therefore d\lambda = -\frac{h}{mv^2}dv = -\frac{h}{pv}dv$$

$$= -\frac{\lambda}{v}dv \text{ [from equation (1)].}$$

35. For a particle in a cubical box write down the energy values for the condition: $n_x + n_y + n_z = 4$ and indicate the level of degeneracy, if any. **[C.U.(H) 2000]**

Soln. : $n_x + n_y + n_z = 4$

Now, $1 + 1 + 2 = 4.$ (1)

Thus energy of the particle in a cubical box, $E = E_{n_x} + E_{n_y} + E_{n_z}$

$$= \frac{h^2}{8ma^2}\left(n_x^2 + n_y^2 + n_z^2\right) = \frac{h^2}{8ma^2}\left(1^2 + 1^2 + 2^2\right) = \frac{6h^2}{8ma^2},$$

where m = mass of the particle, a = edge of the cube.

Again,

n_x	n_y	n_z
1	1	2
2	1	1
1	2	1

Each of the combinations mentioned above corresponds to the same energy $\frac{6h^2}{8ma^2}$. Hence, the level is three-fold degenerate.

36. Show that the function $\psi(x) = \frac{(x^2 - a^2)}{(x - a)}$ is not acceptable for describing the state of a particle in a one-dimensional box ranging from $x = -2a$ to $x = 2a$. **[B.U.(H) 1999]**

Soln. : Given, $\psi(x) = \frac{(x^2 - a^2)}{(x - a)}$.

At $x = -2a$.

$$\psi(x) = \frac{4a^2 - a^2}{-2a - a} = -a.$$

At $x = 2a$,

$$\psi(x) = \frac{4a^2 - a^2}{2a - a} = 3a.$$

So, $\psi(x)$ does not vanish at the boundary. Hence, the given $\psi(x)$ is not acceptable for describing the particle confined in a one-dimensional box ranging from $-2a$ to $2a$.

37. From the definition of linear harmonic oscillator, obtain the expression for the potential energy and hence, construct the Hamiltonian operator for the same. **[C.U.(H) 2001]**

Soln. : A linear restoring force (F) acts on particle executing simple harmonic oscillation which is proportional to its displacement, x, from the centre, namely

$$F = -kx, \quad (1)$$

where k = force constant.

Now, $F(x) = -\frac{\partial V(x)}{\partial x}$, $dV(x) = -F(x)\, dx$ (2)

or, $dV(x) = kx\, dx$ [from equation (1)]. (3)

Integrating equation (3), we get

$V(x) = \frac{1}{2}kx^2$, [since at $x = 0$, $V(x) = 0$] (4)

$V(x)$ = potential energy of the particle.

[p = momentum of the particle, m = mass of the particle.]

Hamiltonian operator for a linear harmonic oscillator, $\hat{H} = \frac{\hat{p}^2}{2m} + \frac{1}{2}k\hat{x}^2$.

Now, $\hat{p} = -i\hbar\frac{\partial}{\partial x}$, $\therefore$ $\hat{p}^2 = -\hbar\frac{\partial^2}{\partial x^2}$.

Thus, $\hat{H} = -\frac{\hbar^2}{2m}\frac{\partial^2}{\partial x^2} + \frac{1}{2}kx^2$.

38. Obtain the energy expression for a particle moving to and fro within a one-dimensional box of length a using de Broglie equation.

Soln. : The energy of a particle moving to and fro within a box of length a,

$$E = \frac{p^2}{2m}, \quad (1)$$

where p is momentum of the particle and m is its mass.

By de Broglie hypothesis every moving object has wave associated with its motion, i.e.,

$$p = \frac{h}{\lambda}, \quad (2)$$

where λ is the wavelength of the wave associated with the particle having momentum p. Again, in one traversal of the region, the particle covers a distance of $2a$.

An integral multiple of the matter wave must fit in the distance covered by the particle in one back and forth motion.

Hence, $2a = n\lambda$.

$$\therefore \quad \lambda = \frac{2a}{n} \quad (3)$$

$$\text{Thus, } E = \frac{p^2}{2m} = \frac{h^2}{2m\lambda^2} \text{ [from equation (2)]}$$

$$= \frac{n^2h^2}{8ma^2} \text{ [from equation (3)]}.$$

39. What is the ground state energy of a 3-dimensional harmonic oscillator ?

Soln. : Energy of the 3-dimensional harmonic oscillator,

$$E_n = E_{n_x} + E_{n_y} + E_{n_z}$$

$$= \left(n_x + \frac{1}{2}\right)h\nu + \left(n_y + \frac{1}{2}\right)h\nu + \left(n_z + \frac{1}{2}\right)h\nu = \left(n_x + n_y + n_z + \frac{3}{2}\right)h\nu$$

$$= \left(n + \frac{3}{2}\right)h\nu.$$

Thus, ground state energy = $E_0 = \frac{3}{2}h\nu$ [since $n = 0$].

40. What is the degeneracy of a level having energy $\frac{7}{2}h\nu$ for a 3-dimensional harmonic oscillator?

Soln. : Energy of the nth level of a 3-dimensional harmonic oscillator,

$$E_n = \left(n_x + n_y + n_z + \frac{3}{2}\right)h\nu = \left(n + \frac{3}{2}\right)h\nu.$$

$$\text{Now,} \quad n + \frac{3}{2} = \frac{7}{2}$$

$$\therefore \quad n = 2.$$

Thus, $n_x + n_y + n_z = 2$

n_x	n_y	n_z
1	1	0
1	0	1
0	1	1
2	0	0
0	0	2
0	2	0

Thus, the given energy level is six-fold degenerate.

41. Comment on zero-point energy of a one-dimensional harmonic oscillator.

Ans. : Energy of a linear harmonic oscillator,

$$E = \frac{p_x^2}{2m} + \frac{1}{2}kx^2,$$

where p_x is momentum of the oscillator and x is displacement from the origin. If E is zero, p_x and x are also zero. It means that there are no uncertainties in both position (Δx) and momentum (Δp_x) and both can be measured simultaneously with an arbitrary precision, i.e., $\Delta x \, \Delta p_x = 0$. But uncertainty principle requires $\Delta x \, \Delta p_x \geq \frac{h}{2}$. So, if Δx is zero, Δp_x has to be infinite and vice versa. Both Δp_x and Δx cannot be zero simultaneously. Thus, zero point energy of the oscillator cannot be zero.

42. Show that E_n of hydrogen atom is n^2-fold degenerate. **[V.U.(H) 1995]**

Soln. : Energy of a hydrogen atom is given by $E_n = -\frac{me^4}{2n^2\hbar^2}$,

where n is the principal quantum number, m is the mass of an electron and e is its charge. The corresponding eigenfunction of energy depends on n, l (azimuthal quantum number), and m magnetic quantum number and is represented as ψ_{nlm}.

Now, for every value of n there are n number of l with minimum value 0 and maximum value $(n - 1)$, and again, for every value of l there are $(2l + 1)$ number of m ranging from $-l$ to $+l$. So, situations could occur when different states corresponding to different l and m but same n (e.g., ψ_{nlm}, $\psi_{nl'm'}$) belong to the same energy state which depends only on n, e.g., for a given n, total number of l is

$$(0, 1, 2, \ldots, (n - 1) =)\ n$$

and total number of m is

$1 + 3 + 5 + \ldots + (2n - 1) = n^2$ (since, for each l, number of $m = 2l + 1$) and hence, E_n of a hydrogen atom is n^2- fold degenerate.

43. What is the degree of degeneracy of Y_{lm} with respect to the eigenvalues of $\hat{L}^2$?

Soln. : $\hat{L}^2 Y_{lm} = l(l+1)\hbar^2 Y_{lm}$

So the eigenvalue depends only on l, not on m. But for each value of l, there are $2l + 1$ number of m and the form of eigenfunction Y_{lm} is governed by l and m. So, situations are likely to occur when several states represented by different Y_{lm} (different in respect to m, for example, Y_{lm} , $Y_{lm'}$, $Y_{lm''}$, etc.) correspond to same eigenvalue which is independent of m.

Hence, the degree of degeneracy of Y_{lm} with respect to $\hat{L}^2$ is $(2l + 1)$.

44. Show that e^{ikx} is an eigenfunction of the operator $\hat{P}_x$. What is the eigenvalue?

Soln. : $\hat{p}_x e^{ikx} = -i\hbar \frac{\partial}{\partial x} e^{ikx} = \hbar k e^{ikx}$.

$\therefore$ eigenvalue is $\hbar k$.

45. Analyse whether the following functions are even or odd or neither :

(a) $\tan x$, (b) e^x, (c) 15, (d) $x \cosh hx$, (e) $2 - 2x$, (f) $(2 + x)(2 - x)$, (g) $x^2 e^{-x^2}$, (h) xe^{-x^2}, (i) xe^{-x}, (j) $x \sin x$, (k) e^{ix}.

Soln. : (a) $f(-x) = \tan(-x) = -\tan x = -f(x)$, odd

(b) $f(-x) = e^{-x} \neq e^x$

$\therefore f(x) = e^x \neq f(-x)$

$\neq -f(-x)$, neither even nor odd.

(c) $-15 \neq 15$, neither even nor odd.

(d) $f(x) = x \cosh x = \frac{1}{2} x (e^x + e^{-x}) = -\frac{1}{2}(-x)(e^{-x} + e^x) = (-x)\cosh(-x)$

$= -f(-x)$. odd.

(e) $f(-x) = 2 - 2(-x) \neq 2 + 2x \neq 2 - 2x$

$\therefore f(x) = 2 - 2x \neq f(-x)$

$\neq -f(-x)$, neither even nor odd.

(f) $f(-x) = (2 - x)(2 + x) = f(x)$, even.

(g) $f(-x) = (-x)^2 e^{-(-x)^2} = x^2 e^{-x^2} = f(x)$, even.

(h) $f(-x) = (-x) e^{-(-x)^2} = -xe^{-x^2} = -f(x)$, odd.

(i) $f(-x) = (-x)e^{-(-x)} = -xe^x \neq xe^{-x}$

$\therefore f(x) = xe^{-x} \neq f(-x)$

$\neq -f(-x)$, neither even nor odd.

(j) $f(-x) = -x\sin(-x) = x \sin x$, even.

(k) $e^{ix} = \cos x + i \sin x \neq \cos x - i \sin x$

$\neq -(\cos x - i \sin x)$ (where $e^{-ix} = \cos x - i \sin x$),

neither even nor odd.

46. Classify these operators into linear or non-linear :

(a) $x^2\dfrac{d^2}{dx^2}$, (b) $(\ )^2$, (c) $\int dx$, (d) exp, (e) $\sum_{x=1}^{n}$, (f) $\sqrt{}$, (g) $\dfrac{d}{dx}$, (h) cos.

Soln. : Let $f(x)$ and $g(x)$ be functions of x, and c be any constant.

(a) $x^2\dfrac{d^2}{dx^2}[cf(x)] = cx^2\dfrac{d^2}{dx^2}f(x)$, and $x^2\dfrac{d^2}{dx^2}[f(x) + g(x)]$

$= x^2\dfrac{d^2}{dx^2}f(x) + x^2\dfrac{d^2}{dx^2}g(x)$, linear.

(b) $(f(x) + g(x))^2 \neq (f(x))^2 + (g(x))^2$

$(cf(x))^2 \neq c(f(x))^2$, non-linear.

(c) $\int (f(x)+g(x))dx = \int f(x)dx + \int g(x)dx$

$\int cf(x)dx = c\int f(x)dx$, linear.

(d) $\exp[f(x) + g(x)] \neq \exp(f(x)) + \exp(g(x))$

$\exp[cf(x)] \neq c\exp[f(x)]$, non linear.

(e) $\sum_{x=1}^{n} cf(x) = c\sum_{x=1}^{n} f(x)$

$\sum_{x=1}^{n} f(x) + g(x) = \sum_{x=1}^{n} f(x) + \sum_{x=1}^{n} g(x)$, linear.

(f) $\sqrt{f(x)+g(x)} \neq \sqrt{f(x)} + \sqrt{g(x)}$

$\sqrt{cf(x)} \neq c\sqrt{f(x)}$, non-linear.

(g) $\dfrac{d}{dx}$ is linear by the same argument as in (a).

(h) $\cos(f(x) + g(x)) \neq \cos f(x) + \cos g(x)$

$\cos(cf(x)) \neq c\cos f(x)$, non-linear.

47. Given a linear operator $\hat{A}\phi = a\phi$, find the value of $e^{\hat{A}}\phi$.

Soln. : $e^{\hat{A}} = I + \hat{A} + \dfrac{\hat{A}^2}{2!} + \dfrac{\hat{A}^3}{3!} + \ldots$

I is identity operator, defined by $I\phi = \phi$ for all ϕ.

Hence, $e^{\hat{A}}\phi = \left(1+\hat{A}+\dfrac{\hat{A}^2}{2!}+\dfrac{\hat{A}^3}{3!}+\ldots\right)\phi$

$= \phi + a\phi + \dfrac{a^2}{2!}\phi + \dfrac{a^3}{3!}\phi + \ldots$

(since, $\hat{A}\phi = a\phi$, $\hat{A}^2\phi = \hat{A}a\phi = a\hat{A}\phi = a^2\phi$, and so on)

$= \left(1+a+\dfrac{a^2}{2!}+\dfrac{a^3}{3!}+\ldots\right)\phi = e^a\phi.$

48. Prove that $i\,[\alpha, \beta]$ is Hermitian, if α and β are Hermitian.

Soln. : $\langle \psi_1 | i[\alpha,\beta] | \psi_2 \rangle = \langle \psi_1 | i(\alpha\beta - \beta\alpha) | \psi_2 \rangle$

$= \langle \psi_1 | i\alpha\beta | \psi_2 \rangle - \langle \psi_1 | i\beta\alpha | \psi_2 \rangle$

$= \langle -i\beta\alpha\psi_1 | \psi_2 \rangle - \langle -i\alpha\beta\psi_1 | \psi_2 \rangle$ (since α and β are Hermitian)

$= \langle -i(\beta\alpha - \alpha\beta)\psi_1 | \psi_2 \rangle = \langle i(\alpha\beta - \beta\alpha)\psi_1 | \psi_2 \rangle = \langle i|\alpha,\beta|\psi_1 | \psi_2 \rangle.$

49. What are the consequences of measurement when (a) two operators commute? (b) two operators do not commute?

Ans. : When two operators commute, it means that the corresponding observables can simultaneously be measured with an absolute precision, i.e., their simultaneous measurements do not have any uncertainties.

When two operators do not commute, the corresponding observables cannot be measured simultaneously and precisely; i.e., their simultaneous measurements are subject to certain uncertainties.

50. (a) Interpret (i) ψ, (ii) $\psi^*\psi\, dx$.

(b) What is node?

Ans. : (a) (i) ψ is a wave function that depends on coordinates and time, and contains all the information about the dynamical properties of the system. It is actually a probability amplitude and can be real or complex.

(ii) $\psi^*\psi$ is the probability of finding a particle between x and $x + dx$ at the position x.

(b) Node is a point after passing through which wave function (ψ) changes sign. At node ψ is zero.

51. Test whether the following functions are acceptable or not within the indicated range :

(i) e^{-x^2} $(-\infty, \infty)$; (ii) $e^{-x}\cos x$ $(0, \infty)$; (iii) $\frac{1}{x}(0, \infty)$; (iv) $\tan x$ $(0, \pi)$; (v) $\sin x$ $(0, \pi)$; (vi) $\frac{\sin x}{x}(0, \infty)$; (vii) $e^{-i\phi}(0, 2\pi)$.

Ans. : (i) e^{-x^2} is acceptable as it is continuous, single-valued, finite, i.e., does not diverge at any point between $-\infty$ to $+\infty$ and its first derivative or slope is also continuous.

(ii) $e^{-x}\cos x$ is acceptable due to the same reasons as in (i).

(iii) $\frac{1}{x}$ is not acceptable since it diverges at 0.

(iv) $\tan x$ is not acceptable, since it diverges at $\frac{\pi}{2}$.

(v) $\sin x$ is acceptable due to the same reasons as in (i).

(vi) $\frac{\sin x}{x}$ is acceptable due to the same reasons as in (i). It is finite at $x = 0$.

(vii) $e^{i\phi}$ is acceptable since it is continuous, single-valued, and finite within the given range of $\phi.(e^{i\phi} = \cos\phi + i\sin\phi)$.

52. Show which of the following functions are acceptable : (i) $e^{-|x|}$ $(-\infty, \infty)$; (ii) $\cos|x|$ $(-2\pi, 2\pi)$; (iii) $\sin|x|$ $(-2\pi, 2\pi)$.

Soln. : (i) $f(x) = e^{-|x|}$

$= e^{-x}$ (when $x \geq 0$)

$= e^{x}$ (when $x \leq 0$)

$f(x)$ is continuous, finite, single-valued.

Now, $y' = \dfrac{d(e^{-|x|})}{dx} = -e^{-x}$ (when $x \geq 0$)

$= e^{x}$ (when $x \leq 0$).

Thus, $e^{-|x|}$ is not acceptable within the given range as its derivative diverges at $x = 0$.

(ii) $f(x) = \cos|x|$,

$= \cos x$ $(x \geq 0)$

$= \cos x$ $(x \leq 0)$

$\therefore \quad y' = \dfrac{d(\cos|x|)}{dx} = -\sin x$ $(x \geq 0)$

$= -\sin x$ $(x \leq 0)$.

$f(x)$ is continuous, finite, single-valued and its derivative is also continuous. Hence, it is acceptable.

(iii) $f(x) = \sin|x|$

$= \sin x$ $(x \geq 0)$

$= -\sin x$ $(x \leq 0)$.

$f(x)$ is continuous, finite, single-valued.

But $y' = \cos x$ $(x \geq 0)$

$= -\cos x$ $(x \leq 0)$.

So, y' is not continuous at $x = 0$.

Hence, $\sin|x|$ is not acceptable.

53. What is the meaning of nodes of wave functions? How do you explain the passage of electron across the node? Do the 'signs' on the two sides of a node make any difference in the electronic charge distribution?

Ans. : Answer to the first part of the question has been given in 50(b).

Electron does not have any well-defined trajectory as a consequence of uncertainty principle. At node its uncertainty in momentum will be infinite by virtue of uncertainty principle as its position is now precisely known. Uncertainty in momentum must be comparable to its momentum in magnitude and it is found at the other side of the node.

Electronic charge distribution is equal to the product of the wavefunctions, viz, $\psi^*\psi$. So the different signs of wavefunction on the two sides of a node do not make any difference in electronic charge distribution.

54. Why are the wave functions associated with different states of a particle in a box orthogonal?

Ans. : Wave functions of a particle in a box are energy eigenfunctions and they are non-degenerate. Non-degenerate eigenfunctions of a Hermitian operator are orthogonal. Hence, wave functions of a particle in a box are orthogonal as Hamiltonian is a Hermitian operator.

55. Functions ψ_1 and ψ_2 are individually normalised and are mutually orthogonal to each other. Normalise $\psi = \psi_1 + 3^{1/2}\ \psi_2$. **[I.I.T.B. M.Sc. Admission Test 2000]**

Soln. : Let $\psi' = N\psi$.

By the problem, $\langle \psi' | \psi' \rangle = 1$

or, $N^2 \langle \psi | \psi \rangle = 1$

or, $N^2 \langle \psi_1 + \sqrt{3}\psi_2 | \psi_1 + \sqrt{3}\psi_2 \rangle = 1$

or, $N^2 \left[\langle \psi_1 | \psi_1 \rangle + \sqrt{3} \langle \psi_2 | \psi_1 \rangle + \sqrt{3} \langle \psi_1 | \psi_2 \rangle + 3 \langle \psi_2 | \psi_2 \rangle \right] = 1$

or, $N^2\ [1 + 3] = 1$ [Given, $\langle \psi_1 | \psi_2 \rangle = \langle \psi_2 | \psi_2 \rangle = 1$ and $\langle \psi_1 | \psi_2 \rangle = \langle \psi_2 | \psi_2 \rangle = 0$]

or, $N^2 = \frac{1}{4}$, $\therefore N = \frac{1}{2}$.

56. The experimental results of the photoelectric effect data for metals Cs and Na are given in the figure below. Explain why the Cs-line is to the left of the Na-line in terms of the atomic structure ideas known to you. **[I.I.T.B. M.Sc. Admission Test 2000]**

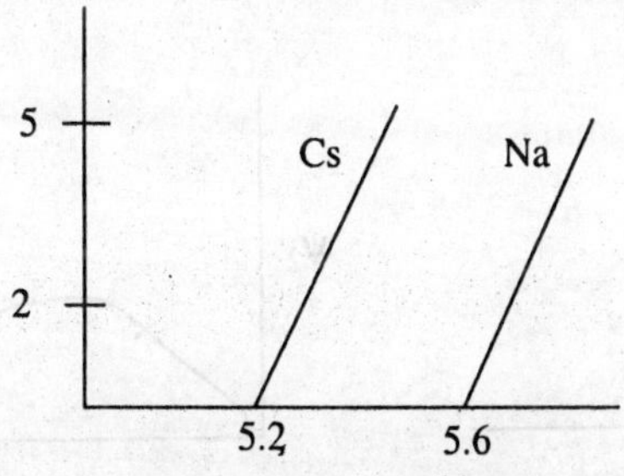

Frequency (in 10^{14} Hz)

Ans. : The given figure shows that the frequency of light required to eject an electron is greater for Na than Cs. It happens as Cs has outermost electrons in 6*s* shell while Na has that in 3*s* shell and screening effect of innermost electrons is more pronounced on 6*s* electron than on 3*s* electron. Hence, the former electron is more loosely bound than the latter and thus, requires less energy to escape attraction to nucleus.

57. A cubic box with each side measuring 10 Å (with zero potential inside and infinite potential outside) contains a system of 4 quantum particles. Find the degeneracy of the lowest energy state of the system. **[C.U. 2003]**

Ans. : Second energy state of cube is 3-fold degenerate. Each energy state can accomodate at most 2 quantum particles. Lowest energy state will accomodate 2 particles. Remaining 2 particles can occupy the 2nd energy state in the following different ways :

$(E_{n_x} E_{n_y} E_{n_z})$	(E_{211})	(E_{112})	(E_{121})
	2	0	0
	0	2	0
	0	0	2
	1	1	0
	0	1	1
	1	0	1

Hence, degeneracy of lowest energy state is six.

58. Prove that commutator of two Hermitian operators is always anti-Hermitian.

Hints : Prove $\left(\hat{A}\hat{B} - \hat{B}\hat{A}\right)$ is anti-Hermitian. Proceed as problem 23.

59. What is radial distribution function?

Ans. : Radial distribution function, $R_{nl}^2(r)r^2$ determines the probability of finding an electron anywhere between two concentric spheres of radii r and $r + dr$, i.e., it gives the probability of finding the electron at a distance r from the nucleus.

60. Identify the following hydrogenic orbitals with given radial parts as (i) $N_1(\beta_2 - r)$ $\exp[-\alpha_2 r]$ and (ii) $N_2 r \exp[-\alpha_3 r]$, where N_i, α_i, β_i are constants. Explain your answer using plots of the orbitals.

[C.U. 2006]

Ans. : Let $\psi_1 = N_1(\beta_2 - r)\, e^{-\alpha_2 r}$, $\psi_2 = N_2 r\, e^{-\alpha_3 r}$.

ψ_1 is zero only at $r = \beta_2$ and after passing through which it changes sign. So it has only one node at $r = \beta_2$, whereas ψ_2 has no node but becomes zero at nucleus.

So, ψ_1 is $2s$ orbital, since for $2s$ orbital number of node $= n - l - 1 = 2 - 0 - 1 = 1$ and s orbital has non-zero amplitude at nucleus and ψ_2 is $2p$ orbital since for $2p$ orbital, number of node $= 2 - 1 - 1 = 0$.

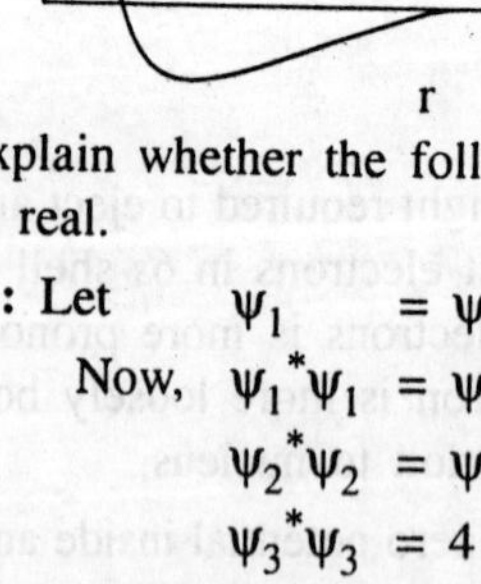

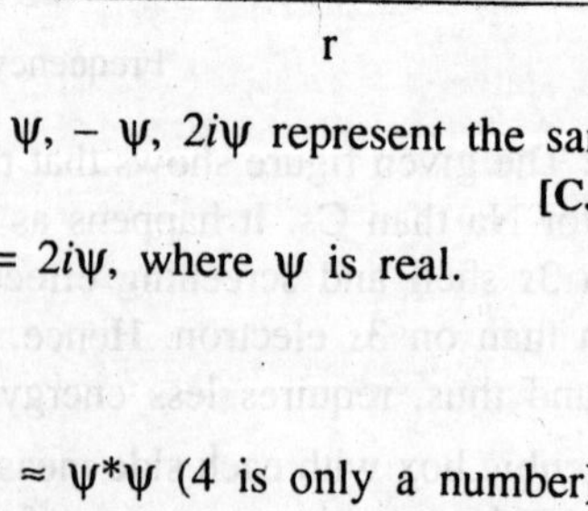

61. Explain whether the following functions ψ, $-\psi$, $2i\psi$ represent the same state, given ψ is real.

[C.U. 2006]

Soln. : Let $\psi_1 = \psi$, $\psi_2 = -\psi$, $\psi_3 = 2i\psi$, where ψ is real.

Now, $\psi_1^*\psi_1 = \psi^*\psi$

$\psi_2^*\psi_2 = \psi^*\psi$

$\psi_3^*\psi_3 = 4\, i^* i \psi^*\psi = 4\psi^*\psi \approx \psi^*\psi$ (4 is only a number)

Probability densities for ψ_1, ψ_2 and ψ_3 are same and so they represent the same state.

62. What is meant by degeneracy? Find the quantum number associated with energy level $\dfrac{17h^2}{8ma^2}$ of a particle in a cubical box and hence, indicate the degree of degeneracy.

[C.U.(H) 2005]

Soln. : If several linearly independent eigenfunctions of an operator correspond to the same eigenvalue, the number of eigenfunctions is the degeneracy with respect to that operator representing some property.

For 2nd part of the problem, proceed as Problem no. 35.

Here, $n_x^2 + n_y^2 + n_z^2 = 17$

or, $2^2 + 2^2 + 3^2 = 17.$

The given energy level is 3-fold degenerate.

63. Determine $[x, p_y]$, $[t, E]$.

Soln. : $[x, p_y]\ \psi(x, y) = xp_y\ \psi(x, y) - p_y x\ \psi(x, y)$

$$= -i\hbar x\frac{\partial\psi(x,y)}{\partial y} + i\hbar x\frac{\partial\psi(x,y)}{\partial y} \quad [p_y = -i\hbar\frac{\partial\psi}{\partial y}]$$

$$= 0.$$

$$E = i\hbar\frac{\partial\psi}{\partial t}$$

$$[t, E]\ \psi(x, t) = tE\ \psi(x, t) - Et\ \psi(x, t)$$

$$= i\hbar t\frac{\partial}{\partial t}\psi(x, t) - i\hbar\frac{\partial}{\partial t}t\ \psi(x, t)$$

$$= i\hbar t\frac{\partial}{\partial t}\psi(x, t) - i\hbar t\frac{\partial}{\partial t}\psi(x, t) - i\hbar\psi(x, t)$$

$$= -i\hbar\psi(x, t)$$

$$[t, E] = -i\hbar.$$

64. Find the average $\langle x\rangle$ of a particle confined in a one-dimensional box. Would this location be right place to find the particle if it is in second energy state.

Hints : Let the particle be moving in a one-dimensional box of length a between 0 to a. $\langle x\rangle = \int_0^a \sin\frac{n\pi x}{a}x\sin\frac{n\pi x}{a}dx = \frac{1}{a}\int_0^a x\left(1-\cos\frac{2n\pi x}{a}\right)dx = \frac{a}{2}.$

i.e., $\langle x\rangle$ is at the centre of the box.

In 2nd state, $\psi = 0$ at the centre of the box. Thus, in second energy state particle will not be found at this location.

65. The wave function of a quantum state of hydrogen atom with principal quantum number $n = 2$ is $\psi_{2lm}(r, \theta, \phi) = \frac{1}{\sqrt{32\pi}}\left(\frac{1}{a_o}\right)^{\frac{3}{2}}\left(2-\frac{r}{a_o}\right)\exp\left(-\frac{r}{2a_o}\right).$

(i) Identify the values of quantum number l and m and hence, the atomic orbital.

(ii) Find where the radial node of the wave function occurs. **[JAM 2011]**

Hints : (i) Since the wave function has no angular part, both the quantum numbers l and m are zero, and the given wavefunction corresponds to $2s$ atomic orbital.

(ii) See Prob. 31 in the section Numerical Problems of this chapter.

66. An atomic orbital is described by the weave function

$$\psi(r) = \frac{1}{\sqrt{\pi a_0^3}} e^{-\left(\frac{r}{a_0}\right)}, \text{ where } a_0 \text{ is the Bohr radius.}$$

Given : $d\tau = r^2 \sin\theta\, dr\, d\theta\, d\phi$ and $\int_0^\infty r^n e^{-\beta r} dr = \frac{n!}{\beta^{n+1}}$ (n is a positive integer).

(a) Identify the atomic orbital and calculate the mean or the average radius of this orbital in terms of a_0.

(b) Calculate the most probable radius (in terms of a_0) at which an electron will be found when it occupies this orbital.

Hints : (a) 1s orbital : See problem 27 for (a) and problem 47 for (b) in numerical section.

Numerical Problems

1. When aluminium is irradiated with light, the maximum energy of ejected electron is 2.3 eV for λ = 2000 Å and 0.223 eV for λ = 3000 Å. Calculate Planck constant and the work function of aluminium. What is threshold frequency for aluminium?

Soln. : Given, energies of ejected electrons :

$$h\nu_1 - W = h\frac{c}{\lambda_1} - W = (2.3 \times 1.6 \times 10^{-19}) \text{ J} \quad (1)$$

$$\text{and } h\frac{c}{\lambda_2} - W = (0.23 \times 1.6 \times 10^{-19}) \text{ J.} \quad (2)$$

Given, λ_1 = 2000 Å and λ_2 = 3000 Å.

$$\text{Hence, } h = \frac{(2.3-0.23)\times 1.6\times 10^{-19}\text{ J}}{c\left(\frac{1}{\lambda_1}-\frac{1}{\lambda_2}\right)}$$

$$= \frac{\left(2.07\times 1.6\times 10^{-19}\right)\text{J}}{\left(2.9979\times 10^{8}\right)\text{m.sec}^{-1}\left(\frac{1}{2000\times 10^{-10}}-\frac{1}{3000\times 10^{-10}}\right)\text{m}^{-1}}$$

$$= \mathbf{6.628 \times 10^{-34}\ J.sec.}$$

$$\text{Thus, work function : } W = h\frac{c}{\lambda_1} - 2.3 \times 1.6 \times 10^{-19}$$

$$= \frac{\left(6.628\times 10^{-34}\right)\text{J.sec}\times\left(2.9979\times 10^{8}\right)\text{m.sec}^{-1}}{\left(2000\times 10^{-10}\right)\text{m}} - (2.3 \times 1.6 \times 10^{-19}) \text{ J}$$

$$= 6.255 \times 10^{-19} \text{ J} = \mathbf{3.91\ eV.}$$

Threshold frequency of aluminium,

$$\nu_0 = \frac{(6.255\times10^{-19})\,\text{J}}{(6.628\times10^{-34})\,\text{J.sec}} \quad [\because h\nu_0 = W]$$

$$= \mathbf{9.437 \times 10^{14}\ Hz.}$$

2. Calculate the de Broglie wavelength of an electron that has been accelerated through a potential difference of 300 volts.

Soln. : $\lambda = \dfrac{h}{\sqrt{2meV}}$, where λ is de Broglie wavelength.

$$= \frac{6.626\times10^{-34}\ \text{J.sec}}{\sqrt{2\times(9.109\times10^{-31})\,\text{kg}\times(1.602\times10^{-19}\times300)\,\text{J}}} = 7.0812\times10^{-11}\ \text{m} = \mathbf{0.70812\ Å.}$$

3. The work function for sodium metal is 1.82 eV. Calculate the threshold frequency for sodium. **[C.U.(H) 2002]**

Soln. : $h\nu_0 = W = 1.82$ eV, where W = work function, ν_0 = threshold frequency.

$$\nu_0 = \frac{(1.82\times1.602\times10^{-19})\,\text{J}}{(6.626\times10^{-34})\,\text{J.sec}} = 4.40\times10^{14}\ \text{sec}^{-1} = \mathbf{4.40\times10^{14}\ Hz.}$$

4. Calculate the wavelength at which a cavity at 3000 °K will radiate most per unit wavelength.

Soln. : By Wien's displacement law, $\lambda_{max} = \dfrac{\text{constant}}{T} = \dfrac{2.898\times10^{-3}\ \text{m}\,°\text{K}}{3000\,°\text{K}}$

$$= 9.660\times10^{-7}\ \text{m} = 9660\ Å.$$

5. At a given temperature, for a black body cavity, λ_{max} = 6000 Å. What will be the λ_{max} if temperature of the cavity is decreased such that the rate of emission of spectral radiation decreases two-fold?

Soln. : According to Stefan-Boltzmann law,

$$R_T = \sigma T^4, \tag{1}$$

where R_T is total emissive power at temperature T, and σ is a constant.

Now, $R_{T'} = \dfrac{R_T}{2}$ at T'.

Here, $R_{T'} = \dfrac{R_T}{2} = \sigma T'^4$. (2)

Dividing equation (1) by equation (2),

$$\left(\frac{T}{T'}\right)^4 = 2, \ \therefore \ \frac{T}{T'} = 2^{\frac{1}{4}}.$$

By Wien's displacement law,

$$\lambda_{max} T = \lambda'_{max} T'.$$

$$\therefore \quad \lambda'_{max} = \lambda_{max}\frac{T}{T'} = 6000\ Å \times 2^{\frac{1}{4}} = \mathbf{7135.24\ Å.}$$

6. 2.14 eV is required to remove an electron from caesium. (a) What is the cut-off wavelength for photoelectric emission from caesium? (b) Will a light of $\lambda = 6000$ Å liberate an electron from caesium?

Soln. : (a) Given, $h\nu_0 = 2.14$ eV,

$$\text{or, } \frac{hc}{\lambda_0} = 2.14 \text{ eV},$$

$$\therefore \quad \lambda_0 = \frac{(6.626\times10^{-34})\text{J.sec}\times(2.9979\times10^{8})\text{m.sec}^{-1}}{(2.14\times1.602\times10^{-19})\text{J}}$$

$$= 5.79418 \times 10^{-7} \text{ m} = \mathbf{5794.18\ Å.}$$

(b) Caesium will not show photoelectric effect with a light of $\lambda = 6000$ Å, since its cut-off wavelength is 5794.18 Å.

7. Potassium surface is irradiated with a radiation of 4000 Å. Work function of potassium is 2.30 eV. What is the kinetic energy of (a) fastest and (b) slowest emitted electrons? (c) What is maximum possible wavelength capable of removing electron from potassium surface? (d) What is stopping potential? (e) How many electrons will liberate per unit time from the surface if the intensity of incident radiation is 1.5 W/m^2?

Soln. : (a) Maximum kinetic energy of ejected electron :

$$\frac{1}{2}mv^2 = E = h\nu - W$$

$$= \frac{(6.626\times10^{-34})\text{ J.sec}\times(2.9979\times10^{8})\text{m.sec}^{-1}}{4000\times10^{-10}\text{ m}} - (2.3\times1.602\times10^{-19})\text{ joule}$$

$$= 1.28142 \times 10^{-19} \text{ J.} \quad (1)$$

(b) Kinetic energy of slowest emitted electron = 0.

(c) Work function = 2.3 eV.

$$h\nu_0 = 2.3 \times 1.602 \times 10^{-19} \text{ J.} \quad (2)$$

Hence, $h\,\dfrac{c}{\lambda_0} = 2.3 \times 1.602 \times 10^{-19}$.

$$\therefore \lambda_0 = \frac{(6.626\times10^{-34})\text{J.sec}\times(2.9979\times10^{8})\text{ m.sec}^{-1}}{(2.3\times1.602\times10^{-19})\text{J}} = 5.39111 \times 10^{-7} \text{ m}$$

$$= 5391.11 \text{ Å.} \quad (3)$$

Hence, maximum possible value of λ that can remove an electron = 5391.11 Å.

(d) If stopping potential is V_0, then

$$eV_0 = 1.28142 \times 10^{-19} \text{ J [from 1]}$$

$$\therefore \quad V_0 = \frac{1.28142\times10^{-19}\text{ J}}{1.602\times10^{-19}\text{C}} = 0.799 \text{ V [J = C × V]}$$

(e) Let the number of liberated photoelectrons be n.

Now, $nh\nu = 1.5$ W/m^2.

$$\therefore \quad n = \frac{1.5\text{ W / m}^2\times(4000\times10^{-10})\text{m}}{(6.626\times10^{-34})\text{J.sec}\times(2.9979\times10^{8})\text{m.sec}^{-1}} = \mathbf{3.0205 \times 10^{18}/m^2.sec.}$$

8. The stopping potential for photoelectrons emitted from a surface irradiated by a light of wavelength $\lambda = 3000$ Å is 1.91 V. When the incident wavelength is changed, the potential is found to be 0.90 V. What is the new wavelength?

Soln. : Given, $\frac{1}{2}mv^2$ = kinetic energy of liberated electron =

$$1.91 \text{ eV} = \frac{hc}{\lambda} - W \text{ (where } \lambda = 3000 \text{ Å) and} \quad (1)$$

$$0.90 \text{ eV} = \frac{hc}{\lambda'} - W \text{ (Let } \lambda \text{ be a new wavelength.)} \quad (2)$$

Subtracting equation (2) from equation (1),

$$1.01 \times 1.602 \times 10^{-19} \text{ J}$$

$$= \left(6.626\times10^{-34}\right) \text{J.sec} \times \left(2.9979\times10^{8}\right) \text{m.sec}^{-1}. \left(\frac{1}{3000\times10^{-10}\text{m}} - \frac{1}{\lambda'}\right)$$

$\therefore \quad \lambda' = 3.9702 \times 10^{-7}$ m = **3970.2 Å.**

9. (a) What is the wavelength associated with a ball of 50 gm moving at a speed of 500 m/sec? (b) Why is the wave nature of ball not revealed through diffraction effects from solid?

Soln. : (a) Wavelength, $\lambda = \frac{h}{p} = \frac{h}{mv} = \frac{6\cdot 626\times10^{-34}\text{J.sec}}{\left(50\times10^{-3}\right)\text{kg}\times500 \text{ m/sec}}$

$= 2.65 \times 10^{-35}$ m = **2.65×10^{-25} Å.**

(b) λ associated with the ball is too much smaller than the interatomic separation ($\approx$ 1Å) in solid to be detected through diffraction pattern.

10. Calculate the de Broglie wavelength of an electron moving with kinetic energy 500 eV.

Soln. : de Broglie wavelength, $\lambda = \frac{h}{p}$

$$= \frac{h}{\sqrt{2m \text{ KE}}} \text{ [KE = kinetic energy]}$$

$$= \frac{\left(6.626\times10^{-34}\right)\text{J.sec}}{\sqrt{2\times\left(9.109\times10^{-31}\right)\text{kg}\times500\,\text{ev}\times\left(1.602\times10^{-19}\right)\text{J/eV}}}$$

$= 5.485 \times 10^{-11}$ m = **0.549 Å.**

11. Calculate the smallest possible uncertainty in the position of an electron moving with a velocity of 3×10^7 m/sec,

Soln. : Uncertainty in position $\Delta x = \frac{h}{4\pi m \Delta v}$

$$= \frac{\left(6.626\times10^{-34}\right)\text{J.sec}}{4\times3.14\times\left(9.109\times10^{-31}\right)\text{kg}\times3\times10^{7}\text{ m.sec}^{-1}} = 1.931 \times 10^{-12} \text{ m} = \mathbf{1.93 \text{ pm.}}$$

12. Find the uncertainty in the momentum of a particle when its position is determined within 0.01 cm. Find also the uncertainty in the velocity of an electron, a proton and an α-particle respectively when they are located within 5×10^{-10} m.

Soln. : Uncertainty in the momentum of the particle,

$$\Delta p = \frac{h}{4\pi\Delta x} = \frac{(6.626 \times 10^{-34})\ \text{J.sec}}{4 \times 3.14 \times 0.01\ \text{cm} \times 10^{-2}\ \text{m / cm}} = \mathbf{5.28 \times 10^{-31}\ kg.m.sec^{-1}}.$$

Uncertainty in velocity of an electron :

$$\Delta v = \frac{h}{4\pi m\Delta x} = \frac{(6.626 \times 10^{-34})\ \text{J.sec}}{4 \times 3.14 \times (9.109 \times 10^{-31})\ \text{kg} \times (5 \times 10^{-10})\ \text{m}} = \mathbf{1.16 \times 10^{5}\ m.sec^{-1}}.$$

Uncertainty in velocity of proton :

$$\Delta v = \frac{(6.626 \times 10^{-34})\ \text{J.sec}}{4 \times 3.14 \times (1.672 \times 10^{-27})\ \text{kg} \times (5 \times 10^{-10})\ \text{m}} = \mathbf{63.104\ m.sec^{-1}}.$$

Uncertainty in the velocity of an α-particle :

$$\Delta v = \frac{(6.626 \times 10^{-34})\ \text{J.sec}}{4 \times 3.14 \times \left(\frac{4 \times 10^{-3}}{6.022 \times 10^{23}}\right) \text{kg} \times (5 \times 10^{-10})\ \text{m}}$$

$$= \mathbf{15.88\ m.sec^{-1}}.$$

13. Show that $f(y) = (16y^4 - 48y^2 + 12)\, e^{-\frac{y^2}{2}}$ is an eigenfunction of the operator $y^2 - \frac{d^2}{dy^2}$. Find the eigenvalue.

Soln. : $\left(y^2 - \frac{d^2}{dy^2}\right)(16y^4 - 48y^2 + 12)\, e^{\frac{y^2}{2}}$

$$= (16y^6 - 48y^4 + 12y^2)\, e^{-\frac{y^2}{2}} - \frac{d^2}{dy^2}(16y^4 - 48y^2 + 12)\, e^{-\frac{y^2}{2}}$$

$$= A - \frac{d^2 B}{dy^2}.$$

$$\frac{dB}{dy} = (-16y^5 + 112y^3 - 108y)\, e^{-\frac{y^2}{2}}$$

$$\frac{d^2 B}{dy^2} = (16y^6 - 192y^4 + 444y^2 - 108)\, e^{-\frac{y^2}{2}}.$$

Now, $A - \frac{d^2 B}{dy^2}$ gives

$$\left(y^2 - \frac{d^2}{dy^2}\right) f(y) = (144y^4 - 432y^2 + 108)\, e^{-\frac{y^2}{2}} = 9f(y).$$

Thus, eigenvalue is **9**.

14. Show that $e^{-\frac{1}{2}x^2}$ is an eigenfunction of the operator $\frac{d^2}{dx^2} - x^2$. Find the eigenvalue.

Soln. : $\left(\frac{d^2}{dx^2} - x^2\right) e^{-\frac{1}{2}x^2}$

$$= \frac{d}{dx}\left(-xe^{-\frac{1}{2}x^2}\right) - x^2 e^{-\frac{1}{2}x^2} = -e^{-\frac{1}{2}x^2} + x^2 e^{-\frac{1}{2}x^2} - x^2 e^{-\frac{1}{2}x^2}$$

$= -e^{-\frac{1}{2}x^2}$. Hence, eigenvalue is **–1.**

15. Polynomials $P_0(x) = a_0$, $P_1(x) = a_1 + b_1x$, $P_2(x) = a_2 + b_2x + c_2x^2$ are the first three members of an orthogonal set of functions in the interval $-1 \leq x \leq 1$. Evaluate the constants a_0, a_1, b_1, etc.

Soln. : $\int_{-1}^{+1} P_0^2(x)dx = 1$

or, $\int_{-1}^{+1} a_0^2 dx = 1$, $\therefore$ $\boldsymbol{a_0 = \frac{1}{\sqrt{2}}}$. (1)

Again, $\int_{-1}^{+1} P_0P_1 dx = 0$

or, $a_0a_1\int_{-1}^{+1} dx + a_0b_1\int_{-1}^{+1} xdx = 0$

or, $a_0a_1 = 0$, $\therefore$ $\boldsymbol{a_1 = 0}$ $[\because a_0 = \frac{1}{\sqrt{2}}]$. (2)

Now, $\int_{-1}^{+1} P_1^2(x)dx = 1$

or, $\int_{-1}^{+1}\left(a_1^2 + b_1^2x^2 + 2a_1b_1x\right)dx = 1$ or, $b_1^2\left(\frac{2}{3}\right) = 1$. [$\because$ from equaton (2), $a_1 = 0$]

$\therefore$ $\boldsymbol{b_1 = \sqrt{\frac{3}{2}}}$ (3)

Again, $\int_{-1}^{+1} P_0P_2 dx = 0$

or, $\int_{-1}^{+1}\left(a_0a_2 + a_0b_2x + a_0c_2x^2\right)dx = 0$ or, $2a_0a_2 = -\frac{2}{3}a_0c_2$.

$\therefore$ $\boldsymbol{a_2 = -\frac{c_2}{3}}$. (4)

Now, $\int_{-1}^{+1} P_1 P_2 dx = 0$

or, $\int_{-1}^{+1} \left(b_1 a_2 x + b_1 b_2 x^2 + b_1 c_2 x^3\right) dx = 0$

or, $b_1 b_2 = 0.$

$\therefore$ $\mathbf{b_2 = 0}$ [$\because$ from equation (3), $b_1 = \sqrt{\frac{3}{2}}$]. (5)

Again, $\int_{-1}^{+1} P_2^2 dx = 1$

or, $\int_{-1}^{+1} a_2^2 dx + \int_{-1}^{+1} 2a_2 c_2 x^2 dx + \int_{-1}^{+1} c_2^2 x^4 dx$ [$\because$ from equation (5), $b_2 = 0$]

$\therefore$ $\mathbf{a_2 = \frac{\sqrt{5}}{2\sqrt{2}}}$.

Thus, $\mathbf{c_2} = -3a_2 = -\mathbf{\frac{3\sqrt{5}}{2\sqrt{2}}}$.

16. The speed of an electron is measured to be 100 m/sec with an uncertainty of 0.1%. Find the minimum uncertainty in locating the position of the electron.

Soln. : v = 100 m/sec

Δv = 0.001 × 100 m/sec = 0.1 m/sec.

$\Delta p = m\Delta v$ = (9.109 × 10^{-31}) kg × 0.1 m/sec = 9.109 × 10^{-32} kg. m/sec.

$$\Delta x = \frac{h}{4\pi\Delta p} = \frac{\left(6.626\times10^{-34}\right)\text{J.sec}}{4\times3.14\times\left(9.109\times10^{-32}\right)\text{kg.m/sec}} = \mathbf{5.792 \times 10^{-4}\ m.}$$

17. Calculate the de Broglie wavelength of an electron travelling at 1.00% speed of light.

[C.U.(H) 2002]

Soln. : Velocity v = (2.9979 × 10^8) m/sec × 10^{-2} = 2.9979 × 10^6 m/sec.

$$\text{Wavelength, } \lambda = \frac{h}{mv} = \frac{\left(6.626\times10^{-34}\right)\text{J.sec}}{\left(9.109\times10^{-31}\right)\text{kg}\times\left(2.9979\times10^{6}\right)\text{m/sec}}$$

$= 2.426 \times 10^{-10}$ m = **2.43 Å.**

18. Average life-time of an excited atom is 10^{-8} sec.

(a) What is the minimum uncertainty in frequency of the radiation emitted by the atom while decaying to ground state?

(b) What is the uncertainty in energy of the excited state?

(c) Calculate the natural width ($\Delta\nu/\nu$) of the spectral lines of sodium atom at λ = 5890 Å.

Soln. : (a) Uncertainty in frequency, $\Delta\nu = \frac{1}{4\pi\Delta t} = \frac{1}{4\times3.14\times10^{-8}\text{ sec}}$

$= \mathbf{7.96 \times 10^6\ sec^{-1}}$.

(b) Uncertainty in energy

$= h\Delta\nu = (6.626 \times 10^{-34})$ J.sec $\times$ 7.96×10^6 sec^{-1}

$= 5.275 \times 10^{-27}$ J = **3.29×10^{-8} eV.**

(c) $\lambda = 5890$ Å $= \dfrac{c}{\nu}$.

$$\therefore \frac{\Delta\nu}{\nu} = \frac{7.96\times 10^6 \text{ sec}^{-1}}{c} = \frac{7.96\times 10^6 \text{ sec}^{-1}\times 5890\times 10^{-10} \text{ m}}{2.9979\times 10^8 \text{ m/sec}} = \mathbf{1.56 \times 10^{-8}}.$$

19. Assume that $\psi(x) = c \sin\left(\dfrac{n\pi x}{L}\right)$ is a wave function of a particle in a box of length L. Verify that it satisfies $H\psi = E\psi$. What is the value of E so obtained? Calculate c from normalization of ψ. **[B.U.(H) 2000]**

Soln. : $$H\psi = -\frac{h^2}{8\pi^2 m}\frac{d^2}{dx^2}\, c \sin\left(\frac{n\pi x}{L}\right)$$

$$= \frac{n^2h^2\pi^2}{8\pi^2 mL^2}\, c \sin\left(\frac{n\pi x}{L}\right) = \frac{n^2h^2}{8mL^2} c \sin\left(\frac{n\pi x}{L}\right) = E\psi.$$

$$\therefore \quad \boldsymbol{E = \frac{n^2h^2}{8mL^2}}.$$

$$c^2\int_0^L \sin^2\frac{n\pi x}{L}dx = 1$$

or, $$\frac{c^2}{2}\int_0^L\left(1-\cos\frac{2n\pi x}{L}\right)dx = 1 \quad [\because \cos 2x = 1 - 2\sin^2 x]$$

or, $$\frac{c^2}{2}\left[(x)_0^L - \left(\frac{\sin 2n\pi x/L}{2n\pi/L}\right)_0^L\right] = 1$$

or, $$c^2 \times \frac{L}{2} = 1 \quad [\text{since } n = 1, 2, 3, \ldots, \text{etc.}, \sin 2n\pi = 0]$$

$$\therefore \quad \boldsymbol{c = \sqrt{\frac{2}{L}}}.$$

20. Consider a particle is moving to and fro in a box of length a.

(a) Find the averge value of p_x, for $n = 1$ state. Comment on result.

(b) If the particle is in its ground state, what is the probability of finding the particle in the range $\frac{a}{3}$ to $\frac{2a}{3}$? Find the same for the particle in its first and second excited states.

Soln. : (a) $$\langle p_x\rangle = \int \psi_1^*\hat{p}_x\psi_1 dx = \frac{2}{a}\int_0^a \sin\frac{\pi x}{a}\left(-i\hbar\frac{\partial}{\partial x}\right)\sin\frac{\pi x}{a}dx$$

[ψ for particle in a box is $\sqrt{\dfrac{2}{a}}\sin\left[\dfrac{n\pi x}{a}\right]$]

$$= -i\hbar\frac{2}{a}.\frac{\pi}{a}\int_0^a \sin\frac{\pi x}{a}\cos\frac{\pi x}{a}dx = -i\frac{\hbar}{a^2}\int_0^a \sin\frac{2\pi x}{a}dx = \frac{i\hbar}{a^2}\frac{\left(\cos\frac{2\pi x}{a}\right)_0^a}{2\pi/a} = \mathbf{0.}$$

Probability of finding positive value of momentum for a particle in a box is exactly same as that of its negative value of momentum and they compensate each other to give zero average momentum.

(b) Probability of finding the particle in the ground state in the range $\frac{a}{3}$ to $\frac{2a}{3}$,

$$\int_{a/3}^{2a/3}\psi_1^*\psi_1 dx = \frac{2}{a}\int_{a/3}^{2a/3}\sin^2\frac{\pi x}{a}dx \text{ (since } \psi_n = \sqrt{\frac{2}{a}}\sin\frac{n\pi x}{a}\text{, here } n = 1)$$

$$= \frac{1}{a}\int_{a/3}^{2a/3}\left(1-\cos\frac{2\pi x}{a}\right)dx = \frac{1}{a}\left[\left(\frac{2a}{3}-\frac{a}{3}\right)-\frac{a}{2\pi}\left(\sin\frac{2\pi x}{a}\right)_{a/3}^{2a/3}\right]$$

$$= \frac{1}{3}-\frac{1}{2\pi}\left(\sin\frac{4\pi}{3}-\sin\frac{2\pi}{3}\right) = \frac{1}{3}+\frac{0\cdot 866}{\pi} = 0.609.$$

Probability of finding the particle in the first excited state within the range $\frac{a}{3}$ to $\frac{2a}{3}$,

$$\int_{a/3}^{2a/3}\psi_2^*\psi_2 dx$$

$$= \frac{2}{a}\int_{a/3}^{2a/3}\sin^2\frac{2\pi x}{a}dx = \frac{1}{a}\int_{a/3}^{2a/3}\left(1-\cos\frac{4\pi x}{a}\right)dx = \frac{1}{3}-\frac{1}{4\pi}\left(\sin\frac{8\pi}{3}-\sin\frac{4\pi}{3}\right) = \mathbf{0.196.}$$

Probability of finding the particle in the 2nd excited state between $\frac{a}{3}$ to $\frac{2a}{3}$.

$$\int_{a/3}^{2a/3}\psi_3^*\psi_3 dx = \frac{2}{a}\int_{a/3}^{2a/3}\sin^2\frac{3\pi x}{a}dx = \frac{1}{a}\int_{a/3}^{2a/3}\left(1-\cos\frac{6\pi x}{a}\right)dx\cdot$$

Proceeding in the same way, probability = **0.333.**

21. The wave function of a certain state of a system is the linear combination $\psi = \sqrt{\frac{1}{4}}\phi_1 + \sqrt{\frac{3}{4}}\phi_2$, where ϕ_1 and ϕ_2 are energy eigenfunctions with distinct eigenvalues E_1 and E_2 respectively. What is the probability that the system energy will be observed with E_1? **[B.U.(H) 1994]**

Soln. : $$\langle E\rangle = \int\psi^* H\psi d\tau = \int\left(\sqrt{\frac{1}{4}}\phi_1^* + \frac{\sqrt{3}}{4}\phi_2^*\right)H\left(\sqrt{\frac{1}{4}}\phi_1 + \sqrt{\frac{3}{4}}\phi_2\right)d\tau$$

$$= \frac{1}{4}\int\phi_1^* H\phi_1 d\tau + \frac{3}{4}\int\phi_2^* H\phi_2 d\tau$$

(since ϕ_1 and ϕ_2 are non-degenerate eigenfunctions of H,

$$\int \phi_1^* H\phi_2 d\tau = \int \phi_2^* H\phi_1 d\tau = 0)$$

$$= \frac{1}{4}E_1 \int \phi_1^* \phi_1 d\tau + \frac{3}{4}E_2 \int \phi_2^* \phi_2 d\tau \text{ (given, } H\phi_1 = E_1\phi_2,\ H\phi_2 = E_2\phi_2)$$

$$= \frac{1}{4}E_1 + \frac{3}{4}E_2 \ (\because \int \phi_1^* \phi_1 d\tau = \int \phi_2^* \phi_2 d\tau = 1).$$

The probability that the system energy will be observed with E_1 is $\frac{1}{4}$.

22. A particle with a mass 6.65×10^{-27} kg is confined in an infinite square well of width a. The energy of the third level is 2×10^{-24} J. Calculate a.

Soln. : $E = \dfrac{n^2h^2}{8ma^2}$

$$\therefore a = \frac{3 \times \left(6.6626 \times 10^{-34}\right) \text{J.sec}}{\sqrt{8 \times \left(6.65 \times 10^{-27}\right) \text{kg} \times \left(2 \times 10^{-24}\right) \text{J}}} \quad (\because n = 3)$$

$= 6.094 \times 10^{-9}$ m = **6.09 nm.**

23. A particle of mass 10^{-6} kg is rolling on the smooth floor of a 1×10^{-4} m wide box with a speed of 3.3313×10^{-3} m/sec. Applying particle in a box problem, calculate the quantum number corresponding to the translational energy of a ball. Will you consider the energy to be quantized in a practical sense?

Soln. : Kinetic energy, $T = \dfrac{1}{2}mv^2 = \dfrac{n^2h^2}{8ma^2}$

or, $n^2 = \dfrac{4m^2v^2a^2}{h^2}$.

$$\therefore \quad n = \frac{2 \times 10^{-6} \text{kg} \times 10^{-4} \text{m} \times 3.3313 \times 10^{-3} \text{m/sec}}{6.626 \times 10^{-34} \text{J.sec}} = \mathbf{1 \times 10^{21}}.$$

$$\Delta E = E_{n+1} - E_n = \frac{(2n+1)h^2}{8ma^2} = \frac{\left(2 \times 10^{21} + 1\right)\left(6.626 \times 10^{-34} \text{J.sec}\right)^2}{8 \times 10^{-6} \text{kg} \times \left(10^{-4} \text{m}\right)^2}$$

$= 1.098 \times 10^{-32}$ J.

Thus, the energy **may not be considered quantized** in a practical sense.

24. An electron is confined to a molecule of length 2 nm. What is the minimum energy and minimum excitation energy from the state? What is the probability of locating the electron between $x = 0$ and $x = 0.5$ nm?

Soln. : Minimum Energy, $E = \dfrac{h^2}{8ma^2} = \dfrac{\left(6.626 \times 10^{-34} \text{J.sec}\right)^2}{8 \times \left(9.109 \times 10^{-31}\right) \text{kg} \times \left(2 \times 10^{-9} \text{m}\right)^2}$

$= \mathbf{1.51 \times 10^{-20}}$ **J.**

Minimum excitation energy, $\Delta E = \dfrac{(2n+1)h^2}{8ma^2}$

$$= \frac{3\times\left(6.626\times10^{-34}\,\text{J.sec}\right)^2}{8\times9.109\times10^{-31}\,\text{kg}\times\left(2\times10^{-9}\,\text{m}\right)^2} = \mathbf{4.52\times10^{-20}\ J.}$$

Probability of locating electron between $x = 0$ and $x = 0.5$ nm,

$$\int_a^b \psi_1^*\psi_1 dx = \frac{2}{l}\int_0^{0.5}\sin\frac{\pi x}{l}\sin\frac{\pi x}{l}dx \quad \left(\text{since } \psi_n = \sqrt{\frac{2}{l}}\sin\frac{n\pi x}{l}\right)$$

$$= \frac{1}{l}\int_0^{0.5}\left(1-\cos\frac{2\pi x}{l}\right)dx = \frac{1}{l}\left[(x)_0^{0.5} - \frac{l}{2\pi}\left(\sin\frac{2\pi x}{l}\right)_0^{0.5}\right]$$

$$= \frac{0.5}{l} - \frac{1}{2\pi} = \mathbf{0.091}\ (\because\ l = 2\text{ nm}).$$

25. Calculate the first excitation energy of a proton confined to a region roughly equal to the diameter of a nucleus (10^{-15} m). Calculate the probability that in its ground state it will be found between $x = 0.25l$ to $x = 0.75l$.

Soln. : First excitation energy of proton :

$$\Delta E = \frac{(2n+1)h^2}{8ma^2}$$

$$= \frac{3\times\left(6.626\times10^{-34}\,\text{J.sec}\right)^2}{8\times1.672\times10^{-27}\,\text{kg}\times\left(10^{-15}\,\text{m}\right)^2}$$

$= 9.847 \times 10^{-11}$ J $= 614 \times 10^6$ eV $=$ **614 MeV.**

Probability of locating the particle within $0.25l$ to $0.75l$,

$$\int_a^b \psi_1^*\psi_1 dx = \frac{2}{l}\int_{0.25l}^{0.75l}\sin^2\frac{\pi x}{l}dx$$

$$= \frac{1}{l}\int_{0.25l}^{0.75l}\left(1-\cos\frac{2\pi x}{l}\right)dx = \frac{1}{l}\left[0.5l - \frac{1}{2\pi}\left(\sin\frac{2\pi x}{l}\right)_{0.25l}^{0.75l}\right] = \mathbf{0.82.}$$

26. If benzene is regarded as a square box of side 0.4 nm containing 6π-electrons, then calculate the wavelength of light to promote a π-electron to the first excited state.

Soln. : Energy of benzene, $E_{gs} = \dfrac{4h^2}{8m_ea^2} + \dfrac{10h^2}{8m_ea^2} + \dfrac{10h^2}{8m_ea^2} = \dfrac{24h^2}{8m_ea^2}.$

Energy of excited state, $E_{ex} = \dfrac{4h^2}{8m_ea^2} + \dfrac{10h^2}{8m_ea^2} + \dfrac{5h^2}{8m_ea^2} + \dfrac{8h^2}{8m_ea^2} = \dfrac{27h^2}{8m_ea^2}.$

$$E_{ex} - E_{gs} = h\nu = \frac{3h^2}{8m_ea^2}$$

or, $\lambda = \dfrac{8cm_e a^2}{3h}$

$$= \frac{8\times\left(2.9979\times10^{8}\right)\text{m.sec}^{-1}\times\left(9.109\times10^{-31}\right)\text{kg}\times\left(0.4\times10^{-9}\,\text{m}\right)^2}{3\times6.626\times10^{-34}\ \text{J.sec}}$$

$= 1.758 \times 10^{-7}$ m = **175.8 nm.**

27. A particle has the wave function $\psi(r) = \left(\dfrac{1}{\pi a_0^3}\right)^{\frac{1}{2}} e^{-\frac{x}{a_0}}$, where a_0 is constant. Find $\langle r \rangle$.

Soln. : $\langle r \rangle = \displaystyle\int_0^\infty \psi^*(r) r \psi(r) 4\pi r^2 dr$

$$= \frac{1}{\pi a_0^3}\int_0^\infty 4\pi r^3 e^{-\frac{2r}{a_0}} dr = \frac{4}{a_0^3}\cdot\frac{3!}{\left(\frac{2}{a_0}\right)^4} = \frac{3}{2}a_0.$$

28. Calculate the zero point energy of a particle of mass 10^{-3} kg constrained to move only along x-axis when it is fixed to a spring which is stretched 10^{-2} metres by a force of 0.1 newton.

Soln. : By Hook's law, force constant $k = \dfrac{F}{-x} = \dfrac{0.1\,\text{N}}{10^{-2}\,\text{m}} = 10$ kg.sec^{-2}.

Frequency $\nu = \dfrac{1}{2\pi}\sqrt{\dfrac{k}{m}}$

$$= \frac{1}{2\pi}\sqrt{\frac{10\,\text{kg.sec}^{-2}}{10^{-3}\,\text{kg}}} = 15.92\ \text{Hz}.$$

Zero point energy, $E_0 = \dfrac{1}{2}h\nu$

$= \dfrac{1}{2}\times\left(6.626\times10^{-34}\right)$J.sec $\times$ 15.92 Hz = **5.27 × 10^{-33} J.**

29. Verify that the function $Ae\left(-\frac{1}{2}Bx^2\right)$ is an eigenfunction of linear harmonic oscillator, where $B = \dfrac{\sqrt{(mk)}}{\hbar}$. **[V.U.(H) 2001]**

Soln. : $H = -\dfrac{h^2}{2m}\dfrac{\partial^2\psi}{\partial x^2} + \dfrac{1}{2}kx^2$. (1)

$\psi = Ae^{-\frac{1}{2}Bx^2}$. (2)

Now $\dfrac{\partial\psi}{\partial x} = \dfrac{\partial}{\partial x}Ae^{-\frac{1}{2}Bx^2} = -ABx\ e^{-\frac{1}{2}Bx^2} = \psi'$ (3)

and $\dfrac{\partial^2\psi}{\partial x^2} = \dfrac{\partial\psi'}{\partial x} = -AB\,e^{-\frac{1}{2}Bx^2} + AB^2x^2\ e^{-\frac{1}{2}Bx^2} = -B(1 - Bx^2)\psi.$ (4)

[putting the value of ψ from equation (2)]

Thus, $H\psi = -\frac{\hbar^2}{2m}\frac{\partial^2\psi}{\partial x^2} + \frac{1}{2}kx^2\psi$

$= \frac{\hbar^2}{2m}B\left(1 - Bx^2\right)\psi + \frac{1}{2}kx^2\psi$ [from equation (4)]

$= \frac{\hbar^2}{2m}.\frac{\sqrt{mk}}{\hbar}\psi - \frac{\hbar^2}{2m}.\frac{mk}{\hbar^2}x^2\psi + \frac{1}{2}kx^2\psi \quad \left(\text{given, } B = \frac{\sqrt{mk}}{\hbar}\right)$

$= \frac{\hbar}{2}\sqrt{\frac{k}{m}}\psi - \frac{1}{2}kx^2\psi + \frac{1}{2}kx^2\psi \;=\; \frac{1}{2}\hbar\sqrt{\frac{k}{m}}\psi$

$= \frac{1}{2}h\nu\psi$ (where $\nu = \frac{1}{2\pi}\sqrt{\frac{k}{\mu}}$ and $\hbar = \frac{h}{2\pi}$)

$= E\psi.$

Hence, $E = \frac{1}{2}h\nu.$

30. The 1s wave function for H atom is $\psi = \left(\pi a_0^3\right)^{-\frac{1}{2}} e^{-\frac{r}{a_0}}$. Calculate the probability of the electron being within a distance of a_0 (Bohr radius) from the nucleus.

[V.U.(H) 2002; C.U.(H) 2005]

Soln. : $P_{1s} = \int_0^{a_0} \psi^*\psi\, d\tau$

$= \int_0^{a_0} \psi^*\psi r^2 dr \int_0^{\pi} \sin\theta\, d\theta \int_0^{2\pi} d\phi \quad [d\tau = r^2\sin\theta\; dr d\theta\; d\phi]$

$= 4\pi \int_0^{a_0} \psi^*\psi r^2\, dr$

$= \frac{4}{a_0^3}\int_0^{a_0} r^2 e^{-\frac{2r}{a_0}} dr \left[\because\; \psi = \left(\pi a_0^3\right)^{-\frac{1}{2}} e^{-\frac{r}{a_0}}\right]$

$$= \frac{4}{a_0^3}\left(\left(\frac{r^2 e^{-\frac{2r}{a_0}}}{-\frac{2}{a_0}}\right)_0^{a_0} - 2\int_0^{a_0} r\frac{e^{-\frac{2r}{a_0}}}{-\frac{2}{a_0}} dr\right)$$

$$= \frac{4}{a_0^3}\left[\frac{a_0^3 e^{-2}}{-2} + a_0\int_0^{a_0} re^{-\frac{2r}{a_0}} dr\right] = -2\, e^{-2} + \frac{4}{a_0^2}\left[\left(\frac{re^{-\frac{2r}{a_0}}}{-\frac{2}{a_0}}\right)_0^{a_0} - \int_0^{a_0}\frac{e^{-\frac{2r}{a_0}}}{-\frac{2}{a_0}} dr\right]$$

$$= -2e^{-2} - 2\,e^{-2} + \frac{2}{a_0}\int_0^{a_0} e^{-\frac{2r}{a_0}}\,dr$$

$$= -2e^{-2} - 2e^{-2} - e^{-2} + 1 = \mathbf{0.3233}.$$

31. The radial wave function for 2*s* orbital of a hydrogen atom is given by $R_{20} = N\left(2-\frac{r}{a_0}\right)e^{-\frac{r}{2a_0}}$, N is constant. (i) Determine the number and location of node (*s*) in 2*s* wave function. (ii) Write down the expression for radial distribution function of a 2*s* electron and sketch the radial distribution curve. **[C.U.(H) 2002]**

Soln. : (i) The number of radial node $= n - l - 1$.

For 2*s* orbital, $n = 2$, $l = 0$.

Hence, the number of node $= 2 - 1 = 1$.

(ii) Radial distribution function for 2s electron,

$$R_{20}^2 r^2 = \left(N\left(2-\frac{r}{a_0}\right)e^{-\frac{r}{2a_0}}\right)^2 r^2,$$

where r = distance from nucleus.

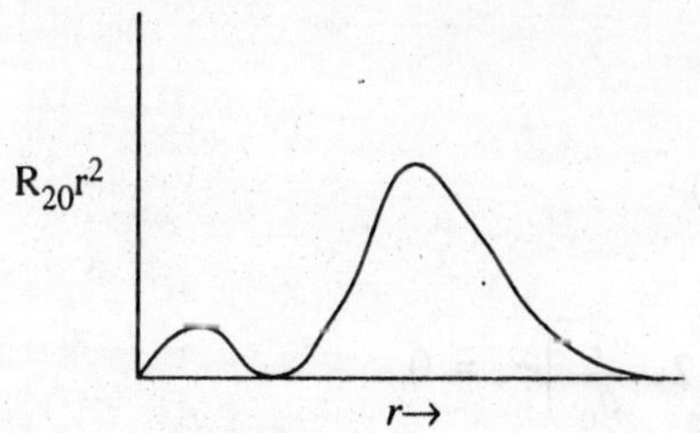

At the location of node, $R_{20} = 0$.

or, $N\left(2-\frac{r}{a_0}\right)e^{-\frac{r}{2a_0}} = 0$

or, $2 - \frac{r}{a_0} = 0$ (since N and $e^{-\frac{r}{2a_0}}$ are not zero)

$\therefore \quad r = 2a_0$.

Hence, node is at $2a_0$ distance from nucleus.

32. Estimate the wavelength of the light absorbed when a π-electron of butadiene is excited from the highest occupied energy level to the lowest vacant energy level. For the sake of simplicity assume that the π-electrons of butadiene move in a one-dimensional box of length 7.0 Å. **[C.U.(H) 2004]**

Soln. : Ground state energy of butadiene,

$$E_{gr} = \frac{2h^2}{8m_e a^2} + \frac{8h^2}{8m_e a^2}.$$

Energy of the excited state,

$$E_{ex} = \frac{2h^2}{8m_e a^2} + \frac{4h^2}{8m_e a^2} + \frac{9h^2}{8m_e a^2}.$$

$$\Delta E = E_{ex} - E_{gr} = h\nu = h\frac{c}{\lambda} = \frac{5h^2}{8m_e a^2}$$

$$\text{or,} \quad \lambda = \frac{8m_e a^2 c}{5h}$$

$$= \frac{8\times(9.109\times10^{-31})\,\text{kg}\times(7\times10^{-10}\,\text{m})^2\times(2.9979\times10^8)\,\text{m.sec}^{-1}}{5\times(6.626\times10^{-34})\,\text{J.sec}}$$

$$= 3.231 \times 10^{-7}\text{m} = \mathbf{3231\ \text{Å}}.$$

33. The 2*s* wave function for H atom is $\psi_{2s} = N\left(2 - \frac{r}{a_0}\right)e^{-\frac{r}{2a_0}}$, where N is a constant, r is the distance from nucleus and a_0 is Bohr radius. Find the distance from the nucleus at which radial probability density is maximum.

Soln. : Radial probability density for 2*s* electron = $\psi^*_{2s}\psi_{2s}r^2 = [R_{20}(r)]^2r^2$.

At the distance where radial probability density is maximum,

$$\frac{d[R_{20}(r)]^2 r^2}{dr} = 0$$

$$\text{or,} \quad \frac{d}{dr}\left[\left(2 - \frac{r}{a_0}\right)^2 e^{-\frac{r}{a_0}} r^2\right] = 0$$

$$\text{or,} \quad -\frac{2}{a_0}r^2 + 2r\left(2 - \frac{r}{a_0}\right) - \frac{1}{a_0}\left(2 - \frac{r}{a_0}\right)r^2 = 0$$

$$\text{or,} \quad -\frac{6r^2}{a_0} + \frac{r^3}{a_0^2} + 4r = 0 \quad \text{or,} \quad \frac{r^2}{a_0^2} - \frac{6r}{a_0} + 4 = 0 \ (\text{since, } r \neq 0)$$

$$\text{or,} \quad r = \frac{\frac{6}{a_0} \pm \sqrt{\frac{36}{a_0^2} - \frac{16}{a_0^2}}}{\frac{2}{a_0^2}} = 3a_0 \pm \sqrt{5}a_0.$$

$$\therefore \quad r_1 = 5.236\ a_0 \text{ and } r_2 = 0.764\ a_0.$$

Radial probability density for ψ_{2s} becomes maximum at **5.236** a_0 and at **0.764** a_0 as well.

34. Evaluate transmission coefficients for (i) electron (ii) proton for penetrating a rectangular potential barrier of height 12 MeV and thickness 10^{-14} m.. Both the particles have total energies 5 MeV.

Soln. : Transmission coefficient,

$$T = \frac{1}{\cosh^2\beta a + \frac{1}{4}\left(\frac{\alpha}{\beta} - \frac{\beta}{\alpha}\right)^2 \sinh^2\beta a}$$

$$\beta = \sqrt{\frac{2mV}{\hbar^2}\left(1-\frac{E}{V}\right)}.$$

$$a = \frac{\sqrt{2mE}}{\hbar}.$$

Here, $a = 10^{-14}$ m, $E = 5$ MeV, $V = 12$ MeV.

(i) For electron,

$$\beta = \frac{2\pi}{6.626\times10^{-34}\text{ J.sec}}\sqrt{2\times(9.109\times10^{-31})\text{kg}\times(7\times10^6)\text{eV}\times(1.602\times10^{-19})\text{J/eV}}$$

$= 1.355 \times 10^{13}$ m^{-1}.

$\therefore$ $T =$ **0.981.**

(ii) For proton,

$$\beta = \frac{2\pi}{(6.626\times10^{-34})\text{J.sec}}\sqrt{2\times(1.672\times10^{-27})\text{kg}\times(7\times10^6)\text{eV}\times(1.602\times10^{-19})\text{J/eV}}$$

$= 5.81 \times 10^{14}$ m .

$\therefore$ $T =$ **3.494×10^{-5}.**

35. Calculate the transmission coefficients of electron tunnelling through a rectangular barrier of height 2 eV and width (i) 2 Å, (ii) 4 Å. Electron has total energy 1 eV.

Soln. : $E = 1$ eV, $V = 2$ eV, $a = 2$ Å $= 2 \times 10^{-10}$ m.

$$\beta = \frac{2\pi\sqrt{2\times(9.109\times10^{-31}\text{ kg})\times1\text{ eV}\times(1.602\times10^{-19})\text{J/eV}}}{6.626\times10^{-34}\text{ J.sec}}$$

$= 5.12 \times 10^9$ m^{-1}.

$\therefore$ transmission coefficient,

$T =$ **0.404.**

When $a = 4$ Å $= 4 \times 10^{-10}$ m^{-1},

$T =$ **0.064.**

36. Normalise the function $e^{im\phi}$.

Soln. : $N^2 \int_0^{2\pi} e^{-im\phi}e^{im\phi}d\phi = 1$ (N = normalisation constant)

or, $N^2 \int_0^{2\pi} d\phi = 1$ or, $N^2.2\pi = 1$.

$$\therefore \quad N = \frac{\mathbf{1}}{\sqrt{\mathbf{2\pi}}}.$$

37. If ψ is an eigenfunction of operator α with eigenvalue a, then $\beta\psi$ is an eigenfunction of α with eigenvalue $\alpha + k$ provided $[\alpha, \beta] = k\beta$.

Soln. : Given, $(\alpha\beta - \beta\alpha)\psi = k\beta\psi$.

$\therefore$ $\alpha\beta\psi = \beta\alpha\psi + k\beta\psi = \beta a\psi + k\beta\psi$ (since $\alpha\psi = a\psi$)

Hence, $\alpha\beta\psi = (a + k)\beta\psi$.

38. An operator $\hat{T}_i$ is defined as $\hat{T}_i \psi(x) = \psi(x + i)$.

(a) Is $\hat{T}_i$ linear? (b) Evaluate : $\left(\hat{T}_1^2 - 3\hat{T}_1 + 2\right) x^2$.

Soln. : (a) If $\hat{T}_i c\psi(x) = c\psi(x + i)$, where c is any constant; $\hat{T}_i$ will be called linear.

(b) $\left(\hat{T}_1^2 - 3\hat{T}_1 + 2\right)x^2$

$$= \hat{T}_1(x + 1)^2 - 3(x + 1)^2 + 2x^2 \quad [\because \hat{T}_i\psi(x) = \psi(x + i)]$$

$$= (x + 2)^2 - 3(x + 1)^2 + 2x^2 = 1 - 2x.$$

39. Calculate the probability of a particle in the right-half of a one-dimensional box in its first excited state. Will all the other states give the same result? Explain.

Soln. : $\psi(x) = \sqrt{\dfrac{2}{a}} \sin \dfrac{2\pi x}{a}$, when $n = 2$

$$\int_0^{\frac{a}{2}} \psi^* \psi dx = \frac{2}{a} \int_0^{\frac{a}{2}} \sin^2 \frac{2\pi x}{a} dx$$

$$= \frac{1}{a} \int_0^{\frac{a}{2}} \left(1 - \cos \frac{4\pi x}{a}\right) dx \quad (\because 2 \sin^2 x = 1 - \cos 2x)$$

$$= \frac{1}{a} \left[\frac{a}{2} - \frac{a}{4\pi}\left(\sin \frac{4\pi x}{a}\right)_0^{\frac{a}{2}}\right] = \frac{1}{2}.$$

The probability of finding the particle is same in the two halves of the box for all the states of the particle.

Function of the particle in a box is the combination of two momentum eigenfunctions.

$$\sin \frac{n\pi x}{a} = \frac{1}{2i}\left(e^{i\frac{n\pi x}{a}} - e^{-i\frac{n\pi x}{a}}\right) \quad (\because e^{ix} = \cos x + i \sin x)$$

$$= \frac{1}{2i}\left(e^{ikx} - e^{-ikx}\right) \text{ (where } k = \frac{n\pi}{a}\text{)}.$$

So, the probability that the particle will possess positive momentum ($\hbar k$) is same as the particle will possess negative momentum ($\hbar k$) and hence, the probability of finding the particle in the left-half is same as the probability of finding the particle in the right-half.

40. Show that the functions of simple harmonic oscillator

$$\psi_0(x) = \frac{1}{\pi^{\frac{1}{4}}} e^{-\frac{x^2}{2}}$$

$$\psi_1(x) = \left(\frac{1}{4\pi}\right)^{\frac{1}{4}} 2xe^{-\frac{x^2}{2}} \text{ are orthogonal.}$$

Soln. : $\displaystyle\int_{-\infty}^{+\infty} \psi_0(x)\psi_1(x)dx = \left(\frac{1}{2\pi}\right)^{1/2} \int_{-\infty}^{+\infty} xe^{-x^2} dx$

$$= \left(\frac{1}{2\pi}\right)^{1/2}\left[\int_0^{+\infty} xe^{-x^2/2}dx + \int_0^{\infty}(-x)e^{-x^2/2}dx\right] = 0.$$

Since the integral is an odd function of x its magnitude from $-\infty$ to $+\infty$ must be zero.

41. A particle of mass m moving in one dimension between $x = a$ and $x = b$ is described by a wave function $\psi = \frac{A}{x}$, where A is normalisation constant. (a) Calculate A, (b) What is the average value of x?

Soln. : (a) $\psi = \frac{A}{x}$

$$\int_a^b \psi^*\psi = 1$$

$$\text{or, } A^2\int_a^b \frac{dx}{x^2} = 1 \quad \text{or, } A^2\left[-\frac{1}{x}\right]_a^b = 1$$

$$\text{or, } A^2\left(\frac{b-a}{ab}\right) = 1, \quad \therefore A = \sqrt{\frac{ab}{b-a}}.$$

$$\text{(b) } \bar{x} = \int_a^b \psi^* x\psi\, dx = \frac{ab}{b-a}\int_a^b \frac{x}{x^2}dx$$

$$= \frac{ab}{b-a}\int_a^b \frac{dx}{x} = \frac{ab}{b-a}\ln\frac{b}{a}.$$

42. The velocity of a rifle bullet is 1000 m/sec. If it weighs 30 gm, calculate the minimum precision in measurement of its position when its momentum is not perturbed by more than one part in a million.

Soln. : $v = 1000$ m/sec

$$\Delta v = 1000 \text{ m/sec} \times \frac{1}{10^6},$$

Hence, $\Delta p = m\Delta v = 30 \times 10^{-3}$ kg $\times 10^{-3}$ m/sec $= 3 \times 10^{-5}$ kg.m/sec.

$$\text{Thus, } \Delta x = \frac{\hbar}{2\Delta p}$$

$$\text{or, } \Delta x = \frac{6.626\times10^{-34}\text{ J.sec}}{4\times\pi\times(3\times10^{-5})\text{kg.m/sec}} = 1.76 \times 10^{-30} \text{ m.}$$

43. Show that the wave functions corresponding to different energy levels for a particle in a box are orthogonal.

Soln. : $\psi_n = \sqrt{\frac{2}{a}}\sin\frac{n\pi x}{a}$ [n = 1, 2, 3, ..., etc.]

$$\psi_m = \sqrt{\frac{2}{a}}\sin\frac{m\pi x}{a} \quad [m = 1, 2, 3, \ldots, \text{etc.}]$$

$$\int_0^a \psi_n\psi_m dx = \frac{2}{a}\int_0^a \sin\frac{n\pi x}{a}\sin\frac{m\pi x}{a} = \frac{1}{a}\int_0^a\left[\cos(n-m)\frac{\pi x}{a}-\cos(n+m)\frac{\pi x}{a}\right]$$

$$= \frac{1}{a}\left[\frac{1}{n-m}\left(\sin(n-m)\frac{\pi x}{a}\right)_0^a - \frac{1}{n+m}\left(\sin(n+m)\frac{\pi x}{a}\right)_0^a\right]$$

$= 0$ [since $\sin 0 = 0$ and $\sin (n - m)\pi = \sin (n + m)\pi = 0$ as n and m are integer numbers]

44. A stream of electrons each having an energy $E = 4$ eV is directed towards a potential barrier of height 5 eV. The width of the barrier is 2.1×10^{-7} cm. Calculate the fraction transmitted.

Hints : $\beta = \frac{1}{\hbar}\sqrt{2m(V-E)}$

$$= \frac{2\pi\sqrt{2\times\left(9.109\times10^{-31}\right)\text{kg}\times(5-4)\text{eV}\times\left(1.602\times10^{-19}\right)\text{J/eV}}}{6.626\times10^{-34}\text{ J.sec}}$$

$= 5.12 \times 10^{-19}\text{ m}^{-1}$.

$\beta a = 5.12 \times 10^{-19} \times 2.1 \times 10^{-9} = 10.76$.

$T = 1.3 \times 10^{-9}$.

45. A particle has the wave funcion $\psi = \sqrt{\frac{2(1+\alpha^2)}{\alpha^2}}\, e^{-|x|} \sin \alpha x$. Calculate $\langle x\rangle$ and $\langle p_x^2\rangle$.

Soln. : $\psi = \sqrt{\frac{2(1+\alpha^2)}{\alpha^2}} e^{-|x|} \sin \alpha x = Ae^{-|x|}\sin \alpha x$, where $A = \sqrt{\frac{2(1+\alpha^2)}{\alpha^2}}$

$\psi_1 = Ae^{-x} \sin \alpha x \qquad 0 \le x \le \infty$

$\psi_2 = Ae^{x} \sin \alpha x \qquad -\infty \le x \le 0$

$$\langle x\rangle = \int_{-\infty}^{+\infty} \psi x \psi\, dx$$

$$= \int_0^{+\infty} \psi_1 x\psi_1 dx + \int_{-\infty}^{0} \psi_2 x \psi_2\, dx$$

$$= A^2\left[\int_0^\infty e^{-2x} x\sin^2 \alpha x\, dx + \int_{-\infty}^0 e^{2x} x \sin^2 \alpha x\, dx\right]$$

$$= A^2\left[\int_0^\infty e^{-2x} x\sin^2 \alpha x\, dx + \int_0^\infty e^{-2x}(-x) \sin^2 \alpha x\, dx\right] = \mathbf{0}.$$

$$\langle p_x^2\rangle = -A^2\hbar^2\left[\int_0^\infty \psi_1 \frac{d^2}{dx^2}\psi_1\,dx + \int_{-\infty}^0 \psi_2 \frac{d^2}{dx^2}\psi_2\,dx\right]$$

$$= -2A^2\hbar^2\left[\int_0^\infty (1-\alpha^2)e^{-2x}\sin^2\alpha x\,dx + \int_0^\infty 2\alpha e^{-2x}\sin\alpha x\cos\alpha x\,dx\right]$$

$$= -2\hbar^2\left[(1-\alpha^2)\frac{1}{2}-1\right] \text{ [integreating two terms by parts]}$$

$$= \boldsymbol{\hbar^2(1+\alpha^2)}.$$

Hints : $\displaystyle\int_0^\infty e^{-2x}\sin^2\alpha x\,dx = \frac{1}{2}\int_0^\infty e^{-2x}(1-\cos 2\alpha x)dx$

$$= \frac{1}{2}\left[-\frac{e^{-2x}}{2} - \frac{e^{-2x}}{4(1+\alpha^2)}(2\alpha\sin 2\alpha x - 2\cos 2\alpha x)\right]_0^\infty$$

$$= \frac{1}{2}\left[\frac{1}{2} - \frac{1}{2(1+\alpha^2)}\right] = \frac{1}{4}.\frac{\alpha^2}{1+\alpha^2}$$

$$\int_0^\infty e^{-2x}\sin\alpha x\cos\alpha x\,dx$$

$$= \frac{1}{2}\int_0^\infty e^{-2x}\sin 2\alpha x\,dx = \frac{1}{2}\frac{\left(e^{-2x}\right)_0^\infty}{4(1+\alpha^2)}\left[-2\sin 2\alpha x - 2\alpha\cos 2\alpha x\right]_0^\infty$$

$$= -\frac{\alpha}{4(1+\alpha^2)}.$$

46. Normalize the function $\psi = e^{-zr/a_0}$.

Soln. : $\displaystyle N^2\int_0^\infty \psi 4\pi r^2\,dr = 1$

or, $\displaystyle N^2 4\pi\int_0^\infty e^{-2zr/a_0} r^2\,dr = 1$

or, $\displaystyle N^2 4\pi.\ \frac{2}{\left(\frac{2z}{a_0}\right)^3} = 1$

or, $\displaystyle N^2 = \left(\frac{z}{a_0}\right)^3.\frac{1}{\pi},\ \therefore N = \left(\frac{1}{\pi}\right)^{1/2}\left(\frac{a_0}{z}\right)^{3/2}.$

47. Is wave function for H atom is $\psi_{1s} = Ne^{-r/a_0}$, where N is a constant, r is a distance from the nucleus and a_0 is Bohr radius. Find the most probable value of r in $1s$ state.

Soln. : Radial probability density $= R_{10}^2 r^2$.

$$\frac{d\left(R^2 r^2\right)}{dr} = 0$$

or, $$\frac{d}{dr}\left(e^{-2r/a_0} r^2\right) = 0$$

or, $$2re^{-2r/a_0} - \frac{2}{a_0} r^2 e^{-2r/a_0} = 0$$

or, $$2re^{-2r/a_0}\left(1 - \frac{r}{a_0}\right) = 0$$

or, $$1 - \frac{r}{a_0} = 0 \text{ (since, } re^{-2r/a_0} \neq 0).$$

$\therefore r = a_0$ = distance from nucleus at which radial probablity density is maximum.

48. A particle is confined within a one-dimensional box of length a. Find the minimum possible energy it will have from uncertainty principle.

Soln. : The maximum uncertainty in position of the particle = length of box within which the particle is confined.

Hence, $\Delta x = a$.

By uncertainty principle, $\Delta x \Delta p_x \geq \frac{\hbar}{2}$.

Hence, minimum uncertainty in momentum of the particle

$$= \Delta p_x = \frac{\hbar}{2\Delta x} = \frac{\hbar}{2a}.$$

Now, energy of the particle confined in a box,

$$E = \frac{\langle p_x^2 \rangle}{2m}.$$

For particle in a box, $\Delta p_x^2 = \langle p_x^2 \rangle$.

Hence, $$E = \frac{\Delta p_x^2}{2m} = \frac{\hbar^2}{8ma^2}.$$

49. Calculate the minimum uncertainty in position for an automobile of mass 500 kg moving with speed 50 ± 0.001 km.hr^{-1}.

Comment on the result [$h = 6.627 \times 10^{-34}$ J.sec]. **[C.U. 2001]**

Soln. : By the problem, $\Delta v_x = 2 \times 0.001$ km.hr^{-1}.

$$\Delta p_x = m\Delta v_x = 500 \text{ kg} \times 0.002 \text{ km.hr}^{-1} = 0.278 \text{ N.sec}.$$

By uncertainty principle, $\Delta x \Delta p_x = \frac{\hbar}{2}$.

$$\therefore \Delta x = \frac{\hbar}{2 \times 0.278} = \frac{6.627 \times 10^{-34}}{4 \times \pi \times 0.278} \text{m} = 1.89 \times 10^{-34} \text{ m}.$$

Δx is too small to be detected. Hence, uncertainty principle is not important for macroscopic bodies like microscopic bodies.

CHAPTER 16

SPECTROSCOPY AND SOME PHYSICAL PROPERTIES

Required Formulae

1. If an angle θ is subtended by two bond moments μ_1 and μ_2, then resultant dipole moment

$$\mu = \sqrt{\mu_1^2 + \mu_2^2 + 2\mu_1.\mu_2 \cos\theta}\cdot$$

If $\mu_1 = \mu_2$, $\mu = 2\mu_1 \cos\dfrac{\theta}{2}$.

2. Molar polarization, $P_m = \dfrac{D-1}{D+2}\dfrac{M}{\rho}$,

where D = dielectric constant
M = molecular weight
ρ = density.

3. Molar refraction, $R_m = \dfrac{n^2-1}{n^2+2}\dfrac{M}{\rho}$,

where n = refractive index.

4. $P_m = \dfrac{4}{3}\pi N_A\left(\alpha_a + \alpha_e + \dfrac{\mu^2}{3KT}\right)$

α_a = atomic polarizability
α_e = electronic polarizability
μ = permanent dipole moment
k = Boltzmann constant
T = temperature
N_A = Avogadro number.

5. Specific rotation, $[\alpha]_\lambda^t = \dfrac{100\alpha_{obs}}{lC}$,

α_{obs} = observed angle of rotation

l = length of polarimeter in dm
C = concentration of substrate in gm per 100 ml of solution.

Molecular rotation $[M]_\lambda^t = \dfrac{[\alpha]_\lambda^t \times M}{100}$, where M = molar mass.

6. For a solution mixture of A and B, molar refraction of solution

$= R_M = X_A R_M(A) + X_B R_M(B)$,

where X_A = mole fraction of A
$R_M(A)$ = molar refraction of A
X_B = mole fraction of B
$R_M(B)$ = molar refraction of B

$$R_M = \frac{n^2-1}{n^2+2}\left(\frac{X_A M_A + X_B M_B}{\rho_{sol}}\right)$$

n = refractive index of solution
M_A = molar mass of A
M_B = molar mass of B.

7. Rotational constant, $B = \dfrac{h}{8\pi^2 Ic}$ cm^{-1} = $\dfrac{h^2}{8\pi^2 I}$ joule = $\dfrac{h}{8\pi^2 I}$ sec^{-1},

where I = moment of inertia = μr^2
c = velocity of light
h = Planck constant, μ = reduced mass
r = bond length.

8. Rotational energy, $E = BJ(J + 1)$ cm^{-1}.

9. Spacing between the lines in rotation spectrum = $2B$.

10. Vibrational energy = $\left(n+\dfrac{1}{2}\right)h\nu$ joule = $\left(n+\dfrac{1}{2}\right)\nu$ sec^{-1}.

11. Vibrational frequency, $\bar{\nu} = \dfrac{1}{2\pi c}\sqrt{\dfrac{k}{\mu}}$ cm^{-1} = $\dfrac{1}{2\pi}\sqrt{\dfrac{k}{\mu}}$ sec^{-1},

where k = force constant.

12. Spacing between lines in vibration spectrum = ν sec^{-1}.

13. Ratio of populations at two energy levels J and J' : $\dfrac{N_J}{N_{J'}} = e^{-(\varepsilon_J - \varepsilon_{J'})/KT}$,

ε_J, $\varepsilon_{J'}$ = energy of the levels J and J'
N_J, $N_{J'}$ = population of the respective energy levels.

If degeneracy of the levels = $2J + 1$, then $N_J / N_{J'} = \dfrac{2J+1}{2J'+1} e^{-(\varepsilon_J - \varepsilon_{J'})/KT}$

Short Questions

1. What is polarizability? What is polarization? Can a molecule with zero dipole moment undergo polarization?

Ans. : Polarizability is the dipole moment induced in molecule by an external electric field of unit strength. If μ_{ind} is the induced dipole moment and E is the strength of the applied field, then $\mu_{ind} \propto E$ or, $\mu_{ind} = \alpha E$, where α is the polarizability.

It is the measure of easiness with which a molecule can be polarized.

Polarization (P) is the dipole moment induced by an external field per unit volume.

A molecule with zero dipole moment can be polarized by the distortion of its electronic distribution and nuclei by an applied field.

2. Why is CCl_4 molecule is non-polar although the C-Cl bond is polar?

[B.U.(H) 2002]

Ans. : Net dipole moment of a molecule is the vector addition of all the bond moments. Now, bond moment of one C-Cl bond is exactly opposite to the vector addition of other three C-Cl bond moments, since CCl_4 has tetrahedrally symmetrical structure.

3. Find out the dimension of dipole moment μ. What is the SI unit of dipole moment? How do the induced polarization and the orientation polarization depend on temperature?

[B.U. 2003]

Ans. : Dimension of dipole-moment : current × time × length = ITL

SI unit of dipole moment μ : C-m (coulomb-metre)

Polarization is induced in both polar and non-polar molecules by the displacement of electronic distribution and nuclei towards opposite ends of the external electric field. This type of polarization is called *distortion polarization* and it is independent of emperature. But polar molecules can undergo another type of polarization, called *orientation polarization*, in addition to distoration polarization. Orientation polarization occurs due to lignments of molecular dipoles along external field and it depends on temperature. As emperature increases, the random orientation of the molecular dipoles also increase and t, in turn, decreases the tendency of these dipoles to orient along field direction, and thus rientation polarization decreases. Thus, for polar molecules induced polarization depends n temperature while for non-polar molecules it is temperature independent.

4. Show that sum of the bond moment of BCl_3 molecule is zero.

Ans. : The shape of BCl_3 molecule is triangular planar :

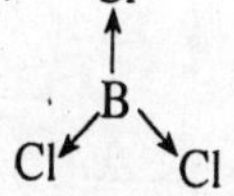

Let the bond moment of each C-Cl bond be μ making an angle of 120° with each her. The resultant moment of two C-Cl bond moments

$$= \mu' = \sqrt{\mu^2 + \mu^2 + 2\mu^2 \cos 120°} = \sqrt{2\mu^2 - \mu^2} = \mu.$$

It is opposite to the remaining one C-Cl bond moment.

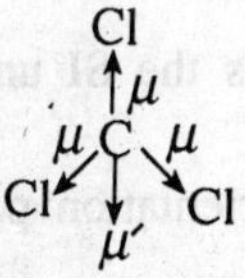

Hence, the sum of the bond moments = $\mu - \mu' = \mu - \mu = 0$.

5. Explain the graphical plot required to determine the dipole moment of a polar ostance from the molar polarization values measured at a number of temperatures.

[B.U. 1997]

Ans. : For polar substance orientation polarizability α_0 is directly proportional to its dipole moment μ in the applied field direction and inversely proportional to temperature T of the system, namely

$$\alpha_0 = \frac{\mu^2}{3KT}, \text{ where } k \text{ is Boltzmann constant.}$$

Total polarizability α_t is sum of distortion (α_d) and orientation (α_o) polarizabilities

$$\alpha_t = \alpha_d + \alpha_o$$

According to Debye equation,

$$P_m = \frac{4}{3}\pi N_A \alpha_t$$

$$\text{or, } P_m = \frac{4}{3}\pi N_A \left(\alpha_d + \frac{\mu^2}{3KT}\right),$$

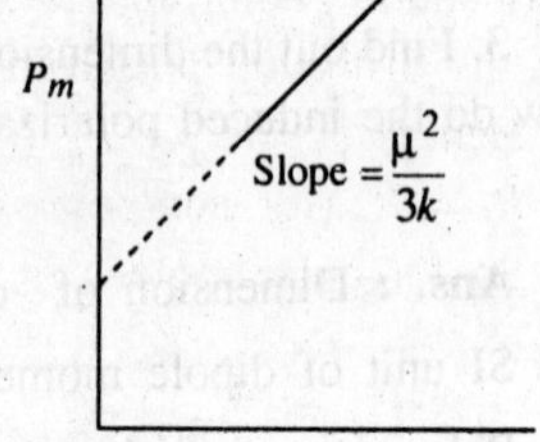

where P_m = molar polarization

N_A = Avogadro number.

Hence, P_m of polar molecules depends on temperature. Measuring a number of differen P_m values at different temperatures and plotting those values against $\frac{1}{T}$, the dipole momen μ can be determined from the slope $\left(\frac{\mu^2}{3k}\right)$.

6. How does molar polarization vary with temperature? How would you determine the dipole moment of a molecule using the variation of molar polarization with temparature?

Ans. : Same as the answer of Q. 5.

7. What is the CGS unit of orientation polarizability?

Ans. : Orientation polarizability, $\alpha_o = \frac{\mu^2}{3kT}$, where μ = permanent dipole moment k = Boltzmann constant, T = temperature.

$$\text{Hence, its unit} = \frac{(\text{esu.cm})^2}{(\text{erg.K}^{-1})\text{K}}$$

$$= \frac{(\text{esu.cm})^2}{\text{erg}} = \frac{\left(\text{dyne}^{\frac{1}{2}}.\text{cm.cm}\right)^2}{\text{dyne.cm}} = \text{cm}^3.$$

8. What is the SI unit of orientation polarization?

Ans. : Orientation polarization, $P_o = \frac{1}{3\varepsilon_0}\left(\frac{\mu^2}{3kT}\right)$,

where ε_0 = permitivity of vacuum.

μ = permanent dipole moment

k = Boltzmann constant

T = temperature in °K.

So, its SI unit $= \dfrac{1}{\left(C^2.N^{-1}.m^{-2}\right)}\dfrac{(C.m)^2}{\left(J.K^{-1}\right)K}$

$$= \frac{C^2.m^2}{\left(C^2.N^{-1}.m^{-2}\right)(N.m)} = m^3.$$

9. Is it possible to distinguish a polar molecule from a non-polar molecule from the temperature variation of molar polarization? [C.U.(H) 1996]

Ans. : In polar molecule sum total of distortion polarization P_d and orientation polarization P_o contribute to the total induced polarization (P_t) while in non-polar molecule only the distortion polarization constitutes the total polarization (P_t).

Hence, $P_t = P_o + P_d$ (for polar molecules).

$P_t = P_d$ (for non-polar molecule).

So for polar molecules molar polarization,

$$P_m = \frac{4}{3}\pi N_A\left(\alpha_d + \alpha_\mu\right) = \frac{4}{3}\pi N_A\left(\alpha_d + \frac{\mu^2}{3kT}\right)$$

where N_A = Avogadro Number

α_d = distortion polarizability

α_μ = orientation polarizability

μ = permanent dipole moment,

k = Boltzmann constant

and T = temperature in kelvin,

and for non-polar molecule,

$$P_m = \frac{4}{3}\pi N_A\alpha_d.$$

α_d is characteristic of the molecule and is independent of temperature, and thus P_m for non-polar molecule is also independent of temperature. But for polar molecules, besides α_d, α_μ also contributes to P_m and α_μ depends on temperature. So, P_m for polar molecule varies with temperature. It decreases with increasing temperature while the same for non-polar molecule remains constant.

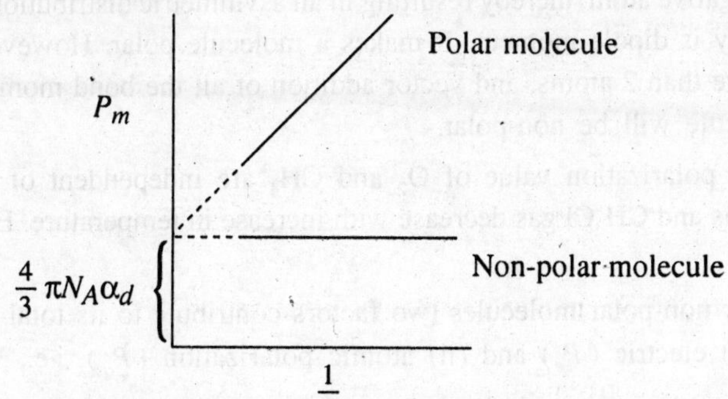

So, observing temperature variation of molar polarization (P_m), it is possible to distinguish polar molecule from non-polar molecule.

10. Show that the polarizability has a dimension of volume? Arrange the following molecules in the order of the decreasing polarizability HCl, HF, HI, HBr.

Ans. : Polarizability, $\alpha = \dfrac{\text{induced dipole moment}}{\text{external electric field}}$

$$= \frac{\mu}{E} = \frac{\text{charge} \times \text{distance}}{\text{charge}/(\text{distance})^2} = \frac{Q \times L}{Q/L^2} = L^3.$$

Hence, polarizability has the dimension of volume.

Polarizability decreases successively from HI to HF in the order HI > HBr >HCl > HF, as the bond length decreases successively from HI to HF in the same order.

11. What will be the nature of total polarization P_t versus $\frac{1}{T}$ for CH_2Cl and C_6H_6? **[B.U. 2003]**

Ans. : CH_2Cl is polar molecule and C_6H_6 is non-polar molecule.

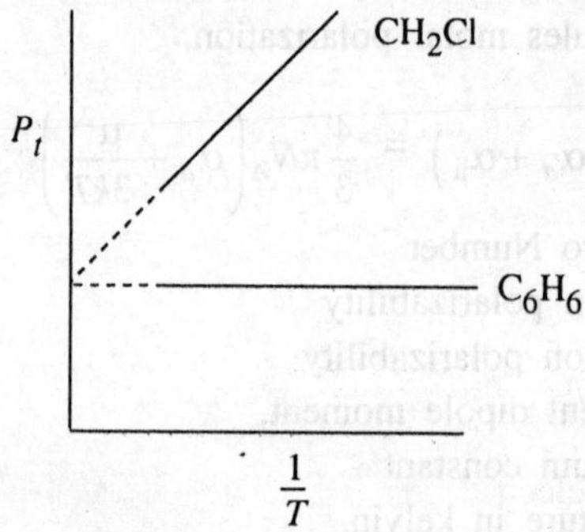

The reason is given in answer to Q. 9.

12. Atoms are non-polar but a molecule may be either polar or non-polar—discuss. **[C.U. 1993]**

Ans. : In isolated atom, negative electronic charge is spherically disposed around the positively charged nucleus. This symmetric charge distribution makes atom non-polar. But a molecule may be homonuclear or heteronuclear. In homonuclear molecule again the symmetric charge distribution makes the atom non-polar. In heteronuclear molecule due to difference in eletronegativity of the atoms the negative charge distribution shifts towards more electronegative atom, thereby resulting in an asymmetric distribution of charge density. This asymmetry is dipole moment. It makes a molecule polar. However, if the molecule consists of more than 2 atoms, and vector addition of all the bond moments becomes null, then the molecule will be non-polar.

13. Molar polarization value of O_2 and CH_4 are independent of temperature while those of HCl gas and CH_3Cl gas decrease with increase in temperature. Explain with proper reason. **[C.U.(H) 2003]**

Ans. : For non-polar molecules two factors contribute to its total molar polarization P_m, namely (i) electric (P_e) and (ii) atomic polarization (P_a), i.e., $P_m = P_e + P_a$.

Electronic and atomic polarization arise due to distortion of electronic distribution and nucleus respectively towards opposite ends of the external electric field. Together they are called distortion polarization (P_d). It is independent of temperature, since it is evident from the definition that the ease with which charge distribution will be distorted depends on the external field and on the molecule itself. O_2 and CH_4 are non-polar molecule and hence their molar polarization is independent of temperature.

In case of polar molecule another type of polarization, called *orientation polarization*, (P_o) is present, i.e., $P_m = P_o + P_e + P_a = P_o + P_d$. Now, the rise in temperature increases random orientation of molecular dipoles, thereby reducing the tendency of the dipoles to orient along the field direction which in turn decreases orientation polarization. In other words, it has an inverse relation with temperature. HCl and CH_3Cl are polar molecules and so their molar polarization decreases with increase in temperature.

14. Total molar polarization is partly temperature dependent and partly temperature independent—why? **[B.U. 1999]**

Ans. : For non-polar molecules total molar polarization is completely temperature independent while for polar molecule it is partly temperature dependent and partly temperature independent.

See Q. 13.

15. Write Clausius-Mosottii equation in CGS and SI systems.

Clausius-Mosottii equation relates the distortion polarizability of the molecule to the dielectric constant of the non-polar substance.

Ans. : CGS System :

$$\frac{D-1}{D+2}\frac{M}{\rho} = P_m = \frac{4}{3}\pi N_A \alpha_d$$

D = dielectric constant of the medium

M = molar mass of the medium

ρ = density of the medium

N_A = Avogadro number

α_d = distortion polarizability

P_m = molar polarization.

SI System :

$$\frac{\left(\frac{\varepsilon}{\varepsilon_0}\right)-1}{\left(\frac{\varepsilon}{\varepsilon_0}\right)+2}\frac{M}{\rho} = P_m = \frac{1}{3\varepsilon_0} N_A \alpha_d$$

ε = permitivity of medium

ε_0 = permitivity of vacuum.

Other symbols signify usual meaning.

16. Why is *p*-dichlorobenzene is non-polar but *p*-dihydroxybenzene is polar?

Ans. : In *p*-dichlorobenzene two Cl atoms are in the same plane of the benzene ring and thus the vector sum of two C-Cl bond moments are zero as they are oppositely directed. But, in the case of *p*-dihydroxybenzene two OH bonds are not coplanar with benzene, rather they are directed at an angle to the benzene ring and thus, it is polar.

p-dichlorobenzene. 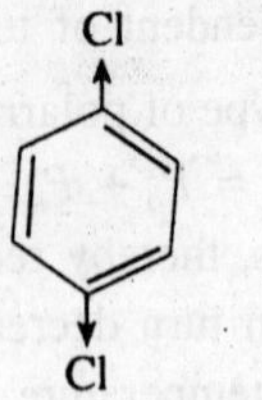 $\mu = 0$

p-dihydroxybenzene. 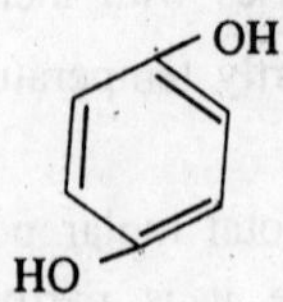 $\mu \neq 0$

17. SO_2 has dipole moment of 1.6 D while CO_2 has zero dipole moment—explain.

Ans. : SO_2 is non-linear molecule and thus, two S-O bond moments are at an angle θ with each other giving rise to net dipole moment which is equal to $2\mu_{SO} \cos\frac{\theta}{2}$. 

CO_2 is planar molecule and its two C-O bond moments are oppositely directed making an angle of $\theta = 180°$ and hence, CO_2 is non-polar. $O \equiv C \equiv O$

18. Three different dichlorobenzenes have the dipole moments (i) 1.48 D, (ii) 0 and (iii) 2.25 D. Identify the compounds.

Ans. : Net dipole moment μ is vector addition of all the bond moments. If μ_1 and μ_2 make an angle θ with each other, then

$$\mu = \sqrt{\mu_1^2 + \mu_2^2 + 2\mu_1\mu_2 \cos\theta},$$

if $\mu_1 = \mu_2$, $\mu = 2\mu_1 \cos\frac{\theta}{2}$,

For *o*-dichlorobenzene, $\theta = 60°$. 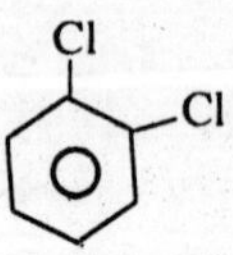

$\therefore \mu = 2\mu_1 \cos 30° = \sqrt{3}\mu_1$.

For *m*-dichlorobenzene, $\theta = 120°$. 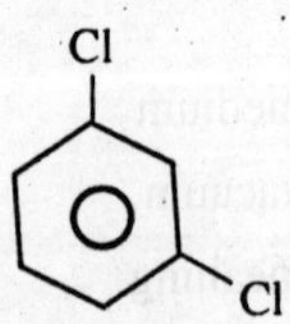

$\therefore \mu = \mu_1$.

And for *p*-dichlorobenzene, $\theta = 180°$. 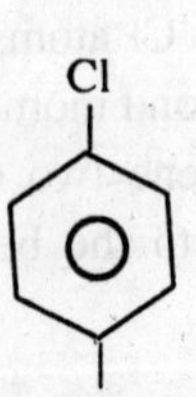

$\therefore \mu = 0$.

Hence, ortho has the highest dipole moment and para has the least. Thus (i) is ortho, (ii) is para and (iii) is meta.

19. What is specific rotation?

Ans. : Specific rotation is defined as the angle of rotation of the plane of linearly polarized light produced by 1 decimetre length of solution containing 1 gram of the optically active substance per ml and is represented as

$$[\alpha]_D^t = \frac{100\alpha_{obs}}{lC}$$

$[\alpha]_D^t$ = specific rotation measured at temperature t using sodium D line.

α_{obs} = observed angle of rotation

l = length (in dm) of the column of solution

C = concentration of the substance in gram per 100 ml of solution.

20. On which factors does the angle of rotation of the plane of linearly polarized light produced by an optically active substance depend?

Ans. : The factors are :

(i) the nature of substance,

(ii) concentration of the solution,

(iii) length of the solution traversed by polarized light,

(iv) nature of solvent,

(v) wavelength of the light used,

(vi) temperature.

21. What are the units of molar polarization and molar refraction?

Ans. : Both molar polarization and molar refraction have the unit cm^3/mol or m^3/mol.

22. Write Debye equation in CGS and SI systems?

Ans. : Debye equation relates the total polarizability, i.e., orientation and distortion polarizability of the polar molecule to the dielectric constant of the molecule.

CGS System :

$$\frac{D-1}{D+2}\frac{M}{\rho} = P_m = \frac{4}{3}\pi N_A\left(\alpha_d + \frac{\mu^2}{3kT}\right),$$

where D = dielectric constant of the substance

M = molar mass of the substance

ρ = density

P_m = molar polarization

N_A = Avagadro number

α_d = distortion polarizability

μ = permanent dipole moment

k = Boltzmann constant

T = temperature.

SI System :

$$\frac{\left(\frac{\varepsilon}{\varepsilon_0}\right)-1}{\left(\frac{\varepsilon}{\varepsilon_0}\right)+2}\frac{M}{\rho} = P_m = \frac{1}{3\varepsilon_0}N_A\left(\alpha_d + \frac{\mu^2}{3kT}\right)$$

ε = permitivity of medium

ε_0 = permitivity of vacuum.

All other symbols signify the usual meaning.

23. The dipole moment of CO_2 is zero but that for H_2O is 1.85 D — explain. **[C.U. 1994]**

Ans. : CO_2 is a linear molecule having the structure O = C = O while H_2O is non-linear having the structure . For further information see Q. 17.

24. What is polarizability volume? What is its unit?

Ans. : Polarizability volume, $\alpha' = \frac{\alpha}{4\pi\varepsilon_0}$,

where α = polarizability of the molecule

ε_0 = permitivity of vacuum.

Unit of $\alpha' = \frac{J^{-1}.C^2.m^2}{J^{-1}.C^2.m^{-1}} = m^3$.

25. What is the unit of polarizability in SI system?

Ans. : Unit of polarizability, $\alpha\left(=\frac{\mu_{ind}}{E}\right)$

$$= \frac{C.m}{N.C^{-1}} = C^2.m.N^{-1} = J^{-1}C^2.m^2.$$

(unit of dipole moment = C.m, unit of E = $N.C^{-1}$)

26. What is Lorentz-Lorenz equation?

Ans. : Lorentz-Lorenz equation relates molar polarization (P_m) to molar refraction (R_m).

Thus, it is written as $P_m = R_m = \frac{n^2-1}{n^2+2}\frac{M}{\rho}$, (1)

where n is the refractive index of the substance, M is its molar mass and ρ is its density.

This relation holds, both for polar or non-polar molecules, if the *alternation* frequency of external field is in optical region (> 10^{15} Hz). Now, at optical region P_m contains only electronic polarization (P_e).

It is written as $P_m = P_e = \frac{4}{3}\pi N_A \alpha_e$, (2)

where N_A is Avogadro number and α_e is electronic polarizability.

Thus, it is evident from equations (1) and (2),

$$R_m = \frac{4}{3}\pi N_A \alpha_e. \quad (3)$$

Equation (3) is also known as Lorentz-Lorenz equation.

27. Why does Maxwell relation $D = n^2$ hold for non-polar molecule but not for polar molecule?

Ans. : For polar molecule dielectric constant (D) depends on electronic polarizability (α_e), atomic polarizability (α_a) and orientation polarizability (α_o) while for non-polar molecule D depends only on α_e and α_a by the following relations,

$$\frac{D-1}{D+2}\frac{M}{\rho} = \frac{4}{3}\pi N_A(\alpha_e + \alpha_0 + \alpha_a) \text{ (for polar molecule)}$$

and $$\frac{D-1}{D+2}\frac{M}{\rho} = \frac{4}{3}\pi N_A(\alpha_e + \alpha_a) \text{ (for non-polar molecule),}$$

where M is molar mass and ρ is the density. Now, D is measured using alternating current of frequency less than 10^6 Hz, when all the three polarizabilities are present. But refractive index (n) is measured employing optical technique at the frequency greater than 10^{15} Hz, when α_o and α_a cease to exist. Refractive index is related to polarizability by the relation $\frac{n^2-1}{n^2+2}\frac{M}{\rho} = \frac{4}{3}N_A\alpha_e$. Since, α_o is altogether absent in case of non-polar molecule and α_a contributes minimal, $D = n^2$ holds for them, but α_o contributes significantly for polar molecule and thus, $D = n^2$ does not hold for them.

28. What is the essential condition for Raman activity of a molecule? What are Stokes and anti-Stokes lines in the Raman spectrum? Mention two important applications of Raman spectroscopy. **[C.U. 1995]**

Ans. : If polarizability of molecule changes with its vibration, the molecule is Raman active and this is the essential condition for the molecule to be Raman active.

Stokes lines in the Raman spectrum correspond to the radiation scattered by the molecule with a frequency lower than that of incident radiation.

Anti-Stokes lines in the Raman spectrum are due to the radiation scattered by the molecule with a frequency higher than that of incident radiation.

Two important applications of Raman spectroscopy are :

(i) determination of bond length of molecule

(ii) determination of structure of molecule.

29. How many normal modes are possible in C_2H_6, C_6H_6, H_2O, HCl and CO_2?

Ans. : The number of normal modes are $3N$–5 for linear molecule and $3N$–6 for non-linear molecule, where N is the number of atoms.

Hence, the number of normal modes are for

C_2H_6 : $3 \times 8 - 6 = 18$

C_6H_6 : $3 \times 12 - 6 = 30$

H_2O : $3 \times 3 - 6 = 3$

HCl : $3 \times 2 - 5 = 1$

CO_2 : $3 \times 3 - 5 = 4$.

30. Why water cannot be used as a solvent in infrared spectroscopy?

Ans. : Water absorbs strongly in infrared region. So, infrared spectroscopy of the sample under investigation cannot be obtained without interference from the absorption of water if water is used as solvent and thus, it cannot be used as a solvent in infrared spectroscopy.

31. What are the essential conditions for a compound to be infrared active and Raman active? Explain with suitable examples. **[C.U.(H) 2002]**

Ans. : If the dipole moment of the molecule changes with vibration, the molecule is infrared active, e.g., H_2O is IR active as its dipole moment changes with vibration, whereas H_2 is not inactive as H_2 is non-polar molecule and during vibration it does not generate moment. Again, although CO_2 is non-polar molecule, during antisymmetric stretching and bending, it generates dipole moment which changes with vibrations and hence, these particular vibrations are only inactive.

If polarizability of the molecules changes with vibration, the molecule is Raman active, e.g., both H_2O and H_2 are Raman active as their polarizability changes with vibration. But only symmetric stretching of CO_2 is Raman active, as polarizability changes only during this type of stretching.

32. Which of the following molecules show (a) rotational spectrum, (b) infrared spectrum, (c) Raman spectrum?

HCl, CS_2, Br_2, CCl_4, C_2H_4

Ans. : HCl will show (i) rotational, (ii) infrared and (iii) Raman spectrum, since it has permanent dipole moment which changes with rotation and vibration and its polarizability also changes with vibration.

CS_2 (S = C = S) will not show rotational spectrum, as it does not possess permanent dipole moment. But, its asymmetric stretching and bending are infrared active as dipole moment changes during these vibrations, and its symmetric stretching is Raman active as this stretching changes its polarizability.

Br_2 will show only Raman spectrum for the same reasons given above.

CCl_4 will show only Raman spectrum for the same reasons.

C_2H_4 will give Raman spectrum and infrared spectrum (only for assymetry stretching and bending).

33. What are Stokes and anti-Stokes lines in Raman spectrum? Why are Stokes lines more intense than anti-Stokes lines?

Ans. : Radiation scattered by molecules with a frequency lower than the frequency (ν) of incident radiation gives rise to Stokes lines. If ν is the frequency of incident radiation, then Stokes lines appear with the frequency ν_m ($\nu_m < \nu$), where ($\nu - \nu_m$) corresponds to the energy absorbed by molecules to go to higher energy state from lower one.

Radiation scattered by molecule with a frequency higher than the frequency (ν) of the incident radiation gives rise to anti-Stokes lines. If ν is the frequency of incident radiation, then anti-Stokes lines appear with frequency ν_m ($\nu_m > \nu$) where ($\nu_m - \nu$) corresponds to energy absorbed by molecules to come down from higher energy level to lower one.

Spectral intensity depends on the popualtion of the molecules of the energy level from which transition occurs. Population of lower energy level is greater than that of higher one at normal temperatures. So Stokes lines are more intense than the anti-Stokes lines.

34. Deduce the expression for the possible transition energies associated with rotation vibration spectra of a diatomic molecule. Hence, show that the vibration-rotation spectra would consist of lines with equal spacing on either side of band origin. **[C.U.(H) 2000]**

Ans. : Vibrational transition from $n = 0$ state to $n = 1$ state is accompanied by rotational transition. Let the rotational level in $n = 0$ state be designated by J and in $n = 1$ state by J'.

Now, rotational-vibrational energy of diatomic molecule (using simple harmonic oscillator and rigid-rotator approximation)

$$= \varepsilon_{rot,vib} = BJ\ (J + 1) + \left(n + \frac{1}{2}\right)\bar{\nu}_o,$$

where $\bar{\nu}_o$ = vibrational frequency in cm^{-1}

B = rotational constant = $\dfrac{h}{8\pi^2 Ic}$ cm^{-1}.

(h = Planck constant, I = moment of inertia = μr^2, μ = reduced mass, r = bond length, c = velocity of light)

So, transition energy in rotation-vibration spectrum of diatomic molecule =

$$\Delta\varepsilon = \varepsilon_{J',n=1} - \varepsilon_{J,n=0}$$

$$= BJ'(J' + 1) + \frac{3}{2}\bar{\nu}_o - BJ(J + 1) - \frac{1}{2}\bar{\nu}_o$$

$= \bar{\nu}_o + B(J' - J)(J' + J + 1)$ (assuming B remains the same for all the vibrational levels).

The rotational selection rule is $\Delta J = \pm 1$.

So, when $\Delta J = +1$, $J' - J = 1$ or, $J' = J + 1$,

then $\Delta\varepsilon = \bar{\nu}_o + 2BJ'$, where $J' = 1, 2, 3, \ldots$, etc. (1)

When $\Delta J = -1$, $J' - J = -1$, $\therefore J = J' + 1$,

then $\Delta\varepsilon = \bar{\nu}_o - 2BJ$, where $J = 1, 2, 3, \ldots$, etc. (2)

Combining (1) and (2),

$$\Delta\varepsilon = \bar{\nu}_o + 2Bm, \text{ where } m = \pm 1, \pm 2, \pm 3, \ldots, \text{ etc.}$$

The frequency $\bar{\nu}_o$ is called *band origin which remains absent in the spectrum.*

Lines to high frequency region (on the right side of band origin) correspond to +ve values of m and lines to low frequency region (on the left side of band origin) corresponds to –ve values of m. So rotational vibrational spectra consists of equally spaced lines with the separation of $2B$ between them on either side of the band origin.

35. Show that the lines in the rotation spectrum of a diatomic molecule are equispaced under rigid rotator approximation. **[B.U.(H) 2004]**

Ans. : Under rigid rotator approximation, rotational energy of diatomic molecule,

$$\varepsilon_{rot} = BJ\ (J + 1)\ cm^{-1},\ J = 0, 1, 2, \ldots,$$

where $B = \dfrac{h}{8\pi^2 Ic}$ = rotational constant.

(h = Planck constant, I = moment of inertia, c = speed of light)

J = rotational quantum number.

Now, under this approximation selection rule for rotational transition is $\Delta J = \pm 1$.

So, for the transition $J = 0 \rightarrow J = 1$,

transition energy, $\Delta\varepsilon = 2B \text{ cm}^{-1}$

and for the transition $J = 1 \rightarrow J = 2$,

$\Delta\varepsilon = 6B - 2B = 4B \text{ cm}^{-1}$, and so on.

So the lines in rotation spectrum are equally spaced with the separation of $2B \text{ cm}^{-1}$ in between them

36. State Franck-Condon principle.

Ans. : Franck-Condon principle states that an electronic transition takes place so rapidly in comparison with the nuclear motion that nuclei maintain their initial position and momentum just after the transition. It happens as the nuclei are much heavier than the electrons. According to this principle electronic transition from lower electronic state to higher one takes place vertically keeping internuclear distance and nuclear momentum intact.

37. Why are the spectral lines broad?

Ans. : Three factors contribute to the broadening of the spectral line, namely,

(i) Doppler broadening, (ii) Collisional broadening, (iii) life-time broadening.

(i) Doppler broadening : Molecules emitting radiation are in ceaseless random motion. So there will be Doppler shift of frequency of radiation. If the molecules approach observer, then the frequency will shift towards higher region and if molecules move away from observer, it will shift towards lower region.

(ii) Collisional broadening : It arises from the collision between the molecules, or between the molecules and the container. If the mean time between the collisions is Δt_{col}, then the resulting collisional broadening is $\Delta E_{col} \approx \dfrac{\hbar}{2\Delta t_{col}}$.

(iii) Life-time broadening : Uncertainty principles is responsible for this broadening. If life-time of the molecule in excited state is Δt, then the resulting broadening is $\Delta E = \dfrac{\hbar}{2\Delta t}$.

38. Determine various degrees of freedom of HCl, CH_3Cl, C_6H_6, H_2CO, CO_2, NH_3, CH_4, H_2O.

Ans. : Total number of degrees of freedom of a molecule consisting of N atoms is $3N$. Of these $3N$ degrees of freedom, 3 degrees of freedom are for translation, 2 (for linear) or 3 (for non-linear) for rotation and the rest constitutes vibrational degrees of freedom.

Molecule	*Translational*	*Rotational*	*Vibrational*	*Total*
HCl	3	2	1	6
CH_3Cl	3	3	9	15
C_6H_6	3	3	30	36
H_2CO	3	3	6	12
CO_2	3	2	4	9
NH_3	3	3	6	12
CH_4	3	3	9	15
H_2O	3	3	3	9

39. Which of the following molecules are microwave active—H_2, HCl, CH_4, CH_3I, H_2O, SF_6?

Ans. : HCl, CH_3I, H_2O are microwave active, since they have permanent dipole moment.

40. Why is laser used in Raman spectrometer?

Ans. : Laser is used as a source of radiation in Raman spectrometer to irradiate the sample, since intensity of scattered radiation is very low and the shifts, measured in frequency, of scattered radiation from incident radiation are quite small. So, highly intense monochromatic radiation from laser is required to make the shifts detectable.

41. Write the differences between harmonic and anharmonic oscillator.

Soln. :

Harmonic Oscillator	***Anharmonic Oscillator***
1. Energy of harmonic oscillator $E_n = \left(n+\frac{1}{2}\right)h\nu$, $n = 1,2,3, \ldots$, etc., where ν = frequency of vibration $= \frac{1}{2\pi}\sqrt{\frac{k}{\mu}}$. n = vibrational quantum number.	1. Energy of anharmonic oscillator $E_n = \left(n+\frac{1}{2}\right)h\nu_e - \left(n+\frac{1}{2}\right)^2 h\nu_e x_e$. where x_e = anharmonicity constant.
2. Energy levels are equispaced. $E_{n+1} - E_n = h\nu$	2. Separation between successive energy levels decreases as n increases. $E_{n+1} - E_n = h\nu_e(1-2x_e(n+1))$
3. Selection rule for transition is $\Delta n = \pm 1$.	3. Selection rule for transition is $\Delta n = \pm 1, \pm 2, \pm 3, \ldots$, etc.

Numerical Problems

1. Calculate the specific rotation of sucrose if a 20% solution of it when placed in a 2-decimetre polarimeter tube shows an angle of rotation of 26.4°. **[C.U.(H) 1994]**

Soln. : Specific rotation of sucrose $= [\alpha]_D^t = \frac{100\alpha_{obs}}{lC} = \frac{100 \times 26.4}{2 \times 20} = \mathbf{66°}$

2. The dipole moment of water molecule is 1.84 D. Calculate the O—H bond moment, the bond angle being 105°. **[C.U.(H) 2001]**

Estimate the % ionic character of the bond, if O—H length is 96 pm.

Soln. : Let μ_{O-H} be the bond moment of OH bond.

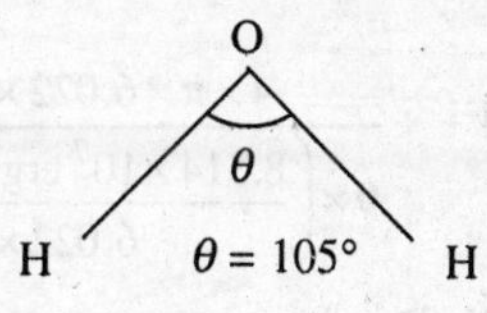

Given, $1.84 \text{ D} = \sqrt{\mu_{OH}^2 + \mu_{OH}^2 + 2\mu_{OH}^2 \cos\theta}$

$= 2\mu_{OH} \cos\frac{\theta}{2} = 2\mu_{OH} \cos 52.5°$

$\therefore \mu_{O-H} = 1.51$ D.

If H_2O is 100% ionic, then bond moment of OH will be,

$\mu_{OH\ ion} = 4.8 \times 10^{-10} \times 96 \times 10^{-10}$ esu.cm

$= 460.8 \times 10^{-20}$ esu.cm (10^{-18} esu.cm = 1D)

$= 4.608$ D.

Hence, % ionic character $= \dfrac{1.478}{4.608} \times 100 = \mathbf{32.07\%}.$

3. Dipole moment for HCl molecule is 1.03 D and internuclear separation is 1.275 Å. Calculate the % ionic character of the bond. **[Pune B.Sc. 1986]**

Soln. : $\mu_{HCl,\ ion} = 4.8 \times 10^{-10} \times 1.275 \times 10^{-8}$ esu.cm

$= 6.12 \times 10^{-18}$ esu.cm $= 6.12$ D.

Hence, % ionic character $= \dfrac{\mu_{obs}}{\mu_{ion}} \times 100 = \dfrac{1.03}{6.12} \times 100 = \mathbf{16.83\%}.$

4. Dielectric constant of a liquid of molecular weight 112 is 4.288. Its refractive index is 1.348 and its density is 1.108 gm/cc at 25 °C. Calculate molar polarization and molar refraction.

Soln. : $D = 4.288$, $n = 1.348$, $M = 112$, $r = 1.108$ gm/cc.

Hence, molar polarization $= P_m = \dfrac{D-1}{D+2}\dfrac{M}{\rho}$

$$= \frac{(4.288-1)}{(4.288+2)} \cdot \frac{112}{1.108} \text{ cc. mole}^{-1} = \mathbf{52.86\ cc.\ mole^{-1}}.$$

Molar refraction $= R_m = \dfrac{n^2-1}{n^2+2} \cdot \dfrac{M}{\rho} = \dfrac{(1.348)^2-1}{(1.348)^2+2} \times \dfrac{112}{1.108} = \mathbf{21.64\ cc.\ mole^{-1}}.$

5. From the data given in the last problem, determine dipole moment of the compound assuming atomic polarization is 5% of electronic polarization.

Soln. : Electronic polarization $= P_e = R_m = 21.64$ cc.mol^{-1}.

Atomic polarization $= P_a = \dfrac{5}{100} \times P_e = 1.082$ cc.mol^{-1}.

Molar polarization $= P_m = 52.86$ cc.mol^{-1}

Now $P_m = P_a + P_e + P_o$

P_o = orientation polarization $= \dfrac{4\pi N_A \mu^2}{9kT}$

Thus, 52.86 cc.mol^{-1}

$$= 1.082 \text{ cc. mol}^{-1} + 21.64 \text{ cc. mol}^{-1} + \frac{4 \times \pi \times 6.022 \times 10^{23} \times \mu^2 (\text{esu.cm})^2}{9 \times \left(\dfrac{8.314 \times 10^7 \text{ erg.K}^{-1}.\text{mole}^{-1}}{6.022 \times 10^{23}}\right) \times 298.15°\text{K}}$$

$\therefore\ \mu = 1.21 \times 10^{-18}$ esu.cm $= 1.21$ D.

6. Dipole moments of $C_6H_5NO_2$, C_6H_5Cl and *p*-nitrotoluene are 3.93 *D*, 1.55 *D* and 4.39 D respectively. Predict the dipole moments of *m*-nitrotolune, *o*-nitrotolune, ortho- and meta-dichlorobenzene and *o*-dinitrobenzene.

Soln. : Given,

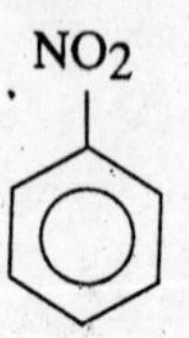
$\mu = 3.93$D

$\mu = 4.39$D

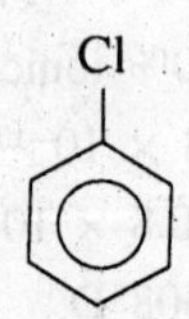
$\mu = 1.55$D

Dipole moment of tolune  $= 4.39\text{ D} - 3.93\text{ D} = 0.46\text{ D.}$

Hence, bond moments of C-Cl, C-NO_2, C-CH_3 are 1.55 D, 3.93 D and 0.46 D respectively.

(a) 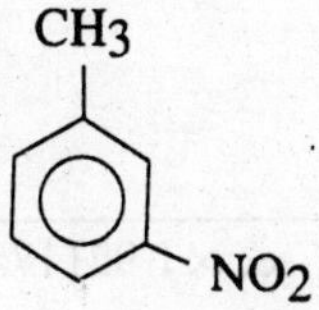

$$\mu = \sqrt{\mu^2_{C-CH_3} + \mu^2_{C-NO_2} + 2\mu_{C-CH_3}\mu_{C-NO_2}\cos\theta}$$

$$= \sqrt{(0.46)^2 + (3.93)^2 + 2\times 0.46\times 3.93\times \cos 120°}$$

$$= 3.72\text{ D.}$$

(b) 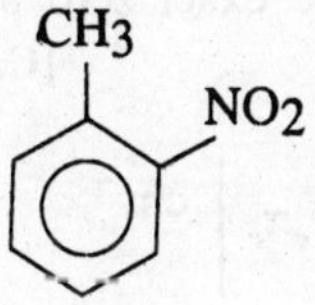

$$\mu = \sqrt{\mu^2_{C-CH_3} + \mu^2_{C-NO_2} + 2\mu_{C-CH_3}\mu_{C-NO_2}\cos 60°}$$

$$= \sqrt{(0.46)^2 + (3.93)^2 + 2\times 0.46\times 3.93\times \cos 60°} = 4.18\text{ D.}$$

(c) 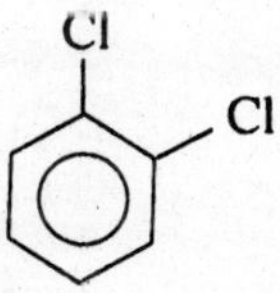

$$\mu = \sqrt{\mu^2_{C-Cl} + \mu^2_{C-Cl} + 2\mu_{C-Cl}\mu_{C-Cl}\cos 560°}$$

$$= 2\,\mu_{C-Cl}\cos 30° = 2\times 1.55\times \cos 30°\text{ D} = 2.68\text{ D.}$$

(d) 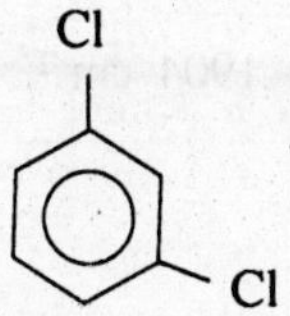

$$\mu = 2\mu_{C-Cl}\cos(120°/2) = 2\mu_{C-Cl}\cos 60° = 1.55\text{ D.}$$

(e) 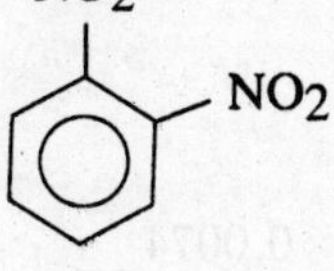

$$\mu = 2\mu_{C-NO_2}\cos(60°/2) = 2\mu_{C-NO_2}\cos 30°$$

$$= 2\times 3.93\times \cos 30°\text{ D} = 6.806\text{ D.}$$

7. In the pure rotational spectrum of CO, $\overline{\nu}_{0\to1}$ = 3.84235 cm^{-1}. Calculate the internuclear distance of the molecule, if μ_{CO} = 11.38365 × 11^{-27} kg? **[C.U.(H) 2003]**

Soln. : Given, $2B$ = 3.84235 cm^{-1}.

$\therefore\ B$ = 1.921175 cm^{-1}.

$$\text{Now, } B = \frac{h}{8\pi^2 Ic} = \frac{h}{8\pi^2 \mu r^2 c}$$

$$\text{or, } r^2 = \frac{h}{8\pi^2 \mu c B}$$

$$= \frac{6.626\times10^{-34}\,\text{J.sec}}{8\times\pi^2\times11.38365\times10^{-27}\,\text{kg}\times2.9979\times10^8\,\text{m / sec}\times1.921175\times100\,\text{cm}^{-1}}$$

= 1.28 × 10^{-20} m.

$\therefore\ r$ = 1.13 × 10^{-10} m = **1.13 Å.**

8. The fundamental and first overtone transition of NO molecule are centred at 1876 cm^{-1} and 3724 cm^{-1} respectively.

Evaluate (i) the equilibrium vibrational frequency, (ii) the exact zero point energy. **[C.U.(H) 2003]**

Soln. : Vibration energy $= E_n = \left[\left(n+\frac{1}{2}\right)\overline{\nu}_e - \left(n+\frac{1}{2}\right)^2 \overline{\nu}_e x_e\right]$ cm^{-1},

where n = vibrational quantum number,

$\overline{\nu}_e$ = equilibrium vibrational frequency in cm^{-1}

x_e = anharmonicity constant.

(i) Given, for transition from n = 0 to n = 1 state,

$$\Delta\varepsilon_{0\to1} = 1876\text{ cm}^{-1} = \overline{\nu}_e - 2\overline{\nu}_e x_e. \quad (1)$$

For tansition from n = 0 to n = 2 state,

$$\Delta\varepsilon_{0\to2} = 3724\text{ cm}^{-1} = 2\overline{\nu}_e - 6\overline{\nu}_e x_e. \quad (2)$$

Multiplying equation (1) by 3, we get

$$5628\text{ cm}^{-1} = 3\overline{\nu}_e - 6\overline{\nu}_e \overline{x}_e. \quad (3)$$

Subtracting equation (2) from equation (3), we get $\overline{\nu}_e$ = 1904 cm^{-1}.

Hence, equilibrium vibrational frequency = **1904 cm^{-1}.**

Now, from equation (1),

$$1876\text{ cm}^{-1} = \overline{\nu}_e(1 - 2x_e)$$

$$\text{or, } 1 - 2x_e = \frac{1876\,\text{cm}^{-1}}{1904\,\text{cm}^{-1}} = 0.9853.$$

$\therefore\ x_e$ = 0.0074.

(ii) The exact zero point energy

$$= E_{n=0} = \frac{1}{2}\overline{\nu}_e - \frac{1}{4}\overline{\nu}_e x_e = \frac{1904}{2}\text{cm}^{-1} - \frac{1904\times0.0074}{4}\text{cm}^{-1}$$

= **948.48 cm^{-1}.**

9. Chlorobenzene has μ = 1.55 D and nitrobenzene μ = 3.80 D. Estimate the dipole moments of meta-dinitrobenzene, ortho-dichlorobenzene and meta-chloronitrobenzene. The observed moments are 3.90 D, 2.25 D and 3.40 D. How would you explain any discrepancy?

[B.U. 1990]

Ans. : Bond moments of C-Cl and C-NO_2 are 1.55 D and 3.80 D respectively.

Hence, 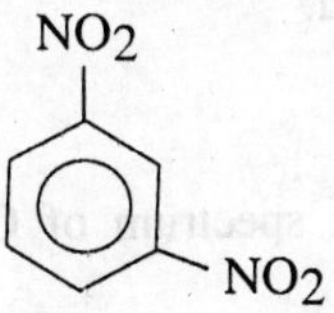

$$\mu = 2\mu_{C-NO_2} \cos\ (120°/2) = 3.8\ D.$$

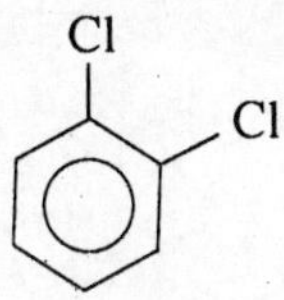

$$\mu = 2\mu_{C-Cl} \cos\ (60°/2) = 2.68\ D.$$

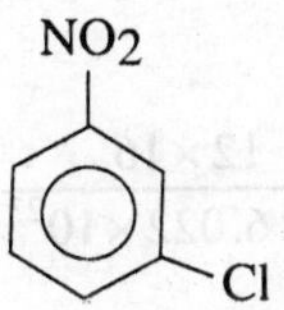

$$\mu = \sqrt{\mu^2_{C-Cl} + \mu^2_{C-NO_2} + 2\mu_{C-Cl}\mu_{C-NO_2} \cos 120°}$$

$$= \sqrt{(1.55)^2 + (3.8)^2 + 2 \times 1.55 \times 3.8 \times \cos 120°\ D} = 3.31D.$$

For *o*-dichlorobenzene calculated value differs much from the observed value. This is due to 'ortho effect' which arises due to close proximity of two groups introducing repulsion which thereby increases the bond angle subtended by two groups.

10. The dipole moment of HCl is 1.03 D and its bond length is 1.27 Å. Calculate (i) the charge on the constituent atoms and (ii) the percentage of ionic character.

[B.U. 1987; V.U. 1996]

Soln. : (i) Dipole moment of HCl = charge on the constituent atom (δe) × bond length

$\therefore$ 1.03 D = $\delta e \times 1.27 \times 10^{-8}$ cm

Therefore, $\delta e = \dfrac{1.03 \times 10^{-18}\ \text{esu.cm}}{1.27 \times 10^{-8}\ \text{cm}}$ = **8.11 × 10^{-11} esu.**

(ii) See Problem no. 3.

11. A compound *AB* having partly ionic and partly covalent characters has the dipole moment 1.03 debye. Show how the per cent ionic character may be calculated on the basis of charge (4.8×10^{-10} esu) separated by equilibrium bond length 1.3 Å. **[B.U. 1985]**

Hints : If *AB* is completely ionic, then $\mu_{cal} = 4.8 \times 10^{-10}$ esu × 1.3 Å.

% of ionic character = $\dfrac{\mu_{cal}}{\mu_{obs}} \times 100$ = **16.51%.**

12. Suppose there are two energy levels and the difference between them is $2kT$. Calculate the ratio of number of molecules in the two levels. **[B.U. 2001]**

Soln. : Let N_i and N_j be the number of molecules in the two levels i and j.

$\varepsilon_j - \varepsilon_i = 2\,kT$

$$\frac{N_j}{N_i} = \frac{e^{-\varepsilon_j/kT}}{e^{-\varepsilon_i/kT}} = e^{-(\varepsilon_j-\varepsilon_i)/kT} = e^{-2} = \mathbf{0.135}.$$

13. The bond angle in H_2S is 97°, while its dipole moment is 0.95 D. Calculate S-H bond moment. **[V.U.(H) 2002]**

Soln. : Let $\mu_{S\text{-}H}$ be the bond moment of S-H bond.

$$\therefore\ 0.95\ \text{D} = 2\mu_{S\text{-}H}\cos\frac{\theta}{2},\ \mu_{S\text{-}H} = \mathbf{0.7168\ D}.$$

14. The spacing between rotational lines in the IR spectrum of CO is 3.86 cm⁻¹. Calculate the bond length of C—O bond. **[V.U.(H) 2002]**

Soln. : Spacing between rotational lines in IR spectrum = $2B$, where B = rotational constant.

Given, $2B = 3.86\ \text{cm}^{-1}$.

$\therefore\ B = 1.93\ \text{cm}^{-1}$

$$B = \frac{h}{8\pi^2 Ic} = 1.93\ \text{cm}^{-1}$$

$$I = \mu r^2$$

$$\mu = \frac{m_o m_c}{m_o + m_c} = \frac{(12\times16)/(6.022\times10^{23})^2}{(12+16)/(6.022\times10^{23})} = \frac{12\times16}{28\times6.022\times10^{23}}$$

$= 1.139 \times 10^{-23}$ gm $= 1.139 \times 10^{-26}$ kg.

$$\therefore\ r = \sqrt{\frac{h}{8\pi^2\mu cB}} = 1.13\times10^{-10}\ \text{m} = \mathbf{1.13\ Å}.$$

15. The molar polarization of a substance is 46 cm³.mol⁻¹ at 25 °C. Calculate the dipole moment of the molecule if the distortion polarizability is 1.808×10^{-23} cm³. **[V.U.(H) 2001]**

Soln. : Molar polarization $= P_m$ = distortion polarization + orientation polarization $= P_d + P_o$.

Given,

$P_m = 46\ \text{cm}^3.\text{mol}^{-1}$, $T = 273.15 + 25 = 298.15$ °K.

$$P_d = \frac{4}{3}\pi N_A\,\alpha_d,\quad \alpha_d = 1.808\times10^{-23}\ \text{cm}^3.$$

$$= \frac{4}{3}\times\pi\times6.022\times10^{23}\times1.808\times10^{-23} = 45.61\ \text{cm}^3.\text{mol}^{-1}.$$

$$P_o = P_m - P_d = 0.39\ \text{cm}^3.\text{mol}^{-1}.$$

$$P_o = \frac{4\pi N_A\mu^2}{9kT}$$

$$\therefore\ \mu = \sqrt{\frac{9P_o kT}{4\pi N_A}} = \frac{3}{2\times N_A}\sqrt{\frac{P_o\times R\times T}{\pi}}$$

$$= \frac{3}{2\times6.022\times10^{23}}\sqrt{\frac{0.39\times8.314\times10^7\times298.15}{\pi}}$$

$= 1.38 \times 10^{-19}$ esu.cm $=$ **0.138 D.**

16. At 27 °C 1.1 (M) lactic acid ($C_3H_6O_3$) solution taken in a polarimeter tube of 10 cm length rotates the plane polarized light by +2.3°. Calculate the molar rotation of lactic acid. **[C.U. 1996]**

Soln. : Molecular weight of $C_3H_6O_3 = M = 90$ gm

$C = 1.1$ (M) lactic acid = 99 gm lactic acid/1000 cc, l = 1dm,α_{obs} = +2.3°.

$$[\alpha]_\lambda^t = \frac{100\alpha_{obs}}{l.C} = \frac{100\times 2.3}{1\times 9.9} = 23.23°$$

$$[M]_\lambda^t = \frac{[\alpha]_\lambda^t \times M}{100} = \frac{23.23\times 90}{100} = \mathbf{20.91°.}$$

17. The refractive index of benzene at 298 K for light of wave length 600 nm is 1.498. The density of benzene is 0.844×10^3 kg.m^{-3}. Estimate the polarization of benzene. **[V.U.(H) 1999]**

Soln. : Benzene is non-polar molecule. So D = n^2 and $P_o = 0$.

Hence, $P_e = \dfrac{n^2-1}{n^2+2}.\dfrac{M}{\rho}$

$n = 1.498$, $M = 78 \times 10^{-3}$ kg

$\rho = 0.844 \times 10^3$ kg.m^{-3}.

Neglecting atomic polarization (P_a) since it is negligibly small,

$$P_e = P_m = \frac{1.498^2-1}{1.498^2+2}\,\frac{78\times 10^{-3}}{0.844\times 10^3}\ \text{m}^3.\text{mole}^{-1}$$

$= 2.71 \times 10^{-5}$ m^3.mole^{-1} = **27.1 cm³.mole⁻¹**.

18. The rotational spectrum of HCl has lines 21.1 cm^{-1} apart. Calculate the moment of inertia and bond length of HCl. **[V.U.(H) 1995]**

Hints : $2B = 21.1$ cm^{-1}, $B = \dfrac{h}{8\pi^2 Ic}$ cm^{-1}

I = moment of inertia,

$I = \mu r^2$, $\mu = \dfrac{m_1m_2}{m_1+m_2}$, r = bond length

$I = 2.653 \times 10^{-40}$ gm.cm², r = **1.28 Å.**

19. The dipole moment of H_2O is 1.84 debye. Calculate H—OH bond angle, given HO bond moment = 1.5 D. **[C.U. 1986]**

Soln. : Let the H—OH bond angle be θ.

Hence, 1.84 D = 2 × 1.5 D × cos $\dfrac{\theta}{2}$, ∴ θ = **104.34°.**

20. The refractive index of gaseous normal alkane C_nH_{2n+2} is 1.00139 at STP. If atomic refractions for H and C are 1.10 cc/mole and 2.42 cc/mole respectively, find the molecular formula of alkane. **[C.U. 1984]**

Soln. : Molar refraction $R_m = \dfrac{n^2-1}{n^2+2}.\dfrac{M}{\rho}$.

Now at STP, molar volume = $\dfrac{M}{\rho}$ = 22400 cc.

$$\therefore \quad R_m = \frac{(1.00139)^2 - 1}{(1.00139)^2 + 2} \times 22400 = 20.75 \text{ cc/mole.}$$

Now atomic refractions are additive and are added to make the whole of molar refraction.

Hence, $20.75 = 2.42\, n + 1.10\, (2n + 2)$

or, $4.62\, n = 18.55, \therefore n = 4.02 \approx 4.$

Hence, molecular formula of alkane C_nH_{2n+2} is $\mathbf{C_4H_{10}}$.

21. The density of $SiHBr_3$ (mol. wt. 269) is 2.690 gm/cc. At 25 °C its refractive index is 1.578 and dielectric constant is 3.570. Estimate its dipole moment in debyes (neglecting atomic polarization). **[C.U. 1999]**

Hints : $P_m = P_e + P_a + P_o$

$$P_a = 0,\ P_e = R_m = \frac{n^2 - 1}{n^2 + 2} \times \frac{M}{\rho}$$

$$P_m = \frac{D - 1}{D + 2} \times \frac{M}{\rho}$$

$$P_o = \frac{4}{3}\pi N_A \times \frac{\mu^2}{3kT}$$

$\mu = \mathbf{0.79\ D.}$

22. The dipole moments of HCl, HBr and HI are 3.44, 2.64 and 1.00 respectively in 10^{-30} coulomb-metre unit. Express them in debye. If equilibrium internuclear distances are 127 pm, 141 pm and 161 pm respectively, find the fractional charges on the halogen atoms and examine whether they are in order of the electronegativities of the halogens. **[B.U. 1999]**

Soln. : 1 coulomb = 3×10^9 esu

So, $\mu_{HCl} = 3.44 \times 10^{-30}$ coulomb.metre $\times 3 \times 10^9$ esu/coulomb $\times 10^2$ cm/metre

$= 1.032 \times 10^{-18}$ esu.cm $= 1.032$ debye

$\mu_{HBr} = 2.64 \times 10^{-30} \times 3 \times 10^{11} = 0.792 \times 10^{-18}$ esu.cm $= 0.792$ debye.

$\mu_{HI} = 1 \times 10^{-30} \times 3 \times 10^{11} = 3 \times 10^{-19}$ esu.cm $= 0.3$ debye.

If r is internuclear distance and δe is fractional charges on atoms, then

$\mu = \delta e \times r.$

For HCl, $r = 127$ pm, $\therefore \delta e = \dfrac{1.032 \times 10^{-18} \text{ esu.cm}}{127 \times 10^{-10} \text{ cm}} = \mathbf{8.12 \times 10^{-11}}$ **esu.**

For HBr, $r = 2.64$ pm, $\therefore \delta e = \dfrac{7.92 \times 10^{-19} \text{ esu.cm}}{141 \times 10^{-10} \text{ cm}} = \mathbf{5.62 \times 10^{-11}}$ esu.

For HI, $r = 161$ pm, $\therefore \delta e = \dfrac{3 \times 10^{-19} \text{ esu.cm}}{161 \times 10^{-10} \text{ cm}} = \mathbf{1.86 \times 10^{-11}}$ **esu.**

They are in the order of decreasing electronegativities of halogens.

23. Given atomic masses of H = 1.0078 amu, Cl = 34.98 amu and the mean internuclear distance of H-Cl is 1.275 Å. Calculate the reduced mass and moment of inertia of HCl. **[Pune B.Sc. 1985]**

Soln. : Reduced mass, $\mu = \dfrac{m_1 m_2}{(m_1 + m_2)} = \dfrac{(M_1 M_2)/N_A^2}{(M_1 + M_2)N_A}$

$$= \frac{1.0078 \times 34.98 \times 10^{-3}}{(1.0078 + 34.98) \times 6.022 \times 10^{23}} \text{ kg} = \mathbf{1.626 \times 10^{-27} \text{ kg.}}$$

Moment of inertia, $I = \mu r^2$ (r = bond length)

$= 1.626 \times 10^{-27}$ kg $\times$ $(1.275 \times 10^{-10}$ m$)^2$

$= \mathbf{2.644 \times 10^{-47}}$ **kg. m²**.

24. The rotational spectrum of HF has lines 41.9 cm^{-1} apart. Calculate the moment of inertia and bond length in HF. **[Delhi B.Sc.(H) 1976]**

Soln. : Given, $2B = 41.9$ cm^{-1} (B = rotational constant)

$B = 20.95$ cm^{-1}

Now, $B = \dfrac{h}{8\pi^2 Ic}$ cm^{-1}.

$\therefore$ moment of inertia, $I = \dfrac{6.626 \times 10^{-34} \text{ J.sec}}{8 \times \pi^2 \times 20.95 \text{ cm}^{-1} \times 2.9979 \times 10^{10} \text{ cm / sec}}$

$= \mathbf{1.34 \times 10^{-47}}$ **kg.m²**.

Now, reduced mass, $\mu = \dfrac{(M_1 M_2)/N_A^2}{(M_1 + M_2)N_A}$

$$= \frac{19 \times 10^{-3}}{20 \times 6.022 \times 10^{23}} \text{ kg} = 1.578 \times 10^{-27} \text{ kg.}$$

Hence, bond length, $r = \sqrt{\dfrac{I}{\mu}}$

$$= \sqrt{\frac{1.34 \times 10^{-47} \text{ kg. m}^2}{1.578 \times 10^{-27} \text{ kg}}} = 9.21 \times 10^{-11} \text{ m} = \mathbf{0.92 \text{ Å.}}$$

25. Calculate the ratio of molecules in rotational level $J = 2$ relative to $J = 0$ for a molecule with $B = 1.2$ cm^{-1} at 300 K.

Soln. : $\dfrac{N_{J=2}}{N_{J=0}} = \dfrac{e^{-E_{J=2}/kT}}{e^{-E_{J=0}/kT}}$

$N_{J=2}$ = population at $J = 2$ level

$N_{J=0}$ = population at $J = 0$ level

$E_{J=2} = BJ(J + 1) = 6B$ cm^{-1}, $E_{J=0} = 0$

or, $\dfrac{N_{J=2}}{N_{J=0}} = e^{-(E_{J=2} - E_{J=0})/kT} = e^{-6Bhc/kT}$

$$= e^{-\frac{6 \times 1.2 \text{ cm}^{-1} \times 2.9979 \times 10^{10} \text{ cm/sec} \times 6.626 \times 10^{-34} \text{ J.sec}}{300 \text{ K} \times 1.38066 \times 10^{-23} \text{ J.K}^{-1}}} = 0.966.$$

But Jth rotational level is $2J + 1$ degenerate, Thus, considering degeneracy

$$\frac{N_{J=2}}{N_{J=0}} = 5 \times e^{-(E_{J=2} - E_{J=0})/kT} = \mathbf{4.83.}$$

26. Spacing between lines in microwave spectra of $H^1 I^{127}$ is 13.2 cm^{-1}. Calculate the bond length of HI **[V.U. 1994]**

Soln. : Given, $2B = 13.2\ cm^{-1}$, $\therefore\ B = 6.6\ cm^{-1}$.

Hence, moment of inertia $I = \dfrac{h}{8\pi^2 Bc}\ cm^{-1}$

$$= \frac{6.626\times10^{-34}\ J.sec}{8\times\pi^2\times6.6\ cm^{-1}\times2.9979\times10^{10}\ cm.sec^{-1}}$$

$$= 4.241 \times 10^{-47}\ kg.m^2.$$

$$[\ \mu = \frac{M_1M_2/N_A^2}{(M_1+M_2)/N_A} = \frac{127\times10^{-3}}{128\times6.022\times10^{23}}\ kg$$

$$= 1.648 \times 10^{-27}\ kg]$$

Thus, bond length $r = \sqrt{\dfrac{I}{\mu}} = \sqrt{\dfrac{4.241\times10^{-47}\ kgm^2}{1.648\times10^{-27}\ kg}}$

$= 1.604 \times 10^{-10}$ m = **1.604 Å.**

27. C-H stretching vibration in organic compounds occur near 2900 cm^{-1}. Approximately what wave number would be the C-D stretching vibration? **[B.U.(H) 1992]**

Soln. : $\bar{\nu} = \dfrac{1}{2\pi c}\sqrt{\dfrac{k}{\mu}}\ cm^{-1} = 2900\ cm^{-1}$

Reduced mass of C-H,

$$\mu = \frac{12\times10^{-3}}{13\times6.022\times10^{23}}\ kg = \mathbf{1.53} \times 10^{-27}\ kg.$$

$$\therefore\quad k = 4\pi^2\mu\bar{\nu}^2c^2$$

$$= 4\pi^2 \times 1.53 \times 10^{-27}\ kg \times (2.9979 \times 10^{10}\ cm/sec \times 2900\ cm^{-1})^2$$

$$= 456.512\ kg.\ sec^{-2}.$$

Now, reduced mass of C-D,

$$\mu' = \frac{12\times2\times10^{-3}}{(12+2)\times6.022\times10^{23}}\ kg = 2.846 \times 10^{-27}\ kg.$$

$$\bar{\nu}^2 = \frac{1}{2\pi c}\sqrt{\frac{k}{\mu'}} = \frac{1}{2\pi\times2.9979\times10^{10}\ cm.sec^{-1}}\sqrt{\frac{456.512\ kg.sec^{-2}}{2.846\times110^{-27}\ kg}} = \mathbf{2126.24\ cm^{-1}}.$$

28. Absorption maxima in the I.R. spectra of H^1Br^{80} was observed as 37.8 nm. Calculate the force constant of the bond between HBr.

Soln. : Reduced mass of HBr, $\mu = \dfrac{80\times1\times10^{-3}}{(80+1)\times6.022\times10^{23}}\ kg$

$= 1.64 \times 10^{-27}$ kg.

$$\lambda = \frac{c}{\nu} = 37.8\ nm = 37.8 \times 10^{-9}\ m.$$

$$\nu = \frac{2.9979\times10^8\ m.sec^{-1}}{37.8\times10^{-9}\ m} = 7.93 \times 10^{15}\ sec^{-1}.$$

$$7.93 \times 10^{15} \text{ sec}^{-1} = \frac{1}{2\pi}\sqrt{\frac{k}{1.64\times10^{-27} \text{ kg}}}.$$

$\therefore$ force constant $k = 4\pi^2 \times (7.93 \times 10^{15} \text{ sec}^{-1})^2 \times 1.64 \times 10^{-27}$ kg = **4.1×10^6 N.m^{-1}.**

29. In I.R. spectrum of H^1Cl^{35}, the centre of fundamental band has a frequency of 8.6×10^{13} Hz. Calculate force constant.

Soln. : Reduced mass of HCl, $\mu = \frac{35\times1}{(35+1)\times6.022\times10^{23}}$ gm = 1.614×10^{-24} gm.

Force consant, $k = 4\pi^2\nu^2\mu = 4\pi^2 \times (8.6 \times 10^{13} \text{ sec}^{-1})^2 \times 1.614 \times 10^{-24}$ gm
= **4.7×10^5 dyne.cm^{-1}.**

30. The vibrational Raman spectrum of chlorine molecule shows a series of Stokes lines separated by 0.98 cm^{-1} and a similar series of anti-Stokes lines. What is the bond length of Cl_2? (At. wt. = 35)

Soln. : Stokes lines and anti-Stokes lines are separated by $4B$ cm^{-1}.

Given, $4B$ = 0.98 cm^{-1}.

$\therefore$ B = 0.245 cm^{-1}.

Reduced mass of Cl_2, $\mu = \frac{35\times35\times10^{-3}}{(35+35)\times6.022\times10^{23}} = 2.91 \times 10^{-26}$ kg.

Now, $B = \frac{h}{8\pi^2 Ic} = \frac{h}{8\pi^2\mu r^2 c}$.

Hence, bond length $r = \sqrt{\frac{h}{8\pi^2 B\mu c}}$

$$= \sqrt{\frac{6.626\times10^{-34} \text{ J.sec}}{8\pi^2\times0.245 \text{ cm}^{-1}\times2.91\times10^{-26} \text{ kg}\times2.9979\times10^{10} \text{ cm.sec}^{-1}}}$$

= 1.98×10^{-10} m = **1.98 Å.**

31. What is the value of J in the most highly populated rotational level of ICl (B = 0.114 cm^{-1}) and CH_4 (B = 5.24 cm^{-1}) at room temperature?

Soln. : J_{max} for ICl

$$= \sqrt{\frac{kT}{2Bhc}} - \frac{1}{2}$$

$$= \sqrt{\frac{1.38066\times10^{-23} \text{ J.K}^{-1}\times298.15 \text{ K}}{2.0114 \text{ cm}^{-1}\times6.626\times10^{-34} \text{ J.sec}\times2.9979\times10^{10} \text{ cm.sec}^{-1}}} - \frac{1}{2}$$

= **30.**

J_{max} for CH_4

$$= \sqrt{\frac{1.38066\times10^{-23} \text{ J.K}^{-1}\times298.15 \text{ K}}{2\times5.24 \text{ cm}^{-1}\times6.626\times10^{-34} \text{ J.sec}\times2.9979\times10^{10} \text{ cm.sec}^{-1}}} - \frac{1}{2} = \mathbf{4.}$$

32. Using 435.8 nm lines mercury as a source of radiation a Raman line was observed at 444.7 nm. What is the Raman shift?

Soln. : λ_{source} = 435.8 nm = 435.8 × 10^{-7} cm..

$\therefore$ $\bar{\nu}_{source} = (\lambda_{source})^2 = 22946.31\ cm^{-1}$.

$\nu_{emitted}$ = 444.7 nm = 444.7 × 10^{-7} cm.

$\therefore$ $\bar{\nu}_{emitted} = (\lambda_{emitted})^{-1} = 22487.07\ cm^{-1}$.

Hence, Raman shift = $\bar{\nu}_{source} - \bar{\nu}_{emitted}$ = **459.24 cm^{-1}**.

33. Irradiation of CCl_4 with 436 nm lines of mercury line gave lines at 439.9 nm, 441.8 nm, 444.6 nm and 450.7 nm respectively in the Raman spectrum. Calculate the shifts in cm^{-1}.

Soln. : λ_{source} = 436 nm = 436 × 10^{-7} cm.

$\bar{\nu}_{source} = (\lambda_{source})^{-1} = 22935.78\ cm^{-1}$.

λ_1 = 439.9 nm = 439.9 × 10^{-7} cm.

$\bar{\nu}_1 = 22732.44\ cm^{-1}$

$\Delta\bar{\nu}_1 = \bar{\nu}_{source} - \bar{\nu}_1$ = **203.34 cm^{-1}**.

λ_2 = 441.8 nm = 441.8 × 10^{-7} cm

$\Delta\bar{\nu}_2 = \bar{\nu}_{source} - \bar{\nu}_2$ = **301.1 cm^{-1}**.

λ_3 = 444.6 nm = 444.6 × 10^{-7} cm

$\Delta\bar{\nu}_3 = \bar{\nu}_{source} - \bar{\nu}_3$ = **443.65 cm^{-1}**.

λ_4 = 450.7 nm = 450.7 × 10^{-7} cm

$\Delta\bar{\nu}_4 = \bar{\nu}_{source} - \bar{\nu}_4$ = **748.07 cm^{-1}**.

34: The wave numbers recorded in the vibrational spectra of the following molecules in their $n = 0$ states are : HD 3627 cm^{-1}, DCl 1990 cm^{-1}, HCl 2885 cm^{-1} and D_2 2990 cm^{-1}. Calculate the energy change of the reaction : DCl + HD = D_2 + HCl.

Soln. : Total vibrational energy of D_2 + HCl,

$$E = \frac{1}{2}hc\left(\bar{\nu}_{D_2} + \bar{\nu}_{HCl}\right)$$

$$= \frac{1}{2} \times 6.626 \times 10^{-34}\ \text{J.sec} \times 2.9979 \times 10^{10}\ \text{cm.sec}^{-1} \times (2885 + 2990)\ \text{cm}^{-1}$$

$$= 5.84 \times 10^{-20}\ \text{J}.$$

Total vibrational energy of (DCl + HD),

$$E' = \frac{1}{2}hc\left(\bar{\nu}_{D_2} + \bar{\nu}_{HCl}\right)$$

$$= \frac{1}{2} \times 6.626 \times 10^{-34}\ \text{J.sec} \times 2.9979 \times 10^{10}\ \text{cm.sec}^{-1} \times (3627 + 1990)\text{cm}^{-1}$$

$$= 5.58 \times 10^{-20}\ \text{J}.$$

Energy change of the reaction,

$$\Delta E = [(E' - E) \times N_A]\ \text{J.mole}^{-1}$$

$= (5.84 - 5.58) \times 10^{-20}\ \text{J} \times 6.022 \times 10^{23}\ \text{mole}^{-1}$ = **1.56 kJ.mole^{-1}**.

35. The hydrogen chloride rotational constant B = 10.59 cm^{-1} and fundamental and first overtone vibration lines are shown at 2885 cm^{-1} and 5679 cm^{-1} respectively. Calculate

(i) rotational energy for the molecule at J = 2.

(ii) zero point energy of the molecule,

(iii) hot band wave numbers. **[Delhi B.Sc.(H) 1982]**

Soln. : (i) Rotational line is given by

$$\bar{\nu} = BJ(J+1)\ \text{cm}^{-1} = 10.59\ \text{cm}^{-1} \times 2 \times 3 = \mathbf{63.54\ cm^{-1}}.$$

(ii) Vibrational energy,

$$\varepsilon_n = \bar{\nu}_e\left(n+\frac{1}{2}\right) - \bar{\nu}_e x_e\left(n+\frac{1}{2}\right)^2 \text{cm}^{-1}.$$

Given, for the transition from $n = 0$ to $n = 1$,

$$\Delta\varepsilon_{0\to1} = 2885\ \text{cm}^{-1} = \bar{\nu}_e - 2\bar{\nu}_e x_e \qquad (1)$$

and for the transition from $n = 0$ to $n = 2$,

$$\Delta\varepsilon_{0\to2} = 5679\ \text{cm}^{-1} = 2\bar{\nu}_e - 6x_e\bar{\nu}_e. \qquad (2)$$

Multiplying equation (1) by 2, we get

$$5770\ \text{cm}^{-1} = 2\bar{\nu}_e - 4x_e\bar{\nu}_e. \qquad (3)$$

Subtracting equation (2) from equation (3), we get

$$2x_e\bar{\nu}_e = 91\ \text{cm}^{-1}. \qquad (4)$$

Adding equation (4) and equation (1),

$$\bar{\nu}_e = (2885 + 91)\ \text{cm}^{-1} = 2976\ \text{cm}^{-1}.$$

Thus, from equation (4), $x_e = \dfrac{91\,\text{cm}^{-1}}{2\times 2976\,\text{cm}^{-1}} = 0.0153.$ (5)

Hence, zero point energy,

$$\varepsilon_{n=0} = \frac{1}{2}\bar{\nu}_e - \left(\frac{1}{2}\right)^2\bar{\nu}_e x_e$$

$$= \frac{1}{2}\bar{\nu}_e\left(1-\frac{1}{2}x_e\right) = \frac{1}{2}\times 2976\,\text{cm}^{-1}\left(1-\frac{1}{2}\times 0.0153\right)$$

$$= \mathbf{1476.62\ cm^{-1}}.$$

(iii) Hot bond wave numbers corresponding to transition from $n = 1$ to $n = 2$,

$$\Delta\varepsilon_{1\to2} = \bar{\nu}_e(1 - 4x_e)\ \text{cm}^{-1}. \qquad (6)$$

Either putting the values of $\bar{\nu}_e$ and x_e in (6) or subtracting (2) from (3), we obtain

$$\Delta\varepsilon_{1\to2} = \mathbf{2794\ cm^{-1}}.$$

36. The C = O bond energy in acetone is 728 kJ.mol^{-1}. Will a light of 250 nm be able to break the bond ? **[B.U. 1999]**

Soln. : Energy of one photon of wavelength 250 nm,

$$E = h\frac{c}{\lambda} = \frac{6.626\times10^{-34}\ \text{J.sec}\times 2.9979\times10^{8}\ \text{m.sec}^{-1}}{250\times10^{-9}\ \text{m}} = 7.946\times10^{-19}\ \text{J}.$$

Energy of one Einstein of photon $= N_A E = 6.022\times10^{23}\times 7.946\times10^{-19}$ J.
$= 478.51$ kJ.

It is less than C = O bond eneregy. So the bond will not break.

37. The refractive index of CCl_4 for sodium *D* line at 20 °C is 1.457 and its density is 1.595 gm/cc. Calculate the molar refraction. If the atomic refraction of carbon be 2.42, calculate atomic refraction of chlorine. **[C.U. 1981]**

Soln. : Molar refraction $= R_m = \dfrac{n^2-1}{n^2+2}\dfrac{M}{\rho}$

n = refractive index = 1.457

M = molecular weight of CCl_4 = 154 gm

ρ = density = 1.595 gm/cc.

$$\text{Thus, } R_m = \frac{(1.457)^2-1}{(1.457)^2+2}\cdot\frac{154\,\text{gm}}{1.595\,\text{gm/cc}} = \mathbf{26.29\ cc.}$$

Atomic refractions are additive, Thus,

26.29 = atomic refraction of carbon + 4 × atomic refraction of chlorine

= 2.42 + 4 R_{Cl}

R_{Cl} = **5.97 cc.**

38. Two optically active substances A and B have specific rotations of +50° and –30° respectively. A mixture containing 6 gm A and 4 gm B is dissolved in water and the solution is made up to 50 cc. Find the angle of rotation for the solution kept in a 5 cm polarimeter tube. **[C.U. 1985]**

Soln. : If C is concentration in gm per 100 cc, l (dm) is length of the tube, and α is angle of rotation, then specific rotation $= [\alpha] = \dfrac{100\alpha}{lC}$.

For A, $[\alpha] = +50°$, l = 5 cm = 0.5 dm, $C_A = \dfrac{12\,\text{gm}}{100\,\text{cc}}$.

$$\therefore 50 = \frac{100\alpha_A}{0.5\times 12},\ \therefore \alpha_A = 3°.$$

For B, $[\alpha] = -30°$, $C_B = \dfrac{8\,\text{gm}}{100\,\text{cc}}$.

$$\therefore -30 = \frac{100\alpha_B}{0.5\times 8},\ \therefore \alpha_B = -1.2°.$$

So, the angle of rotation for the solution $= \alpha_A + \alpha_B = 3° - 1.2° = \mathbf{1.8°}$.

39. What is the frequency of the rotational line shown by a diatomic molecule, having the moment of inertia value 16.5×10^{-40} gm. cm^2, the excited molecule being in the quantum state J = 2? **[B.U. 1983]**

Soln. : According to the selection rule of rotational transition $\Delta J = \pm 1$.

Here J for excited state is 2. Hence, transition occurred from J = 1 state.

Energy for J = 1 state $= E_1 = BJ\ (J+1)\text{cm}^{-1} = 2B\ \text{cm}^{-1}$.

For J = 2 state, $E_2 = B.2\ (2+1) = 6B\ \text{cm}^{-1}$.

So, $\Delta E = E_2 - E_1 = 4B\ \text{cm}^{-1}$.

Hence, wave number of frequency of the rotational line shown by a diatomic molecule

$$= 4B = \frac{4h}{8\pi^2 Ic} = \frac{4\times 6.626\times10^{-27}\,\text{erg.sec}}{8\times\pi^2\times16.5\times10^{-40}\,\text{gm.cm}^2\times2.9979\times10^{10}\,\text{cm/sec}} = \mathbf{6.79\ cm^{-1}}.$$

40. The molar polarization of fluorobenzene vapour is 70.62 cm^3.mole^{-1} at 351 K and 62.47 cm^3.mole^{-1} at 423 K. Calculate the dipole moment of molecule. **[V.U. 2000]**

Soln. : Molar polarization, $P_m = A + \dfrac{B}{T}$.

$$A = \frac{4}{3}\pi N_A(\alpha_e + \alpha_A)$$

$$B = \frac{4}{3}\pi N_A . \frac{\mu^2}{3k}$$

Given, $70.62\ \text{cm}^3.\text{mole}^{-1} = A + \dfrac{B}{351K}$ (1)

and $62.47\ \text{cm}^3.\text{mole}^{-1} = A + \dfrac{B}{423K}$. (2)

Subtracting (2) from (1), we get

$$8.15\ \text{cm}^3.\text{mole}^{-1} = B\left(\frac{1}{351K} - \frac{1}{423K}\right).$$

$\therefore\ B = 16806.32\ \text{cm}^3.\text{K.mole}^{-1}$.

Thus, $16806.32\ \text{cm}^3.\text{K.mole}^{-1} = \dfrac{4}{3}\pi N_A . \dfrac{\mu^2 . N_A}{3R}$

$$\mu^2 = \frac{9\times 16806.32\,\text{cm}^3.\text{K mol}^{-1}\times 8.314\times 10^7\,\text{erg.mol}^{-1}.\text{K}^{-1}}{4\times\pi\times\left(6.022\times 10^{23}\right)^2}$$

$= 2.76 \times 10^{-36}$ esu^2.cm^2.

$\therefore\ \mu = 1.66 \times 10^{-18}$ esu.cm = **1.66 D.**

41. Calculate the moment of inertia of CO molecule having isotopic combination $C^{12}O^{16}$ given $r_{CO} = 1.13$ Å. **[B.U. 1987]**

Soln. : Reduced mass $\mu_{CO} = \dfrac{12\times 16}{(12+16)\times 6.022\times 10^{23}} = 1.14 \times 10^{-23}$ gm

So moment of inertia $= I = \mu r^2$

$= 1.14 \times 10^{-23}$ gm $\times (1.13 \times 10^{-8})^2$ cm^2

= **1.46×10^{-39} gm.cm^2.**

42. The fundamental vibrational frequency of HCl is 2885 cm^{-1}. Assuming HCl and DCl may be treated as simple harmonic oscillator, calculate the fundamental frequency of DCl (atomic mass of chlorine = 35 dalton). **[I.I.T., M.Sc. Admission Test]**

Soln. : Reduced mass of HCl,

$$\mu_{HCl} = \frac{m_H m_{Cl}}{m_H + m_{Cl}} = \frac{(1\times 35)/\left(6.022\times 10^{23}\right)^2}{(1+35)/6.022\times 10^{23}}\text{gm}$$

$$= \frac{35}{36\times 6.022\times 10^{23}}\text{gm} = 1.61 \times 10^{-24}\ \text{gm}.$$

$$\mu_{DCl} = \frac{(2\times 35)}{(2+35)\times 6.022\times 10^{23}}\text{gm} = 3.14 \times 10^{-24}\ \text{gm}.$$

$$\bar{\nu}_{HCl} = 2885\ \text{cm}^{-1} = \frac{1}{2\pi c}\sqrt{\frac{k}{\mu_{HCl}}}$$

$k = (2885\ \text{cm}^{-1})^2 \times (2\pi c)^2\ \mu_{HCl}$.

$$\bar{\nu}_{DCl} = \frac{1}{2\pi c}\sqrt{\frac{k}{\mu_{DCl}}},\ k = \bar{\nu}_{DCl}^2 \times (2\pi c)^2 \times \mu_{DCl}.$$

Thus,

$(2885\ \text{cm}^{-1})^2 \times (2\pi c)^2\ \mu_{HCl} = \bar{\nu}_{DCl}^2 \times (2\pi c)^2 \times \mu_{DCl}$

$$\text{or, } \bar{\nu}_{DCl}^2 = \frac{(2885\ \text{cm}^{-1})^2 \times 1.61 \times 10^{-24}\ \text{gm}}{3.14 \times 10^{-24}\ \text{gm}} = (4267640.84\ \text{cm}^{-1})^2.$$

$\therefore\ \bar{\nu}_{DCl} = \mathbf{2065.83\ cm^{-1}}$.

43. Calculate the reduced mass of $I^{127}\ Cl^{35}$. The bond length of the molecule is 2.32Å. Calculate its moment of inertia. **[B.U. 1993]**

Hints : $I = \mu r^2 = 2.45 \times 10^{-38}$ gm.cm^2.

See Problem no. 41.

44. How many times does a molecule of H^1Cl^{35} rotate per second in the $J = 1$ rotational level? Given, $B = 10.6$ cm^{-1}. **[B.U. 1989]**

Soln. : Classical rotational energy $= \frac{1}{2}I\omega^2$.

Quantum mechanical rotational energy in $J = 1$ state $= BJ\ (J + 1)$ cm^{-1}

$= BhcJ\ (J + 1)$ joule $= 2Bhc$ joule $(J = 1)$

Now, $\frac{1}{2}I\omega^2 = 2Bhc$

$$\text{or, } \omega^2 = \frac{4Bhc}{I}$$

$$= \frac{4Bhc \times 8\pi^2 Bc}{h}\quad \left[\because B = \frac{h}{8\pi^2 Ic},\ \therefore I = \frac{h}{8\pi^2 Bc}\right]$$

$= 32B^2\pi^2c^2$

$\omega = 4B\pi c\sqrt{2} = 4 \times 10.6\ \text{cm}^{-1} \times \pi \times 2.9979 \times 10^{10}\ \text{cm/sec} \times \sqrt{2}$

$= 5.65 \times 10^{12}$ radian.sec^{-1} = $\mathbf{8.99 \times 10^{11}}$ **revolutions.sec^{-1}** (dividing by 2π)

45. A liquid of molecular weight 112 and density 1.108 gm/cc. has D.E constant 4.2 and R.I 1.348. Calculate its molar polarization and molar refraction. **[C.U. 1986]**

Hints : $P_m = \frac{D-1}{D+2}\frac{M}{\rho}$, (52.87 cc.mole^{-1})

$R_m = \frac{n^2-1}{n^2+2}\frac{M}{\rho}$ (21.64 cc.mole^{-1}).

46. The refractive index of benzene at 298 K for light of wavelength 600 nm is 1.498. The density of benzene is 0.844×10^3 kg.m^{-3}. Estimate the polarizability of benzene **[V.U. 199-]**

Soln. : See Problem no. 17.

$P_m = 27.1$ cm^3.mole^{-1} $= \frac{4}{3}\pi\ N_A\ \alpha$ [α = polarizability]

$$\alpha = \frac{27.1\,cm^3.mole^{-1} \times 3}{4 \times \pi \times 6.022 \times 10^{23}} = 1.07 \times 10^{-23}\ cm^3.$$

47. The dipole moment of o-xylene is 0.693 D. Find the dipole moment of tolune. What will be the value in SI units ? **[B.U. 1993]**

Soln. :

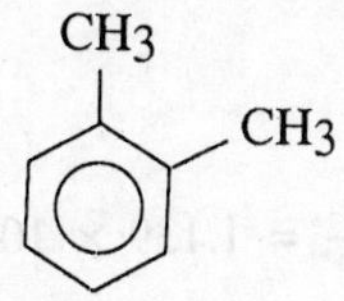

$\mu_{o\text{-xylene}} = 0.693D.$

Let μ be bond moment of C—CH_3.

Given, $0.693 = 2\mu \cos \dfrac{60°}{2}$.

$\therefore\ \mu = 0.40D.$

$1\ D = 10^{-18}$ esu.cm

$1\ C.m = 3 \times 10^9$ esu $\times 10^2$ cm $= 3 \times 10^{11}$ esu.cm.

$$1\ D = \frac{10^{-18}}{3 \times 10^{11}}\ C.m = 3.33 \times 10^{-30}\ \text{coulomb.metre.}$$

$\therefore$ 0.4 D $= 1.332 \times 10^{-30}$ C.m.

48. A liquid of mol. wt. 18 and density 0.995×10^3 kg.m^{-3} has a dielectric constant 78.5 and refractive index 1.383. Calculate the values of its molar polarization, molar refraction and dipole moment neglecting atomic polarization. **[V.U. 1989]**

Hints : Density $= \rho = 0.995 \times 10^3$ kg.m^{-3} $= 0.995$ gm.cm^{-3}

$$P_m = \frac{D-1}{D+2}\frac{M}{\rho}$$

$$R_m = \frac{n^2-1}{n+2}\frac{M}{\rho} = P_e$$

$$P_m = P_o + P_e + P_a \quad [P_a = 0]$$

$$P_o = P_m - P_e = \frac{4}{3}\pi N_A \frac{\mu^2}{3kT} \quad (T = 298.15\ °K)$$

$$\mu = 0.8 \times 10^{-18}\ \text{esu.cm} = 0.8\ D.$$

49. How many revolutions does a molecule $Na^{23}Cl^{35}$ make in $J = 9$ state? The rotational constant of NaCl is 6500 MHz.

Soln. : Classical rotational energy $= \frac{1}{2}I\omega^2$.

Quantum mechanical energy $= hBJ\ (J + 1)$.

Now, $\quad \frac{1}{2}I\omega^2 = hBJ\ (J + 1)$

or, $\omega^2 = \dfrac{2hBJ(J+1)}{I} = 2hBJ\ (J+1) \times \dfrac{8\pi^2 B}{h}$ $\left[B = \dfrac{h}{8\pi^2 I}\ \therefore\ I = \dfrac{h}{8\pi^2 B}\right]$

$= 16\ B^2\pi^2\ J(J + 1).$

$\therefore\ \omega = 4\pi B\sqrt{J(J+1)} = 4\pi B\sqrt{9(9+1)}\quad [J = 9]$

$= 4\pi\ (6500 \times 10^6\ sec^{-1})\sqrt{90} = $ **7.75×10^{11} radian.sec^{-1}**

$=$ **1.23×10^{11} revolutions.sec^{-1}.**

50. In the near I.R. spectrum of CO there is an intense band at 2144 cm^{-1}. Calculate (a) the fundamental vibration frequency of CO, (b) the period of vibration, (c) the force constant, (d) the zero point energy in cal/mole. **[B.U. 1990]**

Soln. : (a) $\bar{\nu}$ = 2144 cm^{-1}, $\nu = \bar{\nu}c$ = **6.43 × 10^{13} sec^{-1}.**

(b) 6.43 × 10^{13} vibrations per sec all executed in one sec. So time for one vibration

$$= \text{period of vibration} = \frac{1}{6.43\times10^{13}} = \mathbf{1.55 \times 10^{-14}\ sec.}$$

(c) $\nu = \frac{1}{2\pi}\sqrt{\frac{k}{\mu}}$, where $\mu = \frac{12\times16}{(12+16)\times6.022\times10^{23}}$ = 1.139 × 10^{-23} gm.

k = 1.86 × 10^6 gm.sec^{-2} = **1.86 × 10^6 dyne.cm^{-1}.**

(d) Zero point energy = $\frac{1}{2}h\nu = \frac{1}{2}\times$ 6.626 × 10^{-34} J. sec × 6.43 × 10^{13} sec^{-1}

= 2.13 × 10^{-20} J = 5.09 × 10^{-21} cal [4.184 J = 1 cal]

Thus, zero point energy per mole

= 5.09 × 10^{-21} × 6.022 × 10^{23} = **3065.19 cal.$mole^{-1}$.**

51. $K^{39}Cl^{35}$ has an intense line at 278.0 cm^{-1}, in far infrared spectrum. Calculate the force constant and the period of vibration of $K^{39}Cl^{35}$.

Soln. : Force constant, $k = 4\pi^2\ \bar{\nu}^2c^2\mu_{KCl}$ dyne.cm^{-1}.

$$\mu_{KCl} = \frac{39\times35}{(39+35)\times6.022\times10^{23}} = 3.063\times10^{-23}\text{ gm}$$

$\bar{\nu}$ = 278.0 cm^{-1}

k = **8.39 × 10^4 dyne.cm^{-1}.**

Period of vibration, $T = \frac{1}{\nu} = \frac{1}{c\bar{\nu}}$ = **1.19 × 10^{-13} sec.**

52. Calculate the refractive index of a mixture of CH_3OH and C_2H_5OH containing 0.75 mole fraction of CH_3OH. Given,

ρ_{CH_3OH} = 0.792 gm.cm^{-3}, $\rho_{C_2H_5OH}$ = 0.789 gm.cm^{-3}

n_{CH_3OH} = 1.328 $n_{C_2H_5OH}$ = 1.3611

$\rho_{mixture}$ = 0.7903 gm.cm^{-3}

where ρ = density, n = refractive index.

Soln. : Molar refraction for $CH_3OH = R_{CH_3OH} = \frac{n^2-1}{n^2+2}\frac{M}{\rho}$

$$= \frac{(1.328)^2-1}{(1.328)^2+2}\times\frac{32}{0.792} = 8.197\text{ cm}^3.$$

Molar refraction for $C_2H_5OH = R_{C_2H_5OH} = \frac{(1.3611)^3-1}{(1.3611)^2+2}\times\frac{46}{0.789}$ = 12.902 cm^3.

Molar refraction for solution = 0.75 × 8.197 + 0.25 × 12.902 = 9.37.

Now, $R_{sol} = \frac{n^2-1}{n^2+2}\times\frac{M_{sol}}{\rho_{sol}}$, where n = refractive index of the solution.

$$\text{Hence, } 9.37 = \frac{n^2-1}{n^2+2} \times \frac{0.75\times 32 + 0.25\times 46}{0.7903}$$

$$\text{or, } \frac{n^2-1}{n^2+2} = 0.2085, \therefore n = \mathbf{1.34.}$$

53. A microwave spectrometer capable of operating only between 60 and 90 cm^{-1} was used to observe rotational spectra of HI and DI. Absorptions were measured as follows for three successive lines :

HI (cm^{-1})	DI (cm^{-1})
64.275	65.070
77.130	71.577
89.985	78.084

Determine the J values between which transition occus for the first line listed above for HI and DI and find the B values for each molecule. Hence, find the atomic wt. of D. (At. wt. of I = 127, H = 1). State the assumption used for calculation. **[C.U. 2006]**

Soln : Both HI and DI are assumed to be rigid rotor. Hence, distance between two adjacent lines in the rotational spectra = $2B$.

By the problem, for HI molecule,

$2B = (77.130 - 64.275)\ cm^{-1}$

$= (89.985 - 77.130)\ cm^{-1} = 12.855\ cm^{-1}$. (1)

$\therefore\ B = 6.4275\ cm^{-1}$. (2)

For DI molecule,

$2B' = (71.577 - 65.070)\ cm^{-1}$

$= (78.084 - 71.577)\ cm^{-1}$

$= 6.507\ cm^{-1}$ (3)

$\therefore\ B' = 3.2535\ cm^{-1}$. (4)

Again, energy separation of two adjacent rotational levels = $2B(J + 1)$.

By the problem, for HI

$2B(J + 1) = 64.275\ cm^{-1}$

or, $12.855\ cm^{-1}\ (J + 1) = 64.275\ cm^{-1}$ [from Eq. (1)]

or, $J + 1 = 5$

$\therefore\ J = 4$.

And for DI $6.507\ (J + 1)\ cm^{-1} = 65.07$ [from Eq. (3)]

or, $J + 1 = 10$

$\therefore\ J = 9$.

Thus, for HI the first line appears due to the transition from $J = 4 \rightarrow J = 5$ and for DI, transition from $J = 9 \rightarrow J = 10$ gives rise to the first line.

$$\text{Now, for HI, } \mu = \frac{m_1 m_2}{m_1 + m_2} = \frac{M_1 M_2/N_A^2}{(M_1 + M_2)/N_A}$$

$$= \frac{127}{128 \times 6.022 \times 10^{23}} \text{ gm} = 1.648 \times 10^{-24} \text{ gm.}$$

Now, $B = \dfrac{h}{8\pi^2\mu r^2 C}$ cm^{-1}.

$$\therefore\ r_{HI}^2 = \frac{6.626\times10^{-27}\ \text{erg.sec}}{8\times\pi^2\times6.4275\ \text{cm}^{-1}\times1.648\times10^{-24}\ \text{gm}\times2.9979\times10^{10}\ \text{cm.sec}^{-1}}$$

$= 2.642 \times 10^{-16}$ cm^2.

Hence, $r_{HI} = 1.625 \times 10^{-8}$ cm.

Bond length remains same within experimental error on isotopic substitution.

Henc, $r_{HI} = r_{DI} = 1.625 \times 10^{-8}$ cm.

$$\therefore\ \mu_{DI} = \mu' = \frac{6.626\times10^{-27}\ \text{erg.sec}}{8\times\pi^2\times3.2535\ \text{cm}^{-1}\times\left(1.625\times10^{-8}\ \text{cm}\right)^2\times2.9979\times10^{10}\ \text{cm.sec}^{-1}}$$

$= 3.258 \times 10^{-24}$ gm.

Let atomic weight of D be x gm.

Thus, $3.258 \times 10^{-24} = \dfrac{127x}{(127+x)\times6.022\times10^{23}}$

or, $1.962 = \dfrac{127x}{127+x}$

or, $\dfrac{1}{127}+\dfrac{1}{x} = \dfrac{1}{1.962}$

$\therefore\ x = 1.99 \approx 2$, i.e., atomic weight of D is **2 gm**.

CHAPTER 17

STATISTICAL THERMODYNAMICS

Required Formulae

1. If ε_i and ε_j are the energies of *i*th energy level with degeneracy g_i and *j*th energy level with degeneracy g_j respectively,

$$\frac{N_i}{N_j} = \frac{g_i e^{-\beta\varepsilon_i}}{g_j e^{-\beta\varepsilon_j}}, \qquad (1)$$

where N_i = population at *i*th energy level

N_j = population at *j*th energy level

$\beta = \frac{1}{kT}$ (k = Boltzmann constant, T = temperature)

and $$N_i = \frac{g_i e^{-\beta\varepsilon_i}}{q}, \; N_j = \frac{g_j e^{-\beta\varepsilon_j}}{q}. \qquad (2)$$

The above relations are known as Boltzmann relation,

where N = total numbers of molecules

q = partition function $= \sum_i g_i e^{-\beta\varepsilon_i}$.

2. If n identical but distinguishable particles are arranged in i energy levels in such a way that ε_1 energy level accomodates n_1, ε_2 accomodates n_2, ε_i accomodates n_i and so on, the total distinct number of ways of realising such distribution,

$$W = \frac{n!}{n_1!n_2!\ldots n_i!} = \prod_i \frac{n!}{n_1!}.$$

If i^{th} energy level is g_i-fold degenerate, $W = \prod_i g_i^{n_i} \frac{n!}{n_i!}$.

Short Questions

1. (a) Consider an isolated system containing N molecules, distributed among the possible non-degenerate energy levels represented by $\varepsilon_1, \varepsilon_2, \ldots$, etc. Assuming the expression for thermodynamic probability, show that under equilibrium conditions number of molecules (N_i) in the level ε_i is given by $N_i = A \exp(-b\varepsilon_i)$, where A and b are constants.

(b) Answer the following questions taking $b = \frac{1}{kT}$:

(i) Show that $N_{i+1} < N_i$ for any finite temperature.

(ii) Obtain A in terms of b and show that it is related to the partition function.

(iii) What is the probability that a molecule has an energy ε_i? **[C.U.(H) 2001]**

Ans. : (a) Let N_1 molecules be in the energy level ε_1, N_2 molecules in the level ε_2 and so on.

$$\sum_i N_i = N,\ i = 1, 2, \ldots, \text{etc.}$$

So, weight of the configuration = thermodynamic probability = $W = \frac{N!}{N_1!N_2!\ldots}$

$$\text{or, } \ln W = \ln \frac{N!}{N_1!N_2!N_3!\ldots}$$

$$= \ln N_i! - \ln (N_1!\ N_2!\ N_3!\ \ldots) = \ln N_i! - (\ln N_1! + \ln N_2! + \ln N_3! + \ldots)$$

$$\text{or, } \ln W = \ln N! - \sum_i \ln N_i!\ \ i = 1, 2, \ldots, \text{etc.}$$

Using Stirling's approximation

$$\ln W = (N \ln N - N) - \sum_i (N_i \ln N_i - N_i)$$

$$= N \ln N - N - \sum_i N_i \ln N_i + N \ \left(\text{since, } \sum_i N_i = N\right)$$

$$= N \ln N - \sum_i N_i \ln N_i = \sum_i N_i \ln N - \sum_i N_i \ln N_i .$$

At equilibrium, the configuration which will dominate can be found out from the maximum value of W which is a function of N_i. At the maximum value of W,

$$d \ln W = 0$$

$$\text{or, } -\Sigma d N_i \ln N_i - \Sigma N_i d \ln N_i = 0 \ \left(\text{since } \sum_i N_i = N = \text{constant}; \sum_i dN_i = 0\right)$$

$$\text{Again, } \Sigma N_i d \ln N_i = \Sigma \frac{N_i}{N_i} d N_i = \Sigma d N_i = 0.$$

$$\therefore \Sigma d N_i \ln N_i = 0.$$

Maximum value of W is subject to two constants, namely,

$$\sum_i N_i \varepsilon_i = E, \text{ where } E = \text{total energy}$$

$$\text{and } \sum_i N_i = N,\ i = 1, 2, \ldots, \text{etc.}$$

$$\text{Now, } \sum_i dN_i = 0 \text{ and } \sum_i \varepsilon_i dN_i = 0,$$

since E and N are constant.

Applying Lagrange's method of multipliers $\sum_i (a + b\varepsilon_i + \ln N_i)\, dN_i = 0$, where a and b are constants.

N_i are all independent.

So, $a + b\varepsilon_i + \ln N_i = 0$ is true for every i.

Thus, $\ln N_i = -a - b\varepsilon_i$.

$N_i = e^{-a} \times e^{-b\varepsilon_i} = A\, e^{-bei}$ (since e^{-a} is constant).

(b) (i) $b = \frac{1}{kT}$. Thus, $N_i = Ae^{-\frac{\varepsilon_i}{kT}}$.

$e^{-\frac{\varepsilon_i}{kT}}$ is a monotonically decreasing function of energy.

Number of molecules in ε_i energy level at temperature $T = N_i = Ae.^{-\frac{\varepsilon_i}{kT}}$.

Number of molecules in ε_{i+1} energy level at temperature $T = N_{i+1} = Ae^{\frac{-e_{i+1}}{kT}}$.

Now, $\varepsilon_i < \varepsilon_{i+1}$.

Thus, $e^{-\varepsilon_i/kT} > e^{-\varepsilon_{i+1}/kT}$ (k and T are constants).

Hence, $N_i > N_{i+1}$ **(Proved).**

(ii) $\sum_i N_i = N$

or, $\sum_i Ae^{-b\varepsilon_1} = N$

or, $A = \frac{N}{\sum_i e^{-b\varepsilon_i}}$.

Partition function $= q = \sum_i e^{-b\varepsilon_i}$.

$\therefore \quad A = \frac{N}{q}$.

(iii) $N_i = \frac{Ne^{-b\varepsilon_i}}{\sum_i e^{-b\varepsilon_i}}$

or, $\frac{N_i}{N}$ = probability that a molecule has an energy $\varepsilon_i = \frac{e^{-b\varepsilon_i}}{\sum_i e^{-b\varepsilon_i}}$.

2. What is thermodynamic probability? **[C.U.(H) 2002]**

Ans. : Thermodynamic probability is the distinct number of ways of realising a ticular distribution.

Let 4 particles (*a, b, c, d*) occupy 2 cells in the following way :

Cell 1 3

Cell 2 1

This can only be done in 4 ways.

Cell 1	*abc*	*abd*	*adc*	*bcd*
Cell 2	*d*	*c*	*b*	*d*

So thermodynamic probability of the above distribution is 4.

3. What is the condition for applying Boltzmann relation?

Ans. : Two following conditions should be ensured for applying Boltzmann relation :

(i) The system must be in thermal equilibrium at temperature *T*.

(ii) The system must contain fixed number of particles.

4. What is the significance of partition function?

Ans. : Partition function $q = \sum_i g_i e^{-\beta \varepsilon_i}$,

where g_i = degeneracy of *i*th energy level,
ε_i = energy of *i*th energy level and
$\beta = \sum_i g_i e^{-\beta \varepsilon_i}$.

q describes how the total energy of the system is partitioned among several energy states at temperature *T*.

At absolute zero, $q = g_0$.

Thus, at $T = 0$, *q* becomes equal to degeneracy of ground state, i.e., all the molecules come down to ground state.

As *T* increases, *q* also increases. So at high temperatures all the energy states of the system are accessible to a molecule.

So, *q* also indicates the number of energy states that are thermally accessible to a molecule at temperature *T*.

5. Which state function of the system is related to the maximum value of thermodynamic probability and how? **[C.U. 2002]**

Ans. : Entropy (*S*) is related to the maximum value of thermodynamic probability *W*.

$S = k \ln W$, k = Boltzmann constant.

If *W* of any distribution of some particles is greater than the others, entropy will also be greater for it.

6. Show that the entropy is a logarithmic function of thermodynamic probability.

Ans. : Entropy, *S*, is the function of thermodynamic probability, *W*.

$\therefore S = f(W)$.

Let a system having entropy *S* and thermodynamic probability *W* be divided into two non-interacting parts with entropies S_1 and S_2 and the corresponding thermodynamic probabilities are W_1 and W_2.

Then, $S = S_1 + S_2 = f(W_1) + f(W_2) = f(W) = f(W_1, W_2)$

$f(W_1, W_2) = f(W_1) + f(W_2)$ suggests that the function is logarithmic.

Thus, $S = a \ln W + b$, where a and b are constants.

7. Show that the relation connection S of a system with the partition function Q is given by $S = Nk_B \ln Q + \frac{E}{T}$,

where E = internal energy of the system,

k_B = Boltzmann constant

N = number of molecules in the system

T = temperature of the system. **[C.U.(H) 2003]**

Ans. : If W is thermodynamic probability, then $W = \frac{N!}{N_1!N_2!N_3!}$,

where $N = N_1 + N_2 + N_3 + \ldots$, N_1, N_2, N_3, ..., etc. are the number of molecules in ε_1, ε_2, ε_3, ... energy levels.

$$\text{Now, entropy} = S = k_B \ln W = k_B\left(\ln N! - \sum_{i=1}^{N} \ln N_i!\right)$$

$$= k_B\left(N\ln N - \sum_i N_i \ln N_i\right) \quad \text{[applying Stirling's approximation]}$$

$$= k_B\left(\sum_i N_i \ln N - \sum_i N_i \ln N_i\right) = -k_B \sum_i N_i \ln\frac{N_i}{N} = -Nk_B \sum_i \frac{N_i}{N} \ln\frac{N_i}{N}$$

$$= -Nk_B \sum \frac{e^{-\beta\varepsilon_i}}{Q} \ln\frac{e^{-\beta\varepsilon_i}}{Q} \quad \left[\text{since, according to Boltzmann relation, } \frac{N_i}{N} = \frac{e^{-\beta\varepsilon_i}}{Q}\right]$$

$$= Nk_B \sum_i \frac{e^{-\beta\varepsilon_i}}{Q}(-\beta\,\varepsilon_i - \ln Q)$$

$$= Nk_B \ln Q + Nk_B \beta \sum \frac{e^{\beta\varepsilon_i}\varepsilon_i}{Q} = Nk_B \ln Q + \frac{E}{T},$$

$$\text{since} \quad \frac{\sum_i e^{\beta\varepsilon_i}\varepsilon_i}{Q} = \bar{E} = \frac{E}{N}$$

8. Consider a system of n molecules, distributed among non-degenerate energy levels represented by ε_0, ε_1, ε_2, ..., etc. Write down the expression for molecular partition function for the system. Show that internal energy (U) of a system can be expressed as

$$U = nkT^2\left(\frac{\partial \ln Q}{\partial T}\right)_V.$$

where k is the Boltzmann constant, T and V being the temperature and volume of the system respectively. **[C.U. 2002]**

Soln. : Partition function, $Q = \sum_i e^{-\beta\varepsilon_i}$ $i = 0, 1, 2, \ldots$, etc.

Average internal energy, $\overline{U} = \dfrac{\sum_i e^{-\beta\varepsilon_i\varepsilon_i}}{\sum e^{-\beta\varepsilon_i}}$.

Now, $\sum e^{-\beta\varepsilon_i\varepsilon_i} = -\sum\left(\dfrac{\partial e^{-\beta\varepsilon_i}}{\partial\beta}\right)_V = -\left(\dfrac{\partial}{\partial\beta}Q\right)_V$

$\beta = \dfrac{1}{kT}$

$\therefore \quad \left(\dfrac{\partial\beta}{\partial T}\right)_V = -\dfrac{1}{kT^2}$.

Thus, $\overline{U} = -\dfrac{1}{Q}\left(\dfrac{\partial Q}{\partial\beta}\right)_V = kT^2\left(\dfrac{\partial \ln Q}{\partial T}\right)_V$.

$\therefore \; U = n\overline{U} = nkT^2\left(\dfrac{\partial \ln Q}{\partial T}\right)_V$.

9. The molecules of a gas have two energy states, zero and ε, and degeneracies g_1 and g_2, respectively. Write down the expression for molecular partition function.

Ans. : Molecular partition function, $Q = g_1 e^{-\beta\varepsilon_1} + g_2 e^{-\beta\varepsilon_2}$

Since, $\varepsilon_1 = 0$, $\varepsilon_2 = \varepsilon$,

$Q = g_1 + g_2 e{-}\beta\varepsilon$ (where $\beta = \dfrac{1}{kT}$, k = Boltzmann constant, T = temperature)

10. What is ensemble?

Ans. : An ensemble is an imaginary collection of large number of systems each having same macroscoping properties but different microscopic properties.

11. How do you define canonical ensemble, grand canonical ensemble and microcanonical ensemble?

Ans. : Canonical ensemble : Each member of the imaginary collection of systems has same *N, V, T,* where *N* is the number of particles of single component, *V* is volume of each member, and *T* is temperature.

Grand canonical : Each member has same μ, *V, T,* where μ is chemical potential.

Microcanonical ensemble : Each member has the same *N, V, E,* where *E* is the energy of each component.

12. If *N* molecules are distributed among the possible non-degenerate energy level ε_1, ε_2, ..., etc. in an isolated system, the number of molecules N_i in the level ε_i can be represented by $N_i = N\dfrac{e^{-\beta\varepsilon_i}}{\sum_i e^{-\beta\varepsilon_i}}$. Find the value of β.

Soln. : Thermodynamic probability $= W = \dfrac{N!}{N_1!N_2!N_3!} = \dfrac{N!}{\sum_i N_i!}$.

Entropy, $S = k \ln W$, where k = Boltzmann's constant.

$$\ln W = N \ln N - \sum_i N_i \ln N_i \text{ (applying Stirling's approximation)}$$

$$= \Sigma N_i \ln N - \sum_i N_i \ln N_i = \Sigma N_i \ln \frac{N_i}{N} = -\Sigma N_i \ln \frac{e^{-\beta\varepsilon_i}}{\sum_i e^{-\beta\varepsilon_i}}$$

$$= -\Sigma N_i(-\beta\varepsilon_i) + \sum_i N_i \ln Q \text{ (where } Q = \sum_i e^{-\beta\varepsilon_i}\text{)}$$

$$= \beta E + N \ln Q$$

(since, $\sum_i N_i\varepsilon_i = E$ = total energy of the system and $\sum_i N_i = N$)

Thus, $S = k\,\beta E + N\,k \ln Q$.

Now, $$\left(\frac{\partial S}{\partial E}\right)_V = k\beta + kE\left(\frac{\partial \beta}{\partial E}\right)_V + Nk\left(\frac{\partial \ln Q}{\partial E}\right)_V$$

$$= k\beta + kE\left(\frac{\partial \beta}{\partial E}\right)_V + Nk\left(\frac{\partial \ln Q}{\partial Q}\right)_V\left(\frac{\partial Q}{\partial \beta}\right)\left(\frac{\partial \beta}{\partial E}\right)_V$$

Now, $Q = \sum_i e^{-\beta\varepsilon_i}$.

Hence, $$\frac{\partial Q}{.\partial \beta} = -\sum_i \varepsilon_i e^{\beta\varepsilon_i}.$$

Thus, $$\left(\frac{\partial S}{\partial E}\right)_V = k\beta + kE\left(\frac{\partial \beta}{\partial E}\right)_V - Nk\frac{\sum e^{-\beta\varepsilon_i\varepsilon_i}}{Q}\left(\frac{\partial \beta}{\partial E}\right)_V$$

$$= k\beta + kE\left(\frac{\partial \beta}{\partial E}\right)_V - Nk\,\overline{E}\left(\frac{\partial Q}{\partial \beta}\right)$$

$$= k\beta \text{ (since } N\overline{E} = E\text{)}. \quad (1)$$

Again, $TdS = dE + PdV$.

$$\therefore \left(\frac{\partial S}{\partial E}\right) = \frac{1}{T}. \quad (2)$$

Equating (1) and (2), $\beta = \dfrac{1}{kT}$.

13. Write down the vibrational partition function.

Ans. : Vibrational energy $= \varepsilon_n = \left(n + \frac{1}{2}\right)h\nu$, where n = 0, 1, 2, ..., etc.

and $\nu = \dfrac{1}{2\pi}\sqrt{\dfrac{k}{\mu}}$ = vibrational frequency,

where k is force constant of the molecule and μ = reduced mass.

Hence, vibrational partition function

$$q_{\text{vib}} = \sum_{n=0}^{\infty} e^{-\beta\varepsilon_n} = \sum_{n} e^{-\beta\left(n+\frac{1}{2}\right)h\nu} = e^{-\beta h\nu/2} \sum_{n=0}^{\infty} e^{-\beta h\nu n}$$

$$= e^{-\beta h\nu/2}\,(1+e^{-\beta h\nu} + e^{-2\beta h\nu} + \ldots) = \frac{e^{-\beta h\nu/2}}{1-e^{-\beta h\nu}} \quad \left(\text{since, } 1 + x + x^2 = \frac{1}{1-x}\right).$$

14. Derive Barometric distribution formula from Boltzmann energy distribution.

Ans. : If N particles are distributed among the possible non-degenerate energy levels ε_1, ε_2, ..., etc, in such a way that N_1 particles are in ε_1, N_2 particles are in ε_2 and so on, the number of particles N_i in ith energy level at certain temperature T can be written as under :

$$\frac{N_i}{N} = \frac{e^{-\beta\varepsilon_i}}{\sum_{i=1}^{N} e^{-\beta\varepsilon_i}}, \qquad (1)$$

where i = 1, 2, ..., N and $\beta = \frac{1}{kT}$.

Equation (1) is known as Boltzmann relation.

So if N_i and N_j are the number of particles in ε_i and ε_j energy levels, then according to Boltzmann relation,

$$\frac{N_i}{N_j} = \frac{e^{-\beta\varepsilon_i}}{e^{-\beta\varepsilon_j}} = e^{-\beta(\varepsilon_i-\varepsilon_j)}. \qquad (2)$$

Equation (2) can describe distribution of particles in atmosphere at temperature T, where gravitational energy varies with altitude.

Let N_o and N_h be the number of particles at the ground level and at altitude h respectively at temperature T. Thus, $\frac{N_h}{N_0} = e^{-\beta(\varepsilon_h-\varepsilon_o)} = e^{-\beta\,\Delta\varepsilon}$.

The difference of energy $\Delta\varepsilon$ represents the potential energy difference of the particles at height h with respect to ground. Thus, $\Delta\varepsilon = mgh$, where m is mass of the particle and g is gravitational force.

$$\text{So, } N = N_o e^{-\beta mgh} = N_o e^{-\frac{mgh}{kT}}. \qquad (3)$$

The above equation is known as barometric distribution formula.

15. Calculate the contribution of a vibrational degrees of freedom to the average energy of a molecule. Comment on its value at high temperature.

Ans. Vibrational partition function, $q_{vib} = \frac{e^{-\beta h\nu/2}}{1-e^{-\beta h\nu}}$,

where $\beta = \frac{1}{kT}$ and energy of nth vibrational level = $\varepsilon_n = \left(n+\frac{1}{2}\right)h\nu$, n = 0, 1, 2.

Now for an assembly of N number of molecules,

$$\langle E_{\text{vib}} \rangle = NkT^2 \frac{d \ln q_{\text{vib}}}{dT}.$$

Here, $N = 1$, Thus,

$$\langle E_{\text{vib}} \rangle = kT^2 \frac{d \ln q_{\text{vib}}}{dT}$$

$$= kT^2 \frac{d}{dT}\left[-\frac{h\nu}{2kT} - \ln\left(1 - e^{-\frac{h\nu}{kT}}\right)\right]$$

$$= kT^2 . \frac{h\nu}{2} . \frac{1}{kT^2} + \frac{e^{-\frac{h\nu}{kT}} . \frac{h\nu}{kT^2} . kT^2}{1 - e^{-\frac{h\nu}{kT}}}$$

$$= \frac{h\nu}{2} + \frac{h\nu}{e^{\beta h\nu} - 1}.$$

If $kT >> h\nu$ (at high temperature),

$$\langle E_{\text{vib}} \rangle = \frac{h\nu}{2} + kT \left(e^{\beta h\nu} \approx 1 + \frac{h\nu}{kT}\right)$$

$$= kT \text{ (since } \frac{h\nu}{2} << kT).$$

16. State the difference between Maxwell-Boltzmann, Bose-Einstein and Fermi-Dirac statistics.

Ans.

Maxwell-Boltzmann statistics	Bose-Einstein statistics	Fermi-Dirac statistics
(a) Particles are distinguishable	(a) Particles are industinguishable	(a) Particles are indistinguishable
(b) Particles do not have spin	(b) Particles have integral spin	(b) Particles have half-integral spin
(c) No symmetry requirement is imposed on the corresponding wave function of the system over interchanging any two particles.	(c) Wave function should be symmetric over interchanging any two particles	(c) Wave function should be antisymmetric over interchanging any two particles.
(d) No restriction over occupancy of an energy state	(d) No restriction.	(d) Energy state can accomodate only two particles with opposite spin.

Numerical Problems

1. (a) Consider 20 molecules divided equally between four non-degenerate energy levels. What is the thermodynamic probability (W) for this distribution? (b) How does the value of W change if one molecule is removed from one level and added to another? **[C.U.(H) 2001]**

Soln. : (a) Number of molecules in each energy level $= \dfrac{20}{4} = 5$.

Total number of molecules = 20.

$$W = \frac{20!}{5!5!5!5!} = \mathbf{1.17 \times 10^{10}}.$$

(b) If one molecule is removed from one level and added to another

$$W_1 = \frac{20!}{5!5!6!4!} = \mathbf{9.78 \times 10^{9}}.$$

2. Suppose a molecule has two energy levels $\varepsilon_1 = 0$ and $\varepsilon_2 = kT$. Calculate (i) the partition function and (ii) ratio of the number of molecules in the two levels [k = Boltzmann constant and T is the temperature in K] **[C.U. 2001]**

Soln. : (i) Partition function $q = e^{-\varepsilon_1/kT} + e^{-\varepsilon_2/kT} = 1 + e^{-1} = \mathbf{1.37}$.

(ii) Let N_1 and N_2 be the number of molecules at ε_1 and ε_2 energy levels respectively.

$$\frac{N_1}{N_2} = \frac{e^{-\varepsilon_1/kT}}{e^{-\varepsilon_2/kT}} = e^{-(\varepsilon_1-\varepsilon_2)/kT} = e \text{ (since, } \varepsilon_1 = 0,\ \varepsilon_2 = kT)$$

$= \mathbf{2.72}$.

3. Consider a system of six distinguishable particles. One of the macrostates has the following distribution of particles :

Energy	0ε	1ε	2ε	3ε	4ε
No. of particles	0	0	2	2	2

Calculate its themodynamic probability. **[C.U. 2002]**

Soln. : Thermodynamic probability, $W = \dfrac{6!}{2!2!2!} = \mathbf{90}$.

4. Calculate the number of ways of arranging 5 different particles among 3 energy ways levels such that one energy level has 1 particle, two have 2 each. **[C.U. 2003]**

Soln. : The total number of ways, W

$$= \frac{5!}{1!2!2!} = \mathbf{30}.$$

5. Calculate the difference in entropy between two macrostates, I and II, where six distinguishable particles are distributed in 3 different energy levels (0, ε and 2ε) in the following manner :

Energy	0	ε	2ε
macrostate I	3	3	0
macrostate II	2	2	2

Soln. : Entropy, $S = k \ln W$,

where W is thermodynamic probability

and k is Baltzmann's constant.

For macrostate 1, $W_1 = \dfrac{6!}{3!3!} = 20.$

For macrostate 2, $W_2 = \dfrac{6!}{2!2!2!} = 90.$

So the difference in entropy between two macrostates, $\Delta S = S_2 - S_1$

$$= k \ln W_2 - k \ln W_1 = k \ln \frac{W_2}{W_1} = k \ln \frac{90}{20}$$

$$= 1.38066 \times 10^{-23} \times 1.504 \text{ J.K}^{-1}$$

$$= \mathbf{2.079 \times 10^{-23} \text{ J.K}^{-1}}.$$

6. Calculate the rotational partition function of a diatomic molecule at high temperature.

Soln. : Rotational energy of Jth level $= E_J = BJ(J + 1)$,

where $B = \dfrac{h^2}{8\pi^2 I}$

h = Planck constant

I = moment of inertia.

Rotational partional function $= q_{\text{rot}}(T) = \displaystyle\sum_{J=0}^{\infty} (2J+1)e^{\beta BJ(J+1)}$, where $\beta = \dfrac{1}{kT}$.

Degeneracy of Jth level $= 2J + 1$.

Since $\dfrac{\beta}{kT}$ is quite small at ordinary temperature for most molecules, the sum can be approximated by an integral. Hence, at high temperature,

$$q_{rot}(T) = \int_0^{\infty} (2J+1)e^{-BJ(J+1)/kT} = \int_0^{\infty} e^{-BJ(J+1)/kT} d\{J(J+1)\} = \frac{kT}{B} = \frac{\mathbf{8\pi^2 IkT}}{\mathbf{h^2}}.$$

7. Find the ratio of population of two states such that $E_b - E_a = kT$. Degeneracy of corresponding energy levels are 3 and 1.

Soln. : Let N_b and N_a be the population of state b and state a.

$g_b = 3,\ g_a = 1$

Hence, $\dfrac{N_a}{N_b} = \dfrac{g_a e^{-E_a/kT}}{g_b e^{-E_b/kT}} = \dfrac{1}{3} e^{-(E_a - E_b)/kT} = \dfrac{e}{3} = \mathbf{0.906.}$

8. The energy difference between ground state and first vibrational state for O_2 is 3.139×10^{-20} J. At what temperature would the 2nd vibrational state have half of the population of the 1st vibrational state?

Soln. : Degeneracy of each vibrational state = 1.

Hence, $\dfrac{N_{n=1}}{N_{n=2}} = 2 = e^{-(E_1 - E_2)/kT} = e^{3.139 \times 10^{-20}/T \times 1.38 \times 10^{-23}}$

or, $\ln 2 = 3.139 \times 10^{-20}/T \times 1.38 \times 10^{-23}$

$\therefore$ T = **3281.61 K.**

9. Two distinguishable particles are distributed among four boxes. How many distributions are possible? Calculate thermodynamic probability of each distribution and comment on most probable distribution. Calculate also the probability of getting any one arrangement.

Soln. : Only two distributions are possible.

(i) 2 in same box (2000)

(ii) 2 in different two boxes with one in each box (1100).

Thermodynamic probability of distribution (2000) = $^4P_1 = 4$.

Thermodynamic probability of distribution (1100) = $^4P_2 = 12$.

So the 2nd distribution is most probable since it has highest thermodynamic probability or microstates.

Total number of arrangement = $2^4 = 16$ (also $12 + 4 = 16$).

So probability of getting any one arrangement = $\dfrac{1}{16}$.

10. Suppose, five letters of the word 'LASER' is written on separate cards. The cards are shuffled and arranged in a row. Calculate the probability of obtaining the word 'LASER'

Soln. : Total number of ways the five letters can be arranged taking one at a time $= \lfloor 5 = 120$.

So probability of obtaining the word 'LASER' = $\dfrac{1}{120}$.

11. Suppose, 2 distinguishable particles are distributed among 3 degenerate energy states. How many arangements are possible? What will happen if the particles are (i) bosons (ii) fermions?

Soln. : Here degeneracy = $g = 3$.

Number of particles = $N = 2$.

Total number of arrangements = $W = g^N = 3^2 = 9$.

ε_1	ε_2	ε_3	$(\varepsilon_1 = \varepsilon_2 = \varepsilon_3)$
AB			
	AB		
		AB	
A	B		
A		B	
	A	B	
	B	A	
B	A		
B	A		

(i) For bosons, $W = \dfrac{(g+N-1)!}{N!(g-1)!} = \dfrac{4!}{2!2!} = 6.$

ε_1	ε_2	ε_3
AA		
	AA	
		AA
A	A	
A		A
	A	A

(ii) For fermions, $W = {}^{g}C_N = {}^{3}C_2 = \dfrac{3!}{2!1!} = 3.$

ε_1	ε_2	ε_3
A	A	
	A	A
A		A

CHAPTER 18

MISCELLANEOUS PROBLEMS

Theoretical Problems

(a) *Answer the following in brief as much as possible :*

1. What is the SI unit of R?

Ans. : The SI unit of R is $J.K^{-1}.mol^{-1}$.

2. What is the SI unit of van der Waals constant a?

Ans. : $J.m^3.mol^{-2}$.

3. What is the relation between bar and torr?

Ans. :
$$1 \text{ atm} = 760 \text{ torr}$$
$$= 1.01325 \text{ bar}$$
$$1 \text{ bar} = \frac{760}{1.01325} \text{ torr}$$
$$= 750.06 \text{ torr.}$$

4. Find the maximum number of degrees of freedom of a one-component system.

Ans. : By phase rule, $P + F = C + 2$,

P = no. of phase

F = degrees of freedom

C = no. of component.

$\therefore \quad F = C - P + 2 = 1 + 2 - P$

$= 3 - P.$

Minimum phase = 1.

$\therefore$ maximum degrees of freedom = 3 − 1 = 2.

5. Find the degrees of freedom of an aqueous solution of sucrose and glucose.

Ans. : By phase rule, $F = C - P + 2$.

Here $C = 2$, $P = 1$

$\therefore \quad F = 2 - 1 + 2 = 3.$

6. How is entropy related to probability?

Ans. : If W is thermodynamic probability, S is entropy and k is Boltzmann constant, which is equal to $\frac{R}{N_A}$ [N_A = Avogadro number],

$S = k \ln W$.

7. Remark whether Q = 0 for a cyclic process.

Ans. : Q is not a state function, so $Q \neq 0$.

8. Remark whether every closed system is isolated.

Ans. : An isolated system cannot exchange both matter and energy with the surrounding. Closed systems exchange heat (energy) with the surrounding. Thus, closed system is not an isolated one.

9. Express 1 J in $kg.m^2.sec^{-2}$.

Ans. : Unit of energy is joule.

1 J = force × distance

Dimension of 1 J = $MLT^{-2} \times L = ML^2T^{-2}$.

So, 1 J = 1 $kg.m^2.sec^{-2}$.

10. Express 1 watt in kcal/day.

Ans. : $1 \text{ watt} = \frac{1\text{ J}}{1\text{ sec}} = \frac{1\text{ J} \times 60}{\text{minute}} = \frac{1\text{ J} \times 60 \times 60 \times 24}{\text{day}} = \frac{86400\text{ J}}{\text{day}}$

$$= \frac{86400}{4.184} \text{ cal/day} = 20650.0956 \text{ cal/day} = 20.65 \text{ kcal/day}.$$

11. Remark whether T_B should be greater than T_C.

Ans. : T_C is the highest temperature at which a gas can be liquefyed by application of pressure alone. T_B is the temperature at which a gas can behave ideally, since then intermolecular attraction and repulsion balances each other. As temperature increases, intermolecular attraction starts becoming negligible. Thus, $T_B > T_C$.

12. Remark whether $\Delta S = 0$ for any process in isolated system.

Ans. : $\Delta S = \frac{Q_{rev}}{T}$ for an isolated system. But for spontaneous changes, the heat change is irreversible, and, $Q_{irrev} = 0$ for an isolated system. But there is no relation between irreversible heat change and entropy change. So ΔS is not zero for any spontaneous change in an isolated system.

13. When should a liquid have zero surface tension?

Ans. : Surface tension, γ, is a surface property. At critical temperature (T_C) and pressure (P_C), there is no surface. Hence, at T_C, $\gamma = 0$.

14. For a distribution $P(u)\,du = Ae^{-u^2}\,du$ with u ($-\infty$ to $+\infty$). Find the most probable value.

Ans. : $u = 0$ [for the most probable speed,

$$\frac{d}{du}(Ae^{-u^2})\,du = 0$$

$$\Rightarrow -2uAe^{-u^2} = 0, \therefore u = 0]$$

15. Define compressibility factor.

If at certain temperature, P_i and P are the pressures of ideal gas and real gas respectively at the same molar volume or at the same pressure, V_i and V are the molar volumes of ideal gas and real gas respectively, then compressibility factor,

$$Z = \frac{P}{P_i} = \frac{V}{V_i} = \frac{P}{\frac{nRT}{V_1}} = \frac{PV_1}{nRT}.$$

16. Remark if the pH range for boric acid-borax buffer lies in the acidic range.

Ans. : This is a buffer of weak acid and weak base. Pk_a of boric acid is 9.27. Its actual buffer range is 8.27 to 10.27. Thus, pH range of the system is alkaline.

17. What is the driving force of osmosis?

Ans. : The driving force of osmosis is the chemical potential. The increase of chemical potential in a solvent-solution system decreases the tendency of the solvent to pass to solution. So the driving force is the chemical potential.

18. Draw a volume-temperature diagram for an isothermal expansion.

Ans. :

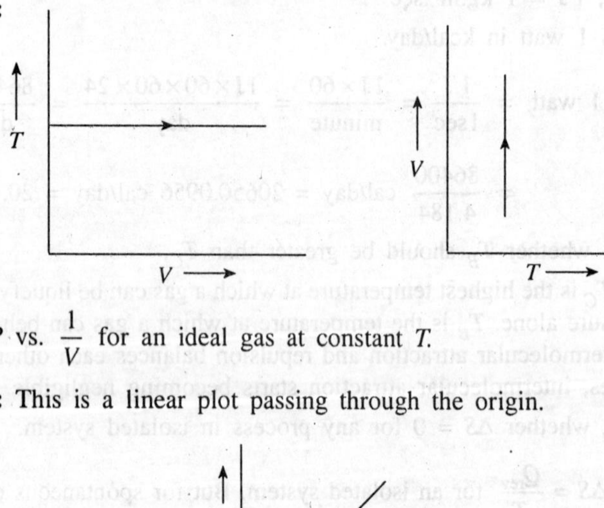

19. Plot P vs. $\frac{1}{V}$ for an ideal gas at constant T.

Ans. : This is a linear plot passing through the origin.

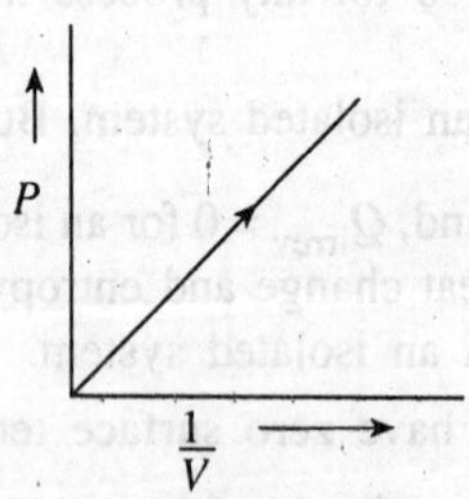

20. What is the basic difference between volume and molar volume?

Ans. : Volume is an extensive property, molar volume is an intensive variable.

21. What physical property does the dimension ML^2T^{-2} correspond to?

Ans. : It corresponds to energy.

22. What is the dimension of viscocity coefficient (η)?

Ans. : $ML^{-1}T^{-1}$ $\qquad [F = \eta A.\frac{dc}{dv}]$

23. Define reduced temperature.

Ans. : If the temperature of a gas in a particular state is T, and T_c is its critical temperature, then $\frac{T}{T_c} = T_r$ = reduced temperature.

24. What kind of average speed is involved in $PV = \frac{1}{3} mnc^2$.

Ans. : The average speed is r.m.s.

25. If x is the length and y is the time, then find the dimension of $\frac{d^3x}{dy^3}$.

Ans. : The dimension is [LT^{-3}], because $\frac{d^3x}{dy^3} = \frac{d}{dy}.\frac{d}{dy}\frac{dx}{\cdot dy}$.

26. Explain : PV is a state function.

Ans. : For state 1, PV is P_1V_1 and for state 2, PV is P_2V_2 irrespective of how the change takes place. Thus, PV is a state function.

27. Explain if Z in $PV = ZnRT$ for a real gas is intensive.

Ans. : By definition, $Z = \frac{PV}{nRT}.\frac{V}{n}$ is molar volume V_m. Thus, $Z = \frac{PV_m}{RT}$. All these quantities are intensive. Therefore, Z is an intensive property (In case of Z, parts are not added to give the whole).

28. Find the degrees of freedom in the reaction mixture $CaCO_3 = CaO + CO_2$.

Ans. : There are three components, $CaCO_3$, CaO and CO_2. But there is an equation relating them, so the no. of components is 2 (3 – 1 = 2).

No. of phases, $P = 3$.

Hence, degrees of freedom, $F = C - P + 2$

$= 2 - 3 + 2 = 1.$

29. Remark whether heat cannot be completely converted into work.

Ans. : Heat can be completely converted into work in a process involved in state change. As for example, isothermal reversible or irreversible expansion of an ideal gas. However, heat cannot be completely converted into work in a cyclic process.

30. Identify X and Y in $\left(\frac{\partial P}{\partial T}\right)_V = \left(\frac{\partial S}{\partial X}\right)_Y$

Ans. : By Maxwell equation we know that, $\left(\frac{\partial P}{\partial T}\right)_V = \left(\frac{\partial S}{\partial V}\right)_T$

Hence, $X = V$ and $Y = T$.

31. Explain whether a pure system with four phases can co-exist.

Ans. : We know that by phase rule, $P + F = C + 2$,

P is the no. of phases,

F is the degree of freedom,

C is the no. of components,

Hence, $F = C - P + 2$.

A pure system means a single component.

$\therefore \quad F = 1 - 4 + 2$

$= -1.$

Hence, four phases cannot co-exist.

32. Given, $Y = aX^2$. How do you get a linear plot?

Ans. : By taking logarithm,

$\log Y = \log a + 2 \log X.$

Thus, plot of log Y vs. log X is linear with 2 as slope and log a as intercep

33. When does solubility decrease with temperature?

Ans. : If dissolution is exothermic, e.g., heat of solution is negative, then solubili will decrease with increase in temperatue.

34. How does K_w change with temperature?

Ans. : K_w is the equilibrium constant of the reaction, $H_2O = H^+ + OH^-$. Th is endothermic and hence, K_w will increase with increase in temperatrure.

35. What pH range will be suitable for acetic acid-acetate buffer?

Ans. : The best buffer region is pH with $(pK_a + 1)$ and $(pK_a - 1)$, where p is the negative logarithm of K_a of acetic acid and $pK_a = 4.75$. Therefore, the p range is 3.75-5.75.

36. For an aqueous solution of varying pH, plot pH vs. pOH at a given temperatur

Ans. : Since pH is varying, taking pH as abcissa (X-axis) and pOH as ordina (Y-axis), the curve is shown below :

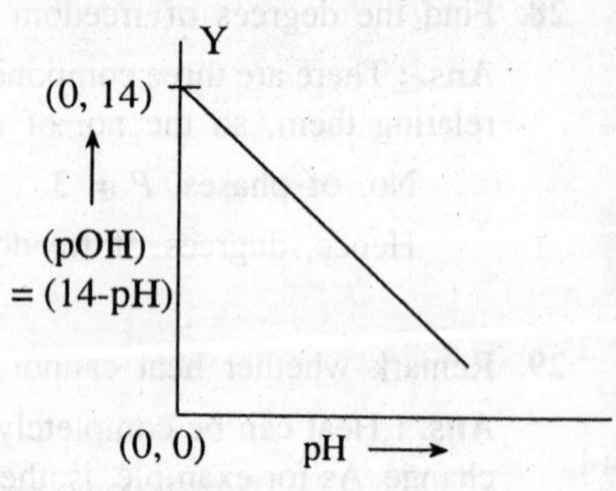

We know that, pH + pOH = 14.

So when pH = 1, pOH = 13.

Hence, it will be a linear plot with 14 as intercept and −1 as slope.

37. Why do not two isothermals intersect each other?

Ans. : Isothermals are plotted in PV-curve in general. Isothermal plot is rectangular hyperbola and other isothermals are parallel to it. Each point in th same isothermal plot are in the given temperature. A plot for P-V for an ide gas is shown below :

Isothermals for an ideal gas at temperatures T_1, T_2 and T_3
$T_3 > T_2 > T_1$.

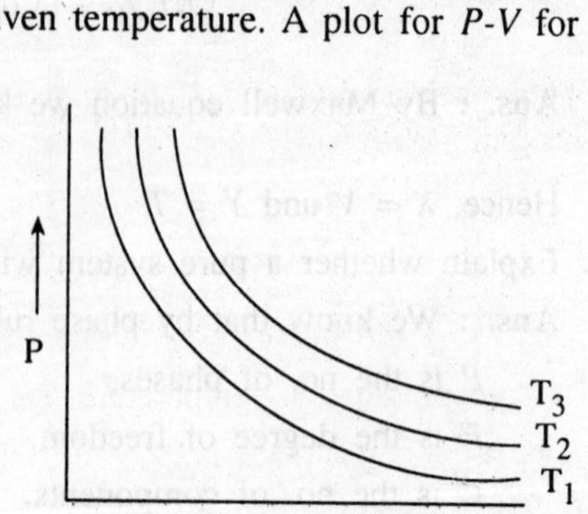

38. Viscosity and conductivity of a gas must be related to another. Why?

Ans. : In viscosity there is momentum transfer—it is transport of momentum. Theremal conductivity is due to transport of K.E. Kinetic energy (K.E.) is $\frac{P^2}{2m}$, where P is momentum. So they must be related.

39. State the principle of corresponding states.

Ans. : The principle of corresponding states predicts that at the same reduced temperature (T_r) and reduced pressure (P_r) all gases have the same reduced volume (V_r). Here reduced temperature (T_r) is $\frac{T}{T_c}$, P_r is $\frac{P}{P_c}$ and V_r is $\frac{V}{V_c}$, where P_c, V_c and T_c are the critical pressure, volume and temperature of the gas respectively and P, V and T are pressure, volume and temperature of the same gas respectively.

40. When is an adiabatic process isoentropic?

Ans. : We know that $\Delta S = \frac{Q_{rev}}{T}$, where ΔS is change in entropy, Q_{rev} is reversible heat change and T is temperature. Thus, for adiabatic reversible change, $\Delta S = 0$, i.e., adiabatic reversible change is isoentropic.

41. In a cryoscopic study of solutions, which concentration unit is preferable—molarity or molality? Why?

Ans. : The unit chosen is molality. Bccause in cryoscopic change, there is a change of temperature and the molality does not depend on temperature (molarity depends on temperature).

42. What is the dimension of enthalpy?

Ans. : Enthalpy has the unit of energy. Thus, its dimension is ML^2T^{-2}.

43. Iodine usually sublimes but it can be melted. How?

Ans. : For the iodine system the triple point is above 1 atm. Naturally solid iodine does not have a normal melting point. It sublimes at 1 atm. But above the triple point which is much above 1 atm, solid iodine can be melted.

44. Should an ideal gas have a viscosity? Why?

Ans. : In case of gases, viscosity is due to momentum transfer by collisions. In ideal gases there can be collisions. So they will have viscosity.

45. Does the B.P. of water change with pressure? How?

Ans. : B.P. of any substance depends on pressure. Dependence is given by Clapeyron equation and Clausius-Clapeyron equation. By Clausius-Clapeyron equation, writing in integrated form,

$$\ln \frac{P_2}{P_1} = \frac{\Delta H}{R} \frac{T_2 - T_1}{T_1 T_2},$$

where P_2 is the pressure at temperature T_2,

P_1 is the pressure at temperature T_1,

ΔH is the change in enthalpy, i.e., latent heat for evaporation,

R is the universal gas constant.

46. What is the dimension of surface tension?

Ans. : Surface tension is the force per unit length.

Thus, the dimension is $MLT^{-2} \times \frac{1}{L} = [MT^{-2}]$.

47. What is Clausius inequality?

Ans. : We know that, $dS = \frac{đQ_{rev}}{T}$ and any $đQ \le đQ_{rev}$.

$\therefore\ dS \ge \frac{đQ}{T}$. This is the Clausius inequality.

$đQ_{rev}$ is reversible heat change and $đQ$ is heat change in spontaneous process.

48. Why is partial molar temperature of a component in a mixture irrelevant?

Ans. : Temperature (T) is an intensive property, and a system at equilibrium has a same temperature in all its parts and in the surrounding. So the partial molar temperature is meaningless.

49. How does chemical potential of a solvent will change on addition of solute?

Ans. : The chemical potential of solvent is stated as $\mu_1^o\ (T, P)$. Addition of solute changes the mole fraction from 1 to X_1, where X_1 is the mole fraction of solvent.

So now $\mu_1 = \mu_1^o(T, P) + RT \ln X_1$, X_1 is a fraction. Hence, $\ln X_1$ is negative and the chemical potential will be reduced.

50. Why is Ostwald dilution law meaningless in case of aqueous HCl solution?

Ans. : Ostwald dilution law is applicable for weak electrolytes BA ionising to B^+ and A^-; $BA \rightleftharpoons B^+ + A^-$ and states that as concentration decreases degree of ionization (α) increases. But HCl in aqueous solution is completely ionized to H^+ and Cl^- at all dilutions and hence, Ostwald dilution law cannot be applied.

51. Why ΔS is not a state function?

Ans. : ΔS is $\int_1^2 dS$ and it does not depend on path, being the same always for the same state change. So dS is an exact differential. State functions characterize a particular state. ΔS does not characterize a particular state and hence, is not a state function.

52. In the titration of NH_4OH with HCl, the pH at the neutralization point is not 7—comment.

Ans. : In the titration of NH_4OH with HCl, at the neutralization point there is NH_4Cl. But NH_4^+ ions undergo hydrolysis and pH becomes < 7 due to presence of some extra H^+ arising from hydrolysis.

$NH_4^{\oplus} + H_2O \rightleftharpoons NH_3 + H_3O^{\oplus}$ ($H^{\oplus}$ is written as $H_3O^{\oplus}$)

53. 1 atm pressure = ? torr?

Ans. : 1 atm = 760 torr = 760 mm of Hg.

54. What is the dimension of relative lowering of vapour pressure?

Ans. : If the V.P. of pure solvent is P^0 and that of solution is P at a particular temperature, then relative lowering is $\frac{P^0 - P}{P^0}$. Thus, relative lowering has no dimension—it is a pure number.

55. How does Carnot cycle look in *T-S* diagram?

Ans. : Carnot cycle looks like a parallelogram. The first change is reversible isothermal, the second change is reversible adiabatic, i.e., isoentropic. The third change is isothermal reversible and the last change is isoentropic.

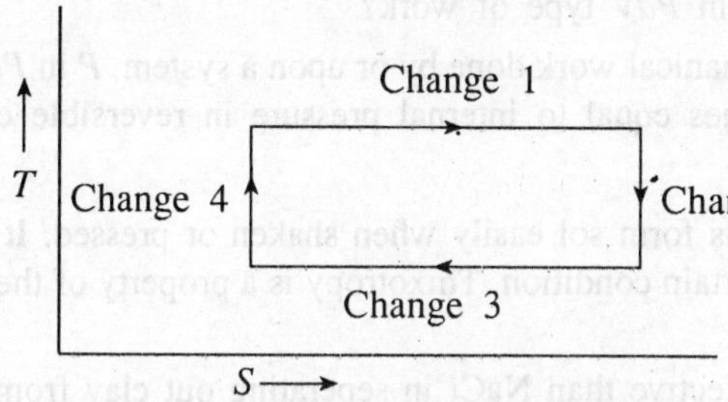

56. What is the S.I. unit of chemical potential?

Ans. : The S.I. unit of chemical potential (μ_i) is J.mol^{-1}.

57. What condition must be satisfied for a binary liquid mixture to show azeotropism?

Ans. : If the binary liquid solution deviates from Raoult's law by increase or decrease vapour pressure (V.P.) and there is a composition at which V.P. becomes maximum or minimum, then an azeotrope results.

58. How does the pH of water depend on temperature?

Ans. : K_w increases as temperature increases. Hence, C_{H^+} increases with rise of temperature. Thus, pH given by $-\log_{10}C_{H^+}$ should decrease.

59. Cite two examples of reversible process.

Ans. : Boiling of a liquid in a closed vessel with movable piston at constant normal *P* and *T*, and similarly melting of a solid at constant normal pressure and temperature are examples of reversible process.

60. Write down any one form of Gibbs-Helmholtz equation.

Ans. : Let the Gibbs free energy be *G*, and *H* be enthalpy, then $G = H + T\left(\frac{\partial G}{\partial T}\right)_P$ is one form of Gibbs-Helmholtz equation.

61. A reaction of the type $A(g) + A(g) \rightarrow A_2(g)$ is expected to be exothermic.—Explain.

Ans. : For a spontaneous change ΔG should be negative. But $\Delta G = \Delta H - T\Delta S$,

where ΔG is the change in Gibbs free energy,

ΔH is the change in enthalpy,

ΔS is the change in entropy.

For the given change ΔS has to be negative. Let it be $-Y$.

$\therefore$ $\Delta G = \Delta H + TY$. Hence, ΔG can be –ve, only if ΔH is –ve. The change is, therefore, exothermic.

62. State the basic law that governs the solvent extraction method.

Ans. : The law is Nernst distribution law which states that if there is a pair of immiscible liquids and a solute is soluble in both the solvents with C_1 being the concentration of solvent 1 and C_2 being the concentration of solvent 2. Then

$\frac{C_1}{C_2} = D$, distribution coefficient, is constant at a constant temperature.

63. What is the dimension of change?

Ans. : Kept to itself the change has no dimension. However, it will have the dimension of property whose change is given.

64. What does P signify in PdV type of work?

Ans. : PdV is the mechanical work done by or upon a system. P in PdV is external pressure which becomes equal to internal pressure in reversible change.

65. What is thixotropy?

Ans. : Thixotropic gels form sol easily when shaken or pressed. It is a property any gel can get at a certain condition. Thixotropy is a property of the initial stages of gelation.

66. Why is alum more effective than NaCl in seperating out clay from river water?

Ans. : Clay is in colloid form and it is precipitated as ions are added. According to Schulze-Hardy rule the precipitating effect of ions increases as the valency of ions increases. Alum consists of Al^{3+}, and NaCl consists of Na^+. Thus, coagulating power of Al^{3+} is much greater. That is why alum is more effective.

67. Write the cell reaction if two equivalent of electricity is passed through the cell—
$H_2(g)$, pt | H^+ (aq), Cl^-(aq), AgCl(s) | Ag(s).

Ans. : Reaction at the R.H.E.,

$$2AgCl \rightleftharpoons 2\,Ag^{\oplus} + 2Cl^-$$
$$2Ag^{\oplus} + 2e \rightleftharpoons 2\,Ag$$

Adding, $2AgCl + 2e \rightleftharpoons 2\,Ag + 2Cl^-$

Reaction at the L.H.E. as R.H.E.,

$$2H^+ + 2e = H_2$$

Now, reaction at L.H.E. as L.H.E., $H_2 = 2H^+ + 2e$

So the cell reaction is

$$2AgCl + H_2 = 2Ag + 2H^+ + 2Cl^-$$

or, $2AgCl + H_2 = 2Ag + 2HCl$.

68. Give example of one reversible and one irreversible cells.

Ans. : Reversible cell, $H_2(g)$ Pt | H^+ (aq), Cl^-(aq), AgCl(s) | Ag(s)

Irreversible cell, Zn | H_2SO_4 | Cu.

69. What is the unit of $\frac{d\ln k}{dt}$ for an nth order reaction?

Ans. : The unit is sec^{-1}.

70. What is the sign of entropy of activation when the transition state is formed by an associative mechanism?

Ans. : It will generally be negative, since the activated particle is one and there are more than one reactant.

71. What is the dimension of $\oint pdq$, where p is the generalized momentum conjugate to the generalized coordinate q?

Ans. : $\oint pdq = nh$. Thus, $\oint pdq$ has the same dimension as h, i.e.,

$ML^2T^{-2} \times T = [ML^2T^{-1}]$

72. Write Gibbs adsorption isotherm.

Ans. : $-\frac{C}{RT} \cdot \frac{d\gamma}{dC} = \Gamma_2$,

where Γ_2 is the surface excess of solute per unit area,
C is the concentration of solute
γ is surface tension of solvent.

73. Calculate the momentum of an electron which is characterized by a de Broglie wavelength of 100 pm.

Ans. : We know that $\lambda = \frac{h}{P}$

or, $P = \frac{h}{\lambda} = \frac{6.625 \times 10^{-34} \text{ J.sec}}{100 \times 10^{-10} \text{ m}} = 6.626 \times 10^{-14}$ kg.m.sec^{-1}.

74. Write an expression of the Hamiltonian operator for a particle in a one dimensional box.

Ans. : $\hat{H} = -\frac{h^2}{8\pi^2 m} \cdot \frac{\partial^2}{\partial x^2}$, since $V = 0$.

75. Give an example of a cationic surfactant.

Ans. : The cationic surfactant may be lauryl amine hydrochloride

$CH_3(CH_2)_{11}\ NH_3^+Cl^-$.

76. Name two experiments in support of (i) wave nature and (ii) particle nature of electron.

Ans. : The experiments of finding *e/m* ratio of an electron support the particle nature of electron. Electron diffraction studies prove the wave nature of electron.

77. Name two experiments in support of (i) wave nature and (ii) particle nature of radiation.

Ans. : The interference of light is an example of wave nature of radiation. Photoelectric effect is one of the example indicating the particle nature of radiation.

78. Arrange boiling, Boyle and critical temperatures (T_b, T_B and T_c) in increasing order with explanation.

Ans. : The increasing order is $T_b < T_c < T_B$.

In the boiling temperature the liquid vaporizes at 1 atm. At the critical temperature the liquid state disappears and at the Boyle temperature the gas behaves as an ideal gas.

79. Plot energy vs. n for the particle in a box in one dimension.

Ans. : We know, $E = \frac{n^2h^2}{8ma^2}$.

For a particular system, $\frac{h^2}{8ma^2}$ is constant (K). So, $\frac{n^2h^2}{8ma^2} = Kn^2$.

Therefore, the plot is parabolic and a plot for n = 1, 2, 3 is shown below :

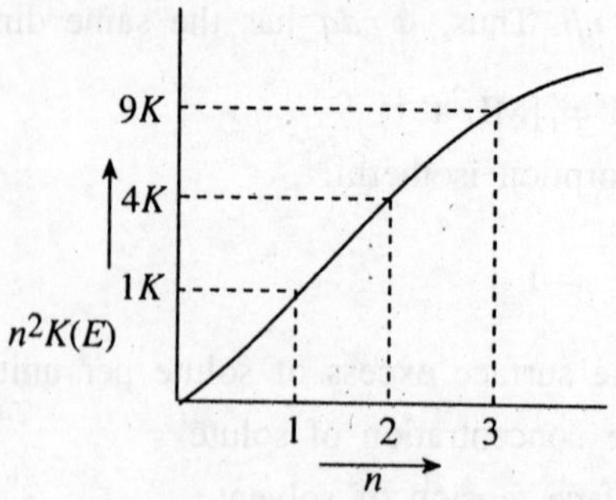

80. Give an example of a photochemical reaction which may be used to measure intensity of radiation.

Ans. : We know, $\varphi = \frac{\text{rate of reaction}}{I_a}$.

$\therefore\ I_a = \frac{\text{rate of reaction}}{\varphi}$.

In the range of radiation (250 – 400) nm the quantum efficiency of the decomposition of uranyl oxalate is 0.57. Hence, measurement of the rate of reaction will lead to I_a calculation.

81. Express mean ionic activity of $Al_2(SO_4)_3$ in terms of ionic activities.

Soln. : $Al_2(SO_4)_3$ ionizes as

$$\underset{C}{Al_2(SO_4)_3} \rightleftharpoons \underset{2C}{2Al^{3+}} + \underset{3C}{3\,SO_4^{-2}}$$

$\therefore$ Ionic activity $a_2 = a^2_{Al^{3+}} \cdot a^3_{SO_4^{-2}}$

$\therefore\ a_\pm = \sqrt[5]{a_2}$

$\therefore\ a_+ = C_+\gamma_+,\ a_- = C_-\gamma_-$

$\therefore\ a_\pm = \sqrt[5]{(2C)^2.\gamma_+^2.(3C)^3.\gamma_-^3} = \sqrt[5]{108C^5.\gamma_\pm^5}$

$\therefore\ a_\pm = (108)^{1/5}.C.\gamma_\pm$

82. Calculate $\frac{\Gamma(5/2)}{\Gamma(3/2)}$.

Soln. : $\Gamma(n+1) = n\Gamma(n)$

$\because\ \Gamma\left(\frac{5}{2}\right) = \frac{3}{2}\Gamma\left(\frac{3}{2}\right) = \frac{3}{2} \times \frac{1}{2}\Gamma(1) = \frac{3}{4}$.

Again, $\Gamma\left(\frac{3}{2}\right) = \frac{1}{2}\Gamma(1) = \frac{1}{2}$.

$$\therefore \quad \frac{\Gamma(5/2)}{\Gamma(3/2)} = \frac{3/4}{1/2} = \frac{3}{2}.$$

83. When does the Bragg reflection condition lead to $\lambda \le d$?

Ans. : We know that Bragg's equation is

$$n\lambda = 2d\sin\theta \quad [\theta = 0°\text{-}90°].$$

When $n = 2$, $\lambda = d\sin\theta$,

$$d = \frac{\lambda}{\sin\theta}.$$

$$\therefore \quad d \ge \lambda \text{ (since, } 0 < \sin\theta < 1).$$

84. For identical experimental conditions the first order Bragg reflection from a plane of cubic crystal comes up at 5.9° and 5.85° respectively at 20 °C and 50 °C. Calculate coefficient of cubic expansion of the solid.

Hints : Coefficient of cubic expansion = $\alpha = \frac{1}{V}\left(\frac{\partial V}{\partial T}\right)_P$

$$\lambda = 2d\sin\theta$$

$$\therefore \quad \frac{dV}{V} = \frac{d_{50}^3}{d_{20}^3} - 1 = \frac{\sin^3\theta_{20}}{\sin^3\theta_{50}} - 1.$$

$$\text{Thus, } \alpha = \frac{1}{30}\left(\frac{\sin^3\theta_{20}}{\sin^3\theta_{50}} - 1\right) = 8.59 \times 10^{-4}\ \text{K}^{-1}.$$

85. Potassium crystallizes in a body centred cubic lattice with cell length 5.2 Å. Determine (i) distance between (a) nearest neighbours and (b) next-nearest neighbours (ii) no. of (a) nearest neighbours and (b) next-nearest neighbours for each K atom and (iii) density of lattice.

Hints :

(i) (a) $b = \frac{\sqrt{3}}{2} \times 5.2 = 4.50$ Å (along body diagonal).

(b) $a = 5.2$ Å (along cell edge).

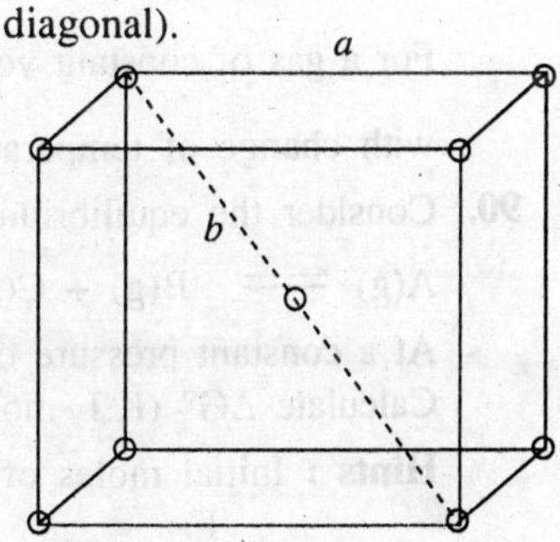

(ii) (a) 8 (at 8 corners).

(b) 6 (along cell edges).

(iii) 0.921 gm.cm^{-3}.

86. Show that $\eta^2\sigma^4/m$ has the dimension of energy where the terms have usual meaning.

Ans. : Dimension of η = ML^{-1}T^{-1}

Dimension of σ = L

Dimension of m = M

$$\text{Thus, dimension of } \frac{\eta^2\sigma^4}{m} = \frac{M^2L^{-2}T^{-2}L^4}{M} = ML^2T^{-2}$$

= dimension of energy.

87. Show that $(PdV + VdP)$ is an exact differential.

Soln. : Let $df = PdV + VdP = MdX + NdY,$

where $M = P,\ dx = dV,\ N = V\ dy = dP.$

If $\left(\frac{\partial M}{\partial y}\right)_x = \left(\frac{\partial N}{\partial x}\right)_y$, then df is an exact differential.

Here, $\left(\frac{\partial P}{\partial P}\right)_V = 1 = \left(\frac{\partial V}{\partial V}\right)_P$

Thus, df is an exact differential.

88. Calculate the fraction of molecules of a gas having energy greater than 50 RT.

Soln. : $\frac{dN_E}{N} = 2\pi\left(\frac{1}{\pi RT}\right)^{3/2} E^{1/2} e^{-E/RT} dE,$

where E = energy per mole.

$$\therefore \quad \frac{N_{>E}}{N} = \int_E^\infty 2\pi\left(\frac{1}{\pi RT}\right)^{3/2} E^{1/2} e^{-E/RT} dE.$$

After integrating, we have

$$\frac{N_{>E}}{N} = 2\left(\frac{E}{\pi RT}\right)^{1/2} e^{-E/RT}$$

$$= 2\left(\frac{50RT}{\pi RT}\right)^{1/2} e^{-50RT/RT} = \mathbf{1.54 \times 10^{-21}}.$$

89. Mean free path is independent of temperature—comment.

Ans. : Mean free path of a gas $= \lambda = \frac{kT}{\sqrt{2}\pi\sigma^2 P}$.

For a gas of constant volume, $\frac{T}{P}$ remains constant. Hence, λ does not change with change of temperature.

90. Consider the equilibrium,

$A(g) \rightleftharpoons B(g) + C(g)$

At a constant pressure of 1 atm. A dissociates to the extent of 50% at 500 K. Calculate $\Delta G°$ (K.J. mol^{-1}) for the reaction. **[JAM 2008]**

Hints : Initial moles of A = a mol.

At equilibrium, moles of A = $\frac{a}{2}$, moles of B = moles of C = $\frac{a}{2}$.

Total number of moles = $\frac{3a}{2}$.

$$K_P = \frac{\left(\frac{a}{2} \times P \Big/ \frac{3a}{2}\right)^2}{\frac{a}{2} P \Big/ \frac{3a}{2}} = \frac{P}{3} = \frac{\mathbf{1}}{\mathbf{3}}.$$

91. Consider the following redox system

$Q + 2H^+ + 2e \rightleftharpoons QH_2$, $E° = 0.699$ V

Calculate pH of the solution at 298 K, if the redox potential of the system is 0.817 V. **[JAM 2008]**

Hints : $E = E° + 0.059$ pH

$E = 0.817$ V, $E° = 0.699$V, pH = 2.

92. A stream of O_2 molecules at 500 K exits from a pin hole in an oven and strikes a slit that selects the molecules travelling in a specific direction. Given that the pressure outside the oven is 2.5×10^{-7} atm, estimate the maximum distance at which slit must be placed from the pin hole, in order to produce a collimated beam of oxygen. (Radius of $O_2 = 1.8 \times 10^{-10}$ m)

Hints : To get a collimated beam, the maximum distance of the slit from the pin hole should be same as mean free path of O_2 molecules.

$$\lambda = \frac{RT}{\sqrt{2}\pi\sigma^2 PN_A} = 0.4733 \text{ m}.$$

93. A photochemical reaction was carried out using a monochromatic radiation (490 nm) of intensity 100 W. When the sample was irradiated for 30 min, 0.3 mol of the reactant was decomposed. Estimate the quantum efficiency. **[JAM 2008]**

Hints : Energy corresponding to 1 einstein of photon of $\lambda = 490$ nm

$$= \frac{N_A hc}{\lambda} = 244088.7121 \text{ J}.$$

$$\text{Energy absorbed in einstein} = \frac{\frac{50}{100}\times 100\times 30\times 60}{244088.7121} = 0.369.$$

$$\therefore\ \phi = \frac{0.3}{0.369} = 0.813.$$

94. Assuming ideal behaviour calculate $\Delta U°$ at STP for the reaction :

$CH_4(g) + 2O_2(g) \longrightarrow CO_2(g) + 2H_2O$ (l), $\Delta H° = -606.9$ kJ.mol^{-1}.

Hints : $\Delta H° = \Delta U° - 2RT$ (for the reaction)

$\Delta U° = -602.358$ kJ.mol^{-1}

95. Find the eigenvalues of the following matrix, given that 2 is one of the eigenvalues.

$$\begin{vmatrix} 1 & 1 & 1 \\ 1 & 0 & -2 \\ 1 & -1 & 1 \end{vmatrix}$$

Soln. : $$\begin{vmatrix} 1-x & 1 & 1 \\ 1 & 0-x & -2 \\ 1 & -1 & 1-x \end{vmatrix} = 0$$

or, $(1 - x)\{(0 - x)(1 - x) - 2\} + 1\{-2 - (1 - x)\} + 1(-1 + x) = 0$

or, $3x + 2x^2 - 6 - x^3 = 0$ or, $(x - 2)(3 - x^2) = 0$.

$\therefore\ x = 2,\ x = \pm\sqrt{3}$.

The eigenvalues are : $2, +\sqrt{3}, -\sqrt{3}$.

96. At 600 K and 200 bar, a 1 : 3 (molar ratio) mixture of A_2 and B_2 reacts to form an equilibrium mixture containing $x_{AB_3} = 0.60$. Assuming ideal gas behaviour, calculate K_p for the reaction :

$A_2(g) + 3B_2(g) \rightleftharpoons 2AB_3(g)$ **[JAM 2007]**

Soln. : Given : $x_{AB_3} = 0.6$.

Thus, $x_{A_2} + x_{B_2} = 0.4$.

Now by the problem, $n_{A_2} : n_{B_2} = 1 : 3$.

Let $n_{A_2} + n_{B_2} + n_{AB_3} = y$.

Thus, $n_{A_2} + n_{B_2} = 0.4\,y$,

Hence, $n_{A_2} = 0.4y \times \frac{1}{4} = 0.1y$, $n_{B_2} = 0.4y \times \frac{3}{4} = 0.3y$.

$\therefore\ X_{A_2} = \frac{n_{A_2}}{y} = 0.1$, and $X_{B_2} = \frac{n_{B_2}}{y} = 0.3$.

Now, $K_p = K_x\,(P)^{\Delta\gamma}$

Here, $\Delta\gamma = -2$

$$\text{Thus, } K_p = \frac{0.6^2}{0.1\times0.33\times\left(\frac{200}{1.01325}\right)^2} = \mathbf{3.42 \times 10^{-3}}.$$

97. A 50 mL 0.05 M solution of Fe(II) is titrated with 0.05M solution of Ce(IV) in the presence of dilute H_2SO_4 at 25 °C. Calculate the equivalence point potential and the equation constant K in terms of log K. [$E^o_{(Fe^{+3}|Fe^{+2})} = +0.75$ V, $E^o_{(Ce^{+4}|Ce^{+3})}$ = 1.45 V] **[JAM 2007]**

Hints : $\Delta G° = -RT \ln K = -nFE° = -2.303\ RT \log K$

$E° = 0.71$ V

$\log K = 11.83$.

98. The vapour pressure of D_2O at 20 °C is 745 mm Hg when 15 gm of a non-volatile compound is dissolved in 200 gm of D_2O, the pressure changes to 730 mm of Hg. Assuming applicability of Rault's law, calculate the molecular weight of the compound. **[JAM 2007]**

Hints : $P = P^o X_{D_2O}$

Here $P = 730$ mm of Hg, $P^o = 745$ mm of Hg and $n_{D_2O} = \frac{200}{200} = 10$.

$$X_{D_2O} = \frac{n_{D_2O}}{n_{D_2O} + n_2} \quad \text{and } W_2 = 15 \text{ gm.}$$

Hence, if M_2 is molecular weight of unknown compound, M_2 = 73gm.

99. The reaction

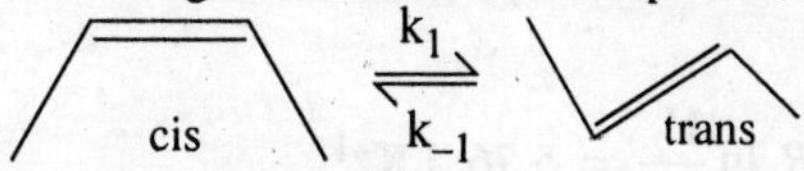

is first order in both directions. At 25 °C, the equilibrium constant (K) of this reaction is 0.4. If 0.115 mol. dm^{-3} of cis isomer is allowed to equilibrate, Calculate the equilibrium concentration of each isomer.

Soln. : Hints : Let at equilibrium concentration of trans-isomer = x mol.dm^{-3}

Then concentration of cis-isomer = $(0.115 - x)$ mol-dm^{-3}.

$$\therefore K = \frac{x}{0.115 - x} = 0.4, \quad \therefore x = 0.033 \text{ mol.dm}^{-3}.$$

100. Consider the 4π electrons in cyclobutadiene to be free particles in a 2-dimensional square box of length 2 Å. Calculate the wavelength of the electronic transition from the HOMO to LUMO. Write down the normalized wave functions for the occupied degenerate states. **[JAM 2007]**

Soln.: For 2 dimensional square box, energy of a state = $E_{n_x n_y} = \left(n_x^2 + n_y^2\right)\dfrac{h^2}{8ml^2}$.

For cyclobutadience, $E_{gr} = E_{11} \times 2 + E_{12} \times 1 + E_{21} \times 1$

$$= \frac{2h^2}{8ml^2} \times 2 + \frac{5h^2}{8ml^2} \times 1 + \frac{5h^2}{8ml^2} \times 1 = \frac{14h^2}{8ml^2}.$$

$$E_{ex} = E_{11} \times 2 + E_{12} \text{ (or } E_{21}) \times 1 + E_{22} \times 1$$

$$= \frac{4h^2}{8ml^2} + \frac{5h^2}{8ml^2} + \frac{8h^2}{8ml^2} = \frac{17h^2}{8ml^2}.$$

$$\Delta E = E_{ex} - E_{gr} = \frac{3h^2}{8ml^2}.$$

Now, $\dfrac{3h^2}{8ml^2} = h\dfrac{c}{\lambda}$.

Here l = 2 Å.

$$\text{Thus, } \frac{3 \times 6.626 \times 10^{-34} \text{ J.sec}}{8 \times (9.109 \times 10^{-34} \text{ kg}) \times (2 \times 10^{-10})^2 \text{ m}^2} = \frac{2.9979 \times 10^8 \text{ m.sec}^{-1}}{\lambda}.$$

$$\therefore \lambda = \frac{2.9979 \times 10^8 \text{ m.sec}^{-1} \times 8 \times (9.109 \times 10^{-31}) \text{kg} \times (2 \times 10^{-10})^2 \text{ m}^2}{3 \times 6.626 \times 10^{-34} \text{ J.sec}}$$

$$= 4.396 \times 10^{-8} \text{m} = 43.96 \text{ nm}.$$

$$\psi_{12} = \frac{2}{l} \sin\frac{\pi x}{l} \sin\frac{2\pi y}{l} \quad (\text{here } l = 2 \text{ Å})$$

$$\psi_{21} = \frac{2}{l} \sin\frac{2\pi x}{l} \sin\frac{\pi y}{l}.$$

101. When an ideal monoatomic gas is expanded from 1.5 bar, 24.8 L and 298 K into an evacuated container, the final volume becomes 49.6 L. Calculate ΔH, ΔS and ΔG for the process.

Hints : This process is isothermal for ideal gas.

$\Delta H = 0$

$$\Delta S = nR \ln \frac{V_2}{V_1} = 5.76 \text{ J.K}^{-1}$$

$\Delta G = -T\Delta S = -1.72$ kJ.

102. The reaction $N_2 + 3H_2 = 2NH_3$ is carried out at 300 K by mixing N_2 and H_2. The standard free energy of formation of NH_3 is –16.4 kJ/mol. After one hour of mixing, the particle pressures of N_2, H_2 and NH_3 are 50 bars, 2 bars and 200 bars respectively. What is the reaction free energy at this stage of the reaction?

Hints : $\Delta G = \Delta G° + RT \ln Q_P$

$$\Delta G° = 2\Delta G^{\circ}_{f,NH_3} - 3\Delta G^{\circ}_{f,H_2} - \Delta G^{\circ}_{f,N_2}$$

$$= -16.4 \times 2 = -32.8 \text{ kJ.}$$

Thus, $\Delta G = -32.8 \times 10^3 + 8.314 \times 300 \ln \frac{(200)^2}{50 \times 2^3} = -21.3$ kJ.

103. One mole of a salt of type MX is dissolved in 1.00 kg of water. The freezing point of the solution is –2.4 °C. Calculate the per cent dissociation of the salt in water.

Hints : $\Delta T_f = ik_f m$, $m = 1$ molal

$$\alpha = \frac{i-1}{\gamma - 1}$$

The per cent dissociation = 29%

104. At 20 °C, the vapour pressure of two pure liquids X and Y which form an ideal solution are 70 torr and 20 torr respectively. If the mole fraction of X in solution is 0.5, find the mole fractions of X and Y in the vapour phase in equilibrium.

Hints : $P = P^{\circ}_X X_S + P^{\circ}_Y Y_S$

Again, $P = P_x + P_y = PX_{vap} + PY_{vap}$.

Given, $P^{\circ}_X = 70$ torr, $P^{\circ}_Y = 20$ torr, $X_S = 0.5$.

Thus, $X_{vap} = 0.78$, $Y_{vap} = 0.22$.

105. The equilibrium constant for the reaction

$$Fe_3O_4 \text{ (s)} + CO \text{ (g)} \rightleftharpoons 3FeO \text{ (s)} + CO_2 \text{ (g)}$$

at 600 °C is 1.00. If a mixture initially consisting of 1 mole of Fe_3O_4, 2 moles of CO, 0.5 mole of FeO and 0.3 mole of CO_2 is heated to 600 °C at constant total pressure of 5 atmospheres, how many moles of each substance would be present at equilibrium? **[JAM 2006]**

Hints : Fe_3O_4 (s) + CO (g) $\rightleftharpoons$ 3FeO (s) + CO_2 (g)

$(a - x) \quad (b - x) \quad (c + 3x) \quad (d + x)$

Let a, b, c and d be initial moles of Fe_3O_4, CO, FeO and CO_2 respectively and x moles of each reactant are converted to products. Since Fe_3O_4 and FeO are solid,

$$K_P = \frac{P_{CO_2}}{P_{CO}} = \frac{X_{CO_2}P}{X_{CO}P} = \frac{X_{CO_2}}{X_{CO}} = K_x,$$

where P is the total pressure.

Now at equilibrium total number of moles = $b - x + x + d = b + d$.

Hence, $X_{CO_2} = \frac{d+x}{b+d}$ and $X_{CO} = \frac{b-x}{b+d}$.

Given that $K_P = 1$, $b = 2$ moles, $d = 0.3$ moles. Thus, $1 = \frac{0.3+x}{2-x}$,

$\therefore$ $x = 0.85$ moles.

Hence at equilibrium,

$n_{Fe_3O_4}$ = (1 – 0.85) moles = 0.15 moles

n_{CO} = 2 – 0.85 = 1.15 moles

n_{FeO} = (0.5 + 0.85) moles = 1.35 moles

n_{CO_2} = (0.3 + 0.85) moles = 1.15 moles.

106. The $Zn^{2+}|Zn$ half cell (E° = –0.762 V) is connected to a $Cu^{2+}|Cu$ half cell (E° = 0.340 V). What is the value of for spontaneous conversion of chemical energy to electrical energy? What is the value of $\log_{10} K$, where K is the equilibrium constant? Use 2.303 RT/F = 0.06 **[JAM 2009]**

Hints : Pt, Zn | Zn^{+2} || Cu^{+2} | Cu, Pt

$$E^\circ_{cell} = E^\circ_{Cu^{+2}/Cu} - E^\circ_{Zn^{+2}/Zn}$$

$$\Delta G^\circ = -nFE^\circ_{cell} = -RT \ln K = -RT\ 2.303 \log K$$

$$\log K = \frac{F}{2.303RT} = \frac{1}{0.06}.$$

107. The limiting ionic molar conductivities of K^{+1} and Cl^- are 73.5 and 76.5 S cm^2 mol^{-1} respectively. If the molar conductivity of 0.1 M KCl solution is 130.0 S cm^2 mol^{-1}, calculate the Kohlrausch's constant for KCl solution. **[JAM 2006]**

Sol. : Hints : $\Lambda = \Lambda^\circ - k\sqrt{c}$

k = kohlrausch's constant

$$\Lambda^\circ = \Lambda^\circ_{K^+} + \Lambda^\circ_{Cl^-}$$

$c = 0.1$, Thus, k = **63.245 S cm² mol**$^{-3/2}$.

108. (a) Given that $\Delta G = -nFE$, derive the expression for the temperature dependence of the cell potential (E) in terms of the change in entropy (ΔS).

(b) For a cell reaction, E (at 25 °C) = 1.26 V, $n = 2$ and $\Delta S = -96.5$ JK^{-1}mol^{-1}. Calculate E at 85 °C.

Hints : ΔG (25 °C) = $-nFE$

$$\left(\frac{\partial \Delta G}{\partial T}\right)_P = -\Delta S$$

ΔG (85 °C) = ΔG(25 °C) – ΔS.60 = $-nFE$ (85 °C)

E (85 °C) = **1.23 volt.**

109. The rate constant k for a second order reaction $P + Q \longrightarrow$ Products is expressed as $\log K = 20 - \dfrac{3000}{T}$, where the concentration is in mol.lit^{-1}, T is in absolute temperature and time is in minutes. The initial concentrations of both the reactants are 0.5 M. Calculate the activation energy and half-life of the reaction at 27 °C (R = 2 cal K^{-1} mol^{-1}) **[JAM 2006]**

Hints : $K = Ae^{-E_a/RT}$

$$\log K = \log A - \frac{E_a}{2.303RT}$$

Given that $\dfrac{E_a}{2.303R} = 3000.$

$\therefore \; E_a = 13818 \text{ cal mol}^{-1}, \quad k = 10^{10} \text{ mol}^{-1}.\text{min}^{-1}.$

Thus, $t_{\frac{1}{2}} = \dfrac{1}{Ka} = 2 \times 10^{-9} \text{ min}^{-1}$

110. (a) A container is partitioned into two compartments, one of which contains 2 moles of He while the other contains 3 moles of Ar. The gases are ideal. The temperature is 300 K and the pressure is 1 bar.

$R = 0.88\ l$ bar mol^{-1} K^{-1}, ln (2/5) = –0.92, ln (3/5) = –0.51.

(i) What is the total Gibbs free energy of the two gases?

(ii) If the partition between the two compartments is removed and the gases are allowed to mix, then what is the Gibbs free energy of the mixture?

(iii) What is the change in enthalpy in this process?

(b) Obtain (i) the molar heat of formation of CH_4(g) and (ii) the average C—H bond energy, to the nearest kilojoule (kJ), from the given data :

	Reaction	ΔH (kJ mol^{-1})
(1)	$CH_4(g) \longrightarrow CH_3(g) + H(g)$	435
(2)	$CH_3(g) \longrightarrow CH_2(g) + H(g)$	444
(3)	$CH_2(g) \longrightarrow CH(g) + H(g)$	444
(4)	$CH(g) \longrightarrow C(g) + H(g)$	339
(5)	$C(\text{graphite}) \longrightarrow C(g)$	717
(6)	$H_2(g) \longrightarrow 2H(g)$	436

[JAM 2009]

Hints : (a) (i) $G = G^o + nRT \ln P$

$P^o = 1$ bar, $P = 1$ bar

$G = G^o.$

(ii) $G_{He} = G^o_{He} + 2RT \ln \dfrac{2}{5} P, \quad G_{Ar} = G^o_{Ar} + 3RT \ln \dfrac{3}{5} P$

$P = 1$ bar

$G_{He} + G_{Ar} = G^o_{He} + G^o_{Ar} - 83.93$ l-bar.

(iii) $\Delta H_{mix} = 0.$

(b) (i) C(graphite) + $2H_2(g)$ = $-CH_4$ $\Delta H_f = -73$ kJ.mole^{-1}.

(ii) Average C—H bond energy = $\frac{73}{4} \approx 18$ kJ.mole^{-1}.

Objective Type Questions

Part A

[*Four choices are given under each of the following questions. Choose the correct answer in each case. Write your answers as* (a), (b), (c) or (d).]

1. The resultant of two vectors of magnitude μ making an angle of θ with respect to each other is :

(a) $2\mu \cos\theta$ (b) $2\mu \cos\theta/2$ (c) $2\mu \sin\theta$ (d) $2\mu \cos 2\theta$.

2. For $A \rightarrow B$, ΔH is given by

(a) $\int C_P dT$ (b) dC_p/dT (c) $C_p \times T$

(d) C_p/T.

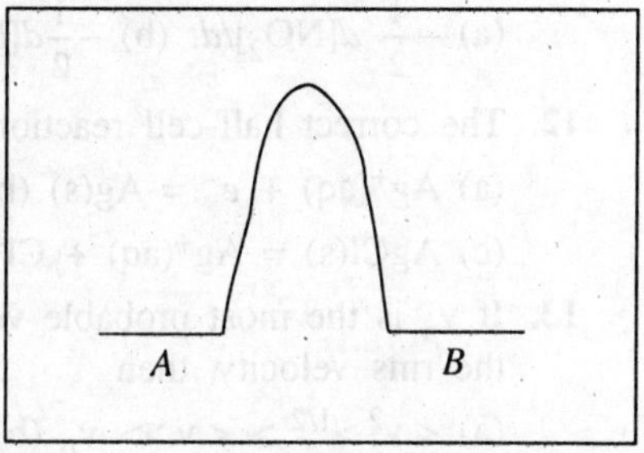

Temperature (*T*)

3. (3N/2)! can be evaluated only when the number N is :

(a) odd (b) prime (c) even (d) without restriction.

4. For 300 MHz, the corresponding wavelength is :

(a) 1 m (b) 2 m (c) 3 m (d) 0.5 m.

5. For an ideal gas undergoing free expansion under adiabatic conditions,

(a) $\Delta U < 0$, $\Delta T < 0$ (b) $\Delta U < 0$, $\Delta T > 0$,

(c) $\Delta U > 0$, $\Delta T > 0$ (d) $\Delta U = 0$, $\Delta T = 0$.

6. For the process $A \rightarrow B$,

(a) $\Delta U = U_B - U_A$ (b) $Q_{total} = Q_B - Q_A$ (c) $W_{total} = W_B - W_A$

(d) $\Delta U = \Delta H + P\Delta V$.

7. The potential $V(r_A, r_B, R) = (e^2/4\pi\varepsilon_o)\,[1/R - 1/r_A - 1/r_B]$, where R is the nucleus-nucleus distance and r_A and r_B are the nucleus-electron distances. The term(s)

(a) $1/R$ is attractive, $1/r_A$ and $1/r_B$ are repulsive

(b) $1/R$ is repulsive, $1/r_A$ and $1/r_B$ are attractive

(c) $1/R$ is attractive, $1/r_A$ and $1/r_B$ are attractive

(d) $1/R$ is repulsive, $1/r_A$ and $1/r_B$ are repulsive.

8. The number of vibrational degrees of freedom in water molecule is

(a) 1 (b) 3 (c) 2 (d) 4.

9. The partition coefficient for the distribution of benzoic acid between H_2O and CCl_4 mixture at a given temperature :

(a) increases with increasing ratio of $H_2O : CCl_4$.
(b) decreases with increasing ratio of $H_2O : CCl_4$
(c) increases with increasing concentration of benzoic acid
(d) remains constant.

10. Given that $X + Y \Leftrightarrow A + B$ (with forward and backward rate constants of k_1 and k_2) and A $\Leftrightarrow C + D$ (with forward and backward rate constants of k_3 and k_4). The equilibrium constant for $X + Y \Leftrightarrow B + C + D$ is
(a) $(k_1/k_2) - (k_3/k_4)$ (b) $(k_1/k_2) \times (k_4/k_3)$ (c) $(k_1/k_2) \times (k_3/k_4)$
(d) $(k_1-k_2)/ (k_3-k_4)$.

11. The rate of the following reaction $2NO(g) + O_2(g) = 2NO_2(g)$ is
(a) $-\frac{1}{2}\, d[NO_2]/dt$ (b) $-\frac{1}{2} d[NO]/dt$ (c) $d[NO]/dt$ (d) $2d[NO]/dt$.

12. The correct half-cell reaction for the electrode Cl^-, AgCl/Ag is :
(a) $Ag^+(aq) + e^- = Ag(s)$ (b) $AgCl(s) + e^- = Ag(s) + Cl^-$ (aq)
(c) $AgCl(s) = Ag^+(aq) + Cl^-(aq)$ (d) $Ag^+(aq) + Cl^-(aq) = AgCl(s)$.

13. If v_p is the most probable velocity, $< v >$ is the mean velocity, and $< v^2 >^{1/2}$ is the rms velocity, then
(a) $< v^2 >^{1/2} > < v >> v_p$ (b) $< v >> < v^2 >^{1/2} > v_p$
(c) $v_p < < v^2 >^{1/2} > < v >$ (d) $v_p = <v^2>^{1/2} = <v>$.

14. The equivalent conductance of 0.1 mol.l^{-1} acetic acid is 39 ohm^{-1}.cm^2.equiv^{-1}. The equivalent conductance of acetic acid at infinite dilution is 390 ohm^{-1}.cm^2. equiv^{-1}. The degree of ionization of 0.1 mol.l^{-1} acetic acid solution is :
(a) 0.2 (b) 0.1 (c) 0.01 (d) 0.5.

15. The amount of electricity required to deposit one equivalent of silver ion from 0.1 molar silver nitrate solution in coulomb.mol^{-1} is :
(a) 96485 (b) 48243 (c) 192270 (d) 9648.

16. The correct quantum numbers for an electron in the $3p_z$ orbital are :
(a) $n = 3,\ l = 1,\ m_l = 0,\ m_s = +1/2$ (b) $n = 3,\ l = 2,\ m_l = 1,\ m_s = -1/2$
(c) $n = 3,\ l = 1,\ m_l = -1,\ m_s = +1/2$ (d) $n = 3,\ l = 0,\ m_l = 1,\ m_s = -1/2$.

17. The entropy of vaporization of toluene is 90 J.K^{-1}.mol^{-1} at its boiling point of 110 °C. The heat of vaporization in kJ.mole^{-1} is :
(a) 9.9 (b) 34.47 (c) 19.8 (d) 0.001.

18. For the reaction,

$$CO(g) + 2H_2(g) \longrightarrow CH_3OH(l)$$

The free energy of formation of CO(g) and $CH_3OH(l)$ are −137 kJ.mol^{-1} and −166 kJ.mol^{-1} respectively at 298 K. This reaction is :
(a) spontaneous as written (b) non-spontaneous as written
(c) spontaneous in the reverse direction (d) spontaneity cannot be predicted.

19. The distribution of probability density calculated from Maxwell-Boltzmann statistics in an arbitrarily chosen direction spreads with
(a) increasing temperature (b) decreasing temperature
(c) temperature independent (d) increasing velocity.

20. The reaction A → B is spontaneous at :

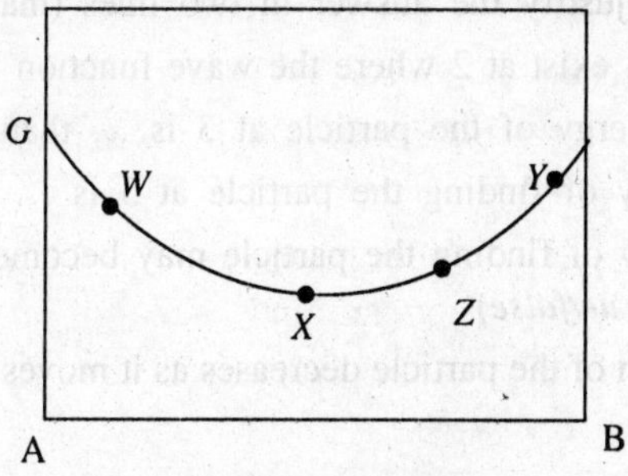

Extent of reaction

(a) W (b) X (c) Y (d) Z.

Part B

[*Answer to the point*]

21. For the reaction $PCl_5(g) = PCl_3(g) + Cl_2(g)$:

(A) Calculate the equilibrium constant using the following data at 25 °C.

Substance	**ΔH° (kJ.mol^{-1})**	**S° (J.K^{-1}.mol^{-1})**
$PCl_3(g)$	–287	312
$PCl_5(g)$	–375	365
$Cl_2(g)$	—	223

(B) Using the result from above, calculate the equilibrium constant of $2PCl_5(g) = 2PCl_3(g) + 2Cl_2(g)$.

22. The observed freezing point depression for an aqueous acetic acid solution of molality = 10^{-3} mol.kg^{-1} is 2×10^{-3} K. If $K_f(H_2O) = 1.86$ K.mol^{-1}.kg, (i) calculate the depression in freezing point, (ii) find the degree of ionization of acetic acid.

23. Starting with the definition of G, show that for a closed system in the absence of non-expansion work,

$V = (\partial G/\partial P)_T$, $S = -(\partial G/\partial T)_P$.

24. Calculate the work done by 2 mol of a monoatomic perfect gas initially at 25 °C, undergoing the following transformations—

(i) isothermal reversible expansion from 10 L to 40 L.

(ii) adiabatic reversible expansion from 10 L to 40 L.

25. (i) Draw the phase diagram of water up to 2 atm and 110 °C.

(ii) Based upon the variation of molar Gibbs function (or chemical potential) with temperature, explain why the melting point of ice decreases with increasing pressure.

26. Calculate the K_{sp} of Hg_2Cl_2; $E°(Hg_2^{\ 2+}/Hg) = 0.79$ V; $E°(Cl^-, Hg_2Cl_2/Hg,) = 0.27$ V.

27. (i) For the reaction $A \rightarrow B$, sketch the variation of the concentration of A with time when (a) it is a zero order, and (b) first order reaction with respect to A.

(ii) Estimate the fraction of molecules with kinetic energy greater than or equal to 100 kJ.mol^{-1} at 300 K.

(iii) Explain in maximum two lines the meaning of frequency factor in the Arrhenius equation.

28. A particle is represented in space with the following wave function. Choose the correct option and justify the answer in two lines (maximum).

(a) The particle ... exist at 2 where the wave function is negative. (*can/cannot*)

(b) The kinetic energy of the particle at 3 is ... than that at 1. (*more/less*)

(c) The probability of finding the particle at 5 is ... than at 1. (*more/less*)

(d) The probability of finding the particle may become zero as it moves along X-direction. (*true/false*)

(e) The momentum of the particle decreases as it moves along X-direction. (*true/false*)

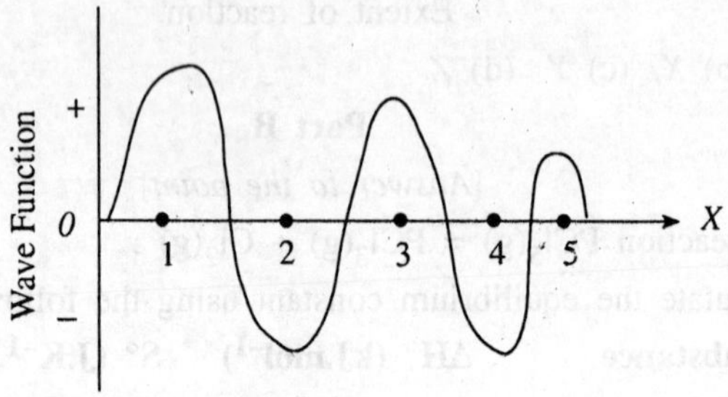

Part C

[*Four choices are given under each of the following questions. Choose the correct answer in each case. Write your answers as (a), (b), (c) or (d).*]

29. Which of the following 0.1 molar solutions has the lowest freezing point?

(a) urea (b) glucose (c) KCl (d) Na_2SO_4.

30. Which of the following gas molecules has the longest free path under the same conditions of temperature and pressure?

(a) O_2 (b) H_2 (c) N_2 (d) Cl_2.

31. An element retains one-tenth of its original radioactivity after 2.303 time units. The half-life in the same unit is :

(a) 2.303 (b) 0.0693 (c) 0.693 (d) 0.2303.

32. The trajectory of a particle in Brownian motion is :

(a) linear (b) zigzag (c) oscillatory (d) curved.

33. The difference, $(\Delta H - \Delta U)$, in the heats of combustion of glucose at constant pressure and constant volume at 25 °C is :

(a) zero (b) $-12RT$ (c) $12RT$ (d) $6RT$.

34. The molar heat capacity at constant volume is independent of temperature for (a) hydrogen gas (b) water vapour (c) helium gas (d) all gases.

35. Identify the exact differential :

(a) dq_{rev} (b) dq_{rev}/T (c) dq_{irrev} (d) dq_{irrev}/T.

36. The acid-catalyzed hydrolysis of an ester is found to obey a pseudo first order kinetics. The corresponding rate constant is independent of :

(a) the temperature

(b) the acid concentration

(c) the ester concentration

(d) none of the above.

37. If the specific rate constant k is to be equal to the pre-exponential factor A in the Arrhenius equation, then

(a) the reaction temperature is very low

(b) the reaction is a catalysed one

(c) the reaction is molecular in nature

(d) the reaction does not need any activation energy.

38. According to the equipartition principle, the average vibrational energy of the linear N_2O molecule at 27 °C is approximately :

(a) 7200 $J.mol^{-1}$ (b) 9600 $J.mol^{-1}$ (c) 4800 $J.mol^{-1}$ (d) 6000 $J.mol^{-1}$.

39. The ionic mobility (μ) and ionic conductivity (λ) are related by the equation

(a) $\mu = \lambda/F$ (b) $\mu = \lambda$ (c) $\mu = n\lambda F$ (d) $\mu = \lambda F$.

40. The conductivity in $ohm^{-1}.m^2.mol^{-1}$ of M^+ ion is 70 and that of A^- is 60. The specific conductance of the saturated solution of MA is 1.3×10^{-6} $ohm^{-1}.m^{-1}$. The solubility of the salt in $mol.l^{-1}$ is :

(a) 10^{-3} (b) 10^{-4} (c) 10^{-5} (d) 10^{-11}.

41. The role of the supporting eletrolyte in polarography is to :

(a) maintain the residual current

(b) increase the diffusion current

(c) suppress the diffusion current

(d) limit the maximum current.

42. For the reaction $N_2O_4(g) \rightleftharpoons 2NO_2(g)$, the degree of dissociation at 50 °C and 1 atm pressure is found to be 0.38. When the reaction is carried out in the presence of an inert gas with a partial pressure of 0.25 atm (total pressure is still 1 atm), the degree of dissociation will :

(a) increase (b) decrease (c) remain unaffected (d) drastically decrease.

43. The first ionization constants of acetic acid, chloroacetic acid, carbonic acid and boric acid are 1.75×10^{-5}, 1.38×10^{-3}, 4.69×10^{-11} and 5.79×10^{-11} respectively. To have the maximum buffer action between pH 4 and 5, one should have a mixture of the following acid and its salt.

(a) chloroacetic acid (b) boric acid (c) carbonic acid (d) acetic acid.

44. Which of the following statements is true for the equivalent conductance of an aqueous solution of KCl?

(a) It is independent of concentration.

(b) It inceases linearly with concentration.

(c) It increases linearly with increase in $[\text{concentration}]^{1/2}$.

(d) It decreases with increase in $[\text{concentration}]^{1/2}$.

45. The enthalpy change for a cell reaction is given by :

(a) $nC_v(T_2-T_1)$ (b) $nF[T(\partial E/\partial T)_P - E]$ (c) $[nFE + (\partial E/\partial T)_P]$

(d) $[T(\partial E/\partial T)_P - E]/nF$.

46. After electrolysis, the pH of an aqueous NaCl solution :
(a) increases (b) decreases (c) remains unaffected (d) is unpredictable.

47. A drop of ink put into a glass of water mixes uniformly with the solvent as time progresses. This is due to the :
(a) gravitational force (b) minimization of potential energy
(c) maximization of entropy (d) osmotic pressure of water.

48. The specific heat of a stable solid can be obtained from :
(a) the Gibbs-Helmholtz equation
(b) the Kirchhoff's equation
(c) the first law of thermodynamics
(d) the second law of thermodynamics.

49. For an adiabatic irreversible expansion of a perfect gas, which of the following relations is correct?
(a) $\Delta T = -P_{ext}\,\Delta V/C_v$ (b) $\Delta T = -P_{ext}\,\Delta V/C_p$
(c) $\Delta T = e^{-nR\ \ln[V(\text{final})/V(\text{initial})]}$ (d) $\Delta T = 0$.

50. For the reaction $N_2(g) + 3H_2(g) \longrightarrow 2NH_3(g)$, $\Delta H° = -46.11\ kJ.mol^{-1}$ at 25 °C. The equilibrium constant of this reaction will increase with :
(a) increase in temperature (b) increase in pressure (c) equilibrium constant does not change by changing temperature or pressure (d) lowering temperature.

51. The reaction in a given direction is spontaneous if the change in Helmholtz Function, A, is negative at :
(a) constant P and T (b) constant V and T
(c) constant H and S (d) constant T and S.

52. The relationship between a (edge length) and r (atomic radius) for a primitive cubic unit cell is :
(a) $a = 2r$ (b) $(2)^{1/2}\ a = 4r$ (c) $(3)^{1/2}\ a = 4r$ (d) $a = 4r$.

53. Ionic solids, in general, are associated with one of the following defects :
(a) interstitial (b) vacancy (c) substitutional (d) Frenkel.

54. How many mirror planes of symmetry are present in CH_3Cl?
(a) none (b) 1 (c) 2 (d) 3.

55. Absorption of molecules in microwave region of the electromagnetic spectrum is due to :
(a) elecronic transitions (b) vibrational transitions
(c) rotational transitions (d) nuclear spin transitions.

56. The attractive part of van der Waals interactions is due to :
(a) ion-dipole interactions (b) dipole-dipole interactions
(c) dipole-induced dipole interactions (d) induced dipole-induced dipole interactions.

57. Number of radial nodes in the $2p_z$ orbital is :
(a) 0 (b) 1 (c) 2 (d) 3.

58. Between two wave functions, the one with larger number of nodes represents the state with :

(a) higher energy (b) lower energy (c) same energy (d) zero energy.

59. An ideal gas kept in a vessel at a temperature T, has :

(a) unequal average velocity along x-, y- and z-directions

(b) zero average veloccity along x-direction

(c) equal non-zero average velocity along x-, y- and z-directions

(d) unequal average speed along x-, y- and z-directions.

60. For a van der Waals gas having constants a and b,

(a) a and b are state functions and state parameters

(b) a and b are state functions but not state parameters

(c) P and T are state functions but not state parameters

(d) P, V and T are all state functions and state parameters.

61. Above 1100 K, iron undergoes a phase transition from a body-centred cubic to another cubic structure accompanied by an increase in density. The new lattice structure is :

(a) simple cubic (b) face centred cubic (c) CsCl type (d) orthorhombic.

62. At the upper critical solution temperature of phenol-water system, at a given pressure the degree of freedom is :

(a) 2 (b) 3 (c) 0 (d) 1.

63. Maximum entropy of mixing is achieved when two ideal gases, X and Y, are mixed in the amounts of :

(a) 1 mole of X and 3 moles of Y (b) 3 moles of X and 1 mole of Y

(c) 2 moles of X and 3 moles of Y (d) 2 moles of X and 2 moles of Y.

64. Two pieces of soft iron have equal mass and are at same temperature and pressure. One of the pieces is placed on a permanent magnet, and the other on a wooden block. The two pieces will have :

(a) same Gibbs free energy (b) different enthalpy

(c) same entropy (d) different entropy.

65. In the reaction, $C_3H_8(g) + 5O_2(g) = 3CO_2(g) + 4H_2O$ (g), 6.6 moles of CO_2 were formed. The extent of reaction, ξ, is :

(a) 2.2 (b) 1.6 (c) 1.3 (d) 6.6.

66. In the relations/statements given below, the correct one is :

(a) $\Delta G = \Delta H + S\Delta T + T\Delta S$ (b) $\Delta S = 0$ for all adiabatic processes.

(c) a solution component at unit activity is in its standard state.

(d) activity and activity coefficient are intensive properties.

67. A liquid solution of two components :

(a) will always freeze entirely at one temperature

(b) will never freeze entirely at one temperature

(c) will have frozen solid phase containing both the components

(d) will never freeze.

68. The molecule whose bond vibration is both IR and Raman active is :
(a) H_2 (b) N_2 (c) HI (d) O_2.

69. Trouton rule states that :
(a) $\Delta H/T_b$ is constant for normal liquids (b) $\Delta H/T_f$ is constant for all liquids (c) $\Delta S > 0$ for spontaneous processes (d) $\Delta G \to 0$ as equilibrium is approached.

70. Two electrodes, $Ag^+ | Ag$, and $Cl^{-1}, AgCl | Ag$ dipped into a solution of aqueous $AgNO_3$ will have :
(a) same E and $E°$ values (b) different E and $E°$ values
(c) same $E°$ value, but different E values (d) different $E°$ values, but same E value.

71. The standard heat of formation at 298 K of :
(a) H_2, D_2 and HD are all equal to zero (b) H_2, D_2, and HD are all non-zero (c) H_2 and D_2 are zero, but that of HD is non-zero (d) HD is zero, but those of H_2 and D_2 are non-zero.

Part D

72. Indicate whether the following statements are true (T) or false (F) :

(a) The molecule-ion O_2^+ is more stable than the neutral molecule O_2.

(b) The species O_2 is diamagnetic.

(c) The $3p_z$ orbital of the hydrogen atom has no radial node.

(d) $\Delta x \, \Delta p_x \geq \dfrac{\hbar}{2}$.

(e) Diamond has a structure of covalent network.

(f) Compared to the infrared transitions, electronic transitions occur at longer wavelength.

(g) H_2 molecule displays a pure rotational spectrum.

(h) If two gases are at the same temperature, their molecules have the same average kinetic energy.

(i) At 600 mm pressure 1 gm of charcoal absorbs about 10 ml of N_2 at 0 °C, 20 ml at –29 °C and 45 ml at –78 °C. These data indicate that adsorption of N_2 on charcoal is accompanied by evolution of heat.

(j) If ΔG_{ads} is zero, then the following adsorption isotherm is obeyed :

$$\theta = \frac{KP}{1+KP}.$$

Part E

73. Choose the correct statement(s).

(a) A hypothetical absolute temperature scale (°A) is defined by assigning 0 °C = 100 °A. The boiling point of water on this scale is :
(i) 100 °A (ii) 136.61 °A (iii) 200 °A.

(b) The following plot for a fixed amount of an ideal gas at constant temperature will not be linear :
(i) P against V (ii) P against $1/V$ (iii) PV against P.

(c) The root mean square speed for a collection of gas molecule at absolute temperature T is :

(i) $(8RT/\pi M)^{1/2}$ (ii) $(3RT/M)^{1/2}$ (iii) $(2RT/M)^{1/2}$.

(d) The height (h) to which a liquid will rise or be depressed in a capillary tube is related to the radius of the tube (r) by :

(i) $h = 2\gamma/\rho rg$ (ii) $h = r\gamma/2\rho g$ (iii) $h = \rho rg/2\gamma$.

(e) The molality of an aqueous solution that has a boiling point elevation of 1 K is :

(i) 0.95 $mol.kg^{-1}$ (ii) 1.95 $mol.kg^{-1}$ (iii) 3.95 $mol.kg^{-1}$

(f) The entropy of a perfect crystal at 0 K is :

(i) > 0 (ii) 0 (iii) < 0.

(g) A change in the free energy is related to the enthalpy change, the entropy change and temperature by :

(i) $\Delta G = \Delta A - T\Delta S$ (ii) $\Delta G = \Delta U - T\Delta S$

(iii) $\Delta G = \Delta H - T\Delta S$ (iv) $\Delta G = \Delta H + T\Delta S$.

(g) In SI units, pressure is expressed in

(i) Pa (ii) bar (iii) atm (iv) torr.

(i) A spontaneous change is always accompanied by

(i) an increase in the entropy of the universe

(ii) a decrease of the entropy of the universe

(iii) an unaltered entropy of the universe.

(j) At equilibrium,

(i) $\Delta G > 0$ (ii) $\Delta G < 0$ (iii) $\Delta G = 0$ (iv) $\Delta G = -RT \ln K$.

Part F

74. State whether the following statements are true (T) or false (F).

(a) In order to minimize liquid junction potentials, one must use a salt bridge containing a salt such that $t_+ = t_-$.

(b) The order of a reaction is always a positive integer.

(c) The standard reduction potential E_1^o of a certain redox couple is greater than E_2^o, the standard reduction potential of another redox couple. Therefore, 1 is a stronger reducing agent than 2.

(d) The dissociation constant of a salt depends upon the solvent used.

(e) The total energy of the universe is variable.

(f) If ΔG has a negative value, the process is not spontaneous.

(g) Entropy of the universe remains constant for a reversible process.

(h) Enthalpy of the universe decreases in an irreversible process.

(i) The change in internal energy is the heat change that takes place at constant volume.

(j) The enthalpy change is the heat change occuring at constant temperature.

Part G

75. Choose the correct statement(s).

(a) An increase in the volume of an ideal gas brings with it :

(i) a decrease in entropy (ii) no change in entropy (iii) an increase in entropy (iv) none of the above.

(b) In homonuclear diatomic bond strength :

(i) increases (ii) decreases (iii) remains the same with increase in bond order.

(c) In heteronuclear diatomic bond length :

(i) increases (ii) decreases (iii) remains the same with increase in bond order.

(d) The probability density for finding the 1*s* electron in a hydrogen atom is maximum at :

(i) $r = 0$ (ii) $r = a_0$ (iii) $r = a_0/2$ (iv) $r = x$.

(e) Compounds of various metals give characteristic colours to a flame. These colours are due to :

(i) nuclear spin transitions (ii) vibrational transitions (iii) electronic transitions.

(f) In fcc metal, the number of atoms per unit cell is

(i) 8 (ii) 14 (iii) 6 (iv) 4.

(g) The rate constants of two reactions, k_1 and k_2, are equal at a certain temperature T. If at a higher temperature T', $k_1' > k_2'$, then

(i) $E_{a.1} < E_{a.2}$ (ii) $E_{a.1} = E_{a.2}$ (iii) $E_{a.1} > E_{a.2}$.

(h) A current of 965 mA flows between two copper electrodes immersed in a solution of $CuSO_4$ in an electrolytic cell for 200 sec. The mass of copper deposited on the cathode is

(i) 0.635 gm (ii) 6.35 gm (iii) 0.0635 gm (iv) 0.00635 gm.

(i) The partition coefficient K of a solute in two immiscible solvents depends on the temperature. This dependence can be represented as :

(i) $K \propto T$ (ii) $K \propto \frac{1}{T}$ (iii) $K \propto T^n$ (iv) $K \propto e^{1/T}$

(v) none of the above.

(j) The unit of rate constant for a zero-order reaction is

(i) sec^{-1} (ii) $mol.litre^{-1}.sec^{-1}$ (iii) $litre.mol^{-1}.sec^{-1}$ (iv) 1.

Part H

76. Choose the most appropriate answer from the alternatives given :

(a) The ionic product of water at 5 °C is 1.9×10^{-13}. The pH of neutral water at this temperature is

(i) 7 (ii) 6.50 (iii) 6.36 (iv) – 6.36.

(b) Cars with catalytic converters must use unleaded fuel because :

(i) it is more fuel efficient (ii) it gives higher mileage (iii) lead poisons the catalyst (iv) lead is harmful for the environment.

(c) The rate of a reaction, $A + B \rightarrow$ products,

remains unchanged when the concentration of A is doubled and B is halved. It is also unchanged when the concentration of A is halved and B is doubled. The rate expression is of the form :
(i) $[A]^2[B]$ (ii) $[A]\ [B]^2$ (iii) $[A]^2[B]^2$ (iv) $[A]^2[B]^{1/2}$.

(d) The ion conductance at infinite dilution at 25 °C of the proton is 349.8 $ohm^{-1}.cm^2$ and the acetate ion is 40.9 ohm^{-1}. Therefore, the equivalent conductance of acetic acid at infinite dilution is :
(i) 390.7 $ohm^{-1}.cm^2$ (ii) 308.9 $ohm^{-1}.cm^2$ (iii) 431.6 $ohm^{-1}.cm^2$ (iv) 268.0 $ohm^{-1}.cm^2$.

(e) If two gases are at the same temperature, their molecules have same :
(i) rms velocity (ii) average speed (iii) average kinetic energy (iv) average potential energy.

(f) The orthorhombic crystal system is :

(i) $a = b = c$	(ii) $a = b \neq c$	(iii) $a \neq b \neq c$	(iv) none of
$\alpha = \beta = \gamma = 90°$	$\alpha = \beta = \gamma = 90°$	$\alpha = \beta = \gamma = 90°$	the above.

(g) A gas is allowed to expand from V_i to V_f against constant external pressure, P_{ex}. The work done by the gas is :
(i) 0 (ii) $P_{ex}(V_f - V_i)$ (iii) $-nRT \ln (V_f / V_i)$ (iv) $nR \ln V_f/V_i$.

(h) Bohr's basic quantum postulate was :
(i) $mvr = nh/2\pi$ (ii) $E = h\nu$ (iii) $c = \nu\lambda$ (iv) all of the above.

(i) A wave function is acceptable, only if it is :
(i) single valued (ii) continuous (iii) square integrable (iv) all of mentioned earlier.

(j) The outermost electron in H_2^- will be in the orbital :
(i) $1\sigma_g$ (ii) $2\sigma_g$ (iii) $1\sigma_u$ (iv) $3\sigma_g$.

Part I

77. State whether the following are True or False :

(a) All real gases can be liquefied at room temperature.

(b) If a strong acid is added to a saturated solution of a salt of a weak acid, more salt will precipitate out.

(c) The equivalent conductance of a strong electrolyte increases with decrease in concentration.

(d) For an elementary reaction the order is same as the molecularity of the reaction.

(e) The extensive use of CFCs (chlorofluorocarbons) is responsible for the Green House Effect.

(f) The number of components in the equilibrium $CaO(s) + CO_2(g) = CaCO_3(s)$ is 2.

(g) The freezing point of water increases with increasing pressure.

(h) The entropy of the system does not change for an adiabatic irreversible change.

(i) If a perfect gas expands against zero external pressure under adiabatic conditions, its temperature falls.

(j) The quantum number l represents the magnitude of the electron's orbital angular momentum.

Part J

Choose the correct answers :

78. Of the following statements, which is not true for an ideal gas?

(a) $(\partial U/\partial V)_T = 0$ (c) $(\partial U/\partial T)_V = 0$

(b) $(\partial U/\partial P)_T = 0$ (d) $(\partial H/\partial P)_T = 0$.

79. The pH of a 10^{-3} M solution of a weak acid HA is 4.0. The dissociation constant of HA is :

(a) 1×10^{-3} (b) 1×10^{-4} (c) 1×10^{-5} (d) 1×10^{-6}.

80. The following reaction is carried out at constant volume and constant temperature :

$$2NO(g) + 2H_2(g) \rightleftharpoons N_2(g) + 2H_2O(g).$$

The equilibrium constant for the above reaction can be calculated by knowing the

(a) initial concentration of all the species and the total final pressure of the system.

(b) initial concentration of all the species.

(c) equilibrium concentration of any three species.

(d) mole fraction of two species at equilibrium and the total final pressure of the system.

81. The criterion for spontaneity and equilibrium of a system is

(a) $(dG)_{TV} \leq 0$ (c) $(dU)_{S,P} \leq 0$

(b) $(dA)_{P,T} \leq 0$ (d) $(dH)_{S,P} \leq 0$.

82. The spontaneous cell reaction and the emf of an electrochemical cell are known at a given temperature. The thermodynamic function (s) which can be determined from this information is/are :

(a) ΔS only (c) ΔH only

(b) ΔG only (d) ΔG, ΔH and ΔS.

83. For one mole of an ideal gas, $(\partial V/\partial P)_T$ is equal to :

(a) $-V/P$ (c) $-RT/P$

(b) RT/P^2 (d) RT/P.

84. The liquid-vapour phase diagram for two miscible liquids A and B is given below. In a mixture of A and B, the mole fraction of A $(X_A) = 0.25$. When the mixture is distilled in a column at 1 atmosphere,

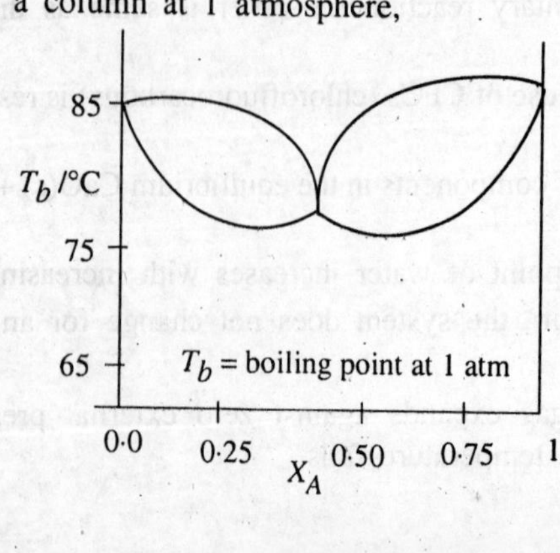

(a) it is possible to separate both A and B in pure form.

(b) it is possible to separate A only in pure form.

(c) it is impossible to separate either A or B in pure form.

(d) it is possible to separate B only in pure form.

85. The concentrations of NaCl, $MgCl_2$ and $FeCl_3$ in a 250 ml aqueous solution are 1 M, 0.5, and 0.0005 M, respectively. The solubility product of AgCl is 2.8×10^{-10}. Half a gram of AgCl is added to the solution. The concentration of Ag^+ in this solution is :

(a) 1.4×10^{-10} M (b) 0.7×10^{-10} M

(c) 2.8×10^{-10} M (d) 1.86×10^{-10} M.

86. The width of the Maxwell-Boltzmann velocity distribution of a gas :

(a) decreases as the temperature increases.

(b) increases as the temperature increases.

(c) remains unchanged with change in temperature.

(d) increases or decreases with temperature depending on the gas.

87. The half-life of a reaction is independent of the initial concentration of the reactants, if :

(a) the reaction is zeroth order, (b) the reaction is first order

(c) the reaction is very fast, (d) the reaction is very slow.

88. In a simple free radical initiated polymerization, the rate of polymerization varies with the initiator concentrations $[I]$ as :

(a) $[I]^0$ (b) $[I]^{1/2}$ (c) $[I]$ (d) $[I]^2$.

89. Which one of the following statements is correct for the equilibrium reaction $H_2(g) + Cl_2(g) \rightleftharpoons 2HCl(g)$ + heat :

(a) increase in temperature shifts the equilibrium to the right

(b) increase in pressure shifts the equilibrium to the left

(c) change in pressure has no noticeable effect on the equilibrium

(d) decrease in pressure shifts the equilibrium to the left.

90. For the following mechanism,

$$NO_2(g) + F_2(g) \xrightarrow[k_1]{} NO_2F(g) + F(g), \text{ slow}$$

$$F(g) + NO_2(g) \xrightarrow[k_2]{} NO_2F(g), \text{ fast}$$

The rate law is given by

(a) rate $= k_1k_2[NO_2]^2$

(b) rate $= k_1[NO_2][F_2]$

(c) rate $= k_2[F][NO_2]$

(d) rate $= k_1[NO_2F]$ $[F]$

91. Four gases, CO_2, SO_2, Cl_2, and H_2O are kept at the same temperature and pressure in separate containers of equal volume. The gas which has the maximum density is

(a) CO_2 (b) SO_2 (c) Cl_2 (d) H_2O.

92. The phase diagram for a pure compound is shown below :

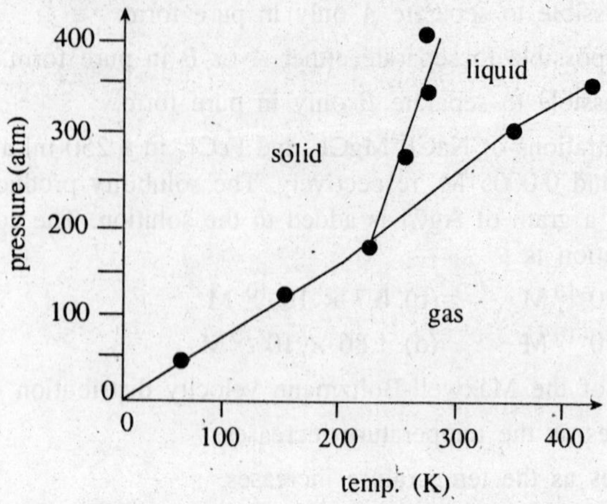

At a temperature greater than 210 K and pressure of 60 atm, the compound exists as :

(a) both solid and gas at equilibrium

(b) vapour

(c) solid

(d) vapour and liquid in equilibrium.

93. The frequency of light of wavelength λ = 500 nm is :

(a) 599.6 Hz (b) 5.996×10^{14} Hz

(c) 59.96×10^{8} Hz (d) 5.926×10^{-14} Hz.

94. A photophysical process which leads to non-radiative decay is called :
(a) fluorescence (b) internal conversion (c) absorption (d) phosphorescence.

95. An allloy of gold and copper crystallizes in a cubic-lattice with gold atoms occupying the corners of the cube and copper atoms the centre of each of the cube faces. The empirical formula of the alloy is :

(a) AuCu (b) Au_2Cu (c) $AuCu_3$ (d) Au_3Cu.

96. Find the radius, r, of the smallest atom that can fit in the centre of an octahedron, made from 6 ligand atoms, each with radius R.

(a) 0.2 R (b) 0.414 R (c) 0.65 R (d) R.

97. The point group symmetry of acetylene is :

(a) C_{2h} (b) C_{2v} (c) $D_{\infty h}$ (d) $C_{\infty v}$.

98. The zero point energy of an electron confined in a 1-dimensional box of length a is :

(a) $h^2/(2\pi^2 ma^2)$ (b) $h^2/(4\pi^2 ma^2)$

(c) $h^2/(8\pi^2 ma^2)$ (d) $h^2/(16\pi^2 ma^2)$.

99. According to the equipartition theorem, the heat capacity of HF at constant volume will be :

(a) $\frac{3R}{2}$ (b) $5R$ (c) $\frac{7R}{2}$ (d) $6R$.

100. The 3d atomic orbital has :

(a) one angular node (b) one radial node

(c) one angular and one radial nodes (d) two angular nodes.

101. The bond length of a homonuclear diatomic molecule can be determined by :

(a) IR spectroscopy (b) Microwave spectroscopy

(c) Raman spectroscopy (d) NMR spectroscopy.

102. What are the coordination numbers of Cs^+ and Cl^- in the caesium chloride structure?

(a) 4 (b) 6 (c) 8 (d) 12.

103. In a diffraction experiment, Bragg reflections are observed for an fcc lattice, only when :

(a) h, k, l values are all odd or all even

(b) h, k, l values are all odd

(c) h, k, l values are all even

(d) $h+k+l$ values are even.

104. According to band theory of solids, conduction occurs in metals because :

(a) valence band is full

(b) band gap has a finite value

(c) the valence band and conduction band overlap

(d) transition occurs from valence band to conduction band

105. The zero point vibrational energy for the deuterium (D_2) molecule is :

(a) lower than that for H_2 (b) higher than that for H_2

(c) equal to that for H_2 (d) equal to zero.

106. The vapour pressure of an aqueous solution of KNO_3 at 100 °C is 750 torr. The activity of water then is :

(a) 0.987 (b) 1.01 (c) 1.00 (d) 0.

107. The maximum work that may be obtained from a chemical reaction is given by the change in the :

(a) enthalpy (b) entropy

(c) Gibbs free-energy (d) internal energy.

108. In the ^1H NMR spectrum, a compound of molecular formula C_4H_6 exhibited only a singlet at δ 1.8 ppm. The compound is :

(a) 1-butyne (b) 2-butyne (c) 1,3-butadiene (d) cyclobutene.

109. The point group of $[PtCl_4]^{2-}$ is :

(a) C_s (b) C_{2v} (c) D_{2d} (d) D_{4h}.

110. By the uncertainty principle, the uncertainty in position (Δs) and the uncertainty in momentum (Δp) of a particle are related as (h = Planck constant) :

(a) $\Delta s.\Delta p \leq h / 4\pi$ (b) $\Delta s.\Delta p \geq h / 4\pi$ (c) $\Delta s.\Delta p = 0$ (d) $\Delta s.\Delta p > 1$.

111. With increasing temperature, the equilibrium constant of a reversible reaction tends towards :

(a) 0 (b) 1 (c) –1 (d) ∞.

112. The absorption of infrared radiation by a molecule is accompanied by a change in :
(a) the zero point energy (b) the dipole moment
(c) the nuclear spin (d) the electronic state.

113. A chemical reaction that is acompanied by a decrease in entropy occurs :
(a) not at all (b) if the enthalpy change is sufficiently negative
(c) only at 0 K (d) if the enthalpy change is sufficiently positive.

114. In the Schrödinger equation, $H\Psi = E\Psi$, Ψ represents :
(a) wavelength (b) wave function (c) probability density (d) potential energy.

115. Suppose, the transformation of A to B occurs by both a reversible first-order reaction and a reversible second-order reaction involving H^+ ion.

$$A \underset{k_2}{\overset{k_1}{\rightleftharpoons}} B \qquad A + H^+ \underset{k_4}{\overset{k_3}{\rightleftharpoons}} B + H^+$$

The relationship between these four rate constants is :
(a) $k_1k_2 = k_3k_4$ (b) $k_1k_3 = k_2k_4$ (c) $k_1k_4 = k_2k_3$.

116. A chemical reaction will be favoured by :
(a) negative values of $\Delta H°$ (b) negative values of $\Delta G°$
(c) positive values of $\Delta S°$ (d) all of the three.

117. If the heats of reaction of the reactions :

$C + O_2(g) = CO_2(g)$, $C + \frac{1}{2}O_2(g) = CO(g)$ and $CO(g) + \frac{1}{2}O_2(g) = CO_2(g)$ are Q, Q_1 and Q_2, respectively, then :
(a) $Q = Q_1Q_2$ (b) $Q = Q_1 + Q_2$ (c) $Q = Q_1 - Q_2$.

118. Same quantity of heat is supplied to three reversible engines working between heat reservoirs with the source temperatures of T_1, T_2, T_3, respectively, and the sink temperature of T K. If $T_1 > T_2 > T_3$, then the maximum work available will be in the order :
(a) $W_1 > W_2 > W_3$ (b) $W_3 > W_2 > W_1$ (c) $W_1 > W_2 < W_3$.

119. The distance between 110 planes of a simple cube crystal with a lattice side of 1.5 Å is :
(a) 1.5 Å (b) 0.75 Å (c) 1.06 Å (d) 0.53 Å.

120. The half-life is independent of initial concentration of the reactant :
(a) only for a zero-order (b) only for a first-order
(c) for both zero and first-orders reaction.

121. Radiative transition from a higher excited singlet electronic state to a lower singlet electronic state is called :
(a) fluorescence (b) phosphorescence (c) internal conversion.

122. According to the equipartition principle the heat capacity at constant volume (C_v) for CO_2 gas is :
(a) $6.5R$ (b) $8R$ (c) $9R$ (d) $8.5R$.

123. The pair that shows critical solution temperature in a composition, temperature phase diagram is :
(a) phenol-water (b) phenol-benzene (c) chloroform-water (d) methanol-water.

124. Under the equilibrium conditions for the reaction $C(s) + CO_2(g) \rightleftharpoons 2CO(g)$, the total pressure is 12 atm. The value of K_p is :

(a) 16 (b) 0.5 (c) 2 (d) 32.

125. The crystal structure which does not have any axis of rotational symmetry is :

(a) triclinic (b) orthorhombic (c) hexagonal (d) cubic.

126. Among the following statements the one that is NOT true for a zero-order reaction is :

(a) $t_{1/2}$ is directly propotional to the initial concentration of the reactant

(b) the time taken for 75% reaction is 1.5 times that of $t_{1/2}$.

(c) the rate constant has the dimension of concentration·time^{-1}.

(d) the concentration of the reactant decreases exponentially with time.

127. The ratio of the most probable velocity of hydrogen to oxygen gas is :

(a) 2.5 (b) 4 (c) 4.5 (d) 16.

128. For an irreversible adiabatic expansion of a perfect gas from volume V_i to V_f the change in entropy of the gas is :

(a) nR ln (V_f/V_i) (b) zero (c) less than zero (d) greater than zero.

129. The latent heat of vaporization of bromine (l) at 59 °C at l atm is 29.76 kJ/mole. The ΔU of this process is :

(a) 127 kJ (b) 29 kJ (c) 27.5 kJ (d) 27 kJ.

130. A solution containing 2.423 gm of a sulphur in 100 gm of naphthalene (m.p. 80.1 °C) gave a freezing point depression of 0.64 °C. The latent heat of fusion of naphthalene is 35.7 cal/gm. The molecular formula of sulphur in the solution is :

(a) S_4 (b) S_2 (c) S_6 (d) S_8.

131. In comparison to the frequency of the IR transition the NMR transition frequency is :

(a) much higher (b) much lower (c) almost same (d) none of these.

132. The microwave active molecule is :

(a) CO_2 (b) HCl (c) p-dinitrobenzene (d) acetylene.

133. As per the uncertainty principle, $\Delta x.\Delta p_y$ =

(a) h (b) zero (c) λ (d) $h/2\pi$.

134. The dipole moment of toluene is m_1 debye. The dipole moment of o-xylene will be :

(a) $3\sqrt{3}m_1$ (b) $\sqrt{3}m_1$ (c) m_1 (d) 0.

135. The non-radiative transition taking place between S_1 and T_1 states is known as :

(a) internal conversion (IC) (b) intersystem crossing (ISC) (c) fluorescence (d) phosphorescence.

136. For a linear molecule of n atoms in two dimensional space, the number of vibrational degrees of freedom is :

(a) $2n - 6$ (b) $2n - 5$ (c) $2n - 3$ (d) $2n - 4$.

137. For the elementary gas phase reaction $CO + NO_2 \longrightarrow CO_2 + NO$, the activation energy is found to be 116 kJ.mol^{-1}. The enthalpies of formation of CO, NO_2, CO_2

and NO are –110 kJ.mol^{-1}, +33 kJ.mol^{-1}, –394 kJ.mol^{-1} and +90 kJ.mol^{-1}, respectively.

The activation energy in kJ.mol^{-1} for the reverse reaction is :

(a) 343 (b) –227 (c) 227 (d) 116.

138. The heat of a chemical reaction in solution may be determined by :

(a) infrared spectroscopy (b) differential thermal analysis

(c) X-ray analysis (d) electron energy loss spectroscopy.

139. The term 'symbol' for the first excited state of the helium atom is :

(a) $^1S_0$ (b) $^3S_1$ (c) $^3P_0$ (d) $^1P_1$.

140. The following diagram shows the heating curve for one mole of a substance that melts at 250 K and boils at 600 K. The horizontal portion of the graph marked 'X' represents :

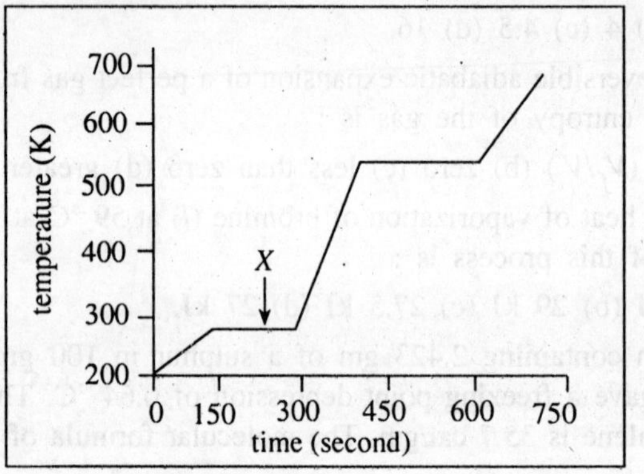

(a) the heat required to bring the substance to its melting point.

(b) the melting of the substance.

(c) the enthalpy of vaporization of the substance.

(d) the heat required to boil the substance.

141. A constant current 200 mA is passed for one minute through a copper coulometer to facilitate the reaction : $Cu^{2+} + 2e^- \longrightarrow Cu^0$. The charge recorded by coulometer is :

(a) 6C (b) 8C (c) 12C (d) 16C.

142. Of the following, the metal that is present in the respiratory protein haemocyanin is :

(a) vanadium (b) iron (c) molybdenum (d) copper.

143. An involatile solute is dissolved in a volatile solvent. Which of the curves below best represents the variation of solvent vapour pressure P with solute mole fraction X_2 :

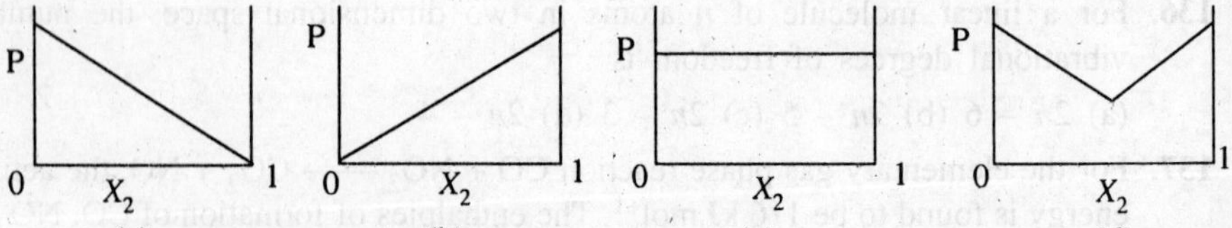

(a) (b) (c) (d)

144. The equivalent conductance and the specific conductance of a saturated AgCl solution are 138.3 $ohm^{-1} \cdot cm^2$ and 1.81×10^{-6} $ohm^{-1} \cdot cm^{-1}$, respectively. The solubility of AgCl in equivalents·$litre^{-1}$ is :

(a) 1.31×10^{-5} (b) 1.31×10^{-4} (c) 1.31×10^{-3} (d) 1.31×10^{-2}.

145. The point groups to which CH_4 and CH_3Cl belong, respectively, are :

(a) T_d and T_d (b) T_d and C_{3v} (c) C_{4v} and C_{3v} (d) C_{3v} and C_{2v}.

146. Choose the correct statement :

(a) Carbon dioxide gas does not show rotational Raman spectrum.

(b) Carbon dioxide gas does not absorb IR radiation.

(c) Carbon dioxide gas does not absorb microwave radiation.

(d) Carbon dioxide gas does not absorb in the UV region.

147. The dipole moment of chlorobenzene is 1.69 D. The dipole moment of *m*-dichlorobenzene is expected to be roughly

(a) 1.69 D (b) $1.69\sqrt{2}$D (c) 0 D (d) $1.69\sqrt{3}$D.

148. The number of nodal surfaces for an atomic orbital having principal quantum number n and azimuthal quantum number l is equal to :

(a) n (b) l (c) $n + l - 1$ (d) $n - l$.

149. When an electron is removed from oxygen molecule to form O_2^+ ion,

(a) the bond length and bond order both increase.

(b) the bond length and bond order both decrease.

(c) the bond length decreases and bond order increases.

(d) the bond length increases and bond order decreases.

150. In an experiment to determine the molecular weight of a substance using the measurement of osmotic pressure, 50% of the compound was in the dimerised form in the solution. If the actual molecular weight was 120, the measured molecular weight would be :

(a) 240 (b) 60 (c) 180 (d) 160.

151. In a kinetics experiment, the observed rate law was $-\dfrac{dC}{dt} = kC^{1/2}$. The units of the rate constant would be :

(a) sec^{-1} (b) $M^{1/2}$ sec^{-1} (c) $M^{-1/2}$ sec^{-1} (d) M^{-1} $sec^{-1/2}$.

152. The number of lines present in the ESR spectrum of methyl radical is :

(a) one (b) four (c) three (d) six.

153. The number of formula units of NaCl present in a unit cell is :

(a) one (b) two (c) three (d) four.

154. Arsenic in germanium is a :

(a) *p*-type conductor (b) *n*-type conductor

(c) ionic conductor (d) metallic conductor.

155. The reduction potentials for $Cu^{2+}(aq) + e \rightarrow Cu^+(aq)$ and $Cu^{2+}(aq) + 2e \longrightarrow Cu(s)$ are 0.15 and 0.34, respectively. Accordingly, the reduction potential for $Cu^+(aq) + e \longrightarrow Cu(s)$ will be :

(a) 0.53 V (b) 0.19 V (c) 0.17 V (d) 0.05 V.

156. Calcium carbonate is heated in a closed container and the equilibrium $CaCO_3(s) \rightleftharpoons CaO(s) + CO_2$ is established. This system has :

(a) 3 phases and 2 independent components

(b) 2 phases and 2 independent components

(c) 2 phases and 3 independent components

(d) 3 phases and 3 independent components.

157. The solubility of Ag_2CrO_4 is 1×10^{-4} M. The solubility product is :

(a) 4×10^{-12} (b) 1×10^{-8} (c) 2×10^{-12} (d) 1×10^{-12}.

158. For a 1*s* atomic orbital, the value of $|\psi|^2$ is :

(a) a minimum at the nucleus

(b) a maximum at a distance of one Bohr from the nucleus

(c) zero at the nucleus

(d) a maximum at the nucleus.

159. A container is divided into two halves by a diaphragm. Both sides of the diaphragm are filled with nitrogen at 1 atm pressure and at 300 K. When the diaphragm is removed, the entropy of mixing ΔS is :

(a) $\Delta S = 0$ (b) $\Delta S > 0$ (c) $\Delta S < 0$ (d) undefined.

160. In a first order reaction, the reactant concentration reduces from 1 mole/litre to 0.25 mole/litre in 100 min. The half-life of the reaction is :

(a) 50 min (b) 75 min (c) 200 min (d) 25 min.

161. When H_2 gas and D_2 gas are at the same temperature, the average speeds of H_2 molecules and D_2 molecules are in the ratio :

(a) $\dfrac{\nu_{H_2}}{\nu_{D_2}} = 1$ (b) $\dfrac{\nu_{H_2}}{\nu_{D_2}} = \sqrt{2}$ (c) $\dfrac{\nu_{H_2}}{\nu_{D_2}} = 2$ (d) $\dfrac{\nu_{H_2}}{\nu_{D_2}} = 1/\sqrt{2}$.

162. The number of molecules of an ideal gas in a volume of 8.2 litres at 380 torr pressure and 27 °C temperature will be :

(a) 1.0×10^{23} (b) 1×10^{22}

(c) 6.02×10^{22} (d) 12.04×10^{23}.

163. The dependence of $\dfrac{P\overline{V}}{RT}$ on temperature and pressure of gaseous nitrogen is shown in the figure. We, therefore, can infer that :

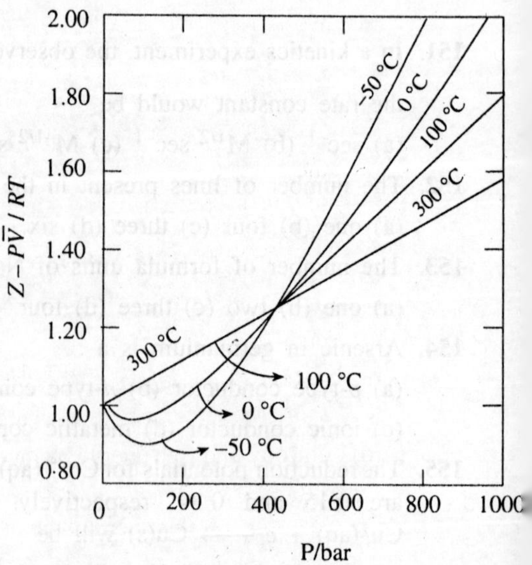

(a) volume occupied by a molecule is less at lower pressure

(b) the molecules feel intermolecular attractions only at large separation

(c) intermolecular attractive force is influenced by temperature
(d) even at zero pressure the molecules experience intermolecular attraction.

164. Crompton effect demonstrates that
(a) electrons have wave-like behaviour
(b) electrons have both wave- and particle-like behaviour
(c) photons have both particle and wave characteristics
(d) photons act like particles.

165. According to classical equipartition principle, the heat capacity of monoatomic solids should be close to (in J/mole/K unit) :
(a) 22.4 (b) 24.1 (c) 24.9 (d) 27.6.

166. Among the following orbitals of a homonuclear diatomic molecule, the bonding one is :
(a) $1\sigma_u$ (b) $2\sigma_u$ (c) $1\pi_g$ (d) $1\pi_u$.

167. Among the following molecules, the shortest bond length is to be found in :
(a) C_2 (b) N_2 (c) O_2 (d) F_2.

168. Bond length of a homonuclear diatomic molecule can be measured using :
(a) microwave absorption spectroscopy
(b) Raman spectroscopy
(c) infrared absorption spectroscopy
(d) electron microscopy.

169. The state of one mole of an ideal monoatomic gas is changed reversibly from (1 lit, 10 bar) to (10 lit, 1 bar) by two paths shown in the figure. The correct statement is :
(a) work done by the gas and change of its internal energy are more for the path-I than for the path-II.
(b) work done for both the paths is same but internal energy changes are different.
(c) work done by the gas for path-II is less but energy change for both the paths is same.
(d) work done by path-I is more but internal energy of the gas in both the states is the same.

170. For the transformation of the system described in Q.169,
(a) the entropy and the free energy of the system will remain unchanged
(b) entropy change of the system will be more for path-I than for path-II
(c) entropy will increase by the same amount for both the paths
(d) entropy and free energy changes cannot be computed based on the given data.

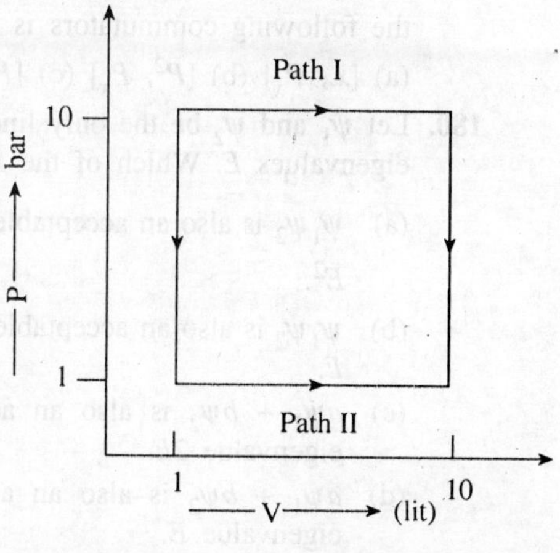

171. For an adiabatic reversible expansion of an ideal gas, the plot of log P versus log V is linear with a slope equal to ($\gamma = C_p/C_v$) :

(a) $1/\gamma$ (b) $-1/\gamma$ (c) γ (d) $-\gamma$.

172. The half-life of a reactant in a chemical reaction for initial concentration 0.1 and 0.4 mol/lit are 200 sec and 50 sec, respectively. The order of the reaction is :

(a) 0 (b) 1 (c) 2 (d) 3.

173. For a bimolecular reaction, $A + B \underset{k_2}{\overset{k_1}{\rightleftharpoons}} AB \xrightarrow{k_2} C$. The product formation is observed to follow the rate law, rate = $(k_1k_2/k_{-1})[A][B]$. Therefore, we can infer that :

(a) $k_1 > k_2 > k_{-1}$ (b) $k_2 > k_1 > k_{-1}$ (c) $k_{-1} >> k_2$ (d) $k_1 < k_{-1}$

174. The pH of a buffer solution containing 4×10^{-3} mole of acetic acid ($pK_a = 4.76$) and 0.4 mole of sodium acetate will be :

(a) 6.76 (b) 4.76 (c) 2.76 (d) 0.76.

175. The law that relates the solubility of a gas to its pressure called :

(a) Raoult's law (b) the distribution law (c) Henry's law (d) Ostwald's law.

176. A crystal having the lattice parameters $a = b \neq c$ and $\alpha = \beta = 90°$:

(a) is monoclinic. (b) is tetragonal or hexagonal.

(c) is orthorhombic. (d) has $\gamma = 90°$.

177. For a semiconductor at 0 K :

(a) there are no states at the Fermi energy

(b) the conduction band is fully filled

(c) the valence and conduction bands overlap

(d) the electrical conductivity is finite.

178. For which of the following systems is a quantum number of zero not allowed?

(a) rigid rotor (b) harmonic oscillator

(c) particle in a box (d) anharmonic oscillator.

179. If x is a position operator and P_x, P_y, P_z are the momentum operators along the x–, y– and z– directions respectively, with $P^2 = P_x^2 + P_y^2 + P_z^2$, then which of the following commutators is non-zero?

(a) $[x, P_x]$ (b) $[P^2, P_x]$ (c) $[P^2, P_z]$ (d) $[x, P_z]$.

180. Let ψ_1 and ψ_2 be the only linearly independent eigenfunctions having identical eigenvalues E. Which of the following statements is true?

(a) $\psi_1\psi_2$ is also an acceptable eigenfunction for the same state with eigenvalue E^2.

(b) $\psi_1\psi_2$ is also an acceptable eigenfunction for the same state with eigenvalue E.

(c) $a\psi_1 + b\psi_2$ is also an acceptable eigenfunction for the same state with eigenvalue $2E$.

(d) $a\psi_1 + b\psi_2$ is also an acceptable eigenfunction for the same state with eigenvalue E.

181. A black body kept at room temperature would emit broad band radiation peaking in the :

(a) infrared, (b) visible (c) radio-frequency (d) X-ray.

182. The graph below shows the constant volume heat capacity C_v for 1 mole of a diatomic molecule in the gaseous state as a function of temperature. R is the gas constant.

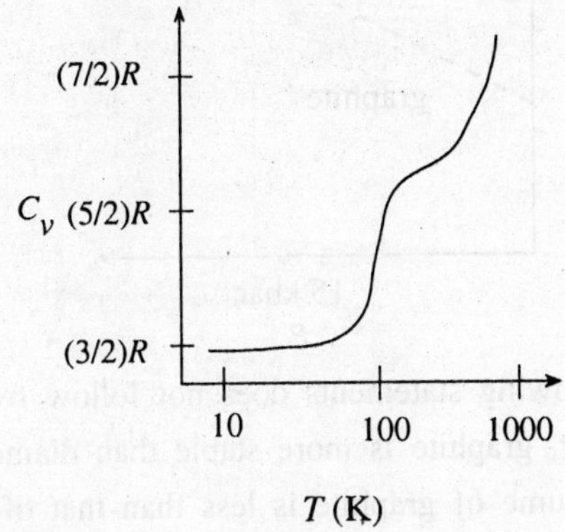

Which of the following statements is not true for the system?

(a) Below 100 K C_v is due largely to translational degrees of freedom.

(b) Vibrational degrees of freedom contribute significantly to C_v only above about 1000 K.

(c) C_v for translation and rotation is $\approx (5/2)\ R$.

(d) Between 10 K and 100 K, rotational degrees contribute most to C_v.

183. The graphs below show the P-V isotherms of an ideal gas at the temperatures T_1 and T_2. In which of these does the dashed line correspond to the process of adiabatic compression between the volume V_1 and V_2?

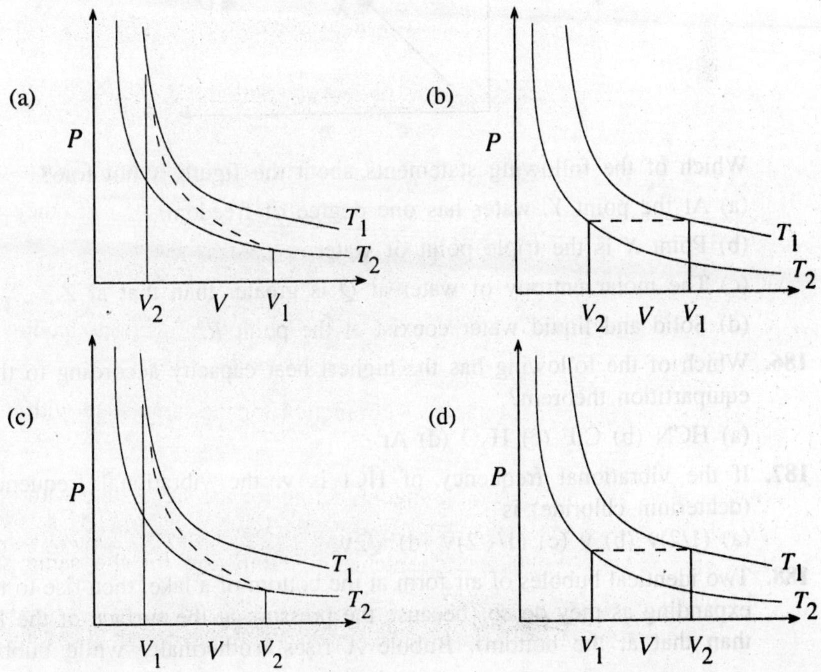

184. The graph below shows the approximate variation of the molar Gibbs free energy G as a function of the pressure P at constant temperature for diamond (solid line) and graphite (dashed line).

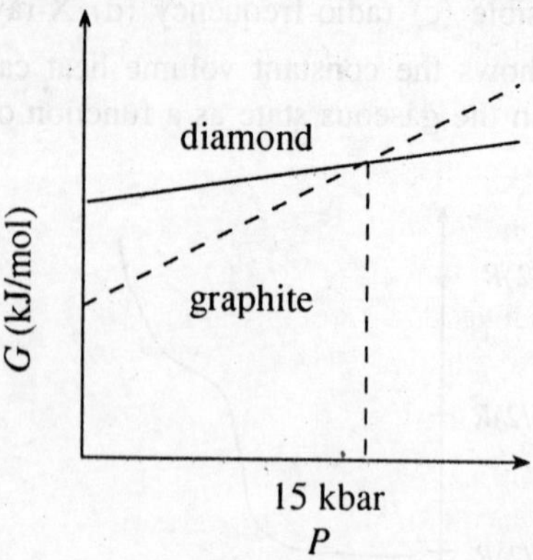

Which of the following statements does not follow from the figure?

(a) Below 15 kbar, graphite is more stable than diamond.

(b) The molar volume of graphite is less than that of diamond.

(c) The molar volume of graphite is more than that of diamond.

(d) Diamond is more stable than graphite above 15 kbar.

185. The graph below shows the P-T phase diagram of pure water :

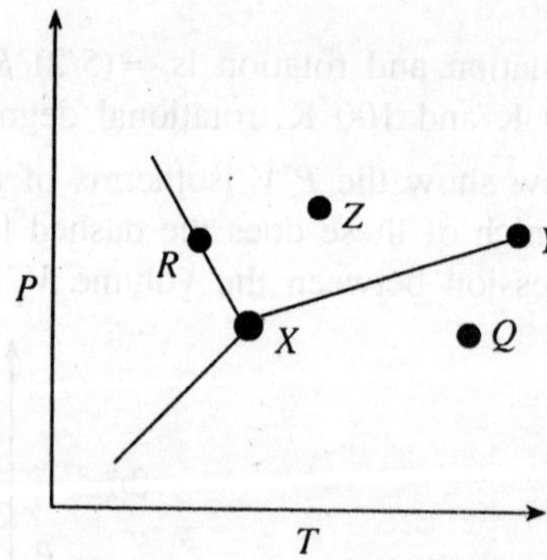

Which of the following statements about the figure is not true?

(a) At the point Y, water has one degree of freedom.

(b) Point X is the triple point of water.

(c) The molar entropy of water at Q is greater than that at Z.

(d) Solid and liquid water coexist at the point R.

186. Which of the following has the highest heat capacity according to the classical equipartition theorem?

(a) HCN (b) ClF (c) H_2O (d) Ar.

187. If the vibrational frequency of HCl is ν, the vibrational frequency of DCl (deuterium chloride) is :

(a) (1/2)ν (b) ν (c) $(1/\sqrt{2})\nu$ (d) $\sqrt{2}\nu$.

188. Two identical bubbles of air form at the bottom of a lake, then rise to the surface expanding as they do so (because the pressure at the surface of the lake is less than that at the bottom). Bubble A rises isothermally while bubble B rises adiabatically. Assuming the air in the bubbles is a monoatomic ideal gas, we can conclude that at the surface :

(a) volume of bubble $A >$ volume of bubble B

(b) volume of bubble $A <$ volume of bubble B

(c) volume of bubble $A =$ volume of bubble B

(d) the volumes are indeterminate.

189. Two different monoatomic ideal gases A and B, each with the same energy U, volume V and mole number N occupies the two halves of a divided chamber. If the partition between the chambers is removed and the gases are allowed to mix, then the total change in entropy is given by :

(a) $Nk_B \ln 2$ (b) $Nk_B \ln (1/2)$ (c) $Nk_B \ln 4$ (d) $(N/2)k_B \ln 2$.

190. A liquid is kept in a closed vessel. If a glass plate of negligible mass and with a small hole is placed on top of the liquid surface, the vapour pressure of the liquid in the vessel would :

(a) increase (b) remain the same (c) decrease (d) be changed in a manner that cannot be predicted.

191. The dissipated work ∂W_d done by a system at pressure P against an external pressure P_{ext} leading to a volume change dV is defined as

$$\partial W_d = (P - P_{ext})\, dV.$$

Which of the following statements is *not* true?

(a) $\partial W_d > 0$, for a spontaneous process.

(b) $\partial W_d > 0$, for a reversible process.

(c) $\partial W_d > 0$, when $P > P_{ext}$.

(d) $\partial W_d > 0$, when $P < P_{ext}$.

192. The emf of the cell Zn | Zn^{2+}(1 M) || Cu^{2+}(1 M) | Cu is 1.1 V. If the standard potential of Zn^{+2} | Zn is –0.78 V, what is the oxidation potential of Cu | Cu^{2+}?

(a) 1.88 V (b) 0.34 V (c) – 0.34 (d) –1.88 V.

193. If a reaction goes to completion in finite time, the order of the reaction is :

(a) zero (b) first (c) second (d) third.

194. According to the equipartition principle the heat capacity at constant volume (C_v) for ethylene gas is :

(a) 11R (b) 1.5R (c) 0.15R (d) 15R.

195. An aqueous solution containing a mole of a non-volatile solute freezes at –0.186 °C. The elevation of boiling point of the same aqueous solution (K_f = 1.86, K_b = 0.512) would be :

(a) 0.186 (b) 0.512 (c) 0.00512 (d) 0.0512.

196. In a system, when the chemical potential of each component is the same for all the phases, the equilibrium is said to be :

(a) metastable (b) composition (c) thermal (d) mechanical.

197. Among the singlet (S), doublet (D) and triplet (T) electronic states, phosphore scence involves transition between :

(a) S and S (b) S and T (c) D and D (d) T and T.

198. ΔH and ΔU for the reaction Fe_2O_3 (s) + $3H_2$(g) → 2Fe(s) + $3H_2O$(l) at constant temperature are related as :

(a) $\Delta H = \Delta U$ (b) $\Delta H = \Delta U + RT$ (c) $\Delta H = \Delta U + 3RT$ (d) $\Delta H = \Delta U - 3RT$.

199. The half-life time for a reaction at initial concentration of 0.1 mol.l^{-1} and 0.4 mol.l^{-1} are 200 sec and 50 sec respectively. The order of the reaction is

(a) 0 (b) 2 (c) 3 (d) 1.

200. On a particular day water boiled at 99.5 °C in Calcutta and 96.1 °C at Darjeeling. Elevation of boiling point of a 0.01 molal sucrose solution on that day was (a) less in Darjeeling (b) equal both in Calcutta and at Darjeeling (c) less in Calcutta.

201. As the frequency of the alternating current used in its measurement is increased, the conductance of an aqueous electrolytic solution :

(a) attains a limiting maximum value (b) remains unchanged (c) decreases linearly.

202. The differential heat of dilution of a solution at any given concentration of the solute, is the difference of enthalpies :

(a) per mole of solute in the solution and in its pure state.

(b) per mole of solvent in the solution and in its pure state.

(c) sum total of the two.

203. The buffer capacity will be maximum at the

(a) half neutralization point of a weak acid by a weak base

(b) half neutralization point of a weak acid by a strong base

(c) full neutralization of a strong acid by a strong base.

204. In a homogeneous reaction $R \rightarrow P$, the time needed for the first 5% conversion of R is independent of its initial concentration. The reaction is :

(a) zero order in R (b) first order in R (c) second order in P.

205. Joule-Thomson coefficient of a van der Waals gas is determined by :

(a) only the pressure correction factor 'a'

(b) only the volume correction factor 'b'

(c) both 'a' and 'b'.

206. A saturated solution of Na_3PO_4, which is completely dissociated to its component ions, contains an excess of the salt. For the system :

(a) P = 2, C = 2, F = 2

(b) P = 3, C = 3, F = 2

(c) P = 3, C = 2, F = 1.

207. One mole of an ideal gas absorbs 300 calories of heat to expand irreversibly from 10 litres of volume at 27 °C to 100 litres of volume at 27 °C. The corresponding change in entropy is :

(a) 4.605 e.u. (b) 1 e.u. (c) 11.1 e.u.

208. Two reactions I and II, have the same values of rate constants at 30 °C. But activation energy of I is greater than that of II ($E_I > E_{II}$). On raising the temperature to 50 °C, the relative increase in rates of the two reactions will be :

(a) less for reaction I (b) more for reaction I (c) the same for both the reactions.

209. For the reaction $N_2(g) + 3H_2(g) \rightleftharpoons 2NH_3(g)$, the thermodynamic equilibrium constant K_a has the unit of :

(a) atmosphere^{-2} (b) mole^{-2}.litre^2 (c) number.

210. For a reversible reaction having $K_a < 1$ and $\Delta H^\circ < 0$ at a given temperature :
(a) ΔG° and ΔS° are both +ve
(b) ΔG° is +ve, ΔS° is –ve
(c) ΔG° is –ve, ΔS° is +ve.

211. As the temperature of a gas increases, the difference in population of gas molecules having most probable velocity and average velocity

(a) decreases, (b) increases, (c) remains the same.

212. Molal boiling point elevation constants of water and ethanol are 0.51 and 1.22 respectively at 1 atmosphere pressure. A 0.01 M solution of urea will have osmotic pressure :

(a) same both in ethanol and in water (b) higher in water (c) higher in ethanol.

213. For a reaction $A + B \rightarrow$ Product, the overall rate of formation of production is doubled when concentration of A only is doubled. However, when that of B only is doubled the rate remains unaffected. The overall order of the reaction is

(a) zero (b) one (c) two.

214. Energy of activation is :

(a) free energy change of overall reaction

(b) potential energy difference between the transition state and the reactant

(c) potential energy difference between the transition state and the product.

215. The molar free energy of a gas is invariant at

(a) constant temperature (b) constant pressure (c) constant temperature and constant pressure.

216. The pH at half-neutralisation point of a weak acid by a strong base is given by :

(a) pH = pK_a (b) pH = $2pK_a$ (c) pH = $\frac{1}{2}pK_a$.

217. When a pure liquid coexists with the vapour and solid, the system is :

(a) bivariant (b) mono-variant (c) non-variant.

218. When one mole of ice is converted to water at 0°C and 1 atm, the work done (l.atm) is :

(a) 1.1×10^{-4} (b) 2.0×10^{-3} (c) 2.0×10^{-4} (d) 1.1×10^{-5}.

219. For a zero order reaction, unit of the rate constant is expressed as :

(a) M^1s^{-1} (b) M^0s^{-1} (c) $M^{-1}s^{-1}$ (d) M^0s^0

220. When 100 gm of water is reversibly heated from 50 °C to 75 °C at 1 atm, the change in entropy (J.K^{-1}) of the universe is :

(a) 0.35 (b) 0.31 (c) 0 (d) 3.1.

221. If the electron were spin $\frac{3}{2}$ particles, instead of spin $\frac{1}{2}$, then the number of electrons that can be accomodated in a level are :

(a) 2 (b) 3 (c) 4 (d) 5.

222. For a particle in a cubic box, the total number of quantum numbers needed to specify its state is :

(a) 1 (b) 2 (c) 3 (d) 9.

223. With increasing pressure, the temperature range over which the liquid state is stable :

(a) decreases (b) increases (c) remains constant

(d) decreases till the critical pressure and then increases.

224. Given that the most probable speed of oxygen gas is 1000 $m.s^{-1}$, the mean/average speed (ms^{-1}) under the same condition is :

(a) 1224 (b) 1128 (c) 886 (d) 816.

225. 1×10^{-6} moles of the enzyme carbonic anhydrase dehydrates H_2CO_3 to produce 0.6 mol of CO_2 per second. The turnover number of the enzyme is :

(a) $N_A \times 6 \times 10^{-5}$ (b) $(1/6) \times 10^{-5}$ (c) $(6 \times 10^5)/N_A$ (d) 6×10^5.

226. For a reaction with rate equation $-\frac{dC}{dt} = KC^2$, C_0 and C are the concentrations of the reactant at time 0 and t respectively. If 10 minutes were required for C_0 to become $C_0/2$, the time required for C_0 to become $C_0/4$ is :

(a) 10 min (b) 20 min (c) 30 min (d) 40 min.

227. For an ideal solution formed by mixing of pure liquids A and B :

(a) $\Delta H_{mixing} = 0$ (b) $\Delta H_{mixing} < 0$ (c) $\Delta H_{mixing} > 0$ (d) $\Delta S_{mixing} = 0$

228. The relationship between the equilibrium constant K_1 for the reaction :

$$CO(g) + \frac{1}{2}O_2(g) \rightleftharpoons CO_2\ (g)$$

and the equilibrium constant K_2 for the reaction :

$2CO(g) + O_2(g) \rightleftharpoons 2CO_2(g)$ is :

(a) $2K_1 = K_2$ (b) $K_1 = K_2^2$ (c) $K_1 = K_2$ (d) $K_1^2 = K_2$.

229. For H-like atoms, the ground state energy is proportional to :

(a) $\frac{\mu}{Z^2}$ (b) $\frac{Z^2}{\mu}$ (c) μZ^2 (d) $\frac{1}{\mu Z^2}$,

where μ is the reduced mass and Z is the nuclear charge.

230. The value of the integral $\int e^{-x} x^2 dx$ is :

(a) $x^2e^{-x} + 2xe^{-x} + 2e^{-x}$ (b) $\frac{-1}{2}(x^2e^{-x} + 2xe^{-x} + 2e^{-x})$

(c) $\frac{1}{2}(x^2e^{-x} + 2xe^{-x} + 2e^{-x})$ (d) $-x^2e^{-x} - 2xe^{-x} - 2e^{-x}$.

231. For the raction $\alpha A \longrightarrow$ products, the plot of $\frac{1}{|A|}$ versus time (t) gives a straight line. The order of the reaction is :

(a) 0 (b) 1 (c) 2 (d) 3.

232. The pH of a solution prepared from 0.005 mole of $Ba(OH)_2$ in 100 cc water is :

(a) 10 (b) 12 (c) 11 (d) 13.

233. For an electron whose x-positional uncertainty is 1×10^{-10} m, the uncertainty in the x-component of the velocity in ms^{-1} will be of the order of :
(Data : $m_e = 9 \times 10^{-31}$ kg, $h = 6.6 \times 10^{-34}$ Js)
(a) 10^6 (b) 10^9 (c) 10^{12} (d) 10^{15}.

234. For the following system in equilibrium $CaCO_3(s) \rightleftharpoons CaO(s) + CO_2(g)$, the number of components (C), phases (P) and degrees of freedom (F), respectively, are :
(a) 2,2,2 (b) 1,3,0 (c) 3,3,2 (d) 2,3,1.

235. Given that $E^o_{Fe^{2+}/Fe} = -0.44$ V and $E^o_{Fe^{3+}/Fe^{2+}} = 0.77$ V, the $E^o_{Fe^{3+}/Fe}$ is :
(a) 1.21 V (b) 0.33 V (c) −0.036 V (d) 0.036 V.

236. The solid-liquid phase diagram for the Mg-Zn system is shown in the adjoining figure where the vertical line at X(Mg) = 0.33 represents the formation of a congruent melting compound $MgZn_2$. The figure is divided into seven regions depending upon the physical state of the system. The composition of the region #6 represents :

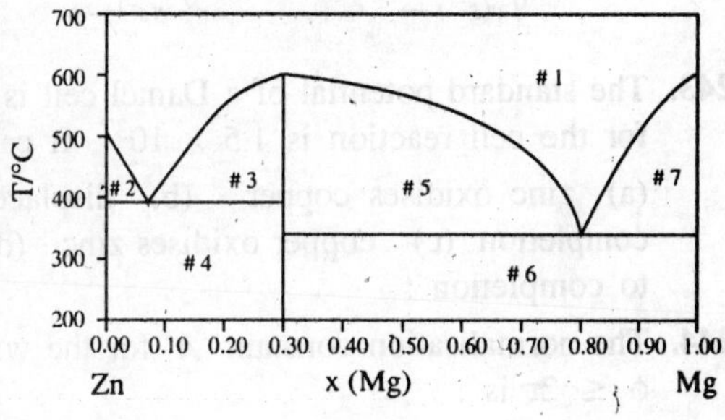

(a) single phase of a solution of Mg and Zn
(b) two-phase region between solid Zn and solid $MgZn_2$
(c) two-phase region between liquid and solid $MgZn_2$
(d) two-phase region between solid Mg and solid $MgZn_2$.

237. In the extraction of metals from their ores, which one of the following reduction methods can bring about a non-spontaneous reduction?
(a) electrolytic reduction (b) reduction by carbon (c) reduction by another metal (d) reduction by hydrogen.

238. The plot of Gibb's free energy G and the extent of a reaction ξ is given for the reaction $A \rightleftharpoons B$. If μ_A and μ_B are the chemical potentials of A and B respectively, the Incorrect statement is

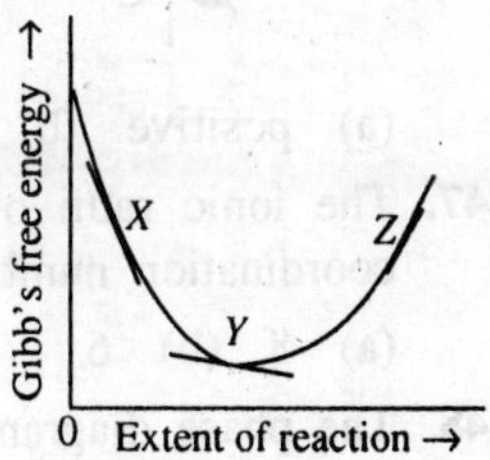

(a) at point X, $\mu_A > \mu_B$
(b) at point Y, $\Delta G = 0$
(c) at point Z, $\mu_A > \mu_B$
(d) at equilibrium, the composition of the reaction mixture can be identified.

239. For a cyclic process performed by an ideal gas, changes in some thermodynamic functions are zero. Indicate the set in which all the functions are zero.
(a) W, ΔU, ΔH, ΔG (b) Q, ΔS, ΔH, ΔA (c) Q, ΔU, ΔS, ΔG (d) ΔU, ΔS, ΔH, ΔA.

240. Which one of the following species is the conjugate base of HO^-?

(a) H_2O (b) O^{2-} (c) O_2 (d) O_2^{2}.

241. An aqueous solution containing 0.01M $FeCl_3$ and 0.06M $HClO_4$ has the same ionic strength as a solution of :

(a) 0.09M NaCl (b) 0.04M Na_2SO_4 (c) 0.06M $CuSO_4$ (d) 0.03M H_3PO_4.

242. Which one of the following figures, showing kinetic energy of the ejected electron versus the frequency (ν) of the incident photon, represents the Einstein's photoelectric effect?

(a) 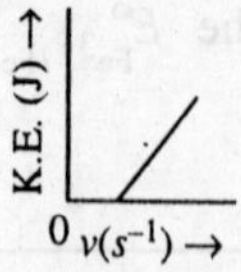

(b) 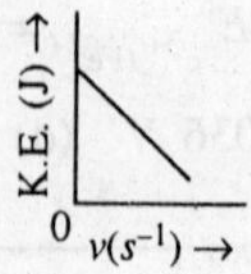

(c) 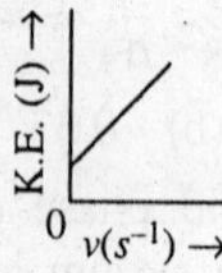

(d) 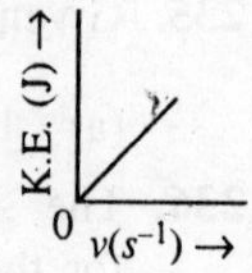

243. The standard potential of a Daniel cell is + 1.10 V and the equilibrium constant for the cell reaction is 1.5×10^{37}. It can be concluded that :

(a) zinc oxidises copper (b) displacement of copper by zinc goes to near completion (c) copper oxidises zinc (d) displacement of zinc by copper goes to completion :

244. The normalisation constant 'A' for the wave function $\psi(\phi) = Ae^{(im\phi)}$ where $0 \le \phi \le 2\pi$ is :

(a) $1/\sqrt{2\pi}$ (b) $\sqrt{2\pi}$ (c) 2π (d) $1/\sqrt{2}$.

245. The pH of a 1.0×10^{-3} M solution of a weak acid HA is 4.0. The acid dissociation constant K_a is :

(a) 1.0×10^{-3} (b) 1.0×10^{-4} (c) 1.0×10^{-5} (d) 2.0×10^{-5}

246. The overlap between the atomic orbitals sketched below is :

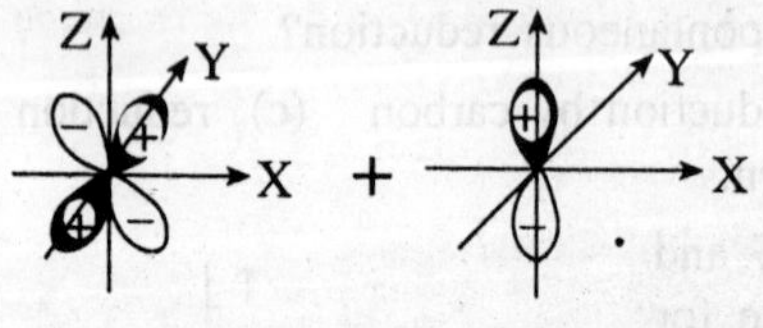

(a) positive (b) negative (c) zero (d) no overlap.

247. The ionic radii of Ca^{2+} and F^- are 100 pm and 133 pm respectively. The coordination number of Ca^{2+} in the ionic solid will be :

(a) 8 (b) 6 (c) 4 (d) 2.

248. The phase diagram of a pure substance is sketched here. The number of degrees of freedom at points P_1, P_2 and P_3, respectively, are :

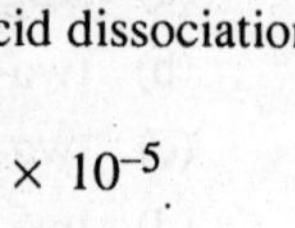

(a) 2, 1, 0 (b) 1, 2, 0 (c) 2, 0, 1 (d) 0, 2, 1.

249. The solubility products (K_{sp}) for three salts MX, MY_2 and MZ_3 are 1×10^{-8}, 4×10^{-9} and 27×10^{-8}, respectively. The solubilities of these salts follow the order :

(a) MX > MY_2 > MZ_3 (b) MZ_3 > MY_2 > MX (c) MZ_3 > MX > MY_2 (d) MY_2 > MX > MZ_3.

250. The temperature (T) dependence of the equilibrium constant (K) of a chemical reaction is correctly described by the following statement :

(a) For an endothermic reaction, the slope of ln K vs. $1/T$ plot is positive.

(b) For an exothermic reaction, K is proportional to T.

(c) For an exothermic reaction, K at a higher temperature is lower than K at a lower temperature

(d) If ΔH is independent of temperature, the change in K with T is smaller at lower temperatures.

251. When the concentration of K^+ across a cell members drops from 0.01 M to 0.001 M, the potential difference across the members is :

(a) 0.0 V (b) 0.0059 V (c) 0.059 V (d) 0.59 V.

252. The statement that is correct for both electrochemical (galvanic) cells and electrolytic cells is :

(a) $\Delta G = -nFE$ (b) free energy decreases in both cells

(c) the cell potentials are temperature independent

(d) chemical energy is converted into electrical energy in both cells.

253. The molar heat capacity at constant volume of a colourless gas is found to be 25 J.mol^{-1}K^{-1} at room temperature. The gas must be :

(a) N_2 (b) O_2 (c) CO_2 (d) SO_2.

254. The wave function for a particle (moving in a ring) is $(2\pi)^{-\frac{1}{2}} \exp(2i\phi)$, where ϕ is the polar angle. The prabability of finding the particle in a small interval $d\phi$ when the value of $\phi = \frac{\pi}{2}$ is :

(a) $d\phi$ (b) $d\phi/2\pi$ (c) $d\phi \exp(i\pi)$ (d) $d\phi \exp(I\pi)/2\pi$.

255. An electric current of 0.965 ampere is passed for 2000 seconds through a solution containing $[Cu(CH_3CN)_4]^+$ and metallic copper is deposited at the cathode. The amount of Cu deposited is :

(a) 0.005 mol (b) 0.01 mol (c) 0.02 mol (d) 0.04 mol.

256. The Maxwell-Boltzmann distribution for molecular speeds is shown in the following figure. In the figure, H is the height of the peak, L is the location of the maximum and W is the width at half-height. As the temperature is decreased,

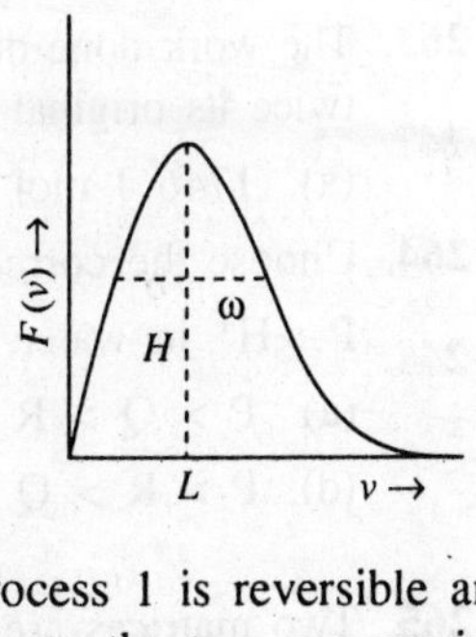

(a) H increases, L decreases and W increases

(b) H increases, L decreases and W decreases

(c) H decreases, L decreases and W increases

(d) H decreases, L decreases and W decreases.

257. A system undergoes two cyclic processes 1 and 2. Process 1 is reversible and process 2 is irreversible. The correct statement relating to the two process is :

(a) ΔS (for process 1) = 0, while ΔS (for process 2) $\neq 0$

(b) $Q_{cyclic} = 0$ for process 1 and $Q_{cyclic} \neq 0$ for process 2

(c) more heat can be converted to work in process 1 than in process 2

(d) more work can be converted to heat in process 1 than in process 2.

258. The molar internal energy of a gas at temperature T is $U_m(T)$. The molar internal energy at $T = 0$ is $U_m(0)$. The correct expression that relates these two with appropriate contributions is :

(a) $U_m(T) = U_m(0) + 3RT$ [linear molecule; translation only]

(b) $U_m(T) = U_m(0) + \frac{5}{2}RT$ [linear molecule; translation and rotation only]

(c) $U_m(T) = U_m(0) + \frac{3}{2}RT$ [nonlinear molecule; translation and rotation only]

(d) $U_m(T) = U_m(0) + RT$ [nonlinear molecule; translation only]

259. If a particle has linear momentum $\vec{p} = -2\vec{i} + \vec{j} + \vec{k}$ at position $\vec{r} = 3\vec{i} - \vec{j} + \vec{k}$, then its angular momentum is :

(a) $\vec{i} + 2\vec{k}$ (b) $-2\vec{i} - 5\vec{j} + \vec{k}$ (c) $5\vec{i} - 2\vec{j}$ (d) $2\vec{i} + 5\vec{j} - \vec{k}$.

260. If ψ is the eigenfunction to the Hamiltonian operator with α as the eigenvalue, then α must be

(a) positive (b) negative (c) an integer (d) real.

261. A quantum mechanical particle of mass m free to rotate on the surface of a sphere of radius r is in the state with energy $\frac{10\hbar^2}{mr^2}$. The degeneracy of this state is :

(a) 20 (b) 10 (c) 9 (d) 4.

262. Choose the INCORRECT statement among the following :

(a) When ideal gases are mixed, the entropy of mixing is always positive.

(b) At equilibrium, the chemical potential of a species is the same in all of the phases of the system.

(c) The total pressure of a mixture of ideal gases is equal to the sum of the partial pressure of each gas in the mixture.

(d) When a gas is allowed to expand, the maximum work is obtained when the process is carried out irreversibly.

263. The work done during the free expansion of one mole of an ideal gas at 27 °C to twice its original volume is (given : RT = 2494 J mol^{-1}, ln 2 = 0.7, log 2 = 0.3) :

(a) 1746 J mol^{-1} (b) −1746 J mol^{-1} (c) zero (d) 748.2 J mol^{-1}

264. Choose the correct order of the diffusion coefficients of the following at 298K.

P : H^+ in water, Q : OH^- in water, R : H_2O in water, S : Sucrose in water,

(a) P > Q > R > S (b) S > R > Q > P (c) S > Q > R > P

(d) P > R > Q > S.

265. Two matrices are given as $X = \begin{pmatrix} 1 & 5 \\ 3 & 7 \end{pmatrix}$ and $Y = \begin{pmatrix} 2 & 4 \\ 6 & 0 \end{pmatrix}$. If X^T is the transpose of X, then what would be X^TY?

(a) $\begin{pmatrix} 20 & 52 \\ 4 & 20 \end{pmatrix}$ (b) $\begin{pmatrix} 20 & 4 \\ 52 & 20 \end{pmatrix}$ (c) $\begin{pmatrix} 32 & 4 \\ 48 & 12 \end{pmatrix}$ (d) $\begin{pmatrix} 44 & 28 \\ 12 & 12 \end{pmatrix}$.

266. Addition of 1.0 gm of a compound to 10 gm of water increases the boiling point by 0.3 °C. The amount of compound needed to propare a 500 ml of 0.1 M solution is (given : assume negligible dissociation or association of the compound, boiling point constant K_b of water = 0.513 K kg mol^{-1}) :
(a) 0.855 gm (b) 17.1 gm (c) 8.55 gm (d) 85.5 gm

267. The molar conductivity of 0.009 M aqueous solution of a weak acid (HA) is 0.005 S m^2 mol^{-1} and the limiting molar conductivity of HA is 0.05 S m^2 mol^{-1} at 298 K. Assuming activity coefficients to be unity, the acid dissociation constant (K_a) or HA at this temperature is
(a) 1×10^{-4} (b) 0.1 (c) 9×10^{-4} (d) 1.1×10^{-5}.

268. The structure of rock salt consists of :
(a) a cubic close-packed array of anions with cations in all the octahedral sites
(b) a cubic close-packed array of cations with anions in all the tetrahedral sites
(c) a hexagonal close-packed array of anions with cations in all the octahedral sites
(d) a cubic close-packed array of anions with cations in all the tetrahedral sites.

269. For an ideal gas the plot that is nonlinear is :
(a) PV vs. T (b) PV vs. P, at constant T (c) P vs. V, at constant T
(d) ln P vs. ln V at constant T.

270. Consider two identical containers, one with 1 mole of H_2 and the other with 1 mole of He. If the root-mean square (RMS) velocities of the two gases are the same then the ratio of the temperatures, $T(H_2)/T(He)$ is :
(a) 1/2 (b) 2 (c) $1/\sqrt{2}$ (d) $\sqrt{2}$.

271. An electron moves around the nucleus in a circular orbit according to the Bohr model. The radial vector $\vec{r}$ and the instantaneous linear momentum vector $\vec{p}$ are shown in the diagram :

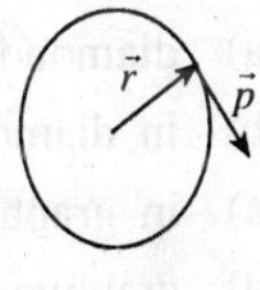

The direction of the angular momentum vector is :
(a) along $\vec{r}$ (b) along $\vec{p}$ (c) opposite to $\vec{p}$
(d) perpendicular to both $\vec{r}$ and $\vec{p}$.

272. X and Y are transformed co-ordinates obtained from p and q as follows :

$$\begin{pmatrix} X \\ Y \end{pmatrix} = \begin{pmatrix} a_1 & a_3 \\ a_2 & a_4 \end{pmatrix} \begin{pmatrix} p \\ q \end{pmatrix}.$$

The correct set of linear equations that represent X and Y are :
(a) $X = a_1p + a_2q$, $Y = a_3p + a_4q$ (b) $X = a_1p + a_3q$, $Y = a_2p + a_4q$ (c) $X = a_2p + a_4q$, $Y = a_1p + a_3q$ (d) $X = a_1p + a_4q$, $Y = a_2p + a_3q$.

273. Which of the following is not a solution of the equation :

$$\frac{d^2x}{dt^2} + \omega^2 x = 0.$$

(a) $x = A \cos \omega t$ (b) $x = A \sin \omega t$ (c) $x = At^2$ (d) $x = A\left(e^{i\omega t} + e^{-i\omega t}\right)$.

274. An electron is found in an orbital with one radial node and two angular nodes. Which orbital the electron is in?

(a) 1*s* (b) 2*p* (c) 3*d* (d) 4*d*.

275. The acceptable valence shell electronic arrangement is :

(a) 2*s*: [↑↓] 2*p*: [↑↓] [↑] [ ] (b) 2*s*: [↑↑] 2*p*: [↑] [↑] [↑]

(c) 2*s*: [↑↓] 2*p*: [↑] [↑] [↑] (d) 2*s*: [↑↓] 2*p*: [↑] [↓] [↑]

276. If K_{sp} is the solubility product of a sparingly soluble salt A_3X_2, then its solubility is

(a) $(K_{sp}/108)^{1/3}$ (b) $(K_{sp})^{1/5}$ (c) $(K_{sp}/72)^{1/5}$ (d) $(K_{sp})^{1/2}$.

277. For the formation of B from A, heat liberated is 20 kJ mol^{-1}. If the activation energy for the reaction B $\longrightarrow$ A is 100 kJ mol^{-1}, then the activation energy (in kJ mol^{-1}) for the reaction A $\longrightarrow$ B is :

(a) 120 (b) 100 (c) 80 (d) 60.

278. For the reaction A + B $\longrightarrow$ Z, the concentration of Z at time *t* is given by $[Z] = [A]_{t=0}(1 - e^{-kt}) + [Z]_{t=0}$, where *k* is the rate constant. The rate law is :

(a) $-\dfrac{d[Z]}{dt} = k[A]$ (b) $\dfrac{d[Z]}{dt} = k[A]$ (c) $\dfrac{d[Z]}{dt} = k[A]$ (d) $\dfrac{d[Z]}{dt} = k[A][B]$.

279. Among the following, the incorrect statement is :

(a) diamond and graphite are two allotropes of carbon

(b) in diamond, each carbon is sp^3 hybridized

(c) in graphite, each carbon is sp^2 hybridized

(d) graphite shows high electrical conductivity in one direction only.

280. The pH of a 1×10^{-8} M HCl solution is close to :

(a) 8.0 (b) 7.1 (c) 6.9 (d) 6.0.

281. The indicator phenolphthalein changes colour at pH ~ 9. This indicator is not suitable for accurate determination of the end point in the titration of :

(a) CH_3COOH with NaOH (b) HCl with NH_4OH

(c) HCl with NaOH (d) HCl with KOH.

282. Based on the principle of equipartition of energy, the molar heat capacity of CO_2 at constant volume $C_{v,m}$ is :

(a) 3.5*R* (b) 6*R* (c) 6.5*R* (d) 9*R*.

283. One mole of a van der Waals gas undergoes reversible isothermal transformation from an initial volume V_1 to a final volume V_2. The expression for the work done is :

(a) $RT \ln \dfrac{V_2}{V_1} + a(V_2 - V_1)$ (b) $-RT \ln \dfrac{V_2 - b}{V_1 - b} + a\left(\dfrac{1}{V_1} - \dfrac{1}{V_2}\right)$

(c) $RT \ln \dfrac{P_2}{P_1}$ (d) $RT \ln \dfrac{V_2 - b}{V_1 - b} - a\left(\dfrac{1}{V_1} - \dfrac{1}{V_2}\right)$.

284. The scalar product of two vectors u and v, where $u = 2i + 3j - 5k$ and $v = i + j + 3k$, is :

(a) -10 (b) $2i + 3j - 15k$ (c) $3i + 4j - 2k$ (d) 10.

285. The minimum concentration of silver ions that is required to start the precipitation of Ag_2S ($K_{sp} = 1 \times 10^{-51}$) in a 0.1 M solution of S^{2-} is :

(a) 1×10^{-49} M (b) 1×10^{-50} M (c) 1×10^{-26} M (d) 1×10^{-25} M.

286. Identify the correct statement regarding Einstein's photoelectric effect :

(a) The number of electrons ejected depends on the wavelength of incident radiation.

(b) Electron ejection can occur at any wavelength of incident radiation.

(c) The number of electrons ejected at a given incident wavelength depends on the intensity of the radiation.

(d) The kinetic energy of the ejected electrons is independent of the wavelength of incident radiation.

287. The hydrolysis constant (K_h) of NH_4Cl is 5.6×10^{-10}. The concentration of H_3O^+ in a 0.1 M solution of NH_4Cl at equilibrium is :

(a) $\sqrt{5.6 \times 10^{-11}}$ (b) $\sqrt{5.6 \times 10^{-10}}$ (c) 5.6×10^{-10} (d) 2.8×10^{-5}.

288. The acid dissociation constant (K_a) for HCOOH, CH_3COOH, $CH_2ClCOOH$ and HCN at 25 °C are 1.8×10^{-4}, 1.8×10^{-5}, 1.4×10^{-3} and 4.8×10^{-10}, respectively. The acid that gives highest pH at the equivalence point when 0.2 M solution of each acid is titrated with a 0.2 M solution of sodium hydroxide is :

(a) HCOOH (b) CH_3COOH (c) $CH_2ClCOOH$ (d) HCN.

289. For an ideal gas undergoing reversible Carnot cycle, the plot of enthalpy (H) versus entropy (S) is :

(a)
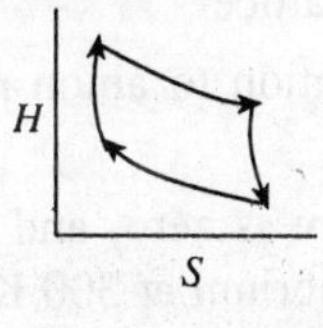

(b)
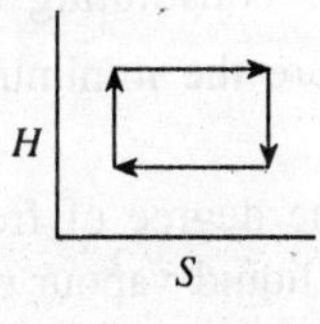

(c)
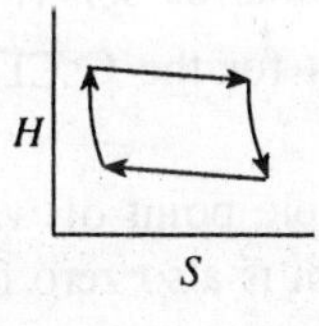

(d)
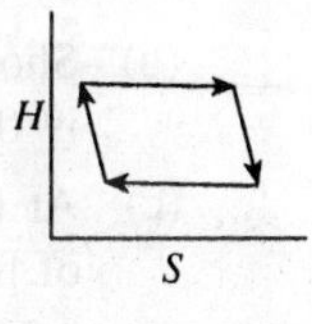

Short Questions

1. Devise a cell in which the following reaction takes place. Write down the half-cell reactions. $2H_2(g) + O_2(g) = 2H_2O(l)$ [**Ans.** $H_2 \mid OH^- \mid O_2$]

2. Calculate the minimum pressure needed to convert diamond to graphite at 298 K. Given, densities of diamond and graphite remain constant at 3.52 gm.cm^{-3} and 2.25 gm.cm^{-3} respectively, and ΔG_f° of diamond is 2.90 kJ per mole at 298 K. (1.51×10^9 Pa)

3. (a) An ideal gas at 300 K and 1.0 atm pressure was passed at a rate of 100 ml/sec through a tube which is thermally insulated from outside, but fitted inside with a heater which dissipates heat at a rate of 0.5 J per second. The gas flows into the atmosphere at a temperature of 30.1 K. Calculate the heat capacity ratio, γ, of the gas. [**Ans.** 1.4]

4. (a) For a noncatalytic reaction, $A \rightarrow B$, the conversion is linear with time. The reaction goes to completion in 4 hours at 300 K and in 2 hours at 315 K. Find out the order of the reaction, and the energy of activation.

[**Ans.** 36.31 kJ]

(b) Two glass capillary tubes, one with a square bore and the other with a circular bore of same cross-sectional area, were dipped into water. Show analytically in which case there will be higher capillary rise. [**Ans.** square]

5. When an excited nucleus decays, it emits gamma-rays. If the half-life of radioactive decay is 69.3×10^{-14} second, what is the uncertainty in the energy of the emitted gamma-rays?

[**Ans.** 3.804×10^{-23} J]

6. Calculate the probability that a particle in a one-dimensional box of side a, is to be found between 0 and $a/2$. [**Ans.** $\frac{1}{2}$]

7. (a) Calculate the unit cell dimensions of NaCl if its density is 2.17 kg.m^{-3}.

[**Ans.** a = 5.75 nm]

(b) For a reaction $2A + B \rightarrow C$, if the concentration of A is doubled the reaction rate increases by a factor of four. If the concentration of B is tripled, the reaction rate increases by a factor of nine. What is the order of the reaction with respect to A and with respect to B? When B is present in large excess and the initial concentration of A is 0.1 moles/litre, the concentration of A decreases to 0.05 moles/litre after 5 seconds. What is the rate constant of the reaction? What is the reaction rate?

[**Ans.** $r = kC_A^2 C_B^2$, k = 2 l.mol^{-1}.sec^{-1}]

(c) Write the Schrödinger equation for the electronic wave function of H_2.

8. (a) Show that in a simple cubic structure only about 52% of the space is filled by the atoms or spherical ions constituting the lattice.

(b) Show that for the CsCl structure the minimum cation to anion radius ratio is 0.732.

(c) At the triple point of water, the degree of freedom is zero, and the degree of freedom is also zero for the liquid-vapour equilibrium at 300 K—explain.

(d) It is easier to break chemical bonds at higher temperatures—justify.

(e) All vibrations of CO_2 are not infrared active—explain.

9. (a) Is it possible to measure precisely and simultaneously the x-coordinate and the y-component of the linear momentum of an electron? Justify in brief.

(b) Functions : ψ_1 and ψ_2 are individually normalized and are mutually orthogonal to each other. Normalize $\psi = \psi_1 + 3^{1/2}\psi_2$.

(c) The experimental results of the photoelectric effect data for metals Cs and Na are given in the figure. Explain why the Cs line is to the left of the Na line in terms of the atomic structure ideas known to you.

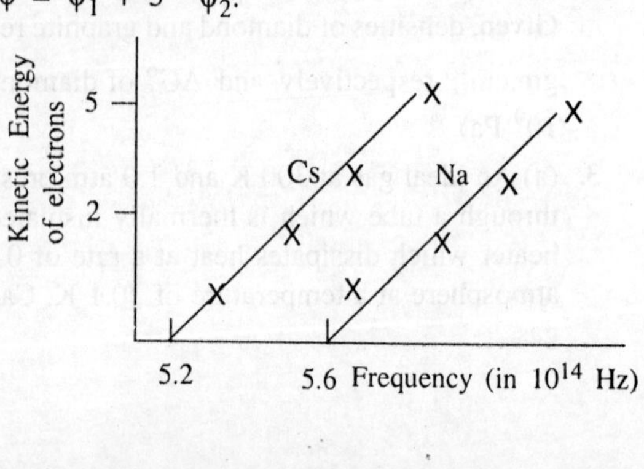

(d) Consider the folllowing wave function for a particle in a box.

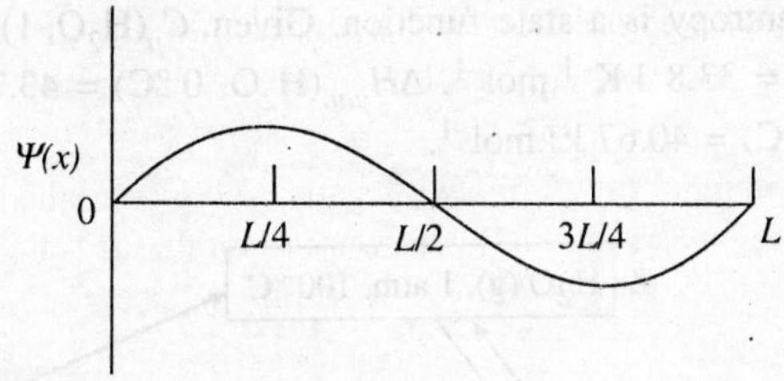

At what positions is the likelihood of finding the particle maximum? At what positions is it minimum? [**Ans.** max $\frac{L}{4}, \frac{3L}{4}$ min $= \frac{L}{2}$]

10. (a) At what compositions corresponding to the points (X, Y or Z) is the reaction $A \rightarrow B$ not spontaneous? Why?

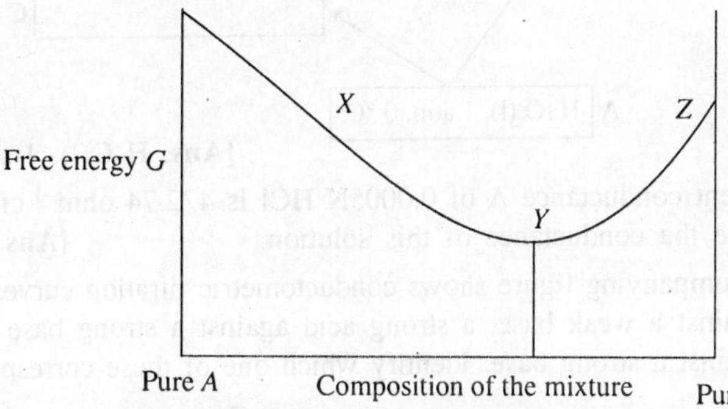

(b) For a given substance, which line of Figure A below belongs to the gaseous phase and which line belongs to the liquid phase? Why? (A = gas, B = liquid)

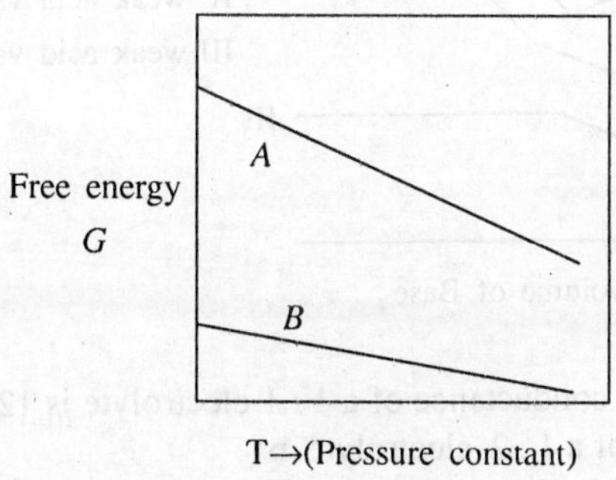

Figure A

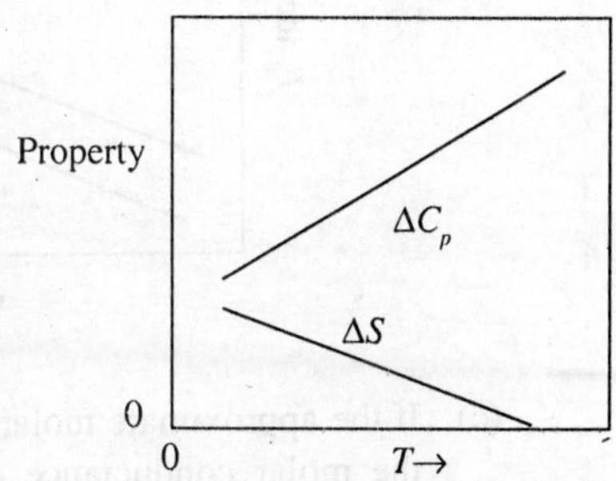

Figure B

(c) For the reaction $A + B = C$, K is the thermodynamic equilibrium constant. If $(\partial K/\partial P)_T = x$, what values can x take and why? [**Ans.** $\frac{1}{P}$]

(d) For a reaction $X_2 + Y_2 = 2XY$, the dependence of the change in heat capacity and entropy change of the reaction on temperature are shown in the Figure B above. Are the reactants stabilized by heating? Rationalize.

(e) Fill in the box C in the following. Calculate the entropy change for each step and verify that entropy is a state function. Given, $C_p(H_2O, 1) = 75.5\ J.K^{-1}.mol^{-1}$. $C_p(H_2O, g) = 33.8\ J.K^{-1}.mol^{-1}$. $\Delta H_{vap}(H_2O, 0\ °C) = 43.773\ kJ.mol^{-1}$. $\Delta H_{vap}(H_2O, 100\ °C) = 40.67\ kJ.mol^{-1}$.

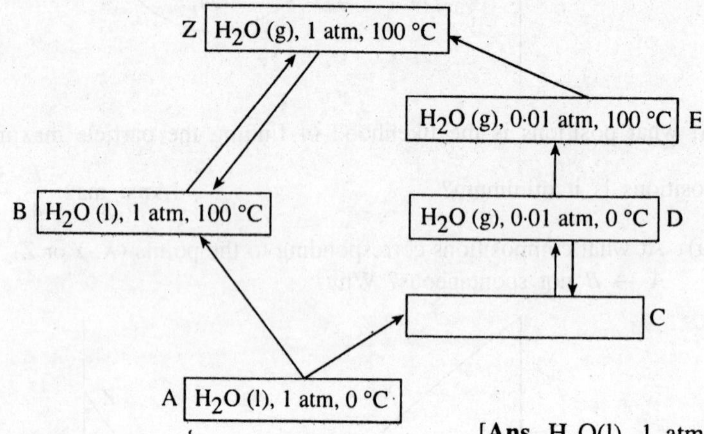

[**Ans.** H_2O(l), 1 atm, 100 °C]

11. (a) Equivalent conductance Λ of 0.0005N HCl is 422.74 $ohm^{-1}.cm^2$ at 25 °C. Calculate the conductance of this solution. [**Ans.** 2.11×10^{-4}]

(b) The accompanying figure shows conductometric titration curves for a weak acid against a weak base, a strong acid against a strong base and a weak acid against a strong base. Identify which one of these corresponds to I, II and III.

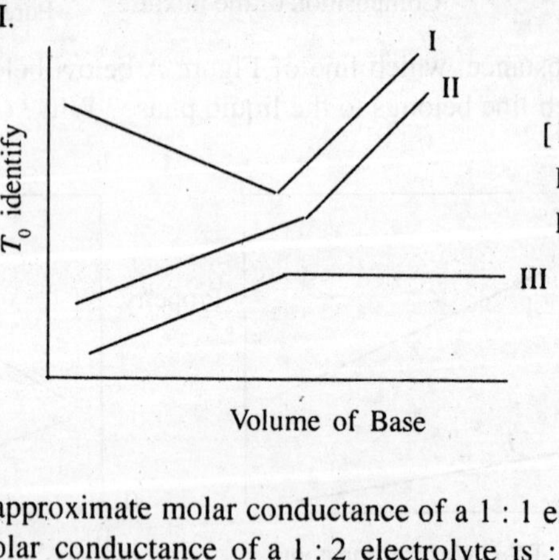

[I strong acid vs. strong base
II weak acid vs. strong base
III weak acid vs. weak base]

(c) If the approximate molar conductance of a 1 : 1 electrolyte is 120 $ohm^{-1}cm^2$, the molar conductance of a 1 : 2 electrolyte is :
(i) 120 $ohm^{-1}.cm^2$ (ii) 180 $ohm^{-1}.cm^2$ or (iii) 240 $ohm^{-1}.cm^2$.

(d) For a zero order reaction, the half-life depends on the logarithm of the initial concentration. Is this statement correct? Justify your answer.

12. (a) Calculate the constant pressure molar heat capacity of hydrogen gas from the data given below :

For the reaction, $H_2(g) + \frac{1}{2}O_2(g) \longrightarrow H_2O(g)$

$\Delta H°(25\ °C) = -241.8\ kJ.mol^{-1}$, $\Delta H°(100\ °C) = -242.6\ kJ.mol^{-1}$. At constant pressure, molar heat capacity of $O_2(g) = 29.37\ J.K^{-1}.mol^{-2}$, molar heat capacity of $H_2O(g)$ = 33.58 $J.K^{-1}.mol^{-1}$. Assume that the heat capacities are independent of temperature. [**Ans.** 29.56 $J.K^{-1}.mol^{-1}$]

(b) For the reaction, $N_2(g) + 2H_2(g) \rightarrow 2NH_3(g)$, $\Delta G° = -33$ $kJ.mol^{-1}$ at 25 °C. If the partial pressures of $N_2(g)$, $H_2(g)$, and $NH_3(g)$ are 3 bar, 1 bar and 4 bar respectively, in which direction the reaction will process? [**Ans.** forward]

Using the data given above calculate the standard Gibbs function of formation of NH_3? [**Ans.** –33 $kJ.mol^{-1}$]

13. (a) A gas at 250 K and 15 atm has a molar volume 12% smaller than that calculated from the perfect gas law. Calculate (a) the compression factor (b) the molar volume of the gas. (c) Which forces are dominating—attractive or repulsive? [**Ans.** (a) 0.88 (b) 1.203 (c) attractive]

(b) The change in the Gibbs function of a certain constant pressure process was found to fit the expression. $\Delta G = -92.6 + 40.3/T$,

where ΔG and T are in k.J and K respectively. Calculate the value of ΔS for the process at 25 °C. [**Ans.** 0.4533 $J.K^{-1}$]

(c) A monoatomic perfect gas is expanded to twice its volume and simultaneously heated to twice its initial temperature. Calculate the change in its entropy. [**Ans.** 14.41 $J.K^{-1}.mol^{-1}$]

(d) The following data was recorded during the potentiometric titration of 25.0 ml of ammonium iron (II) sulphate solution with 0.1095 M cerium (IV) sulphate solution using platinum and saturated calomel electrodes.

Ce^{+4} solution added, ml	E, mV	$\Delta E/\Delta V$, $mVml^{-1}$
22·00	523	
22·50	543	40
22·60	550	70
22·70	557	70
22·80	565	80
22·90	575	100
23·00	590	150
23·10	620	300
23·20	860	2400
23·30	915	550
23·40	944	290
23·50	958	140
24·00	986	56

Calculate the concentration of the ammonium iron(II) sulphate solution. [**Ans.** 0.203 M]

14. The normal boiling point of a binary solution of A and B with mole fraction $x_A = 0.6589$ is 88 °C. At this temperature the vapour pressures of pure A and B are 957 and 379.5 torr, respectively. (a) Is this solution ideal? (b) What is the composition of the vapour above the solution?

[**Ans.** (a) ideal, (b) $X_A = 0.83$, $X_B = 0.17$]

15. A monolayer of N_2 molecule is adsorbed on the surface of 1.00 gm of an Fe/Al_2O_3 catalyst at 77 K. Upon warming the N_2 gas occupies 2.86 cm^3 at 0 °C and 760 torr. What is the surface area of the catalyst if the effective area of a N_2 molecule is 0.167 nm^2? [**Ans.** 12.83 m^2]

16. (a) Calculate the resistance of a conductivity cell containing a solution with specific conductance, $\kappa = 0.001\ ohm^{-1}.cm^{-1}$, and electrodes 1 cm × 2 cm are held 0.5 cm apart. [**Ans.** 250 ohms]

(b) The ionic radii of Mg^{2+} ion is 65 pm and O^{2-} ion is 140 pm. Find the coordination number in MgO.

[**Ans.** around each Mg^{+2} ion there are 6 Cl^{-1} ions and vice versa]

17. Using the following thermochemical equations,

$Ag(s) \longrightarrow Ag^+(aq) + e \qquad \Delta H = 105.6$ kJ

$\frac{1}{2} Cl_2(g) + e \longrightarrow Cl^-(aq) \qquad \Delta H = -167.2$ kJ

$AgCl(s) \longrightarrow Ag^+(aq) + Cl^-(aq) \; \Delta H = 65.5$ kJ

predict the effect of raising temperature on the equilibrium constant of the reaction given below :

$Ag(s) + \frac{1}{2}Cl_2(g) \longrightarrow AgCl(s)$. [**Ans.** yield will decrease]

18. (a) The emf of the cell

$Ag(s)|AgCl(satd.), KCl(0.05\ mol.dm^{-3})|\ AgNO_3\ (0.1\ mol.dm^{-3})|Ag(s)$ is 0.431 V at 298 K. The mean activity coefficient of KCl is 0.817 and that of $AgNO_3$ is 0.723. Calculate the solubility product of AgCl at 25 °C.

[**Ans.** 1.46×10^{-10}]

(b) The conversion of A to B and C goes through the following mechanism :

$$2A \underset{k_2}{\overset{k_1}{\rightleftharpoons}} I \underset{k_4}{\overset{k_3}{\rightleftharpoons}} B + C.$$

Show that the equilibrium constant (K) for the overall reaction is : $K = \frac{k_1 k_3}{k_2 k_4}$.

(c) Express how the equilibrium of the reaction :

$N_2 + 3H_2 \rightleftharpoons 2NH_3$

will be affected if inert gas Ar is added to the system at constant pressure and constant volume. [**Ans.** yield decreases, no effect]

19. (a) Using quantization condition for the de Broglie wavelength on a ring, derive the expression for rotational energy levels of a rigid homonuclear diatomic rotor of bond length $2r$.

(b) The fundamental vibrational frequency of HCl is 2885 cm^{-1}. Assuming that HCl and DCl may be treated as simple harmonic oscillator, calculate the fundamental frequency of DCl (atomic mass of chlorine = 35 dalton).

[**Ans.** 2066 cm^{-1}]

(c) Find the condition under which the function $f(x) = a \sin \alpha x + b \cos \alpha x$ can be eigenfunction of $-id/dx$. **[Ans.** $a = b$]

20. If 4 moles of a MX_2 salt in 1 kg of water raises the boiling point of water by 3.2 K, calculate the degree of dissociation of MX_2 in the solution. (for water, $K_b = 0.5$ K kg mol^{-1}) **[Ans.** $\alpha = 0.3$] **[JAM 2011]**

21. Equilibrium constant for a reaction doubles as the temperature is increased from 300 K to 600 K. Calculate the standard reaction enthalpy (in kJ mol^{-1}) assuming it to be constant in this temperature range. (given : $R = 8.3$ JK^{-1} mol^{-1}, ln 2 = 0.7) **[Ans.** 34.86 J] **[JAM 2010]**

22. A 50 mL solution of 0.1 M monoprotic acid ($K_a = 1 \times 10^{-6}$ at 298 K) is titrated with 0.1 M NaOH at 298 K. Calculate the [H^+] of the solution after the addition of 50 mL of NaOH at this temperature. (given : $K_w = 1 \times 10^{-14}$ at 298 K) **[Ans.** 8.85] **[JAM 2010]**

23. One mole of an ideal gas initially at 300 K and at a pressure of 10 atm undergoes adiabatic expansion

(i) reversibly and

(ii) irreversibly against a constant external pressure of 2 atm until the final pressure becomes equal to the external pressure.

Calculate ΔS_{system} for (i) and (ii). For (ii), express the final answer in terms of R.

Given : Molar heat capacity at constant volume $C_{v,m} = \dfrac{3R}{2}$.

[Ans. (i) = 0, (ii) = 3.67 R] **[JAM 2011]**

24. For the following equilibrium at 300 °C,

N_2O_4 (g) $\rightleftharpoons$ $2NO_2$ (g).

Calculate K_p when N_2O_4 is 30% dissociated and the total pressure is 2 bar. **[Ans.** 0.79] **[JAM 2011]**

Short Questions without Answers

1. Define fine structure of spectral lines.
2. The frequency of light incident on a metal surface is doubled. What will be the effect on kinetic energy of emitted photoelectrons?
3. Write the dimension of angular momentum from its definition.
4. Write down the action integral for angular motion of electron in an atom.
5. Arrange the following molecules in order of increasing polarisability : HCl, HF, HI, HBr.
6. Name the important physical quantity for a diatomic molecule, which is obtained from its rotational constant '*B*'.
7. For hydrogen 1s-state, $|\psi|^2$—shows maximum at the origin—state true or false.
8. Write down the values of *L*, *S* and *J* for the $2P_{3/2}$ term.
9. What is the main reason behind splitting of the atomic spectral lines into fine structrue?
10. Is the photoelectric work function of a metal the same as its ionization potential?
11. Why does not the lowest energy state in H atom have $n = 0$?
12. What is the potential of the half-cell H_2(Pt) (0.01 atm) | $H^+(a = 0.1)$ at 298 K?
13. Give the cell representation of any standard cell.
14. Name an electrode which can be used to determine Cl^- ion concentration of a NaCl solution.
15. Draw the differential plot for potentiometric titration of HCl vs. NaOH.
16. Express mean ionic activity of $Al_2(SO_4)_3$ in terms of ionic activities.
17. Suggest a suitable linear plot for variation of Freundlich isotherm.
18. Draw schematically an energy profile diagram for a uncatalysed reaction.
19. State Walden's rule.
20. Plot schematically μ_{JT} vs. $^1/_T$ for a van der Waals gas.
21. What is the dimension of $\frac{d\gamma}{d\ln C}$, where γ and *C* respectively are surface tension and concentration.
22. Write down the dimension of *R*.
23. Draw log *k* vs. $\sqrt{\mu}$ plot for inversion of sucrose catalysed by HCl.
24. Calculate C_v for H_2 according to the equipartition theorem.
25. Write down Schrödinger time dependent equation.
26. Draw the wave function for the 2nd quantum state of a particle in a one-dimensional box ($0 \leq x \leq a$).
27. Which one of the electronic transitions $n \rightarrow \pi^*$ or $\pi \rightarrow \pi^*$, will appear at longer wavelength?
28. Write the dimension of $i\hbar\frac{\partial}{\partial\theta}$ [θ is angle].
29. Give an example of a photosensitized reaction.
30. Find a suitable eigenfunction of the operator $\frac{\partial}{\partial x}$.

31. What is the dimension of entropy?

32. Remark on the statement "solubility increases with temperature".

33. Define an ideal solution. .

34. What condition is imposed on $\oint dx$ for x to be a state function?

35. What is the relation between eV and J?

36. Show that $(PdV - VdP)$ is not an exact differential.

37. Identify X and Y in Henry law : $X \propto Y$.

38. Remark if any reversible isothermal process is isoenthalpic.

39. Explain whether the partition co-efficient is independent of temperature.

40. What is meant by "critical solution temperature"?

41. Cite an example of an isothermal-isobaric process.

42. How many phases can generally exist for a 1-component system at STP?

43. What is the unit of rate constant for a 2nd order reaction?

44. What is the dimension of surface tension?

45. Define "triple point".

46. For a certain equilibrium $\Delta G^\circ = 0$, find the equilibrium constant.

47. Define chemical potential. What is its dimension?

48. Entropy of an isolated system is constant—criticize.

49. Draw curves to show the pressure dependence of compressibility factor (Z) for H_2 and He.

50. The solubility product of a sparingly soluble salt MA_3 is 2.7×10^{-7}. Find its solubility assuming complete dissociation.

51. Indicate a simple experimental device by which pure HD can be distinguished from an equimolecular mixture of H_2 and D_2 (D = heavy hydrozen).

52. Justify/criticise whether $[\partial(G/T) / \partial(^1/_T)]_P$ is a state function.

53. What does 'zero' in 'zero-point energy' signify?

54. Arrange in order of increasing translational frequencies for electronic, NMR and vibrational spectra.

55. Show that the function $ae^{x/a}$ is an eigenfunction of the operator $\dfrac{d}{dx}$. What is the eigenvalue?

56. Arrive the dimension for 'energy × time'.

57. State whether the threshold frequency in photoelectric phenomena is a characteristic of light used or that of the material employed.

58. Why is CCl_4 molecule non-polar although the C—Cl bond is polar?

59. The electronic and vibrational transitions produce the electronic and vibrational spectra. What are the transitions responsible for NMR spectrum?

60. Write an equation which describes the variation of Å with temperature.

61. Write Boltzmann distribution formula for the case of degenerate energy levels.

62. The rate of reaction is 1.0×10^{-5} mol.dm^{-3}.s^{-1}. Calculate the rate in molecule cm^{-3}.s^{-1} unit.

63. Give an example of a photosensitized reaction which is of immense important to mankind.

64. Define zeta potential.

65. Represent the Weston Cadmium cell using the usual convention.

66. Which of the two solutions, aq. solution of NaCl or an alcoholic solution of NaCl of same conentration will show more non-ideality and why?

67. If pH = 7 − (log x)/2, find K_w of pure water.

68. Explain if 'a' in van der Waals equation follows additivity.

69. How does the photoelectric work function depend on the frequency of light?

70. Expand the operator $\left(\frac{d}{dx}+x\right)^2$.

71. Compare the dimensions of action and angular momentum.

72. How is the stationary Bohr orbit of an atom defined in terms of de Broglie waves?

73. Verify whether the operator Δ^2 is linear.

74. How does the energy associated with a wave function vary with the number of nodes?

75. What is the value of quantum yield for a primary process in photochemistry?

76. Should an ideal gas have viscosity? Why?

77. What is Clausius inequality?

78. A reaction of the type $A(g) + A(g) \rightarrow A_2(g)$ is expected to be exothermic.—Explain.

79. What condition must be satisfied for a binary liquid mixture to show azerotropism?

80. I_2 usually sublimes, but it can be melted.—Explain.

81. How can you depict Boyle's law by a straight line plot other than PV vs. V (or P)?

82. (i) Draw the P-T phase diagram of water. (ii) Label the different regions in this diagram. (iii) On the diagram, show the liquid-vapour equilibrium for a dilute solution of NaCl, with the help of a dashed curve. **[JAM 2009]**

83. The temperature dependence of the Gibbs free energy G is :

$$\left(\frac{\partial(G/T)}{\partial T}\right)_P = -\frac{H}{T^2}.$$

Obtain the expression for the temperature dependence of the equilibrium constant K. Given that $\Delta H° = A + BT$ (where A and B are constants). **[JAM 2009]**

84. The Maxwell probability distribution of molecular speeds for a gas is :

$$F(v)dv = 4\pi v^2\left(\frac{m}{2\pi kT}\right)^{3/2} \exp\left(-\frac{mv^2}{2kT}\right)dv,$$

where v is the speed, m the mass of a gas molecule and k the Boltzmann constant.

(i) Use $F(v)$ to show that the most probable speed u is given by the expression

$$u = \sqrt{\left(\frac{2RT}{M}\right)}.$$

(ii) Use $R = 8\ JK^{-1}\ mol^{-1}$ in the above expression to calculate the u_{mp} for $CH_4(g)$ at 127 °C. **[JAM 2011]**

Answers to Objective Type Questions

Part A

1. (b)
2. (a)
3. (c)
4. (a)
5. (d)
6. (a)
7. (b)
8. (b)
9. (d)
10. (c)
11. (b)
12. (b)
13. (a)
14. (b)
15. (a)
16. (a)
17. (b)
18. (a)
19. (a)
20. (a)

Part B

21. (A) 2.9×10^{-7}
    (B) 8.41×10^{-14}
22. 0.08
24. (i) 6.87 kJ
    (ii) 4.48 kJ
26. 2.36×10^{-18}
27. 2.76×10^{-17}
28. (a) can
    (b) more
    (c) less
    (d) true
    (e) false

Part C

29. (d)
30. (b)
31. (c)
32. (b)
33. (a)
34. (c) (assuming it (to be an ideal gas)
35. (b)
36. (c)
37. (d)
38. (b)
39. (a)
40. (c)
41. (a)
42. (a)
43. (d)
44. (d)
45. (b)
46. (a)
47. (c)
48. (c)
49. (a)
50. (d)
51. (b)
52. (a)
53. (d)
54. (d)
55. (c)
56. (d)
57. (a)
58. (a)
59. (b)
60. (c)
61. (b)
62. (c)
63. (c)
64. (d)
65. (a)
66. (d)
67. (c)
68. (c)
69. (a)
70. (b)
71. (c)

Part D

72. (a) F
    (b) F
    (c) F
    (d) T
    (e) T
    (f) F
    (g) F
    (h) T
    (i) T
    (j) F

Part E

73. (a) (ii)
    (b) (i)
    (c) (ii)
    (d) (i)
    (e) (ii)
    (f) (ii)
    (g) (iii)
    (h) (i)
    (i) (i)
    (j) (iii)

Part F

74. (a) T
    (b) F
    (c) F
    (d) T
    (e) F
    (f) F
    (g) T
    (h) F
    (i) T
    (j) F

Part G

75. (a) (iii)
    (b) (i)
    (c) (ii)
    (d) (ii)

(e) (iii)
(f) (iv)
(g) (iii)
(h) (iii)
(i) (v)
(j) (ii)

Part H

76. (a) (iii)
(b) (iv)
(c) (iii)
(d) (i)
(e) (iii)
(f) (iii)
(g) (ii)
(h) (i)
(i) (iv)
(j) (iii)

Part I

77. (a) F
(b) F
(c) T
(d) T
(e) F
(f) T
(g) F
(h) F
(i) F
(j) F

Part J

78. (c)
79. (c)
80. (a)
81. (d)
82. (b)
83. (a)
84. (d)
85. (a)
86. (b)
87. (b)
88. (b)
89. (c)
90. (b)
91. (b)
92. (b)
93. (b)
94. (b)
95. (a)
96. (b)
97. (c)
98. (c)
99. (c)
100. (d)
101. (c)
102. (c)
103. (a)
104. (c)
105. (a)
106. (a)
107. (c)
108. (b)
109. (d)
110. (b)
111. (b)
112. (b)
113. (b)
114. (b)
115. (c)
116. (b)
117. (b)
118. (a)
119. (c)
120. (b)
121. (a)
122. (a)
123. (a)
124. (a)
125. (a)
126. (d)
127. (b)
128. (d)
129. (d)
130. (d)
131. (b)
132. (b)
133. (b)
134. (b)
135. (b)
136. (c)
137. (a)
138. (b)
139. (b)
140. (b)
141. (c)
142. (d)
143. (a)
144. (a)
145. (b)
146. (c)
147. (a)
148. (b)
149. (c)
150. (d)
151. (b)
152. (b)
153. (d)
154. (b)
155. (a)
156. (a)
157. (a)
158. (d)
159. (a)
160. (a)
161. (b)
162. (a)
163. (c)
164. (d)
165. (c)
166. (d)
167. (b)
168. (b)
169. (c)
170. (c)

171. (d)
172. (c)
173. (c)
174. (a)
175. (c)
176. (b)
177. (a)
178. (c)
179. (a)
180. (d)
181. (a)
182. (d)
183. (a)
184. (c)
185. (a)
186. (a)
187. (c)
188. (a)
189. (c)
190. (b)
191. (b)
192. (c)
193. (a)
194. (d)
195. (d)
196. (b)
197. (b)
198. (d)
199. (a)
200. (a)
201. (a)
202. (a)
203. (b)
204. (b)
205. (c)
206. (a)
207. (a)
208. (b)
209. (c)
210. (b)
211. (b)
212. (c)
213. (b)
214. (b)
215. (c)
216. (a)
217. (c)
218. (b)
219. (a)
220. (c)
221. (c)
222. (c)
223. (a)
224. (b)
225. (d)
226. (c)
227. (a)
228. (d)
229. (c)
230. (d)
231. (c)
232. (d)
233. (a)
234. (d)
235. (c)
236. (d)
237. (a)
238. (c)
239. (d)
240. (b)
241. (b)
242. (a)
243. (c)
244. (a)
245. (c)
246. (a)
247. (b)
248. (a)
249. (c)
250. (c)
251. (b)
252. (b)
253. (d)
254. (b)
255. (c)
256. (b)
257. (c)
258. (b)
259. (b)
260. (d)
261. (c)
262. (d)
263. (c)
264. (a)
265. (b)
266. (c)
267. (a)
268. (a)
269. (c)
270. (a)
271. (d)
272. (b)
273. (c)
274. (d)
275. (c)
276. (a)
277. (a)
278. (b)
279. (d)
280. (c)
281. (a)
282. (c)
283. (b)
284. (a)
285. (d)
286. (c)
287. (a)
288. (d)
289. (b)